# Arthropod Relationships

## The Systematics Association Special Volume Series

*Series Editor*

David M. John

Department of Botany, The Natural History Museum, Cromwell Road, London SW7 5BD, UK

The Systematics Association provides a forum for discussing systematic problems and integrating new information from cytogenics, ecology and other specific fields into taxonomic concepts and activities. It has achieved great success since the Association was founded in 1937 by promoting major meetings covering all areas of biology and palaeontology, supporting systematic research and training courses through the award of grants, production of a membership newsletter and publication of review volumes by its publisher Chapman & Hall. Its membership is open to both amateurs and professional scientists in all branches of biology who are entitled to purchase its volumes at a discounted price.

The first of the Systematics Association's publications, *The New Systematics,* edited by its then president Sir Julian Huxley, was a classic work. Over 50 volumes have now been published in the Association's 'Special Volume' series often in rapidly expanding areas of science where a modern synthesis is required. Its *modus operandi* is to encourage leading exponents to organise a symposium with a view to publishing a multi-authored volume in its series based upon the meeting. The Association also publishes volumes that are not linked to meetings in its 'Volume' series.

Anyone wishing to know more about the Systematics Association and its volume series are invited to contact the series editor. Further information about the Systematics Association can also be found on the website

http://www.thomson.com/systematic/SA/main.html

**Forthcoming titles in the series:**

Evolutionary Relationships Among Protozoa
Coombs *et al.*

Advances in Plant Molecular Systematics
Bateman *et al.*

Homology in Systematics
R.W. Scotland and R.T. Pennington

The Flagellates
B. Leadbeater and J. Green

Other Systematics Association publications are listed after the index for this volume.

The Systematics Association Special Volume Series 55

# Arthropod Relationships

Edited by

R.A. Fortey
*Department of Palaeontology, The Natural History Museum, London, UK*

and

R.H. Thomas
*Department of Zoology, The Natural History Museum, London, UK*

SPRINGER-SCIENCE+BUSINESS MEDIA, B. V.

First edition 1998


Originally published by The Systematics Association in 1998
Softcover reprint of the hardcover 1st edition 1998

Typeset in 10/12pt Times by Saxon Graphics Ltd, Derby

ISBN 978-94-010-6057-8 ISBN 978-94-011-4904-4 (eBook)
DOI 10.1007/978-94-011-4904-4



A catalogue record for this book is available from the British Library

Library of Congress Catalog Card Number: 97–68214

∞ Printed on acid-free text paper, manufactured in accordance with ANSI/NISO Z39.48-1992 (Permanence of Paper).

To the memory of Sidnie Manton, 20 years on from the publication of *The Arthropoda.*

# CONTENTS

# CONTRIBUTORS

L.G. ABELE
Office of the Provost, Florida State University, Tallahassee, FL 32306-3020, USA
email: labele@aa.fsu.edu

J.P. BACON
Sussex Centre for Neuroscience, School of Biological Sciences, University of Sussex, Brighton BN1 9QG, UK
email: J.P.Bacon@sussex.ac.uk

G. BOXSHALL
Department of Zoology, The Natural History Museum, Cromwell Road, London, SW7 5BD, UK
email: G.Boxshall@nhm.ac.uk

D.E.G. BRIGGS
Department of Geology, Wills Memorial Building, University of Bristol, Queen's Road, Bristol, BS8 1RJ, UK
email: D.E.G.Briggs@bristol.ac.uk

G. BUDD
Department of Historical Geology and Palaeontology, Institute of Earth Sciences, Norbyvägen 22, S-75236 Uppsala, Sweden
email: Graham.Budd@pal.uu.se

R.A. DEWEL
Department of Biology, Appalachian State University, Boone, NC 28608, USA
email: dewelra@appstate.edu

W.C. DEWEL
Department of Biology, Appalachian State University, Boone, NC 28608, USA
email: dewelwc@appstate.edu

M.H. DICK
Department of Biology, Middlebury College, Middlebury, Vermont 05753, USA
email: dick@midd-unix.middlebury.edu

W. DOHLE
Institut für Zoologie, Freie Universität Berlin, Königin-Luise-Str. 1–3, D-14195 Berlin, Germany

J.A. DUNLOP
Department of Earth Sciences, University of Manchester, Oxford Road, Manchester,M13 9PL, UK
email: jdunlop@fs1.ge.man.ac.uk

D.J. EERNISSE
Department of Biology MH282, California State University, Fullerton, CA 92834, USA
email: DEernisse@fullerton.edu

M.J. EMERSON (Deceased)
late of the San Diego Natural History Museum

R.A. FORTEY
Department of Palaeontology, The Natural History Museum, Cromwell Road, London, SW7 5BD, UK
email: R.Fortey@nhm.ac.uk

G. FRYER
Institute of Environmental and Biological Sciences, University of Lancaster, Lancaster LA1 4QY, UK

H. HAMILTON
Department of Integrative Biology, Museum of Paleontology, University of California, Berkeley, CA 94720, USA

V. HYPSA
Faculty of Biological Sciences, Branisovská 31, 370 05 Ceské Budejivce, Czech Republic

O. KRAUS
Zoologisches Institut und Zoologisches Museum, Universität Hamburg, Martin-Luther-King-Platz 3, D-20146 Hamburg, Germany

N.P. KRISTENSEN
Zoological Museum, University of Copenhagen, Universitetsparken 15, DK-2100 Copenhagen Ø, Denmark
email: npkristens@zmuc.ku.dk

J. KUKALOVÁ-PECK
Department of Earth Sciences, Carleton University, 1125 Colonel By Drive, Ottawa, K1S 5B6 Canada
email: jkpeck@ccs.carleton.ca

K.J. MÜLLER
Institüt für Paläontologie, Universität Bonn, Nussallee 8, D-53115 Bonn, Germany

C. NIELSEN
Zoological Museum, University of Copenhagen, Universitetsparken 15, DK-2100 Copenhagen Ø, Denmark
email: cnielsen@zmuc.ku.dk

D.-E. NILSSON
Department of Zoology, University of Lund, Helgonavägen 3, S-223 62 Lund, Sweden
email: dan-e.nilsson@zool.lu.se

D. OSORIO
Biological Sciences, Sussex University, Brighton, BN1 9QG, UK
email: D.Osorio@sussex.ac.uk

A.P. RASNITSYN
Arthropoda Laboratory, Paleontological Institute, Russian Academy of Science, Profsoyuznaya Str. 123, Moscow,117647, Russia
email: rasna@glas.net.ru

G. SCHOLTZ
Institut für Biologie, Vergleichende Zoologie Humboldt-Universität zu Berlin, Philippstr. 13, D-10115 Berlin, Germany
email: gerhard=scholtz@rz.hu-berlin.de

F.R. SCHRAM
Institute for Systematics and Population Biology, University of Amsterdam, Post Box 94766, NL–100 GT Amsterdam, The Netherlands
email: schram@bio.uva.nl

P.A. SELDEN
Department of Earth Sciences, University of Manchester, Oxford Road, Manchester, M13 9PL, UK
email: paul.selden@man.ac.uk

W.A. SHEAR
Department of Biology, Hampden-Sydney College, Hampden-Sydney, VA 23943, USA
email: BillS@tiger.hsc.edu

T. SPEARS
Department of Biological Sciences, Florida State University, Tallahassee, FL 32306-2043, USA
email: spears@bio.fsu.edu

R.H. THOMAS
Department of Zoology, The Natural History Museum, Cromwell Road, London, SW7 5BD, UK
email: r.thomas@nhm.ac.uk

J.W. VALENTINE
Department of Integrative Biology, Museum of Paleontology, University of California, Berkeley, CA 94720, USA
email: jwv@ucmp1.berkeley.edu

M. VLÁŠKOVÁ
Faculty of Biological Sciences, Branišovská 31, 370 05 České Budejovce, Czech Republic
email: misa@paru.cas.cz

D. WALOSSEK
Sektion für Biosystematische Dokumentation, University of Ulm, Liststraβe 3, D-89079 Ulm, Germany
email: dieter.walossek@biologie.uni-ulm.de

W.C. WHEELER
Department of Invertebrates, American Museum of Natural History, Central Park West at 79th Street, New York, NY 10024-5192, USA
email: wheeler@amnh.org

P.M. WHITINGTON
Department of Zoology, University of New England, Armidale, NSW, 2351, Australia
email: pwhiting@metz.une.edu.au

R. WILLMANN
II. Zoologisches Institut der Universität, Georg-August Universität-Göttingen, Berliner Straβe 28, D-37073 Göttingen, Germany

M.A. WILLS
Department of Geology, Wills Memorial Building, University of Bristol, Queen's Road, Bristol BS8 1RJ, UK
email: m.a.wills@bristol.ac.uk

J. ZRZAVÝ
Faculty of Biological Sciences, Branišovská 31, 370 05 České Budejovce, Czech Republic
email: zrzavy@entu.cas.cz

# PREFACE

Arthropods – insects, crustaceans, myriapods and arachnids – are the most speciose of all animal groups, and have probably been so for hundreds of millions of years. Their importance in every ecosystem – terrestrial and marine – is not in question. Yet very little has been agreed about how they achieved their pre-eminence. The evolutionary pathways which led to their current diversity are still the subject of controversy, despite the fact that many of the questions of descent have been debated for more than a century. We judged that the time was ripe for a new attempt at a synthesis of current views about arthropod phylogeny, the first since Gupta (1979). There have been many new advances in both methodology and technology since then, not least the impact of cladistics in the former, and molecular biology in the latter. Yet the classical approaches of comparative morphology, embryology, and palaeontology have been far from static – for example, there have been more discoveries of exciting new fossils in the past 20 years than ever before. The way forward seemed to be to bring together the scattered specialists to discuss a modern overview on arthropod relationships. Hence, we invited the leading figures in the various disciplines of arthropodology to a Symposium at the Natural History Museum, London, 17–19 April 1996. We were generously supported by the Linnean Society of London, the Systematics Association, The Royal Society and the Natural History Museum in making the meeting a practical proposition, and to these bodies we tender our thanks. This book is the outcome of three days of mutual education.

We have taken a novel approach to publication of data. Many of the papers include cladistic analyses based upon character matrices – both morphological and molecular. We have published the results of the analyses in the book, but the data are stored and available on a website page. They can therefore be directly downloaded by other investigators, and manipulated as they choose. We trust that this will further aid the utility of the book.

The papers are arranged in such a way as to focus progressively from general problems to issues involving particular groups. The place of the Arthropoda within the animal phyla is considered first, particularly the relationship to annelids; there are reviews of the morphological, developmental and molecular data bearing on the unsolved problems. Most contributors favour arthropod monophyly, although Fryer bids us look carefully at some of the supposed homologies. Next, cladistic treatments of the arthropods as a whole – from several theoretical standpoints – provide a summary of the issues dealt with in subsequent chapters. The reader will be able to judge whether combining all phylogenetic data sources in such analyses is more persuasive than trying to identify uniquely derived apomorphies as a basis for the definition of groups. The important place of the tardigrades within the arthropod stem-lineage is shown by Dewel and Dewel. That Cambrian fossils offer unique evidence in fleshing out the early history of arthropods is illustrated with reference to new Lower Cambrian animals from Greenland, and the celebrated 'Orsten' arthropods from Sweden and elsewhere. It is clear that some of the main lineages of arthropod radiation had already been distinguished by the early Cambrian. There is recent evidence that the antiquity of the main branches in the group may be even older (Fortey *et al.*, 1996; Wray *et al.*, 1996). No doubt this antiquity exacerbates the problems in unequivocally recognizing fundamental branching events.

Even within each major arthropod group there are problems in recognizing the relationships between major taxa. Three different attitudes to such problems in Crustacea are exemplified by the morphological, molecular and 'key character' approaches, respectively. They yield interesting differences. The fossil history of myriapods and chelicerates is reviewed. The reality or otherwise of the 'Uniramia' (myriapods plus insects) is discussed by several authors; recent reconsideration of morphological evidence finds more support for an insect–crustacean relationship than has been claimed in recent years (also Averof and Cohen, 1997). This is one example where molecular phylogenies have led to a re-evaluation of classical evidence. Within the section on hexapods, the reader will find clear accounts of the several schools of thought regarding the most fundamental subdivisions of the greatest group of arthropods. The four papers on this topic provide an object lesson in the problems of interpretation of characters in unscrambling important, but ancient events. On the one hand it is undoubtedly true that, as Kukalovà-Peck claims, there are morphologies preserved in fossils which no longer survive, and which must surely bear on the phylogeny. On the other hand, the interpretation of the fossil characters has proved difficult, and there are cases where ambiguous morphology has been placed at the service of a favoured theory, with the attendant dangers of circularity.

Finally, embryological and physiological evidence is brought into the argument. There are examples in the sensory 'wiring' across arthropod groups of such striking similarity that it seems difficult to comprehend that they may have evolved in parallel.

If there is a general message to be drawn from this compilation, it is that all fields of endeavour have much to contribute to an understanding of arthropod relationships. It would be unwise to neglect the broad range of discoveries in favour of those from a narrow specialization alone. It will be clear that there is, as yet, no 'right' answer about arthropod phylogeny; but the journey towards it gets increasingly interesting.

*R.A.F. and R.H.T.*

## REFERENCES


Averof, M. and Cohen, S.M. (1997) Evolutionary origin of insect wings from ancestral gills. *Nature,* **385**, 627–30.

Fortey, R.A., Briggs, D.E.G. and Wills, M.A. (1996) The Cambrian evolutionary 'explosion': decoupling cladogenesis from morphological disparity. *Biological Journal of the Linnean Society,* **57**, 13–33.

Gupta, A.P. (1979) *Arthropod Phylogeny,* Van Nostrand Reinhold Co., New York.

Wray, G.A., Levinton, J.S. and Shapiro, L.H. (1996) Molecular evidence for deep Precambrian divergences among metazoan phyla. *Science*, 274, 568–73.

# 1 Body plans, phyla and arthropods

*J.W. Valentine and H. Hamilton*

*Department of Integrative Biology, Museum of Paleontology, University of California, Berkeley, CA 94720, USA*
*email: jwv@ucmp1.berkeley.edu*
*Department of Integrative Biology, Museum of Paleontology, University of California, Berkeley, CA 94720, USA*

## 1.1 INTRODUCTION

Phyla are, traditionally, Linnean taxa, each of which can be characterized by a distinctive assemblage of morphological features. The early systematists who attempted to classify animals at the highest level were clearly struck by the unique morphological themes that differentiated the body plans of major groups, eventually termed phyla. The body plans could not be linked by series of intermediate organisms, living or fossil, nor can they today. Although the body plans are unique, the features that characterized them are not. Rather, they are polythetic (Beckner, 1968); that is, each feature may be found in more than one phylum, or may be absent from a member of the phylum; it is the assemblage that is unique. A number of attempts have been made to express the fundamental commonality exhibited by members of phyla – words such as archetype, Bauplan and phylotype spring to mind. Here we wish briefly to review such concepts, and to suggest which are more informative in framing a definition of phyla, with particular reference to the Arthropoda. We use the term **body plan** for the assemblage of morphological features that is found among members of a higher taxon, without implying any particular taxonomic, evolutionary or morphological viewpoint.

## 1.2 CONCEPTS OF THE PHYLUM

### 1.2.1 BAUPLAN AND GROUNDPLAN

The term Bauplan (Woodger, 1945) has been widely used to signify the body plan of phyla. At times it has been used as a neutral term, much as we use the term body plan. It is also commonly employed in a rather Cuvieran manner; a particularly cogent evaluation of this use of the term is given by Brusca and Brusca (1990). For them, a Bauplan refers not only to the basic architectural scheme that is displayed in the morphological commonalities of a taxon, but also to the structure of the individual organ systems and organs that are found to function appropriately within that particular architecture. Thus, the term embraces both structure and function and implies constraints on the extent to which variations may occur within the architecture of phyla, presumably limited by functional coadaptations among the components of the various Baupläne. Such a concept thus embraces both the morphological unity and disparity observed within a phylum, but does not refer directly to an ancestor. **Groundplan** is sometimes used instead of Bauplan by writers of English, for either of the meanings mentioned.

### 1.2.2 ARCHETYPE

The concept of the archetype, developed in detail by Owen (1846, 1848), stretches back at least to Goethe and to Geoffroy Saint-Hilaire (Appel, 1987). Geoffroy in particular regarded what became known as the archetype as an abstraction, not as an actual body plan, but as a construct that displayed a unity of composition that could be found among the taxa of a particular group. The uniting principle was that of 'connections' – the relations between the morphological components of the various taxa, which became one of the principal bases of homology. This abstraction or 'ideal' applied whether or not there were functional similarities among the morphological parts. Thus, archetypes did not refer to ancestors, though subsequently used in that sense by some workers.

### 1.2.3 HYPOTHETICAL ANCESTOR

It is clear that any group of allied taxa has a common ancestor that must have displayed most of its common features, and that the taxa have inherited at least many of those characteristics from that ancestor (Figure 1.1). The common ancestor is thus not hypothetical in the sense that its existence is in doubt, but rather in the sense that the particular ancestor that is postulated is an hypothesis. By sorting out derived from ancestral features in its descendant taxa, a common ancestor may be reconstructed with respect to homologous features. When the morphology of an arthropod ancestor has been hypothesized it has tended to appear

*Arthropod Relationships*, Systematics Association Special Volume Series 55, Edited by R.A. Fortey and R.H. Thomas. Published in 1997 by Chapman & Hall, London. ISBN 0 412 75420 7

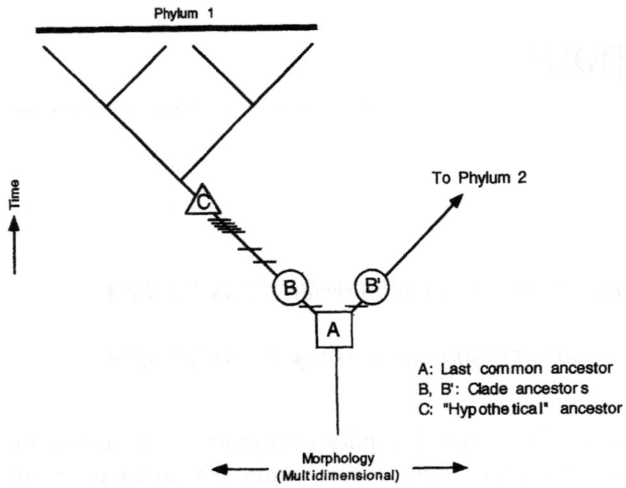


**Figure 1.1** Some key events in the origin of a phylum. A: the last common ancestor (box) of two phyla, 1 and 2. B, B′: the clade ancestors (circles) of each of the phyla, which are sister species and daughters of the last common ancestor. Note that all three of these species, the parent and two daughters, have the same body plans, and indeed may be virtually indistinguishable morphologically, or if distinctive, may not vary in any way that is related to the body plans of phylum 1 or 2. C: the 'hypothetical' ancestor of phylum 1 (triangle). This is the first species with an assemblage of distinctive characters sufficient to be considered as the first representative of the body plan of phylum 1. If the lineage is evolving slowly the choice of this form may be somewhat arbitrary, while if there is a relatively rapid rise of several of the novel characters as the body plan becomes fully established, the choice may be quite constrained and obvious.

rather annelid-like, with a long series of similar segments bearing similar legs (e.g. Snodgrass, 1938).

### 1.2.4 PHYLOTYPE

Allied but distinctive animals, such as are grouped within a phylum or a class, were once thought to be most morphologically similar early in development, and to diverge during subsequent ontogenetic stages (Von Baer's law). However, progressive divergence is far from universal; disparate early developmental stages of allied taxa commonly converge during the course of early ontogeny, only to diverge once more as the adult body types are formed (Seidel, 1960; Sander, 1983; Wray and Raff, 1991; Strathmann, 1993). The stage of maximum similarity has been termed the **phylotypic stage** (Sander, 1983), and may usually coincide with the final positioning of the morphogenetic fields of adult tissues and organs (see review in Hall, 1996).

The phylotypic stage is by no means identical even in the conphyletic taxa that share it, and the stage as represented in any given living member of a phylum is unlikely to be identical with a stage in the ontogeny of the founding ancestor of the phylum. However, it is likely, though not certain, that the phylotypic stage in a given living form will more closely resemble an ontogenetic stage of the phylum's ancestor than will any other stage, as it appears to be the least diverged from the ancestral ontogeny. The features held in common among the phylotypic stages displayed by various members of a phylum may thus help in reconstructing a similar stage in an ancestral ontogeny, just as other features may in reconstructing hypothetical adult ancestors. A reconstructed phylotypic stage is similarly hypothetical.

### 1.2.5 CLADE ANCESTOR

The last common ancestor of two clades, such as sister phyla, is the species that gave rise to daughter species from which each of the phyla has respectively descended (Figure 1.1). As each of the daughter species founded a clade, they are termed clade ancestors here; in effect they are the first members of taxa. The parental last common ancestral species stem will share a common body plan with these sisters. The sisters themselves are not likely to be much more different than any two average sister species today – and indeed the sisters might even be sibling species. Furthermore, any morphological differences that the sisters do display may well have nothing to do with the morphological changes that eventually occur to create the distinctive body plans of their descendants – and this is probably by far the most usual case. Thus, the descendant body plans are derived with respect to both their clade ancestors and their last common ancestor (except in the special case when one body plan remains essentially unchanged, probably a rare case with phyla). If taxa are construed as being founded by their clade ancestor, the morphological concepts of the taxa are lost. Phyla construed in this way can no longer be defined or identified morphologically, since a clade ancestor will not usually possess the body plan that characterizes the Linnean phylum that descends from it, but will have the body plan of some members of its ancestral lineage and of its sister phylum. Identification of the nature of the clade ancestor, however, is clearly of considerable importance in any attempt to understand the evolution of the body plan of a phylum.

### 1.2.6 GENOME ANCESTOR

As molecular tools provide an ever-increasing understanding of the genetic basis of morphological patterning, characterization of the genomic elements that produce the body plans of major taxa is becoming a realistic possibility. Slack *et al.* (1993) have suggested a genomic definition for the kingdom Metazoa. The definitive features are the presence of a series of genes in the *Hox* cluster, and of some related homeobox genes, that are termed the **zootype** and are hypothesized to characterize the kingdom. It is not yet clear whether the type cluster was actually assembled at the very onset of the

Metazoa or evolved later, but *Hox* cluster-type genes have been identified in phyla with primitive body plans such as sponges (Degnan *et al.*, 1995) and cnidarians (Schummer *et al.*, 1992; Miller and Miles, 1993; Shenk and Steele, 1993) and appear to be present in all metazoans. Furthermore, each phylum (and indeed each class) that has been appropriately investigated to date contains a unique *Hox* cluster (for reviews see Akam *et al.*, 1994; Ruddle *et al.*, 1994; Valentine *et al.*, 1996; Holland and Garcia-Fernàndez, 1996; and see also Abele, 1997; Akam, 1997; Carroll, 1997; and Dick; 1997, all this volume). As pointed out by Slack *et al.* (1993), the phylotype should represent an ontogenetic stage when the genome responsible for patterning a body plan is first fully expressed. These genes might be representative of the full ancestral genome associated with the phylum body plan, a concept related to the phylotype and analogous to the zootype of the kingdom. To the extent that the ancestral genome must be reconstructed, it too is hypothetical.

### 1.2.7 DEVELOPMENTAL ANCESTOR

The basic features that underlie the development of body plans are not simply genomes but are developmental fields, fields of patterning that are first found in the egg and produce cells whose fates and positions are progressively specified during morphogenesis so as to produce a given polarity, region, organ or other structure as development unfolds (Wolpert, 1977; Davidson, 1993; Gilbert, 1994; Gilbert *et al.*, 1996). The fields involve not only portions of the genome but also myriad signalling relays and boundary determinants operative at the cellular or tissue levels. Thus, an ancestor can be viewed, not just as so much genetic information or as a combination of some particular somatic features, but as including a developmental process involving them both. Using a developmental ancestor stresses the fact that, after all, the Metazoa owe their very existence to their ability to produce differentiated cells that form a level in a somatic hierarchy. Arguably, the best choice for the ancestor of a phylum should be made at the first appearance (in phylogeny) of the developmental system that underlies its definitive body plan. A developmental definition of an ancestor invokes an entire array of data that we do possess or can determine – morphological data from the fossil record, comparative anatomical data, and all the exquisitely detailed molecular genetic information now coming to light from the living descendants or allies of the ancestral form. Synthesizing and applying such data should produce both the best theoretical and the most practical guide to the concept of a phylum.

## 1.3 ARTHROPODS AND BODY PLANS

### 1.3.1 ARTHROPOD ANCESTRY

It is commonly hypothesized that arthropods have descended from an ancestor with an annelid-like body plan via the process of 'arthropodization' – the acquisition of features that arthropods have that annelids do not, such as a sclerotized cuticle and jointed appendages, plus the loss of those things that annelids have and arthropods do not, such as a eucoelomic hydrostatic skeleton. A clade that includes both arthropods and annelids, often as sisters (e.g. Davidson *et al.*, 1995), is usually termed the Articulata and is supported by a supposed homology of segmentation (and sometimes of other features) in both phyla (Figure 1.2(A)). However, small subunit ribosomal RNA (SSU rRNA) sequences among living representatives of various phyla point to a strikingly different history for the annelid and arthropod body plans (Figures 1.2(B) and 1.3, and references therein). The Arthropoda and the Annelida are members of sister protostome clades that each include a number of other phyla, that is, other body plans. Support is weak for some branchpoints but seems strong for the protostome–deuterostome branch and also for the branch leading to the separate clades containing arthropods and annelids, a branch that is also consistent with analyses of mitochondrial DNA rearrangements (Boore *et al.*, 1995). The sequence of appearance of the major synapomorphies associated with the best-supported branches that lead to bilaterian body plans (Figure 1.3) suggests that the last common ancestor of arthropods and annelids was unsegmented (that is, lacked a set of serially repeated, correlated organs or organ systems that comprise the regular morphological regions termed segments). If the SSU rRNA tree is correct, an unsegmented form branched off between the last common arthropod–annelid ancestor and each of those phyla.

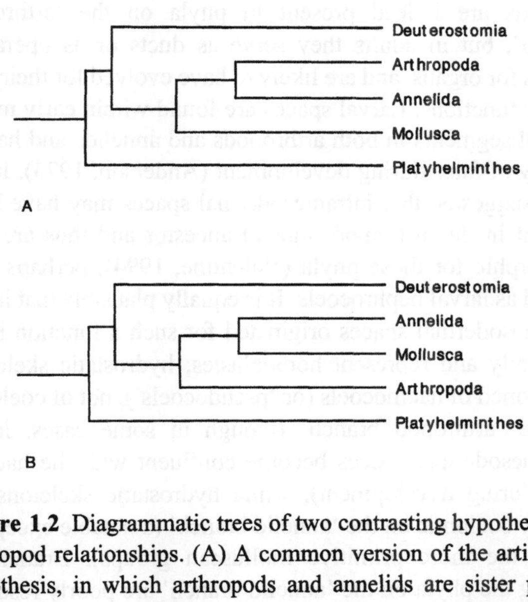


**Figure 1.2** Diagrammatic trees of two contrasting hypotheses of arthropod relationships. (A) A common version of the articulate hypothesis, in which arthropods and annelids are sister phyla. (B) A version of the phylogeny suggested by SSU rRNA sequence comparisons, in which arthropods and annelids belong to sister protostome clades but are not themselves sisters (see also Figure 1.3).

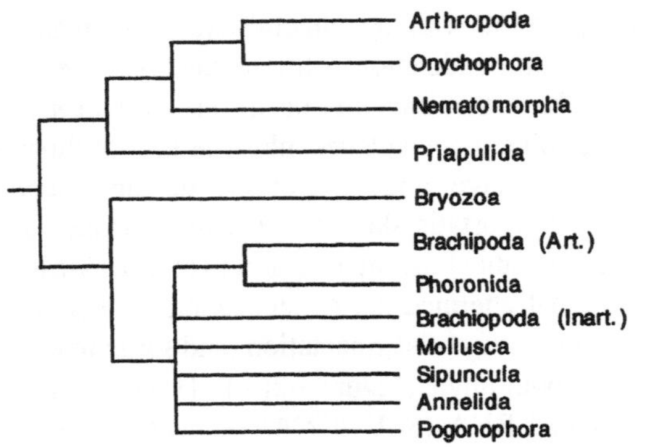


**Figure 1.3** Branching topology of some protostome phyla as interpreted from SSU rRNA sequence comparisons. Distances along or between branches are not to any scale. Although the branching pattern is based on analyses by several methods, not all of these taxa were included in any single analysis. (Chiefly after studies by Lake, 1990; Turbeville *et al.*, 1991; Halanych *et al.*, 1995; Winnepenninckx *et al.*, 1995a,b.)

It is possible that the last common arthropod–annelid ancestor was seriated, that is, possessed some serially repeated organs (but not a series of regularized morphological regions) to service an elongate body, and had a blood vascular system (BVS), with locomotion by pedal or peristaltic waves supported by a haemocoel (Valentine, 1989, 1994). However, it is possible that it was much simpler, without a haemal system at all, resembling in grade a primitive flatworm. At any rate that ancestor almost certainly did not possess a perivisceral coelom. Body spaces that qualify as coeloms are indeed present in phyla on the 'arthropod branch', but in adults they serve as ducts or as operating spaces for organs, and are likely to have evolved for their particular functions. Larval spaces are found within early mesodermal segments in both arthropods and annelids and have a variety of fates during development (Anderson, 1973). It has been suggested that intramesodermal spaces may have been present in the arthropod–annelid ancestor and thus are plesiomorphic for those phyla (Valentine, 1994); perhaps they served as larval nephrocoels. It is equally plausible that larval intramesodermal spaces originated for such a function independently and represent homoplasies; hydrostatic skeletons are formed of haemocoels (or 'pseudocoels'), not of coeloms, on the 'arthropod branch' (though in some cases, larval intramesodermal spaces become confluent with the haemocoel during development), while hydrostatic skeletons are largely coelomic on the 'annelid branch' (a notable exception being the more primitive molluscan groups). Branchings among the phyla on the 'annelid branch' are poorly resolved by SSU rRNA data and little more can be said from either morphological or molecular evidence at this time (although it would be morphologically parsimonious if the molluscs prove to have the deepest branch in the clade that includes annelids).

The general architecture of the last body plan ancestral to the arthropods may be reconstructed if an arthropod sister phylum is identified. Features such as the presence of limbs and other serial structures, a haemocoel as a hydrostatic skeleton, and a moulted cuticle, suggest either tardigrades or onychophorans as the living arthropod sister phylum, and some SSU rRNA evidence also places these clades close to the arthropods (Wheeler *et al.*, 1993; Garey *et al.*, 1996; Giribet *et al.*, 1996). In either of these cases, the common features listed above, perhaps together with a BVS that has been lost in the minute tardigrades, would represent plesiomorphies for arthropods.

What criterion can serve as a reasonable indication that features requisite to an arthropod body plan had appeared? In nearly all definitions of the phylum, the jointing of the carapace of both the body and of appendages is stressed. It is plausible that jointing is a solution to the problem of maintaining flexibility of movement in a body that is becoming encased in an increasingly rigid exoskeleton (Snodgrass, 1938). The rise of arthropod-style segmentation, which involves the body wall and the somatic musculature, the appendages and their musculature, and the innervation and vascularization of both sorts of muscles and other serially repeated tissues, probably was associated with the evolution of jointing. These developments depended upon the relatively rigid exoskeletal elements, to which muscles could be attached, and which could act as levers in many cases. Again, the fluid skeleton that provides body turgor, and that antagonizes muscle contractions in some cases, is a haemocoel; there is no hint that a coelom was ever involved in this segmental locomotory system.

It seems doubtful that a segmented exoskeletal arrangement would have arisen *de novo* in the arthropod ancestor unless there was a preceding serial arrangement of some organs, and unless locomotion was enhanced as skeletonization proceeded. The last common ancestor of the Arthropoda and its sister phylum is therefore likely to have had a series of paired unjointed locomotory appendages, with the requisite supporting systems. The rigid, jointed exoskeleton may have arisen from a periodically moulted cuticle of this segmented clade ancestor. Such a form could also have given rise to the Onychophora and Tardigrada. The key morphological feature in recognizing the arthropod body plan, then, may be a sclerotized jointed integument that is a functional exoskeleton, thus underlying the evolution of the polythetic complex characterizing the arthropod body plan.

### 1.3.2 BODY PLANS AND SYNAPOMORPHIES

In Figure 1.4, the arthropods are placed within the context of a tree of metazoan phyla (plus the presumably ancestral choanoflagellates) among which SSU rRNA sequences have been compared. This topology represents a subjective

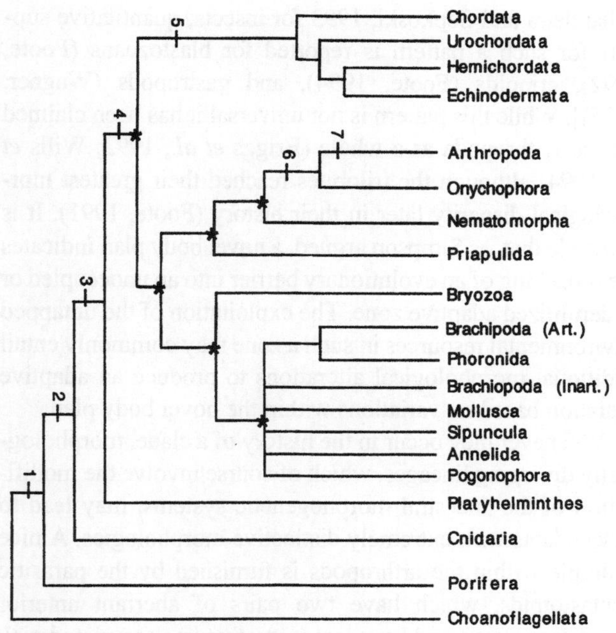


**Figure 1.4** Branching topology of some metazoan phyla based on SSU rRNA sequence comparisons, with some of the major design features that are probably associated as synapomorphies with various clades, indicated by numbers. Branchpoints for which there are no obvious synapomorphies relating to body plans are starred. Distances along or between branches are not to any scale. Although the branching pattern is based on analyses by several methods, not all of these taxa were included in any single analysis. Design features: 1, extracellular matrix, multicellularity; 2, gastrulation; 3, cellular mesoderm; 4, haemal fluid, either in vessels or in a haemocoel; 5, radial cleavage and some other deuterostome developmental characters, such as enterocoely; 6, locomotory appendages; 7, sclerotization of cuticle. (Branching pattern chiefly after studies by Field *et al.*, 1988; Lake, 1990; Turbeville *et al.*, 1991, 1994; Wainright *et al.*, 1993; Wada and Satoh, 1994; Halanych, 1995; Halanych *et al.*, 1995; Winnepenninckx *et al.*, 1995a,b; Garey *et al.*, 1996; Giribet *et al.*, 1996.)

consensus of sorts; some of the branchpoints are poorly supported. For the deeper branches, up to the protosome–deuterostome common ancestor, an orderly set of nested synapomorphies may be inferred that is associated with the body plans of a series of primitive phyla (Figure 1.4). For some of the shallower branches, however, especially those involved in the protostome–deuterostome split and in the early diversification of the major protostome clades (see the starred branchpoints in Figure 1.4), clear body plan synapomorphies are lacking, although each of the living phyla that have descended from clades originating at those branchpoints can be characterized by numerous apomorphies. This situation does not seem to be an artefact of the particular topology figured; rearranging the phyla in some other scheme also fails to create clear, nested synapomorphies that are related to body plans. This lack of significant synapomorphies is a common outcome of transforming higher taxa, defined as branches in a tree, into taxa placed within ranks of a hierarchy; these structures are fundamentally dissimilar (Valentine and May, 1996).

The branchpoint between the 'arthropod' and 'annelid' clades in Figure 1.4 is the point at which their last common ancestor gave rise to their clade ancestors, which were sister species. There is certain to have been some apomorphies in at least one of the sisters – some genetic differences at least, and probably also some external morphological differences. It is also certain that they shared a common body plan. Similarly, the last common ancestor between arthropods and nematomorphs probably exhibited features that were derived with respect to the last common arthropod–annelid ancestor, but that those features produced a distinct new body plan is not at all clear; indeed it seems possible that all of the branchpoints that are starred in Figure 1.4 may have shared a common body plan: that of the haemocoelic and possibly seriated vermiform organism mentioned above. The important problems in the evolutionary morphogenesis of phyla descending from ancestors with common body plans, then, is not how they are interrelated – the branching pattern may even have been random with respect to their later body plans or their later grades of organization – but how they evolved from the clade ancestor to the developmental ancestor. On present evidence it is possible, for example, that the annelid clade ancestor branched from the last common annelid–arthropod ancestor more deeply than did the molluscan clade ancestor, yet that the molluscs have evolved very little from their ancestral body plan while the annelids are highly derived. We do not know that this is true, but any such patterns would vitiate a search within the 'annelid clade' for nested synapomorphies that are related to body plan morphology. Appropriate synapomorphies should be found within the molecular sequences of the genomes, however.

In short, how well the body plans of phyla can be represented as a progressive set of synapomorphies, nested like Russian dolls, depends on the history of body plan diversification. So long as novel body plans represent successive grades of organization, as from sponges to cnidarians to flatworms, or a progressive trend along a theme, as is assumed for onychophorans to arthropods, a nested pattern appears. However, if several novel body plans arise from lineages with a common ancestral body plan, there need be no synapomorphies at all that are defining at the level of the higher taxon. The continuing difficulty encountered in attempting to discover by cladistic methods a morphologically based metazoan tree at the level of the phylum, a Linnean concept, is at least partly owing to this problem.

### 1.3.3 MORPHOLOGICAL DIVERSITY

The wide array of morphological types now known from the early fossil record that appear to be arthropods but that cannot be placed within the morphological bounds displayed by living classes has been extensively reviewed and discussed (e.g. see chapters in this volume). Continuing evaluation of the relationships of these forms has produced a long series of suggestions and some extensive morphologically based cladograms (Wills *et al.*, 1994, 1995; Waggoner, 1996) which tend to identify a few major clades that are euarthropods together with a few allied groups that include living tardigrades and onychophorans and some assemblages of extinct forms (Figure 1.5(A)). The topology of the major clades in such a tree is rather similar to their topology in the latest SSU rRNA trees (Figure 1.5(B)). If the arthropods are defined by the establishment of sclerotization and its morphological consequences, the euarthropod taxa, however they are interrelated, evidently form a phylum.

Higher taxa that are entirely extinct and that fall morphologically at the margin of the euarthropods remain difficult to assign. It is unclear whether *Anomalocaris* and *Opabinia* evolved independently, or whether their probable segmentation and the sclerotized, jointed appendages of *Anomalocaris* imply that they belong within the Euarthropoda but have lost some features, perhaps as adaptations to pelagic life, that characterize other taxa in that clade.

The fossil record is replete with examples of the early radiation of arrays of modifications within newly arisen body plans [stressed by Simpson, 1944, with data chiefly for vertebrates; see also Valentine, 1969 for marine invertebrates, and Labandeira and Sepkoski, 1993 for insects; quantitative support for such a pattern is reported for blastozoans (Foote, 1992), crinoids (Foote, 1994), and gastropods (Wagner, 1995)]. While this pattern is not universal it has been claimed for the Arthropoda as a whole (Briggs *et al.*, 1992; Wills *et al.*, 1994) although the trilobites reached their greatest morphological disparity later in their history (Foote, 1991). It is plausible that, as Simpson argued, a novel body plan indicates the broaching of an evolutionary barrier into an unoccupied or underutilized adaptive zone. The exploitation of the untapped environmental resources in such a zone may commonly entail additional morphological alterations to produce an adaptive radiation based on variations within the novel body plan.

Whenever they occur in the history of a clade, morphologically diverging lineages, which of course involve the modification of the ancestral morphogenetic systems, may lead to descendants with extremely distinctive morphologies. A nice example within the arthropods is furnished by the parasitic pentastomids, which have two pairs of aberrant anterior appendages only, and have lost many features associated with a segmented body plan (Storch, 1993). As a result, pentastomids have sometimes been considered as a phylum of their own, although some workers developed evidence that they are aberrant crustaceans modified for a parasitic lifestyle (e.g. Wingstrand, 1972), an hypothesis later supported by SSU rRNA evidence (Abele *et al.*, 1989). If pentastomids had fallen taxonomically at the margin of living arthropods they might have been generally accepted as a phylum of their own, but their evident history of descent from a crustacean inclines even Linnean systematists to include them in that taxon; their modifications are therefore regarded as 'secondary'. It seems that possession of most elements of a given body plan is not strictly necessary for membership in a phylum. Just how well the pentastomid genome retains the characteristics of the ancestor of the arthropods becomes an important question, as yet unanswered.

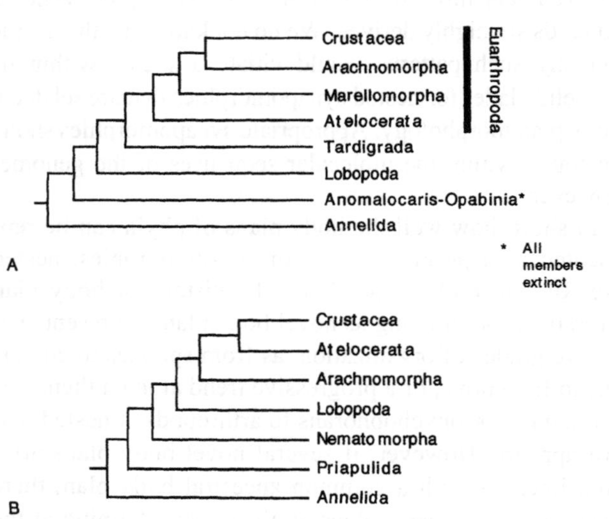


**Figure 1.5** Two views of the phylogenetic relations among major arthropod groups. (A) Topology based on morphology (after Wills *et al.*, 1995). (B) Topology suggested by interpretation of SSU rRNA sequences. (Chiefly after studies by Turbeville *et al.*, 1991; Wheeler *et al.*, 1993; Winnepenninckx *et al.*, 1995b.)

### 1.3.4 DEVELOPMENTAL BASIS OF MORPHOLOGICAL DIVERSITY

It has commonly been suggested that the vast morphological breadth within the phylum Arthropoda is a consequence of the modular nature of the body plan (Brusca and Brusca, 1990). By varying the numbers of body segments and their patterns of tagmosis, by specialization of some segments, and by the specialization of appendages (also exploiting their segmented nature), the functions of segments or tagmata become highly differentiated and large numbers of morphological sub-plans can be and have been generated, some of which have been modified into as large an array of variants as can be found within the body plans of most other entire phyla. Thus, questions arise as to what sort of genetic underpinning has permitted such astonishing morphological diversification, and whether it is a genetic peculiarity of arthropods.

Metazoan development involves the differentiation of distinctive cell types (from two or so to hundreds) in particular locations at particular times during ontogeny. As the entire genome is present in all cells (with minor exceptions such as gametes), these developmental events are orchestrated by a control system that operates by expressing a fraction of the genes in any given cell, and repressing the others. Presumably, body plans were evolved by selection among alternate heritable variations in development; differential success of alternate developmental variations, expressed in extended reproductive success, determined the direction of morphological change. As the early metazoans evolved increasingly complex morphologies, there must have been a concomitant elaboration of regulatory gene functions – preceded by duplication and divergence of the regulatory genes. The patterns of expression and repression naturally mirrored the morphological differentiation of the body plans; if strong morphological differences were present along the anteroposterior (AP) body axis, it follows that there were strong AP differences in gene expression and repression, mediated by appropriate control systems.

Molecular studies of development have demonstrated that many of the genes involved in metazoan patterning and morphogenesis are very ancient, certainly antedating the protostome–deuterostome branching. As the early metazoan radiations unfolded, those ancient genes were used in regulating the development of the disparate body plans that were evolved. The requirement for definitive positioning of different, proliferating cell types gave rise to modularities of control. For example, in early bilaterians, some regulatory genes must have mediated cell type patterns along the AP axis. The subsequent evolution of increased morphological differentiation along this axis, including the eventual rise of complex serial structures in some lineages, was evidently accompanied by elaboration of selector and other downstream regulatory gene functions, also expressed serially, and positioned by the original AP patterning system. The patterning genes thus stand at the base of trees of gene interactions, the control signals forming sets of nested modularities. Such a system is extremely flexible, presumably capable of responding to nearly any morphological change that is accessible to organisms with a given body plan.

The arthropods, while not necessarily possessing any special developmental system, represent an almost perfect marriage of a flexibly modular body plan, capable of producing a vast number of morphological variations, with a flexibly modular developmental system. There are other organisms constructed on modular body plans that have not enjoyed quite such bountiful morphological variety. Annelids are an obvious example, and lobopodians, though displaying significantly greater variety in the Cambrian than at present, seem to have had even less morphological diversity than annelids. We suspect that the arthropodan diversity is a consequence of their exoskeleton, with its jointing in support of both body and limbs, carrying the modularity of the body plan to an extreme not available to soft-bodied forms. If this is correct, it underscores the importance of the arthropod-style exoskeleton as a synapomorphy of the phylum.

## 1.4 CONCLUSIONS

In attempting to define a concept of the phylum we have been led to consider the ideas that have been suggested for defining a body plan. It seems desirable to base the definition on a polythetic assemblage of features. A Cuvieran-style Bauplan is a useful pedagogic concept but the boundaries of phyla are in fact not constrained by its features – the limits to body plan modification are not clear, especially as features may be lost, as when forms become parasitic or miniaturized. The archetype does not seem useful in a world wherein evolution has occurred and ideal design motifs never existed. Phylotypes are not ancestral stages (and may be lacking in some groups). Clade ancestors may usually lack the body plan of the descendent phyla. The concept of a genomic ancestor, sometimes said to be based on the information required of a body plan, in fact fails to include the abundance of epigenetic information present in developmental pathways.

Perhaps the best definition of a phylum combines the hypothetical ancestor, the most useful morphological concept, with the developmental ancestor, the most useful theoretical concept. As the developmental ancestor includes both the genome and the inherited inductive pathways that are necessary to generate the morphology of the hypothetical ancestor, it effectively integrates these concepts. The phylum then becomes the monophyletic clade founded by this developmental ancestor. The euarthropods appear to represent a phylum in this sense.

## ACKNOWLEDGEMENTS

We thank J. Craine, Department of Integrative Biology, University of California, Berkeley, for help with graphics. Supported in part by NSF grants EAR-9317247 and EAR-9317372. Contribution No. 1655, Museum of Paleontology, University of California, Berkeley.

## REFERENCES

Abele, L.G., Kim, W. and Felgenhauer, B.E. (1989) Molecular evidence for inclusion of the phylum Pentastomida in the Crustacea. *Molecular Biology and Evolution*, **6**, 685–91.

Akam, M., *et al.* (eds) (1994) *The Evolution of Developmental Mechanisms*. The Company of Biologists Ltd., Cambridge.

Anderson, D.T. (1973) *Embryology and Phylogeny in Annelids and Arthropods*, Pergamon Press, Oxford.

Appel, T.A. (1987) *The Cuvier–Geoffroy Debate: French Biology in the Decades before Darwin*, Oxford University Press, New York and Oxford.

Beckner, M. (1968) *The Biological Way of Thought*, University of California Press, Berkeley and Los Angeles.

Boore, J.L., Collins, T.M., Stanton, D., Daehler, L.L. and Brown, W.M. (1995) Deducing the pattern of arthropod phylogeny from mitochondrial DNA arrangements. *Nature*, **376**, 163–5.

Briggs, D.E.G., Fortey, R.A. and Wills, M.A. (1992) Morphological disparity in the Cambrian. *Science*, **256**, 1670–3.

Brusca, R.C. and Brusca, G.J. (1990) *Invertebrates*, Sinauer Associates, Sunderland, Massachusetts.

Davidson, E.H. (1993) Later embryogenesis: regulatory circuitry in morphogenetic fields. *Development*, **118**, 665–90.

Davidson, E.H., Peterson, K.J. and Cameron, R.A. (1995) Origin of bilaterian body plans: evolution of developmental regulatory mechanisms. *Science*, **270**, 1319–25.

Degnan, B.M., Degnan, S.M., Giusti, A. and Morse, D.E. (1995) A *Hox/Hom* homeobox gene in sponges. *Gene*, **155**, 175–7.

Field, K.G., Olsen, G.J., Lane, D.J., Giovannoni, S.J., Ghiselin, M.T., Raff, E.C., Pace, N.R. and Raff, R.A. (1988) Molecular phylogeny of the animal kingdom. *Science*, **239**, 748–53.

Foote, M. (1991) Morphologic patterns of diversification: examples from trilobites. *Palaeontology*, **34**, 461–85.

Foote, M. (1992) Paleozoic record of morphologic diversity in blastozoan echinoderms. *Proceedings of the National Academy of Sciences USA*, **89**, 7325–9.

Foote, M. (1994) Morphological disparity in Ordovician–Devonian crinoids and the early saturation of morphological space. *Paleobiology*, **20**, 320–44.

Garey, J.R., Krotec, M., Nelson, D.R. and Brooks, J. (1996) Molecular analysis supports a tardigrade–arthropod association. *Invertebrate Biology*, **115**, 79–88.

Gilbert, S.F. (1994) *Developmental Biology*, Sinauer Associates, Sunderland, Massachusetts.

Gilbert, S.F., Opitz, J.M. and Raff, R.A. (1996) Resynthesizing evolutionary and developmental biology. *Developmental Biology*, **173**, 357–72.

Giribet, G., Carranza, S., Bagũnà, J., Riutort, M. and Ribera, C. (1996) First molecular evidence for the existence of a Tardigrada + Arthropoda clade. *Molecular Biology and Evolution*, **13**, 76–84.

Halanych, K.M. (1995) The phylogenetic position of the pterobranch hemichordates based on 18S rDNA sequence data. *Molecular Phylogenetics and Evolution*, **4**, 72–6.

Halanych, K.M., Bacheller, J.D., Aguinaldo, A.M.A., Liva, S.M., Hillis, D.M. and Lake, J.A. (1995) Evidence from 18S ribosomal DNA that the lophophorates are protostome animals. *Science*, **267**, 1641–3.

Hall, B.K. (1996) *Bauplane*, phylotypic stages, and constraint; why there are so few types of animals. *Evolutionary Biology*, **29**, 215–61.

Holland, P.W.H. and Garcia-Fernàndez, J. (1996) *Hox* genes and chordate evolution. *Developmental Biology*, **173**, 382–95.

Labandeira, C.C. and Sepkoski, J.J., Jr (1993) Insect diversity in the fossil record. *Science*, **261**, 310–15.

Lake, J.A. (1990) Origin of the Metazoa. *Proceedings of the National Academy of Sciences USA*, **87**, 763–6.

Miller, D.J. and Miles, A. (1993) Homeobox genes and the zootype. *Nature*, **365**, 215–16.

Owen, R. (1846) Report on the archetype and homologies of the vertebrate skeleton. *Report of the British Association for the Advancement of Science* (*Southampton Meeting*), 169–340.

Owen, R. (1848) *On the Archetype and Homologies of the Vertebrate Skeleton*, J. Van Voorst, London.

Ruddle, F.H., Bartels, J.L., Bentley, K.L., Kappen, C., Murtha, M.T. and Pendleton, J.W. (1994) Evolution of *HOX* genes. *Annual Review of Genetics*, **28**, 423–42.

Sander, K. (1983) The evolution of patterning mechanisms: gleanings from insect embryogenesis and spermatogenesis, in *Development and Evolution* (eds B.C. Goodwin, N. Holder and C.C. Wylie), Cambridge University Press, Cambridge, pp. 127–59.

Schummer, M., Scheurlen, I., Schaller C. and Gaillot, B. (1992) HOM/HOX homeobox genes are present in hydra (*Chlorohydra viridissima*) and are differentially expressed during regeneration. *EMBO Journal*, **11**, 1815–23.

Seidel, F. (1960) Körpergrundgestalt und Keimstruktur. Eine Erörterung über die bei phylogenetischen Überlegungen. *Zoologischer Anzeiger*, **164**, 245–305.

Shenk, M.A. and Steele, R.E. (1993) Homeobox genes and epithelial patterning in *Hydra*, in *Molecular Basis of Morphogenesis* (ed. M. Bernfield), Wiley-Liss, New York, pp. 231–40.

Simpson, G.G. (1944) *Tempo and Mode in Evolution*, Columbia University Press, New York.

Slack, J.M.W., Holland, P.W.H. and Graham, C.F. (1993) The zootype and the phylotypic stage. *Nature*, **361**, 490–2.

Snodgrass, R.E. (1938) Evolution of the Annelida, Onychophora, and Arthropoda. *Smithsonian Miscellaneous Collections*, **97** (6), 1–159.

Storch, V. (1993) Pentastomida, in *Microscopic Anatomy of Invertebrates, Volume 12, Onychophora, Chilopoda and Lesser Protostomata* (eds. F.W. Harrison and M.E. Rice), Wiley-Liss, New York, pp. 115–42.

Strathmann, R.R. (1993) Hypotheses on the origins of marine larvae. *Annual Review of Ecology and Systematics*, **24**, 89–117.

Turbeville, J.M., Pfeifer, D.M., Field, K.G. and Raff, R.A. (1991) The phylogenetic status of arthropods, as inferred from 18S rRNA sequences. *Molecular Biology and Evolution*, **8**, 669–86.

Turbeville, J.M., Schulz, J.R. and Raff, R.A. (1994) Deuterostome phylogeny and the sister group of the chordates: evidence from molecules and morphology. *Molecular Biology and Evolution*, **11**, 648–55.

Valentine, J.W. (1969) Patterns of taxonomic and ecological structure of the shelf benthos during Phanerozoic time. *Palaeontology*, **12**, 684–709.

Valentine, J.W. (1989) Bilaterians of the Precambrian–Cambrian transition and the annelid–arthropod relationship. *Proceedings of the National Academy of Sciences USA*, **86**, 2272–5.

Valentine, J.W. (1994) Late Precambrian bilaterians: grades and clades. *Proceedings of the National Academy of Sciences USA*, **91**, 6751–7.

Valentine, J.W. and May, C.L. (1996) Hierarchies in biology and paleontology. *Paleobiology*, **22**, 23–33.

Valentine, J.W., Erwin, D.H. and Jablonski, D. (1996) Developmental evolution of metazoan body plans: the fossil evidence. *Developmental Biology*, **173**, 373–81.

Wada, H. and Satoh, N. (1994) Details of the evolutionary history from invertebrates to vertebrates as deduced from the sequences of 18S rDNA. *Proceedings of the National Academy of Sciences USA*, **91**, 1801–4.

Waggoner, B.M. (1996) Phylogenetic hypotheses of the relationships of arthropods to Precambrian and Cambrian problematic fossil taxa. *Systematic Biology*, **45**, 190–222.

Wagner, P.J. (1995) Testing evolutionary constraint hypotheses with early Paleozoic gastropods. *Paleobiology*, **21**, 248–72.

Wainright, P.O., Hinkle, G., Sogin, M.L. and Stickel, S.K. (1993) Monophyletic origins of the metazoa: an evolutionary link with fungi. *Science*, **260**, 340–2.

Wheeler, W.C., Cartwright, P. and Hayashi, C.Y. (1993) Arthropod phylogeny: a combined approach. *Cladistics*, **9**, 1–39.

Wills, M.A., Briggs, D.E.G. and Fortey, R.A. (1994) Disparity as an evolutionary index: a comparison of Cambrian and Recent arthropods. *Paleobiology*, **20**, 93–130.

Wills, M.A., Briggs, D.E.G., Fortey, R.A. and Wilkinson, M. (1995) The significance of fossils in understanding arthropod evolution. *Verhandlung Deutsche Zoologische Gesellschaft*, **88**, 203–15.

Wingstrand, K.G. (1972) Comparative spermatology of a pentastomid, *Raillietiella hemidactyli*, and a branchiuran crustacean, *Argulus foliaceus*, with a discussion of pentastomid relationships. *Biologiske Skrifter. Kongelige Danske Videnskabernes Selskabs*, **19**, 1–72.

Winnepenninckx, B., Backeljau, T. and DeWachter, R. (1995a) Phylogeny of protostome worms derived from 18S rRNA sequences. *Molecular Biology and Evolution*, **12**, 641–9.

Winnepenninckx, B., Backeljau, T., Mackey, L.Y., Brooks, J.M., DeWachter, R., Kumar, S. and Garey, J.R. (1995b) 18S rRNA data indicate that Aschelminthes are polyphyletic in origin and consist of at least three distinct clades. *Molecular Biology and Evolution*, **12**, 1132–7.

Wolpert, L. (1977) *The Development of Pattern and Form in Animals*, Carolina Biological, Burlington, North Carolina.

Woodger, J.H. (1945) On biological transformations, in *Essays on Growth and Form Presented to D'Arcy Wentworth Thompson* (eds. W.E. Le Gros Clark and P.G. Medawar), Cambridge University Press, Cambridge, pp. 95–120.

Wray, G.A. and Raff, R.A. (1991) The evolution of developmental strategy in marine invertebrates. *Trends in Ecology and Evolution*, **6**, 45–50.

# 2 The phylogenetic position of the Arthropoda

*C. Nielsen*

*Zoological Museum, University of Copenhagen, Universitetsparken 15, DK–2100 Copenhagen Ø, Denmark*
*email: cnielsen@zmuc.ku.dk*

## 2.1 INTRODUCTION

The arthropods have been treated as one group of animals for a long time. Linnaeus called them Insecta both in the 1st and the 10th editions of *Systema Naturae* (1735, 1758), and although a number of authors from the first half of the 19th century treated crustaceans, arachnids and insects as separate groups under a common heading, the name Arthropoda was not introduced until Siebold (1848). The first phylogenetic classification of the Animal Kingdom was presented by Haeckel (1866), who in his famous book *Generelle Morphologie der Organismen* drew the first phylogenetic tree and introduced the word 'monophyletic' to describe the nature of the 19 phyla he recognized. However, his classification of lower categories was not based on the same principle. This early attempt at creating a phylogenetic classification of the animals was not followed up during the next century, and textbook and handbook classifications have continued to resemble Linnaeus' 'pre-evolutionary' system. Only Ax (1995) and Nielsen (1995) have attempted to create cladistic arrangements of the phyla.

Several authors, for example Sundberg and Pleijel (1994) and Ax (1995), have pointed out that categories such as phyla and classes are arbitrarily fixed, at least as long as an absolute geological age cannot be assigned to each category. Ax (1995) has taken the full consequence of this, and his classification is without categories above the species (and genus). However, this appears impractical in routine use, and a compromise using monophyletic groups and a number of arbitrarily defined, but well-established categories will probably remain in use for many years.

Few authors have questioned the monophyly of the phylum Arthropoda (see below), and its position in the 'classical' group Protostomia, closely related to the Annelida, has mostly been taken for granted. It is only during the past few decades that we have seen attempts to disrupt these groups. In the following discussion, the definitions of the phyla follow those given by Nielsen (1995); references have been kept to a minimum, and information about the original sources of the characters can be found in Nielsen (1995). The phylogenetic interrelationships taken as the starting point of the discussion are shown in Figure 2.1.

## 2.2 ARTHROPODA AS A MONOPHYLETIC GROUP

Most authors have treated the Arthropoda as a natural or monophyletic group. Almost 60 years ago, Snodgrass (1938) summarized his opinion about the interrelationships of annelids, onychophorans, arthropods, and most of the major arthropod groups in a diagram which is very similar to those seen in many modern textbooks, and which is congruent with the tree in Figure 2.1 (the tardigrades were not considered). However, Tiegs and Manton (1958), Manton (1977) and Anderson (1973) developed a 'polyphyletic' interpretation of the arthropods, because they found that the articulation and musculature of the legs of the various 'arthropod' groups were so different from each other that they could not be derived from a common ancestor which was also an arthropod. Tiegs and Manton (1958, p. 322) speculated that the common ancestor of arthropods and onychoporans could have been an annelid from which a number of arthropod groups had evolved; this would make the Arthropoda a grade rather than a clade. Manton (1972, 1977) more explicitly united Onychophora, Myriapoda and Insecta in the group Uniramia and regarded this and Crustacea, Trilobita and Chelicerata as phyla which had reached the arthropod grade independently. Anderson (1973) based much of his argument on embryological fate maps and proposed that Uniramia was derived from a clitellate-type annelid whereas the origin of the other 'arthropod' groups from non-annelid ancestors appeared more obscure. Fryer (1992, 1996) came out strongly in favour of arthropod polyphyly, but the essence of his argumentation is that 'arthropodization', i.e. the evolution of articulated limbs, may have taken place a number of times. The ancestors of the several arthropod lineages were described as various 'non-annelid worms with a haemocoel' (the chitinous cuticle was not discussed). The onychophorans were removed from the 'Uniramia Vera'

*Arthropod Relationships*, Systematics Association Special Volume Series 55, Edited by R.A. Fortey and R.H. Thomas. Published in 1997 by Chapman & Hall, London. ISBN 0 412 75420 7

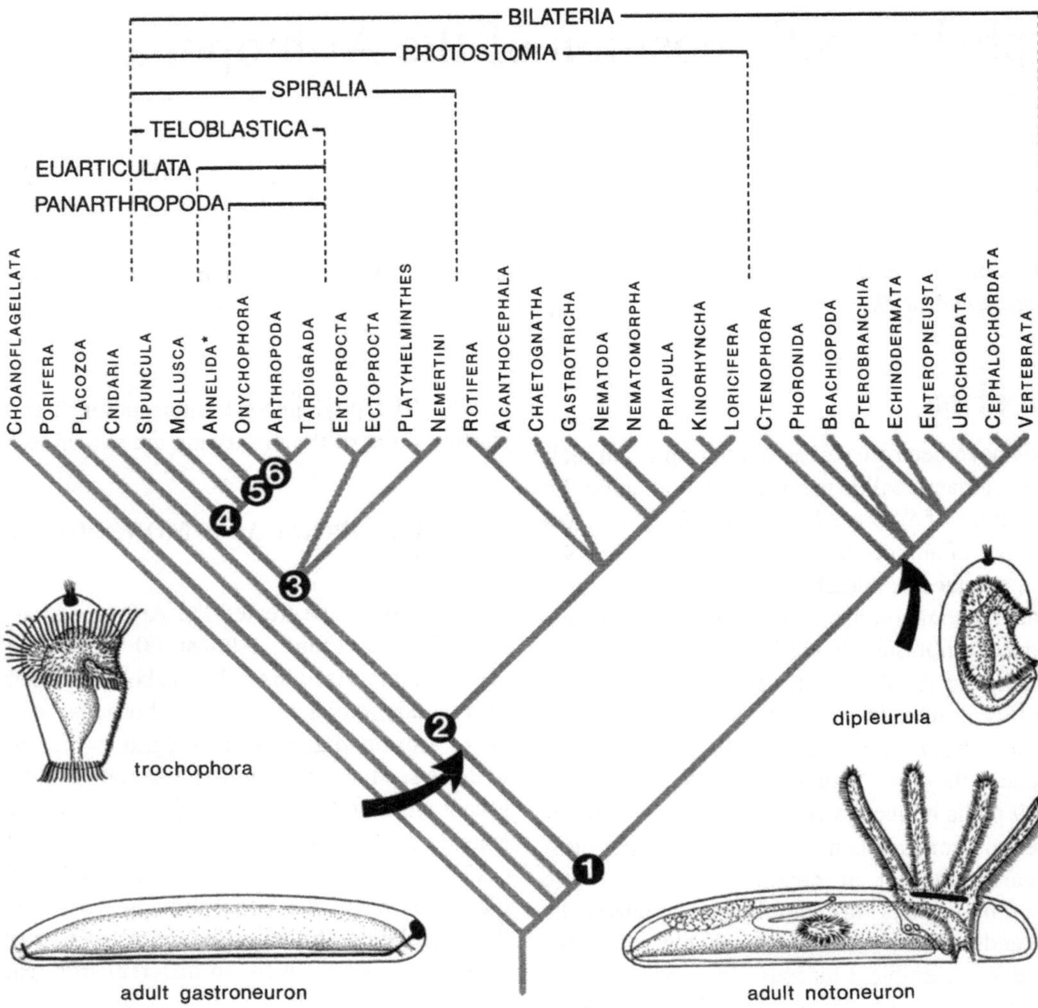


**Figure 2.1** Interrelationships of the animal phyla recognized in the discussion; the phylum Annelida includes Myzostomida, Pogonophora (incl. Vestimentifera), Lobatocerebridae, Gnathostomulida and Echiura. The following nodes are discussed in the text: Node 1. Bilateria: the split between Protostomia and Deuterostomia (the Ctenophora are not considered); Node 2. Protostomia: the split between Spiralia and Aschelminthes; Node 3. Spiralia: the interrelationships between the main spiralian groups are only briefly mentioned: Node 4. Teloblastica (= the coelomate protostomes): the interrelationships of Sipuncula, Mollusca and Euarticulata; Node 5. Euarticulata: the split between Annelida and Panarthropoda; Node 6. Panarthropoda: the interrelationships of Onychophora, Tardigrada and Arthropoda (= Euarthropoda). (Modified from Nielsen, 1994.)

(= Tracheata) but not discussed further; the tardigrades were not discussed at all. Fryer's diagram (1996, Figure 1) shows a number of 'arthropod' groups and their the non-arthropodized ancestors as the monophyletic sister group of the annelids, so in this case the 'polyphyly' is just a question of semantics.

However, all these authors disregarded the morphological evidence of, for example, the α-chitin-containing cuticle of the panarthropods in contrast to the collagenous cuticle of the other spiralians, including the annelids. It appears that their ideas are now giving way to a revival of the more traditional view of Arthropoda and Panarthropoda as monophyletic groups; this is supported by modern analyses of morphological data, also in combination with molecular data (Wheeler *et al.*, 1993).

The phylum Arthropoda can be seen as a monophyletic group characterized by: (i) their chitinous cuticle, appendages with claws, and mixocoel with metanephridia and ostiate heart (see below), characters shared with Onychophora and to some extent Tardigrada; and (ii) the unique segmented exoskeleton, articulated appendages with intrinsic musculature between individual joints, metanephridia without cilia, and probably also by a cephalon comprising a number of fused segments. The following sections discuss the position of the arthropods within supraphyletic groups of various levels.

## 2.3 BILATERIA: PROTOSTOMIA AND DEUTEROSTOMIA (FIGURE 2.1, NODE 1)

The arthropods are clearly bilateral animals, and the monophyly of the supraphyletic taxon Bilateria seems to be universally accepted (the inclusion of the Ctenophora is disputed). Bilateral symmetry is clearly recognizable in all protostomes and deuterostomes, but it has been questioned whether this feature had evolved in the common ancestor or independently in the ancestors of the two groups.

Protostomes typically show a blastopore which becomes divided into mouth and anus by fusion of the lateral blastopore lips. Their central nervous system (Figure 2.2) typically consists of an apical brain and a pair of ventral longitudinal nerves, which become fused in some types, and which develops along the lateral blastopore lips. These two characters are obviously coupled, so although the actual fusion of the blastopore lips has only been observed directly in a few representatives of annelids, onychophorans and nematodes, the embryological origin of the ventral nerve cords from a pair of longitudinal, ventral areas between mouth and anus can be observed in most species of protostomes, including insects with superficial cleavage and blastoderm.

Deuterostomes generally have a blastopore which becomes the adult anus while the mouth develops as a new opening. Their central nervous system develops from a shorter or longer dorsal epithelial area, and the apical organ of the larva is discarded (questionable in the chordates, see Lacalli *et al.*, 1994).

It therefore appears that the ventral areas of the two groups are not homologous: that of the protostomes is the elongated blastopore, while that of the deuterostomes is an area between apical organ and blastopore.

However, the old idea that the ventral side of annelids and insects corresponds to the dorsal side of vertebrates (Geoffroy Saint-Hilaire, 1822) has gained new interest because similar *Hox* genes appear to be involved in the patterning of the body plans of these apparently quite dissimilar groups (Nübler-Jung and Arendt, 1994; Lacalli, 1996). At first, this seems quite incompatible with the different orientation and origin of nervous systems in protostomes and deuterostomes, but it is possible that a further development of Lacalli's ideas can reconcile the two views.

If an advanced gastraea was creeping on the bottom, gathering detritus with the cilia around the blastopore (Figure 2.3), it could develop bilateral symmetry by fixing a ventral side with a nervous concentration comprising the apical organ, a frontal zone between apical organ and blastopore, and a blastoporal zone surrounding the blastopore. Two evolutionary directions could then lead to the two main bilaterian groups: Protostomia, which acquired a much elongated blastoporal zone giving rise to the ventral nerve cord(s), the frontal nervous zone becoming quite short; and Deuterostomia, in which the frontal zone became elongated, developed a new mouth surrounded by a new ciliary system, the blastoporal zone becoming insignificant. This idea links the 'neural' sides of protostomes and deuterostomes, as indicated by the homeobox genes, but it is still difficult to explain how the deuterostomes came to creep on the morphologically ventral side, which faced away from the substratum in the early ancestor. It is possible that the early deuterostome ancestors passed through a holoplanktonic stage, but this does not seem to be supported by observations of living forms.

Other differences between protostomes and deuterostomes include different larval types: trochophore types in protostomes and dipleurula types in deuterostomes (see Figure 2.1), and different origin and morphology of coeloms. Although the larval characters give strong support to the separation of protostomes and deuterostomes they will not be discussed here because the panarthropods lack ciliated primary larvae. The coelomate protostomes (Teloblastica: Figure 2.1, node 4) have varying numbers of coelomic sacs which develop in mesodermal bands originating from mesoteloblasts derived from the 4d-cell in the spiral cleavage or as ectomesoderm (never from the endoderm). Multiciliate cells are the main type in the ciliated epithelia. Conversely, the deuterostomes have three pairs of coeloms, protocoel, mesocoel and metacoel, with the mesoderm originating from the wall of the archenteron. The three pairs of coelomic sacs are easily visible in the non-chordate deuterostomes, and various vestiges of the sacs or associated structures can be identified in the chordates. The mesocoel has long tubular channels extending into the ciliated tentacles of phoronids, brachiopods and pterobranchs, and into the tube feet of the echinoderms; these phyla show only monociliate cells. Thus, the two types of coeloms show very different origin and structure, and it is very difficult to interpret them as homologous, i.e. that one type should be derived from the other.

## 2.4 PROTOSTOMIA: SPIRALIA AND ASCHELMINTHES (FIGURE 2.1, NODE 2)

The phylogeny of the protostomes suggested in Figure 2.1 shows Spiralia and Aschelminthes as sister groups, but the monophyly of the latter group is not strongly supported and was not substantiated by analyses using parsimony programs such as PAUP (Nielsen *et al.*, 1996). However, the monophyly of the Spiralia was substantiated, with Ectoprocta as the phylum with the most weakly supported position.

The characteristic spiral cleavage pattern with alternatingly oblique mitotic spindles and the origin of organs such

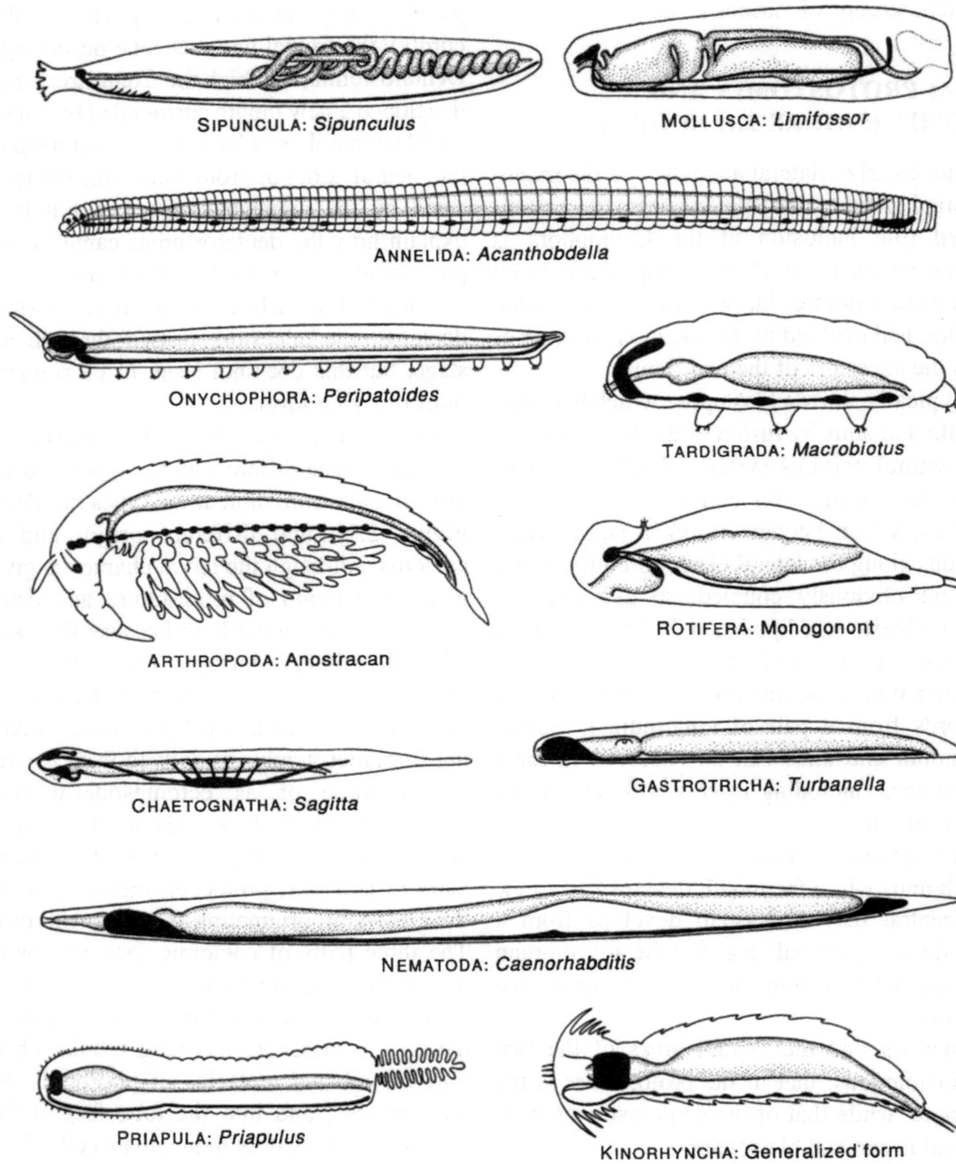


**Figure 2.2** Protostomian nervous systems. Representatives of most protostomian phyla are drawn in left view and the brain with the central nervous system shown in black. (From Nielsen, 1994.)

as prototroch and mesoderm from blastomeres of identical position has been reported in representatives of Platyhelminthes, Nemertini, Entoprocta, Sipuncula, Mollusca and Annelida. The cleavage of ectoproct eggs can perhaps be interpreted as spiral, but further investigations are needed. The 'characteristic' pattern is usually found in species with small eggs and planktotrophic development. However, there is a considerable variation in cleavage patterns within most phyla, and large groups, such as cephalopods, lack any trace of spiral cleavage; most authors interpret this as specializations from the ancestral pattern of the phylum, for example in connection with large amounts of yolk. Not only the spiral pattern but also general protostomian characteristics such as the fate of the blastopore and the origin of the mesoderm can be modified both in single species or large taxa, so that the absence of a spiralian or protostomian character in a group cannot be taken as a proof that the taxon does not belong to Spiralia or Protostomia. A good example is the prosobranch *Viviparus*, which has a blastopore which becomes the adult anus and mesoderm which apparently originates exclusively from ectomesoderm.

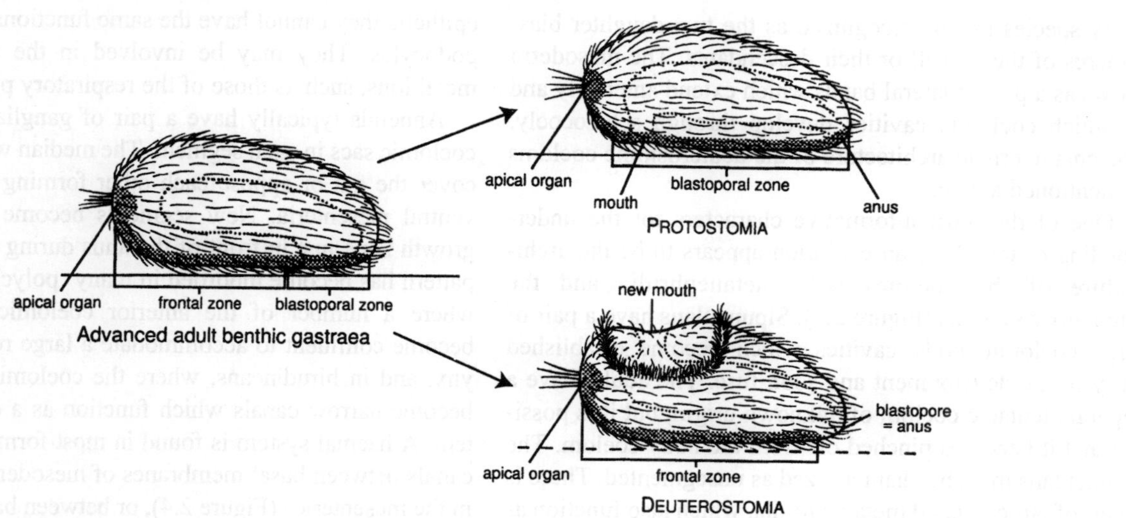


**Figure 2.3** The evolution of early protostome and deuterostome ancestors from a bilaterally symmetrical ancestor of the gastraea type. The nervous systems are shaded.

Arthropod cleavage has been interpreted in various ways; none of the species studied shows an unequivocal spiral pattern, and primary ciliated larvae are not known in any arthropod. Some cirripedes with rather small eggs have total cleavage and a four-cell stage with three smaller cells, which could be interpreted as the A–C cells of the spiral cleavage, and a larger D-cell. The following cleavages do not show a clear spiral pattern, and the fates of the blastomeres are not clearly spiral either. It is perhaps best to conclude that the cleavage of these cirripedes may be interpreted as spiral, but that the pattern is not so characteristic that it can be used as a certain character in a cladistic analysis. The following section will nevertheless show that arthropods share a number of highly complicated characters with other coelomate protostomes, so the uncertainty about the cleavage pattern does not disrupt the argument.

## 2.5 SPIRALIA: TELOBLASTICA (= COELOMATE PROTOSTOMES) AND THEIR ALLIES (FIGURE 2.1, NODE 3)

The interrelationships of the phyla here assigned to the Spiralia are not fully resolved. There is one clearly monophyletic group, Teloblastica, characterized by teloblastic growth of the mesoderm from 4d-cells (not seen in arthropods), coelomic cavities formed through schizocoely, and metanephridia. (It should be emphasized that the larvae of many teloblasticans have protonephridia, like those of the non-coelomate protostomes, and that the metanephridia/gonoducts are adult characters.) The remaining four phyla assigned to Spiralia in Figure 2.1, Platyhelminthes, Nemertini, Entoprocta, and Ectoprocta, are among the phyla which show the least stability in the cladistic analyses by Nielsen *et al.* (1996), but none of them has coelomic cavities with metanephridia.

The entoprocts seem to be the phylum with the most easily assessed characters: their cleavage is clearly spiral with prototroch cells originating from the characteristic blastomeres and mesoderm originating from the 4d-cell, and the larvae are almost schematic trochophores; there is no coelom and the adults have protonephridia. It seems clear that the entoprocts are neotenic with a highly shortened longitudinal axis, and therefore with no longitudinal ventral nerve cord, related to their sessile habits.

The phylogenetic position of the ectoprocts is quite enigmatic, but it can at least be stated that the morphological characters believed to support close relationships with brachiopods and phoronids have turned out to be misinterpreted. At the present stage of knowledge, they probably can best be fitted onto the phylogenetic tree as the sister group of the entoprocts because of similarities in metamorphoses of representatives of the two groups. More detailed investigations, especially of embryology and larval morphology, could perhaps reveal more specific similarities with the entoprocts.

Some platyhelminths and nemertines have spiral cleavage, but their larval types are not easily comparable to trochophores. They will not be discussed further here.

## 2.6 TELOBLASTICA: SIPUNCULA, MOLLUSCA AND EUARTICULATA (FIGURE 2.1, NODE 4)

The coelomate protostomes (called Teloblastica by Nielsen, 1995) share a number of advanced characters which set them apart from the non-coelomate protostomes and also from the deuterostomes, which are all coelomate. The mesoderm differentiates from a pair of teloblast cells, which in many species can be recognized as the two daughter blastomeres of the 4d-cell or their descendants. The mesoderm grows as a pair of lateral bands which extend anteriorly and in which coelomic cavities develop through schizocoely. The characteristic architecture of the deuterostome coeloms is mentioned above.

One of the most informative characters for the understanding of teloblastican evolution appears to be the architecture of the coeloms with metanephridia and the circulatory systems (Figure 2.4). Sipunculans have a pair of large coelomic body cavities which become established early in the development and soon fuse. The adults have a separate tentacle coelom of unknown origin, but it is possible that it becomes pinched off from the body coelom. The sipunculans must be characterized as unsegmented. There is a pair of large ciliated metanephridia, which also function as gonoducts (Figure 2.4). There appears to be no ultrafiltration of primary urine, and the sipunculans lack any trace of a haemal system (i.e. a fluid-filled circulatory system of canals lined by basal membranes, as opposed to a coelom which is lined by an epithelium). Circulation may be furthered by specializations of the tentacle coelom or by coelomic canals in the body wall (Ruppert and Rice, 1995). The tentacle coelom of some species have a long side branch which extends along the descending branch of the gut. This structure is found, for example, in species which bore into hard substrates, and the gas exchange with the surroundings is probably mainly restricted to the tentacle crown. Their tentacle coelom may contain red haemerythrin and the vessel superficially resembles a blood vessel.

Molluscs have the coelomic cavities restricted to the pericardium. There are strong indications of eight segments in the musculature of monoplacophorans and polyplacophorans, and if this is interpreted as true segmentation, it reflects a fixed number of segments, with no addition of new segments from a posterior growth zone. There is typically a haemal system with a dorsal heart, surrounded by the pericardium, and various more or less well defined blood vessels and lacunae. The pericardium is a coelomic sac whose inner wall forms the musculature of the heart; it is connected through a pair of ciliated ducts to a pair of 'kidneys', which are in many cases connected with the gonads (Figure 2.4). The blood vessels leading from the ctenidia form atria or auricles in the pericardium, and their wall (which is a part of the pericardial epithelium) has extensive podocyte areas where primary urine is formed through ultrafiltration from the blood to the pericardial cavity (Andrews and Taylor, 1988). The pericardium is thus a coelomic cavity with primary urine, which becomes modified through a metanephridium, called the ciliated duct, and its duct (the 'kidney'). The rhogocytes reviewed by Haszprunar (1996) are solitary or clustered cells with podocyte-like surface areas. They occur in various positions in the haemocoel or embedded in connective tissue, but since they do not form epithelia they cannot have the same functions as the normal podocytes. They may be involved in the metabolism of metal ions, such as those of the respiratory pigments.

Annelids typically have a pair of ganglia and a pair of coelomic sacs in each segment. The median walls of the sacs cover the gut or appose each other forming the dorsal and ventral mesenteria. New segments become added from a growth zone just in front of the anus during ontogeny. This pattern has become modified in many 'polychaete' families, where a number of the anterior coelomic cavities have become confluent to accommodate a large retractable pharynx, and in hirudineans, where the coelomic cavities have become narrow canals which function as a circulatory system. A haemal system is found in most forms; it consists of canals between basal membranes of mesoderm, for example in the mesenteries (Figure 2.4), or between basal membranes of peritoneum and gut. The large longitudinal vessels have areas with podocytes where primary urine is formed from the blood through ultrafiltration. Many polychaetes have a pair of ciliated metanephridia in each segment, which modify the primary urine and often also function as gonoducts.

Onychophorans, arthropods and tardigrades are conspicuously segmented, with limbs on most segments and a pair of ganglia in all segments. Embryos of some onychophorans and arthropods show the typical development of coelomic cavities in lateral mesodermal bands and the addition of new segments from a posterior growth zone. It appears that each of these sacs pinches off one small compartment connected with a coelomoduct, and that each of these units becomes an excretory organ, a salivary gland, or connected with the gonads. The coelomic compartment of the excretory organs comprises areas with podocytes where primary urine can be formed through ultrafiltration from the mixocoel (the sacculi; see Figure 2.4); the coelomoduct, which is ciliated in onychophorans and without cilia in arthropods, functions as a kidney (and as a gonoduct in the gonads) and is clearly a modified metanephridium. The remaining part of the coelomic sacs becomes more diffuse and their cavity fuses with the primary body cavity to form a mixocoel, with the ostiate heart derived from coelomic walls. The embryology of tardigrades is very imperfectly known, but it can be stated that earlier reports of formation of mesodermal sacs through enterocoely were based on misinterpretations. The panarthropods are discussed in more detail below, but the characters of coelom and excretory organs of onychophorans and arthropods are clearly modifications of the structures seen in annelids.

The morphology and ontogeny of coeloms and nephridia in the teloblastican phyla indicate a transformation series (Figure 2.4). The sipunculans have no haemal system and the metanephridia modify the coelomic fluid directly without an ultrafiltration system; they also function as gonoducts, and this may give support to the old idea that coeloms evolved as gonads (see for example Goodrich, 1946).

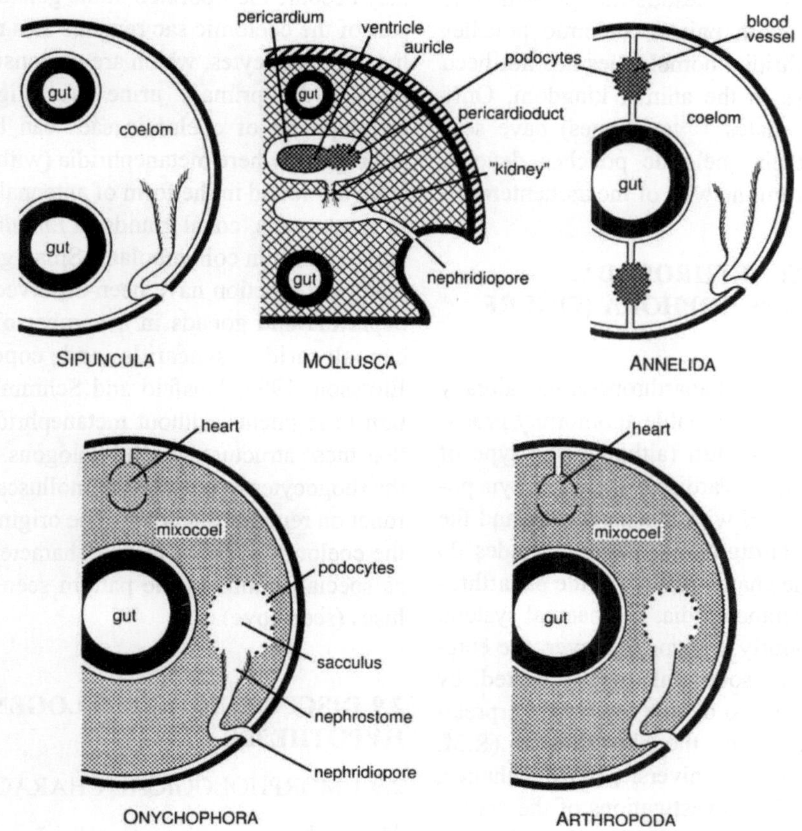


**Figure 2.4** Coelom, metanephridia and haemal systems in the coelomate protostomes (Teloblastica). The mollusc is based on Andrews and Taylor (1988) and Ruppert and Barnes (1994).

Molluscs, annelids and panarthropods have haemal systems which are involved both in circulation and excretion. Molluscs and annelids have ultrafiltration of primary urine from blood vessels to large coelomic cavities and metanephridia which modify the coelomic fluid to urine. Panarthropods have extensive mixocoels and ultrafiltration is restricted to the sacculi, which surround narrow coelomic spaces. The metanephridia function as nephridia in certain segments and as gonads in others; they have retained the cilia in onychophorans, but the cilia have been lost in the arthropods. This is in full agreement with the phylogeny shown in Figure 2.1, node 4.

## 2.7 EUARTICULATA: ANNELIDA AND PANARTHROPODA (FIGURE 2.1, NODE 5)

The addition of segments with coelomic compartments from a posterior teloblastic growth zone during an extended period of the individuals' life is a characteristic only known from annelids, onychophorans and arthropods. Kinorhynchs, such as *Paracentrophyes* (Neuhaus, 1995), hatch with 11 segments and add two segments from a zone anterior to the posteriormost segment during a series of moults, but coelomic compartments, metanephridia or teloblasts have never been reported (the early development is unknown)(see further discussion below).

The addition of new segments from the zone just in front of the pygidium is a conspicuous feature of many adult annelids, but families such as Pectinariidae and Sternaspidae and all the hirudineans attain a species-specific number of segments at an early stage.

Onychophoran embryos add mesoderm from a posterior growth zone, and mesodermal compartments with coelomic sacs become separated from the anterior side of the mesodermal bands, so that the adult number of segments is reached before hatching.

Arthropod segmentation shows much variation. Many crustaceans with pelagic larvae hatch as nauplius larvae with three pairs of appendages, and segments with or without appendages are then added from the growth zone in front of the telson through a number of moults until the fixed adult number is reached.

The tardigrades have a fixed number of segments; their development is poorly known.

The addition from a posterior mesodermal growth zone of well-defined segments with paired coelomic pouches with metanephridia or nephridial homologues has not been encountered anywhere else in the animal kingdom. Only notochordates (cephalochordates + vertebrates) have segmented mesoderm, but their coelomic pouches develop through enterocoely, i.e. from the wall of the archenteron.

## 2.8 PANARTHROPODA: ARTHROPODA, TARDIGRADA AND OCYCHOPHORA (FIGURE 2.1, NODE 6)

The monophyletic nature of the Panarthropoda has already been mentioned above, the most reliable apomorphy probably being the cuticle with α-chitin (although the type of chitin has not been studied in Tardigrada). Other synapomorphies include the mixocoel with metanephridia and the ostiate heart (not seen in tardigrades). The tardigrades do indeed lack a number of the characteristics of the panarthropods, such as mixocoel, metanephridia, and haemal system, and their embryology is poorly known. However, the enterocoelous formation of mesodermal sacs reported by Marcus (1929) has turned out to de due to a misinterpretation, probably of precursors of the claw glands (R.M. Kristensen, Zoological Museum, University of Copenhagen, personal communication). The investigations of the central nervous system by Dewel and Dewel (1996, 1997, this volume) clearly indicate that the tardigrades must be interpreted as the sister group of Arthropoda, with Onychophora as the first outgroup.

A chitinous cuticle which is moulted periodically is also found in three aschelminth phyla, viz. Priapula, Kinorhyncha and Loricifera (Neuhaus *et al.*, in press) which are now by most authors regarded as a monophyletic group [Cephalorhyncha (Nielsen, 1995) or Scalidophora (Ehlers *et al.*, 1996), see also Neuhaus (1994)]; this is discussed further below.

The ontogeny of mesoderm and coeloms in onychophorans and arthropods has been studied by several authors, and it appears that a pair of coelomic sacs develop in the ventral part of each segment, and expand to cover almost the whole of the ectoderm and the gut, but that the walls of the sacs break up and give rise to the heart, the musculature, and in certain segments to nephridia/gonads. Onychophorans show the development quite clearly (Snodgrass, 1938; new investigations using modern microscopical methods are needed): Each coelomic sac, which is connected to a ciliated coelomoduct, breaks up and a small part becomes a sacculus surrounding the internal opening of the coelomoduct while the remaining part becomes individual muscles; parts of the coelomic space become confluent with the haemal space and a mixocoel is formed. In segments with a gonad the coelomoduct becomes the gonoduct, and coelomic compartments from a number of segments may become incorporated in the gonad. In other segments, a part of the coelomic sac remains as a metanephridial sacculus with podocytes, which are responsible for the ultrafiltration of the primary urine (see Figure 2.4). A similar development of coelomic sacs can be observed in some arthropods, where metanephridia (without cilia in the funnel area) are found in the form of antennal and maxillary glands in crustaceans, coxal glands in *Limulus* and arachnids, and labial glands in collembolans. Small groups of podocytes of unknown function have been observed in segments without nephridia and gonads in a number of crustaceans such as cephalocarids, syncarids and copepods (Hessler and Elofsson, 1995; Hosfeld and Schminke, 1997); their position in segments without metanephridial structures suggest that these structures are homologous. These cells resemble the rhogocytes reported from molluscs (see above), but their function remains unknown. The origin and differentiation of the coelomic sacs are highly characteristic and can be seen as specializations of the pattern seen in annelids and molluscs (see above).

## 2.9 DISCUSSION OF PHYLOGENETIC HYPOTHESES

### 2.9.1 MORPHOLOGICAL CHARACTERS

The phylogeny outlined above is based exclusively on morphological characters and on cladistic reasoning without the use of computer programs. The analyses of Nielsen *et al.* (1996) used parsimony programs such as Hennig 86 and PAUP on a data matrix extracted from Nielsen (1995) and came to results compatible with most details of the tree shown in Figure 2.1.

Numerous classifications differing to various extents from these results are found in articles and textbooks, but few are based on cladistic methodology; the discussion will concentrate on those relating to the coelomate protostomes, and especially to the groups Annelida and Arthropoda.

The first computer-generated analysis including most of the animal phyla was that of Brusca and Brusca (1990); it excluded the aschelminths, but the phylogenetic tree showed relationships between coelomate protostomes almost identical to those in Figure 2.1. Meglitsch and Schram (1991), Schram (1991) and Schram and Ellis (1994) obtained trees with the aschelminths as the sister group of the remaining bilaterians and monophyletic groups corresponding to Teloblastica and Panarthropoda; however, the panarthropods came out as sister group of annelids + sipunculans + molluscs. Eernisse *et al.* (1992) analysed a selection of animal phyla in an attempt to elucidate the positions of annelids and arthropods. They concluded that the panarthropods are monophyletic with kinorhynchs and nematodes as the two first outgroups, and that annelids + molluscs + sipunculans (eutrochozoans; the name had been coined earlier by Ghiselin (1988) on the basis of combined studies of morphology and molecules), together with nemertines and flatworms form another clade. Unfortunately, their list of characters show several ambiguities and the data matrix shows a considerable number of errors, so the results must be regarded with suspicion. Also Ruppert and Barnes (1994, Figure 20.6) mention the kinorhynchs as a possible sister group of the arthropods, and the moulting of a chitinous, segmented cuticle is of course an obvious similarity between the two groups. However, there is nothing to indicate that kinorhynchs are descended from ancestors with spiral cleavage, teloblastic mesoderm, coelom, and metanephridia, so they can hardly be interpreted as 'degenerate' panarthropods. The other possibility for a sister group relationship implies that the type of mesoderm formation, coelomic compartments with metanephridia, and haemal systems of annelids and arthropods should represent homoplasies, which appears equally unlikely.

Conway Morris and Peel (1995) presented the surprising idea that molluscs, annelids and brachiopods are closely related, based on a supposed homology of sclerites of *Halkieria*, spicules or hairs of chitons and chaetae (setae) of annelids and brachiopods; they placed the arthropods outside of this lineage. However, it should be noted that the spicules or hairs of the chitons are grooved chitinous structures formed from a number of cells arranged in a horseshoe and containing a medulla with sensory cells (Leise, 1988). They bear little resemblance to the chaetae of annelids and molluscs, which are each formed by one characteristic chaetoblast (Storch and Welsch 1972; O'Clair and Cloney 1974). The similarities between halkieriid sclerites and annelid, especially amphinomid, chaetae is not supported by their calcification; the last reference cited by Conway Morris and Peel (1995, p. 351; Schroeder, 1984, p. 304) directly states that 'despite frequent assertions that the chaetae are calcareous, ... no data on their composition appear to have been published'. There seem to be no other observations supporting Conway Morris and Peel's phylogeny.

Fossil forms have generally not been discussed above because much of the phylogeny is based on characters from embryology, nervous systems and other structures which can only be studied in living forms. However, it should be mentioned that the terrestrial onychophorans appear quite distant in time from their presumed aquatic ancestors, but that this gap seems to be bridged satisfactorily by fossil, aquatic lobopodians (Hou and Bergström, 1995).

## 2.9.2 MOLECULAR CHARACTERS

Comparisons of gene sequences of ribosomal RNA/DNA and mitochondrial DNA were introduced as measures of genetic propinquity of taxa about a decade ago. The expectations have been very high because these characters are present in all metazoan cells and have been presumed to be complex enough to make convergence unlikely. A number of contrasting phylogenetic trees have been published, and most of the authors have stressed that the topologies presented for the basal portion of the tree are not strongly supported. However, many zoologists and palaeontologists have rushed to accept at least some of the new results. It is very difficult to compare the results from the several papers involved because very different selections of species are used. The magnitude of the uncertainty about the results can be illustrated by three examples.

1. A monophyletic group 'Diploblastica' comprising poriferans, *Trichoplax*, cnidarians, and ctenophores is shown in the trees of for example, Philippe *et al.* (1994), Winnepenninckx *et al.* (1995b) and Garey *et al.* (1996), whereas topologies resembling the more conventional one shown in Figure 2.1 are presented for example by Wainright *et al.* (1993), Rodrigo *et al.* (1994) and Halanych (1996). No morphological analysis supports the 'Diploblastica'.
2. Bilateria (excluding Ctenophora) is presented as monophyletic in almost all studies, but there is much disagreement about the first pair of sister groups: no papers show the traditional Protostomia–Deuterostomia split. The nematodes are presented as the sister group of the remaining bilaterians by Winnepenninckx *et al.* (1995b) and Garey *et al.* (1996), whereas the platyhelminths are the sister group in the studies of Ghiselin (1988) and Mackey *et al.* (1996); only very few papers include sequences from both nematodes and flatworms.
3. The interrelationships of arthropods and the other coelomate protostomes depicted in Figure 2.1 are not shown in any of the molecular trees. Arthropods are shown as the sister group of an assemblage consisting of molluscs and annelids + sipunculans + brachiopods (resembling the Eutrochozoa, see above) by Ghiselin (1988), and Turbeville *et al.* (1991) also show the arthropods as sister group of annelids + molluscs (one species of each). Adoutte and Philippe (1993) show arthropods as the sister group of molluscs, annelids and a brachiopod, but neither annelids nor molluscs are monophyletic on the tree. Winnepenninckx *et al.* (1995a) show the arthropods as the sister group of deuterostomes plus many of the traditional protostomian phyla, whereas Winnepenninckx *et al.* (1995b) show arthropods as the sister group of nematomorphs + priapulans and this assemblage as sister group of most of the other traditional protostomes.

An alternative approach, viz. comparisons of the arrangement of genes along the mtDNA, has revealed unexpectedly large variations within phyla compared with among phyla (Boore and Brown, 1994), so these investigations have not –

so far – contributed to our understanding of animal phylogeny. Other large molecules present in all eukaryotes have been sequenced, e.g. elongation factor-1α, but the number of species investigated is small and the resolution at the level under discussion is low (Kobayashi *et al.*, 1996).

The inevitable conclusion is that the molecular data have so far **not** provided any unequivocal answers to questions about the early evolution of metazoans.

The early animal (metazoan) diversification is usually believed to have taken place during a quite short period, about 540–520 million years ago, during the so-called Cambrian evolutionary explosion (but see Fortey *et al.*, 1996). This is supported by the absence of recognizable fossils belonging to phyla still in existence in older deposits and the apparently sudden appearance of representatives of many modern phyla in the early Cambrian. The question is of course whether the analyses of various macromolecules can resolve the speciation events that took place during a supposedly short period. Philippe *et al.* (1994) analysed the reliability of results obtained by the use of the available 18S rRNA database and came to the conclusion that the length of the molecule restricts the reliable resolving power to about 40 million years, and that the combined 18S and 28S rDNA dataset improves the resolution to about 19 million years. This must definitely be borne in mind when molecular phylogenies are assessed.

Finally, it should be mentioned that homeobox genes and their expressions are likely to possess information both about the early diversification of the animals (Slack *et al.*, 1993) and about the relationships at more detailed systematic levels (see for example Popadic *et al.*, 1996).

## 2.10 CONCLUSIONS

The phylogenetic position of the Arthropoda can be inferred through analyses of their morphological and embryological characters, and the supraphyletic groups Panarthropoda (Figure 2.1, node 6), Euarticulata (Figure 2.1, node 5) and Teloblastica (Figure 2.1, node 4) are based on a series of well-studied characters. No sufficiently documented alternatives have been found. In addition, the position of the Teloblastica as a group of Spiralia, and its position within the Protostomia, seem to be well supported, despite lack of agreement in the literature.

Phylogenetic trees built on the basis of morphological characters provide predictions which can be checked, so that the stability of the hypotheses can be tested. So far, the molecular analyses have **not** given us the unambiguous phylogenies that many systematists had expected, and trees built on molecular sequence data give no information about the morphology of ancestral forms. At this point, it appears that molecular data may provide ideas which can inspire new morphological work, but that it is too early to replace the morphology-based cladistic classification with a 'molecular tree'.

ACKNOWLEDGEMENTS

My colleague Dr Danny Eibye-Jacobsen is thanked for constructive criticism and linguistic help.

## REFERENCES

Adoutte, A. and Philippe, H. (1993) The major lines of metazoan evolution: summary of traditional evidence and lessons from ribosomal RNA sequence analysis, in *Comparative Molecular Neurobiology* (ed. Y. Pichon), Birkhäuser, Basel, pp. 1–30.

Anderson, D.T. (1973) *Embryology and Phylogeny of Annelids and Arthropods*, International Series of Monographs in Pure and Applied Biology, Zoology 50, Pergamon Press, Oxford.

Andrews, E.B. and Taylor, P.M. (1988) Fine structure, mechanism of heart function and haemodynamics in the prosobranch mollusc *Littorina littorea* (L.). *Journal of Comparative Physiology B*, **158**, 247–62.

Ax, P. (1995) *Das System der Metazoa*, vol. 1, Gustav Fischer, Stuttgart.

Boore, J.L. and Brown, W.M. (1994) Mitochondrial genomes and the phylogeny of the mollusks. *The Nautilus*, Supplement 2, 61–78.

Brusca, R.C. and Brusca, G.J. (1990) *Invertebrates*, Sinauer Associates, Sunderland.

Conway Morris, S. and Peel, J.S. (1995) Articulated halkieriids from the Lower Cambrian of North Greenland and their role in early protostome evolution. *Philosophical Transactions of the Royal Society B*, **347**, 305–58.

Dewel, R.A. and Dewel, W.C. (1996) The brain of *Echiniscus viridissimus* Peterfi, 1956 (Heterotardigrada): a key to understanding the phylogenetic position of tardigrades and the evolution of the arthropod head. *Zoological Journal of the Linnean Society*, **116**, 35–49.

Eernisse, D.J., Albert, J.S. and Andersen, F.E. (1992) Annelida and Arthropoda are not sister taxa: A phylogenetic analysis of spiralian metazoan morphology. *Systematic Biology*, **41**, 305–30.

Ehlers, U., Ahlrichs, W., Lemburg, C. and Schmidt-Rhaesa, (1996) Phylogenetic systematization of the Nemathelminthes (Aschelminthes). *Verhandlungen der Deutschen zoologischen Gesellschaft*, **89**, 8.

Fortey, R.A., Briggs, D.E.G. and Wills, M.A. (1996) The Cambrian evolutionary 'explosion': decoupling cladogenesis from morphological disparity. *Biological Journal of the Linnean Society*, 57, 13–33.

Fryer, G. (1992) The origin of the Crustacea. *Acta Zoologica (Stockholm)*, **73**, 273–86.

Fryer, G. (1996) Reflections on arthropod evolution. *Biological Journal of the Linnean Society*, **58**, 1–55.

Garey, J.R., Krotec, M., Nelson, D.R. and Brooks, J. (1996) Molecular analysis supports a tardigrade–arthropod association. *Invertebrate Biology*, **115**, 79–88.

Geoffroy Saint-Hilaire, E. (1822) Considérations générales sur la vertèbre. *Mémoires du Muséum d'Histoire Naturelle*, **9**, 89–119.

Ghiselin, M.T. (1988) The origin of molluscs in the light of molecular evidence. *Oxford Surveys in Evolutionary Biology*, **5**, 66–99.

Goodrich, E.S. (1946) The study of nephridia and genital ducts since 1895. *Quarterly Journal of Microscopical Science*, N.S., **86**, 113–392.

Haeckel, E. (1866) *Generelle Morphologie der Organismen. 2 vols*, Georg Reimer, Berlin.

Halanych, K.M. (1996) Convergence in the feeding apparatuses of lophophorates and pterobranch hemichordates revealed by 18S rDNA: an interpretation. *Biological Bulletin* (*Woods Hole*), **190**, 1–5.

Haszprunar, G. (1996) The molluscan rhogocyte (pore-cell, Blasenzelle, cellule nucale), and its significance for ideas on nephridial evolution. *Journal of Molluscan Studies*, **62**, 185–211.

Hessler, R.H. and Elofsson, R. (1995) Segmental podocytic excretory glands in the thorax of *Hutchinsoniella macracantha* (Cephalocarida). *Journal of Crustacean Biology*, **15**, 61–9.

Hosfeld, B. and Schminke, H.K. (1997) Discovery of segmental extranephridial podocytes in Harpacticoida (Copepoda) and Bathynellacea (Syncarida). *Journal of Crustacean Biology*, **17**, 13–20.

Hou, X. and Bergström, J. (1995) Cambrian lobopodians – ancestors of extant onychophorans? *Zoological Journal of the Linnean Society*, **114**, 3–19.

Kobayashi, M., Wada, H. and Satoh, N. (1996) Early evolution of the Metazoa and phylogenetic status of diploblasts as inferred from amino acid sequence of elongation factor-1α. *Molecular Phylogenetics and Evolution*, **5**, 414–22.

Lacalli, T.C. (1996) Dorsoventral axis inversion: a phylogenetic perspective. *BioEssays*, **18**, 251–4.

Lacalli, T.C., Holland, N.D. and West, J.E. (1994) Landmarks in the anterior central nervous system of amphioxus larvae. *Philosophical Transactions of the Royal Society B*, **344**, 165–85.

Leise, E.M. (1988) Sensory organs in the hairy girdles of some mopaliid chitons. *American Malacological Bulletin*, **6**, 141–51.

Linnaeus, C. (1735) *Systema Naturæ sive Regna Tria Naturæ systematice proposita per Classæ, Ordines, Genera, & Species*, Theod. Haack, Lugdunum Batavorum.

Linnaeus, C. (1758) *Systema Naturæ, 10th edition. 10 vols*, Laurentius Salvius, Stockholm.

Mackey, L.Y., Winnepenninckx, B., De Wachter, R., Backeljau, T., Emschermann, P. and Garey, J.R. (1996) 18S rRNA suggests that Entoprocta are protostomes, unrelated to Ectoprocta. *Journal of Molecular Evolution*, **42**, 552–9.

Manton, S.M. (1972) The evolution of arthropodan locomotory mechanisms. Part 10. Locomotory habits, morphology and evolution of the hexapod classes. *Zoological Journal of the Linnean Society*, **51**, 203–400.

Manton, S.M. (1977) *The Arthropoda. Habits, Functional Morphology, and Evolution*, Oxford University Press, Oxford.

Marcus, E. (1929) Zur Embryologie der Tardigraden. *Zoologische Jahrbücher, Anatomie*, **50**, 333–84.

Meglitsch, P.A. and Schram, F.R. (1991) *Invertebrate Zoology*, Oxford University Press, New York.

Neuhaus, B. (1994) Ultrastructure of alimentary canal and body cavity, ground pattern, and phylogenetic relationships of the Kinorhyncha. *Microfauna Marina*, **9**, 61–156.

Neuhaus, B. (1995) Postembryonic development of *Paracentrophyes praedictus* (Homalorhagida): neoteny questionable among the Kinorhyncha. *Zoologica Scripta*, **24**, 179–92.

Neuhaus, B., Kristensen, R.M. and Peters, W. Ultrastructure of the cuticle of Loricifera and electron microscopical demonstration of chitin using gold-labelled wheat germ agglutinin. *Acta Zoologica* (*Stockholm*) (in press)

Nielsen, C. (1994) Larval and adult characters in animal phylogeny. *American Zoologist*, **34**, 492–501.

Nielsen, C. (1995) *Animal Evolution: Interrelationships of the Living Phyla*, Oxford University Press, Oxford.

Nielsen, C., Scharff, N. and Eibye-Jacobsen, D. (1996) Cladistic analyses of the animal kingdom. *Biological Journal of the Linnean Society*, **57**, 385–410.

Nübler-Jung, K. and Arendt, D. (1994) Is ventral in insects dorsal in vertebrates? *Roux's Archives of Developmental Biology*, **203**, 357–66.

O'Clair, R.M. and Cloney, R.A. (1974) Patterns of morphogenesis mediated by dynamic microvilli: chaetogenesis in *Nereis vexillosa*. *Cell and Tissue Research*, **151**, 141–57.

Philippe, H., Chenuil, A. and Adoutte, A. (1994) Can the Cambrian explosion be inferred through molecular phylogeny? *Development*, 1994 suppl., 15–25.

Popadic, A., Rusch, D., Peterson, M., Rogers, B.T. and Kaufman, T.C. (1996) Origin of the arthropod mandible. *Nature*, **380**, 395.

Rodrigo, A.G., Bergquist, P.R. and Bergquist, P.L. (1994) Inadequate support for an evolutionary link between the Metazoa and the Fungi. *Systematic Biology*, **43**, 578–84.

Ruppert, E.E. and Barnes, R.D. (1994) *Invertebrate Zoology*, 6th edn, Saunders College Publishing, Fort Worth.

Ruppert, E.E. and Rice, M.E. (1995) Functional organization of dermal coelomic canals in *Sipunculus nudus* with a discussion of respiratory designs in sipunculans. *Invertebrate Biology*, **114**, 51–63.

Schram, F.R. (1991) Cladistic analysis of metazoan phyla and the placement of fossil problematica, in *The Early Evolution of Metazoa and the Significance of Problematic Taxa* (eds A.M. Simonetta and S. Conway Morris), Cambridge University Press, Cambridge, pp. 35–46.

Schram, F.R. and Ellis, W.E. (1994) Metazoan relationships: a rebuttal. *Cladistics*, **10**, 331–7.

Schroeder, P.C. (1984) Chaetae, in *Biology of the Integument*, vol. 1 (eds J. Bereiter-Hahn, A.G. Matoltsy and K.S. Richards), Springer, Berlin, pp. 297–309.

Siebold, C.T. (1848) *Lehrbuch der vergleichenden Anatomie der wirbellosen Thiere* (vol.1 of C. T. von Siebold & H. Stannius: Lehrbuch der vergleichenden Anatomie), v. Veit & Co., Berlin.

Slack, J.W.M., Holland, P.W.H. and Graham, C.F. (1993) The zootype and the phylotypic stage. *Nature*, **361**, 490–2.

Snodgrass, R.E. (1938) Evolution of the Annelida, Onychophora, and Arthropoda. *Smithsonian Miscellaneous Collections*, **97**(6), 1–159.

Storch, V. and Welsch, U. (1972) Über Bau und Entstehung der Mantelstacheln von *Lingula unguis* L. (Brachiopoda). *Zeitschrift für wissenschaftliche Zoologie*, **183**, 181–9.

Sundberg, P. and Pleijel, F. (1994) Phylogenetic classification and the definition of taxon names. *Zoologica Scripta*, **23**, 19–25.

Tiegs, O.W. and Manton, S.M. (1958) The evolution of the Arthropoda. *Biological Reviews*, **33**, 255–337.

Turbeville, J.M., Pfeifer, D.M., Field, K.G. and Raff, R.A. (1991) The phylogenetic status of arthropods, as inferred from 18S rRNA sequences. *Molecular Biology and Evolution*, **8**, 669–86.

Wainright, P.O., Hinkle, G., Sogin, M.L. and Stickel, S.K. (1993) Monophyletic origins of the Metazoa: an evolutionary link with the Fungi. *Science*, **260**, 340–2.

Wheeler, W.C., Cartwright, P. and Hayashi, C.Y. (1993) Arthropod phylogeny: a combined approach. *Cladistics*, **9**, 1–39.

Winnepenninckx, B., Backeljau, T. and De Wachter, R. (1995a) Phylogeny of protostome worms derived from 18S rRNA sequences. *Molecular Biology and Evolution*, **12**, 641–9.

Winnepenninckx, B., Backeljau, T., Mackey, L.Y., Brooks, J.M., De Wachter, R., Kumar, S. and Garey, J.R. (1995b) 18S rRNA data indicate that Aschelminthes are polyphyletic in origin and consist of at least three distinct clades. *Molecular Biology and Evolution*, **12**, 1132–7.

# 3 A defence of arthropod polyphyly

*G. Fryer*

*Institute of Environmental and Biological Sciences, University of Lancaster, Lancaster, LA1 4QY, UK*

*'Appearance should not be mistaken for truth'.*

Charlotte Brontë (under the *nom de plume* of Currer Bell).
Preface to second edition of *Jane Eyre* (1848)

## 3.1 INTRODUCTION

It has become fashionable to assume – it is certainly not proven – that arthropods are monophyletic. Supporters of this assumption include exponents of several disciplines. Furthermore, some maintain that the only correct approach to the problems of arthropod phylogeny is via cladistics (presumably supported by DNA sequence data for extant forms). One consequence of this is that some of the most important contributions to our understanding of the functional morphology, embryology and evolution of arthropods made during this century by Manton and Anderson, which adduce much evidence in favour of polyphyly, have tended to be ignored (see summaries in Anderson, 1973; Manton, 1977, and Manton and Anderson, 1979). Indeed, because they do not openly embrace cladistic principles, Manton's arguments have been said to have 'a philosophical flaw' (Shear, 1992), to be based on a fallacious concept of phylogenetic evidence (Kristensen, 1975), and, like the work of Anderson, even to be 'old-fashioned Lamarckism' (Tuxen, 1980), though all these commentators acknowledge the valuable contributions of these workers. Recently Tautz *et al.* (1994), who note the need for comparative embryological studies on arthropods, and incidentally cite Anderson's work several times, say that he made 'an extensive effort in this direction', but qualify this by adding 'Unfortunately his studies were strongly influenced by Manton's theory of a polyphyletic origin of the arthropods and some of his inferences have to be treated with caution'. They were not. They were factual studies on the embryology of various annelids and arthropods that could hardly have been influenced by any theory.

Tautz *et al.* go on to say that there is now a consensus based on morphological and molecular data that arthropods have a common ancestor – citing just three papers that, to say the least, are questionable at the factual level and are hardly a secure base for such a claim. Two of them analyse molecular data and are far from being in agreement. Such papers seldom are. The data used in one of them (Ballard *et al.*, 1992) have subsequently been interpreted in a very different manner simply by re-aligning the sequences (Wägele and Stanjek, 1995). It is the assertion that there is a consensus that arthropods are monophyletic that I would treat with caution.

As for the belief that only the cladistic approach is valid, Heywood's (1974) comment is apposite – 'in systematics no one approach, technique, philosophy or set of data has primacy'. Nor do all condemn the Mantonian approach, least of all Anderson, whose extensive embryological studies seemed to him (and who better to judge?) to be in harmony with her conclusions derived from detailed comparative studies on the functional morphology of a wide range of arthropods. As Ridley (1986) says, 'What matters is whether her method is in principle valid. The answer is that it is'. Likewise, Willmer (1990), who makes use of Manton's findings, comes down firmly in favour of arthropod polyphyly.

The claims of arthropod polyphyly are perhaps less radical than some monophylists seem to believe. They are simply that arthropodization occurred more than once and that convergence has been both widespread and detailed. The latter is indisputable. Many shared arthropodan characters, such as a segmented body, locomotion by means of paired appendages, an extensive haemocoel, and a pair of ventral nerve cords with an anterior ganglionic concentration were, I believe, inherited from pre-arthropod ancestors. It is appropriate to add that I believe that the trunk musculature of the precursors of arthropods never proceeded to the condition found in annelids whose muscles are not only

*Arthropod Relationships*, Systematics Association Special Volume Series 55, Edited by R.A. Fortey and R.H. Thomas. Published in 1997 by Chapman & Hall, London. ISBN 0 412 75420 7

histologically different from those of arthropods but followed a different evolutionary direction. In arthropod evolution there is a simple choice of alternatives. Did the most speciose and most diverse of all groups of animals arise as a result of a relatively simple, never to be repeated, event, or was this event repeated several times, perhaps millions of years apart? The possibility of making the transition may have existed for a period as long as the entire history of the dinosaurs or angiosperm plants. It need not all have happened in the Cambrian explosion. Even if populations were small there would be countless billions of opportunities. Furthermore as we shall see, initially that event need not even have included the acquisition of segmented trunk limbs. Let us look at some of the evidence.

## 3.2 EVIDENCE FOR POLYPHYLY

### 3.2.1 MORPHOLOGY

#### (a) Eyes

The object shown in Figure 3.1 is immediately recognisable as an ommatidium of a compound eye. It is not, however, from the eye of an arthropod but from the compound eye of the sabellid polychaete *Dasychone conspersa*. Figure 3.2 is an ommatidium of another sabellid polychaete *Sabella melanostigma* which has up to 240 such eyes on its tentacular crown (Nilsson, 1994) and is just one of many tubiculous polychaetes that have compound eyes. Of slightly different design, but clearly an ommatidium, is that of *Barbatia cancellaria*, an arcacean bivalve mollusc, on whose mantle edge compound eyes occur in large numbers, as they do in *Arca*. As do those of arthropods, the ommatidia of these animals display considerable diversity. Among sabellid polychaetes some species have one-celled, others two-celled and others three-celled ommatidia, while those of the bivalve *Barbatia* are multi-celled.

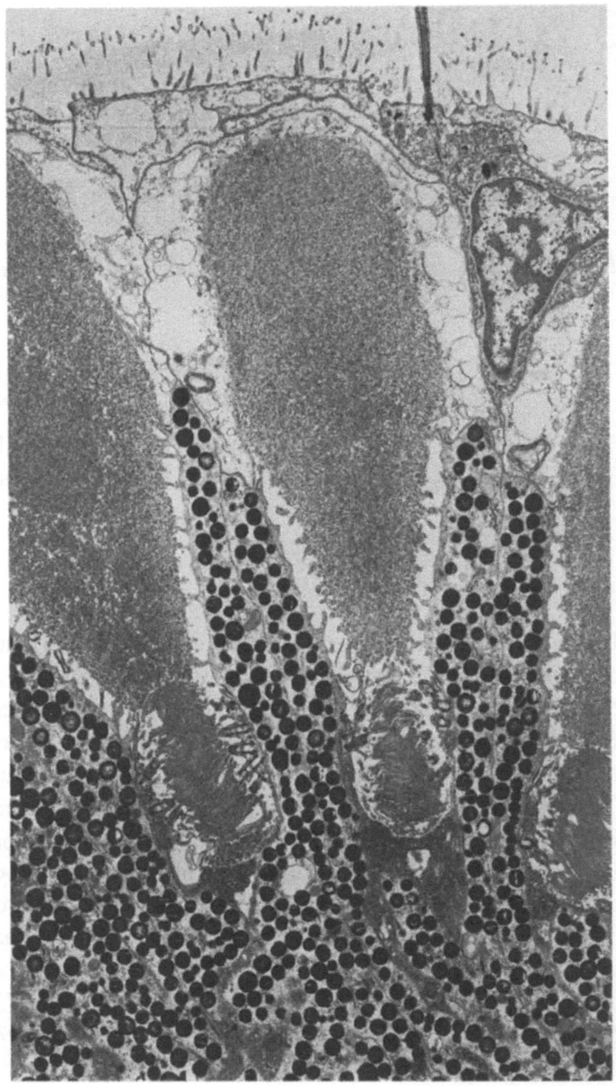

**Figure 3.1** Electron micrograph of an ommatidium of the compound eye of the sabellid polychaete *Dasychone conspersa*. (From a photograph supplied by D.-E. Nilsson.)

The presence of compound eyes in two non-arthropod phyla presents severe problems for the monophyletic hypothesis. It raises the awful possibility that they may also have evolved more than once in arthropods. This seems less improbable than their evolution in very distantly related animals. If this is conceded, the door is open for cases of convergence less spectacular than that of compound eyes, of which of course there are already many well-proven examples. Compound eyes are one of the hallmarks of arthropods and are widely regarded as prime evidence of their monophyly. Paulus (1979) for example argues that the detailed similarity of insect and crustacean ommatidia is so great that 'it cannot be due to convergence' and that these ommatidia evolved only once. As ommatidia occur in the compound eyes of molluscs and polychaetes, where they clearly evolved independently, this is wrong. Furthermore, the compound eyes of pterygote insects are more similar to those of advanced decapod crustaceans, than to those of primitive branchiopods (which also differ from decapods in the nature of their optic ganglia). At the very least this reveals remarkable convergence within arthropods or, as I suggest is the case, reflects the independent evolution of such eyes in different arthropod groups. Even Paulus has to concede that the eyes of the chilopod *Scutigera*, the only extant myriapod with compound eyes, are distinctive, and is driven to suggest that early chilopods may have had ommatidia similar to those of insects, lost them, and then developed secondarily faceted eyes from ocelli. This contorted argument essentially admits their independent evolution. The simplest explanation is that the eyes of *Scutigera* arose independently

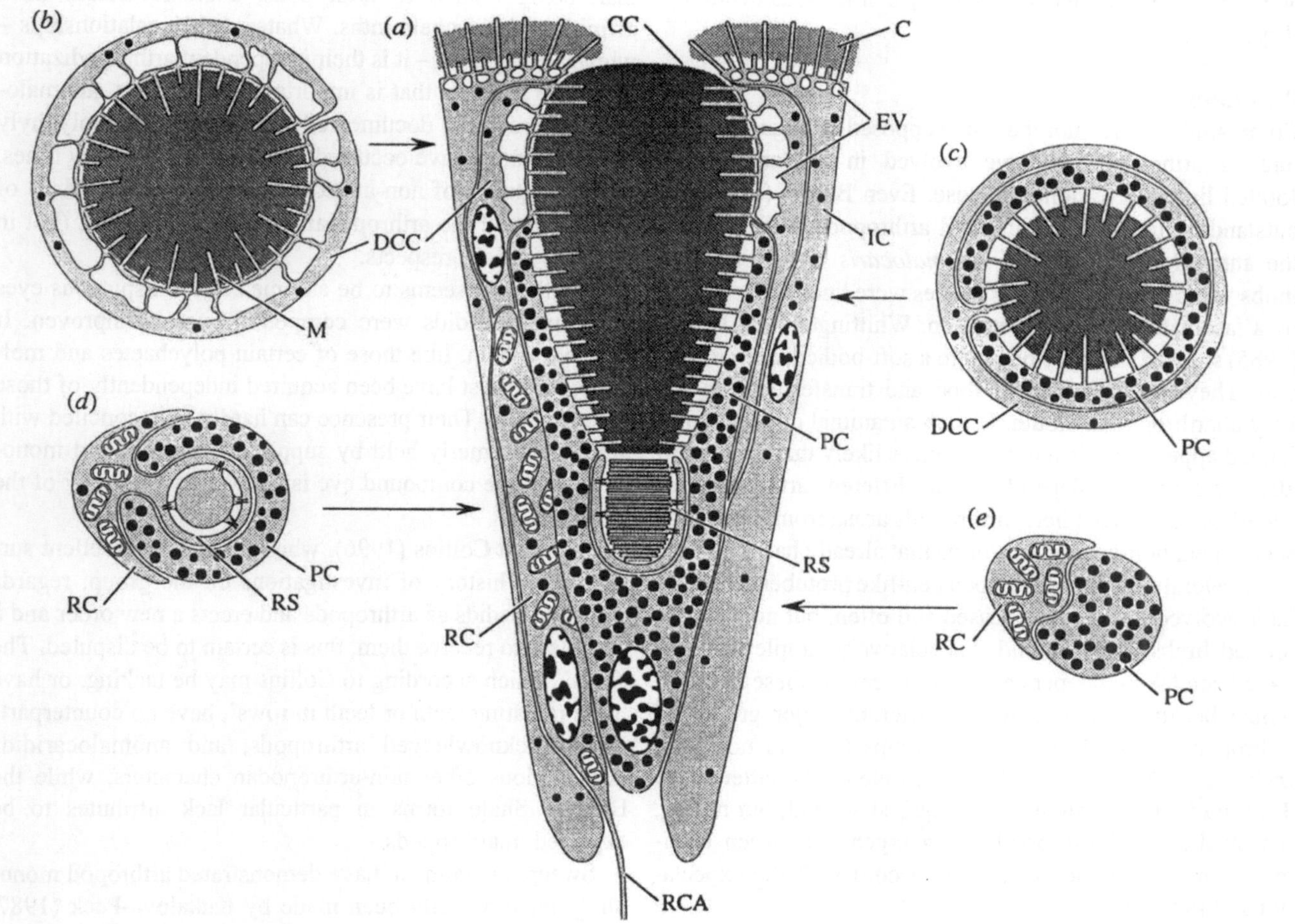


**Figure 3.2** Ommatidium of the sabellid polychaete *Sabella melanostigma* (semi-diagrammatic). (a–e) Sections at different levels through the ommatidium. C, cuticle; CC, crystalline cone; DCC, distal cone cell; EV, empty vesicles; IC, interstitial cell; M, microvilli; PC, pigment cell; RC, receptor cell with ciliary receptive segment (RS) and axon (RCA). (From Nilsson, 1994, with permission.)

in relation to fast running, carnivorous habits, in a member of a group that otherwise lacks them. The compound eyes of sabellid polychaetes (which Nilsson (1994) believes may have arisen independently several times), of arcacean bivalves, and of arthropods are undeniably of independent origin and provide excellent examples of convergence in fine details.

What is more, some of the earliest arthropods had compound eyes more than 530 Myr ago. Such eyes evolved quickly, so why should they not have done so more than once? Arthropods have produced millions of species and astronomical numbers of individuals. Indeed, the stringent demand of monophyly that a single ancestral arthropod acquired compound eyes and handed them on to all existing (and extinct) arthropods scarcely withstands scrutiny. On the contrary, there are good reasons for believing that compound eyes were acquired independently several, perhaps many, times in different lineages (Fryer, 1996), which explains the distinctive types found in *Limulus*, *Scutigera* and certain insects and the otherwise strange distribution of compound eyes among arthropods. A case in point is that of the myodocopines, the only extant ostracods that have compound eyes. There is no evidence that they inherited these from earlier ostracods (Fryer, 1996, Figure 4) and Parker (1995) has recently produced evidence of their independent evolution. The basal stocks of some arthropod lineages appear to have been (or are) eyeless.

Camera-type eyes have evolved many times in the animal kingdom. Spiders and scorpions have such eyes, which is difficult to explain if their ancestors had compound eyes. Nilsson and Pelger (1994) calculate that camera-type eyes could evolve very quickly (from a light-sensitive patch to a focused lens eye in less than half a million years): the same arguments hold for compound eyes. Monophyly also has difficulty in explaining why some Cambrian arthropods were eyeless. It is difficult to believe that compound eyes were acquired, and then lost, so early in arthropod evolution (Fryer, 1996).

**(b) Limbs**

Compound eyes are not the only supposedly diagnostic feature of arthropods to have evolved in non-arthropods. Jointed limbs have done likewise. Even Briggs (1979), an outstanding investigator of fossil arthropods, believed that the anterior appendages of *Anomalocaris* were arthropod limbs when only detached examples were known. However, by a fascinating piece of detection, Whittington and Briggs (1985) showed that they belong to a soft-bodied, non-arthropod. They evidently caught food and transferred it to the very unarthropodan mouth. If such an animal could acquire jointed appendages is it not even more likely that they could also have arisen independently in different arthropod lineages? As envisaged here, arthropods arose from a group of segmented, non-annelidan worms that already had a haemocoel, several of which developed peg-like protuberances that later evolved into chitin-encased and often, but not always, jointed limbs. Why should this relatively simple step not have been taken independently by several of these animals, which became ancestral to the different major groups of arthropods, as well as to other groups that did not quite make the arthropod grade? An example of the latter is the Tardigrada that I now believe acquired several, but not all, the attributes of arthropods. Convergence between tardigrades and arthropods is fully in accord with the expectations of polyphyly.

That segmentation of appendages arose more than once in arthropods is indeed demonstrable. Most extant branchiopod crustaceans retain unsegmented trunk limbs, and have done so since Cambrian times. This is a primitive attribute but one that has proved exceedingly adaptable, especially in the Anomopoda. In two derived orders, however, the Onychopoda and the Haplopoda, which have developed grasping limbs for seizing their prey, a task for which unsegmented limbs are unsuitable, segmented limbs have arisen. The history of branchiopods suggests very strongly that this happened independently of the evolution of jointed trunk limbs in other crustaceans.

Anomalocaridids provide a parallel example involving non-arthropods. While some Chengjiang anomalocaridids are now known to have had segmented limbs (Hou *et al.*, 1995), there is no evidence for such in the somewhat younger but geographically separated Burgess Shale forms. Some, but not all, of the latter had rigid, segmentally arranged, transverse, ventral ribs, interpreted by Collins (1996) as supports for the lateral lobes. Significantly, although Hou *et al.* provide indisputable evidence of segmented trunk limbs in *Parapeytoia yunnanensis*, they regard this as an independent adoption of arthropodization and give reasons for so doing. As some forms lack such limbs this could hardly be otherwise. Hou *et al.* suggest that, rather than being related to arthropods, anomalocaridids have affinities with aschelminths. Whatever their relationships – which are arguable – it is their independent arthropodization in certain respects that is important here. In fact, anomalocaridids provide a documented example of what polyphyly maintains must have occurred several, perhaps many times, namely a group of non-arthropods some representatives of which crossed the arthropodan boundary in several (but in this case not all) respects.

Whether, as seems to be assumed, the conspicuous eyes of anomalocaridids were compound eyes is unproven. If they were, then, like those of certain polychaetes and molluscs, they must have been acquired independently of those of arthropods. Their presence can hardly be reconciled with the belief formerly held by supporters of arthropod monophyly that the compound eye is a key autapomorphy of the group.

Although Collins (1996), who presents an excellent survey of the history of investigations on the group, regards anomalocaridids as arthropods and erects a new order and a new class to receive them, this is certain to be disputed. The 'jaws', which according to Collins may be lacking, or have 'both radiating teeth or teeth in rows', have no counterparts among acknowledged arthropods, and anomalocaridids have various other non-arthropodan characters, while the Burgess Shale forms in particular lack attributes to be expected in arthropods.

Sweeping claims to have demonstrated arthropod monophyly have recently been made by Kukalová-Peck (1987, 1991, 1992) who claims that all arthropod jaws have a five-segmented coxa, that the groundplan of the arthropod leg has 11 segments, that early arthropod legs were polyramous, and that the labrum is derived from a pair of limbs. All these claims are demonstrably wrong (Fryer, 1996), but two brief illustrations are in order to avoid the charge of casual dismissal. The idea of a five-segmented coxa is based on the interpretation of grooves and ridges in fossils of the specialized mandible of a Permian pterygote insect. Abundant contrary evidence is available. For example, branchiopod crustaceans, whose mandibular development has been studied in detail, unambiguously show that the coxa consists of a single segment throughout. Moreover, this has been the case since the Cambrian as Walossek (1993) has shown in his remarkable study of the development of *Rehbachiella* whose phosphatized remains enabled him to study a long series of instars by scanning electron microscopy. The same is true of the Cambrian *Bredocaris*, a putative maxillopodan (Müller and Walossek, 1988), just as it is of copepods. As for the labrum, far from being derived from segmented limbs, in crustaceans it arises, as Manton (1934) showed for *Nebalia*, as unsegmented ectoderm anterior to the incipient stomodaeum. A glance at Anderson's (1973) survey of the embryology of a wide range of arthropods confirms that it is never derived from a pair of limbs. Its convergent evolution in different groups of arthropods as a functional necessity is easy to envisage. As for the groundplan of the arthropod leg having 11 segments, both what is perhaps the most primitive Cambrian arthropod (*Sarotrocercus*) and the most primitive extant arthropods (the Branchiopoda – that date back to the Cambrian) have unsegmented trunk limbs, as the polyphyletic hypothesis would expect.

**(c) Cambrian Problematica**

In recent years, excellent work has been done on early fossil arthropods. Naturally the relationships of these animals, which include the so-called Problematica, have been sought. Whittington, a leader in this field, was driven to the conclusion that the Cambrian Problematica are polyphyletic. More recently, the tendency to treat arthropods as monophyletic has meant that the Cambrian forms must be included in this clade and exercises have been carried out which attempt to recognize relationships not only among these animals but between them and extant forms. Two excellent palaeontologists, Briggs and Fortey (1989), have indeed claimed that the fossils support the hypothesis of monophyly by 'filling in' some of the intermediate steps, that 'steps involved in moving from one taxon to the next identified by cladistic analysis do not require implausible novelties', and that the relationships are revealed by a reasonable series of synapomorphies and require no recourse to polyphyly. The results of these authors and of Briggs *et al.* (1992) and Wills *et al.* (1994) are given in a series of cladograms. It has also been claimed that the number of body plans is fewer than had been thought. The greater the number of body plans, the more difficult it is to support monophyly. These exercises are unconvincing. The use of cladograms presupposes monophyly, and confidence is not increased by changes in suggested affinities in successive cladograms. For example, *Martinssonia*, originally grouped with a number of alleged 'arachnomorphs', alliance with which surely involves 'implausible novelties' was later more reasonably located among crustaceans and their supposed allies. The latest cladogram differs from previous offerings for various reasons (it includes a somewhat different array of organisms, employs different characters, a different method, and a different outgroup). Its producers are honest enough not to 'offer it as a definitive solution to the problem of arthropod relationships'. The most recent placement of *Sarotrocercus* by these authors exemplifies my reservations. It is regarded as sister to *Yohoia*, of uncertain affinities, and *Sanctacaris* is regarded as its next nearest relative. In an earlier version it appeared near *Limulus* and a eurypterid. All placements seem equally curious. Irrespective of the characters used in the exercise, several of which seem irrelevant (e.g. cuticle smooth, no tracheae, no carapace adductor muscles – there is no carapace) or for which *Sarotrocercus* is scored incorrectly (e.g. it is said to have a labrum but no such structure is mentioned in Whittington's (1981) description), I can see no reason for placing *Sarotrocercus* nearer to *Yohoia* or *Sanctacaris* than to any of several other Cambrian arthropods. The anterior tagma of *Sarotrocercus* embraces only two segments, the rudimentary trunk limbs are uniramous, unsegmented and undifferentiated among themselves, and the sole 'cephalic' limb posterior to the anteriormost appendage is identical to the trunk series. In contrast, the anterior tagma of the eyeless *Yohoia* is of four segments, bears three pairs of uniramous walking limbs, and anterior to these, a pair of grasping appendages. Each of the 10 segments of what can be described as the 'thoracic' region bears flap-like, presumably swimming, appendages, of which the first pair at least may have been biramous, with an inner ramus that conceivably acted in concert with the legs of the anterior tagma. There is also a short three-segmented 'abdominal' region bearing a simple 'telson' (Whittington, 1974). *Sanctacaris* has an anterior tagma of six segments, its limbs are highly differentiated, biramous, and differ much between tagmata, the five anteriormost being grasping appendages, the more posterior perhaps similar to the equivalent series of *Yohoia* (Briggs and Collins, 1988).

In collaboration with Wilkinson these authors have even more recently (Wills *et al.*, 1995) suggested yet different affinities. *Sarotrocercus* is now adjudged to be closest to *Habelia* and *Leonchoilia* – two very different animals that are very different from *Yohoia* and *Sanctacaris*. *Habelia* was a benthic crawler with a three-segmented cephalon, whose trunk segments had lateral epimeres. It had a pair of uniramous antenna-like appendages, two further pairs of imperfectly known head appendages, and six pairs of multisegmented, uniramous walking limbs. *Leonchoilia* had a three-segmented cephalon which bore a pair of large uniramous 'great appendages' of unusual design, and two pairs of biramous appendages which, like the 11 pairs of trunk limbs, had a jointed leg-like inner, and a large flap-like outer, branch that probably served as a gill, for propulsion, or both. Both animals were blind. It is difficult to see why *Sarotrocercus* should be thought to be close to either.

Apart from the fact that these are all arthropods, there seems to be nothing to indicate any phyletic relationships between *Sarotrocercus* and any of these very different animals, and no computer programme can make it otherwise. As Wägele (1995) has pointed out, the output of computer cladistics should be regarded as mere statistical results which need biological interpretation. In *Sarotrocercus* we have in fact a creature as similar to the earliest arthropods as perhaps we shall ever get. It is small (length to ca. 11 mm), displays an extremely simple level of tagmosis and a low degree of cephalization, its trunk appendages are still at a stage essentially the same as those to be expected in the worm-like ancestors of arthropods, the first cephalic pair are simply segmented derivatives of pre-arthropod sensory tentacles, the body armour little more than a series of dorsal scutes, and it lacks mouthparts and a labrum. It is easy to visualize this sort of animal evolving several times from segmented worms that began to develop peg-like appendages. Its trunk limbs could hardly differ more from Kukalová-Peck's (1992) suggested polyramous limb with an 11-segmented telopodite. As for its affinities, one could, for example, make a case for regarding it as a very primitive uniramian. If the eyes are indeed compound (which is unproven) it is clear that such could evolve in extremely primitive arthropods.

### (d) Cladistic limitations

Cladistics is not always the logical objective exercise that some believe it to be, as a very relevant example concerning what some regard as the key feature of the so-called Mandibulata shows. Weygoldt (1979) considers that 'the most persuasive character demonstrating the synapomorphous nature of the mandibles' in crustaceans and tracheates is their position, but Walossek and Müller (1992) believe that 'the homologous position of the third head appendage in itself is simply a plesiomorphic character'. Again, as Shear (1997, this volume) has pointed out, of the four alleged synapomorphies of the Myriapoda cited in a recent well-respected textbook, three are wrong.

### (e) Coordinated character changes

In criticizing Manton's approach, Tuxen (1980) believes that consideration of a number of isolated characters may approach their totality without being what he calls 'the rather vague' integrated ensemble. This seems over-optimistic and cannot gainsay the fact that individual characters inevitably have a lower information content than a complex (see also Wägele, 1995). Examples illuminate this debate.

The stream-frequenting atyid prawn *Atya innocous* has highly specialized chelipeds with which it sweeps up particulate food as it walks forward. It can also stand facing the current, spread its cheliped bristles, and hold them together, thus making an efficient filtering basket that catches drifting particles. Chelipeds are generally pincer-like appendages so those of *Atya* are very specialized and we are confronted by the problem of how they evolved, which they did in concert with a remarkable change in behaviour from that of a more generalized ancestor. Here, comparative studies in functional morphology are far more informative than a study of isolated characters.

Having caught its food, *Atya* has to ingest it. For this it has elaborate mouthparts each of which is specialized for dealing with fine particles which they strip from the chelipeds. Again, the best way of understanding how these evolved is by comparison with more generalized mouthparts, including those of *Xiphocaris*, the most primitive atyid. Ingested food is treated further in the gastric mill. Here, the tearing and crushing ossicles of many decapods would be useless: rakes and filters and a specialized gland filter for final treatment are what is needed. All these attributes are part of a coordinated whole, and I would find it more enlightening to compare such a complex of characters with its equivalent in one or more other animals than to be presented with a matrix of numbers, sometimes concerning characters whose significance I did not understand, that I could ask a computer to interpret. Evolution is the transformation of one, often complex, mechanism into another, maintaining functional continuity throughout, and Manton (1964) was surely right to attempt to trace such transformations, as she did so remarkably in, for example, her studies on arthropod mandibles.

Another example is provided by anomopod branchiopods. Although primitive, these tiny animals are complex in structure and display enormous diversity, not least in the specialization of the trunk limbs in relation to food handling and sometimes also to locomotion. Several families are involved in their adaptive radiation. Some anomopods, like *Peracantha*, scrape detritus into a cage made up of the filters of four pairs of limbs. Others, like *Alonopsis*, employ fewer filters in a similar device. Some species do not filter at all but collect food mechanically, as does *Lathonura* (Macrothricidae). *Pseudochydorus* (Chydoridae) has become a scavenger and uses homologues of filters to pitchfork dead bodies of other small crustaceans to the mouthparts, a device from which I believe arose that of *Anchistropus*, which is a parasite on hydras from which it tears chunks of living tissue. The benthic *Ilyocryptus* is perhaps the most perfect filter feeder in the animal kingdom, while *Daphnia* has so expanded and perfected the filters of just two pairs of limbs that it has emancipated itself from the bottom and is able to subsist on material filtered from open water (Fryer, 1968, 1974, 1991).

Comparative studies enable us, legitimately I believe, to suggest the trunk limb arrangement of primitive anomopods (Fryer, 1995) which lived perhaps in Palaeozoic times, and thereby to make a small contribution to our understanding of evolution. Comparisons with more primitive branchiopods are also possible.

The evolution of anomopods has involved coordinated changes in many characters within these complexes. The apparatus that enables the super-specialized *Graptoleberis* to glide over surfaces like a minute snail involves the carapace, labrum, post-abdomen, antennae and three pairs of trunk limbs, and is itself an integral element of the feeding mechanism by means of which the trunk limbs scrape particulate matter from the surfaces over which the animal glides, manipulate it, and pass it to the mouthparts. If any one component of this elaborate mechanical system fails to play its part, the entire mechanism is rendered inoperative. Evolution has here, as always, been concerned with a complex of coordinated structures; not with isolated characters, consideration of which can tell us little about the processes involved.

In considering the evolution and affinities of the Anomopoda are we to ignore such information and the vast differences between anomopods and other branchiopod orders with which some have associated them? Or should we accept that these animals are part of a monophyletic group that includes also five very different orders simply because they are all reputed to share a single apomorphy of very dubious validity, namely an ill-defined bivalved carapace?

A final point on morphology. The evolution of different body plans is easy to appreciate if arthropods are polyphyletic. It presents serious problems for monophyly. Suppose **the** reputed ancestral arthropod adopted, even at a primitive level, a particular body plan involving a specific kind of tagmosis and a particular arrangement of limbs. Stabilizing selection would tend to weed out fundamental deviations from this plan as most would be functionally inefficient. Thus, the evolution of fundamentally different body plans would be very difficult. Trilobites provide an excellent example of such constraining selection. Throughout their long history from the Lower Cambrian to the Permian their body plan remained remarkably stereotyped. This is in striking contrast to the diversification of arthropod body plans, especially during the Cambrian radiation when the tempo of arthropod evolution was particularly brisk. If it be argued that radiation began before much differentiation had been achieved, why not go back a little further and accept that this could have happened among pre-arthropods? After all, the early stages of arthropodization involved far fewer changes than did the evolution of different body plans.

### 3.2.2 PHYSIOLOGY AND RELATED MATTERS

An argument for monophyly is that similar moulting hormones are found in different groups of arthropods. This is equally understandable if arthropods are polyphyletic. Similar problems are often solved in similar ways in different animals: ecdysone also controls the moult cycle of nematodes. All investigated animals use the carotenoid retinene$_1$ or the very similar retinene$_2$ (also called retinal$_1$ and retinal$_2$) as the chromophore of their visual pigments. Crustaceans and insects do so, but this is no indicator of monophyly: so do cephalopods and vertebrates with their very different eyes. It seems that, if vision is to be possible, one of these substances must be present. Another example is the production of light which has evolved independently in crustaceans (several times) and in insects, and in each case involves luciferin and luciferase – as it does in other animals and even plants, because this is the way that light can be produced. It is not an indicator of relationships. Nor is it surprising that similar molecules are sometimes produced at similar sites: these are often the most convenient. What is more significant is when they are produced from different sites as when ecdysteroids are secreted by the Y-organ of crustaceans but by the prothoracic gland of insects. As if to reinforce the point, while both glands are controlled by neuropeptide hormones, this is done in different ways in crustaceans and insects, which points to independent evolution and not to monophyly. There are comparable examples elsewhere in the animal kingdom. Who would claim that mammals and flamingos are monophyletic because both produce milk of similar composition to nourish their young, one from mammary glands, the other from glands in the upper digestive tract, whose secretions are in both cases regulated by the hormone prolactin?

Several other complex molecules with similar functions, though not necessarily identical, are produced by different arthropods, such as haemoglobin by various insects and crustaceans, silk by insects, symphylans, spiders and certain crustaceans, and venoms by insects, spiders, scorpions and centipedes. Silks are particularly informative. These fibroproteins are composite materials that generally consist of a protein matrix in which are embedded protein crystals. Although those of spiders and the silkworm moth, *Bombyx mori*, differ in crystal composition and spatial organization they are basically similar in chemistry and molecular structure. Indeed a spider can produce silks of different composition and with different mechanical properties, and the dragline thread of *Nephila* is of different construction and appears to consist of a structured fibril tube surrounding a thin core (Vollrath *et al.*, 1996).

Silks are secreted by labial glands, Malpighian tubules, accessory tubes of the genital organs or dermal glands in different insects. In larval Lepidoptera there is a single pair of long thin glands: in spiders there are large numbers of glands of several kinds. In each case the walls of the glands consist of a single layer of cells but the detailed cytology is profoundly different. Nevertheless, both exhibit powerful cytoplasmic phosphatase activity localized along the inner border of the cells next to the gland lumen (Bradfield, 1951), and both produce silk. Silks of insects and spiders often have similar uses, e.g. protecting defenceless early stages of development, or even the making of snares in web-spinning caddis and chironomid larvae as well as in spiders. Their independent evolution is almost as remarkable as that of compound eyes, perhaps more so than that of jointed limbs, and like both, indisputable.

None of these similarities can be claimed as an indicator of monophyly, nor can the tracheae that carry oxygen to deep-lying tissues in a variety of arthropods. This indeed is a much more elaborate example of convergence (as which it is universally acknowledged) than the acquisition of even complex molecules by different groups. The chitin from which the tracheae, like the cuticle of arthropods, are made was once deemed to be an indicator of monophyly but is now know to occur in many invertebrates, and even in fungi. It is just the ideal substance from which to make an exoskeleton. The ways in which it is hardened, or reinforced, have also clearly been acquired several times and simply reflect its remarkable suitability for its job.

As convergence is so frequent and supports polyphyly, and not monophyly, the balance must inevitably be in favour of the former, especially as all arthropod lineages would inherit basic physiological processes from their pre-arthropod ancestors.

### 3.2.3 DEVELOPMENTAL GENETICS

Recent advances in developmental genetics have been enormous, and include excellent work on arthropods. The tendency has been to interpret results as indicating monophyly, but a polyphyletic interpretation seems equally convincing. One outcome of this work has been the demonstration that homologous clusters of developmental genes occur in, for example, crustaceans and hexapods, and Averof and Akam (1995) have even shown that while three *Hox* genes specify distinct segment types within the thorax and abdomen of hexapods, they are expressed as largely overlapping domains in the anterior trunk region of the branchiopod crustacean *Artemia*. The immediate conclusion that this is dramatic evidence of monophyly with some divergence evaporates when it becomes known that these genes not only originated before either the Crustacea or the Hexapoda but that they have homologues even among vertebrates. Indeed, comparisons among arthropods suggest that most *Hox* gene duplications preceded diversification of the major arthropod groups (Akam, 1995). Another ancient gene whose wide distribution includes both arthropods and vertebrates is *engrailed*. If arthropods are polyphyletic one would of course **expect** genetic similarities of this sort as all lineages arose from a group of related pre-arthropods which would to a large degree share the same genetic endowment. I suggest that by playing this down proponents of monophyly make polyphyly appear unjustifiably outlandish.

Here, it is appropriate to draw attention to the enormous gap in the fossil record between the earliest hexapods and the Crustacea, with which monophylists would have them share a common ancestry. The separation is fully in keeping with Manton's (1972) belief that hexapods arose independently in the uniramian lineage from soft-bodied ancestors. Their five-fold origin is in effect a muted repeat of the Cambrian explosion made possible by the origin of a land flora that, directly or indirectly, provided food and living conditions, and the opportunity to exploit the gaits and cephalic feeding mechanisms of the small uniramians, or their precursors, that had no such opportunities during their many million-year sojourn in the sea.

### 3.2.4 MOLECULAR SEQUENCE ANALYSIS

Molecular sequence analysis has been hailed as the final arbiter on questions of arthropod relationships but, while its value is not denied, its contributions in this sphere have been confusing and contradictory. Conclusions drawn from studies made to date, differ according to the number and type of taxa involved in the analyses (usually very few), the molecule selected, the method employed, choice of outgroup and so on. The lack of reliability of sequence data *per se* in the inference of phylogenies has been brilliantly expounded by Wägele and Wetzel (1994) who, among other things, show that 'surprising results' can be not merely illogical but nonsense, that the method is no more objective than study of a multicellular structure, which has a much higher information content than a typical sequence, that dendrograms should not be taken as serious reconstructions of phylogeny.

This is not to deny the extremely valuable contribution of sequence analysis which can help to resolve relationships and relatively recent events within defined groups, such as the Crustacea or components thereof, though it sometimes suggests affinities that are clearly nonsense, and has been less successful in comparing long distinct major groups, whatever their ancestry. Many of the events that concern us here took place in early Palaeozoic times, and some of them occurred rapidly in the Cambrian when different body plans probably arose in less, perhaps much less, than 15 Myr. Philippe *et al.* (1994) conclude that the 18S rRNA database cannot confidently resolve cladogenetic events separated by less than about 40 Myr, and that, given a good fossil record and well-dated rocks, palaeontology performs better than molecular methods in the resolution of diversification over short time scales.

Molecular methods may in fact be unable to adjudicate on the question of arthropod monophyly versus polyphyly. They have no means of telling whether events of which they can trace echoes hundreds of millions of years later, took place before or after arthropodization was achieved (Fryer, 1996). If arthropods are polyphyletic as envisaged here, one would expect them to show similarities at the molecular level because, not long before they independently, and several times, achieved the arthropod grade, they had a common ancestor.

## 3.3 CONVERGENCE

Convergence is the scourge of monophyly. The animal kingdom abounds with examples, from the level of single organs and structures, such as compound eyes, tracheae, wings, gills, statocysts, chitinous structures and complex molecules, all of which include arthropods; such similar mechanisms as rolling up in conglobating millipedes, isopods and trilobites (and in the armadillo *Tolypeutes*, pangolins, hedgehogs and others), to whole faunas, such as the marsupials of Australasia vis à vis eutherian mammals elsewhere. The latter comparison includes amazingly detailed similarities in moles, mice and wolves, and now it seems that an element of the Australian avifauna may provide an equally remarkable example. *Petrochromis* of Tanganyika and *Petrotilapia* of Malawi not only bear a remarkable overall similarity but have a complex and otherwise unique dentition that, specialized as it is, has arisen independently in each lake in thes eichlid fishes.

Filters, which are very complex structures, and cannot function unless they evolve in concert with a means of drawing water through them, and some means of removing what is filtered, have certainly evolved several times in the Crustacea. They also evolved remarkably early. Filters amazingly similar to those of advanced extant branchiopods have been found in early Cambrian deposits (Butterfield, 1994) but it is not known from what animal they came. The exopterygote Thysanoptera have achieved the holometabolous condition independently of endopterygote insects, superposition eyes have evolved many times in arthropods, whatever their evolutionary history, and nearly every feature of the Artiodactyla and Perissodactyla has evolved independently. Examples could be multiplied almost indefinitely.

New examples continue to be revealed: for example, the otic tympanum of tetrapods has evidently arisen independently at least three times (Lombard and Bolt, 1979; Clack, 1993) and similarities in the auditory region of elephants and seacows, long thought to be homologies and synapomorphies were arrived at independently (Court, 1994). Anaerobic ciliates with hydrogenosomes (organelles that produce hydrogen via the activity of hydrogenase) have arisen at least four times from aerobic ancestors (Embley *et al.*, 1995). In the light of these and many other examples, why should the acquisition of scutes and simple, chitin-encased, movable appendages (not necessarily segmented) by several members of a group of segmented worms seem improbable? I find it inconceivable that it did not happen. After all, as Hinton (1955) showed, appendages of this sort – prolegs in the Diptera – have evolved independently at least 27 times.

## 3.4 A CHALLENGE TO MONOPHYLY

To convince biologists conversant with the structure, ways of life and diversity of arthropods, that this vast assemblage is monophyletic, proponents of this view will have to do more than cite contradictory sequence data, demonstrate the sharing of ancient genes (which are sometimes fully in accord or accord better, with polyphyly), or produce cladograms based on the juggling of sometimes questionable matrices by a computer. They will have to describe the postulated Ur-arthropod, provide details of the nature and arrangement of its appendages, and explain how it gave rise, via functional intermediates, to the plethora of strikingly differently body plans, with their different locomotory and food collecting devices, that arose rapidly in Cambrian times, and to modern forms. Compound eyes, jointed limbs, tracheae, other structures, and complex molecules have undoubtedly arisen more than once, sometimes very quickly, so, if they agree that the ancestral arthropod arose from a segmented worm-like predecessor, they will have to explain why the simple basic steps of arthropodization could not also have been taken several times, and offer more convincing evidence than has yet been produced that this did not happen. They have to show why, if the ur-arthropod had compound eyes, so many primitive arthropods lack them and why others acquired elaborate camera-type eyes, and explain the distribution of compound eyes among the various groups, which seems not to be in agreement with what monophyly would expect. They have to show how different patterns of embryological development diverged from a single, defined, ancestral type; how crustaceans, trilobites hexapods and xiphosurans came to have different larval forms (bearing in mind that crustaceans already had a nauplius and trilobites a protaspis in Cambrian times); and explain the hiatus of more than 100 Myr between early crustaceans and the earliest known hexapods with which some of them believe they share a common origin. One could add other demands but these are enough to be going on with. By contrast, many of the problems of arthropod evolution that present themselves if the group is deemed to be monophyletic, disappear or are rendered more intelligible if it is in fact polyphyletic. Evidence for polyphyly is indeed plentiful and, I believe, convincing.

ACKNOWLEDGEMENTS

I am much indebted to Derek Briggs who generously provided both information and literature on anomalocaridids, and to Dan Nilsson for the print from which Figure 3.1 was made.

'*Whenever you present the actual simple truth, it is, somehow, always denounced as a lie'.*

Charlotte Brontë, *Shirley* (p. 632) 1849.

## REFERENCES

Akam, M. (1995) *Hox* genes and the evolution of diverse body plans. *Philosophical Transactions of the Royal Society of London B*, **349**, 313–19.

Anderson, D.T. (1973) *Embryology and Phylogeny in Annelids and Arthropods*, Pergamon, Oxford.

Averof, M. and Akam, M. (1995) *Hox* genes and the diversification of insect and crustacean body plans. *Nature*, **376**, 420–3.

Ballard, J.W.O., Olsen, G.J., Faith, D.P., Odgers, W.A., Rowell, D.M. and Atkinson, P.W. (1992) Evidence from 12S ribosomal RNA sequences that onychophorans are modified arthropods. *Science*, **258**, 1345–8.

Bradfield, J.R.G. (1951) Phosphatases and nucleic acids in silk glands: cytochemical aspects of fibrillar protein secretion. *Quarterly Journal of Microscopic Science*, **92**, 87–112.

Briggs, D.E.G. (1979) Anomalocaris, the largest known Cambrian arthropod. *Palaeontology*, **22**, 631–64.

Briggs, D.E.G. and Collins, D. (1988) A Middle Cambrian chelicerate from Mount Stephen, British Columbia. *Palaeontology*, **31**, 779–98.

Briggs, D.E.G. and Fortey, R. A. (1989) The early radiation and relationships of the major arthropod groups. *Science*, **246**, 241–3.

Briggs, D.E.G., Fortey, R.A. and Wills, M.A. (1992) Morphological disparity in the Cambrian. *Science*, **256**, 1670–3.

Butterfield, N.J. (1994) Burgess shale-type fossils from a Lower Cambrian shallow-shelf sequence in northwest Canada. *Nature*, **369**, 477–9.

Clack, J.A. (1993) Homologies in the fossil record: the middle ear as a test case. *Acta Biotheretica*, **41**, 391–409.

Collins, D. (1996) The 'evolution' of *Anomalocaris* and its classification in the arthropod class Dinocarida (*nov.*) and order Radiodonta (*nov.*). *Journal of Paleontology*, **70**, 280–3.

Court, N. (1994) The periotic of *Moeritherium* (Mammalia, Proboscidea): homology or homoplasy in the ear region of Tethytheria McKenna, 1975? *Zoological Journal of the Linnean Society*, **112**, 13–28.

Embley, T.M., Finlay, B.J., Dyal, P.L., Hirt, R.P., Wilkinson, M. and Williams, A.G. (1995) Multiple origins of anaerobic ciliates with hydrogenosomes within the radiation of aerobic ciliates. *Proceedings of the Royal Society of London, B*, **262**, 87–93.

Fryer, G. (1968) Evolution and adaptive radiation in the Chydoridae. (Crustacea: Cladocera): a study in comparative functional morphology and ecology. *Philosophical Transactions of the Royal Society of London B*, **254**, 221–385.

Fryer, G. (1974) Evolution and adaptive radiation in the Macrothricidae (Crustacea: Cladocera): a study in comparative functional morphology and ecology. *Philosophical Transactions of the Royal Society of London B*, **269**, 137–274.

Fryer, G. (1991) Functional morphology and the adaptive radiation of the Daphniidae (Branchiopoda: Anomopoda). *Philosophical Transactions of the Royal Society of London B*, **331**, 1–99.

Fryer, G. (1995) Phylogeny and adaptive radiation within the Anomopoda: a preliminary exploration. *Hydrobiologia*, **307**, 57–68.

Fryer, G. (1996) Reflections on arthropod evolution. *Biological Journal of the Linnean Society*, **58**, 1–55.

Heywood, V.H. (1974) Systematics – the stone of Sisyphus. *Biological Journal of the Linnean Society*, **6**, 169–78.

Hinton, H.E. (1955) On the structure, function and distribution of the prolegs of the Panorpoidea with a criticism of the Berlese–Imms theory. *Transactions of the Royal Entomological Society of London*, **106**, 455–75.

Hou, X-G., Bergström, J. and Ahlberg, P. (1995) *Anomalocaris* and other large animals in the Lower Cambrian Chengjiang fauna of southwest China. *GFF*, **117**, 163–83.

Kristensen, N.P. (1975) The phylogeny of hexapod 'orders'. A critical review of recent accounts. *Zeitschrift für zoologische Systematische und Evolutionforschung*, **13**, 1–44.

Kukalová-Peck, J. (1987) New Carboniferous Diplura, Monura and Thysanura, the hexapod groundplan, and the role of thoracic side lobes in the origin of wings (Insecta). *Canadian Journal of Zoology*, **65**, 2327–45.

Kukalová-Peck, J. (1991) Fossil history and the evolution of hexapod structures, in *The insects of Australia* (2nd ed.) (eds I.D. Nauman and CSIRO), Melbourne University Press, Melbourne, pp. 141–79.

Kukalová-Peck, J. (1992) The 'Uniramia' do not exist: the groundplan of the Pterygota as revealed by Permian Diaphanopterodea from Russia. (Insecta: Palaeodictyopteroidea). *Canadian Journal of Zoology*, **70**, 236–55.

Lombard, R.E. and Bolt, J.R. (1979) Evolution of the tetrapod ear: an analysis and reinterpretation. *Biological Journal of the Linnean Society*, **11**, 19–76.

Manton, S.M. (1934) On the embryology of the crustacean *Nebalia bipes*. *Philosophical Transactions of the Royal Society of London B*, **223**, 163–238.

Manton, S.M. (1964) Mandibular mechanisms and the evolution of arthropods. *Philosophical Transactions of the Royal Society of London B*, **247**, 1–183.

Manton, S.M. (1972) The evolution of arthropodan locomotory mechanisms. Part 10. Locomotory habits, morphology and evolution of the hexapod classes *Zoological Journal of the Linnean Society*, **51**, 203–400.

Manton, S.M. (1977) *The Arthropoda*, Clarendon Press, Oxford.

Manton, S.M. and Anderson, D.T. (1979) Polyphyly and the evolution of arthropods, in *The Origin of Major Invertebrate Groups* (ed M.R. House), Systematics Association Special Volume. No. 12, pp. 269–321.

Müller, K.J. and Walossek, D. (1988) External morphology and larval development of the Upper Cambrian maxillopod *Bredocaris admirabilis*. *Fossils and Strata*, **23**, 1–70.

Nilsson, D.E. (1994) Eyes as optical alarm systems in fan worms and ark clams. *Philosophical Transactions of the Royal Society of London B*, 346, 195–212.

Nilsson, D.E. and Pelger. S. (1994) A pessimistic estimate of the time required for an eye to evolve. *Proceedings of the Royal Society of London B*, **256**, 53–8.

Parker, A.R. (1995) Discovery of functional iridescence and its coevolution with eyes in the phylogeny of the Ostracoda (Crustacea). *Proceedings of the Royal Society of London B*, **262**, 349–55.

Paulus, H.F. (1979) Eye structure and the monophyly of the Arthropoda, in *Arthropod Phylogeny* (ed. A.P. Gupta), Van Nostrand Reinhold, New York, pp. 299–383.

Philippe, H., Chenuil, A. and Adoutte, A. (1994) Can the Cambrian explosion by inferred through molecular biology? *Development 1994 Supplement*, 15–25.

Ridley, M. (1986) *Evolution and Classification*, Longman, London.

Shear, W.A. (1992) End of the 'Uniramia' taxon. *Nature*, 359, 477–8.

Tautz, D., Friedrich, M. and Schröder, R. (1994) Insect embryogenesis – what is ancestral and what is derived? *Development 1994 Supplement*, 193–9.

Tuxen, S.V. (1980) The phylogeny of Apterygota and on phylogeny in general. *Bollettino di Zoologia*, **47** (suppl.), 27–34.

Vollrath, F., Holtet, T., Thøgersen, H.C. and Frische, S. (1996) Structural organisation of spider silk. *Proceedings of the Royal Society of London B*, 263, 147–51.

Wägele, J.W. (1995) On the information content of characters in comparative morphology and molecular systematics. *Journal of Zoological Systematics and Evolutionary Research*, **33**, 42–7.

Wägele, J.W. and Stanjek, G. (1995) Arthropod phylogeny inferred from partial 12 S rRNA revisited: monophyly of the Tracheata depends on sequence alignment. *Journal of Zoological Systematics and Evolutionary Research*, **33**, 75–80.

Wägele, J.W. and Wetzel, R. (1994) Nucleic acid sequence data are not *per se* reliable for inference of phylogenies. *Journal of Natural History*, **28**, 749–61.

Walossek, D. (1993) The Upper Cambrian *Rehbachiella* and the phylogeny of the Branchiopoda and Crustacea. *Fossils and Strata*, **32**, 1–202.

Walossek, D. and Müller, K.J. (1992) 'The Alum Shale Window' – contribution of 'Orsten' arthropods to the phylogeny of Crustacea. *Acta Zoologica*, **73**, 305–12.

Weygoldt, P. (1979) Significance of later embryonic stages and head development in arthropod phylogeny, in *Arthropod Phylogeny* (ed. A.P. Gupta), Van Nostrand Reinhold, New York, pp. 107–35.

Whittington, H.B. (1974) *Yohoia* Walcott and *Plenocaris* n. gen., Arthropods from the Burgess Shale, Middle Cambrian, British Columbia. *Geological Survey of Canada Bulletin*, **231**, 1–27.

Whittington, H.B. (1981) Rare arthropods from the Burgess Shale, Middle Cambrian, British Columbia. *Philosophical Transactions of the Royal Society of London B*, **292**, 329–57.

Whittington, H.B. and Briggs, D.E.G. (1985) The largest Cambrian animal, *Anomalocaris*. Burgess Shale, British Columbia. *Philosophical Transactions of the Royal Society of London B*, **309**, 569–609.

Willmer, P. (1990) *Invertebrate Relationships*, Cambridge University Press, Cambridge.

Wills, M.A., Briggs, D.E.G. and Fortey, R.A. (1994) Disparity as an evolutionary index: a comparison of Cambrian and recent arthropods. *Paleobiology*, 20, 93–130.

Wills, M.A., Briggs, D.E.G., Fortey, R.A. and Wilkinson, M. (1995) The significance of fossils in understanding arthropod evolution. *Verhandlungen der Deutschen zoologischen Gesellschaft*, **88**, 203–15.

# 4 *Hox* genes and annelid–arthropod relationships

*M.H. Dick*

***Department of Biology, Middlebury College, Middlebury, Vermont 05753, USA***
***email: dick@midd-unix.middlebury.edu***

## 4.1 INTRODUCTION

The title of this paper is the same as the provisional title of the topic upon which I was asked to speak at this symposium. I can think of no better title, nor can I think of a worse one. For the phrasing of the title implies that annelids have some special relationship to arthropods, such as that formalized by Cuvier nearly 200 years ago with his Phylum Articulata. It also might imply that homeobox genes can either provide insights into how overtly tagmatized arthropods evolved from a presumably homonomous, annelid-like ancestor, a scenario advanced by Snodgrass (1938), or at least furnish data pertinent to resolving annelid–arthropod relationships. Upon closer examination, an *a priori* juxtaposition of annelids, arthropods, and homeoboxes has no logical justification. There is evidence that annelids and arthropods are not sister groups (Tiegs and Manton, 1958; Eernisse *et al.*, 1992; Raff, 1994). Furthermore, the utility of homeobox genes as a source of systematic characters is an open question.

I admit that the topic initially seemed perfectly logical to me before I reflected on the widespread ambivalence over exactly what homeobox genes mean for systematics. Homeobox genes have been considered as an extraordinary source of data applicable to both phylogeny reconstruction and understanding how body plans diverged (Marx, 1992). The purpose of this paper is to discuss whether these are realistic expectations and how they might be met. Throughout, I use annelid–arthropod relationships as a point of focus. I first introduce homeobox genes, and outline the relevance of *Hox* genes to systematics. I then address the utility of *Hox* and some other homeobox genes as a source of characters for phylogeny reconstruction. Finally, I discuss axial position and segmental diversification in setate annelids. Most homeobox gene expression studies in annelids have been conducted in leeches, which share with arthropods a stereotyped segment number and axial specification early in ontogeny. Many setate annelids maintain a dynamic body axis, capable of respecification throughout ontogeny. Gene expression studies in setate annelids may provide clues to the roles of segmentation and *Hox* genes in the evolution of arthropod tagmatization.

## 4.2 RELEVANCE OF HOX GENES TO SYSTEMATICS

Homeobox genes were originally discovered as the genes underlying homeotic mutations in the fruitfly, mutations that transform one segment into another (Lawrence, 1992; Gehring, 1994). Homeobox genes are defined by containing a conserved 180 base pair (bp) region, the homeobox, which encodes a correspondingly conserved amino acid motif, the homeodomain (Bürglin, 1994). Protein products of these genes bind DNA in a sequence-specific manner and function as transcriptional regulators.

Many classes of homeobox genes are now known. Among these, the *Drosophila* homeotic genes and their homologues in other animals are collectively known as the *Hox* genes (Akam *et al.*, 1994). *Hox* genes are ubiquitous among metazoans (Slack *et al.*, 1993; Bürglin, 1994). Not only are their homeobox sequences conserved, but their chromosomal organization correlates with body regions over which they exert developmental control. That is, in all organisms in which they have been mapped, *Hox* genes are aligned in linear arrays on the chromosome, and their order is generally colinear with anterior boundaries of expression domains along body axes during development (McGinnis and Krumlauf, 1992).

*Hox* clusters are believed to have arisen through lateral duplications beginning with an ancestral homeobox-containing *Hox* gene (Kappen *et al.*, 1989; McGinnis and Krumlauf, 1992), and gene order within clusters was presumably maintained through regulatory interactions (Lewis, 1992). However, there is considerable variation in this basic organization. Clusters vary in size; for example, that of the nematode *Caenorhabditis elegans* contains only four *Hox* genes (Wang *et al.*, 1993), whereas the longest vertebrate cluster contains 11 (Holland, 1992). Clusters themselves were duplicated in some lineages, e.g. chordates (Ruddle *et al.*, 1994) and xiphosurans (Cartwright *et al.*, 1993). Gene copies have been gained or lost from clusters, either before or after cluster duplication events. This variation in *Hox* gene organization provides several levels of potentially informative characters for phylogeny reconstruction.

*Arthropod Relationships*, Systematics Association Special Volume Series 55, Edited by R.A. Fortey and R.H. Thomas. Published in 1997 by Chapman & Hall, London. ISBN 0 412 75420 7

The roles of *Hox* genes in specification of axial positional information and determination of cell fate (McGinnis and Krumlauf, 1992) are important for the establishment of axial structures which vary within and between metazoan phyla. Comparative data on *Hox* genes are thus also pertinent to understanding how body plans diverged. It must be stressed, however, that to avoid circularity, these data can be used to study evolutionary process only when they are interpreted in the context of a phylogeny derived independently of them, or reconstructed by incorporating them into a 'total evidence' approach.

## 4.3 HOX GENES AS A SOURCE OF SYSTEMATIC CHARACTERS

### 4.3.1 SEQUENCE DATA

Homeoboxes are attractive for phylogeny reconstruction because their sequences are unambiguously alignable across the Metazoa. Furthermore, gene trees constructed for the purpose of identifying homeobox fragments (Cartwright *et al.*, 1993) or full-length homeodomain sequences (Balavoine and Telford, 1995) have contained considerable resolution, suggesting that homeobox sequences will be phylogenetically informative.

There are, however, a number of disadvantages in using homeobox genes for systematics. While it is easy to isolate *Hox* fragments using degenerate primers, sequence conservation of these genes decays rapidly outside the homeobox. Techniques such as library screening, RT–PCR, RACE, or inverse PCR must be employed to obtain the entire homeobox sequence, and considerable effort must thus be spent for only approximately 180 bp of sequence. Other classes of homeobox genes which show conserved regions outside the homeobox (Figure 5 in Bürglin, 1994) may prove more useful. The *engrailed* gene has conserved coding regions close to the homeobox which should allow amplification of approximately 300 bp across coelomate taxa. Holland and Williams (1990) and Wray *et al.* (1995) have utilized primer sets which amplify 181–189 bp of the more variable sequence within this region from vertebrates and molluscs, respectively.

Given the rigours of obtaining complete homeobox sequences, it seems counterproductive to target them as opposed to other types of genes for which much longer sequences can be amplified and cloned with equivalent effort. However, sequences of *Hox* and other homeobox gene fragments are rapidly accumulating in the literature from PCR-based surveys of these genes in various animals. An approach that needs to be explored is concatenation of homeobox fragments into longer sequences as input for analysis. The 5E/3F primer set (Pendleton *et al.*, 1993) amplifies an 82 bp fragment from *Hox* genes of paralogue groups 1–9, as well as *caudal* and *Chox-7* (*Ovx-1*) orthologues, from a broad range of phyla. Concatenation of any three, four or five fragments which could be obtained across a set of species representative of higher taxa of interest would yield perfectly aligned input sequences of 246, 328 or 410 bp, respectively, for phylogenetic analysis.

Another drawback of PCR-based surveys is negative data. Surveys of annelids have failed to detect orthologues of genes expected to be present (e.g. *pb* from *Stylaria* and *Abd-B* from *Ctenodrilus*) and it is uncertain whether these genes have been lost, or the method simply failed to detect them. Loss of the genes would be of interest in its own right, but either possibility would leave long stretches of missing data in concatenated sequences. Clearly, there is a need for optimization of surveys by PCR; theoretical considerations for doing so have been explored (Wagner *et al.*, 1993; Misof and Wagner, 1996).

Determining homology is yet another problem in using *Hox* genes as a source of sequence data. For phylogeny reconstruction, only orthologues rather than paralogues can be used; inclusion of paralogues in an analysis may lead to a correct gene tree but a spurious species tree (Fitch, 1970; Hillis and Moritz, 1990; Cartwright *et al.*, 1993). Genes from some *Hox* paralogue groups (lab/*Hox*1, pb/*Hox*2, *Hox*3 and Abd-B/*Hox*9) can be reliably identified across phyla on the basis of homeodomain fragments alone; others (Dfd/*Hox*4, Scr/*Hox*5, Antp/*Hox*6, Ubx/*Hox*7, and abd-A/*Hox*8) may require the entire homeodomain and sometimes flanking sequence for identification. An additional complication is that some *Hox* genes appear to have undergone independent duplications in particular lineages. Two copies of the *labial* gene, for example, were detected by PCR in the oligochaete *Stylaria lacustris* (Snow and Buss, 1994). However, if this duplication proved to be restricted to oligochaetes, or to *Stylaria*, either copy could be used in a higher-level phylogeny including *Stylaria* as a representative oligochaete, because the gene tree would still reflect the species phylogeny.

The current major impediment to exploring the use of *Hox* or other homeobox sequences for phylogeny reconstruction is the patchy distribution of available data, which have tended to accumulate incidentally from studies of developmental mechanisms. Within protostomes, existing data are highly skewed towards arthropods, and within arthropods towards insects. The remedy for paucity of data, of course, is to get more data. However, before targeting *Hox* or other homeobox genes, we need to assess whether they are going to be informative. No one to date has explored rigorously the properties of homeobox sequences for phylogeny reconstruction. An assessment of their utility should now be possible using the vertebrate data, for which complete homeobox sequences are available from numerous taxa, where orthology can often be established, and a well-supported existing phylogeny can be used as a control. The *engrailed* homeobox is another possibility.

#### 4.3.2 GENE STRUCTURE

Characters related to gene structure, such as introns or additional, conserved functional motifs, may prove systematically informative, but as with homeobox sequence data, their utilization is hampered by patchy distribution of existing data and the technical rigours of gathering additional data. Determination of gene structure outside the homeobox generally requires cumbersome isolation and sequencing of genomic and cDNA clones. Some *Hox* and other homeobox genes contain introns within the homeobox (Figure 22 in Bürglin, 1994), and the phylogenetic distribution of some of these introns might be determined by PCR. Variation in both the number of homeodomains and number of zinc-fingers in ZF class homeobox genes (Bürglin, 1994) might similarly be explored.

#### 4.3.3 PRESENCE OR ABSENCE OF ORTHOLOGUES

A more tractable, and potentially useful, type of character is the presence or absence of particular orthologues, indicative of gene duplication or deletion events. Relevant data can be gathered by PCR. An example pertinent to arthropod relationships is the timing of the duplication event which gave rise to *abd-A* and *Ubx* (Valentine *et al.*, 1996). The discovery of a single related gene, *Lox2*, in leeches (Wysocka-Diller *et al.*, 1989; Shankland *et al.*, 1991) suggested that this duplication might have occurred in arthropods after their divergence from annelids. However, two genes related to *Ubx* and *abd-A* are now known from the leech *Hirudo medicinalis* (*Lox 2* and *Lox 4*; Wong *et al.*, 1995), and orthologues of these leech genes have been found in the polychaete *Ctenodrilus serratus* (*CTs-Lox2* and *CTsx-2*, respectively; Dick and Buss, 1994). A branching diagram (Wong *et al.*, 1995) provides evidence that independent duplications led to *Ubx*/*abd-A* in arthropods and *Lox2*/*Lox4* in annelids. Here, some pertinent data exist from molluscs. A survey of *Aplysia californica* (F.H. Ruddle, personal communication) has detected putative orthologues of both *Lox2* and *Lox4* which are identical in derived amino acid sequence to the *CTs-Lox2* and *CTsx-2* fragments from *Ctenodrilus*. A duplication event leading to *Lox2* and *Lox4* in annelids apparently preceded the divergence of annelids and molluscs. Wong *et al.* (1995) referred cursorily to yet another gene, *Lox15* from *Hirudo*, which shows extensive similarity to *Lox2* and *Lox4*. Thus, either of these may have duplicated again in annelids.

While the distinctive *Lox2* homeodomain appears to be a synapomorphy for annelids and molluscs, data from a broader range of protostome taxa will be required to confirm this. Such data could be gathered from appropriate taxa by using the *Hox*5E (Pendleton *et al.*, 1993) primer in conjunction with another degenerate primer specific for the conserved 3′ flanking region characteristic of these homeoboxes.

#### 4.3.4 GENOMIC ORGANIZATION

Characters of this type include cluster number, gene order, gene spacing, and direction of transcription. Such data are generally not trivial to gather. The exception is cluster number, which with reasonable care can be inferred using PCR (Cartwright *et al.*, 1993; Misof and Wagner, 1996). Inferences of cluster number from estimated gene complement are based on the assumption that a cluster organization of *Hox* genes is conserved across the Metazoa. This assumption is in turn based on the broad phylogenetic distribution of the few taxa in which clusters have been genetically or physically mapped, e.g. the vertebrates *Mus* and *Homo*, the cephalochordate *Branchiostoma*, the insects *Drosophila* and *Tribolium*, and the nematode *Caenorhabditis*. One piece of evidence suggests that a cluster exists in cnidarians (Miller and Miles, 1993). Cluster number has not yet provided much information pertinent to arthropod relationships, because a single cluster has been inferred or mapped in representative molluscs (Degnan and Morse, 1993; F.H. Ruddle, personal communication), annelids (Shankland *et al.*, 1991; Dick and Buss, 1994; Snow and Buss, 1994), crustaceans (Averof and Akam, 1993) and hexapods. A notable exception within arthropods is *Limulus*, in which four clusters have been inferred (Cartwright *et al.*, 1993). It may be premature to rule out cluster number as a useful systematic character.

#### 4.3.5 GENE EXPRESSION

The use of homeobox gene expression patterns as systematic characters is problematic, and the problems stem from the need to treat characters at the proper level of homology, or synapomorphy (Dickinson, 1995). Consider the use of *Hox* genes in elucidating protostome relationships. *Hox* genes have a general role among metazoans in specification of axial positional information (McGinnis and Krumlauf, 1992), and this role is thus plesiomorphic for protostomes. Within a particular phylum, the *Hox* genes will be involved in patterning not only axial structures common to protostomes, such as the CNS, but also axial structures peculiar to the phylum. In the former case, expression patterns will be plesiomorphic, and in the latter case they are likely to be autapomorphic for the phylum. Similarly, the segment polarity gene *engrailed* is expressed in a segmentally iterated stripe at the posterior margin of segmental anlagen in leeches, insects, and crustaceans (Patel *et al.*, 1989; Wedeen and Weisblat, 1991; Ramírez *et al.*, 1995). While this expression is indicative of homology of metameres, this homology likely extends to a metameric organization primitive among protostomes.

In a beautiful example of the necessity of considering the appropriate level of homology in interpreting gene expression studies, Panganiban *et al.* (1997) have shown that expression of the *Distal-less* (*Dll*) gene and its orthologues in developing appendages of annelids and arthropods, as well as oncyophorans and deuterostomes, reflects a primitive genetic pathway underlying the formation of appendages and other body wall outgrowths. This pathway was likely utilized by the last common ancestor of protostomes and deuterostomes in patterning a body wall outgrowth, perhaps of sensory or locomotory function. Correlation of *Dll* expression with coelomate appendages represents homology of a patterning system, but not necessarily homology of appendages.

Comparative expression of segmentation and *Hox* genes is better known among arthropods than any other phylum (Akam, 1995), and more limited expression data exist for segmentation (Wedeen and Weisblat, 1991; Savage and Shankland, 1996) and *Hox* genes in annelids. Studies of *Hox* genes in annelids have largely been confined to leeches. The primary role of *Hox* genes in *Hirudo* and *Helobdella* appears to be specification of neuronal identities in the CNS (Shankland *et al.*, 1991; Nardelli-Haefliger and Shankland, 1992; Aisemberg and Macagno, 1994; Wong *et al.*, 1995; Berezovskii and Shankland, 1996), although expression is known from other tissues (Nardelli-Haefliger and Shankland, 1992). *Hox* genes are also known to control segment-specific neuronal differences in *Drosophila* (Coe and Scott, 1988). They are expressed in a broad range of other tissues (McGinnis and Krumlauf, 1992), but the role upon which most attention has been focused is specification of overt segmental identity via expression in the embryonic and larval epidermis.

Unfortunately, comparisons of *Hox* expression patterns between leeches and arthropods alone tell us nothing about annelid–arthropod relationships. Between the two, a role in specification of overt segmental (appendage) identities has been demonstrated only for arthropods. Even if *Hox* genes were found to determine, for example, differences between thoracic and abdominal parapodia in some annelids, this would not be evidence of homology between parapodia and jointed appendages, but only of a conserved axial patterning system. Nor can these comparisons tell us much about evolutionary transitions. We cannot *a priori* assume an annelid–arthropod sister group relationship, accept the idea that the common ancestor of arthropods was a homonomously segmented organism much like an annelid, and consider expression patterns in arthropods as derived. There is evidence that annelids and arthropods are not sister groups (Eernisse *et al.*, 1992). Even if they proved to be, interpretation of expression patterns would be meaningful only in the context of a protostome phylogeny. Comparable expression studies are lacking from the Mollusca, to say nothing of other minor protostome phyla such as echiurans and sipunculans.

## 4.4 AXIAL POSITION AND SEGMENTAL DIFFERENTIATION IN SETATE ANNELIDS

Among arthropods, patterns of tagmatization are of fundamental systematic importance, and these patterns are regulated by segmentation and *Hox* genes. It was initially suggested by Lewis (1978) that segmental diversification in arthropods was driven by *Hox* gene duplications. This hypothesis has been falsified (Akam *et al.*, 1994), partly on the basis of evidence that annelids have *Hox* clusters as extensive as those of arthropods (Nardelli-Haefliger and Shankland, 1992; Dick and Buss, 1994; Snow and Buss, 1994).

The question arises as to the role of all these *Hox* genes in the many annelids which do not show overt segmental differentiation. It is hard to avoid Snodgrass' (1938) legacy, which in the present context leads us to expect that *Hox* genes should play similar roles in patterning overtly homonomous annelids and primitively homonomous arthropods. A corollary is that both the regionalization of positional information and linkage of position to specification of overt segmental identity occurred independently in several arthropod lineages. Here I argue that many overtly homonomous annelids are essentially tagmatized (covert tagmatization), and that there is a continuum in how axial positional information is manifested in annelids, from regional differences in regeneration capability to specification of overt segmental identities. It may be that covert tagmatization was a preadaptation for overt tagmatization in both annelids and arthropods.

Beklemishev (1969) discussed structural heteronomy in annelids, both internal and external. Axial position in oligochaetes and polychaetes is reflected by different capabilities of various regions of the body to regenerate the full complement of missing segments following amputation. Older literature on annelid regeneration provides insights into this dynamic aspect of the body axis in setate annelids.

The syllid polychaete *Autolytus pictus* reaches a length of up to 80 similar setigerous segments. Okada (1929) found axial differences in regeneration ability shown in Table 4.1.

**Table 4.1** Axial differences in completeness of the anterior regenerate in the syllid polychaete *Autolytus pictus* (Data from Okada, 1929.)

| *Position of amputation at setigerous segment* | *Extent of regenerate* |
|---|---|
| 1–6 | Head* + all segments |
| 7 | Head + 3–7 segments |
| 8 | Head + 3–8 segments |
| 9–13 | Head + 4 segments |
| 14–42 | Head + 0–1 segments; no pharynx |
| 42–pygidium | No regeneration |

*Head = prostomium + peristomium.

Worms cut anywhere from segments 1 to 6 replace the head region (prostomium and peristomium) and the correct number of missing segments; worms cut at segment 7 replace the head region and as few as three, but up to the correct number, of missing segments. From segment 14 posteriorly to segment 42, the worm replaces at most the head region and one setigerous segment. Segment 14 has further significance. Before spawning, *A. pictus* produces a 'stolon,' an individual specialized for gamete production and swimming, by metamorphosis of existing segments. The stolon head always differentiates from segment 14. Similar axial differences in extent of regeneration are also known for oligochaetes (Goss, 1969). Studies of this type demonstrate that axial position is accurately maintained in adult worms, that it is regionalized, and that it can be re-established following amputation.

Other studies clearly demonstrate that positional information is linked to segmental differentiation. Some polychaetes such as fanworms are overtly tagmatized into a head region (prostomium and tentacle-bearing collar segment), thorax, and abdomen, with thoracic and abdominal segments differing in the arrangements of setae and ciliary tracts. Wherever it is cut along the body, the fanworm *Sabella pavonina* regenerates only the head region and one setigerous segment of thoracic form (reviewed by Berrill, 1952). If bisected at the level of the abdomen, *S. pavonina* regenerates this typical complement from the posterior piece, which, however, now has fewer than the normal number of 5–11 thoracic segments. To re-establish the proper thorax length, external regulation occurs in which an appropriate number of abdominal segments undergo metamorphosis to thoracic segments.

Internal regulation of segment-specific structures is also known. The oligochaete *Lumbriculus variegatus*, for example, regenerates no more than eight segments anteriorly, regardless of where amputation occurs, then reorganizes internal structures such as vascular commissures and nephridia in the following 10 segments of the old worm to re-establish the organization of a normal worm (Berrill, 1952).

After elegantly reviewing regeneration experiments in polychaetes and oligochaetes, Goss (1969) wrote, '... segments [in annelids] are not identical like the links in a chain. Each one of them has a measure of individuality and occupies a specific position in relation to all the others ... we cannot escape the conclusion that completeness in a regenerate is an expression of qualitative factors operating in segment differentiation.' To the modern developmental geneticist, Goss' mention of qualitative factors evokes the possibility of segmentation or *Hox* gene involvement (Buss and Dick, 1992).

What little is known of the roles of *Hox* genes in annelids comes from an exclusive focus on leeches, largely due to interest in these organisms by neurobiologists. Leeches are highly specialized, and share several traits with arthropods. Both have a stereotyped number of adult segments (although some arthropods show inconstancy in segment number; Beklemishev, 1969), lack the capacity for axial regeneration, and pattern the A/P axis only once, during early ontogeny. These traits are not characteristic of other annelids. Polychaetes tend to have indeterminate growth and exhibit considerable capacities for axial regeneration; many of them utilize fragmentation or stolonization in asexual propagation or spawning (Berrill, 1952; Herlant-Meewis, 1964; Schroeder and Hermans, 1975). Oligochaetes have a more-or-less stereotyped number of adult segments, but are capable of axial regeneration; some also undergo fragmentation or stolonization.

A distinction can be made between a static axis, which is laid down once early in ontogeny, and a dynamic axis which remains capable of respecification throughout ontogeny. Most coelomates, including those in which segmentation and *Hox* gene expression studies have been conducted, have a static axis. Leeches, arthropods, and molluscs have all foregone a dynamic axis in favour of useful novelties. Leeches have developed a posterior sucker, which prohibits maintenance of a posterior growth zone (the sucker itself represents an overt tagma involving seven fused segments). Arthropods have hardened the body wall with an exoskeleton and linked growth to moulting, making axial regeneration prohibitively slow. Most molluscs have encased the axis in a rigid shell; postembryonic growth is not possible at the posterior end, but is accomplished by peripheral addition of new material by the mantle. All three groups have adopted haemocoelic circulation to a greater or lesser extent, with a concomitant loss of aspects of internal segmental individuality.

Nemerteans and many platyhelminths are capable of regeneration and fragmentation (Berrill, 1952). It is possible that, among bilateral metazoans, use of axial positional information was primitively a dynamic process throughout the life of the organism, rather than one geared toward a static endpoint through embryogenesis. Alternatively, the dynamic axis of setate annelids may itself be a novelty, as useful as any of those which prohibit maintenance of such an axis. This is an important distinction. Selective forces acting on patterning genes for a primary role early in ontogeny may be different from those for a role in laying down the initial axis and then maintaining it throughout ontogeny.

Whether or not *Hox* genes were primitively used in patterning a static versus a dynamic axis has a bearing on the role they played in early metazoan evolution. In setate annelids, both the regionalization of axial positional information, and the manner in which it is manifested as segmental identity, appear to be quite variable. This suggests caution in inferring the homology of tagmata among arthropods on the basis of gene expression patterns. A shift early in arthropod evolution from a dynamic, variable axis to a static axis might have preserved or 'frozen' some of this variability as different patterns of tagmatization.

Additional gene expression studies are clearly needed from across the range of protostome phyla better to understand the evolutionary roles of segmentation and *Hox* genes in the radiation of metazoans. Setate annelids are of particular interest. They use positional information along a dynamic axis in a diversity of ways, and this should provide a wealth of 'natural experiments' pertinent to this goal.

## 4.5 CONCLUSIONS

Homeobox genes, and particularly the *Hox* genes, have attracted interest as a source of data both for phylogeny reconstruction and for understanding the divergence and radiation of metazoan body plans. Annelid–arthropod relationships provide a point of focus for this paper to discuss expectations and how, or if, they might be met. *Hox* genes provide several types of characters potentially useful for systematics: sequence data; gene structure; presence or absence or orthologues; genomic organization, including cluster number; and expression patterns. These characters have varying degrees of suitability as data for phylogeny reconstruction. The relevance of homeobox gene expression studies in annelids to understanding tagmatization in arthropods is limited by the fact that most studies have been upon leeches, which are atypical of other annelids. They share with arthropods a stereotyped number of segments, inability to regenerate, and specification of the body axis once, early in ontogeny. Setate annelids (polychaetes and oligochaetes) use axial positional information dynamically throughout ontogeny in processes of continuous growth, regeneration and fragmentation. Many overtly homonomous annelids display covert tagmatization, with axial position reflected in differential regeneration capability. The suggestion is advanced that a primitively regionalized, dynamic, variable body axis might have served as a preadaptation for overt tagmatization in both annelids and arthropods. A switch from a dynamic to a static axis early in arthropod evolution might explain differences in patterns of tagmatization among arthropod subphyla.

ACKNOWLEDGEMENTS

I thank Lisa Nagy, Leo Buss, Neil Blackstone, and Keith Magni for useful and insightful comments on various drafts of the manuscript. This work was supported by the Middlebury College Faculty Professional Development Fund.

## REFERENCES

Aisemberg, G.O. and Macagno, E.R. (1994) *Lox1*, an *Antennapedia*-class homeobox gene, is expressed during leech gangliogenesis in both transient and stable central neurons. *Developmental Biology*, **161**, 455–65.

Akam, M. (1995) Hox genes and the evolution of diverse body plans. *Philosophical Transactions of the Royal Society of London Series B Biological Sciences*, **349**, 313–19.

Akam, M., Averof, M., Castelli-Gair, J., Dawes, R., Falciani, F. and Ferrier, D. (1994) The evolving role of Hox genes in arthropods. *Development*, Supplement, 209–15.

Averof, M. and Akam, M. (1993) HOM/Hox genes of *Artemia*: implications for the origin of the insect and crustacean body plans. *Current Biology*, **3**, 73–8.

Balavoine, G. and Telford, M.J. (1995) Identification of planarian homeobox sequences indicates the antiquity of most Hox/homeotic gene subclasses. *Proceedings of the National Academy of Sciences of the USA*, **92**, 7227–31.

Beklemishev, W.N. (1969) *Principles of Comparative Anatomy of Invertebrates, Volume 1, Promorphology*, University of Chicago Press, Chicago.

Berezovskii, V.K. and Shankland, M. (1996) Segmental diversification of an identified leech neuron correlates with the segmental domain in which it expresses *Lox2*, a member of the *Hox* gene family. *Journal of Neurobiology*, **29**, 319–29.

Berrill, N.J. (1952) Regeneration and budding in worms. *Biological Review*, **27**, 401–38.

Bürglin, T.R. (1994) A comprehensive classification of homeobox genes, in *Guidebook to the Homeobox Genes* (ed D. Duboule), Oxford University Press, Oxford, pp. 27–71.

Buss, L.W. and Dick, M.H. (1992) The middle ground of biology: themes in the evolution of development, in *Molds, Molecules, and Metazoa* (eds P.R. Grant and H.S. Horn), Princeton University Press, Princeton, pp. 77–97.

Cartwright, P., Dick, M. and Buss, L.W. (1993) HOM/Hox type homeoboxes in the chelicerate *Limulus polyphemus*. *Molecular Phylogenetics and Evolution*, **2**, 185–92.

Coe, C.Q. and Scott, M.P. (1988) Segmentation and homeotic gene function in the developing nervous system of *Drosophila*. *Trends in Neurosciences*, **3**, 101–6.

Degnan, B.M. and Morse, D.E. (1993) Identification of eight homeobox-containing transcripts expressed during larval development and at metamorphosis in the gastropod mollusc *Haliotis rufescens*. *Molecular Marine Biology and Biotechnology*, **2**, 1–9.

Dick, M.H. and Buss, L.W. (1994) A PCR-based survey of homeobox genes in *Ctenodrilus serratus* (Annelida: Polychaeta). *Molecular Phylogenetics and Evolution*, **3**, 146–58.

Dickinson, W.J. (1995) Molecules and morphology: where's the homology? *Trends in Genetics*, **11**, 119–21.

Eernisse, D.J., Albert, J.S. and Anderson, F.E. (1992) Annelida and Arthropoda are not sister taxa: a phylogenetic analysis of spiralian metazoan morphology. *Systematic Biology*, **41**, 305–30.

Fitch, W.M. (1970) Distinguishing homologous from analogous proteins. *Systematic Zoology*, **19**, 99–113.

Gehring, W.J. (1994) A history of the homeobox, in *Guidebook to the Homeobox Genes* (ed D. Duboule), Oxford University Press, Oxford, pp. 3–10.

Goss, R.J. (1969) *Principles of Regeneration*, Academic Press, New York.

Herlant-Meewis, H. (1964) Regeneration in annelids, in *Advances in Morphogenesis*, Volume 4 (eds M. Abercrombie and J. Brachet), Academic Press, New York, pp. 155–215.

Hillis, D.M. and Moritz, C. (1990) *Molecular Systematics*, Sinauer Associates, Sunderland.

Holland, P. (1992) Homeobox genes in vertebrate evolution. *BioEssays*, **14**, 267–73.

Holland, P.W.H. and Williams, N.A. (1990) Conservation of *engrailed*-like homeobox sequences during vertebrate evolution. *FEBS Letters*, **277**, 250–2.

Kappen, C., Schughart, K. and Ruddle, F.H. (1989) Two steps in the evolution of Antennapedia-class vertebrate homeobox genes. *Proceedings of the National Academy of Sciences of the USA*, **86**, 5459–63.

Lawrence, P.A. (1992) *The Making of a Fly*, Blackwell Scientific Publications, London.

Lewis, E.B. (1978) A gene complex controlling segmentation in *Drosophila*. *Nature*, **276**, 565–70.

Lewis, E.B. (1992) Clusters of master control genes regulate the development of higher organisms. *Journal of the American Medical Association*, **267**, 1524–31.

Marx, J. (1992) Homeobox genes go evolutionary. *Science*, **255**, 399–401.

McGinnis, W. and Krumlauf, R. (1992) Homeobox genes and axial patterning. *Cell*, **68**, 283–302.

Miller, D.J. and Miles, A. (1993) Homeobox genes and the zootype. *Nature*, **365**, 215–16.

Misof, B.Y. and Wagner, G.P. (1996) Evidence for four *Hox* clusters in the teleost, *Fundulus heteroclitus*. *Molecular Phylogenetics and Evolution*, **5**, 309–22.

Nardelli-Haefliger, D. and Shankland, M. (1992) *Lox2*, a putative leech segment identity gene, is expressed in the same segmental domain in different stem cell lineages. *Development*, **116**, 697–710.

Okada, Y.K. (1929) Regeneration and fragmentation in the syllidian polychaetes. *Roux' Archiv für Entwicklungsmechanik*, **45**, 602–85.

Panganiban, G., Irvine, S.M., Lowe, C., Roehl, H., Corley, L.S., Sherbon, B., Grenier, J.K., Fallon, J.F., Kimble, J., Walker, M., Wray, G.A., Swalla, B.J., Martindale, M.Q., and Carroll, S.B. (1997) The origin and evolution of animal appendages. *Proceedings of the National Academy of Sciences of the USA*, **94**, 5162–6.

Patel, N.H., Kornberg, T.B. and Goodman, C.S. (1989) Expression of *engrailed* during segmentation in grasshopper and crayfish. *Development*, **107**, 201–12.

Pendleton, J.W., Nagai, B., Murtha, M.T. and Ruddle, F.H. (1993) Expansion of the *Hox* gene family and the evolution of chordates. *Proceedings of the National Academy of Sciences of the USA*, **90**, 6300–4.

Raff, R.A. (1994) Developmental mechanisms in the evolution of animal form: origins and evolvability of body plans, in *Early Life on Earth* (ed. S. Bengtson), Columbia University Press, New York, pp. 489–500.

Ramírez, F-A., Wedeen, C.J., Stuart, D.K., Lans, D. and Weisblat, D.A. (1995) Identification of a neurogenic sublineage required for CNS segmentation in an annelid. *Development*, **121**, 2091–7.

Ruddle, F.H., Bentley, K.L., Murtha, M.T. and Risch, N. (1994) Gene loss and gain in the evolution of the vertebrates. *Development*, Supplement, 155–61.

Savage, R.M. and Shankland, M. (1996) Identification and characterization of a *hunchback* orthologue, *Lzf2*, and its expression during leech embryogenesis. *Developmental Biology*, **175**, 205–17.

Schroeder, P.C. and Hermans, C.O. (1975) Annelida: Polychaeta, in *Reproduction of Marine Invertebrates, Volume III, Annelids and Echiurans* (eds A.C. Giese and J.S. Pearse), Academic Press, New York, pp. 1–213.

Shankland, M., Martindale, M.Q., Nardelli-Haefliger, D., Baxter, E. and Price, D.J. (1991) Origin of segmental identity in the development of the leech nervous system. *Development*, Supplement **2**, 29–38.

Slack, J.M.W., Holland, P.W.H. and Graham, C.F. (1993) The zootype and the phylotypic stage. *Nature*, **361**, 490–2.

Snodgrass, R.E. (1938) Evolution of the Annelida, Onycophora, and Arthropoda. *Smithsonian Miscellaneous Collections*, **97**(6), 1–159.

Snow, P. and Buss, L.W. (1994) HOM/Hox-type homeoboxes from *Stylaria lacustris* (Annelida: Oligochaeta). *Molecular Phylogenetics and Evolution*, **4**, 360–4.

Tiegs, O.W. and Manton, S.M. (1958) The evolution of the Arthropoda. *Biological Reviews of the Cambridge Philosophical Society*, **33**, 255–337.

Valentine, J.W., Erwin, D.H. and Jablonski, D. (1996) Developmental evolution of metazoan body plans: the fossil evidence. *Developmental Biology*, **173**, 373–81.

Wagner, A., N. Blackstone, P. Cartwright, M. Dick, B. Misof, P. Snow, G. P. Wagner, J. Bartels, M. Murtha and Pendelton, J. (1993) Surveys of gene families using polymerase chain reaction: PCR selection and PCR drift. *Systematic Biology*, **43**, 250–61.

Wang, B.B., Müller-Immergluck, M.M., Austin, J., Robinson, N.T., Chisholm, A. and Kenyon, C. (1993) A homeotic gene cluster patterns the anteroposterior body axis of *C. elegans*. *Cell*, **74**, 29–42.

Wedeen, C.J. and Weisblat, D.A. (1991) Segmental expression of an *engrailed*-class gene during early development and neurogenesis in an annelid. *Development*, **113**, 805–14.

Wong, V.Y., Aisemberg, G.O., Gan, W.-B. and Macagno, E.R. (1995) The leech homeobox gene *Lox4* may determine segmental differentiation of identified neurons. *The Journal of Neuroscience*, **15**, 5551–9.

Wray, C.G., Jacobs, D.K., Kostriken, R., Vogler, A.P., Baker, R. and DeSalle, R. (1995) Homologues of the *engrailed* gene from five molluscan classes. *FEBS Letters*, **365**, 71–4.

Wysocka-Diller, J.W., Aisemberg, G.O., Baumgarten, M., Levine, M. and Macagno, E.R. (1989) Characterization of a homologue of bithorax-complex genes in the leech *Hirudo medicinalis*. *Nature*, **341**, 760–3.

# 5 Arthropod and annelid relationships re-examined

*D.J. Eernisse*

*Department of Biology MH282, California State University, Fullerton, CA 92834, USA*
*email: DEernisse@fullerton.edu*

## 5.1 INTRODUCTION

The prevailing view of two centuries recognizes annelid worms as the sister taxon of arthropods. However, recent studies have suggested that there are other animal groups nearer to annelids; under this view arthropods do not belong to the clade, Eutrochozoa Ghiselin, 1988, comprising annelids, molluscs, and several other protostome phyla. This recent work is based on both morphology (Eernisse *et al.*, 1992; Schram and Ellis, 1995) and molecular sequence comparisons; the latter is based on several gene regions, including 18S rRNA (Field *et al.*, 1988; Ghiselin, 1988; Patterson, 1989; Lake, 1990; Turbeville *et al.*, 1991; Ruitort *et al.*, 1993; Valentine, 1994; Halanych *et al.*, 1995; Winnepenninckx *et al.*, 1995a,b; Giribet *et al.*, 1996), mitochondrial 12S rRNA (Ballard *et al.*, 1992), and the two largest subunits of RNA polymerase II (Sidow and Thomas, 1994). However, Rouse and Fauchald (1995) continued to find support for the conventional grouping of annelids and arthropods as sister taxa, termed Articulata Cuvier, 1817, in their cladistic analysis based on 13 morphological characters.

The Eutrochozoa hypothesis challenges the interpretation of segmentation arising as a shared evolutionary novelty restricted to the hypothesized clade of 'articulate' animals: annelids and arthropods. Conventionally, segmentation is restricted to Articulata, a clade that would also almost certainly include other 'segmented' animals such as the arthropod-like onychophorans, tardigrades, and possibly the annelid-like pogonophorans. On the other hand, if annelids are more distantly related to arthropods, this requires that segmentation in annelids and arthropods arose homoplastically. Teloblastic segment addition in annelids and arthropods could be entirely convergent, or parallel in its evolution from a more general plesiomorphic metamerism. It might even be a reversal, with unsegmented animals such as molluscs descended from segmented ancestors. These different possible optimizations of segmentation as a homoplasy suggest some intriguing possible reinterpretations of patterns of evolution in early bilaterian animals. For example, segmentation in vertebrates, which is normally considered to have evolved independently from that in annelids and arthropods, might be reconsidered as an expression of a deeper metamerism shared by all three groups and present in their common ancestor. Evidently, some lineages within this clade lost apparent traces of this basic metameric body plan.

It is clearly important to establish the correct outgroup for the arthropods. Researchers have often established polarity for variation in arthropod traits on the assumption that the ancestor of arthropods was annelid-like. However, if the more proximal outgroup were different this would evidently change our understanding of character evolution in arthropods. Outgroup identification has a pivotal role in phylogenetic analysis and later optimization of character change in an ingroup of interest – the best outgroups are the closest (Maddison *et al.*, 1984). Many of the numerous studies seeking to unravel the biology and palaeontology of arthropods or annelids would be open to new interpretations should a living group other than annelids become the probable arthropod sister taxon. For example, there is widespread use of arthropods as model organisms in fields such as medical and molecular biology. A changed view could be especially relevant to recent discoveries that genes, including linked clusters of *Hox* genes, have a likely homologous role in pattern formation for all bilaterian animals (Carroll, 1995; De Robertis and Sasai, 1996).

If some grouping of annelids is not sister taxon, or paraphyletic, to Arthropoda *s.l.*, then what is? There is little consensus. Studies that do not support the conventional Articulata hypothesis disagree about which other members of Bilateria most recently share a common ancestor with arthropods. Rouse and Fauchald (1995) are among those who do not accept the evidence for Eutrochozoa. They consider the lack of viable alternative hypotheses as a primary reason to question results based on sequence comparison. They are proponents of Articulata, which they define (de Queiroz and Gauthier, 1990, 1992) as the clade stemming from the first ancestor to show segmentation (i.e. repetition of homologous body structures derived by teloblastic growth) and longitudinal musculature broken

*Arthropod Relationships*, Systematics Association Special Volume Series 55, Edited by R.A. Fortey and R.H. Thomas. Published in 1997 by Chapman & Hall, London. ISBN 0 412 75420 7

into bands. I follow this definition here. Rouse and Fauchald and others (Nielsen, 1995; Haszprunar, 1996) hypothesize that segmentation is a shared derived condition uniting annelids with arthropods (plus onycophorans and tardigrades) and annelids (plus pogonophorans and sometimes also echiurans) to the exclusion of other members of a more inclusive clade of metazoans with bilateral symmetry (Bilateria).

Rouse and Fauchald's (1995) notion of Articulata is compatible with a variety of previous views because it does not require arthropod monophyly. Arthropods have usually been regarded as derived from an exoskeleton-bearing ancestor that was more or less closely related to annelids (e.g. Boudreaux, 1979), but a common variation is to have two or more 'arthropod' lineages arising independently from anything ranging from an annelid-like worm to a segmented haemocoel-bearing ancestor quite unlike modern annelids, but presumably with a segmented condition traceable to the common ancestor of annelids and arthropods (Tiegs and Manton, 1958; Manton, 1977; Anderson, 1981; Willmer, 1990; Fryer, 1996). All of these views are consistent with Articulata as construed here.

I will attempt to identify the sister group of a broad clade derived from the most recent common ancestor of arthropods, tardigrades, and onychophorans (Nielsen, 1995; Dewel and Dewel, 1996). This group is referred to as Arthropoda *sensu lato*. I am less concerned with interrelationships of groups within Arthropoda such as whether onychophorans or tardigrades (or both) might be arthropod subgroups (Schram, 1991; Ballard *et al.*, 1992; Wheeler *et al.*, 1993; Garey *et al.*, 1996a; Giribet *et al.*, 1996; Moon and Kim, 1996) and whether myriapods or crustaceans are the more proximal sister taxon of insects (Boore *et al.*, 1995; Friedrich and Tautz, 1995; Budd, 1996b). The members of Arthropoda *s.l.* include an ever-expanding assemblage of Cambrian arthropods or near arthropods whose fossils are the most widely studied of any Cambrian fossil group (Briggs *et al.*, 1993; Budd, 1996a; Fortey *et al.*, 1996; Waggoner, 1996).

Some version of Articulata is depicted in nearly all current zoology textbooks (Brusca and Brusca, 1990; review by Eernisse *et al.*, 1992) including many examples of diagrams (Barnes, 1987) that depict arthropods arising from an annelid ancestor. Articulata is often assumed in studies making broad comparisons among animals (Davidson *et al.*, 1995). Even some authors employing cladistic methods have implicitly (Wheeler *et al.*, 1993; Wheeler, 1995) or explicitly (Haszprunar, 1996) assumed the monophyly of Articulata in their analyses, or have favoured this hypothesis despite lack of evident support in their cladistic analyses (Nielsen *et al.*, 1996). The early results of Meglitsch and Schram (1991) in favour of Articulata were demonstrated to be unresolved upon reanalysis (Eernisse *et al.*, 1992), and became consistent with Eutrochozoa when some corrections to the original data were made (Schram and Ellis, 1995).

This study reports one set of new analyses featuring the most widely sampled gene for metazoans, 18S rRNA. There are now over 300 metazoan 18S rRNA sequences available, but most authors have incorporated only a small fraction of these available sequences into their analyses. Nearly all published comparisons on metazoan 18S rRNA have been based on fewer than 20 sequences. The danger is that this may be too few sequences to root accurately the ingroup of interest. For example, based on simulation studies, Sanderson (1996) suggested that a minimum of 40 taxa are needed for identifying the root node of a large clade. Lecointre *et al.* (1993) reached similar conclusions for a real data set composed of 31 taxa. They concluded that inferences obtained from subsets of 4, 8, 16 or 24 species were not congruent with the 31-species result.

Philippe *et al.* (1994) did not explicitly consider the effects of taxon sampling in a more recent article on the 'Cambrian explosion' of metazoan radiations. Based on bootstrap replicates of a resampled data set of 55 or 69 18S rRNA sequences clustered using the neighbour-joining algorithm, Philippe *et al.* reached the conclusion that 18S rRNA sequences were unlikely to resolve cladogenetic events involved with the radiation of animal phyla, even if much more data of the same sort was available. While their demonstration may extend to their particular data set, would a more taxonomically diverse sample of sequences still be unlikely to resolve these relationships?

Herein, up to 103 sequences were analysed simultaneously. Three selected subsets of 28 or 29 sequences, and one subset of 66 sequences, were analysed separately in order to test whether inferences would have differed without inclusion of so many sequences. The 18S rRNA gene was selected for comparison because it is presently the only gene region for which numerous sequences exist that are relevant to testing competing support for the Eutrochozoa and Articulata hypotheses. Tests of these hypotheses must minimally include at least one representative and comparable sequence for each of three critical taxa, arthropods, annelids, molluscs, and must also have at least one outgroup representative. This requirement alone eliminates most genes or gene regions sampled for any single group from present consideration. Also, it is likely that a relatively broad taxonomic sampling will be necessary to accurately infer high-level metazoan relationships. While protein-coding genes could hold great promise for inferring metazoan relationships, the most important reason to use the 18S rRNA gene is its broad taxonomic representation (at least one complete sequence is available for nearly all animal phyla). It is the only gene that currently satisfies Sanderson's (1996) recommendation to include at least 40 ingroup sequences to confidently infer the root node. Simulations also support the notion that the accuracy of trees found by parsimony algorithms can be expected to increase as more taxa are included (Hillis, 1996; but see Kim, 1996).

There are additional advantages that have contributed to the popularity of the 18S rRNA gene for use in estimation of phylogeny. It is likely homologous with other small subunit ('16S-like') ribosomal RNAs found in all living organisms, and because it has variable regions interspersed by less variable regions, it has demonstrated utility for resolving from ancient (e.g. all life) to recent (e.g. within species group) divergences (Sogin, 1991). Invariant regions allow the use of 'universal' oligonucleotide primers, which enable broad taxonomic sampling. The 18S rRNA gene is thought to be relatively free from confounding problems of gene duplication that could lead to inadvertent comparison of paralogous gene copies, despite it being in multiple tandemly repeated units within the genome. This is because the 18S rRNA gene exhibits concerted evolution such that intraspecific variation is slight (Hillis and Dixon, 1991). Its increasingly well understood secondary structure (Kjer, 1995; Halanych, 1996b) is potentially useful in phylogenetic studies.

## 5.2 METHODS

This study is based on a large manual alignment of metazoan and outgroup 18S rRNA sequences obtained from GenBank, starting with the partial sequences reported by Field *et al.* (1988) and expanded in subsequent years and repeatedly analysed as additional sequences have become available. The present analyses are based on a working alignment restricted to 188 mostly entire 18S rRNA sequences, derived from a larger alignment of more than 300 complete or partial sequences. These analyses include a relatively complete analysis based on 103 selected sequences, as well as additional analyses that test how sensitive results might be to pruning of sequences.

Of the 85 of the 188 sequences excluded from analyses, over half were considered to be taxonomically redundant, over 20 were sequences made available by the kind cooperation of Dr G.-S. Min (personal communication) but have been excluded here because they are still awaiting publication. The unpublished sequences are taxonomically diverse and had an important role in improving the alignment despite their exclusion from presented analyses. A remaining few sequences were excluded because they were considered problematic (e.g. suspected of misidentification). Details of those 103 sequences included in the present analyses are available on the world web site: http:\www.systassoc.chapmanhall.com. These details include their accession number, name used in the cladograms figured here, abbreviated classification, and sequence length. The 53 associated citations are too numerous to cite individually but full citations are available from GenBank, as referenced by accession number, and are provided as part of the documentation in the available data file (see below).

The 188 mostly entire 18S rRNA sequences were aligned manually with format conversions and error checking performed using software by Eernisse (1992, 1995). A secondary structure model, modified from Neefs *et al.* (1990), was maintained with the alignment and utilized to a limited extent to aid with the alignment. Visual inspection of alignment ambiguities was performed by eye as assisted by colour sequence editor software by Eernisse (1992, 1995; also unpublished results) and Swofford and Eernisse (unpublished results), without which the visual inspection and manipulations performed would have been substantially more difficult. Sites with remaining alignment ambiguities were identified by eye and excluded from all analyses reported here in order to avoid possible systematic errors. Altogether, 922 out a total of 3008 sites were excluded. Considering the large number of sequences to be aligned, the manual methods were found to be more feasible than explored algorithm-based methods and, judging from visual inspection of attempts to use alignment algorithms for parts of this data set, manual methods appeared to lead to more believable results with substantially fewer, and better placed, gaps.

Phylogenetic analyses were performed with the program, PAUP* 4.0d47 (provided by D.L. Swofford, Smithsonian Institution), using parsimony as the search criterion. Because there was no compelling reason to do otherwise for this data set, all sites were weighted equally (Eernisse *et al.*, 1992; Eernisse and Kluge, 1993; Sullivan, 1996). Use of PAUP*'s exact ('branch-and-bound') algorithm was not feasible given the large number of sequences analysed (Swofford *et al.*, 1996). Instead, searches employed PAUP*'s most thorough tree bisection–reconnection branch swapping heuristic search algorithm (Swofford and Beagle, 1993) in conjunction with 20 (or 10 for the 103-sequence analysis) random addition sequences in order to more effectively find islands of most parsimonious trees (Maddison, 1991). Gaps were considered as missing data in these analyses.

All analyses were based on the same alignment and included sites, differing only in which sequences were included. The first analysis (Analysis 1) compared all 103 sequences listed on http:\www.systassoc.chapmanhall.com so that taxonomic sampling was relatively more dense than in previous published studies, with multiple outgroups to the ingroup, Bilateria. The second analysis (Analysis 2) included 66 of the same 103 sequences and had only a single non-bilaterian outgroup, a cnidarian. The remaining three analyses (Analyses 3–5) were based on three data sets, each comprising 28 or 29, mostly different, 18S rRNA sequences. These also included the same cnidarian outgroup as in Analysis 2. The sequences selected for Analyses 3 to 5 were selected to be approximately comparable with each other, and to represent sampled diversity within Bilateria as fully as possible with the fewer numbers of sequences. The sequences represented in Analyses 1 to 5 are indicated by their presence in the resulting cladograms (Figures 5.1–5.4).

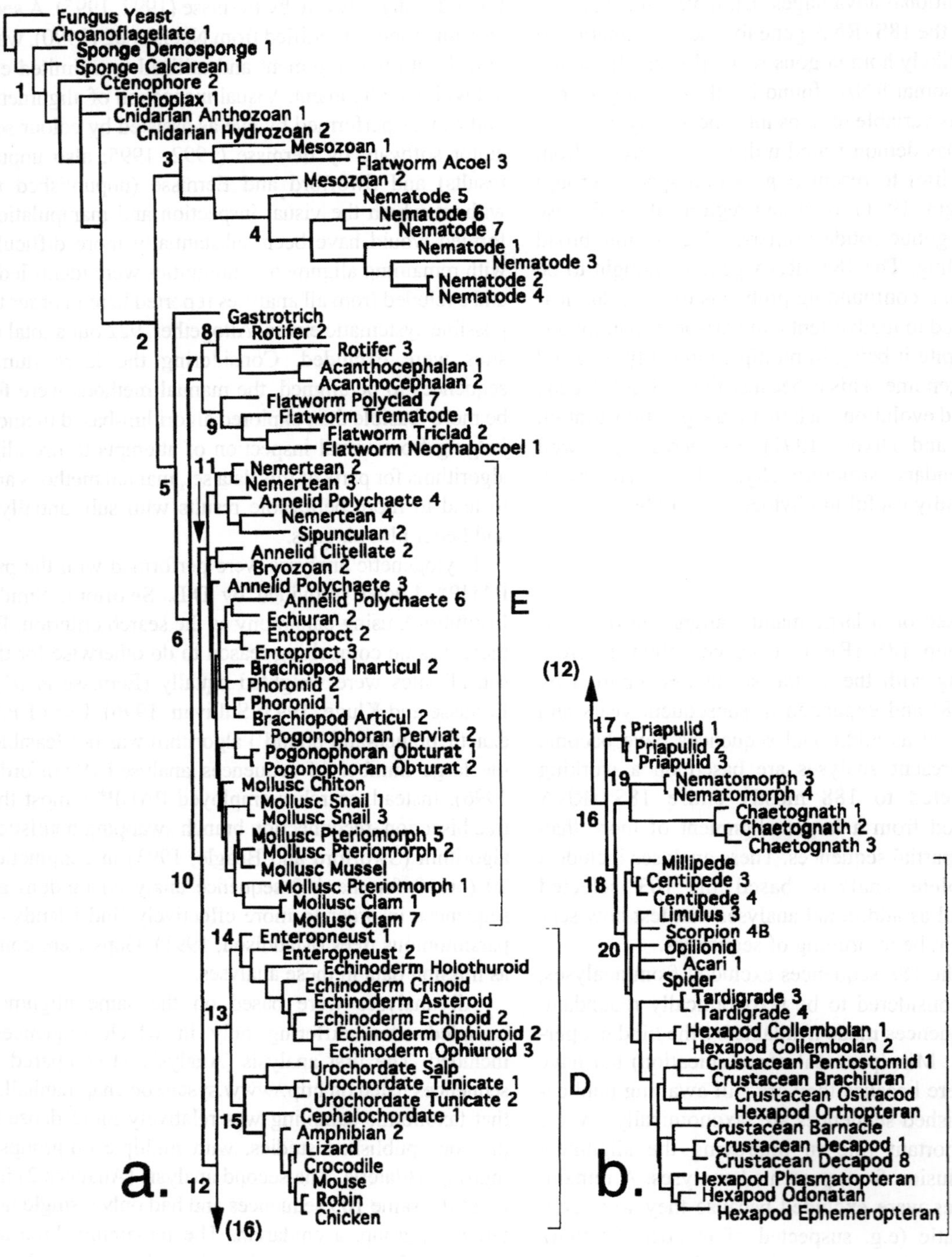


**Figure 5.1** (a,b) 18S rRNA parsimony strict consensus phylogram summary of 20 minimum length trees (length = 7860, C.I. excluding uninformative characters = 0.2477; see Analysis 1, Table 5.2 for additional details). Horizontal branch length is proportional to the relative number of synapomorphies supporting particular internal or terminal branches. Numbers at nodes correspond to numbered clades in text. Figure 5.1(a) joins to 5.1(b) by connecting arrows (clade 13 and 16 are sister taxa, joined at node 12). For a simplified version, see Figure 5.3(b). Corresponding scientific names and sequence information are provided on the www site. Abbreviations: E, Eutrochozoa; D, Deuterostomia; A, Arthropoda *s.l.* (see text).

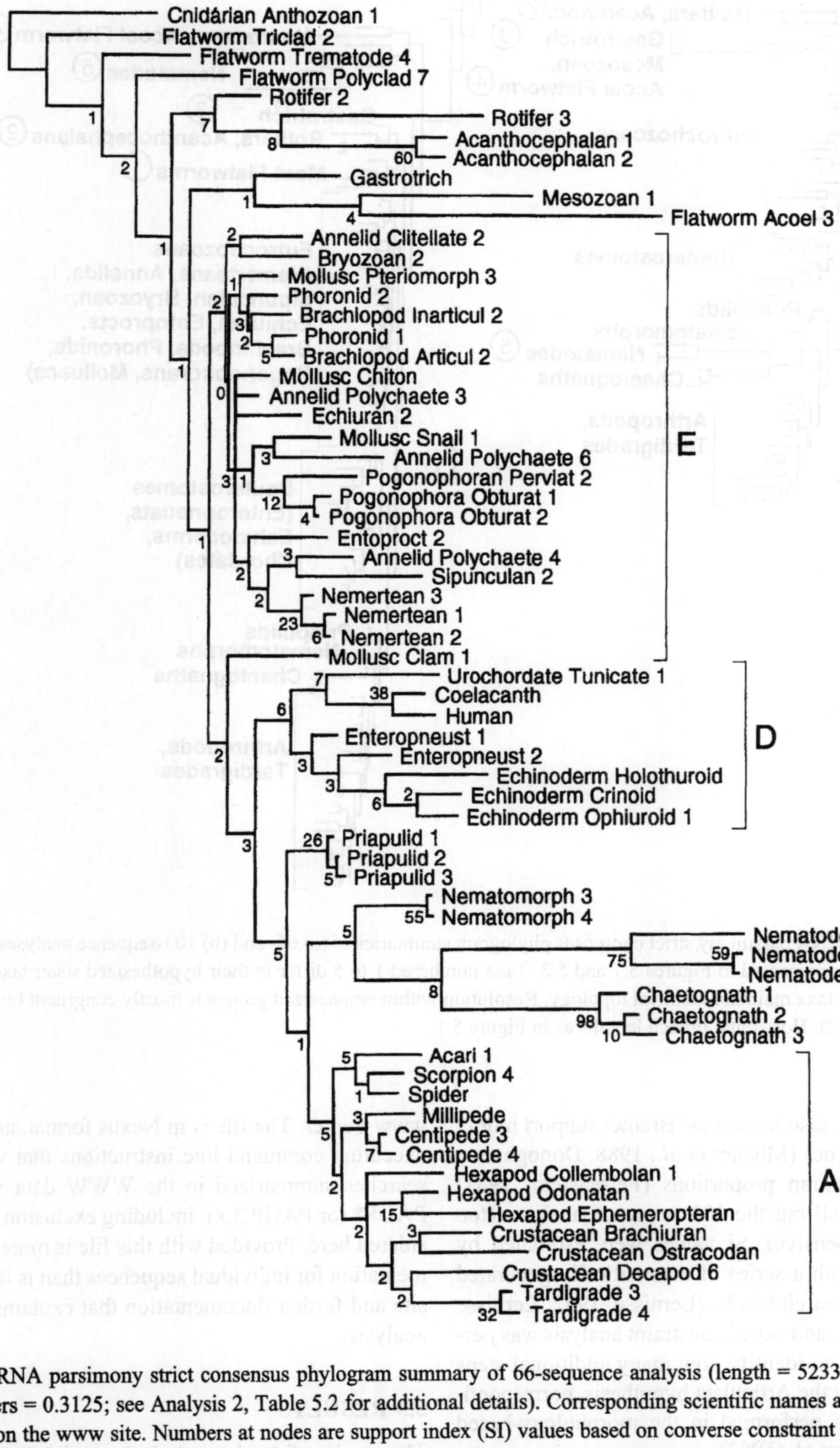


**Figure 5.2** 18S rRNA parsimony strict consensus phylogram summary of 66-sequence analysis (length = 5233, C.I. excluding uninformative characters = 0.3125; see Analysis 2, Table 5.2 for additional details). Corresponding scientific names and sequence information are provided on the www site. Numbers at nodes are support index (SI) values based on converse constraint searches with PAUP* (see text). For a simplified version, see Figure 5.3(a). Abbreviations and horizontal branch lengths as in Figure 5.1.

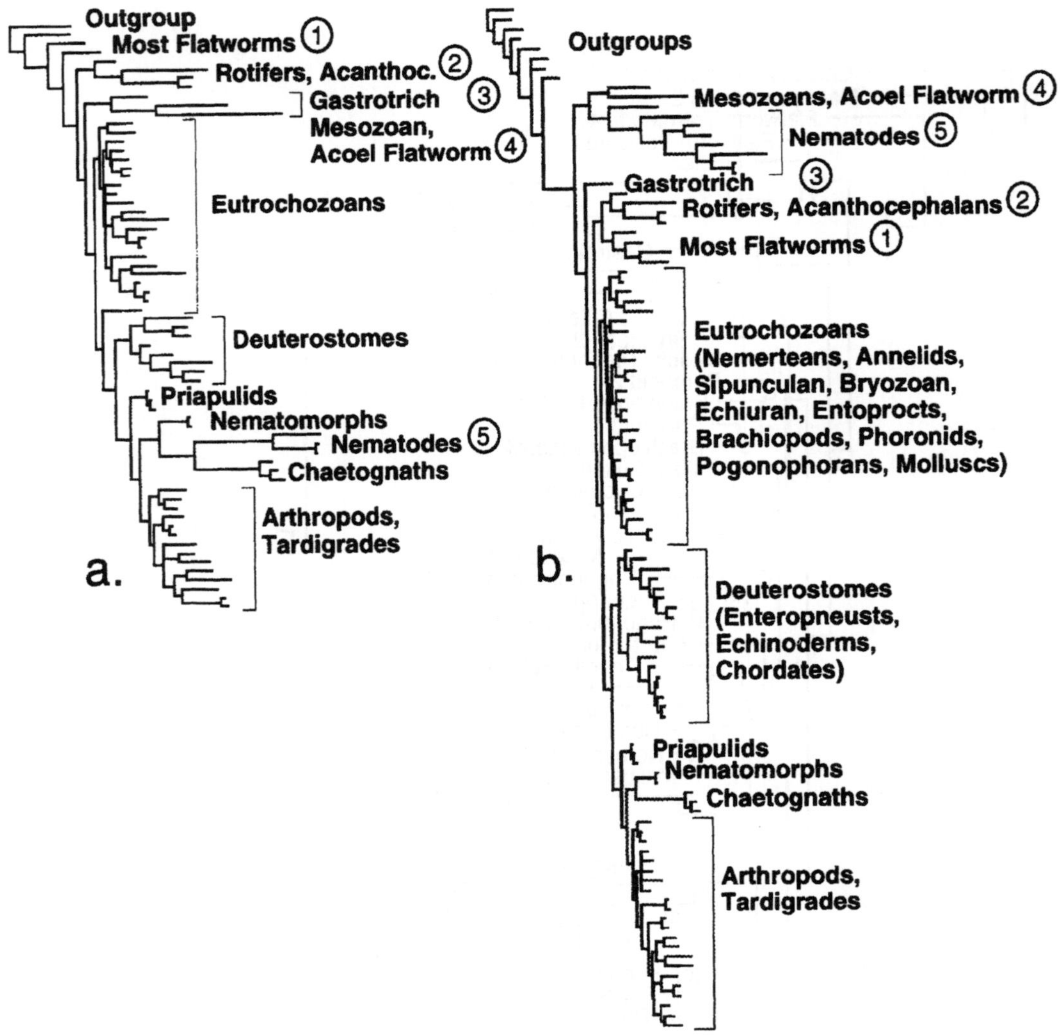


**Figure 5.3(a,b)** 18S rRNA parsimony strict consensus phylogram summaries of (a) 66- and (b) 103-sequence analyses. Simplified summaries based on results presented in Figures 5.1 and 5.2. Taxa numbered 1 to 5 differ in their hypothesized sister taxon relationship in (a) and (b). Remaining taxa maintain identical topology. Resolution within emphasized groups is mostly congruent but differences exist (see Figures 5.1 and 5.2). Horizontal branch lengths as in Figure 5.1.

Support index (SI; also known as 'Bremer support index' or 'decay index') values (Mishler *et al.*, 1988; Donoghue *et al.*, 1992) and bootstrap proportions (Felsenstein, 1988) were calculated for all but the 103 sequence analysis (too computationally expensive). SI values were estimated by executing PAUP* with a series of automatically generated converse constraint search blocks (Eernisse, 1992; Eernisse and Kluge, 1993). An additional constraint analysis was performed with PAUP* to identify how many additional steps would be required by the Articulata hypothesis, corresponding with an analysis performed in the morphology-based analysis of Eernisse *et al.* (1992).

A data file containing the 18S rRNA alignment of 103 sequence is available from the Systematics Association www Server. The file is in Nexus format, and includes self-executing command-line instructions that will replicate all searches summarized in the WWW data when used with PAUP* (or PAUP 3.x), including exclusion of sites not considered here. Provided with this file is more complete documentation for individual sequences than is listed at the www site and further documentation that explains details of each analysis.

## 5.3 RESULTS

The results of Analyses 1, 2, 4a, and 5 (www site) are presented in Figures 5.1–5.4, all based on the same aligned data set of 18S rRNA with the same sites excluded due to

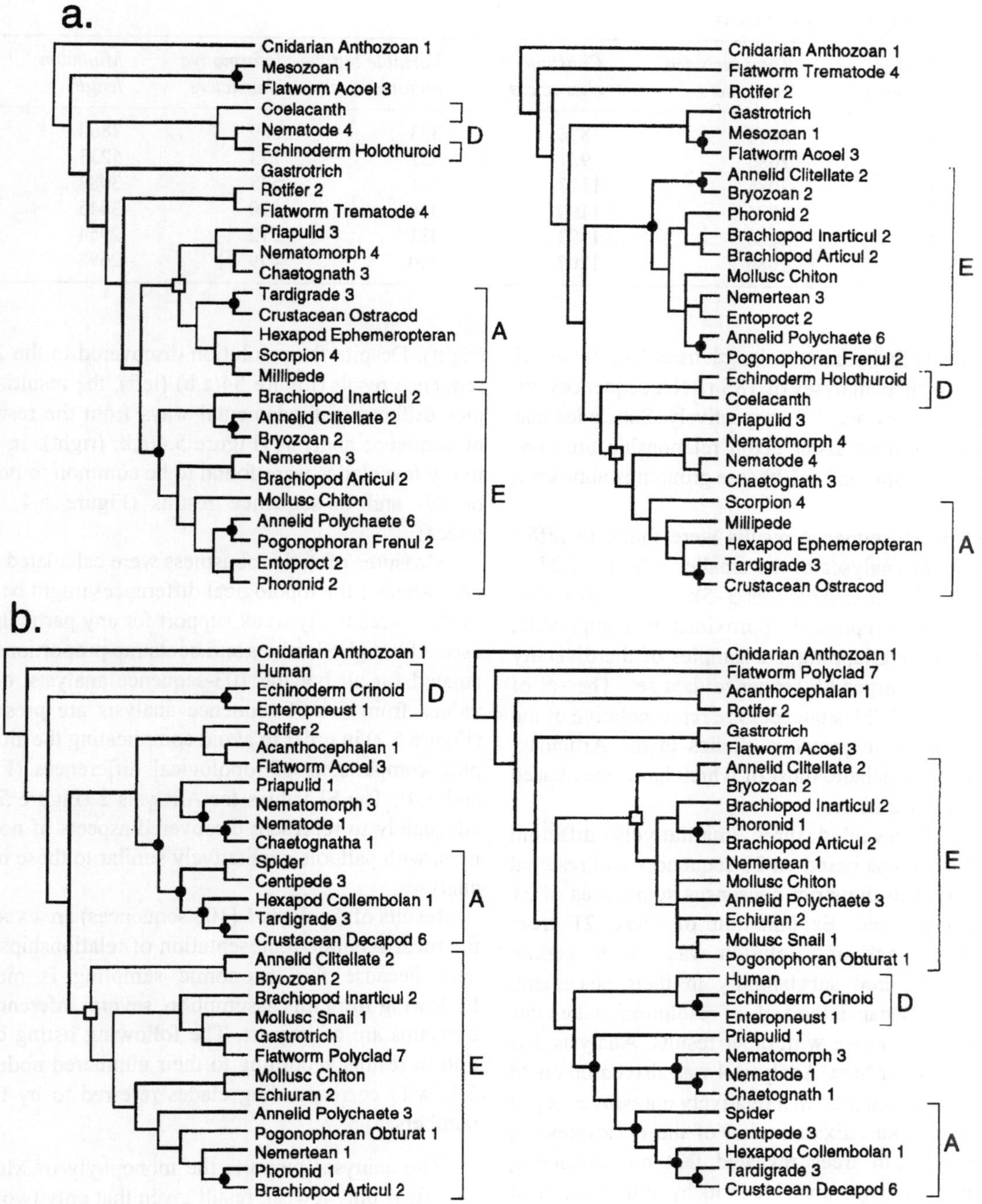


**Figure 5.4(a,b)** 18S rRNA parsimony strict consensus cladogram summaries. (a) Analysis 4a (27 sequences, Table 5.1) on left compared with pruned result of Analysis 2 (66 sequences, Table 5.2 and Figure 5.2) on right. (b) Analysis 5 (29 sequences, Table 5.1) on left compared with pruned result of Analysis 2 (66 sequences, Table 5.2 and Figure 5.2) on right. Bullets at nodes indicate clades that are identical between left and right cladograms. Open boxes are large clades that differ only in one or two included taxa in comparison of left and right figures.

alignment ambiguities, and all analysed with the parsimony criterion and equal weighting. Pertinent aspects of the character data base and parsimony search results are listed in Table 5.1.

All three analyses supported Eutrochozoa, not Articulata. Based on constraint searches with all 103 sequences included, an additional 27 steps were required if arthropods (including tardigrades) and annelids were constrained to be sister taxa (L = 7887).

Important differences were found in the respective topologies of the 103-, 66-, and 27- to 29-sequence analyses (Figures 5.1–5.4), even though all of these analyses produced

**Table 5.1** Summary of 18S rRNA analysis

| *Analysis no.* | *No. of sequences* | *Total included characters* | *Constant characters* | *Variable but uniformative* | *Informative characters* | *Minimum length* | *Number of trees* |
|---|---|---|---|---|---|---|---|
| 1 | 103 | 2086 | 824 | 393 | 869 | 7860 | 20 |
| 2 | 66 | 2086 | 981 | 332 | 773 | 5233 | 12 |
| 3 | 28 | 2086 | 1127 | 421 | 538 | 3082 | 21 |
| 4 | 28 | 2086 | 1190 | 376 | 520 | 3045 | 5 |
| 4a | 27 | 2086 | 1193 | 381 | 512 | 2954 | 1 |
| 5 | 29 | 2086 | 1207 | 400 | 479 | 2698 | 1 |

results of relatively few minimum-length trees. The results of 103- and 66-sequence analyses of 18S rRNA sequences are depicted in Figures 5.1 and 5.2, respectively. Subclades that differed in hypothesized sister taxon relationship are contrasted in Figures 5.3(a) and 5.3(b) (see groupings numbered 1 to 5).

Likewise, the 66-sequence results were found to differ from the results of analyses of the smaller subsets of 27 to 29 sequences (Table 5.1, Analyses 3–5). The smaller data sets were chosen to represent approximately comparable, but only partially overlapping, subsamples of the diversity represented by the larger 66 sequence data set. The set of analyses based on 27–29 sequences are representative of the taxonomic sampling in previous studies in the Articulata versus Eutrochozoa debate, most of which have been based on fewer than 20 sequences.

Analyses 3–5 revealed three qualitatively different results. Analysis 3 was based on 28 sequences and resulted in little resolution in the strict consensus summaries of 21 minimum length trees. Examination of these 21 trees revealed that the lack of resolution was due to certain sequences that differed substantially in their placement, from tree to tree. Because the lack of resolution makes this result difficult to compare with other results, Analysis 3 is not considered further here. Analysis 4 was also based on 28 sequences, and also resulted in a relatively unresolved result in the strict consensus. Examination of the corresponding five minimum length trees revealed that one sequence, 'Clam 1' (www site), was unstable in its hypothesized sister taxon relationship and this was the reason for little apparent resolution in the consensus, despite fewer minimum length trees than in Analysis 3. Removal of this sequence and reanalysis of the remaining 27 sequences (Table 5.1, Analysis 4a) resulted in a single most parsimonious tree, and this result is depicted in Figure 5.4(a) (left). Finally, Analysis 5 was based on 29 sequences, and resulted in a completely resolved result (Figure 5.4(b) (left).

To compare the results of Analyses 4a and 5 (Figure 5.4(a,b) (left) with those of the 66-sequence result (Analysis 2, Figure 5.2), sequences that were not present in the 27- or 29-sequence result were pruned from the Analysis 2 result to yield trees with the same sequences (Figure 5.4(a,b) (right). Despite the resolution discovered in the 27- or 29-sequence result (Figure 54(a,b) (left), the resulting topologies differed in fundamental ways from the results of the 66-sequence analysis (Figure 5.4(a,b, (right). In fact, relatively few clades were found to be common to both the 27- or 29- and 66-sequence results (Figure 5.4, bullets at nodes).

Measures of node robustness were calculated to investigate whether the topological differences might be explained as due to relatively weak support for any particular hypothesis. Although SI values and bootstrap proportions were calculated for all but the 103-sequence analysis, only the SI values from the 66-sequence analysis are presented here (Figure 5.2) in order to avoid complicating the already complex comparisons of topological differences (Figures 5.3 and 5.4). The SI values for Analysis 2 (Figure 5.2) appear adequately to represent discovered aspects of node robustness, with patterns qualitatively similar to those in the other analyses.

Results of Analysis 1 (103 sequences) are expected to be the most accurate representation of relationships estimated here because the taxonomic sampling is more dense. Following from this assumption, several inferences of relationships are of interest. The following listing calls attention to results according to their numbered node in Figure 5.1, with corresponding clades referred to by these same numbers.

1. The analysis supports the monophyly of Metazoa, but this is only a weak result given that only two functional outgroups, representing fungi and choanoflagellates, are included.
2. The monophyly of the ingroup, Bilateria, is strongly supported. This is apparent from the long length of the internal branch separating this clade from other metazoans (see also Adoutte and Philippe, 1993; Philippe *et al.*, 1994; Raff *et al.*, 1994). In contrast, the curious topology supported for the outgroups (e.g. paraphyly of sponges, ctenophores as basal to a paraphyletic cnidarian assemblage) might be due to relatively short internal branches separating these taxa, suggesting that these relationships might change with more dense taxonomic sampling.

3. A clade comprising of two (paraphyletic) mesozoans, the only acoel flatworm represented, and nematodes is sister taxon to other bilaterian metazoans (see also Katayama *et al.*, 1993, 1995; Pawlowski *et al.*, 1996). This placement of nematodes differs substantially from that supported in Analysis 2 (Figure 5.2).
4. The seven sampled nematodes are supported as a monophyletic grouping. Analyses of fewer metazoan sequences typically result in a polyphyletic resolution of nematodes, probably because these tend to be pronounced 'long branch' sequences.
5. Clade 5 is a broad grouping of all bilaterian animals except those in Clade 3. Note that the presence of the remaining flatworms in Clade 5, but not the acoel flatworm, suggests that Platyhelminthes is polyphyletic. Further sampling of acoels is needed.
6. Gastrotrichs are supported as sister taxon to the remaining members of Clade 5. However, note that a very different resolution is supported in the 66-sequence result, though with SI = 1, indicating relatively little support for any placement at present. It is interesting that the basal position of gastrotrichs and rotifers could support the suggestion by Davidson *et al.* (1995) and others that the ancestor of Bilateria may have been a micrometazoan.
7. A clade comprised of rotifers, acanthocephalans, and the remaining flatworms is sister taxon to the remaining members of Clade 6.
8. A clade of acanthocephalans plus rotifers is supported as in recent studies (Winnepenninckx *et al.*, 1995b; Wallace *et al.*, 1996), but the clade of acanthocephalans is sister taxon to Rotifer 3, a bdelloid rotifer. Garey *et al.* (1996b) have recently proposed on similar evidence that Bdelloidea plus Acanthocephala comprise a hitherto unrecognized subclade of Rotifera they term Lemniscea.
9. All represented flatworms except the acoel form a group that is sister taxon to Clade 8.
10. The evidence for Clade 10 suggests support for the notion of Coelomata, or a single origin for a true coelom. The only complications of this interpretation are that it would require the subsequent loss of a true coelom in entoprocts, priapulids, and nematomorphs, which all have been considered to have a pseudocoelom.
11. Clade 11 is a large superphylum group referred to here as Eutrochozoa. Within this clade are two subclades, one composed of molluscs and pogonophorans, and its sister taxon including all annelids, entoprocts, ectoprocts, brachiopods, phoronids, nemerteans, sipunculans and echiurans (Ghiselin, 1988; Turbeville *et al.*, 1992; Carlson, 1995; Halanych *et al.*, 1995; Winnepenninckx, 1995a). There is little support for Halanych *et al.*'s (1995) suggestion that ectoprocts are sister taxon to other eutrochozoans (Conway Morris, 1995; Conway Morris *et al.*, 1996; Halanych, 1996a). Brachiopods plus phoronids is supported as a eutrochozoan subclade, but the other lophophorate group, ectoprocts, is not closely associated (cf. Brusca and Brusca, 1990; Nielsen, 1995).
12. The remaining coelomate bilaterians in Clade 12 together are sister taxon of Eutrochozoa (Clade 11). Differing from the results of some recent studies (Philippe *et al.*, 1994; Winnepenninckx *et al.*, 1995a,b), this grouping of arthropods with deuterostomes renders the conventional grouping of 'protostomes' (arthropods plus eutrochozoans) polyphyletic.
13. The monophyly of Deuterostomia (Clade 13) is supported.
14. A clade comprising enteropneusts (paraphyletic) and echinoderms is sister taxon to remaining deuterostomes, the chordates (see also Turbeville *et al.*, 1994).
15. The monophyly of Chordata is supported, but not the conventional hypothesis that its sister taxon is Hemichordata, which includes enteropneusts (Clade 14).
16. Clade 16 is the sister taxon of deuterostomes (Clade 14). All members of Clade 16 have a chitinous cuticle that is moulted. Chaetognaths are the only members of this clade with only a partial covering of their body by this cuticle (G. Shinn, personal communication); the rest have their entirely covered by a chitinous cuticle.
17. Priapulids form a monophyletic group, but note that all three priapulids sampled have been identified as members of the same species (www site).
18. Clade 18 includes the remaining members of Clade 16, which are sister taxon to priapulids.
19. A clade comprised of nematomorphs and chaetognaths as sister taxon of remaining members of Clade 18 is supported. This is consistent with results reported by Halanych (1996b) except for the exclusion of nematodes from this clade (Winnepenninckx *et al.*, 1995b).
20. Monophyly of arthropods is supported only if tardigrades are included. Resolution within Clade 20 supports myriapods in a basal position, chelicerates form the next branch, then tardigrades as sister taxon of a clade composed of hexapods and crustaceans. Support is evident for pentastomids being crustaceans, as previously suggested by sperm morphology (Wingstrand, 1972) and 18S rRNA studies (Abele *et al.*, 1989). These results are also consistent with recent suggestions (Friedrich and Tautz, 1995; Telford and Thomas, 1995; Budd, 1996b; Popadić *et al.*, 1996) that crustaceans, not myriapods, are closest relatives of hexapods.

## 5.4 DISCUSSION

In contrast to the Articulata hypothesis, some authors have found support for an alternative grouping of annelids with other phyla, most notably molluscs, to the exclusion of arthropods. The present set of analyses further corroborates this suggestion. Following Ghiselin (1988), this clade is here referred to as Eutrochozoa, considered to be essentially equivalent and senior synonym to Lophotrochozoa Halanych *et al.*, 1995. Previous support for Eutrochozoa has been based on morphology (Eernisse *et al.*, 1992) and all published sequence analyses that have included arthropods, annelids, molluscs, and outgroups. These include previously cited comparisons of 18S rRNA-, mitochondrial 12S rRNA, and RNA polymerase II-coding sequences. Despite this support from molecular results, there may be some valid cause for doubt. Most previous studies have been based on relatively few sequences, so that results might be an artefact of inadequate taxonomic sampling. There are doubts about the suitability of 18S rRNA to resolve this particular burst of divergence among metazoans. The morphological support for Eutrochozoa has also been questioned.

### 5.4.1 THE IMPORTANCE OF BROAD TAXON SAMPLING FOR MOLECULAR STUDIES

There is much interest in the historical reconstruction of evolutionary events leading to the 'Big Bang' of bilaterian metazoan radiation. The critical period of body plan diversification has been claimed to coincide with a logistic pattern of increase in the first appearance of diverse animal fossils, especially pronounced about 530 to 525 Myr midway through the Early Cambrian (Bowring *et al.*, 1993). Others have argued that the actual radiation could have occurred much earlier if the presumed ancestor and its divergent descendants were micrometazoans, too small to leave fossils (Valentine, 1989, 1996; Davidson *et al.*, 1995; Fortey *et al.*, 1996; Wray *et al.*, 1996). However ancient these cladogenic events were, recent attempts to resolve the branching pattern have employed comparisons of 18S rRNA sequences.

Previous authors have especially focused on the extent of node robustness to address the important question of how prone the results are to change, given more data from the same universe of characters (e.g. bootstrapping), or else they have explored what near-minimum length trees of competing topology might exist (e.g. support- or decay-index searches). As already mentioned, previous studies have supported Eutrochozoa, not Articulata. However, all previous studies that have included the requisite representatives of arthropods, annelids, molluscs, and appropriate outgroups minimally necessary for testing Articulata versus Eutrochozoa have generally included fewer than 20 sequences in their analyses. Inadequate taxonomic sampling has been shown to be a confounding factor in groups for which more extensive sequence databases exist (Doyle and Donoghue, 1987; Gauthier *et al.*, 1988; Lecointre *et al.*, 1993). More extensive taxonomic sampling can also lead to the resolution of branching patterns that were not stable with fewer taxa sampled. This is an important point because some authors have gone so far as to conclude that a particular gene region cannot resolve a particular branching order when, in fact, their results could change with more extensive taxonomic sampling.

Philippe *et al.* (1994) were pessimistic about resolving ancient metazoan branching patterns with 18S rRNA. Their doubts came from application of a test (Lecointre *et al.*, 1994) of the proportion of neighbour joining trees reconstructed from bootstrap replicates (Felsenstein, 1988) applied to a data set of 55 selected or 69 total aligned 18S rRNA sequences. Their test led them to conclude that, given their data set, the radiation of bilaterian metazoans is likely to be remain unresolved even if substantially more sequence data of the same type were to become available. However, these authors did not consider whether this same lack of resolution would exist as taxonomic sampling became more complete. The present results suggest differences in topology become evident as more sequences are included in the analysis. The question now becomes: is the estimate becoming more accurate in the sense characterized by Hillis (1995), or simply different?

It is impossible to prove that topological results of the most inclusive, 103-sequence, analysis are more accurate with respect to estimating the true genealogy than analyses presented herein or in prior studies based on fewer sequences, although several aspects of these results are consistent with this interpretation. First, the 103-sequence result was the only one considered here that grouped sequences regarded as members of the same phyla, Mollusca and Nematoda, into monophyletic assemblages. Second, the less inclusive analyses support a greater number of groupings at odds with conventional estimates of animal relationships, for example, placing most flatworms as sister taxon to other bilaterians, a large group also including acoel flatworms. Third, a frequent empirical observation is that denser taxonomic sampling provides a more accurate overall resolution. One expects better resolution because with sparse taxonomic sampling the most rapidly evolving, divergent 'long branch' lineages tend to attract. Breaking up long branches improves groupings resulting from historical resemblance and avoids spurious groupings due to homoplasy. Importantly, this improvement should coincide with filling in taxonomic sampling gaps when extremes of diversity have already been sampled, and not necessarily while previously unsampled diversity is added to an analysis (Hillis, 1996). Fourth, recent simulation (Hillis, 1996; Sanderson, 1996) and empirical (Lecointre *et al.*, 1994) studies have shown that approximately 40 sequences are necessary before the root node for an ingroup can be accurately determined (Sanderson, 1996). The matter is far from settled, however, because at least one theory-based study has reached the opposite conclusion, specifically, for those situations where 'long branch' taxa are added at the tips of other long branches (Kim, 1996).

Besides pointing to new interpretations of metazoan phylogeny, this study has sobering implications for previous studies of higher-level metazoan phylogeny, most of which are based on 20 or fewer sequences. Even those with more sequences are biased toward sampling representatives of a single phylum (e.g. arthropods by Ballard *et al.*, 1992). Phylogenetic estimates derived from three roughly comparable 27 to 29 taxon assemblages (Analyses 3–5) exhibited either lack of resolution or important topological variation when those results were contrasted with those from more inclusive analyses (Analyses 1–2). Even more telling, taxon sampling appears to still influence topological results when 66- and 103-sequence results are compared (Figure 5.3).

The inclusion of only a few selected subset analyses illustrates the general problem of limited species sampling (this is not an exhaustive consideration of the problems). This is also not a problem limited to 18S rRNA. Related analyses (D.J. Eernisse, in preparation) were performed based on relatively few (e.g. 10) sequences, but with multiple genes combined into a massive single data set of more than 22 000 aligned sites, and also analysed according to separate gene data sets. These analyses have revealed dramatic cases where ingroup topology was found to be highly sensitive to the choice of outgroup(s). Based on the present study, it can be predicted that problems are expected to persist even when more than 20 sequences are available for those genes.

### 5.4.2 DO THE MOLECULES AGREE WITH MORPHOLOGY?

Eernisse *et al.*'s (1992) analysis of 138 morphological characters was the first morphological study to discover robust support of Eutrochozoa. Depending on the exact composition of Articulata, at least nine and as many as 17 additional steps would be required by constraining Articulata to be a clade, given their matrix and equal weighting. However, Nielsen *et al.* (1996), Haszprunar (1996), and especially Rouse and Fauchald (1995) have criticized particular codings of characters in the Eernisse *et al.* matrix. Moreover, Schram and Ellis (1995) called attention to the possibility that inclusion of other (mostly pseudocoelomate) taxa, not considered by Eernisse *et al.*, could have influenced their result. In order to test whether the support for Eutrochozoa was due to miscoding of characters or exclusion of taxa, Eernisse and colleagues (unpublished results) performed a reanalysis of these morphological data. Characters were recoded to reflect published and unpublished criticisms, and characters were scored for 11 additional phyla bringing the total to 30 phyla (coded as 38 terminal taxa). Reanalysis showed some lack of resolution compared with the original result but support for Eutrochozoa remained robust, and was not an artefact of miscoding of characters or exclusion of taxa.

Besides challenging particular coding, Rouse and Fauchald (1995) considered whole sets of characters in Eernissse *et al.*'s (1992) matrix as invalid because of suspected linkage. By claiming linkage, Rouse and Fauchald are implying that some characters were scored as separate when they were really manifestations of the same character. They rejected nearly all the 138 characters in the Eernisse *et al.* matrix, and based their analysis on 13 total characters. However, Eernisse *et al.* (1992, p. 309) were explicit in 'accepting at face value the independence of characters with unique distributions across taxa'. In other words, two characters with different distributions were assumed as independent, because they did not exactly co-occur in the sampled taxa. As in Hennig's Auxiliary Principle, which states that one should never assume convergent or parallel evolution, instead assuming homology in the absence of contrary evidence (Hennig, 1966), character linkage is an ever-present possibility (like homoplasy). Eernisse *et al.* advocated making fewer assumptions about character linkage before making the analysis. This leaves open the possibility that linked homoplasies could be revealed by a character congruence approach. Linked homologies are more difficult to characterize because synapomorphies can accumulate in a lineage whether or not they are linked. If cladogenesis occurred between the evolution of novelties then it might be possible to distinguish linked versus unlinked synapomorphies with additional taxonomic sampling. At worst, unrecognized character linkage is equivalent to weighting a character twice or more what it should be. Another potential problem is that linked characters that are, in fact, mutually exclusive could be optimized as both 'present' in those taxa scored as missing (but see Pleijd, 1995). While character linkage could remain a problem, it is doubtful that an alternative analysis based on only 13 characters (Rouse and Fauchald, 1995) is sufficient to satisfy proponents of the Articulata hypothesis.

The results here suggest that arthropods may share close ancestry with animals that happen to have an entire (Tardigrada, Nematomorpha, Priapulida) or partial (Chaetognatha) generally chitinous exoskeleton that is moulted; annelids do not have such a moulted chitinous exoskeleton. This suggests that zoologists have placed undue emphasis on teloblastic segment addition. Potential sister taxa such as nematomorphs, chaetognaths, and priapulids (plus perhaps the kinorhynchs for which 18S rRNA sequences are not yet available) assume a greater importance, not as models of arthropod ancestors but as groups that have diverged from a common ancestor to Clade 16 in the 103-sequence result (Figure 5.1(b)).

While it will eventually be desirable to combine morphology with this and other molecular data sets (Eernisse and Kluge, 1993), combined analyses must await a more extensive reconsideration of morphology (D.J. Eernisse *et al.*, unpublished results). A possible advantage of not including morphological characters here is the freedom from potential bias that inappropriate character coding, however slight or extreme, might have had on the present results.

### 5.4.3 PROTOSTOMES AND DEUTEROSTOMES?

Some striking aspects of the 66- and 103-sequence results suggest mostly unexplored possibilities for continued study.

In particular, the support for Eutrochozoa appears to be strong enough that the reliable resolution of three main coelomate bilaterian clades (the eutrochozoans, the deuterostomes, and the arthropods plus other animals with a moulted chitinous exoskeleton) is more pressing. These analyses support the hypothesis that deuterostomes, not eutrochozoans, are sister taxon to the extended arthropod clade. While support for this hypothesis is admittedly weak, or even contradicted by other studies (Philippe *et al.*, 1994; Winnepenninckx *et al.*, 1995a,b), there is little to gain from division of bilaterian animals into a 'protostome versus deuterostome' dichotomy. Deuterostomes appear to be a monophyletic assemblage, but protostomes are perhaps best regarded as a paraphyletic group which, if rediagnosed to be monophyletic, should be considered as synonymous with 'Coelomata', a taxon which includes some members with a derived loss of the eucoelomate condition. The weakly supported deuterostome–arthropod association also has important implications for consideration of features common to these groups, for example, the recent fly–mouse comparisons of the *Hox* gene complex (Carroll, 1995).

## 5.5 CONCLUSIONS

We can conclude the following points:

1. Analyses based on 103 metazoan 18S ribosomal RNA sequences, or selections of these sequences, do not support a close arthropod–annelid relationship. Rather, annelids are supported as closer to various unsegmented coelomate bilaterians including molluscs, together comprising a clade termed Eutrochozoa (from the trochophore larval stage common in these groups).
2. These results agree with recent studies based on sequence comparison or morphology, and suggest that segmentation is not a shared-derived feature peculiar to the conventionally hypothesized arthropod + annelid clade, termed Articulata.
3. By analysing subselections of the 103 sequences, it was discovered that the extent of taxon sampling can dramatically influence topological results. This is important because virtually all published prior studies have been based on fewer than 20 sequences.
4. There is intriguing support for animals with moulted chitinous exoskeletons (nematomorphs, chaetognaths and priapulids) as the most proximal living outgroups to arthropods.

## REFERENCES


Abele, L.G., Kim, W. and Felgenhauer, B.E. (1989) Molecular evidence for inclusion of the phylum Pentastomida in the Crustacea. *Molecular Biology and Evolution*, **6**, 685–91.

Adoutte, A. and Philippe, H. (1993) The major lines of metazoan evolution: summary of traditional evidence and lessons from ribosomal RNA sequence analysis, in *Comparative Molecular Neurobiology* (ed Y. Pichon), Birkhäuser Verlag, Basel, Switzerland, pp. 1–30.

Anderson, D.T. (1981) Origins and relationships among the animal phyla. *Proceedings of the Linnean Society of New South Wales*, **106**, 151–66.

Ballard, J.W.O., Olsen, G.J., Faith, D.P., Odgers, W.A., Rowell, D.M. and Atkinson, P.W. (1992) Evidence from 12S ribosomal RNA sequences that onychophorans are modified arthropods. *Science*, **258**, 1345–8.

Barnes, R.D. (1987) *Invertebrate Zoology*, 5th edn, Saunders College Publishing, Orlando.

Boore, J.L., Collins, T.M., Stanton, D., Daehler, L.L. and Brown, W.M. (1995) Deducing the pattern of arthropod phylogeny from mitochondrial DNA rearrangements. *Nature*, **376**, 163–5.

Boudreaux, H.P. (1979) *Arthropod Phylogeny with Special Reference to Insects*, Wiley, Inc., New York.

Bowring, S.A., Grotzinger, J.P., Isachsen, C.E., Knoll, A.H., Pelechaty, S.M. and Kolosov, P. (1993) Calibrating rates of Early Cambrian evolution. *Science*, **261**, 1293–8.

Briggs, D.E.G., Fortey, R.A. and Wills, M.A. (1993) How big was the Cambrian evolutionary explosion? A taxonomic and morphological comparison of Cambrian and Recent arthropods, in *Evolutionary Patterns and Processes* (eds D.R. Lees and D. Edwards), Linnean Society Symposium Series, **14**, 33–44.

Brusca, R.C. and Brusca, G.J. (1990) *Invertebrates*, Sinauer Associates, Sunderland, Massachusetts.

Budd, G.E. (1996a) The morphology of *Opabinia regalis* and the reconstruction of the arthropod stem group. *Lethaia*, **29**, 1–14.

Budd, G.E. (1996b) Progress and problems in arthropod phylogeny. *Trends in Ecology and Evolution*, **11**, 356–8.

Carlson, S.J. (1995) Phylogenetic relationships among extant brachiopods. *Cladistics*, **11**, 131–97.

Carroll, S.B. (1995) Homeotic genes and the evolution of arthropods and chordates. *Nature*, **376**, 479–85.

Conway Morris, S. (1995) Nailing the lophophorates. *Nature*, **375**, 365–6.

Conway Morris, S., Cohen, B.L., Gawthrop, A.B., Cavalier-Smith, T. and Winnepenninckx, B. (1996) Lophophorate phylogeny. *Science*, **272**, 282.

Cuvier, G. (1817) *Le règne animal distribuè d'après son organisation, pour servir de base à l'histoire naturelle des animaux et d'introduction à l'anatomie comparée*, Vol. 2, Paris.

Davidson, E.H., Peterson, K.J. and Cameron, R.A. (1995) Origin of bilaterian body plans: Evolution of developmental regulatory mechanisms. *Science*, **270**, 1319–25.

De Robertis, E.M. and Sasai, Y. (1996) A common plan for dorsoventral patterning in Bilateria. *Nature*, **380**, 37–40.

de Queiroz, K. and Gauthier, J. (1990) Phylogeny as a central principle in taxonomy: phylogenetic definitions of taxon names. *Systematic Zoology*, **39**, 307–22.

de Queiroz, K. and Gauthier, J. (1992) Phylogenetic taxonomy. *Annual Review of Ecology and Systematics*, **23**, 449–80.

Dewel, R.A. and W.C. Dewel (1996) The brain of *Echiniscus viridissimus* Peterfi, 1956 (Heterotardigrada): a key to understanding the phylogenetic position of tardigrades and the evolution of the arthropod head. *Zoological Society of the Linnean Society*, **116**, 35–49.

Donoghue, M.J., Olmstead, R.G., Smith, J.F. and Palmer, J.D. (1992) Phylogenetic relationships of Dipscales based on *rbc*L sequences. *Annals of the Missouri Botanical Garden*, **79**, 249–65.

Doyle, J.A. and Donoghue, M.J. (1987) The importance of fossils in elucidating seed plant phylogeny and macroevolution. *Review of Palaeobotany and Palynology*, **50**, 63–95.

Eernisse, D.J. (1992) DNA Translator and Aligner: HyperCard utilities to aid phylogenetic analysis. *Computer Applications in the Biosciences*, **8**, 177–84.

Eernisse, D.J. (1995) DNA Stacks, Version 1.1, Software package available from the author.

Eernisse, D. J. and Kluge, A.G. (1993) Taxonomic congruence versus total evidence, and the phylogeny of amniotes inferred from fossils, molecules and morphology. *Molecular Biology and Evolution*, **10**, 1170–95.

Eernisse, D. J., Albert, J. S. and Anderson, F.E. (1992) Annelida and Arthropoda are not sister taxa: a phylogenetic analysis of spiralian metazoan morphology. *Systematic Biology*, **41**, 305–30.

Felsenstein, J. (1988) Phylogenies from molecular sequences: inference and reliability. *Annual Review of Genetics*, **22**, 521–65.

Field, K.G., Olsen, G.J., Lane, D.J., Giovannoni, S.J., Ghiselin, M.T., Raff, E.C., Pace, N.R. and Raff, R.A. (1988) Molecular phylogeny of the animal kingdom. *Science*, **239**, 748–53.

Fortey, R.A., Briggs, D.E.G. and Wills, M.A. (1996) The Cambrian evolutionary 'explosion': decoupling cladogenesis from morphological disparity. *Biological Journal of the Linnean Society*, **57**, 13–33.

Friedrich, M. and Tautz, D. (1995) Ribosomal DNA phylogeny of the major extant arthropod classes and the evolution of myriapods. *Nature*, **376**, 165–7.

Fryer, G. (1996) Reflections on arthropod evolution. *Biological Journal of the Linnean Society*, **58**, 1–55.

Garey, J.R., Krotec, M., Nelson, D.R. and Brooks, J. (1996a) Molecular analysis supports a tardigrade–arthropod association. *Invertebrate Biology*, **115**, 79–88.

Garey, J.R., Near, T.J., Nonnemacher, M.R. and Nadler, S.A. (1996b) Molecular evidence for Acanthocephala as a sub-taxon of Rotifera. *Journal of Molecular Evolution*, **43**, 287–92.

Gauthier, J., Kluge, A.G. and Rose, T. (1988) Amniote phylogeny and the importance of fossils. *Cladistics*, **4**, 105–209.

Ghiselin, M.T. (1988) The origin of molluscs in the light of molecular evidence. *Oxford Surveys in Evolutionary Biology*, **5**, 66–95.

Giribet, G., Carranza, S., Baguñá, J., Riutort, M. and Ribera, C. (1996) First molecular evidence for the existence of a Tardigrada + Arthropoda clade. *Molecular Biology and Evolution*, **13**, 76–84.

Halanych, K.M. (1996a) Response to Conway Morris *et al. Science*, **272**, 283.

Halanych, K.M. (1996b) Testing hypotheses of chaetognath origins: long branches revealed by 18S ribosomal DNA. *Systematic Biology*, **45**, 223–46.

Halanych, K.M., Bacheller, J.D., Aguinaldo, A.M., Liva, S.M., Hillis, D.M. and Lake, J.A. (1995) Evidence from 18S ribosomal DNA that the lophophorates are protostome animals, *Science*, **267**, 1641–3.

Haszprunar, G. (1996) The Mollusca: Coelomate turbellarians or mesenchymate annelids?, in *Origin and Evolutionary Radiation of the Mollusca* (ed J. Taylor), Centenary Symposium of the Malacological Society of London, Oxford University Press, New York, pp. 3–28.

Hillis, D.M. (1995) Approaches for assessing phylogenetic accuracy. *Systematic Biology*, **44**, 3–16.

Hillis, D.M. (1996) Inferring complex phylogenies. *Nature*, **383**, 130–1.

Hillis, D.M. and Dixon, M.T. (1991) Ribosomal DNA: molecular evolution and phylogenetic inference. *Quarterly Review of Biology*, **66**, 411–53.

Katayama, T., Yamamoto, M., Wada, H. and Satoh, N. (1993) Phylogenetic position of acoel turbellarians inferred from partial 18S rDNA sequences. *Zoological Sciences*, **10**, 529–36.

Katayama, T., Wada, H., Furuya, H., Satoh, N. and Yamamoto, M. (1995) Phylogenetic position of the dicyemid Mesozoa inferred from 18S rDNA sequences. *Biological Bulletin*, **189**, 81–90.

Kim, J. (1996) General inconsistency conditions for maximum parsimony: effects of branch lengths and increasing numbers of taxa. *Systematic Biology*, **45**, 363–74.

Kjer, K.M. (1995) Use of rRNA secondary structure in phylogenetic studies to identify homologous positions: an example of alignment and data presentation from the frogs. *Molecular Phylogenetics and Evolution*, **4**, 314–30.

Lake, J.A. (1990) Origin of the metazoa. *Proceedings of the National Academy of Science, USA*, **87**, 763–6.

Lecointre, G., Philippe, H., Van Le, H.L. and Le Guyader, H. (1993) Species sampling has a major impact on phylogenetic inference. *Molecular Phylogenetics and Evolution*, **2**, 205–24.

Lecointre, G., Philippe, H., Van Le, H. L. and Le Guyader, H. (1994) How many nucleotides are required to resolve a phylogenetic problem? The use of a new statistical method applicable to available sequences. *Molecular Phylogenetics and Evolution*, **3**, 292–309.

Maddison, D.R. (1991) The discovery and importance of multiple islands of most-parsimonious trees. *Systematic Zoology*, **40**, 315–28.

Maddison, W.P., Donoghue, M.J. and Maddison, D.R. (1984) Outgroup analysis and parsimony. *Systematic Zoology*, **33**, 83–103.

Meglitsch, P. and Schram, F.R. (1991) *Invertebrate Zoology*, 3rd edn, Oxford University Press, New York.

Mishler, B.D., Bremer, K., Humphries, C.J. and Churchill, S.P. (1988) The use of nucleic acid sequence data in phylogenetic reconstruction, *Taxon*, **37**, 391–5.

Moon, S. Y. and Kim, W. (1996) Phylogenetic position of the Tardigrada based on the 18S ribosomal RNA gene sequences. *Zoological Journal of the Linnean Society*, **116**, 61–9.

Neefs, J.-M., Van de Peer, Y., Hendriks, L. and De Wachter, R. (1990) Compilation of small ribosomal subunit RNA sequences. *Nucleic Acids Research*, **18**, 2237–317.

Nielsen, C. (1995) *Animal Evolution: Interrelationships of the Living Phyla*, Oxford University Press, Oxford.

Nielsen, C., Scharff, N. and Eibye-Jacobsen, D. (1996) Cladistic analyses of the animal kingdom. *Biological Journal of the Linnean Society*, **57**, 385–410.

Patterson, C. (1989) Phylogenetic relationships of major groups: conclusions and prospects, in *The Hierarchy of Life* (eds B.

Fernholm, K. Bremer, and H. Jörnvall), Elsevier Science Publishers, B.V. (Biomedical Division), Amsterdam, pp. 471–88.

Pawlowski, J., Montoya-Burgos, J.-I., Fahrni, J.F., Wüest, J. and Zaninetti, L. (1996) Origin of the Mesozoa inferred from 18S rRNA gene sequences. *Molecular Biology and Evolution*, **13**, 1128–32.

Philippe, H., Chenuil, A. and Adoutte, A. (1994) Can the Cambrian explosion be inferred through molecular phylogeny? *Development 1994*, Supplement, 15–25.

Pleijel, F. (1995) On character coding for phylogeny reconstruction. *Cladistics*, **11**, 309–15.

Popadić, A., Rusch, D., Peterson, M., Rogers, B.T. and Kaufman, T.C. (1996) Origin of the arthropod mandible. *Nature*, **380**, 395.

Raff, R.A., Marshall, C.R. and Turbeville, J.M. (1994) Using DNA sequences to unravel the Cambrian radiation of the animal phyla. *Annual Review of Ecology and Systematics*, **25**, 351–75.

Rouse, G.W. and Fauchald, K. (1995) The articulation of annelids. *Zoologica Scripta*, **24**, 269–301.

Ruitort, M., Field, K.G., Raff, R.A. and Baguñá, J. (1993) 18S rRNA sequences and phylogeny of Platyhelminthes. *Biochemical Systematics and Ecology*, **21**, 71–7.

Sanderson, M.J. (1996) How many taxa must be sampled to identify the root node of a large clade? *Systematic Biology*, **45**, 168–73.

Schram, F.R. (1991) Cladistic analysis of metazoan phyla and the placement of fossil problematica, in *The Early Evolution of Metazoa and the Significance of Problematic Taxa* (eds A. Simonetta and S. Conway Morris), Cambridge University Press, Cambridge, pp. 35–46.

Schram, F.R. and Ellis, W.N. (1995) Metazoan relationships: a rebuttal. *Cladistics*, **10**, 331–7.

Sidow, A. and Thomas, W.K. (1994) A molecular evolutionary framework for eukaryotic model organisms. *Current Biology*, **4**, 596–603.

Sogin, M.L. (1991) Early evolution and the origin of eukaryotes. *Current Biology*, **1**, 457–63.

Sullivan, J. (1996) Combining data with different distributions of among-site rate variation. *Systematic Biology*, **45**, 375–80.

Swofford, D.L. and Begle, D.P. (1993) PAUP: Phylogenetic Analysis Using Parsimony, version 3.1. User's manual for computer program by D.L. Swofford. Illinois Natural History Survey, Champaign, Illinois.

Swofford, D.L., Olsen, G.J., Waddell, P.J. and Hillis, D.M. (1996) Phylogenetic inference, in *Molecular Systematics*, 2nd edn (eds D.M. Hillis, C. Moritz, and B.K. Mable), Sinauer, Sunderland, Massachusetts, pp. 407–514.

Telford, M.J. and Thomas, R.H. (1995) Demise of the Atelocerata? *Nature*, **376**, 123–4.

Turbeville, J.M., Pfeiffer, M., Field, K.G. and Raff, R.A. (1991) The phylogenetic status of arthropods, as inferred from 18S rRNA sequences. *Molecular Biology and Evolution*, **8**, 669–86.

Turbeville, J.M., Field, K.G. and Raff, R.A. (1992) Phylogenetic position of phylum Nemertini, inferred from 18S rRNA sequences: molecular data as a test of morphological character homology. *Molecular Biology and Evolution*, **9**, 235–49.

Turbeville, J.M., Schulz, J.R. and Raff, R.A. (1994) Deuterostome phylogeny and the sister group of the chordates: evidence from molecules and morphology. *Molecular Biology and Evolution*, **11**, 648–55.

Valentine, J.W. (1989) Bilaterians of the Precambrian–Cambrian transition and the annelid–arthropod relationship. *Proceedings of the National Academy of Science, USA*, **86**, 2272–5.

Valentine, J.W. (1994) Late Precambrian bilaterians: grades and clades. *Proceedings of the National Academy of Science, USA*, **91**, 6751–7.

Valentine, J.W. (1996) Developmental evolution of metazoan body plans: the fossil evidence. *Developmental Biology*, **173**, 373–81.

Waggoner, B.M. (1996) Phylogenetic hypotheses of the relationships of arthropods to Precambrian and Cambrian problematic fossil taxa. *Systematic Biology*, **45**, 190–222.

Wallace, R.L., Ricci, C. and Melone, G. (1996) A cladistic analysis of pseudocoelomate (aschelminth) morphology. *Invertebrate Biology*, **115**, 104–12.

Wheeler, W.C. (1995) Sequence alignment, parameter sensitivity, and the phylogenetic analysis of molecular data. *Systematic Biology*, **44**, 321–31.

Wheeler, W.C., Cartwright, P. and Hayashi, C.Y. (1993) Arthropod phylogeny: a combined approach. *Cladistics*, **9**, 1–39.

Willmer, P. (1990) *Invertebrate Relationships: Patterns in Animal Evolution*, Cambridge University Press, Cambridge.

Wingstrand, K.G. (1972) Comparative spermatology of a pentastomid, *Raillietiella hemidactyli*, and a branchiuran crustacean, *Argulus foliaceus*, with a discussion of a pentastomid relationships. *Kongelige Danske Videnskabernes Selskab Biologiske Skrifter*, **19**, 1–72.

Winnepenninckx, B., Backeljau, T. and De Wachter, R. (1995a) Phylogeny of protostome worms derived from 18S rRNA sequences, *Molecular Biology and Evolution*, **12**, 641–9.

Winnepenninckx, B., Backeljau, T., Mackey, L.Y., Brooks, J.M., De Wachter, R., Kumar, S. and Garey, J.R. (1995b) 18S rRNA data indicate that aschelminthes are polyphyletic in origin and consist of at least three distinct clades. *Molecular Biology and Evolution*, **12**, 1132–7.

Wray, G.A., Levinton, J.S. and Shapiro, L.H. (1996) Molecular evidence for deep Precambrian divergences among metazoan phyla. *Science*, **274**, 568–73.

# 6 Evolutionary correlates of arthropod tagmosis: scrambled legs

*M.A. Wills, D.E.G. Briggs and R.A. Fortey*

*Department of Geology, Wills Memorial Building, University of Bristol, Queen's Road, Bristol BS8 1RJ, UK*
*email: m.a.wills@bristol.ac.uk*

*Department of Geology, Wills Memorial Building, University of Bristol, Queen's Road, Bristol, BS8 1RJ, UK*
*email: D.E.G.Briggs@bristol.ac.uk*

*Department of Palaeontology, The Natural History Museum, Cromwell Road, London, SW7 5BD, UK*
*email: R.Fortey@nhm.ac.uk*

## 6.1 INTRODUCTION

In 1974, John Cisne made two predictions concerning tagmosis in arthropods.

1. The degree of appendage differentiation should correlate positively with measures of overall morphological specialization. Hence, taxa which diverge greatly from the ancestral condition would be expected to be more highly tagmatized.
2. The degree of tagmosis for higher arthropod taxa should reflect the complexity of their ecological role (but not necessarily reflect the ecological specialization of species within those groups).

Here, we investigate some aspects of these claims for representative fossil and Recent arthropod taxa. Although characters coding for the specialization of specific appendages have been included in previous cladistic and phenetic data bases (Briggs *et al.*, 1992a, 1993; Wills *et al.*, 1994, 1995), the grouping of appendages into specific tagmata has not been considered directly (because in many cases identical patterns cannot be considered homologous).

Emerson and Schram (1997, this volume: see also Schram and Emerson, 1991) have demonstrated the phylogenetic utility of recognizing common 'fields and nodes' across higher arthropod taxa (Arthropod Pattern Theory; APT). Nodes are regions along the body at which tagma boundaries, gonopores or other distinctive features tend to arise. The occurrence of the same nodes in different taxa (even if the same features *per se* are not expressed) almost certainly has a genetic basis (Averof and Akam, 1993), and the coding of these synapomorphies in a cladistic context offers much for our understanding of the deeper branches in arthropod evolution. Tagmosis as measured by Cisne (1974), by contrast, does not score regions of appendage differentiation as homologous (although it may include common boundaries that are). Rather it seeks to provide an index of the degree of appendage specialization and the division of labour among them.

Here we present disparity and tagmosis indices for Cambrian and Recent taxa included in our most recent compilation of morphological data (Wills *et al.*, 1997).

## 6.2 AN INDEX OF APPENDAGE SPECIALIZATION

Cisne (1974) measured the degree of limb tagmosis of a large number of aquatic arthropod orders using the Brillouin expression (Brillouin, 1962), an equation first conceived for general applications in information theory.

The coefficient of limb tagmosis ($h$) is given by:

$$h = \frac{1}{N \ln 2} \ln \frac{N!}{N_1! N_2! \ldots N_i!}$$

where $N$ is the total number of limb pairs, including a caudal furca (where present), and $N_i$ is the number of limb pairs of the $i$th type. The calculation can be simplified by omitting tagmata composed of just a single limb-pair (since 1! = 1), although these appendage pairs must be counted in the total number of pairs ($N$). In the theoretically least tagmatized arthropods, where all limb pairs are of the same morphology, the expression takes on a value of zero. In the most highly tagmatized forms, where each limb pair is of a different morphology, the value approximates to the log to the base two of the total number of limb pairs (or the natural log of the total number of limb pairs, divided by the natural log of two). The value of $h$ is therefore sensitive to the total number of limb pairs, the number of limb types (or

*Arthropod Relationships*, Systematics Association Special Volume Series 55, Edited by R.A. Fortey and R.H. Thomas. Published in 1997 by Chapman & Hall, London. ISBN 0 412 75420 7

tagmata) and the relative number of limbs of each type (or the distribution of limbs among tagmata). All of these properties are desirable.

Brillouin values are based on the specialization and morphology of appendage types, irrespective of the fusion or otherwise of the somites bearing them. Thus, for example, the second to twelfth appendages of *Molaria* are all treated as part of the same grouping, despite the fact that the second to fourth of these belong to the cephalon, and the fifth to twelfth belong to the trunk. Hence, tagmatization measures of this type cannot be deduced directly from the characters coded in the morphological data base. In addition, none of the indices takes account of the form of morphological appendage specialization, nor the order in which tagmata of different sizes occur down the length of the body. Morphologically and phylogenetically very different taxa can therefore yield identical values for $h$. However, the intention is to measure the **degree** and not the form of appendage specialization in different groups. The decision as to how much differentiation is necessary between limbs in order to recognize them as members of a different tagma is inevitably somewhat subjective, but the treatment here is as internally consistent as possible. Several limb formulae differ from those used by Cisne (1974), either because new evidence from more detailed study has become available (in the case of fossils) or because of a different interpretation of morphology. A list of limb formulae is given in Appendix 6A.

## 6.3 DO RECENT ARTHROPODS HAVE MORE HIGHLY DIFFERENTIATED APPENDAGES THAN THEIR CAMBRIAN COUNTERPARTS?

Cisne (1974) recognized a logistic increase in the mean value of limb specialization (and by inference, ecological specialization) for aquatic, free-living arthropods over the period from the Cambrian to the Recent. This has been tested again by coding the 25 Recent and 37 fossil taxa used in the phylogeny of Wills *et al.* (1997). This includes many arthropods not considered by Cisne, and incorporates new information and interpretations of others.

The mean number of limb pairs (however differentiated) per taxon is greater in Recent (22.25) than Cambrian (16.13) arthropods, although this difference is not statistically significant (t-test). Moreover, the mean number of different appendage **types** is approaching twice the value in Recent (7.46) as compared with Cambrian forms (4.08), a highly significant difference (46 d.f. (as in all these tests), $t = 5.55$, $P = 0.000$). This could be attributable in part to the greater number of appendages in Recent forms (more latitude for specialization). However, the ratio of the number of appendage types to the number of appendage pairs is itself much greater in Recent (0.51) than Cambrian (0.32) taxa, again corresponding to a highly significant difference ($t = 2.76$, $P = 0.008$). Hence, Recent arthropods not only have a greater number of different types of appendages, but a given appendage is more likely to be differentiated from its immediate neighbours. Although the mean number of appendages incorporated into the cephalon for Cambrian taxa (3.88) is nearly 80% that for Recent taxa (4.88), the difference is highly significant ($t = 3.16$, $P = 0.003$).

The mean Brillouin value for Cambrian taxa (0.92) is less than two-thirds that for Recent taxa (1.55), but the difference is highly significant ($t = 3.82$, $P = 0.000$). Hence, while the disparity of Cambrian and Recent forms is comparable (Wills *et al.*, 1994 and below), the level of appendage specialization for Cambrian arthropods is significantly lower than that for their Recent counterparts. Cisne (1974) predicted that the degree of tagmosis for higher taxa should reflect the complexity of their ecological role. Interactions in Cambrian ecosystems were almost certainly less complex than those of the present day (Valentine, 1973; Bergström, 1991). Moreover, Cambrian fossils are drawn from a limited number of ecosystems (e.g. the Burgess Shale is a mid-slope, muddy-bottomed community), while Recent taxa are sampled from all possible environments. Therefore, less differentiation of appendages for specific trophic, locomotory, respiratory and other functions would be anticipated among Cambrian genera.

## 6.4 WHICH ARTHROPOD GROUPS ARE MOST DISPARATE?

In a series of papers (Briggs *et al.*, 1992a,b, 1993; Wills *et al.*, 1994), we sought to compare levels of disparity (the range of body plan design) in Cambrian and Recent arthropods using several phenetic indices. A more extensive database incorporating other Palaeozoic taxa coded for 97 shared characters has since been compiled (Wills *et al.*, 1997). These genera have been ordinated on to principal coordinates following Wills *et al.* (1994). The first three coordinates encompassed 38% of the variance in the data set (Figure 6.1), while 90% of the variance was captured by the first 25 axes (used as the basis for all disparity measures quoted here). Comparisons here are predominantly based on sum of range measures (Wills *et al.*, 1994), although some other values are given in Table 6.1. As in previous studies, Cambrian and Recent taxa occupy comparable amounts of morphospace on all measures. The phylogeny in Figure 6.2 (after Wills *et al.*, 1997) includes pie charts for several major clades, illustrating their disparity as a fraction of that for all arthropods. These measures are not additive in the sense of Foote (1993), but rather illustrate the total extent of morphological diversity for component groups.

The arachnomorphs are by far the most disparate of the major arthropod clades, even when this comparatively large group is rarefied down to a sample size of four (the number of taxa in the smallest group compared here) (Figure 6.3). The relatively high disparity of the arachnomorphs is unsurpris-

**Table 6.1** A variety of disparity indices for several arthropod groups based on the scores of taxa on the first 25 principal coordinates

| *Group* | *No. of genera* | *Sum of ranges* | *Product of ranges* ($\sqrt[25]{}$) | *Sum of variances* | *Product of variances* ($\sqrt[25]{}$) |
|---|---|---|---|---|---|
| Atelocerata | 10 | 33.861 | 1.271 | 5.042 | 0.158 |
| Hexapoda | 6 | 24.924 | 0.911 | 3.481 | 0.095 |
| Myriapoda | 4 | 15.809 | 0.570 | 1.851 | 0.051 |
| Marrellomorpha | 4 | 24.796 | 0.866 | 4.721 | 0.114 |
| Arachnomorpha | 20 | 58.015 | 2.247 | 10.143 | 0.347 |
| Chelicerata + Agla. + Chel. | 6 | 35.036 | 1.302 | 6.629 | 0.201 |
| Chelicerata | 4 | 25.586 | 0.963 | 4.511 | 0.143 |
| Trilobita | 4 | 27.536 | 1.004 | 5.697 | 0.158 |
| Crustacea | 22 | 50.185 | 1.892 | 7.600 | 0.233 |
| Malacostraca – Canadaspidida | 4 | 23.866 | 0.811 | 4.901 | 0.101 |
| Maxillopods – Ostracoda | 4 | 18.865 | 0.701 | 2.381 | 0.072 |
| Orsten genera | 4 | 23.589 | 0.841 | 4.179 | 0.109 |
| Basal six genera | 6 | 33.468 | 1.279 | 6.475 | 0.208 |

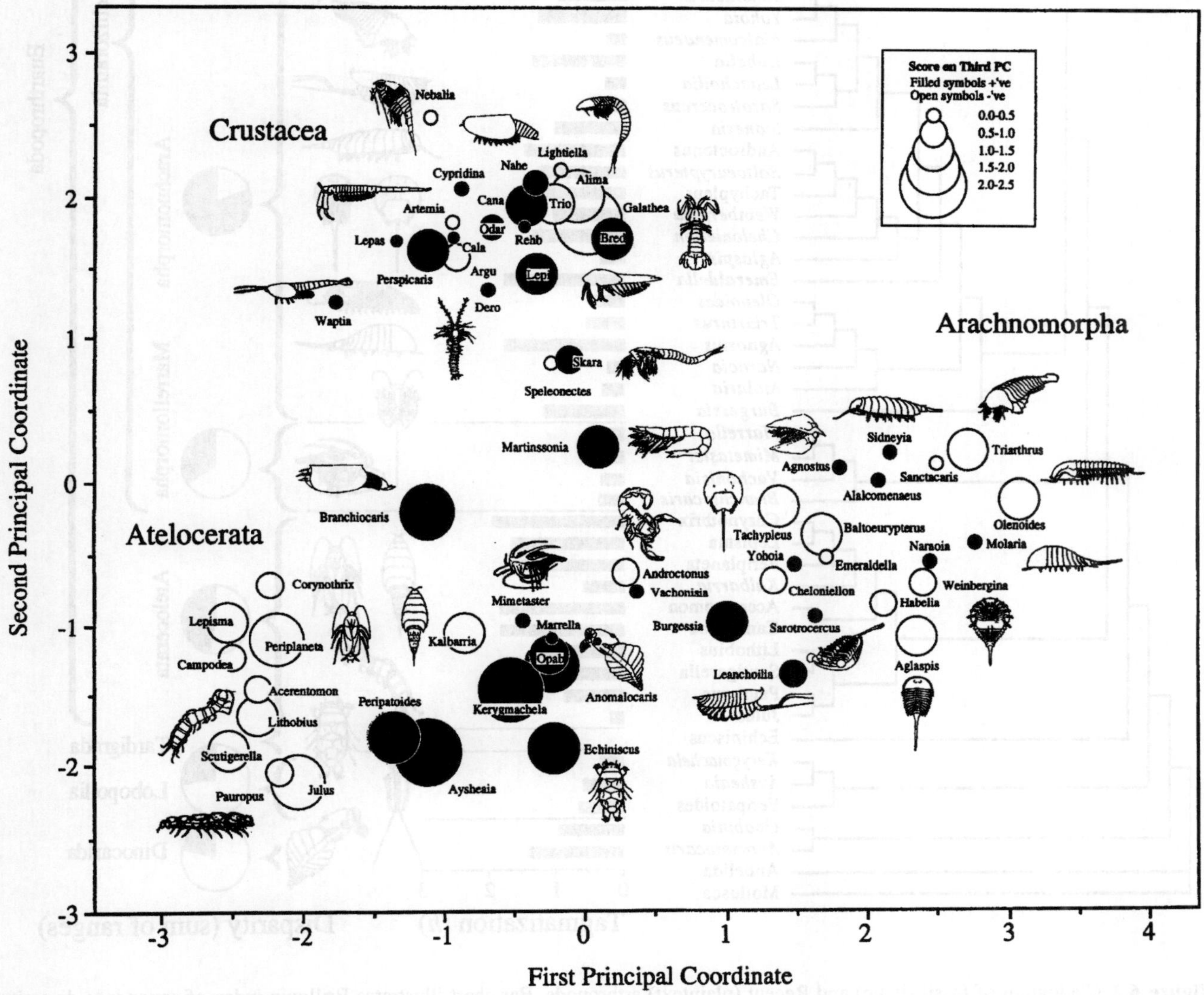


**Figure 6.1** Principal coordinates analysis of fossil and recent arthropods. Taxa have been plotted with respect to the first three coordinates (third axis in and out of page), which together encompass 38% of the variance in the data set.

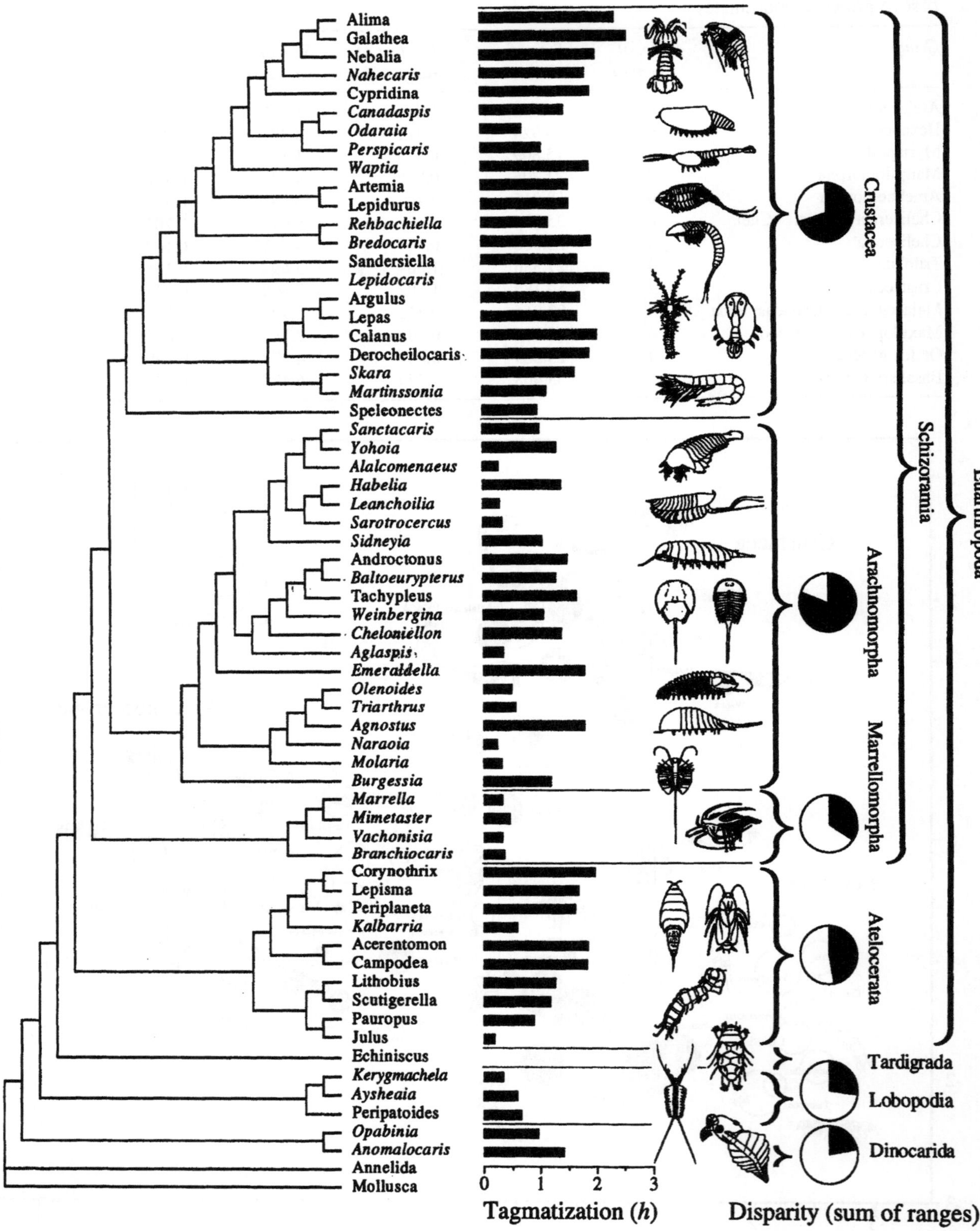


**Figure 6.2** Cladogram of fossil (italic) and Recent (plaintext) arthropods. Bar chart illustrates Brillouin index of appendage diversity ($h$) for each taxon. Pie charts indicate disparity of selected major clades (sum of ranges on first 25 principal coordinates as a fraction of the value for all arthropods). (After Wills *et al.*, 1997.)

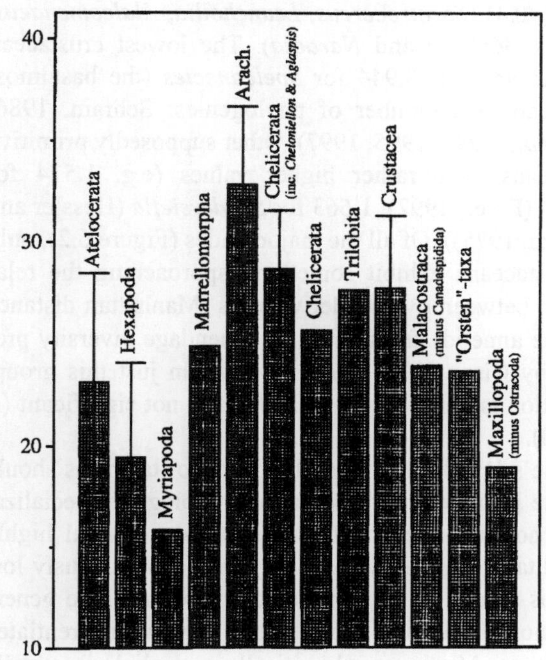


**Figure 6.3** Rarefied disparity of selected major groups. Values expressed as the mean sum of ranges on the first 25 PCs for 200 randomized draws (without replacement) of four taxa (the smallest group considered). Upper and lower 95% confidence intervals included where $n>4$.

ing, since it contains two distinctive clades formally recognized as classes (Trilobita and Chelicerata) in addition to a dozen or more problematic fossils. When corrections are made for sample-size differences (Figure 6.3), the four trilobites coded are almost as highly dispersed in morphospace as the average sample of four taxa drawn from all arachnomorphs. The trilobites alone, therefore, are a highly disparate clade [e.g. compare the body plans of *Agnostus* (Müller and Walossek, 1987) with *Emuella* (Pocock, 1970)]. So, too, are the chelicerates (depending upon how this group is defined – see Wills, 1996; Wills *et al.*, 1997) (calculated both with and without *Cheloniellon* and *Aglaspis* here).

The Crustacea have a lower disparity than the arachnomorphs (sum of ranges = 86%), which becomes slightly more marked at lower sample sizes (Figure 6.3) (crustaceans cluster tighter than arachnomorphs towards the centre of the morphospace they occupy (Wills *et al.*, 1994); see Figure 6.1). When crustaceans and arachnomorphs are rarefied to all possible sample sizes (Figure 6.4), each remains outside the 90% confidence interval for the other over most of its range. Four crustaceans sampled at random from within the morphospace have an average disparity equivalent to that of the trilobites (Figure 6.3). The four malacostracan taxa alone have a disparity almost half that of all the crustaceans taken together (rising to 79% with a correction for the sample-size difference: Figure 6.3). Orsten taxa (although not a clade here) have a disparity comparable with that of the Malacostraca, while the maxillopods (*sans* Ostracoda) constitute a much more cohesive grouping (36% of the Crustacea as a whole).

The ten atelocerates occupy just 58% of the morphospace embraced by the arachnomorphs, and 68% of that taken up by the crustaceans. All three groups remain distinct when rarefied (Figure 6.4), and at a sample size of four, the 90% error bars for the arachnomorphs and crustaceans do not overlap. All atelocerates are built with relatively little variation in body plan. The hexapods show an average disparity comparable with that of the maxillopods when corrected for sample-size differences, while the myriapods are the most conservative of all the groups examined.

The marrellomorphs (ranked cladistically alongside the arachnomorphs plus the crustaceans; Wills *et al.*, 1994, 1995, 1997) are more disparate than the myriapods, and have a level comparable with that of the hexapods, malacostracans and chelicerates. With sample-size corrections, they are more disparate than all atelocerates taken together, and occupy a sizeable fraction (85%) of the amount of morphospace filled by crustaceans. This is wholly in accordance with their status as a major taxon within the arthropods.

## 6.5 IS THERE A RELATIONSHIP BETWEEN APPENDAGE SPECIALIZATION AND OVERALL MORPHOLOGICAL DERIVEDNESS?

The first prediction of Cisne (1974) was that degree of appendage specialization should correlate positively with

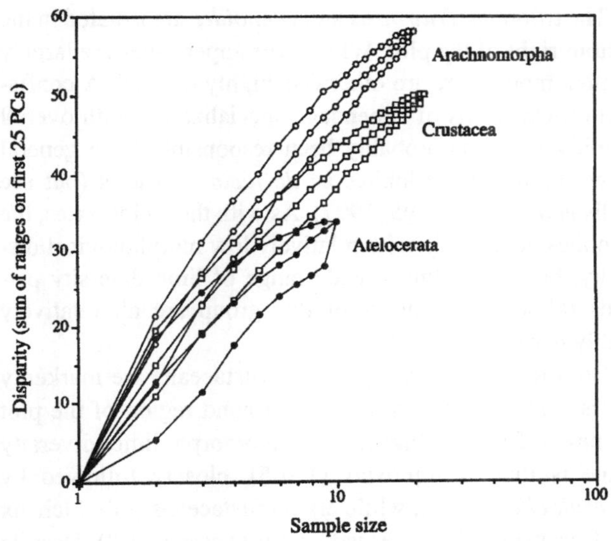


**Figure 6.4** Rarefaction of disparity measures for three major clades over all sample sizes. Values based on 200 random draws of taxa without replacement. Approximate upper and lower 95% confidence intervals shown.

overall morphological derivedness. Manhattan distance from the annelid would be expected to correlate with the degree of tagmatization, since phenetic distance will indicate overall morphological derivedness, regardless of phylogenetic position or the implied number of character state changes between a taxon and the root. However, correlation is very low ($r^2 = 0.045$), and not significant ($P = 0.099$). Correlations for Cambrian and Recent taxa alone are even lower ($r^2 = 0.010$ and 0.011 respectively).

The most tagmatized genus, *Galathea* ($h = 2.574$), is also the third most phenetically distant from the annelid (Manhattan distance 54). However, the arthropod **most** distant from the annelid (*Olenoides*, 60) is among the least tagmatized ($h = 0.476$). The use of a patristic measure of derivedness (measured in character state changes on the cladogram; Figure 6.2) produced an equally poor overall correlation.

A few broad trends do emerge from the plots. The lobopods, among the least phenetically derived taxa, also exhibit some of the lowest indices of limb diversity. The Cambrian genera *Anomalocaris* and *Opabinia* (less derived than any euarthropods) show higher, though below-average Brillouin values. The atelocerates, while all relatively close to the primitive morphology (*Periplaneta* has the maximum Manhattan distance of 36) show a wide range of limb diversity values. *Julus* has the lowest value of all the arthropods (0.165), while the highest values belong to the hexapods (e.g. *Campodea*, 1.797). Hence, even within a relatively well-constrained clade of comparatively low morphological disparity, the variation in the degree of limb tagmosis (range 1.632) is a large proportion (68%) of that observed in the arthropods as a whole (range 2.409).

The trilobites *Triarthrus* and *Olenoides* are isolated in the bottom right of the plot. While their appendages are largely homonomous, they are otherwise highly derived. A conflation of low levels of appendage specialization with overall primitiveness has probably been responsible for a general view holding the trilobites as *de facto* ancestors (but see Wills *et al.*, 1994, 1995, 1997). As with the atelocerates, the trilobites are a clade with relatively low morphological disparity, but subtending a wide range of limb diversity patterns (although members of this group are all relatively highly derived).

Both the arachnomorphs and crustaceans are markedly interspersed, except in the far left-hand region of the plot (Figure 6.5). The highest arachnomorph limb diversity value is that of *Agnostus* (1.765), closely followed by *Emeraldella* (1.761), while eight crustaceans (of which six are Recent) reach higher values (in excess of 1.8). Despite the higher disparity of the arachnomorph clade in comparison with the crustaceans, the range of Brillouin values for the arachnomorphs is lower (although the variance in these values is greater). Several arachnomorphs have very low values (<0.4) (*Sarotrocercus*, *Leanchoilia*, *Alalcomenaeus*, *Aglaspis*, *Molaria* and *Naraoia*). The lowest crustacean Brillouin value is 0.944 for *Speleonectes* (the basalmost crustacean in a number of phylogenies: Schram, 1986; Wills *et al.*, 1994, 1995, 1997). Other supposedly primitive crustaceans have rather higher values (e.g. 1.514 for *Artemia* (Fryer, 1992), 1.663 for *Sandersiella* (Hessler and Newman, 1975)). Of all the major clades (Figure 6.2), only the crustaceans exhibit something approaching the relationship between overall derivedness (Manhattan distance from the annelid) and indices of appendage diversity predicted by Cisne (1974), but even within just this group, correlation is very low ($r^2 = 0.207$) and not significant ($P = 0.710$).

Cisne's (1974) claim that the degree of tagmosis should correlate positively with overall morphological specialization is not supported by the present data. Several highly derived taxa (e.g. many trilobites) have conspicuously low tagmosis values, while other much more primitive genera (e.g. anomalocaridids) have more highly differentiated appendages. Moreover, phenetically and phylogenetically very proximate taxa often display markedly different tagmosis patterns and values (e.g. compare notostracan and cladoceran branchiopods, julid and glomerid diplopods, olenellid and agnostid trilobites). There is no clear pattern in $h$-values when these are plotted on the cladogram (beyond those observed for major clades) (Figure 6.1), and similar patterns of tagmosis have arisen independently in distantly related taxa (see Appendix 6A).

## 6.6 IS THERE A RELATIONSHIP BETWEEN APPENDAGE SPECIALIZATION VALUES AND DISPARITY?

Although we would not necessarily expect more disparate groups or clades to exhibit higher overall tagmosis values (there is actually a slightly negative correlation for the groups identified in Figure 6.3; $r^2 = 0.042$), we might anticipate a greater range or variance in $h$ values in more disparate clades. The range of Brillouin indices for the clades and other groups identified in Figure 6.3 has been plotted against disparity (sum of ranges on first 25 principal coordinates, uncorrected for sample-size differences) to investigate this (Figure 6.6). Some relationship emerges ($r^2 = 0.391$, $P = 0.030$) (although the groups considered are not mutually exclusive). The marrellomorphs show the lowest range in Brillouin values ($h$) (see also Figure 6.2), but have a disparity equal to or greater than several groups of comparable size but much greater range in $h$ (e.g. Myriapoda and Malacostraca). The hexapod clade has a disparity comparable to that of the marrellomorphs, but among the highest range in Brillouin values (although this is much lower if the euthycarcinoid is removed).

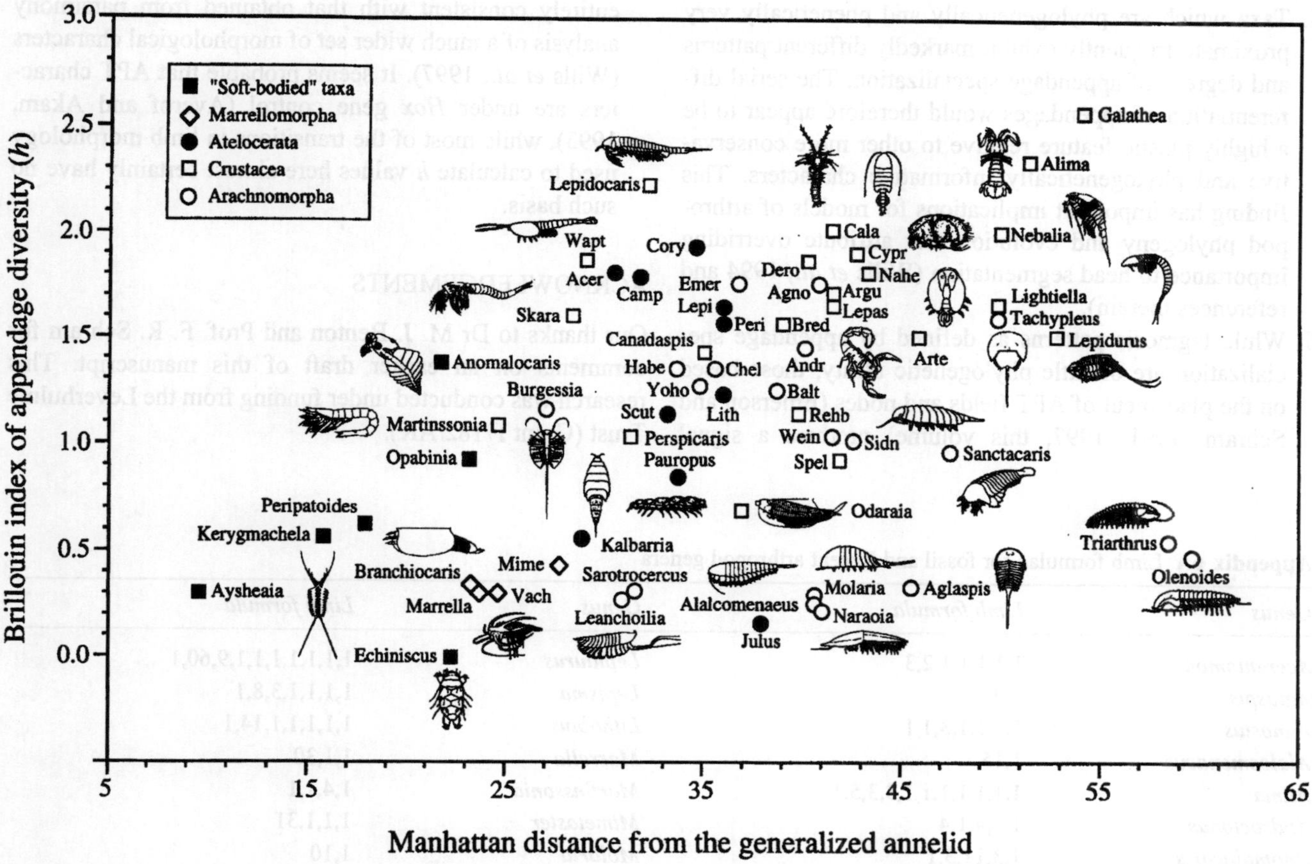


**Figure 6.5** Scatterplot of Brillouin index of appendage diversity against Manhattan distance from the generalized annelid for fossil and Recent arthropods. (Derived from the database of Wills *et al.*, 1997.)

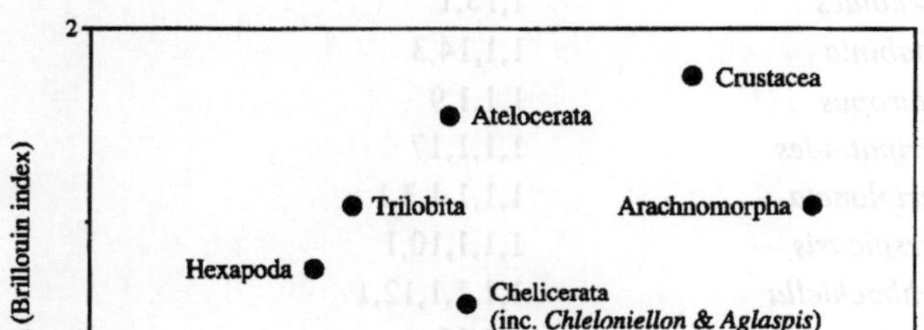


**Figure 6.6** Scatterplot of the range in Brillouin values against disparity (sum of ranges on first 25 PCs) for groups from Figure 6.3.

## 6.7 CONCLUSIONS

1. On average, Recent taxa exhibit significantly more specialization of their appendages than their Cambrian counterparts. Higher levels of appendage specialization were predicted by Cisne (1974) as a reflection of the complexity of the ecological role of higher taxa. This relationship appears to hold for the comparison of Cambrian and Recent groups.
2. While tagmosis is significantly more complex in Recent than Cambrian arthropods, levels of morphological disparity are comparable. Hence, while Cambrian arthropods are as morphologically diverse as Recent taxa, their ecological diversity is lower.
3. The arachnomorphs (containing the trilobites and chelicerates) are the most disparate major arthropod clade, closely followed by the crustaceans. The atelocerates are much more constrained morphologically.
4. There is no simple relationship between overall derivedness (measured as Manhattan distance from the annelid) and the degree of limb tagmosis, whether for arthropods taken as a whole, for just Cambrian or Recent representatives, or within major clades.

5. Taxa which are phylogenetically and phenetically very proximate frequently exhibit markedly different patterns and degrees of appendage specialization. The serial differentiation of appendages would therefore appear to be a highly plastic feature relative to other more conservative and phylogenetically informative characters. This finding has important implications for models of arthropod phylogeny and evolution that attribute overriding importance to head segmentation (Wills *et al.*, 1994 and references therein).
6. While tagmosis patterns as defined by appendage specialization are of little phylogenetic utility, those based on the placement of APT fields and nodes (Emerson and Schram, 1991, 1997, this volume) produce a signal entirely consistent with that obtained from parsimony analysis of a much wider set of morphological characters (Wills *et al.*, 1997). It seems probable that APT characters are under *Hox* gene control (Averof and Akam, 1993), while most of the transitions in limb morphology used to calculate *h* values here almost certainly have no such basis.

ACKNOWLEDGEMENTS

Our thanks to Dr M. J. Benton and Prof. F. R. Schram for comments on an earlier draft of this manuscript. This research was conducted under funding from the Leverhulme Trust (Grant F/182/AK).

**Appendix 6A** Limb formulae for fossil and Recent arthropod genera

| *Genus* | *Limb formula* | *Genus* | *Limb formula* |
|---|---|---|---|
| *Acerentomon* | 1,1,1,1,1,2,3 | *Lepidurus* | 1,1,1,1,1,1,1,9,60,1 |
| *Aglaspis* | 1,9 | *Lepisma* | 1,1,1,1,3,8,1 |
| *Agnostus* | 1,1,1,1,3,1,1 | *Lithobius* | 1,1,1,1,1,14,1 |
| *Alalcomenaeus* | 1,15 | *Marrella* | 1,1,30 |
| *Alima* | 1,1,1,1,1,1,1,3,3,5,1 | *Martinssonia* | 1,4,1,1 |
| *Androctonus* | 1,1,4,1,4 | *Mimetaster* | 1,1,1,31 |
| *Anomalocaris* | 1,3,11,3,1 | *Molaria* | 1,10 |
| *Argulus* | 1,1,1,1,1,4,1 | *Nahecaris* | 1,1,1,1,1,8,5,1 |
| *Artemia* | 1,1,1,1,1,11,1,1 | *Naraoia* | 1,18 |
| *Aysheaia* | 1,10 | *Nebalia* | 1,1,1,1,1,8,4,2,1 |
| *Baltoeurypterus* | 1,4,1,5 | *Odaraia* | 2,1,1,1,30 |
| *Branchiocaris* | 1,1,46,1 | *Olenoides* | 1,15,1 |
| *Bredocaris* | 1,1,1,1,7,1,1 | *Opabinia* | 1,1,14,3 |
| *Burgessia* | 1,3,7,1 | *Pauropus* | 1,1,1,9 |
| *Calanus* | 1,1,1,1,1,1,4,1,1 | *Peripatoides* | 1,1,1,17 |
| *Campodea* | 1,1,1,1,2,1,7,1 | *Periplaneta* | 1,1,1,1,3,1 |
| *Canadaspis* | 1,1,1,2,8,1 | *Perspicaris* | 1,1,1,10,1 |
| *Cheloniellon* | 1,1,4,8,1 | *Rehbachiella* | 1,1,1,1,12,1 |
| *Corynothrix* | 1,1,1,1,1,2,1,1 | *Sanctacaris* | 5,1,10 |
| *Cypridina* | 1,1,1,1,1,1,1,1 | *Sandersiella* | 1,1,1,1,8,1,1,1 |
| *Derocheilocaris* | 1,1,1,1,1,1,4,1 | *Sarotrocercus* | 1,10 |
| *Echiniscus* | 4 | *Scutigerella* | 1,1,1,1,12,1 |
| *Emeraldella* | 1,1,2,2,6,6 | *Sidneyia* | 1,4,5,1 |
| *Galathea* | 1,1,1,1,1,1,1,1,1,4,1,1,1 | *Skara* | 1,1,1,2,1,1 |
| *Habelia* | 1,2,6,5 | *Speleonectes* | 1,1,1,1,1,1,31,1 |
| *Julus* | 1,1,1,124 | *Tachypleus* | 1,4,1,1,1,5 |
| *Kalbarria* | 1,1,11 | *Triarthrus* | 1,3,27 |
| *Kerygmachela* | 1,11,1 | *Vachonisia* | 1,1,2,75 |
| *Leanchoilia* | 1,13 | *Waptia* | 1,1,1,1,1,4,4,1 |
| *Lepas* | 1,1,1,1,1,1,5,1 | *Weinbergina* | 1,6,6 |
| *Lepidocaris* | 1,1,1,1,3,5,1,1 | *Yohoia* | 1,3,1,9 |

## REFERENCES


Averof, M. and Akam, M. (1993) HOM/*Hox* genes of *Artemia*: implications of insect and crustacean body plans. *Current Biology*, **3**, 73–8.

Bergström, J. (1991) Metazoan evolution around the Precambrian–Cambrian transition, in *The Early Evolution of Metazoa and the Significance of Problematic Taxa* (eds A.M. Simonetta and S. Conway Morris), Cambridge University Press, Cambridge, pp. 25–34.

Briggs, D.E.G., Fortey, R.A. and Wills, M.A. (1992a) Morphological disparity in the Cambrian. *Science*, **256**, 1670–3.

Briggs, D.E.G., Fortey, R.A. and Wills, M.A. (1992b) Cambrian and Recent morphological disparity. (Response to Foote and Gould, and Lees). *Science*, **258**, 1817–18.

Briggs, D.E.G., Fortey, R.A. and Wills, M.A. (1993) How big was the Cambrian explosion? A taxonomic and morphologic comparison of Cambrian and Recent arthropods, in *Evolutionary Patterns and Processes* (eds. D.R. Lees and D. Edwards), Linnean Society Symposium Series, Linnean Society of London, pp. 33–44.

Brillouin, L. (1962) *Science and Information Theory*, 2nd edn, Academic Press, New York.

Cisne, J. L. (1974) Evolution of the world fauna of aquatic free-living arthropods. *Evolution*, **28**, 337–66.

Foote, M. (1993) Contributions of individual taxa to overall morphological disparity. *Paleobiology*, **19**, 403–19.

Fryer, G. (1992) The origin of the Crustacea. *Acta Zoologica*, **73**, 273–86.

Hessler, R.R. and Newman, W.A. (1975) A trilobitomorph origin for the Crustacea. *Fossils and Strata*, **4**, 437–59.

Müller, K.J. and Walossek, D. (1987) Morphology, ontogeny, and life habit of *Agnostus pisiformis* from the Upper Cambrian of Sweden. *Fossils and Strata*, **19**, 1–124.

Pocock, K.J. (1970) The Emuellidae: a new family of trilobites from the Lower Cambrian of South Australia. *Palaeontology*, **13**, 522–6.

Schram, F.R. (1986) *Crustacea*, Oxford University Press, Oxford.

Schram, F.R. and Emerson, M.J. (1991) Arthropod Pattern Theory, a new approach to arthropod phylogeny. *Memoirs of the Queensland Museum*, **31**, 1–18.

Valentine, J.W. (1973) *Evolutionary Paleoecology of the Marine Biosphere*, Prentice Hall, New Jersey.

Wills, M.A. (1996) Classification of the arthropod *Fuxianhuia*. *Science*, **272**, 746–8.

Wills, M.A., Briggs, D.E.G. and Fortey, R.A. (1994) Disparity as an evolutionary index. A comparison of Cambrian and Recent arthropods. *Paleobiology*, **20**, 93–130.

Wills, M.A., Briggs, D.E.G., Fortey, R.A. and Wilkinson, M. (1995) The significance of fossils in understanding arthropod evolution. *Verhandlungen der Deutschen Zoologischen Gesellschaft*, **88**, 203–15.

Wills, M.A., Briggs, D.E.G., Fortey, R.A., Wilkinson, M. and Sneath, P.H.A. (1997) An arthropod phylogeny based on fossil and Recent taxa, in *Arthropod Fossils and Phylogeny* (ed. G.E. Edgecombe), Columbia University Press, New York.

# 7 Theories, patterns, and reality: game plan for arthropod phylogeny

*M.J. Emerson*[†] *and F.R. Schram*

*Late of the San Diego Natural History Museum*
*Institute for Systematics and Population Biology, University of Amsterdam, Post Box 94766, NL-100 GT, Amsterdam, Netherlands*
*email: schram@bio.uva.nl*

## 7.1 INTRODUCTION

In a series of papers some years ago (Emerson and Schram, 1990a,b; Schram and Emerson, 1991) we proposed Arthropod Pattern Theory (APT). Before Michael Emerson's death (22 March, 1990), he and I had planned to perform a cladistic analysis of the fossil and recent arthropods using APT features. He prepared a preliminary data set, and we actually ran some analyses. A series of circumstances, however, intervened after the preparation of Schram and Emerson (1991), and time and opportunity only now have presented themselves to allow one of us (F.R.S.) to take this project up again. This still remains, nevertheless, very much a joint effort, since Michael had prepared before his death extensive notes and observations concerning arthropod body plans, and these have formed the foundation of this analysis.

## 7.2 IMPLICATIONS OF RECENT RESEARCH FOR APT

Although the Late Mississippian fossil crustacean, *Tesnusocaris goldichi* Brooks, 1955, formed the immediate focus of our attention in the late 1980s (culminating in Emerson and Schram, 1991) along with its peculiar arrangement of trunk limbs (Emerson and Schram, 1990b), a more or less fully developed Arthropod Pattern Theory emerged only in Schram and Emerson (1991). Essentially the theory advances three hypotheses.

### 7.2.1 HYPOTHESIS 1

The first hypothesis states that 'the biramous limb of Crustacea (and probably all arthropods bearing such) evolved by means of basal fusion of duplopodous, uniramous limbs'. This arose originally out of our work on *Tesnusocaris* with its peculiar arrangement of trunk limbs. We furthermore believed (Schram and Emerson, 1991) that there was at least one other duplopodous animal in the fossil record, viz., *Branchiocaris pretiosa* (Resser), although our interpretation stands in contrast to that found in Briggs (1976). We were encouraged, however, to consider the possibility of fusing two leg bases to form a single biramous limb by the conclusion of Itô (1989) that the copepod basis evolved from the fusion of articles or parts of articles in the exopod and endopod of a limb not unlike that seen in living remipedes.

Subsequent to our posing this idea (Emerson and Schram, 1990b), the issue of duplopody has surfaced in the work of others. Ferrari (1995), in his re-examination of Claus' ideas about copepod mouthparts, believed that certain aspects of APT might support the concepts of Claus. Ferrari outlined some genetic tests with regard to patterns of engrailed expression in the head segments of copepods that might confirm or disprove APT predictions.

Quite independent of Ferrari's work, Panganiban *et al.* (1995) studied the expression of the gene distal-less (*Dll*) in the crustaceans *Artemia* and *Mysidopsis*. They found that in the ontogeny of head limbs, a single anlagen appears that later splits to form the characteristic biramous type. However, in trunk limbs of *Mysidopsis*, they noted that '*Dll* expression rapidly resolves to two clusters of cells per half segment each of which begins to form a proximodistal outgrowth' (p. 1364). Each abdominal limb also developed from two fields, but these appeared sequentially rather than together as in the thoracopods. Panganiban and her colleagues concluded that since the same gene, *Dll*, is involved in limb formation of both insects and crustaceans there had to be identical modes of origin for both groups. This suggestion, however, seems to ignore the fact that the double limb anlagen is exactly what APT would predict, two fields of cells that later merge to form a single limb. Furthermore,

[†] Deceased

*Arthropod Relationships*, Systematics Association Special Volume Series 55, Edited by R.A. Fortey and R.H. Thomas. Published in 1997 by Chapman & Hall, London. ISBN 0 412 75420 7

the single anlagen noted by the geneticists in the crustacean head recalls the observations of Ferrari (1995) relevant to uniramous mouthparts and suggests that perhaps APT should be modified to allow duplopody only in the trunk. However, Panganiban *et al.* (1995) accepted at face value the duplopody of *Tesnusocaris* and remark that this condition could have been a derived one that might be explained by the 'modulation of the dorsoventral positioning of groups of *Dll*-expressing cells' (p. 1365).

### 7.2.2 HYPOTHESIS 2

The second hypothesis of APT states that 'a uniramian diplosegment, or two monomeres, is homologous to a single crustacean (and by extension, other biramian arthropods) body segment, or duplosegment'. Schram and Emerson (1991) noted several features of arthropod comparative anatomy and ontogeny that could be explained by such a radical hypothesis. Since then, Spiridinov (1993) examined some peculiar features in the sixth abdominal segment and tail fan of Euphausiacea and interpreted these too as vestiges of a terminal duplosegment. Furthermore, Scholtz (1995) in his exploration of *engrailed* expression in the freshwater crayfish, *Cherax destructor*, noted nine segmental anlagen in the abdomen, i.e. three extra anlagen behind that of the sixth pleomere. Scholtz did not mention APT at all, and he would almost certainly reject this observation here; however, it is interesting to note that the extra segmental anlagen occur as multiples of two. The seventh pair of anlagen, which has been noted in many malacostracan forms (for further references, see Schram, 1986; or Schram and Emerson, 1991), fuses into the sixth set of ganglionic anlagen and these together form the sixth abdominal duplomere. Thus, Scholtz's eighth and ninth pairs of anlagen could then represent the remnants of a seventh duplomere.

However, Zrzavý and Stys (1994) reanalysed the duplosegment hypothesis and pointed out that the parasegmental compartmentalization in the ontogeny of the adult arthropod segment is virtually identical in insects and crustaceans, and no remnant of an APT duplomere can be recognized. This is indeed a potentially valid criticism of APT. Nevertheless, their arguments are somewhat circular in that they claim that a close relationship between crustaceans and hexapods would then require that the similarity in ontogeny represents an apomorphy and thus reflects true homology of segments, but this presupposes that there **is** a close relationship between these two groups to begin with. They do raise, and perhaps prematurely discard, another alternative hypothesis, viz. that the insectan and crustacean segments are **both** duplomerous. At any rate, the prime comparison of segment formation between uniramian or atelocerate forms and crustaceans has heretofore focused not on insects, but rather on various conditions as seen in myriapods, groups about which we know all too little concerning their ontogeny.

### 7.2.3 HYPOTHESIS 3

The APT third hypothesis focuses on issues of body plan homologies and states that 'suites of segments in arthropods evolve as units, with the transition of tagmata, the location of gonopores and anus, and body termination occurring at specific points along the body that are shared between disparate groups'. This hypothesis later formed the basis for Minelli and Schram (1994) to recognize the phenomenon of positional homology. They drew attention to points along bodies that were homologous between groups, not because of the specific structures that occurred at these points, but because these points or 'hot spots' appeared to share some more fundamental, underlying genetic control.

Sundberg (1995) employed APT to evaluate positional homologies in Cambrian trilobites. His analysis entailed a statistical overview of the occurrence of structural changes along the length of the bodies of these trilobites in an attempt to test whether these occurrences corresponded to APT nodes. He concluded that APT explanations do hold true to a certain degree – not to a level of absolute certainty, but unquestionably at a frequency beyond the realm of pure chance.

Akam (1995) and Averof and Akam (1995), in an interesting set of papers, explored the expression of *Hox* genes in insect and crustacean (*Artemia*) body plans. Their papers did not take up APT, but the patterns they describe and illustrate do in fact lend themselves to an APT interpretation. Akam (1995) believed that the expression of *abdA* indicated that the insect abdomen and the crustacean thorax are homologous. Averof and Akam (1995) added consideration of *Ubx* and *abdB* to this scheme that essentially coincided with the *abdA* research. Under the tenets of the traditional scheme of universal arthropod segment homology (Figure 7.1(b,c)) the *abdA* and *Ubx* expression would begin considerably posterior along the body in insects as opposed to where they begin in crustaceans. On the other hand, under an APT interpretation of segment homology (Figure 7.1(a,b)), the expression of the genes *Antp, abda* and *Ubx* begins at corresponding points in the anterior regions of the *Drosophila* and *Artemia* bodies. To express this another way in the language of APT, the *abdA* gene serves as 'a post node 1/pre-gonopore activator' and the *abdB* gene as 'a controller of a gonopore hotspot'. These genes, therefore, appear to serve the roles that Minelli and Schram (1994) predicted (suggested in their principle 8), viz., that the establishment of anatomical landmarks (hotspots) have priority in morphogenesis.

It is not possible to suggest phylogenetic relationships based on the genetic analysis of only two taxa, *Drosophila* and *Artemia*. Until we have genetic data for Hox expression in myriapods, many groups of crustaceans, and in the cheliceriforms we really have no basis for determining which patterns of expression are plesiomorphic and which are apomorphic. Furthermore, while we have a modicum of

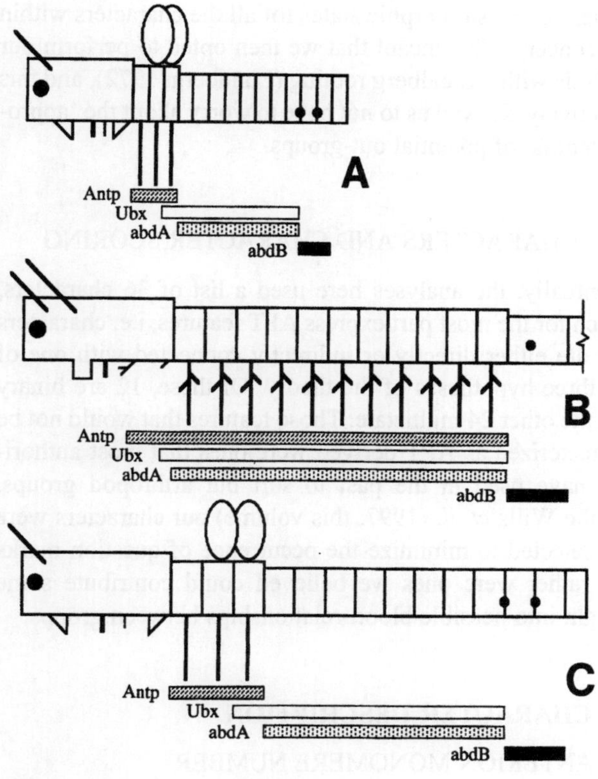


**Figure 7.1** Diagrammatic comparison of arthropod body plans with the domains of Hox gene expression. (A) APT interpretation of insect body plan, in which two adjacent monosegments are homologous to one crustacean duplosegment. (B) Body plan of the crustacean *Artemia*, post-genital segments deleted. (C) Traditional interpretation of insect body plan, in which any segment is homologous to a segment in crustaceans. Comparing (A) with (B) produces an APT interpretation of insect and crustacean types; note the similarity of Hox gene expression that is narrowly staggered at their anterior aspects. Comparison of (B) with (C) produces a traditional interpretation of crustacean and insect types; note the widely different staggering of Hox gene expression in the anterior end as opposed to the similarity of Hox gene expression at their posterior aspects in regards to termination of Ubx and abdA and expression of abdB. *Abdominal B* may function in the genetic determination of the 'gonopore hotspot' in the sense of Minelli and Schram (1994).

information concerning gene expression in segment formation, we have comparatively little knowledge about the patterns of expression of those genes that are known to control whole regions of segments, such as gap genes (e.g., see Nüsslein-Volhard and Wieschaus, 1980; Lawrence, 1992). This latter is an aspect of development that would provide crucial insight into the generation and evolution of major Baupläne within the arthropods.

It seems clear from the above brief review that opinions differ as to whether APT can serve as an effective organizing principle in regards to considerations of arthropod phylogeny. So it now becomes crucial to determine just what the APT hypotheses can tell us about the relationships of arthropod groups.

## 7.3 CLADISTIC METHODS

### 7.3.1 TAXA

#### (a) Choice of taxa

It is not the intention to present here an exhaustive phylogenetic analysis of *all* possible taxa of arthropods. Nevertheless, we wanted to include a reasonable sample of various body plan types from all major groups of living and fossil arthropods. There are two reasons for this. The first of these obviously connects with trying to obtain a broadly based assessment of the relationships of arthropod body plans. To do this we needed to include a reasonable sample of arthropod diversity. Second, we also wanted a check on the efficacy of our characters. By including a representative array of the established arthropod groups, we would hope to see some of those groups re-emerging from our analysis.

#### (b) Monophyly of higher taxa

This last point is important. Not to get such key taxa emerging as monophyletic clades would mean that either the assumptions about arthropod phylogeny of the past two centuries were entirely wrong, or the APT characters we employed have little phylogenetic signal in them. Although we trusted that the emergence of 'familiar' higher taxa might form the basis for assessing the usefulness of our own characters, we did not necessarily demand that such occur. One could assume that if clades emerged from the analysis that were consistently present – as measured by a majority rule or strict consensus – then those monophyletic clades might have a high chance of actually reflecting a real relationship. Conversely, under the assumptions implicit in the APT hypotheses, we could also infer that clades detected in this manner, even if they brought together what have traditionally been viewed as widely separate groups, might reflect derivation from some shared ground pattern. Likewise, separating traditionally aligned groups might reflect lack of derivation from a shared ground pattern.

#### (c) Fossil taxa

Practical considerations ensue in the choice of taxa in this regard (Novacek, 1992). Most of the phylogenetically interesting arthropod taxa are fossils. However, analysis of such long-extinct creatures entails trying to deal with both missing information (e.g. it is quite difficult, but not necessarily impossible, to obtain knowledge about the location of gonopores in fossil taxa) and/or inappropriate queries from the data base (e.g. characters that deal with conditions of appendages in an animal, like *Opabinia*, that appear to have fins and no true appendages). Rather than insert every conceivable fossil arthropod into the matrix, we chose to focus instead on fossils for which we have a fair degree of information about limbs and body plan details. Most of these in fact, e.g. Burgess Shale arthropods and the Cambrian Orsten from Sweden, have figured centrally in speculations about arthropod phylogeny. Inclusion of additional extinct forms would have increased the number of question marks in the matrix and thus only served to increase the total number of equally parsimonious trees (and consequently the level of uncertainty) and not necessarily increased the information content of the analysis.

In one case, however, we deliberately scored *Canadaspis perfecta* in two different ways. Briggs (1978) concluded that *Canadaspis* possessed the body plan of a phyllocarid crustacean. Dahl (1984), on the other hand, questioned many of Briggs' assumptions. In our database, we scored a taxon, *Canadaspis*, according to the conclusions of Briggs and another taxon, we called 'canadaspoid', to take into account some of the arguments of Dahl. This seemed necessary to us, since a cladistic analysis of a traditional morphologic data base of fossil and recent crustaceomorphs (Schram and Hof, in press) placed *Canadaspis* among the stem group taxa and not within the Crustacea *sensu stricto*.

### 7.3.2 CHOICE OF OUT-GROUP

In several preliminary analyses of APT over the years, both of us experimented with different options as regards to out-group. For example, early runs with small test data sets utilized onychophorans. Tardigrades were also considered. It is ordinarily best to avoid *a priori* assumptions when scoring or polarizing features, but this is an area currently of intense discussion and debate (Wägele, 1992). However, it is our experience that only the simultaneous analysis of the entire data set affords useful phylogenetic information about character status and polarity, rather than determining these aspects before the analysis. In other words, we believe it is better to allow the relationships of the taxa implicit in the entire data set to reveal the status of the characters in a theory-free context, rather than to interpret the characters in a theory-bound manner that forces the phylogeny of the taxa.

Having stated this, and in spite of this basic belief, we recognized from the beginning that APT cannot be disguised in any way as being a totally theory-free set of assumptions. The problem then of trying to pick an out-group for our analysis is that the chosen out-group may in fact have little or nothing to do with APT assumptions. For example, in a theory that deals with limb evolution and body region homologies, to force an analysis rooted in annelids and/or molluscs – much in the manner of Wills *et al.* (1995) – is to ignore some important points. In the end, we developed a hypothetical animal, which one of us (M.J.E.) termed the 'gular worm'. This could be viewed as a paper creature possessing plesiomorphic states for all the characters within our concern. This meant that we then opted to perform our analysis with a Lundberg rooting (Lundberg, 1972), and this effectively allowed us to not have to worry about the 'appropriateness' of potential out-groups.

### 7.3.3 CHARACTERS AND CHARACTER SCORING

Eventually, the analyses here used a list of 36 characters, which for the most part express APT features, i.e. characters that are either directly or indirectly connected with one of the three hypotheses of the theory. Of these, 12 are binary and the other 24 multistate. Those features that would not be characterized as APT derived were ones that most authorities have used in the past to sort out arthropod groups. Unlike Wills *et al.* (1997, this volume) our characters were not selected to minimize the occurrence of question marks but rather were ones we believed could contribute some insight into possible blood-relationships between groups.

## 7.4 CHARACTER DESCRIPTION

### 01. ANTERION MONOMERE NUMBER

state 0 = no more than 2
state 1 = with up to 4
state 2 = with up to 6
state 3 = with up to 8
state 4 = with up to 10
state 5 = with up to 12
state 6 = with up to 14
state 7 = with up to 16
state 8 = with up to 18

This feature gauges the extent of the anterior-most region of the body (we employ a neutral term 'anterion' to designate this area), whether it be a 'pro-cephalon', true head, or prosoma. This assumes that this region has a distinct function(s) and morphology from the trunk region that immediately follows.

### 02. NODE 1 EXPRESSION

state 0 = not expressed
state 1 = expressed

This is one of the key APT features. Its expression *per se* is not keyed to a particular anatomical manifestation, but simply records that some 'anatomical event' does occur in the segments in question, whether that event is a distinctive appendage, location of a gonopore, transition of body region. As such, the node's expression is a manifestation of some underlying genetic control of this area as a 'hotspot', which we believe is fundamentally different from any anatomical structures (and their own genetic controls) that may occur at the hot spot.

### 03. HEAD STATE

state 0 = head present
state 1 = prosoma present

There is a generally conceded difference between what constitutes a head and a prosoma. Heads have antennae, prosomas do not; heads have their posterior limbs modified as mouthparts, prosomas have their posterior limbs pediform in structure; heads tend to have true jaws, prosomas do not but may exhibit gnathobases on all or most of the limbs behind the mouth. Now, whether heads precede prosomas in arthropod evolution, or whether each of these is a unique anatomical condition is not clear at this time. In this respect, it would be interesting to know the expression of genes like *Antennapedia* in the prosoma of horseshoe crabs and arachnids.

### 04. ANTERION CONDITION

state 0 = with monomeres only
state 1 = with duplomeres

Another key APT character. It is conceivable that this feature should be modified in light of the suggestions of Ferrari (1995) discussed earlier, which would imply that the head never develops duplomeres. However, since our goal is to evaluate APT as an organizing principle and not evaluate any number of alternatives, we did not explore these other options.

### 05. MOUTH LOCATION

state 0 = anterior
state 1 = within the first 2 monomeres
state 2 = within the second 2 monomeres

This feature deals with the location of the mouth in regards to the anterior segmentation. Thus, the mouth can be placed in a position that one can consider as truly anterior, or within the 'acronal' region, or in the region just posterior to the acron. This feature is not strictly speaking an APT character.

### 06. MOUTH ORIENTATION

state 0 = anteriorly directed
state 1 = ventrally directed
state 2 = posteriorly directed

This feature deals with the location of the mouth in relation to the body axes. Thus, the mouth can be oriented anteriorly, directed ventrally, or deflected posteriorly. This character also makes indirect reference to the form of the foregut, i.e. straight, bent at more or less a right angle to the body axis in its anterior aspect, or U-shaped. This also is not strictly speaking an APT character. This feature impinges on the issue of what constitutes a true labrum as opposed to a hypostome (Walossek and Müller, 1990), issues which we cannot include in this paper.

### 07. FIRST HEAD LIMB STATE

state 0 = uniramous
state 1 = biramous
state 2 = absent

The sensory antennae(ules) serve as an important diagnostic feature of arthropods and near arthropods. These are typically uniramous, with some exceptions (e.g. biramy in some crustaceans, absence in cheliceriforms). There may be some evidence emerging (see the contribution of Dewel and Dewel, 1996 and this volume) that may require that arthropod workers as a group be careful in distinguishing between true antennae, that is a limb of the acronal region, and frontal appendages, as seen for example in anostracan crustaceans, tardigrades, and *Anomalocaris*. However, this is only a recently made suggestion, and we have maintained for the time being the traditional assessment of the anterior anatomy of *Anomalocaris* (see our character 8).

### 08. FIRST HEAD LIMB CONDITION

state 0 = antenniform
state 1 = absent
state 2 = otherwise developed

Although there is some unavoidable overlap between this character and that of character 7 (in particular to the condition seen in cheliceriforms), this feature deals more specifically with the alternative conditions that one can find to the first limb serving as an antenna. While one might be inclined to see a direct connection between the raptorial limb of *Anomalocaris* and arthropod antenna, the connection between the lobate bodies one sees in *Yohoia* and the anterior of other arthropods is not at all clear. However, the suggestions of Dewel and Dewel (1997, this volume) cannot be discarded.

### 09. SECOND MONOLIMB OF HEAD

state 0 = podium
state 1 = labrum
state 2 = mandible
state 3 = antenniform (part)
state 4 = fang or cheliceriform (part)
state 5 = pediform (part)
state 6 = antennulate

At this point it becomes necessary to distinguish between what appendages may develop on a monosegment and what may result from the fusion of these monosegment limbs when a duplosegment is formed. Thus, APT maintains that the mandibles of insects and myriapods cannot be equated to the whole mandible of a crustacean or the chelae of a chelicerate or pycnogonid. This results from the consequences of the first and second hypotheses of APT.

10. THIRD MONOLIMB OF HEAD

state 0 = podium
state 1 = antenniform (part)
state 2 = cheliform (part)
state 3 = pediform (part)
state 5 = absent

See comments on character 9.

11. FOURTH AND FIFTH MONOLIMBS

state 0 = podia
state 1 = mandible + maxilla
state 2 = gnathobasic mandible with palp
state 3 = gnathobasic mandible without palp
state 4 = pediform
state 5 = pedipalp

See comments on character 9. In this case, we do not need to treat the fourth and fifth monolimbs as distinct elements but treat them as linked merely for simplicity's sake.

12. TRUNK STATE

state 0 = with monomeres
state 1 = with some kind of segment pairing
state 2 = with diplomeres
state 3 = with duplomeres

This is the central feature of the second hypothesis of APT. State 1 refers to the condition where there is some underlying genetic pair rule manifestation during development that may not be readily obvious in the gross anatomical state.

13. TRUNK LIMBS

state 0 = uniramous
state 1 = duplopodous
state 2 = biramous
state 3 = polyramous

This figures as the central feature of the first hypothesis of APT. Character state 3 is a reference to a special multi-branched, largely unarticulated 'corm' – branchiopodium or phyllopodium – characteristic of certain kinds of crustaceans, which emerged from the analysis of Schram (1986) and which he denoted as the Phyllopoda.

14. CONDITION OF TRUNK LIMBS LATERALLY

state 0 = unadorned laterally
state 1 = with epipodites

This feature, along with character 14, is not exactly a feature directly related to APT. However, the development of lateral or medial elements on limbs seems to only occur on duplopodous appendages and not on uniramous ones (Kukalová-Peck notwithstanding, 1997, this volume).

15. CONDITION OF TRUNK LIMBS MEDIALLY

state 0 = unadorned medially
state 1 = with endites
See comments under character 14.

16. DUPLOMERE 9 EXPRESSION

state 0 = not expressed
state 1 = expressed

This is a key APT feature. It grows, however, out of a curious phenomenon (see also character 23). Generally it is the nodes that are the hotspots for anatomical 'events'. The Fields are generally regions of structural uniformity. However, within the second Field there is a duplomere that does act as a hotspot, and it is the only one that does so. It therefore, according to Minelli and Schram (1994), must have some kind of underlying genetic control just as the nodes themselves that is distinct from the genetic controls of anatomical structures or events that may occur there.

17. NODE 2 EXPRESSION

state 0 = not expressed
state 1 = expressed

This is one of the prime APT hotspots. Again, its underlying genetic control should be treated as different from what specific structures or anatomical events that may occur at within this node.

18. ANTERION–TRUNK TRANSITION

state 0 = within the first Field
state 1 = within node 1
state 2 = within duplomere 9

This forms a fine example for what we believe are complementary yet different genetic constraints. This feature is not the same as character 2 or character 16. We believe that the events that occur at the hotspots are separate issues from the existence of the hotspots themselves. In our view, the basic architecture of the body plan is laid down first. Only after the evolution of a genetically determined regionalized body do transitions from one tagma to another, location of gonopores, or terminations of bodies evolve at these spots.

19. TRUNK FORMAT

state 0 = single undifferentiated region
state 1 = thorax–abdominal transition in node 3
state 2 = thorax–abdominal transition in duplomere 16
state 3 = thorax–abdominal transition in node 2
state 4 = thorax–abdominal transition in duplomere 9
state 5 = thorax–abdominal transition in monosegments 13–15
state 6 = thorax–abdominal transition in node 1

This is a character that illustrates the importance of the APT homologies derivable from hypothesis 3. Anatomical events, like shifts from thorax to abdomen, do not occur anywhere along the arthropod body but largely within the confines of these specific hotspots only.

### 20. (POST)ABDOMINAL LIMBS

state 0 = no separate abdomen or postabdomen
state 1 = present
state 2 = posteriorly absent
state 3 = completely absent

This is indirectly an APT feature, dealing with largely limb expression within nodes or Fields.

### 21. FEMALE GONOPORE LOCATION

state 0 = terminal
state 1 = node 3
state 2 = duplomere 16
state 3 = node 2
state 4 = duplomere 9
state 5 = node 1

Again, the manifestation of gonopore location at the APT hotspots would not be expected if such events were due entirely to chance.

### 22. MALE GONOPORE LOCATION

state 0 = terminal
state 1 = node 3
state 2 = duplomere 16
state 3 = node 2
state 4 = duplomere 9
state 5 = node 1

See comment on character 21.

### 23. DUPLOMERE 16 EXPRESSION

state 0 = not expressed
state 1 = expressed

Like character 16, this is a subsidiary hotspot that occurs in the midst of a Field that otherwise is devoid of 'anatomical activity'. What structural events that may actually occur here must be under genetic controls other than those that strictly control the hotspot itself.

### 24. NODE 3 EXPRESSION

state 0 = not expressed
state 1 = expressed

The most posterior of the prime APT hotspots.

### 25. HEAD PROTECTION

state 0 = segments not fused to form a distinct head shield
state 1 = modest head shield present
state 2 = head shield laterally expanded
state 3 = head shield wrapped around the sides of the cephalon

This is a character not especially related to APT *per se*. This feature, and the one that follows hereon, have been employed by almost all authorities in helping to define and distinguish between major arthropod body plans.

### 26. STATE OF CARAPACE

state 0 = no carapace
state 1 = developed as a shield extending back over the trunk
state 2 = bivalved

We have kept this character deliberately simple. Other authors (Schram, 1986) have often striven to recognize even more variants in carapace expression, such as length of the carapace, whether it is hinged or not, and so forth. In this analysis, we believed that it was not necessary to get that specific about anatomical variation, since we were interested in the major underlying pattern of arthropod phylogeny and not particularly in sorting terminal taxa.

### 27. EYES

state 0 = simple
state 1 = absent
state 2 = compound and sessile
state 3 = compound and stalked

### 28. ANUS LOCATION

state 0 = terminal
state 1 = base of telson
state 2 = on penultimate segment

### 29. BODY EXTENSIONS

state 0 = fins
state 1 = true limbs

The distinguishing of fins and limbs as well as lobopods and telopods may be breaking down (see several contributions in this volume). We retain this traditional, but non-APT, feature in its more typical interpretation.

### 30. LIMB STATE (ENDO-)

state 0 = lobopods
state 1 = 10 or more arthropodites
state 2 = 8–9 arthropodites
state 3 = 6–7 arthropodites
state 4 = 4–5 arthropodites
state 5 = 2–3 arthropodites

This is a problematic feature. Many authors have made arguments pro and con the homologies of individual leg segments (e.g. see Kukalová-Peck, 1997, or Boxshall, 1997, both this volume). While recognizing that there is much of value in many of these arguments, we also believe that arthropod researchers are nowhere near a consensus concerning segment homologies. The pattern that emerges from considering this feature as formulated here is one of decreasing segments numbers in the limb through time.

31. STATE OF OUTER RAMUS (EXO-)

state 0 = not present
state 1 = whole and unarticulated
state 2 = articulated
state 3 = lost

This too is a vexing issue. One might legitimately argue that one should make no distinction between character states '0' and '3.' However, it seems clear that often times there is evidence from embryology or comparative anatomy that allows one to interpret whether the lack of an outer branch is primary or secondary (e.g. in the case of *Gammarus* or the cheliceriforms).

32. CAUDAL RAMI

state 0 = none
state 1 = short
state 2 = long

33. TAIL

state 0 = no tail fan
state 1 = telson and uropods

This is a non-APT feature useful for the most part to define malacostracan crustaceans.

34. BODY TERMINATION

state 0 = in fourth Field
state 1 = in node 3
state 2 = in duplomere 16
state 3 = in node 2
state 4 = in duplomere 9
state 5 = in node 1

This character is related to the basic APT homologies derived from hypothesis 3.

35. 'UPPER LIP'

state 0 = none
state 1 = lip fold flat
state 2 = lip fold fleshy
state 3 = labrum as fused limbs

Although there is some overlap between this feature and character 9, this feature is dealing with the alternative forms of 'labrum', i.e. those not related to appendage fusion as postulated in insects and myriapods, but which arise as folds of the ventral surface of the head in front of the mouth. As such it is a feature related to recognizing crustaceomorphs. Here too, the arguments of Walossek and Müller (1990) may have relevance.

36. PARAGNATHS

state 0 = none
state 1 = present

This of course is a non-APT feature, but another related to distinguishing crustaceomorphs.

## 7.5 PARSIMONY ANALYSIS

### 7.5.1 METHODS EMPLOYED

The basic matrix Table 7.1 (http://www.systassoc.chapman-hall.com) was analysed with PAUP 3.1.1 (Swofford, 1993). Due to the size of the data set, we ran only heuristic searches. In the initial run through the data, we employed random stepwise addition with the MULPARS option deactivated to obtain a core of trees for further analysis. We then saved these trees and reanalysed the data a second time, specifying the maximum tree length and reactivating MULPARS.

Initially, the data set was run unordered and unweighted. In an attempt to explore how much more resolution we could induce in the trees, we then went back and ordered certain multistate features, viz., characters 1, 5, 7, 8, 12, 13, 18–22, and 34. Subsequently, we deleted characters 21 and 22 in order to try and minimize the occurrence of question marks in the matrix. Finally, we removed the fossil taxa to analyse only the relationships among the living taxa. After all analyses, we derived strict and 50% majority rule trees to evaluate the information content in the resultant 'forests'.

Further examination of the results occurred in MacClade (Maddison and Maddison, 1992). The trees files were carried over into the tree display option allowing study of character distributions, and rearrangement of branches to compare alternative hypotheses of branch relationships with regard to character distributions and tree statistics.

### 7.5.2 RESULTS OF THE ANALYSIS

All analyses, except the one in which we deleted the fossils and which produced only 256 trees, overflowed the memory capacity of our computer between 14 300 and 14 400 trees. This occurred after swapping on thousands of trees. Such a huge number of trees is not unexpected when utilizing a data base which contains so many fossil forms. Despite the fact that when the overflow occurred there still remained several thousand more trees to swap on, we had confidence in our results because we obtained similar, albeit not identical, phylogenetic patterns when we repeated the analyses more than once. These results where perceived as somewhat

**Table 7.1** Basic data matrix of APT characters and a representative array of living and fossil (†) arthropod taxa. See text for details concerning the characters.

| *Taxon* | 12345 | 1<br>67890 | 11111<br>12345 | 11112<br>67890 | 22222<br>12345 | 22223<br>67890 | 33333<br>12345 | 3<br>6 |
|---|---|---|---|---|---|---|---|---|
| ancestor | 00000 | 000?? | ?0??? | 00000 | 00000 | 000?? | ?0000 | 0 |
| Peripatopsis | 20001 | 10021 | 00000 | 00000 | 00000 | 00010 | 00000 | 0 |
| Aysheia† | 00000 | 00000 | 00000 | 00000 | ??0?0 | 01010 | 00050 | 0 |
| Sottyxerxes† | 20002 | 100?? | 12000 | 00013 | ??012 | 03111 | 0000? | ? |
| Schramixerxes† | 20002 | 100?? | 12000 | 11043 | ????2 | 03111 | 0003? | 0 |
| Branchiocaris† | 10002 | 1003? | 03100 | 00000 | ??001 | 21013 | 11000 | 0 |
| Marella† | 00002 | 20032 | 43200 | 00000 | ??002 | 11013 | 20001 | 0 |
| Odaraia† | 200?2 | 20032 | 33200 | 10000 | ??001 | 23011 | 12001 | 0 |
| Canadaspis† | 21012 | 20032 | 33200 | 01133 | ??011 | 23011 | 10011 | ? |
| canadaspoid† | 10002 | 20035 | 00010 | 01053 | ????1 | 23011 | 00031 | ? |
| Yohoia† | 41012 | 02243 | 43000 | 00123 | ??1?2 | 0?114 | 30010 | 0 |
| Burgessia† | 31012 | 10054 | 43200 | 01100 | ????2 | 11113 | 10031 | 0 |
| Sidneyia† | 41012 | 20054 | 43201 | 01133 | ????0 | 03013 | 10130 | 0 |
| Sanctacaris† | 70112 | 121?? | 43201 | 10201 | ??012 | 02113 | 2001? | 0 |
| Triarthrus† | 31012 | 20054 | 43201 | 00100 | ??012 | 02013 | 20001 | 0 |
| Olenoides† | 31012 | 20054 | 43201 | 01131 | ??012 | 02013 | 22011 | 0 |
| Naraoia† | 31012 | 20054 | 43201 | 00100 | ??012 | 01013 | 20011 | 0 |
| Agnostus† | 31012 | 20054 | 43201 | 11161 | ????2 | 01113 | 20030 | 0 |
| Opabinia† | 0?0?1 | 2???? | ????? | ??00 | ????0 | 0300? | ?00?0 | 0 |
| Anomalocaris† | ??0?1 | 102?? | ????? | ??00 | ????3 | 0300? | ?00?0 | 0 |
| Emeradella† | 51012 | ?0054 | 43211 | 00113 | ??012 | 01113 | 20011 | 0 |
| Leanchoilea† | 31012 | 12132 | 43200 | 00100 | ??1?3 | 01111 | 10020 | 0 |
| Molaria† | 31012 | 20054 | 43200 | 01143 | ????2 | 01213 | 1003? | ? |
| Alalcomenaeus† | 31012 | 20054 | 43201 | 00100 | ??1?3 | 0311? | 1002? | ? |
| Limulus | 70112 | 12143 | 53200 | 10201 | 441?2 | 02113 | 10021 | 0 |
| Pynogonum | 30112 | 02143 | 53000 | 1?1?? | 44??1 | 00112 | 30040 | 0 |
| Hughmilleria† | 70112 | 12143 | 53000 | 10223 | 44112 | 00113 | 30010 | 0 |
| Bilobosternia | 70112 | 12143 | 53000 | 10223 | 44112 | 00113 | 30010 | 0 |
| Thelyphonus | 60112 | 12143 | 53000 | 10233 | 44111 | 00113 | 30010 | 0 |
| Liphisteus | 60112 | 12143 | 53000 | 11203 | 44011 | 00013 | 30010 | 0 |
| Hyalomma | 60112 | 02143 | 53000 | 11203 | 44??1 | 01012 | 30030 | 0 |
| Lasionectes | 51012 | 21032 | 33200 | 01100 | 31012 | 01014 | 21002 | 1 |
| Tesnusocaris† | 51012 | 21032 | 33100 | 0?100 | ????2 | 02012 | 22002 | ? |
| Gammarus | 51012 | 11032 | 23010 | 01131 | 33013 | 02013 | 30111 | 1 |
| Anaspides | 51012 | 11032 | 23210 | 01131 | 33013 | 03013 | 20101 | 1 |
| Hutchinsoniella | 41012 | 20032 | 23311 | 01133 | 33102 | 01013 | 22002 | 1 |
| Lepas | 41012 | 20232 | 23200 | 01133 | 53??? | 21011 | 20032 | 1 |
| Calanus | 51012 | 10032 | 23200 | 01133 | 331?3 | 00014 | 22021 | 1 |
| Penaeus | 71012 | 11032 | 23211 | 01131 | 11011 | 13013 | 20111 | 1 |
| Derocheilocaris | 41012 | 20032 | 23000 | 11133 | 441?1 | 01016 | 31022 | 1 |
| Bredocaris† | 41012 | 20054 | 43201 | 01133 | ????3 | 12013 | 11032 | 0 |
| Rhebachiella† | 41012 | 20054 | 43201 | 00103 | ??113 | 12012 | 11012 | 0 |
| Martinsonia† | 41012 | 20054 | 43200 | 01102 | ????3 | 01113 | 21032 | 0 |
| Lophogaster | 51012 | 11032 | 23200 | 01131 | 33013 | 13113 | 20111 | 1 |
| Nebalia | 41012 | 21032 | 23310 | 01132 | 33013 | 23016 | 11012 | 1 |
| Lepidocaris† | 41012 | 20032 | 33201 | 00123 | ??101 | 01015 | 11002 | ? |
| Lynceus | 41012 | 20032 | 33311 | 00100 | 221?1 | 22016 | 11012 | 1 |
| Branchinecta | 41012 | 20032 | 33311 | 00123 | 22100 | 03016 | 11002 | 1 |
| Scutigera | 31002 | 10015 | 12000 | 01100 | 00??1 | 02013 | 02033 | 0 |
| Geophilus | 31002 | 10015 | 10000 | 00100 | 00001 | 02013 | 00003 | 0 |
| Lithobius | 31002 | 10015 | 12000 | 01100 | 00??1 | 02014 | 00033 | 0 |
| Arctobolus | 31002 | 10015 | 12000 | 00000 | 55000 | 02014 | 00003 | 0 |
| Polyxenus | 21002 | 10015 | 12000 | 01000 | 55??0 | 02014 | 00033 | 0 |
| Acerentulus | 31002 | 10015 | 10000 | 01063 | 00??1 | 01014 | 00033 | 0 |
| Periplaneta | 31002 | 10015 | 11000 | 11063 | 40??1 | 02014 | 00033 | 0 |
| Heterojapyx | 310?2 | 10015 | 10000 | 11063 | 44??1 | 01014 | 02033 | 0 |
| Pauropus | 21002 | 11015 | 12000 | 1?000 | 55??0 | 01014 | 00043 | 0 |
| Scutigerella | 31002 | 10015 | 10000 | 01100 | 55??0 | 01014 | 00033 | 0 |

counterintuitive at first – how could so many trees contain any phylogenetic pattern? Surprisingly, however, even the strict consensus trees contained a lot of structure (Figure 7.2) and for the most part further resolution appeared in the 50% majority rule trees (Figure 7.3).

Exploration of the results revealed that this was due in large part to many of the APT characters determining the deep bifurcations in the tree. It is clear, however, that the array of APT characters used here has little to tell us about most of the terminal bifurcations, except some of those that occur lower in the tree and thus more influenced by the deep branches. What is most interesting is that some distinct differences in character distribution patterns prevail with regard to the various APT hypotheses, which indicate that not all these hypotheses are equally effective. These analyses below are plotted on, and discussed in light of, the 50% majority rule tree of the partially ordered analysis.

A summary of the tree statistics is found in Table 7.1. In all the trees, it became immediately clear that certain traditionally recognized groups, such as Trilobita and Crustacea, did not always appear as distinct clades. Consequently, the partially ordered trees were then exported into MacClade and rearranged to produce a fully resolved tree with monophyletic clades of these taxa. This variant on an 'ideal' cladogram (Figure 7.4), however, proved to be 31 steps, almost 10%, longer than those of the base analysis.

## 7.6 DISCUSSION

### 7.6.1 APT HYPOTHESES

Character 13 (the nature of the trunk limbs) is the central feature for the first APT hypothesis. As one can see in Figure 7.5, the important APT character state of duplopody has some limited value in determining the tree branchings, but only two taxa display this feature, viz., the fossil remipede crustacean *Tesnusocaris* and the Cambrian Burgess Shale problematicum *Branchiocaris*. The location of *Branchiocaris* (Figure 7.3(b)) does provide some APT confirmation, but *Tesnusocaris* exhibits its duplopody as an autapomorphy. In the unordered analysis (Figure 7.3(a)), both these taxa exhibit duplopody as an autapomorphy. While we did record an intermediate stage of duplopody between a grade of primary uniramy and a clade of biramy, it would appear that the occurrence of duplopods, given the current state of our knowledge about fossil arthropod limbs, could in fact be equally well explained as an autapomorphy manifested through variations in timing of the *distalless* gene expression (as suggested by Panganiban *et al.*, 1995).

The other character states of feature 13, viz. possession of biramous and polyramous limbs, are not particularly APT-oriented. These character states do, nevertheless, serve to provide some structure to the basic tree since biramy helps determine a 'schizoramian' clade (in the sense of Hessler and Newman, 1975) that contains crustaceans, trilobites, and many Burgess creatures (though biramy is lost for the most part among the cheliceriforms), and polyramy (a distinctive limb type with multiple branches and generally devoid of distinct articles) that serves to define the branchiopods and phyllopods (Schram, 1986).

So, despite the interesting developmental genetic data concerning the *distalless* gene expression alluded to above, we must conclude from the cladistic analysis that the first APT hypothesis in which the biramous limb of arthropods forms from the basal fusion of two sets of uniramous limbs, is not completely effective as an organizing principle.

The second APT hypothesis, which deals with the formation of arthropod segments, is closely linked with the first hypothesis dealing with arthropod appendages. The critical multistate feature in this regard is character 12. The pair-rule genes that mark segmental pairing are known only in some insects. Until more information emerges concerning the genetic control of segments within the myriapods, we cannot effectively evaluate the role of these genes in arthropod somite patterning. However, the issue of diplosegments is clear with or without APT, and this definitely helps determine deep tree structure. Indeed, in the 50% majority rule tree it would appear that one could utilize this feature as a defining one for arthropods as a whole (with alternative patterns existing among various uniramian or atelocerate groups (Figure 7.6(a)). In turn, the appearance of diplosegments also clearly defines a clade that emerges above the paraphyletic Atelocerata. The issues are less clear with regard to this feature, however, when the Variant on the

**Table 7.1** Summary of tree statistics for the series of PAUP and MacClade analyses performed on the data matrix (website http://www.sytassoc.chapmanhall.com).

| *Method* | *Total number of trees* | *Length* | *CI* | *HI* | *RI* | *RC* |
|---|---|---|---|---|---|---|
| Unordered | 14 400 | 298 | 0.329 | 0.671 | 0.713 | 0.234 |
| Partially ordered | 14 300 | 383 | 0.256 | 0.704 | 0.704 | 0.180 |
| Characters 21 and 22 deleted | 14 300 | 346 | 0.254 | 0.746 | .709 | 0.180 |
| Fossils deleted | 256 | 164 | 0.500 | 0.500 | 0.766 | 0.383 |
| Variant on the ideal | From MacClade | 415 | 0.240 | 0.760 | 0.670 | 0.160 |

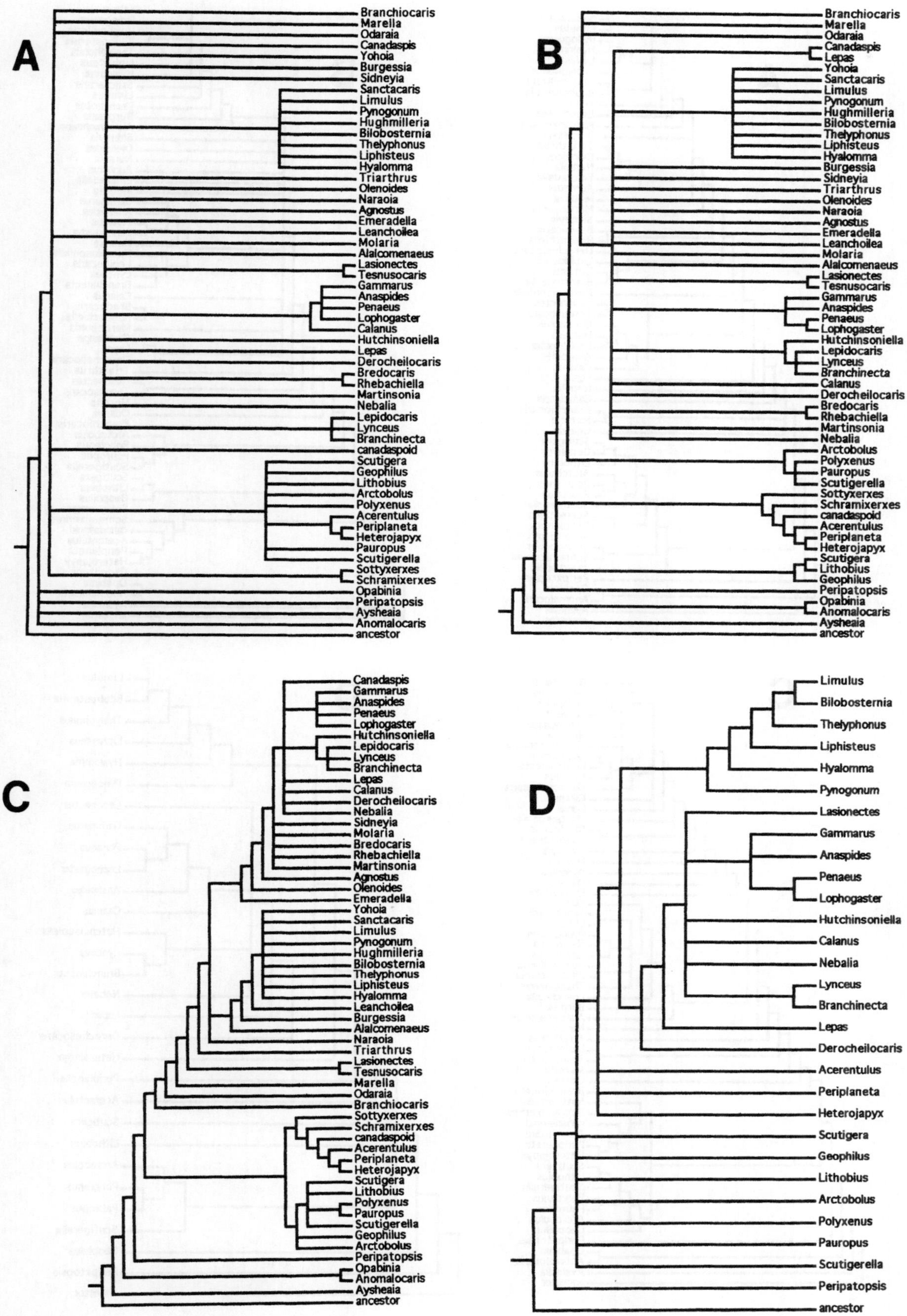


**Figure 7.2** Strict consensus trees of APT analyses. (A) Use of all taxa and all characters unordered. (B) Use of all taxa with a partial ordering of the data (see text for details). (C) As in (B) but with characters 21 and 22 deleted. (D) As in (B) but with all fossil taxa deleted.

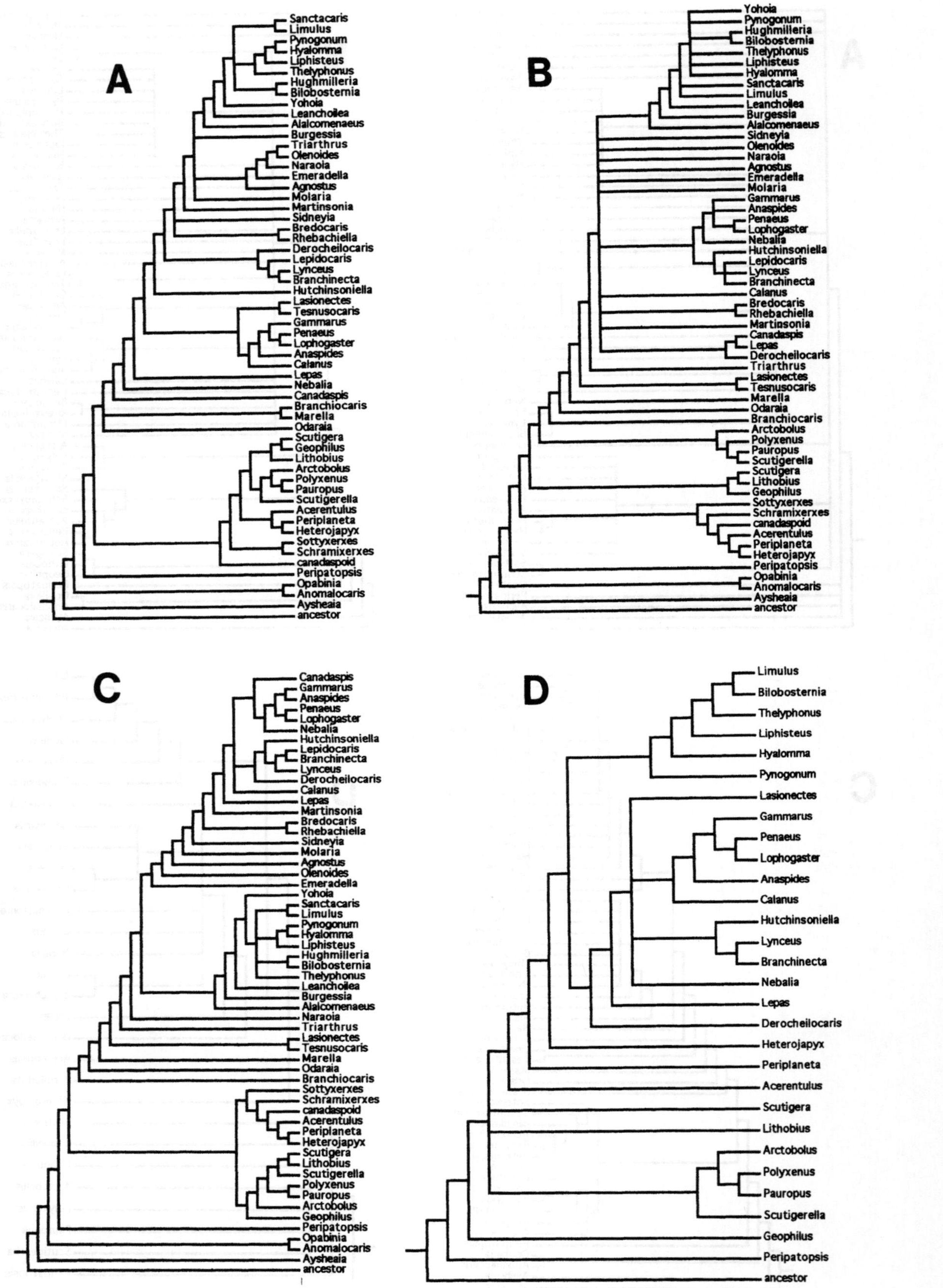


**Figure 7.3** A 50% majority rule consensus trees of APT analyses. (A) Use of all taxa and all characters unordered. (B) Use of all taxa with a partial ordering of the data (see text for details). (C) As in (B) but with characters 21 and 22 deleted. (D) As in (B) but with all fossil taxa deleted.

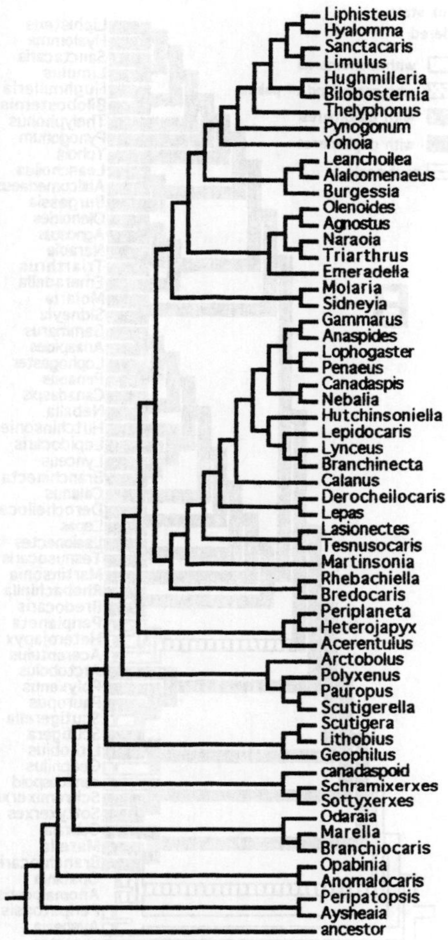


**Figure 7.4** Variant on the ideal tree, produced by manipulating MacClade branches derived from the partially ordered analysis (see Table 7.1) in order to get the shortest tree that had clear monophyletic and/or adjacent paraphyletic groupings of traditionally recognized major groups of arthropods. This tree has 415 steps, i.e. 32 steps longer than the most parsimonious trees derived from the analysis. Our manipulation revealed that a paraphyletic Atelocerata produced a tree significantly shorter than a monophyletic clade, as did the location of a distinct marellomorph clade (rather than a paraphyletic series) between anomalocaridoids and the atelocerates rather than further up in the tree.

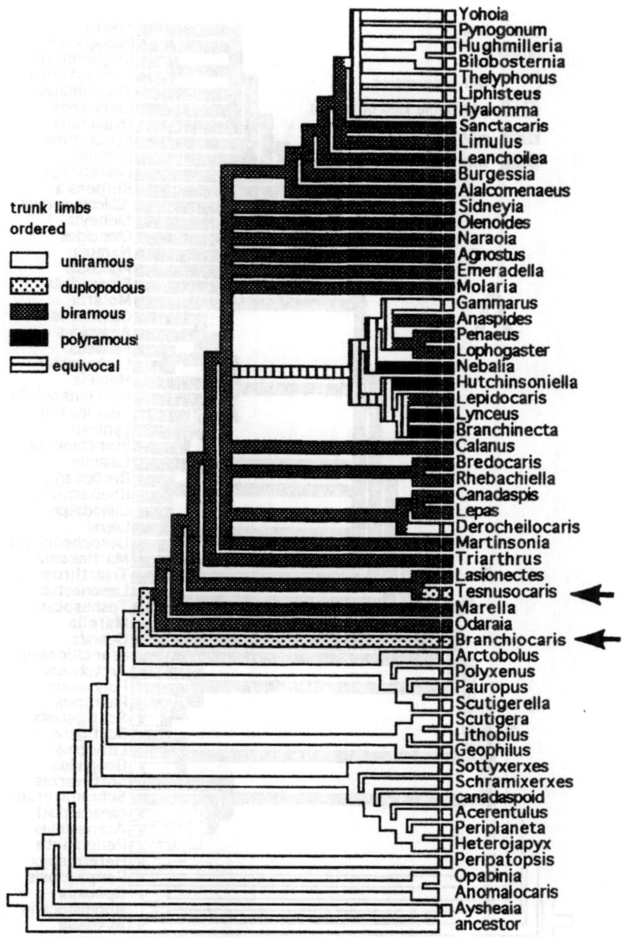


**Figure 7.5** Distribution of states of character 13 (nature of trunk limbs) on the partially ordered 50% majority rule tree. Arrows highlight the occurrence of duplopodous limbs. Biramous limbs define a schizoramian clade that includes not only a monophyletic Cheliceriformes characterized by a 'reversal' to uniramous limbs, but also a possible (in this tree) paraphyletic phyllopod Crustacea characterized by polyramous limbs.

Ideal tree is used (Figure 7.6(b)) where it would appear that duplosegmentation marks two widely separated clades.

While it might seem at first glance that the efficacy of the second hypothesis is possibly confirmed, when the total array of characters is optimized, we can see that many features co-occur at these two bifurcations (Figure 7.6(a)). At point 'a', where diplomery appears, we also note the co-occurrence of characters 5(2), 9(1), 10(5), 11(1), and 27(2 and 3). Likewise, at point 'b' on Figure 7.6(a), we also record the appearance of characters 9(3), 13(1), 25(1), 31(1), and 35(3).

When we combine the results of the cladistic analysis with the ambiguous evidence from other sources alluded to above, we can conclude that while the second APT hypothesis is useful in determining the deep bifurcations in the tree, they are not the only features to operate at those points. The second hypothesis is still viable as an organizing principle, but in light of some of the genetic evidence on segment formation alluded to above the traditional views of segment formation in arthropods could yet prevail.

The third APT hypothesis, which deals with points of homology along the length of the arthropod body, would appear to be very effective indeed. In the original formulation of Schram and Emerson (1991) the third hypothesis did not appear in our minds as important as the first two hypotheses. Yet Gould (1991) claimed that 'when we try to

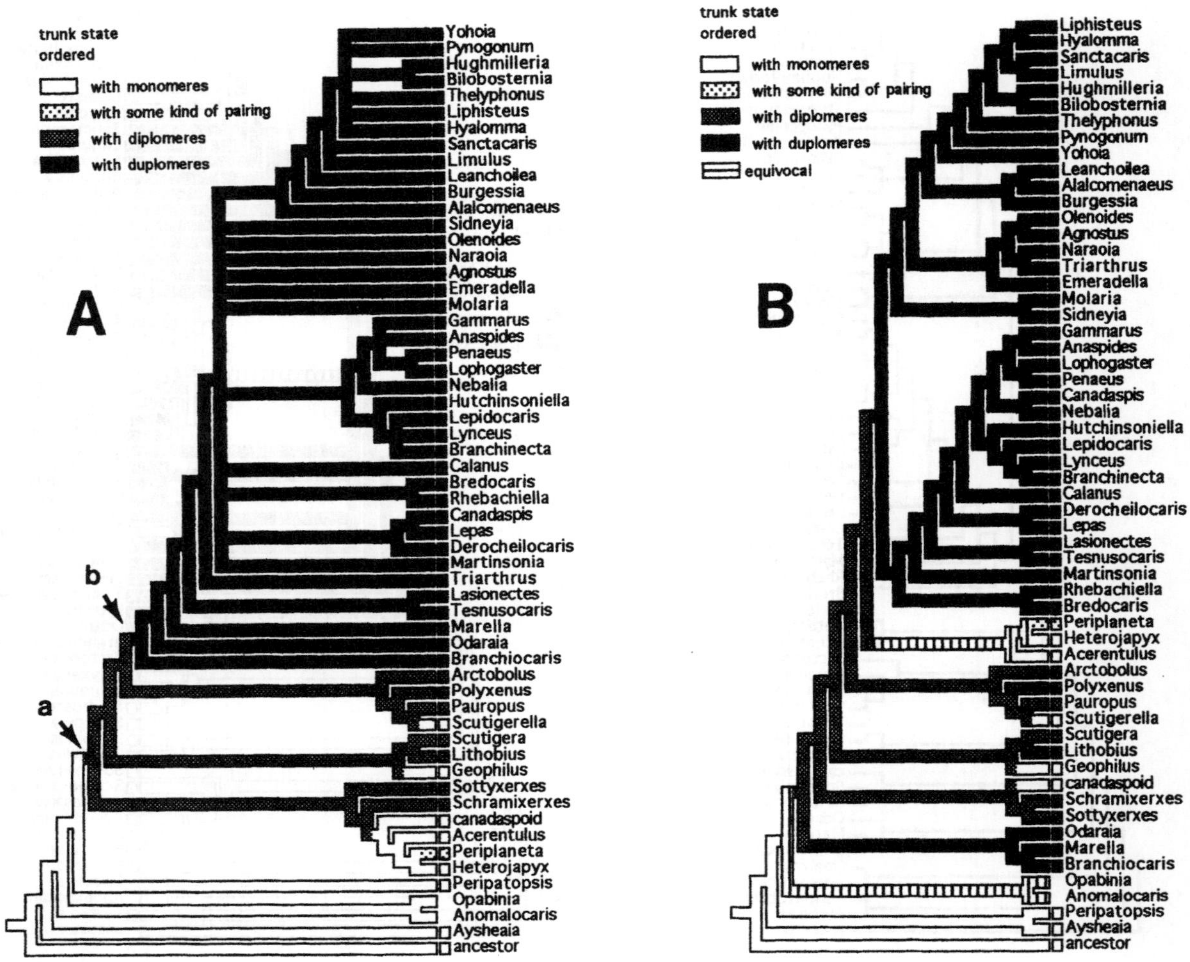


**Figure 7.6** Distribution of states of character 12 (nature of trunk segmentation). (A) Distribution charted on the partially ordered 50% majority rule tree. At point **a** (the appearance of diplomery) co-occur characters 5(2), 9(1), 10(5), 11(1), and 27 (2/3) while at point **b** (the appearance of duplomery) co-occur characters 9(3), 13(1), 25(1), 31(1), and 35(3). (B) Distribution charted on the variant on an ideal tree.

compare organisms of markedly different body plans, it is not feasible to map one form on another'. The third hypothesis of APT maintains that claims like that of Gould are simply not true. It may appear on the surface difficult to find homologous points between animals so different in gross anatomy as scorpions, crabs, and cockroaches. Nevertheless, if we act to minimize or remove those features of anatomy that are purely idiosyncratic to specific species, then the underlying features of the Baupläne can be compared directly and a kind of syncretic mapping of one plan on another achieved. This was essentially the approach of Starabogatov (1991), albeit on a more limited scale. It is in fact the homologous points that APT derived from such comparisons that provide much of the structure to the basic trees seen in Figures 7.2 and 7.3.

We can illustrate this by examining the distribution of characters 2, 16, and 24, which deal with the occurrences of nodes 1, 2, and 3 (Figure 7.7). We can see from this that nodes 1 and 3 provide considerable structure to the basic tree. Node 1 (Figure 7.7(a)) demarcates hexapod, myriapod, and higher arthropod clades with loss of the node's expression among the marellomorphs and for the most part all the cheliceriforms. Node 3 (Figure 7.7(c)) occurs in the long-bodied euthycarcinoids and the crustaceomorph and arachnomorph arthropods with the loss of expression in the clade containing branchiopods and cephalocarids. Node 2 has a more dispersed occurrence (Figure 7.7(b)), and its presence (or in one case loss – among branchiopods) characterizes a number of widely separated clades.

Similar patterns hold true for other features related to hypothesis 3, all of which are not illustrated here for lack of space. For example, character 16, the expression of duplomere 9, highlights a number of taxa as an autapomorphy (viz., *Derocheilocaris*, *Agnostus*, *Odaraia*, *Pauropus*, *Schramixerxes*, and the small clade of *Heterojapyx* and *Periplaneta*). However, it is a good synapomorphy of the

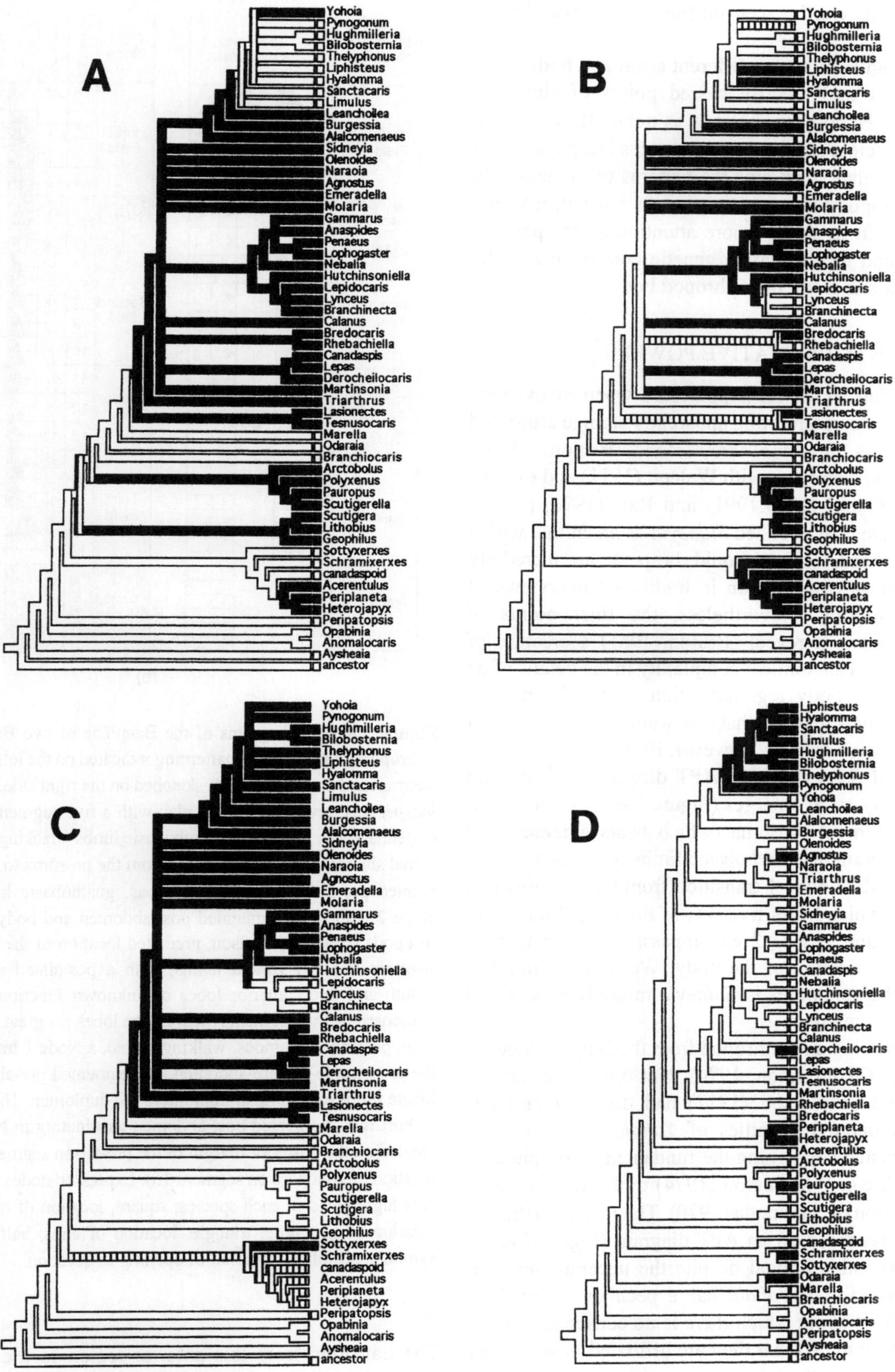


**Figure 7.7** Distribution of some of the character expressions important to the third hypothesis of APT. (A) Character 2 (expression of node 1). (B) Character 17 (expression of node 2). (C) Character 24 (expression of node 3). (D) Character 16 (expression of duplomere 9). See text for details.

cheliceriforms as can be seen on the variant tree (Figure 7.7(d)).

Thus, we can in fact map different arthropod body plans one on another and identify shared points of similarity between them. Hypothesis 3 stands as quite effective as an organizing principle and appears to express Baupläne under genetic control that constrains the patterns of branching in the basic arthropod phylogenetic tree. As a result, it would definitely benefit us to pay more attention to comparative morphology and to elucidating genetic control of specific points and regions along the arthropod body.

## 7.6.2 APT'S INTERPRETATIVE POWERS

We can see a further use for hypothesis 3 when we examine its effectiveness in interpreting apparently strange arthropod fossils, such as *Sidneyia inexpectans* (Bruton, 1981). Several authors, beginning with Walcott (1911) and extending down to Gould (1989,1991) and Raff (1996, p. 128, Figure 4.5) have characterized *Sidneyia* as an animal with a one-segment head, which would bespeak an incredibly primitive state of organization in traditional overviews of arthropod evolution. Nevertheless, the finer points of anatomy of *Sidneyia* reveal (Bruton, 1981) a creature of great subtlety and anatomical complexity in the details of its appendage form, body regionalization, and tail fan morphology, structural features that one would not expect in an allegedly primitive animal. However, if we render the basic morphology of *Sidneyia* in an APT diagram (Schram and Emerson, 1991), what emerges (Figure 7.8(a)) is an animal with a five-segment prosoma (which bears antennae and four sets of uniramous gnathobasic limbs but which lacks a prosomal shield), a node 1 transition from the prosoma to a preabdomen (which bears five sets of biramous limbs with gnathobases), and a node 2 encompassing a short postabdomen and termination of the body. We might venture to predict that the gonopores in *Sidneyia* might be discovered on the eleventh segment.

Another problematic arthropod from the Burgess Shale is *Yohoia tenuis* (Whittington, 1974). Not only have authors since the time of Walcott (1912) had differing opinions about the taxonomic affinities of *Yohoia*, but also little agreement existed concerning the fundamental morphology of the animal, e.g. Whittington (1974) took exception with the interpretation of Simonetta (1970). The reconstruction of Whittington, rendered on an APT diagram (Figure 7.8(b)), reveals a reasonable animal despite the uncertainties concerning the head end. *Yohoia* has a peculiar set of three lobes at the front about which there is no consensus. In this case, we have assumed that they are affiliated in some way with the anterior-most acronal region along with an anteriorly oriented mouth. The peculiar great appendage seems to be just posterior to this area and together with the following three 'walking' limbs form a head or short prosomal area.

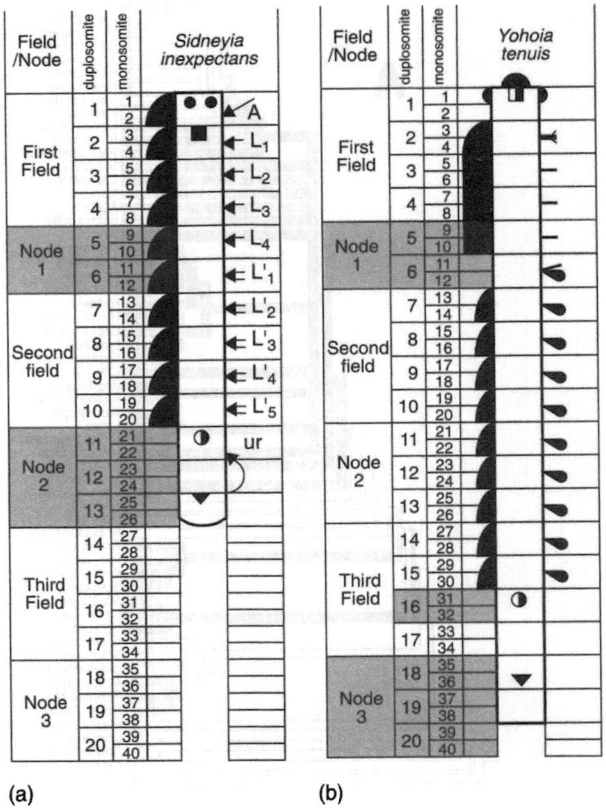


**Figure 7.8** APT diagrams of the Baupläne of two Burgess Shale arthropods, with segment patterning indicated on the left side of each diagram and appendage form sketched on the right side. (a) *Sidneyia inexpectans*, revealing a body plan with a five-segmented prosoma (antenna + four uniramous, gnathobasic limbs – lacking a fused prosomal shield), a Node 1 transition from the prosoma to the five-segmented pre-abdomen (with biramous, gnathobasic limbs), and a Node 2 with three-segmented post-abdomen and body termination (uropods flanking the telson, predicted location of the gonopore on segment 11). (b) *Yohoia tenuis*, with a postulated anterior-most mouth and three anterior lobes of unknown function, a prosoma composed of five segments (bearing the lobes + a great appendage + three possibly uniramous, walking limbs), a Node 1 transition from the prosoma to the pre-abdomen, 10-segmented pre-abdomen with lobate limbs extending to duplomere 16, duplomere 16 transition to a short four-segmented post-abdomen terminating in Node 3 (anus located on the posterior margin of the pre-telson segment, predicted location of gonopore on segment 16). Expressed nodes and duplomeres highlighted in each species; square, location of mouth; circle, location of gonopore; triangle, location of anus; half black/white symbols, postulated location of opening in question.

The transition from this anterior region to the trunk then occurs in node 1. The 10-segmented preabdomen extends without interruption from node 1 to duplomere 16 (with no expression of node 2). At duplomere 16 a short postabdomen begins with the body termination occurring in node

3. We again predict that gonopores in *Yohoia* might be located on duplomere 16.

There is no opportunity here to map the body plans of further fossil arthropods using these APT diagrams. Notwithstanding statements to the contrary, APT can afford us an effective way to understand the possible Baupläne homologies between apparently disparate and strange forms (see also Schram and Emerson, 1991).

### 7.6.3 COMPARISONS WITH OTHER ANALYSES

The only other cladistic analysis that includes a wide array of fossil and recent arthropod forms is that derived from the ongoing work of Wills and his associates (Wills *et al.*, 1995 and references therein). These workers used a character matrix that, at least in its original conception, utilized features that would yield a matrix with few question marks. As such, while they employed characters that can be easily determined for almost all the forms in question, these features are not necessarily ones that have obtained wide recognition and use in earlier literature dealing with the phylogeny of arthropods. Wills' data set, however, had a primary use for evaluating the multidimensional morphospace occupied by fossil arthropods in comparison with recent forms, and thus directly took up the challenge of Gould (1991) to quantify the measure of morphological disparity. Nevertheless, in the course of this work, a cladogram of arthropods resulted (Figure 7.9) that bears interesting comparisons with the cladograms we obtained here. Wills and colleagues rooted their tree to annelids and molluscs; even so parallels exist to our Figures 7.2 and 7.3. *Anomalocaris* and *Opabinia* occur near the base of the tree with onychophorans nearby. They obtained a single atelocerate clade, which we sometimes also obtained (Figures 7.2(c) and 7.3(c)). The euthycarcinoids appear among the atelocerates in both our analyses, and these agree essentially with the analysis of euthycarcinoids obtained by McNamara and Trewin (1993). A distinct marellomorph clade emerges well down in the tree for Wills, as we also found (Figure 7.4), although our analysis also places *Odaraia* in proximity to *Branchiocaris* and *Marrella* (Figures 7.2 and 7.3). We also found distinct crustaceomorph and arachnomorph clades, but the placement of Burgess animals differs in details between their analysis and ours. [Note, however, that the scope of variation of all possible sister group relationships among these taxa can be quite significant, as shown in the monographic treatment of Wills (1994).]

These morphology-based trees can differ significantly from trees derived from molecular data sets or from arrays of taxa that include a minimum number of fossils (Wheeler *et al.*, 1993). It seems clear to us that, regardless of the characters utilized, analyses of arthropod phylogeny that use extensive arrays of fossils have more in common with each other than those that concentrate on only modern forms (Glenner *et al.*, 1995). This suggests that the fossils act as a unique source of information that, if ignored, should seriously call into question the results obtained from such analyses.

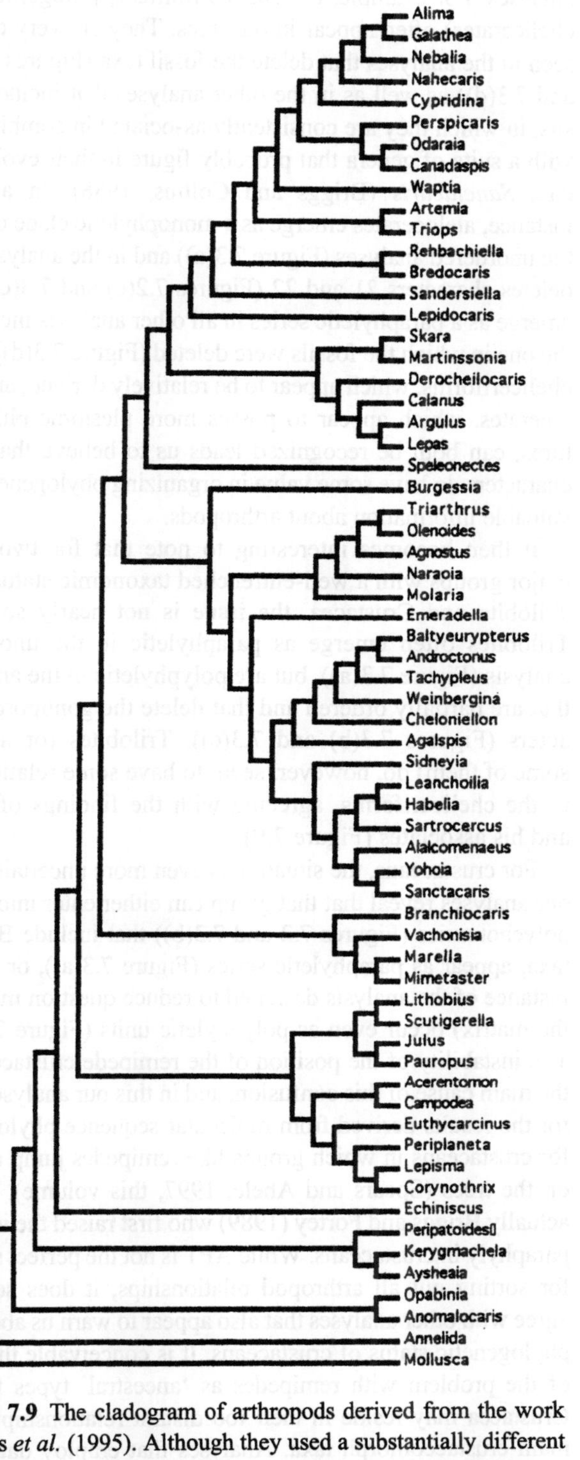


**Figure 7.9** The cladogram of arthropods derived from the work of Wills *et al.* (1995). Although they used a substantially different character set and a more diverse array of taxa, there are still common points of agreement with the results expressed in Figures 7.2 and 7.3.

### 7.6.4 POLYPHYLY VERSUS MONOPHYLY OF WELL-ENTRENCHED GROUPS

We stated above that the appearance of historically entrenched or readily recognisable groups might tell us something about the nature of the data we explore with cladistic analyses. For example, the cheliceriforms (pycnogonids and chelicerates) often appear in our trees. They are very clearly seen in the analyses that delete the fossil taxa (Figures 7.2(d) and 7.3(d)) as well as in the other analyses that include fossils, in which they are consistently associated in combination with a suite of genera that probably figure in their evolution, e.g., *Sanctacaris* (Briggs and Collins, 1988). In another instance, atelocerates emerge as a monophyletic clade only in the unordered analyses (Figure 7.3(a)) and in the analysis that deletes characters 21 and 22 (Figures 7.2(c) and 7.3(c)), but emerge as a paraphyletic series in all other analyses including the one in which the fossils were deleted (Figure 7.3(d)). That cheliceriforms, which appear to be relatively derived, and atelocerates, which appear to posses more plesiomorphic features, can both be recognized leads us to believe that APT characters do have some value in organizing phylogenetically valuable information about arthropods.

It then becomes interesting to note that for two other major groups with a well-entrenched taxonomic status, viz., Trilobita and Crustacea, the issue is not nearly so clear. Trilobites often emerge as paraphyletic in the unordered analysis (Figure 7.3(a)), but are polyphyletic in the analyses that are partially ordered and that delete the gonopore characters (Figures 7.3(b) and 7.3(c)). Trilobites (or at least some of them) do, however, seem to have some relationship to the cheliceriforms, agreeing with the findings of Wills and his associates (Figure 7.9).

For crustaceans, the situation is even more uncertain since our analyses reveal that that group can either enter into grand polychotomies (Figures 7.2 and 7.3(b)) that include Burgess taxa, appear as paraphyletic series (Figure 7.3(a)), or (in the instance of the analysis designed to reduce question marks in the matrix) occur even as polyphyletic units (Figure 7.3(c)). The instability of the position of the remipede crustaceans is the main cause of this confusion, and in this our analyses mirror the results derived from molecular sequence phylogenies for crustaceans in which groups like remipedes jump around on the trees (Spears and Abele, 1997, this volume). It was actually Briggs and Fortey (1989) who first raised the issue of paraphyly in crustaceans. While APT is not the perfect system for sorting out all arthropod relationships, it does seem to agree with other analyses that also appear to warn us about the phylogenetic status of crustaceans. It is conceivable that part of the problem with remipedes as 'ancestral' types for the Crustacea may reside in their too distant relationship to the main crustaceomorph taxa. Analyses that employ data relevant to crustaceomorphs alone (Walossek and Müller, 1990; Schram and Hof, in press) give a coherent picture of the phylogenetic structure of crown and stem group crustaceans, but when larger and more extensive data bases are employed (Wills, 1994) the issue becomes more clouded. We still have much to learn about arthropod phylogeny.

### 7.6.5 THE EFFICACY OF THEORIES

Finally, we must make some comments on the nature and uses of theories such as APT. Earlier, we reviewed the current literature relevant to APT, and if a common criticism can be distilled from the papers that challenge the theory (Zrzavý and Stys, 1994), or by extension even from those that ignore it (Averof and Akam, 1995; Scholtz, 1995), it is that APT is not needed because there are other (more traditional) explanations of these phenomena.

Such a conclusion implies two things. First, if one explanation prevails, there is no reason to search for any other. Second, theories can be proved or disproved in the search for truth, i.e. an objective ultimate reality that in fact exists. We can probably safely advance the viewpoint that these two principles are pillars of past and present phylogenetic speculations. Yet such viewpoints are not the only ones that prevail in the world of scientific investigation.

Wenner (1989) present a spectrum of philosophical viewpoints (Figure 7.10) outlining two main approaches to doing science. One asserts that there is objective truth and that we can arrive at knowledge of it. We can do this by seeking either to verify our hypotheses with data that agree with them (the approach of Carnap and the logical positivists), or to falsify our hypotheses with data that disproves them (the approach of Popper). Opposed to these 'realists' is a school of 'relativists' who assert that we can never know the truth but that we can organize old, and search for

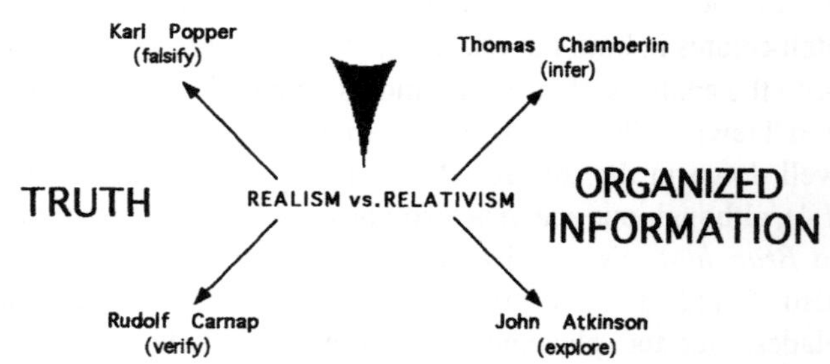


**Figure 7.10** Scheme to suggest the relationships among different approaches to scientific activity. Those methods on the left are directed at discovering some real and ultimate truth and can either proceed by striving to verify hypotheses (Carnap's school) or attempting to falsify them (Popper's school). Those methods on the right are directed at organizing useful knowledge and can proceed either by a rather creative exploration (in the sense of Atkinson) or by a more structured use of multiple alternative hypotheses (in the sense of Chamberlin). (Chart modified from that of Wenner, 1989.)

new, knowledge. This organization can occur in an unstructured, almost artistic way, in which, as Atkinson suggests, we gain insight from the 'creation' of phenomena in our minds as we explore nature and convert others to our viewpoint. In contrast to this approach, we might also work in a more structured way, as Chamberlin put forth, and apply multiple alternative hypotheses to organize the knowledge at hand and see which hypotheses are more effective in relating observations to each other.

While Wenner believes that there can be value to any of these approaches, clearly the most effective methods are those in which the application of multiple hypotheses (or as he prefers, 'strong inference') prevails. Fields of science that utilize the method of Chamberlin are among the most rapidly advancing, e.g. particle physics and molecular biology, while fields that focus on verifying or falsifying some perceived truth are among the most static, e.g. ecology and synthetic evolutionary biology. We might say that rather than viewing arthropodologists as being at a disadvantage with having to deal with contending and contentious hypotheses concerning arthropod phylogeny, we should actually foster and encourage such pluralism.

## 7.7 CONCLUSIONS

We have performed a series of cladistic analyses of APT. The evaluations of the three hypotheses of APT, as outlined by Schram and Emerson (1991), appear to have differing levels of effectiveness. The first hypothesis concerning the formation of the biramous limb is unproven since the available information on duplopody among arthropods is too limited and other equally likely explanations for duplopody occur. The second hypothesis concerning segment formation is still viable and in combination with other APT features contributes to the deep bifurcations in the arthropod tree, but the traditional viewpoint of universal arthropod segment homology may in the end prove more likely. Whether or not duplomery actually occurred seems irrelevant to determining the main framework of arthropod relationships. The third hypothesis appears to form a powerful tool for comparing diverse arthropod body plans. Inclusion of large numbers of fossils in future analyses of arthropod phylogeny may raise levels of uncertainty about the cladograms obtained, but such inclusion is necessary to ensure that important knowledge about arthropod form and function is not overlooked in erecting hypotheses of phylogeny. Doubts may exist concerning the monophyly of certain well-established arthropod groups such as crustaceans and trilobites.

## ACKNOWLEDGEMENTS

The second author would like to thank Dominique and Sharon Emerson, Michael's sisters, for their support and encouragement over the years to continue with this project. In addition, Dominique Emerson and Chris Wagner extended financial help to complete this article and participate in the conference in London. Cees Hof and Prof. Adrian Wenner provided much useful discussion, and Cees Hof assisted greatly with the graphics and computing. I also want to thank Drs Martin Christoffersen, Jens Høeg, Ronald Sluys, and an anonymous reviewer for making constructive comments and suggestions for improvement on the manuscript.

## REFERENCES

Akam, M. (1995) *Hox* genes and the evolution of diverse body plans. *Philosophical Transactions of the Royal Society, London, B*, **349**, 313–19.

Averof, M. and Akam, M. (1995) *Hox* genes and the diversification of insect and crustacean body plans. *Nature*, **376**, 420–3.

Briggs, D.E.G. (1976) The arthropod *Branchiocaris* n. gen., Middle Cambrian, Burgess Shale, British Columbia. *Geological Survey of Canada Bulletin*, **264**, 1–29.

Briggs, D.E.G. (1978) The morphology, mode of life, and affinities of *Canadaspis perfecta* (Crustacean: Phyllocarida), Middle Cambrian, Burgess Shale, British Columbia. *Philosophical Transactions of the Royal Society, London, B*, **281**, 439–87.

Briggs, D.E.G. and Collins, D. (1988) A Middle Cambrian chelicerate from Mount Stephen, British Columbia. *Palaeontology*, **31**, 779–98.

Briggs, D.E.G. and Fortey, R.A. (1989) The early radiation and relationships of the major arthropod groups. *Science*, **246**, 241–3.

Bruton, D.L. (1981) The arthropod *Sidneyia inexpectans*, Middle Cambrian, Burgess Shale, British Columbia. *Philosophical Transactions of the Royal Society, London, B*, **295**, 619–56.

Dahl, E. (1984) The subclass Phyllocarida (Crustacea) and the status of some early fossils; a neontologist's view. *Videnskabelige Meddelelser fra Dansk naturhistorisk Forening*, **145**, 61–76.

Dewel, R.C. and Dewel, W.C. (1996) The brain of *Echiniscus viridissimus* Peterfi, 1956 (Heterotardigrada): a key to understanding the phylogenetic position of tardigrades and the evolution of the tardigrade head. *Zoological Journal of the Linnean Society*, **116**, 35–49.

Emerson, M.J. and Schram, F.R. (1990a) A novel hypothesis for the origin of biramous limbs in arthropods, in *Arthropod Paleobiology: Short Courses in Paleontology. No. 3* (ed. D.G. Mikulic), University of Tennessee, Knoxville, pp. 157–76.

Emerson, M.J. and Schram, F.R. (1990b) The origin of crustacean biramous appendages and the evolution of Arthropoda. *Science*, **250**, 667–9.

Emerson, M.J. and Schram, F.R. (1991) Remipedia. Part 2. Paleontology. *Proceedings of the San Diego Society of Natural History*, **7**, 1–52.

Ferrari, F. (1995) Identity of distal segments of the maxilla 2 and the maxillipede in copepods: new teeth for Carl Claus' old saw. *Crustaceana*, **68**, 103–10.

Glenner, H., Grygier, M.J., Høeg, J.T., Jensen, P.G. and Schram, F.R. (1995) Cladistic analysis of the Cirripedia Thoracica. *Zoological Journal of the Linnean Society of London*, **114**, 365–404.

Gould, S.J. (1989) *Wonderful Life*, Norton, New York.

Gould, S J. (1991) The disparity of the Burgess Shale arthropod fauna and the limits of cladistic analysis: why we must strive to quantify morphospace. *Paleobiology*, **17**, 411–23.

Hessler, R.R. and Newman, W.A. (1975) A trilobitomorph origin for the Crustacea. *Fossils and Strata*, **4**, 437–59.

Itô, T. (1989) Origin of the basis in copepod limbs, with reference to remipedean and cephalocarid limbs. *Journal of Crustacean Biology*, **9**, 85–103.

Lawrence, P.A. (1992) *The Making of a Fly*, Blackwell Scientific, Oxford.

Lundberg, J.G. (1972) Wagner networks and ancestors. *Systematic Biology*, **21**, 398–413.

Maddison, W.P. and Maddison, D.R. (1992) *MacClade: Analysis of phylogeny and character evolution*, ver. 3.01 [Software included], Sinauer Assoc., Sunderland, Mass.

McNamara, K.J. and Trewin, N.H. (1993) A euthycarcinoid arthropod from the Silurian of Western Australia. *Palaeontology*, **36**, 319–35.

Minelli, A. and Schram, F.R. (1994) Owen revisited: a reappraisal of morphology in evolutionary biology. *Bijdragen tot de Dierkunde*, **64**, 65–74.

Novacek, M.J. (1992) Fossils, topologies, missing data, and the higher level phylogeny of eutherian mammals. *Systematic Biology*, **41**, 58–73.

Nüsslein-Volhard, C. and Wieschaus, E. (1980) Mutations effecting segment number and polarity in *Drosophila*, *Nature*, **287**, 13–19.

Panganiban, G., Sebring, A., Nagy, L. and Carroll, S. (1995) The development of crustacean limbs and the evolution of arthropods. *Science*, **270**, 1363–6.

Raff, R.A. (1996) *The Shape of Life*, University of Chicago Press, Chicago.

Scholtz, G. (1995) Expression of *engrailed* gene reveals 9 putative segment-anlagen in the embryonic pleon of the freshwater crayfish *Cherax destructor*. *Biological Bulletin*, **188**, 157–65.

Schram, F.R. (1986) *Crustacea*, Oxford University Press, New York.

Schram, F.R. and Emerson, M.J. (1991) Arthropod Pattern Theory: a new approach to arthropod phylogeny. *Memoirs of the Queensland Museum*, **31**, 1–18.

Schram, F.R. and Hof, C.H.J. Fossils and the interrelationships of major crustacean groups, in *Arthropod Fossils and Phylogeny* (ed G. Edgecomb), Columbia University Press, New York (in press).

Simonetta, A.M. (1970) Studies on non-trilobite arthropods of the Burgess Shale (Middle Cambrian). *Palaeontographica Italiana*, **66**, 35–45.

Spiridinov, V.A. (1993) The preanal plate and spine of Euphausiacea, their variability and homology (with particular reference to the austral species of the genus *Euphausia* Dana). *Arthropoda Selecta*, **1**, 3–15.

Starabogatov, Ya.I. (1991) The systematics and phylogeny of the lower chelicerates (a morphological analysis of the Paleozoic groups). *Paleontological Journal*, for 1990, 2–13.

Sundberg, F.A. (1995) Arthropod pattern theory and Cambrian trilobites. *Bijdragen tot de Dierkunde*, **64**, 193–213.

Swofford, D.L. (1993) *PAUP: phylogenetic analysis using parsimony*, Version 3.1. Computer program distributed by Illinois State Natural History Survey, Champaign, Illinois.

Wägele, J.W. (1992) Review of methodological problems of 'computer cladistics' exemplified with a case study on isopod phylogeny (Crustacea: Isopoda). *Zeitschrift für zoologishe Systematiek und Evolutions-forschung*, **32**, 81–107.

Walcott, C.D. (1911) Cambrian geology and paleontology. II. Middle Cambrian Merostomata. *Smithsonian Miscellaneous Collections*, **57**, 17–40.

Walcott, C.D. (1912) Cambrian geology and paleontology. II. Middle Cambrian Branchiopoda, Malacostraca, Trilobita, Merostomata. *Smithsonian Miscellaneous Collections*, **57**, 145–248.

Walossek, D. and Müller K. (1990) Upper Cambrian stem-lineage crustaceans and their bearing upon the monophyletic origin of Crustacea and the position of *Agnostus*. *Lethaia*, **23**, 409–27.

Wenner, A.M. (1989) Concept-centered versus organism-centered biology. *American Zoologist*, **29**, 1177–97.

Wheeler, W.C., Cartwright, P. and Hayashi, Y. (1993) Arthropod phylogeny: a combined approach. *Cladistics*, **9**, 1–39.

Whittington, H.B. (1974) *Yohoia* Walcott and *Plenocaris* n. gen., arthropods from the Burgess Shale, Middle Cambrian, British Columbia. *Geological Survey of Canada, Bulletin*, **231**, 1–27, plates 1–18.

Wills, M.A. (1994) *The Cambrian Radiation and the Recognition of Higher Taxa*, PhD Thesis, University of Bristol.

Wills, M.A., Briggs, D.E.G., Fortey, R.A. and Wilkinson, M. (1995) The significance of fossils in understanding arthropod evolution. *Verhandlungen Deutsches zoologisches Gesellschaft*, **88**, 203–15.

Zrzavý, J. and Stys, P. (1994) Origin of the crustacean schizoramous limb: a reanalysis of the duplosegment hypothesis. *Journal of Evolutionary Biology*, **7**, 743–56.

# 8 Sampling, groundplans, total evidence and the systematics of arthropods

*W.C. Wheeler*

*Department of Invertebrates, American Museum of Natural History, Central Park West at 79th Street, New York, NY 10024-5192, USA*
*email: wheeler@amnh.org*

## 8.1 INTRODUCTION

The outline of arthropod relationships was clearly and firmly established by Snodgrass (1938) (Figure 8.1). All work on these taxa since then concerns the support for, and discussion of the basic groups he delineated. Although the efforts of Tiegs, Manton, and Anderson (Tiegs and Manton, 1958; Manton, 1964, 1973, 1979; Anderson, 1979) to incorporate functional morphology and observational embryology diverted discussion from Snodgrass' basic principles, the field has returned to the apportionment of variation so productive in the past.

Since Snodgrass, arthropod systematics has seen two fundamental advances: synapomorphy and DNA. Technical innovation has presented molecular genetic data in immense quantity, and the theoretical advances of Hennig (1966) have offered the framework for their interpretation. Although the cladistic paradigm allows (some might say requires) simultaneous analysis of morphological and molecular data, this combination of evidence is rarely attempted

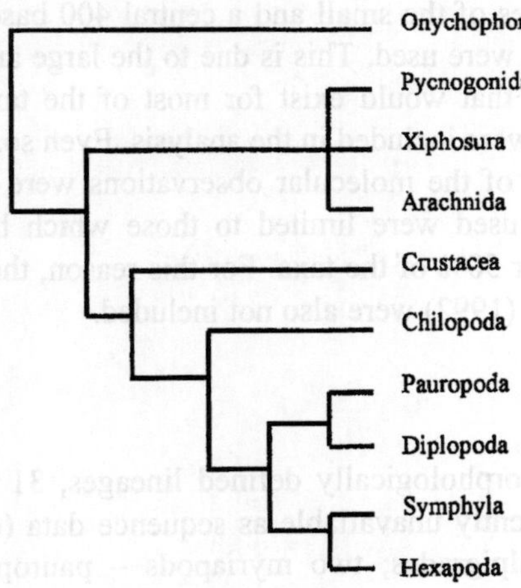


**Figure 8.1** Phylogeny of the extant arthropods. (After Snodgrass, 1938.)

(Wheeler *et al.*, 1993). This is due, in part, to the sampling problems of molecular studies (reviewed by Wheeler, 1997) and the use of groundplans and single-character analysis in morphological work (see papers of Walossek and Boxshall, 1997, this volume).

The discussion presented here is based on two analytical notions. First, that large, diverse samples of taxa are better able to recover the phylogenetic pattern of higher taxa; and second, that diverse types of information (characters) offer more robust indicators of phylogeny than single systems or sources of data. This is the kernel of the 'total evidence' approach (Kluge, 1989).

Although 'total' evidence is something of a misnomer, the concept – that all evidence currently available be used simultaneously – is hard to deny. This does not mean or imply that no new data could be gathered which would overturn the results, just that, for now, this is the best we can do. Hence data from hard and soft-part anatomy, behaviour, development, molecular sequence and gene organization are included in my analysis. If we combine information from behaviour, anatomy and development, it is difficult to see why we should exclude molecular characters from the data set (Kraus and Kraus, 1994). It seems illogical to reserve or segregate organismal variants *a priori*, because we cannot know which features are informative and congruent without simultaneous analysis. Lastly, although not examined here, there is the question of accommodating our knowledge of extinct taxa with molecular systematics. Unless data are combined, the overwhelming majority of creatures which have ever lived – the fossil ones – will be excluded from integration with living taxa.

Another motivation for my analysis comes from desire to employ better samples of lower taxa to arrange higher groups. The fundamental questions of arthropod phylogenetics concern the interrelationships of four lineages: chelicerates, crustaceans, myriapods and hexapods. Of these, the monophyly of the Myriapoda is most frequently questioned. Each of these lineages has been divided into constituent lower taxa

*Arthropod Relationships*, Systematics Association Special Volume Series 55, Edited by R.A. Fortey and R.H. Thomas. Published in 1997 by Chapman & Hall, London. ISBN 0 412 75420 7

('orders'), which though not equivalent in any sense, reflect the cladistic diversity of the groups. Where possible, samples have been used from each of these and relevant outgroups. This should help to augment the quality of groundplan estimates for higher taxa through cladistic sampling.

Although most recent analyses have accepted the monophyly of the Arthropoda and even the basic split between chelicerates on one side and mandibulates (crustaceans, myriapods and hexapods) on the other, argument persists within the Mandibulata. Recent molecular work (Field *et al.*, 1988; Turbeville *et al.*, 1991; Freidrich and Tautz, 1995; Garey *et al.*, 1996; Giribet *et al.*, 1996) (Figure 8.2) has pointed to a Hexapoda + Crustacea grouping as opposed to the more traditional Tracheata (Hexapoda + Myriapoda). As pointed out earlier (Wheeler *et al.*, 1993), molecular data point to the Crustacea + Hexapoda group while morphological analysis offers near uniform support for Tracheata (but see Dohle, 1997, this volume). Some morphological analyses even present the 'Myriapoda' as paraphyletic with respect to the hexapods showing the Labiata as Hexapoda grouped with the Symphyla, Pauropoda and Diplopoda to the exclusion of the Chilopoda (Pocock, 1893; Snodgrass, 1938; Kraus and Kraus, 1994). The status and sister group relations of the myriapods form the main thrust of the following analysis.

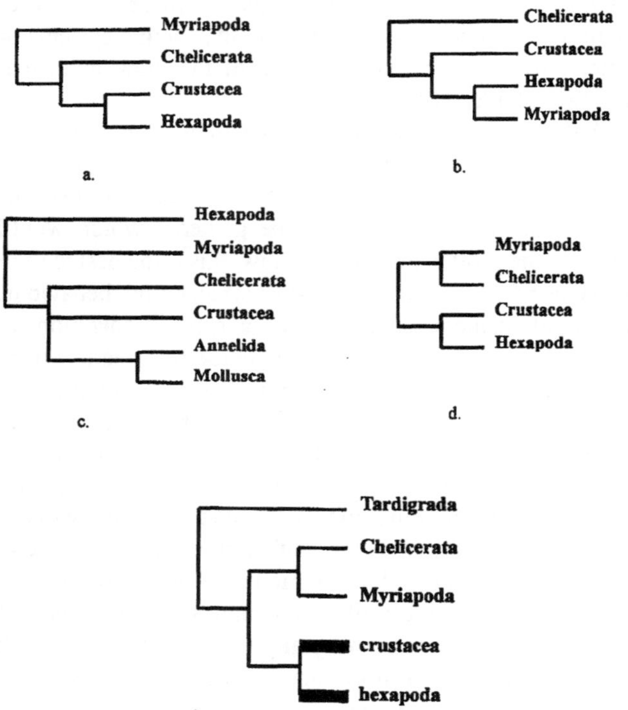


**Figure 8.2** Molecular phylogenies of the Arthropoda. (a) Field *et al.* (1988); (b) Abele *et al.* (1989); (c) Lake (1990); (d) Turbeville *et al.* (1991); and (e) Giribet *et al.* (1996). The cladogram of Garey *et al.* (1996) resembles that of Giribet but contains no myriapodan sample and places a branchiopod in an unresolved clade with the hexapod and crustacean taxa. Taxa in lower case (and with thicker lines) are not monophyletic.

## 8.2 THE DATA SET

### 8.2.1 CHARACTERS

In attempting to include as much data as possible, characters were garnered from both morphological and molecular sources. The analysis of non-sequence data from variants among and between higher taxa resulted in 90 defined lineages (Table 8.1). These include five outgroup (Mollusca, Polychaeta, Clitellata, Onychophora and Tardigrada), 13 chelicerate, 35 crustacean, four myriapod and 33 hexapod taxa. These lineages were defined by variation in the 552 non-sequence characters. Of these, 121 concerned relationships among arthropod taxa at the highest level (from a variety of sources), 96 concerned chelicerate interrelationships [mainly derived from the work of Weygolt and Paulus (1979), Yoshikura (1975) and Schultz (1990)], 248 bore on the hexapod orders [from a variety of sources, mainly Hennig (1981), Kristensen (1995) and Boudreaux (1979)] and 87 concerned the crustaceans (entirely from Emerson and Schram, 1997, this volume). For these 90 lineages, 45% of the entries were missing or inapplicable – an unfortunately high number. Most of these are due to the inapplicability of ingroup variation characters in non-ingroup taxa (e.g. wing venation in worms).

The molecular characters are drawn from three sources (see data on web site). The first are mitochondrial gene order characters of Boore *et al.* (1995) which are included in the non-sequence data (i.e. morphology, behaviour, etc.). The other molecular sources are the small (18S) and large (28S) ribosomal subunit DNAs. Although the entirety of each locus has been sequenced for many taxa, only the middle 1200 bases of the small and a central 400 bases of the large subunit were used. This is due to the large amount of missing data that would exist for most of the taxa if the entire genes were included in the analysis. Even so, approximately 23% of the molecular observations were missing. The regions used were limited to those which had been sequenced for 50% of the taxa. For this reason, the data of Ballard *et al.* (1992) were also not included.

### 8.2.2 TAXA

Of the 90 morphologically defined lineages, 31 of these taxa are currently unavailable as sequence data (one chelicerate – palpigrades; two myriapods – pauropods and symphylans; and 26 crustacean lineages; Table 8.1). Most of the remaining lineages are represented by two sequenced taxa resulting in 136 terminal taxa for analysis. Those taxa

**Table 8.1** Taxa used in the study

| *Higher group* | *Taxon* | *18S rDNA* | *28S rDNA* |
|---|---|---|---|
| Mollusca | | | |
| Cephalopoda | *Loligo pealei* | Wheeler | ND |
| Polyplacophora | *Lepidochiton cavernae* | Wheeler | ND |
| Annelida | | | |
| Polycheata | *Glycera* sp. | Wheeler | ND |
| Oligocheata | *Lumbricus terrestris* | Wheeler | ND |
| | *Tubifex* sp. | Freidrich | Freidrich |
| Hirudinea | *Haemopis marmorata* | Wheeler | ND |
| Onychophora | | | |
| Peripatoidae | *Peripatus trinitatis* | Wheeler | ND |
| Peripatopsidae | *Peripatoides novozealandia* | Wheeler | ND |
| Tardigrada | *Macrobiotus hufelandi* | Giribet | ND |
| Chelicerata | | | |
| Pycnogonida | *Anoplodactylus portus* | Wheeler | Hayashi |
| | *Anoplodactylus lentus* | Hayashi | Hayashi |
| | *Colosendeis* sp. | Hayashi | Hayashi |
| Xiphosura | *Limulus polyphemus* | Wheeler | Hayashi |
| Scorpiones | *Centruroides hentsii* | Wheeler | Hayashi |
| | *Androctonus australis* | Hayashi | Hayashi |
| | *Hadrurus arizonensis* | Hayashi | Hayashi |
| | *Paruroctonus meaensis* | Hayashi | Hayashi |
| Araneae | *Peucetia viridans* | Wheeler | Hayashi |
| | *Gea heptagon* | Hayashi | Hayashi |
| | *Erypelma californica* | Freidrich | Freidrich |
| | *Thelechoris striatipes* | Hayashi | Hayashi |
| | *Heptathelia kimurai* | Hayashi | Hayashi |
| | *Liphistius bristowei* | Hayashi | Hayashi |
| Palpigrada | Morphology only | ND | ND |
| Psuedoscorpiones | *Americhenernes* sp. | Hayashi | Hayashi |
| Solifugae | *Chanbria regalis* | Hayashi | Hayashi |
| Opiliones | *Vonones ornata* | Hayashi | Hayashi |
| | *Leiobunum* sp. | Hayashi | Hayashi |
| Acari | *Amblyomma americanum* | Hayashi | Hayashi |
| | *Rhiphicephalus sanguineus* | Hayashi | Hayashi |
| | *Tetranychus urticae* | Hayashi | Hayashi |
| Ricinulei | Ricinoididae (juvenile) | Hayashi | Hayashi |
| Amblypygi | Amblypygid sp. | Hayashi | Hayashi |
| Thelyphonida | *Mastogoproctus giganteus* | Wheeler | Hayashi |
| Schizomida | *Trithyreus pentapeltis* | Hayashi | Hayashi |
| Crustacea | | | |
| Nectiopoda | Morphology only | ND | ND |
| Stomatopoda | Morphology only | ND | ND |
| Anaspidacea | Morphology only | ND | ND |
| Bathynellacea | Morphology only | ND | ND |
| Lophogastrida | Morphology only | ND | ND |
| Mysida | Morphology only | ND | ND |
| Mictacea | Morphology only | ND | ND |
| Isopoda | Morphology only | ND | ND |
| Amphipoda | Morphology only | ND | ND |
| Cumacea | *Diastylis* sp. | Kim | ND |
| Tanaidacea | Morphology only | ND | ND |
| Spelaeogriphacea | Morphology only | ND | ND |
| Thermosbaenacea | Morphology only | ND | ND |
| Euphausiacea | Morphology only | ND | ND |
| Amphionidacea | Morphology only | ND | ND |
| Dendrobranchiata | Morphology only | ND | ND |
| Caridea | Morphology only | ND | ND |
| Euzygida | Morphology only | ND | ND |

**Table 8.1** (continued)

| *Higher group* | *Taxon* | *18S rDNA* | *28S rDNA* |
|---|---|---|---|
| Reptantia | *Callinectes* sp. | Wheeler | Hayashi |
| | *Procambarus leonensis* | Spears | ND |
| Leptostraca | Morphology only | ND | ND |
| Cephalocarida | Morphology only | ND | ND |
| Notostraca | Morphology only | ND | ND |
| Anostraca | *Artemia salina* | Nelles | Freidrich |
| | *Branchinecta packardi* | Spears | ND |
| Conchostraca | Morphology only | ND | ND |
| Cladocera | *Bosmina longirostris* | Kim | ND |
| Ostracoda | Podocopid sp. | Spears | ND |
| | *Stenocypris major* | Kim | ND |
| Mystacocarida | Morphology only | ND | ND |
| Branchiura | *Argulus nobilis* | Abele | ND |
| | *Porocephalus crotali* | Spears | ND |
| Tantulocarida | Morphology only | ND | ND |
| Copepoda | *Calanus pacificus* | Spears | ND |
| Rhizocephala | Morphology only | ND | ND |
| Ascothoracida | Morphology only | ND | ND |
| Acrothoracica | *Trypetesa lampas* | Spears | ND |
| Thoracica | *Balanus* sp. | Wheeler | Hayashi |
| | *Calantica villosa* | Spears | ND |
| | *Octolasmis lowei* | Spears | ND |
| Facetotecta | Morphology only | ND | ND |
| Myriapoda | | | |
| Chilopoda | *Scutigera coleoptrata* | Wheeler | Hayashi |
| | *Lithobius* sp. | Freidrich | Freidrich |
| Diplopoda | *Spirobolus* sp. | Wheeler | Hayashi |
| | *Polyxenus* sp. | Freidrich | Freidrich |
| | *Megaphyllum* sp. | Freidrich | Freidrich |
| Pauropoda | Morphology only | ND | ND |
| Symphyla | Morphology only | ND | ND |
| Hexapoda | | | |
| Collembola | *Psuedochorutes* | Freidrich | Freidrich |
| | *Podura aquatica* | Carpenter | Carpenter |
| Protura | *Nipponentomon* sp. | Carpenter | Carpenter |
| Diplura | *Metajapyx* sp. | Carpenter | Carpenter |
| | *Campodea tillyardi* | Carpenter | Carpenter |
| Archeognatha | *Petrobius brevistylus* | Freidrich | Freidrich |
| | *Trigoniopthalmus alternatus* | Whiting | Whiting |
| Zygentoma | *Lepisma* sp. | Carpenter | Carpenter |
| | *Thermobius domestics* | Carpenter | Carpenter |
| Ephemeroptera | *Stenonema* sp. | Carpenter | Carpenter |
| | *Ephemerella* sp. | Whiting | Whiting |
| Odonata | *Libellula pulchella* | Wheeler | Whiting |
| | *Calopteryx* sp. | Carpenter | Carpenter |
| Plecoptera | *Megarcys stigmata* | Whiting | Whiting |
| | *Cultus decisus* | Whiting | Whiting |
| Embiidina | *Oligotoma saundersii* | Whiting | Whiting |
| | *Clothoda* sp. | Carpenter | Carpenter |
| Grylloblatta | *Grylloblatta* sp. | Carpenter | Carpenter |
| Dermaptera | *Forficula auricularia* | Carpenter | Carpenter |
| | *Labia* sp. | Carpenter | Carpenter |
| | *Labidura riparia* | Whiting | Whiting |
| Isoptera | *Reticulotermes virginiana* | Carpenter | Carpenter |
| Blattaria | *Blaberus* sp. | Carpenter | Carpenter |
| Mantodea | *Mantis religiosa* | Wheeler | Whiting |
| Orthoptera | *Ceuthophilus* sp. | Carpenter | Carpenter |
| | *Melanoplus* sp. | Whiting | Whiting |

| Higher group | Taxon | 18S rDNA | 28S rDNA |
|---|---|---|---|
| Phasmida | *Timema californica* | Carpenter | Carpenter |
| | *Phyllium* sp. | Carpenter | Carpenter |
| Pthiraptera | *Dennyus hirudensis* | Whiting | Whiting |
| Thysanoptera | *Taeniothrips inconsequens* | Whiting | Whiting |
| Psocodea | *Cerastipsocus venosus* | Wheeler | Whiting |
| Hemiptera | *Saldula pallipes* | Wheeler | Whiting |
| | *Buenoa* sp. | Wheeler | Whiting |
| Coleoptera | *Priacma serrata* | Whiting | Whiting |
| | *Calpocaccus posticatus* | Whiting | Whiting |
| Neuroptera | *Lolomyia texana* | Whiting | Whiting |
| Megaloptera | *Corydalus cognatus* | Whiting | Whiting |
| Raphidiodea | *Agulla* sp. | Whiting | Whiting |
| Hymenoptera | *Hemitaxonus* sp. | Whiting | Whiting |
| | *Ophion* sp. | Whiting | Whiting |
| Lepidoptera | *Papilio troilus* | Wheeler | Whiting |
| | *Galleria mellonella* | Whiting | Whiting |
| Trichoptera | *Leptocerus* sp. | Whiting | Whiting |
| | *Pycnopsyche* sp. | Whiting | Whiting |
| Mecoptera | *Nannochorista neotropica* | Carpenter | Carpenter |
| | *Boreus coloradensis* | Whiting | Whiting |
| Siphonaptera | *Ctenocephalides canis* | Whiting | Whiting |
| | *Hystrichopsylla schefferi* | Whiting | Whiting |
| Strepsiptera | *Crawfordia n.* sp | Whiting | Whiting |
| | *Xenos pecki* | Whiting | Whiting |
| Diptera | *Laphria* sp. | Whiting | Whiting |
| | *Tipula* sp. | Whiting | Whiting |

Abele= Abele *et al.* (1989); Giribet = Giribet *et al.* (1996); Hendriks = Hendriks *et al.* (1988); Freidrich = Freidrich and Tautz (1995); Hayashi = Wheeler and Hayashi (unpublished); Kim= Kim *et al.* (1993); Nelles = Nelles *et al.* (1984); Sharp = Sharp and Li (1987); Spears= Spears *et al.* (1994); Tautz = Tautz *et al.* (1988); Wheeler = Wheeler *et al.* (1993); Whiting = Whiting *et al.* (in press); ND = no data; Carpenter = Wheeler, Whiting, Wheeler, and Carpenter (in press).

without sequence data were placed on the basis of morphology alone with the molecular data coded as missing. This resulted in an overall level of missing data of approximately 29%.

## 8.3 ANALYSIS

### 8.3.1 MORPHOLOGICAL

Morphological characters were analysed using Goloboff's (1995) parsimony-based NONA (version 1.1). These searches used 'tbr' branch swapping on 50 random addition sequences.

### 8.3.2 MOLECULAR

Phylogenetic analysis of molecular sequence data (18S and 28S rDNA) were performed via direct optimization of sequences (Wheeler, 1996), without the intermediate step of multiple alignment, using MALIGN (Wheeler and Gladstein, 1992, version 2.7 on a dedicated cluster of workstations). As with the morphological data, 'tbr' type branch swapping was employed and 50 random addition sequences attempted. For this analysis, an insertion–deletion cost of 2 : 1 was used and a transversion : transition ratio of 2 : 1. These values, though somewhat arbitrary, have been shown to optimize character congruence in other arthropod studies (Wheeler, 1995, 1997). Insertion–deletion events were treated independently and included as phylogenetic information (Wheeler, 1993). Other investigators (Friedrich and Tautz, 1995) have used similar parameter values – though rarely gaps. The choice of these parameters, however, can affect the outcome of phylogenetic analysis (Wheeler, 1995), hence the robustness of these results awaits further appraisal.

### 8.3.3 TOTAL EVIDENCE

When the morphological and molecular data were combined to create 'total evidence' cladograms, morphological character transformations were assigned the same weight as insertion–deletion events. Otherwise, all weighting was equal, in other words, morphological (552) and molecular

(~1400) characters were employed without regard to source. The combined data were analysed in the same manner as described for the molecular data alone.

## 8.4 RESULTS

Phylogenetic analysis of morphological (non-sequence) characters yielded 87 most parsimonious cladograms of length 1204 with a C.I. of 0.55 and an R.I. of 0.85 (Figure 8.3). The molecular (18S and 28S) data alone produced a single tree at weighted length 10599 (Figure 8.4). Combined data yielded a single tree at 16079 weighted steps (Figure 8.5) The most parsimonious cladogram forced to link crustaceans and hexapods to the exclusion of myriapods had a length of 16167 steps – 88 steps longer (0.55 %; Figure 8.6). The comparison of the individual morphological and molecular analyses to the combined data produces 4.13% additional homoplasy (ILD of Mickevich and Farris, 1981), showing a low level of character incongruence between the main sources of data.

## 8.5 CONCLUSIONS

The most salient conclusion from this study is that as far as these data are concerned, the Tracheata are monophyletic as are the Labiata = (Hexapoda + ((Diplopoda + Pauropoda) + Symphyla)) with the myriapods relegated to paraphyly. There are three factors which bear on the confidence which can be placed on this result: analytical robustness, missing data, and missing – that is, extinct – taxa.

The robustness of these results is unknown. The analysis performed here is based on a specific set of assumptions which include an insertion–deletion cost of twice that of transversions, a transversion cost twice that of transitions, and non-sequence character change equal in cost to insertion–deletion events. Although these values are similar to those used in other studies (Freidrich and Tautz, 1995; Wheeler, 1995), the consistency of phylogenetic results under varying parameter values is unknown, but may be important. This is especially pertinent given the small differential in support between the Tracheata scheme and Crustacea + Hexapoda.

Missing data may have an insidious effect on phylogenetic analysis (Nixon and Davis, 1991; Platnick, 1991). In situations of ambiguity or high levels of missing data, these defects are unpredictable. Although additional sequencing effort will remove some missing values, most of the non-sequence missing values cannot be established. This is because many 'missing' values are inapplicabilities, that is, no corresponding feature or attribute can be identified in a taxon. For instance, cheliceral features in myriapods or wing-vein characters in Onychophora can never be appropriately coded. However, given that these features do not, in general, affect the relative placement

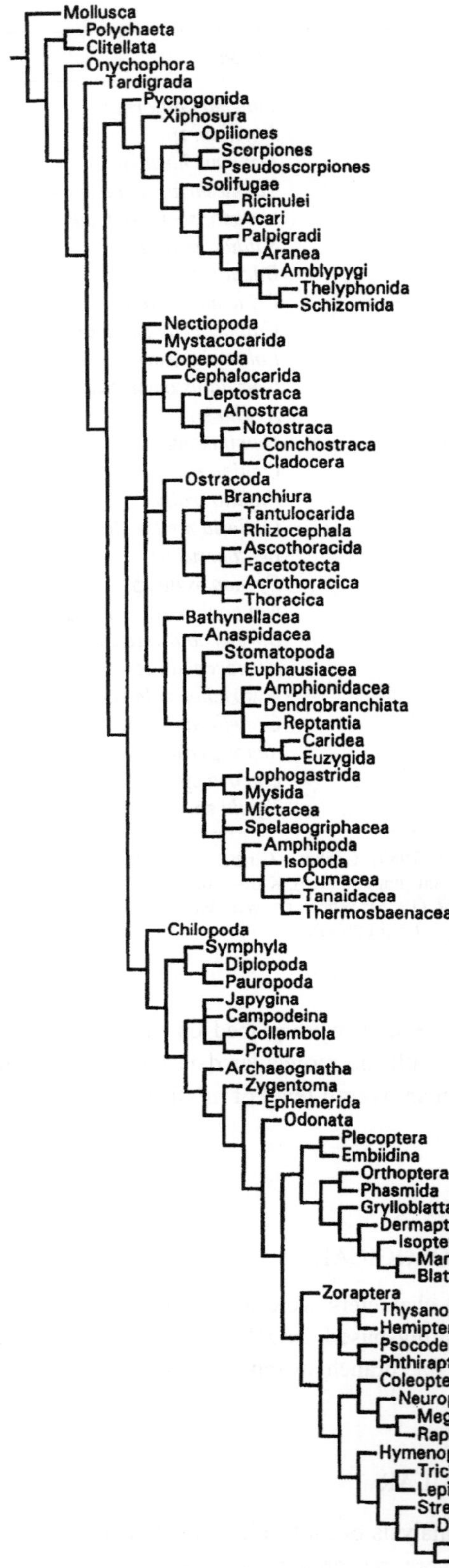


**Figure 8.3** Cladogram of arthropod lineages based on the 552 non-sequence characters of tables published on web site. There were 87 equally parsimonious representations of the 90 taxa found at a length of 1204, a C.I. of 0.55, and an R.I. of 0.85.

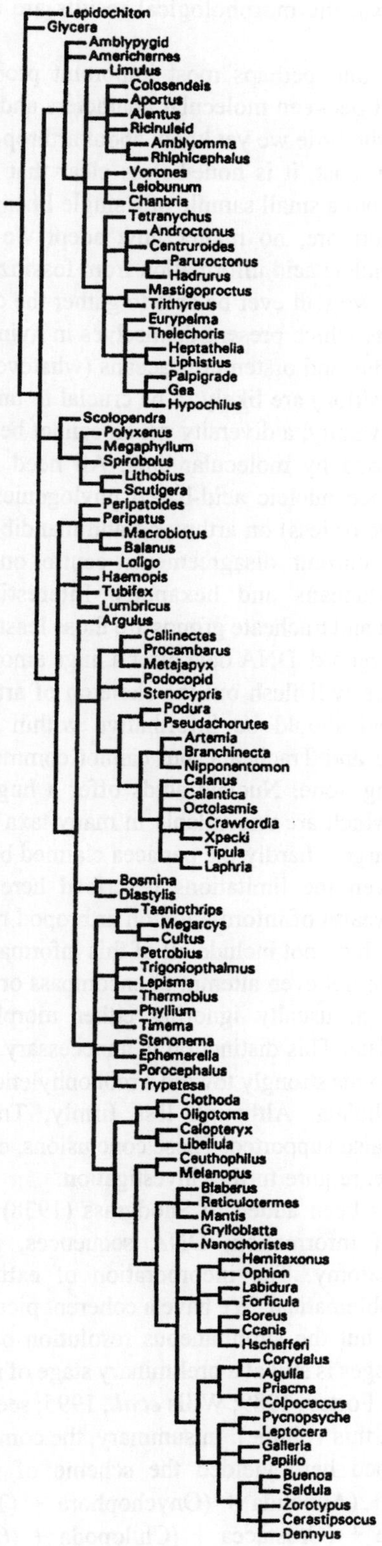


**Figure 8.4** Cladogram of sampled arthropod lineages based solely on molecular sequence information. The cladogram of 106 taxa is based on approximately 1000 bp of 18S rDNA and 350 bp of the 28S rDNA. The total weighted length is 10 599 weighted steps, given insertion–deletion events weighted twice transversions and transversions twice transitions.

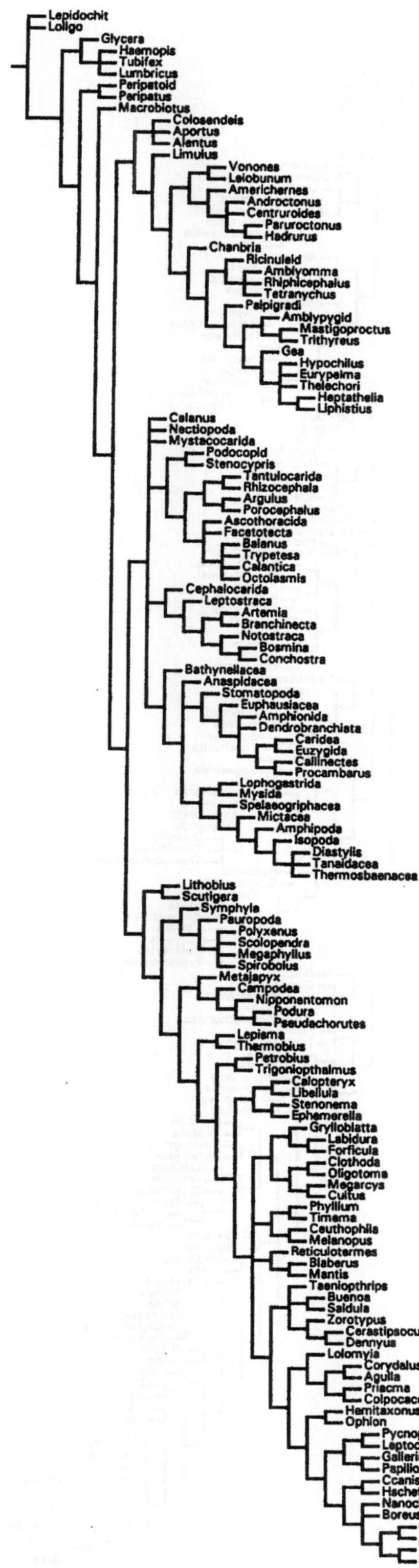


**Figure 8.5** Total evidence cladogram of arthropod lineages. Combined data for 136 terminals yielded a single cladogram at 16 079 weighted steps. Non-sequence changes were weighted equally with insertion deletion events. Other weights were as in Figure 8.5

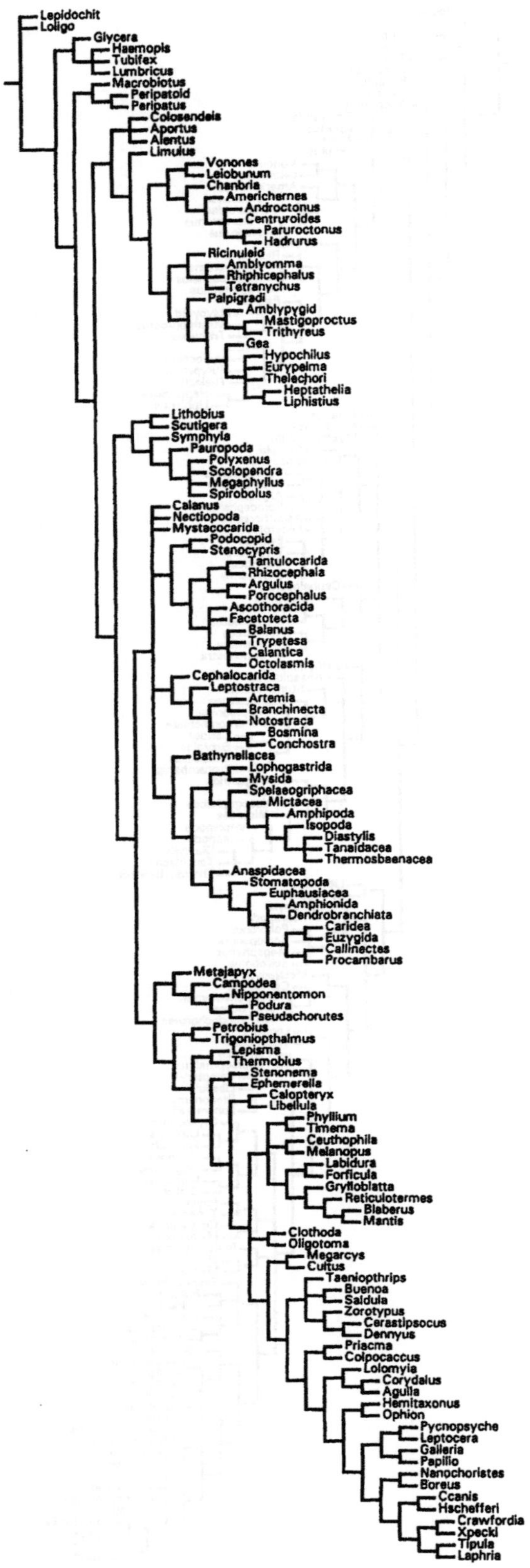


**Figure 8.6** Cladogram of arthropod lineages with Hexapoda + Myriapoda. This most parsimonious cladogram forced to contain Hexapoda + Crustacea has a length of 16 167 weighted steps. Analysis as in Figure 8.5.

of higher taxa, the morphological results are most likely stable.

The final and perhaps most important problem is the estrangement between molecular characters and Palaeozoic taxa. Given the little we yet know about arthropod diversity in the distant past, it is nonetheless clear that crown chelicerates are but a small sample of a single lineage of arachnates. Furthermore, no matter how adept we become at extracting nucleic acid information from fossilized samples, it is unlikely we will ever be able to gather the quantities of sequence data which present themselves in living creatures. Anomalocarids and orsten crustaceans (whatever their phylogenetic position) are likely to be crucial to understanding arthropod diversity; a diversity which cannot be seen, much less understood by molecular data. All need not be lost, however, since nucleic acid-based phylogenies have converged (more or less) on arthropod and mandibulate monophyly. The current disagreements centre on myriapods versus crustaceans and hexapods. Interestingly, basal mandibulate and tracheate groups are those least represented in the fossil record. DNA data offer a huge amount of information which will flesh out the skeleton of arthropod systematics, and should be informative within Chelicerata, Mandibulata and Tracheata, but cannot comment on basal lineages long gone. Nucleic acids offer a huge wealth of characters which are unavailable in many taxa – inapplicability writ large – hardly the panacea claimed by some.

Even given the limitations described here, these data reflect the wealth of information on arthropod relationships. Studies which do not include **all** of this information are limited. They do not even attempt to encompass or explain natural variation, usually ignoring either morphological or molecular data. This distinction is unnecessary. The sum of these data points strongly toward a monophyletic Arthropoda and Mandibulata. Although less firmly, Tracheata and Labiata are also supported. These conclusions, especially the labiate clade, require further investigation.

What has been added to Snodgrass (1938) is a greater diversity of information, DNA sequences, internal and external anatomy. The incorporation of extinct lineages remains problematical. We have a coherent picture of extant arthropods, but the simultaneous resolution of extant and extinct lineages is still at a preliminary stage of investigation (Briggs and Fortey, 1989; Wills *et al.*, 1995; see also Zrzavý *et al.*, 1997, this volume). In summary, the combined analysis performed here yielded the scheme of relationships (Mollusca + (Annelida + (Onychophora + (Tardigrada + (Chelicerata + (Crustacea + (Chilopoda + ((Symphyla + (Pauropoda + Diplopoda)) + Hexapoda)))))))).

## ACKNOWLEDGEMENTS

I would like to acknowledge the great input of a number of people who contributed to construction and collection of these data, especially Gregory Edgcombe, Norman Platnick, James Carpenter, and Alan Harvey for discussion of morphological characters and Aloyisius Philips, Cheryl Hayashi, Michael Whiting, and Gonzalo Giribet for use of their unpublished sequences. As always, the numerous errors are my own.

## REFERENCES


Abele, L.G., Kim, W. and Felgenhauer, B.E. (1989) Molecular evidence for the inclusion of the phylum Pentastomida in the Crustacea. *Molecular Biology and Evolution*, **6**, 685–91.

Anderson, D.T. (1979) Embryos, fate maps, and the phylogeny of arthropods, in *Arthropod Phylogeny* (ed. A.P. Gupta) Van Nostrand, New York, pp. 59–106.

Ballard, J.W.O., Olsen, G.J., Faith, D.P., Odgers, W.A., Rowell, D.M. and Atkinson, P.W. (1992) Evidence from 12S ribosomal RNA sequences that onychophorans are modified arthropods. *Science*, **258**, 1345–8.

Boore, J.L., Collins, T.M., Stanton, D., Daehler, L.L. and Brown, W.M. (1995) Deducing the pattern of arthropod phylogeny from mitochondrial rearrangements. *Nature*, **376**, 163–5.

Boudreaux, H.B. (1979) *Arthropod Phylogeny with Special Reference to Insects*, Wiley and Sons.

Briggs, D.E.G. and Fortey, R.A. (1989) The early radiation and relationships of the major arthropod groups. *Science*, **246**, 241–3.

Field, K.G., Olsen, G.J., Lane, D.J., Giovannoni, S.J., Ghiselen, M.T., Raff, E.C., Pace, N.R. and Raff, R.A. (1988) Molecular phylogeny of the animal kingdom. *Science*, **239**, 748–53.

Friedrich, M. and Tautz, D. (1995) Ribosomal DNA phylogeny of the major extant arthropod classes and the evolution of the myriapods. *Nature*, **376**, 165–7.

Garey, J.R., Krotec, M., Nelson, D.R. and Brooks, J. (1996) Molecular analysis supports a tardigrade–arthropod association. *Invertebrate Biology*, **115**, 79–88.

Giribet, G., Carranza, S., Baguñá, J., Riutort, M. and Ribera, C. (1996) First molecular evidence for the existence of a Tardigrada + Arthropoda clade. *Molecular Biology and Evolution*, **13**, 76–84.

Goloboff, P. (1995) NONA. Program and Documentation. Version 1.1

Hendriks, L., De Baere, R., Van Broeckhoven, C. and De Wachter, R. (1988) Primary and a secondary structure of the 18S ribosomal RNA of the insect species *Tenebrio molitor. Federation of European Biochemical Societies*, **232**, 115–20.

Hennig, W. (1981) *Insect Phylogeny*, John Wiley and Sons, New York. (Translation of Hennig, 1969.)

Kim, W., Yoon, S.M. and Kim, J. (1993) The 18S ribosomal RNA gene of a crustacean branchiopod *Bosmina longirostris*: comparison with another branchiopod *Artemia salina. Nucleic Acids Research*, **21**, 3583.

Kluge, A. (1989) A concern for evidence and a phylogenetic hypothesis for relationships among *Epicrates* (Boidae, Serpentes). *Systematic Zoology*, **38**, 1–25.

Kraus, O. and Kraus, M. (1994) Phylogentic system of Tracheata (Mandibulata): on 'Myriapoda'–Insecta interrelationships, phylogenetic age and primary ecological niches. *Verhandeln naturwissenschaft*, **34**, 5–31.

Kristensen, N.P. (1995) Forty years' insect phylogenetic systematics. *Zoologische Bieträge, Neue Folge*, **36**, 83–124.

Lake, J.A. (1990) Origin of the Metazoa. *Proceedings of the National Academy of Sciences, USA*, **87**, 763–6.

Manton, S.M. (1964) Mandibular mechanisms and the evolution of arthropods. *Philosophical Transactions of the Royal Society, London, B*, **247**, 1–183.

Manton, S.M. (1973) Arthropod phylogeny – a modern synthesis. *Journal of Zoology*, **171**, 111–30.

Manton, S.M. (1977) *The Arthropoda. Habits, Functional Morphology and Evolution*, Clarendon Press, Oxford.

Manton, S.M. (1979) Functional morphology and the evolution of the hexapod classes, in *Arthropod Phylogeny* (ed. A.P. Gupta), Van Nostrand, New York, pp 387–466.

Mickevich, M.F. and Farris, S.J. (1981) The implications of congruence in *Menidia. Systematic Zoology*, **30**, 351–70.

Nelles, L., Fang, B.-L., Volckaert, G., Vandenberghe, A. and De Wachter, R. (1984) Nucleotide sequence of a crustacean 18S ribosomal RNA gene and secondary structure of eukaryotic small subunit ribosomal RNAs. *Nucleic Acids Research*, **14**, 2345–64.

Nixon, K.N. and Davis, J.I. (1991) Polymorphic taxa, missing values, and cladistic analysis. *Cladistics*, **7**, 233–41.

Platnick, N. (1991) On missing entries in cladistic analysis. *Cladistics*, **7**, 337–43.

Pocock, R.J. (1893) On the classification of the tracheate Arthropoda. *Zoologischer Anzeiger*, **16**, 271–5.

Sharp, P.M. and Li, W.-H. (1987) Molecular evolution of ubiquitin genes. *Trends in Ecology and Evolution*, **2**, 328–32.

Shultz, J.W. (1990) Evolutionary morphology and phylogeny of Arachnida. *Cladistics*, **6**, 1–38.

Snodgrass, R.E. (1938) Evolution of the Annelida, Onychophora and Arthropoda. *Smithsonian Miscellaneous Collections*, **97**, 1–159.

Spears,T., Abele,L.G. and Applegate, M.A. (1994) Phylogenetic study of cirripedes and selected relatives (Thecostraca) based on 18S rDNA sequence analysis. *Journal of Crustacean Biology*, **14**, 641–56.

Tautz D., Hancock, J.M., Webb, D.A., Tautz, C. and Dover, G.A. (1988) Complete sequence of the rRNA genes in *Drosophila melanogaster. Molecular Biology and Evolution*, **5**, 366–76.

Tiegs, O.W. and Manton, S.M. (1958) The evolution of the Arthropoda. *Biological Reviews*, **33**, 255–337.

Turbeville, J.M., Pfeifer, D.M., Field, K.G. and Raff, R.A. (1991) The phylogenetic status of the arthropods, as inferred from 18S rRNA sequences. *Molecular Biology and Evolution*, **8**, 669–702.

Weygoldt, P. and Paulus, H.F. (1979) Untersuchungen zur Morphologie, Taxonomie und Phylogenie der Chelicerata. II. Cladogramme und die Enfaltung der Chelicerata. *Zietschrift für Zoolische Systematik und Evolution-Forschung*, **17**, 117–200.

Wheeler, W.C. (1993) The triangle inequality and character analysis. *Molecular Biology and Evolution*, **10**, 707–12.

Wheeler, W.C. (1995) Sequence alignment, parameter sensitivity, and the phylogenetic analysis of molecular data. *Systematic Biology*, **44**, 321–32.

Wheeler, W.C. (1996) Optimization Alignment: the end of multiple sequence alignment in phylogenetics?, *Cladistics* (in press).

Wheeler, W.C. (1997) Molecular systematics and arthropods, in *Arthropod Fossils and Phylogeny* (ed. G.E. Edgecombe), Columbia University Press, New York (in press).

Wheeler, W.C. and Gladstein, D.L. (1992) MALIGN ver. 2.7. Program and Documentation. New York, New York.

Wheeler, W.C., Cartwright, P. and Hayashi, C. (1993) Arthropod phylogenetics: a total evidence approach. *Cladistics*, **9**, 1–39.

Whiting, M.F., Carpenter, J.C., Wheeler, Q.D. and Wheeler, W.C. (1996) The Strepsiptera problem: phylogeny of the holometabolous insect orders inferred from 18S and 28S ribosomal DNA sequences and morphology. *Systematic Biology* (in press).

Wills, M.A., Briggs, D.E.G., Fortey, R.A. and Wilkinson, M. (1995) The significance of fossils in understanding arthropod evolution. *Verhandlungen der Deutschen Zoologischen Gesellschaft*, 88, 203–15.

Yoshikura, M. (1975) Comparative embryology and phylogeny of Arachnida. *Kumamoto Journal of Sciences Biology*, **12**, 71–142.

# 9 Arthropod phylogeny: taxonomic congruence, total evidence and conditional combination approaches to morphological and molecular data sets

*J. Zrzavý, V. Hypša and M. Vlášková*

*Faculty of Biological Sciences, Branišovská 31, 370 05 České Budějovice, Czech Republic*
*email: zrzavy@entu.cas.cz*
*Faculty of Biological Sciences, Branišovská 31, 370 05 České Budějovice, Czech Republic*
*Faculty of Biological Sciences, Branišovská 31, 370 05 České Budějovice, Czech Republic*

## 9.1 INTRODUCTION

Both molecular and morphological data can be used for constructing branching diagrams indicative of phylogeny. The question 'Is it feasible to combine different data sets into a single data matrix in phylogenetic reconstruction?' will be discussed in this paper, using the phylogeny of extant arthropods as an illustration. Three possible answers to this question have been formulated (reviewed in Huelsenback *et al.*, 1996) – the data should be combined either **never** (taxonomic congruence), or **always** (total evidence), or **under some circumstances** (conditional combination).

The taxonomic congruence (TC) approach involves partitioning evidence into separate data subsets, seeking the best-fitting hypothesis for each subset, and deriving a consensus of those topologies. Alternatively, the total evidence (TE) approach aims to find the best-fitting phylogenetic hypothesis for an unpartitioned set of synapomorphies. Supporters of TE emphasize that the different data subsets being analysed are equally weighted, but their constituent characters are unequally weighted when the sizes of the subsets differ, and that partitioning evidence into classes is artificial. On the other hand, adherents of TC stress that the repeating patterns (visualized by consensus trees) are what is being sought.

Naturally, the TC approach does not distinguish between cases in which sampling error is responsible for the different phylogenetic trees (as trees estimated from each data partition are based on a smaller sample of characters than in the combined analysis), and cases in which the conflicting estimates stem from fundamentally different evolutionary mechanisms underlying the data. Only if the different trees result from stochastic variation within small data samples should combining the data actually improve the estimate.

Recently, a different approach, conditional data combination (CC), was presented (de Queiroz, 1993; Huelsenback *et al.*, 1996). This approach proposes that data partitions should be subjected to a statistical test of homogeneity. Heterogeneous data partitions are those that result in significantly different estimates of phylogeny when analysed separately. If the difference is not significant, then the data should be combined.

Here, morphological and molecular data sets concerning 27 arthropod higher taxa are compared. Among different proposed tests for data heterogeneity, a bootstrap analysis was selected as proposed by de Queiroz (1993). If there is high support for conflicting clades, then the data are not combined before analysis, but consensus techniques are used instead to summarize those portions of the separate trees that agree. All clades with bootstrap support lower than some arbitrary threshold value are compressed to unresolved polytomies and a combinable-components ('semistrict') consensus tree is derived from these reduced trees. We then ask what the differences are between trees derived by TE, TC, and the present modification of CC approaches (Figure 9.1).

## 9.2 ARTHROPOD PHYLOGENY

### 9.2.1 MORPHOLOGICAL HYPOTHESES

Morphologists have never reached general agreement on the phylogeny of arthropods (for higher taxa nomenclature, see Table 9.1). Three basic hypotheses are considered here (Štys and Zrzavý, 1994):

1. Arthropods are monophyletic and split into Arachnomorpha and Mandibulata (Boudreaux, 1979; Weygoldt, 1986; Wägele, 1993; Wheeler *et al.*, 1993).
2. Arthropods are monophyletic and split into Atelocerata and Schizoramia (Cisne, 1974; Briggs *et al.*, 1992; Budd, 1993; Wills *et al.*, 1994, 1995).

*Arthropod Relationships*, Systematics Association Special Volume Series 55, Edited by R.A. Fortey and R.H. Thomas. Published in 1997 by Chapman & Hall, London. ISBN 0 412 75420 7

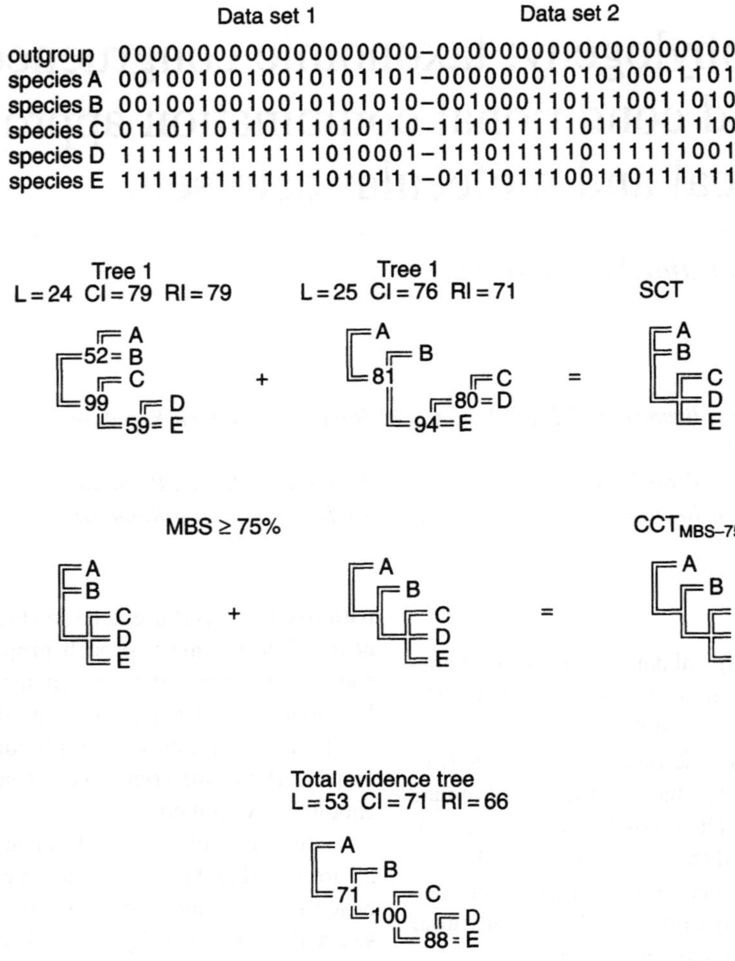


**Figure 9.1** Schematic showing the methodology used in this paper. MBS, minimal bootstrap value (numbers on nodes), SCT, strict consensus tree, $CCCT_{MBS\text{-}75}$, combinable-components consensus tree (including only clades of MBS ⩾ 75%).

3. Arthropods are polyphyletic and should be considered several independent phyla, viz. Uniramia, Crustacea and 'Chelicerata' (= either Arachnomorpha or Euchelicerata), or Uniramia and Schizoramia (Anderson, 1973; Manton, 1977; Bergström, 1979; Budd, 1993).

### 9.2.2 MOLECULAR HYPOTHESES

To date, molecular cladistic studies agree that the extant arthropods are monophyletic (Field *et al.*, 1988; Turbeville *et al.*, 1991; Ballard *et al.*, 1992; Adoutte and Philippe, 1993; Wheeler *et al.*, 1993; Boore *et al.*, 1995; Friedrich and Tautz, 1995; Giribet *et al.*, 1996) and that they include the Pentastomida (as modified crustaceans: Abele *et al.*, 1989, 1992; Giribet *et al.*, 1996). One analysis includes the Onychophora as an arthropod ingroup (Ballard *et al.*, 1992), but according to Wheeler *et al.* (1993) the Euarthropoda and Onychophora are probable sister groups, and Giribet *et al.* (1996) question arthropod affinities of the Onychophora. Molecular cladists have not obtained the Atelocerata as a monophyletic group; they usually agree that the Crustacea and Hexapoda form a clade, sometimes with crustaceans paraphyletic with respect to either monophyletic or polyphyletic hexapods (Adoutte and Philippe, 1993; Averof and Akam, 1995b; Friedrich and Tautz, 1995; Telford and Thomas, 1995; Giribet *et al.*, 1996). The myriapods then represent a more basal branch, either a sister group of all other Euarthropoda (Field *et al.*, 1988; Turbeville *et al.*, 1991; Ballard *et al.*, 1992), or of the Euchelicerata (Friedrich and Tautz, 1995; Giribet *et al.*, 1996).

Wägele and Stanjek (1995) have been able to produce a traditional tree topology (including Euarthropoda, Mandibulata, and Atelocerata) after application of different alignment parameters to the mitochondrial 12S rRNA data published by Ballard *et al.* (1992). However, the poorly resolved strict consensus of their fundamental trees is compatible with both the traditional textbook topology and Ballard *et al.*'s (1992) unorthodox tree.

**Table 9.1** Nomenclature of arthropod higher taxa.

| *Taxon* | *Author* | *Composition* |
|---|---|---|
| Arachnomorpha | Heider, 1913 | Pycnogonida – Euchelicerata |
| Atelocerata | Heymons, 1901 | Myriapoda–Hexapoda |
| Branchiopoda | Latreille, 1817 | Anostraca–Diplostraca– Notostraca |
| Cormogonida | **New name** | Euchelicerata–Mandibulata (*kormos* stem, trunk, *gonos* = sex; indicating the usual position of their gonopores on trunk segments, unlike the Pycnogonida with gonopores situated on free leg bases) |
| Crustacea | Pennant, 1777 | Remipedia–Eucrustacea |
| Diplura | Börner, 1904 | Campodeina–Japygina |
| Euarthropoda | Cuenot, 1949 | Arachnomorpha–Mandibulata |
| Euchelicerata | Weygoldt, 1986 | Xiphosura–Arachnida |
| Eucrustacea | Zimmer, 1926 | Maxillopoda–Thoracopoda |
| Hexapoda | Latreille, 1825 | Parainsecta–Diplura–Ectognatha |
| Ichthyostraca | **New name** | Branchiura–Pentastomida (*ichthys* fish; indicating their unique parasitism on primarily aquatic fish-like chordates or their direct descendants) |
| Malacostraca | Latreille, 1806 | Eumalacostraca–Leptostraca |
| Mandibulata | Snodgrass, 1938 | Myriapoda–Hexapoda–Crustacea |
| Maxillopoda | Dahl, 1956 | Mystacocarida–Copepoda–Thecostraca– Tantulocarida–Oligostraca |
| Myriapoda | Latreille, 1796 | Chilopoda–Progoneata |
| Oligostraca | **New name** | Ostracoda–Ichthyostraca (*oligos* few; indicating reduced number of their true body segments) |
| Panarthropoda | Nielsen, 1995 | Tardigrada–Onychophora– Euarthropoda |
| Pancrustacea | Zravý and Štys, 1977 | Hexapoda–Crustacea (indicating the possible crustacean origin of hexapods; even if the true crustaceans are monophyletic the hexapod ancestor was probably typologically crustacean–like). |
| Parainsecta | Kukalová–Peck, 1987 | Collembola–Protura |
| Progoneata | Pocock, 1893 | Symphyla–Pauropoda–Diplopoda |
| Schizoramia | Hessler and Newman, 1975 | Arachnomorpha–Crustacea |
| Thoracopoda | Hessler, 1975 | Malacostraca–Cephalocarida– Branchiopoda |
| Uniramia | Manton, 1972 | Onychophora–Atelocerata |

### 9.2.3 TOTAL EVIDENCE HYPOTHESIS

There is little agreement between morphologists and molecular biologists. Wheeler has attempted a synthesis of morphological and molecular phylogenetic signals (Wheeler *et al.*, 1993; see also Wheeler, 1994). They used a TE approach with morphological (100 characters) and molecular data (18S rDNA and polyubiquitin sequences), and obtained a mostly traditionally ordered tree (including Euarthropoda, Arachnomorpha, Mandibulata and Atelocerata). Interestingly, the TE data set provides the same tree as the morphological data set alone. The 18S rDNA data set yields the euarthropod clade as an unresolved polytomy (Pycnogonida, Euchelicerata, Crustacea, Hexapoda, Myriapoda), the polyubiquitin (UBQ) data set alone gives obviously absurd results, and the combined 18S rDNA–UBQ data set provides a tree differing from the TE tree only in the pattern of mandibulate interrelationships (viz. Myriapoda versus Pancrustacea). Successive approximation character weighting (Farris, 1969) applied to the combined 18S rDNA–UBQ data set alters the tree by pulling the Pycnogonida to the base of the arthropods.

## 9.3 MATERIAL AND METHODS

### 9.3.1 TERMINAL TAXA (OTUs)

Twenty-seven terminal taxa (operational taxonomic units, OTUs) were selected so as to be monophyletic with reasonable credibility, and to cover the basic morphological/ecological diversity of the arthropods: Anostraca (ANO), Arachnida (ARA), Branchiura (BRA), Campodeina (CAM = Diplura part.), Cephalocarida (CEP), Chilopoda (CHI), Collembola (COL), Copepoda (COP), Diplopoda (DIP), Diplostraca (DIS = Conchostraca + Cladocera), Ectognatha (ECT = Insecta s.str.), Eumalacostraca (EUM), Japygina (JAP = Diplura part.), Leptostraca (LEP = Phyllocarida), Mystacocarida (MYS), Notostraca (NOT), Onychophora (ONY), Ostracoda (OST), Pauropoda (PAU), Pentastomida (PEN), Protura (PRO), Pycnogonida (PYC), Remipedia (REM), Symphyla (SYM), Tantulocarida (TAN), Thecostraca (THE = Cirripedia s.lat.), and Xiphosura (XIP). However, monophyly of some of these OTUs is disputable, which may cause distortion of results (Arachnida, Thecostraca, perhaps also Ostracoda and Japygina).

### 9.3.2 DATA

The morphological data set includes 305 structural, developmental, physiological, neurobiological, spermatological, and ecological characters extracted from the literature (see http://www.systassoc.chapmanhall.com). Several novel characters concerning the arrangement of mitochondrial genes (Boore *et al.*, 1995) and the activity of genes involved in body segmentation, tagmatization, and limb formation (Whitington *et al.*, 1991; Averof and Akam, 1995a,b; Panganiban *et al.*, 1995; Popadić *et al.*, 1996) were included. We included all characters used for phylogenetic analyses by previous authors, even if their methodology was not cladistic, e.g. characters concerning locomotory and jaw-operation patterns (Manton, 1977) and fate-map organization (Anderson, 1973).

Sequences from six different molecules were obtained from the GenBank and from Wheeler *et al.* (1993). A summary of the sequences used is in Table 9.2 (see http://www.systassoc.chapmanhall.com for complete data). The homologous sequences were roughly aligned and spliced using CLUSTALV (Higgins *et al.*, 1992) and then more precisely aligned by MALIGN (Wheeler and Gladstein, 1992) (gap : change cost ratio, 3 : 1, transversion : transition ratio, 2 : 1). The regions that could not be aligned confidently were excluded from analysis. After excluding non-informative (invariant and autapomorphic) positions, presence of a gap was treated as a fifth character state. If gaps contained more than one position, the additional gap sites were treated as missing entries. Sequences for each OTU were arranged tandemly. The OTUs do not correspond in morphological and molecular data: the former were selected for higher clades' hypothetical groundplans and the latter for individual species. We computed a consensus molecular sequence for all species within an OTU, despite the risk of circularity involved in this method depending on previous phylogenetic hypotheses not only about the monophyly of the OTU but also about its inner cladistic structure. For example, the consensus sequence of true insects (Ectognatha) is not simply a set of character states shared by all studied insect species but rather a set of character states shared by at least one pterygote insect with its 'apterygote' outgroup (either Zygentoma or Archaeognatha). Where no well-established phylogenetic hypothesis is available, the consensus sequence of an OTU includes only nucleotides shared by all its species.

**Table 9.2** Number of species analysed concerning individual molecular sequences

| | *12S* | *16S* | *28S* | *18S* | *5.8S* | *UBQ* |
|---|---|---|---|---|---|---|
| | | | *Outgroups* | | | |
| Chordata | 1/GB | 1/GB | 1/GB | 1/GB | 1/GB | – |
| Mollusca | – | 1/GB | 1/GB | 1/GB | 1/GB | 2/W |
| Annelida | 1/GB | – | 1/GB | 1/GB | – | 3/W |
| | | | *Ingroups* | | | |
| Onychophora | 3/GB | – | – | 2/W | – | 2/W |
| Pycnogonida | – | – | – | 1/W | – | 1/W |
| Xiphosura | – | 2/GB | 1/GB | 1/GB | – | 1/W |
| Arachnida | 1/GB | 3/GB | 4/GB | 3/GB | 1/GB | 4/W |
| Anostraca | 1/GB | 1/GB | 1/GB | 2/GB | 1/GB | – |
| Eumalacostraca | 1/GB | 1/GB | 1/GB | 7/GB | – | 1/W |
| Thecostraca | 2/GB | – | – | 6/GB | – | 1/W |
| Branchiura | – | – | – | 1/GB | – | – |
| Pentastomida | – | – | – | 1/GB | – | – |
| Ostracoda | – | – | – | 1/GB | – | – |
| Chilopoda | 2/GB | – | 1/GB | 1/GB | – | 1/W |
| Diplopoda | – | – | 2/GB | 2/GB | – | 1/W |
| Collembola | – | – | 1/GB | 1/GB | – | – |
| Ectognatha | 4/GB | 3/GB | 6/GB | 7/GBW | 4/GB | 8/W |

12S, mt 12S rDNA, 16S, mt 16S rDNA, 28S, 28S rDNA, 18S, 18S rDNA, 5.8S, 5.8S rDNA, UBQ, polyubiquitin; GB, sequence obtained from the GenBank, W, sequence obtained from Wheeler *et al.* (1993). (see http://www.systassoc.chapmanhall.com for the complete list of species)

### 9.3.3 CLADISTIC ANALYSES

Maximum parsimony analyses were performed with molecular characters treated as unordered, the transversion : transition ratio set to either 1 : 1 or 2 : 1, and morphological characters were treated either as unordered or ordered. Several different outgroups were used with the morphological data set: 'og1' includes the hypothetical most plesiomorphic states of all characters ('?' when disputable). The other outgroups represent consensuses of character states in presumably primitive arthropods: og2 = og1-ONY; og3 = og1-ONY-XIP-REM-CEP-CHI; og4 = og1-ONY-XIP-REM-CEP-ANO-PEN-CHI. The molecular outgroup is a consensus of all the included non-arthropod sequences.

Five basic cladistic analyses were performed on: (i) the morphological data set of the 14 'molecular' OTUs (Figure 9.2(a)); (ii) the molecular data set of the 14 OTUs (Figure 9.2(b)); (iii) the whole data set of the 14 OTUs (Figure 9.2(c)); (iv) the morphological data set of all 27 OTUs (Figure 9.3(a)); and (v) the whole data set of the 27 OTUs (Figure 9.3(b)). The Hennig86 computer program (Farris, 1988) was used for all cladistic analyses ('mh*; bb*' option, the heuristic tree search involving the construction of multiple trees and branch-swapping). Successive approximation character weighting (SACW; Farris, 1969) was used for determination of the best supported tree from sets of original trees.

### 9.3.4 BOOTSTRAP ANALYSIS

The bootstrap values used here are the proportions of trees within a replicate showing a clade, averaged across all replicates. Replicates are produced by randomly sampling the data with replacement. In the minimal bootstrap (MBS) analysis, a

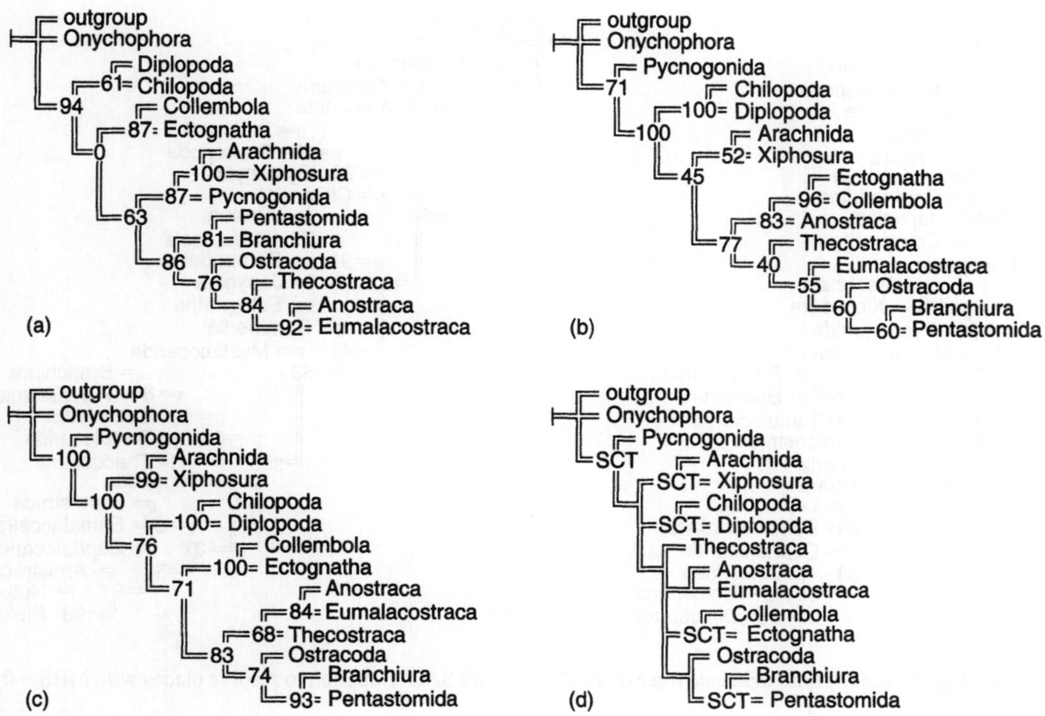


**Figure 9.2** (a) Molecular, (b) morphological and (c) TE trees of the 14 'molecular' OTUs. (MBS values are indicated, SACW applied to resolve clades with MBS = 0). (d) A tree containing all components which appear in at least one CCCT (SCT — components which appear in SCT as well)

strict consensus tree (SCT) is produced from all MEPTs within a replicate, and only the clades that are included in the SCT are considered supported by this replicate. The HEYJOE programme (Siddall, 1995; Internet: 'zoo.toronto.edu') was used ('mh*; bb*; n' option; 1000 replications).

### 9.3.5 CONSENSUS TREES

Two types of consensus trees were used to summarize sets of trees derived from morphological versus molecular data sets. The strict consensus tree (SCT) is derived from combining only those components from a set of trees that appear in all of the fundamental trees. The combinable-components consensus tree (CCCT) is derived by including those components from a set of trees that are not contradictory (Figure 9.1).

In the following analyses all components with MBS values less than a given arbitrarily chosen value (e.g. MBS < 70%) are compressed to unresolved polytomies, and a CCCT ('$CCCT_{MBS-70}$') is computed from the reduced morphological and molecular trees. Two 'relative' MBS values were also used: only clades belonging to one-third (or to two-thirds) of best-supported components are retained, and the reduced trees are then combined by CCCT ('$CCCT_{MBS-1/3}$', '$CCCT_{MBS-2/3}$').

## 9.4 RESULTS

### 9.4.1 THE MOLECULAR DATA SET

The analysis of purely molecular data (transversion: transition ratio 2:1) provided a tree including the Euarthropoda, Myriapoda, Euchelicerata and Hexapoda, but several unorthodox arrangements were obvious, viz. 'Cormogonida', Euchelicerata–Pancrustacea, and Pancrustacea (Figure 9.2(b); Table 9.3). If the transversion : transition ratio is set to 1 : 1, the resulting tree has the same topology as that above, with one exception: the eumalacostracans are removed to a more basal position and become a sister group of all other Pancrustacea.

### 9.4.2 THE MORPHOLOGICAL DATA SET

The unordered morphological analysis performed separately for the 'molecular' OTUs (to be directly compared with the above molecular tree) provides trees that include monophyletic Euarthropoda, Hexapoda, Myriapoda and Schizoramia (Figure 9.2(a)). The interrelationships among myriapods, hexapods and schizoramians are unresolved. When SACW is used, hexapods group with schizoramians in a single resulting tree. If characters are treated as ordered, a monophyletic Atelocerata is included.

The purely morphological analysis of all 27 OTUs

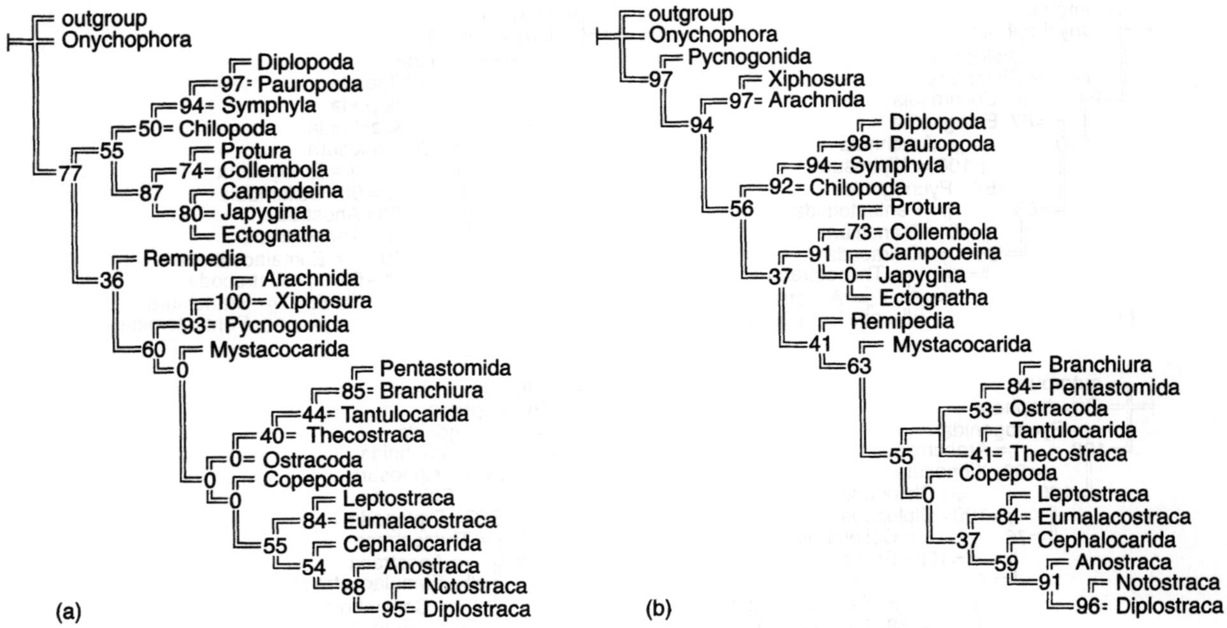


**Figure 9.3** (a) Morphological and (b) total evidence trees of all OTUs (with SACW applied to resolve clades with MBS = 0). MBS values indicated.

provides very similar results, regardless whether ordered or unordered options are used. They do disagree on schizoramian interrelationships as they are resolved after the application of SACW: REM-(Arachnomorpha-(MYS-((OST-(THE-(TAN-(BRA-PEN))))-(COP-Thoracopoda)))) in the unordered analysis (Figure 9.3(a)); REM-((Arachnomorpha-MYS)-(Thoracopoda-(COP-(OST-THE-TAN-(BRA-PEN )))))) in the ordered analysis. Thus, they agree on the monophyly of the Schizoramia, Arachnomorpha–Eucrustacea, Arachnomorpha, Eucrustacea less Mystacocarida, and Ostracoda–(Thecostraca–Tantulocarida–Ichthyostraca).

Both unordered purely morphological analyses of 14 and 27 OTUs are similar to each other except for crustacean relationships (Table 9.3). Even within the purely morphological data set there are considerable differences in the number of 'missing entries' per OTU, and the high proportion of missing entries may cause some artificial groupings in the tree. All poorly known and/or highly diverse OTUs, and OTUs of unstable position were then analysed separately: a cladistic analysis was performed for each such OTU using only characters positively known for it. In these partial analyses pycnogonids group with Euchelicerata, remipedes with Eucrustacea, cephalocarids with Branchiopoda, notostracans with Diplostraca, mystacocarids with Copepoda–Thoracopoda–Thecostraca–Tantulocarida–Oligostraca, thecostracans with Tantulocarida–Ichthyostraca, tantulocarids with Ichthyostraca, copepods with Thoracopoda, ostracods with Thecostraca–Tantulocarida–Ichthyostraca, centipedes with Progoneata, proturans with Collembola, and japygines within an unresolved trichotomy with campodeines and true insects (even after application of SACW). All these positions are compatible or even identical with those within the original morphological tree (Figure 9.3(a)), except for the problematic relative positions of Thecostraca, Tantulocarida, Ostracoda and Ichthyostraca.

### 9.4.3 COMBINED MOLECULAR AND MORPHOLOGICAL RELATIONSHIPS

The molecular and morphological trees are very different and are well-supported by their basic data, as indicated by high MBS values for competing Cormogonida and Arachnomorpha clades (Figure 9.2(a,b)). This suggests that combining both data sets into a single data matrix is not legitimate. On the other hand, the SCT of both trees (Figure 9.2(d): 'STC') includes only a few clades.

The TE analysis of the 14 'molecular' OTUs provides a tree including both 'molecular' and 'morphological' clades (Figure 9.2(c)). A single clade, the Mandibulata, appears in the TE trees but is not included in any of the fundamental trees.

The topology of a $CCCT_{MBS}$ of the molecular and morphological trees from which the least supported clades were excluded depends on which clades are considered 'less supported'. $CCCT_{MBS\text{-}40}$ to $CCCT_{MBS\text{-}70}$ are identical to the SCT; in $CCCT_{MBS\text{-}75}$ and $CCCT_{MBS\text{-}80}$, the Pancrustacea appears as additional well-supported component; in $CCCT_{MBS\text{-}90}$,

**Table 9.3** Presence of clades in individual arthropod trees

| | 'Molecular' OTUs[a] | | | | | All OTUs[a] | |
|---|---|---|---|---|---|---|---|
| *Tree* | *morp* | *mol* | *TC* | *CC* | *TE* | *morp* | *TE* |
| Number of OTUs | 14 | 14 | 14 | 14 | 14 | 27 | 27 |
| Number of characters | 305[b] | 616 | – | – | 921 | 305[b] | 921 |
| Number of MEPTs | 2 | 1 | – | – | 1 | 6 | 5 |
| L | 593 | 1,148 | – | – | 1,771 | 855 | 2,040 |
| CI | 0.59 | 0.63 | – | – | 0.61 | 0.45 | 0.54 |
| RI | 0.51 | 0.45 | – | – | 0.45 | 0.57 | 0.50 |
| *Presence of a clade* | | | | | | | |
| ANO-Hexapoda[c] | 0 | 1 | ? | 0 | 0 | 0 | 0 |
| Arachnomorpha | 1 | 0 | ? | 0 | 0 | 1 | 0 |
| Atelocerata | ? | 0 | ? | ? | 0 | 1 | 0 |
| *Cormogonida | 0 | 1 | ? | (1) | 1 | 0 | 1 |
| *Crustacea | 1 | 0 | ? | ? | 1 | 0 | 1 |
| *Euarthropoda | 1 | 1 | 1 | 1 | 1 | 1 | 1 |
| *Euchelicerata | 1 | 1 | 1 | 1 | 1 | 1 | 1 |
| Euchel.–Pancrustacea | 0 | 1 | ? | ? | 0 | 0 | 0 |
| EUM–Oligostraca[d] | 0 | 1 | ? | 0 | 0 | 0 | 0 |
| *Hexapoda | 1 | 1 | 1 | 1 | 1 | 1 | 1 |
| Hexapoda–Schizoramia | 1 | 0 | ? | 0 | 0 | 0 | 0 |
| *Ichthyostraca | 1 | 1 | 1 | 1 | 1 | 1 | 1 |
| *Mandibulata | 0 | 0 | ? | ? | 1 | 0 | 1 |
| *Myriapoda | 1 | 1 | 1 | 1 | 1 | 1 | 1 |
| *Oligostraca | 0 | 1 | ? | (1) | 1 | 0 | 1 |
| OST–THE–Thoracopoda[e] | 1 | 0 | ? | 0 | 0 | 0 | 0 |
| *Pancrustacea | 0 | 1 | ? | 1 | 1 | 0 | 1 |
| Schizoramia | 1 | 0 | ? | 0 | 0 | 1 | 0 |
| THE–EUM–Oligostraca[de] | 0 | 1 | ? | 0 | 0 | 0 | 0 |
| THE–Thoracopoda[e] | 1 | 0 | ? | ? | 1 | 0 | 0 |
| *Thoracopoda | 1 | 0 | ? | (1) | 1 | 1 | 1 |
| *Branchiopoda | – | – | – | – | – | 1 | 1 |
| CAM–JAP–ECT | – | – | – | – | – | 1 | ? |
| *CEP–Branchiopoda | – | – | – | – | – | 1 | 1 |
| *COP–Thor.–THE–TAN–Olig. | – | – | – | – | – | ? | 1 |
| COP–Thoracopoda | – | – | – | – | – | ? | ? |
| *PAU–DIP | – | – | – | – | – | 1 | 1 |
| *Eucrustacea | – | – | – | – | – | ? | 1 |
| *Malacostraca | – | – | – | – | – | 1 | 1 |
| *NOT–DIS | – | – | – | – | – | 1 | 1 |
| *Parainsecta | – | – | – | – | – | 1 | 1 |
| *Progoneata | – | – | – | – | – | 1 | 1 |
| TAN–Ichthyostraca | – | – | – | – | – | 1 | 0 |
| *THE–TAN | – | – | – | – | – | 0 | 1 |
| THE–TAN–Ichthyostraca | – | – | – | – | – | 1 | 0 |
| *THE–TAN–Oligostraca | – | – | – | – | – | ? | 1 |
| **Number of resolved clades** | 12 | 12 | 5 | 9 | 12 | 19 | 22 |

morp, a tree derived from a morphological data set; mol, a tree derived from a molecular data set; TC, strict consensus of morphological and molecular trees (taxonomic congruence approach); CC, combinable–components consensus of reduced morphological and molecular trees (conditional combination approach); TE, a tree derived from a combined moprhological-molecular data set (total evidence approach); 0, absent; 1, present, (1), rarely present, ?, presence possible but not positively indicated (unresolved); –, inapplicable; OTU, operational taxonomic unit ('terminal taxon'); MEPT, maximum equally parsimonious tree; L, length of MEPT(s); CI, consistency index of MEPT(s); RI, retention index of MEPT(s). Clades supported by all, or at least not contradicted by any, combined morphological–molecular analyses (TE, SCT, CCCT) are asterisked.

[a] 'All OTUs' are Anostraca+, Arachnida+, Branchiura+, Campodeina, Cephalocarida, Chilopoda+, Collembola+, Copepoda, Diplopoda+, Diplostraca, Ectognatha+, Eumalacostraca+, Japygina, Leptostraca, Mystacocarida, Notostraca, Onychophora+, Ostracoda+, Pauropoda, Pentastomida+, Protura, Pycnogonida+, Remipedia, Symphyla, Tantulocarida, Thecostraca+, and Xiphosura+; 'molecular OTUs' are marked by '+' in the above list;

[b] The list of morphological characters and the data matrix are available from http://www.systassoc.chapmanhall.com;

[c] NOT-DIS and possibly also CEP should probably be included;

[d] LEP should probably be included;

[e] TAN should probably be included.

Ichthyostraca are absent while Cormogonida and Thoracopoda appear. In the CCCTs based on 'relative MBS' values, four (Euarthropoda, Myriapoda, Euchelicerata, Hexapoda) and seven clades (the above plus Pancrustacea, Oligostraca, Ichthyostraca) are distinguished in $CCCT_{MBS-1/3}$ and $CCCT_{MBS-2/3}$, respectively. A tree containing all components which appear in at least one $CCCT_{MBS}$ is shown in Figure 9.2(d).

Application of the TE approach to the combined data set of all 27 OTUs provides trees whose topology is compatible with that of the TE tree of the 14 'molecular' OTUs (Figure 9.3(b)).

## 9.5 DISCUSSION: RELATIONSHIPS OF ARTHROPOD CLADES

The clades derived from individual analyses are summarized in Table 9.3. Purely morphological and purely molecular analyses provide rather dissimilar trees (Figures 9.2(a) and 9.3(a) versus Figure 9.2(b)). The TE trees (Figures 9.2(c) and 9.3(b)) include components derived from both data sets, e.g. 'molecular' (Cormogonida, Pancrustacea, Oligostraca) and 'morphological' (Thoracopoda). In general, both purely morphological trees of 14 and 27 OTUs are quite similar one to another (Figures 9.2(b) and 9.3(a)), as are both TE trees (Figures 9.2(c) and 9.3(b)). Only the robustness of a monophyletic Atelocerata (in morphological trees) and the crustacean interrelationships (in both morphological and combined trees) appear to be strongly affected by the list of analysed OTUs.

The Euarthropoda, Euchelicerata, Myriapoda, Hexapoda and Ichthyostraca represent the most strongly supported clades, shared by all the trees; the Cormogonida, Mandibulata, Pancrustacea, Crustacea, Oligostraca and Thoracopoda represent additional clades that are supported, or at least not contradicted, by any combined analysis (SCT, CCCT, TE).

### 9.5.1 MYRIAPODS AND HEXAPODS

The Myriapoda have always been found monophyletic in these analyses. The alternative hypotheses of close relationships of the Symphyla to centipedes within monophyletic myriapods (Boudreaux, 1979), or of progoneates (or symphylans) to hexapods (Sharov, 1966; Dohle, 1988; Kraus and Kraus, 1994), are supported by neither morphological nor molecular data. However, the hexapod affinity of the Symphyla cannot be excluded until more molecular information is available (G. Giribet, personal communication).

The monophyly of Hexapoda is not falsified by the present analysis, however, it has been questioned by some morphologists (reviewed by Štys and Zrzavý, 1994) and molecular biologists [Giribet *et al.*, 1996: (ECT-ANO)-(PEN-(COL-(THE-EUM)))]. The hexapods seem to be primarily subdivided into Parainsecta and CAM-JAP-ECT clade ('Insecta' according to Kukalová-Peck, 1991). Monophyly of the latter clade is disputable, and its inner structure remains unresolved. The alternative hypotheses (either Campodeina or Campodeina–Japygina as sister groups of the Parainsecta) (Štys and Zrzavý, 1994; Kristensen, 1995) are not supported by the present study.

### 9.5.2 ATELOCERATA VERSUS PANCRUSTACEA

Monophyly of the Atelocerata, however supported by 'important' morphological characters, such as reduction of 2nd antennae, presence of the tentorium, Tömösváry's organ, and Malpighian tubules (Boudreaux, 1979; Weygoldt, 1986; Kukalová-Peck, 1991; Wägele, 1993), seems hardly defensible when confronted with the molecular data (but see Wheeler *et al.*, 1993; Wägele and Stanjek, 1995, for different opinions).

The alternative concept of the Pancrustacea seems supported by most molecular papers (Field *et al.*, 1988; Turbeville *et al.*, 1991; Ballard *et al.*, 1992; Averof and Akam, 1995b; Boore *et al.*, 1995; Friedrich and Tautz, 1995; Telford and Thomas, 1995; Giribet *et al.*, 1996). The possible morphological autapomorphies of pancrustaceans are insufficiently known from the comparative viewpoint; they concern eye ultrastructure (Paulus, 1979; Osorio and Bacon, 1994; Osorio *et al.*, 1995), neurogenic pattern-forming processes (Whitington *et al.*, 1991; Osorio *et al.*, 1995), and suppression of the distal mandibular segments (Popadić *et al.*, 1996).

### 9.5.3 PANCRUSTACEAN RELATIONSHIPS

The cladistic structure of Pancrustacea is uncertain and the crustaceans may well be paraphyletic with respect to the hexapods, a hypothesis neither contradicted nor strongly supported by the present analysis.

The monophyletic Branchiopoda may group together with cephalocarids and, more basally and less certainly, with the undoubtedly monophyletic Malacostraca. The alternative hypothesis supposing branchiopod–cephalocarid affinities of the Leptostraca (Schram, 1986) is supported neither by the present data set, nor by the molecular data of Spears and Abele (1997, this volume). However, the interrelationships of branchiopods, cephalocarids and malacostracans remain questionable – note the competing molecular hypothesis about monophyly of Branchiopoda–Hexapoda (Adoutte and Philippe, 1993; Friedrich and Tautz, 1995) (Figure 9.2(b)) or of Branchiopoda–Hexapoda part. (Giribet *et al.*, 1996). The monophyly of a branchiopod–hexapod clade is supported by the apparently homologous body tagmosis and position of genital segments (Averof and Akam, 1995a; Zrzavý and Štys, 1997).

Monophyly of the Maxillopoda is very improbable

(Abele *et al.*, 1992; Giribet *et al.*, 1996), and relationships among individual 'maxillopodan' clades are uncertain; several greatly incompatible morphological hypotheses on maxillopod relationships are reviewed by Huys *et al.* (1993). Only Ichthyostraca, Oligostraca, and perhaps Thecostraca–Tantulocarida may be monophyletic groups. The thecostracan–tantulocarid clade is potentially supported by the shared sexually dimorphic position of gonopores (Huys *et al.*, 1993; Schram and Høeg, 1995). The monophyly of the oligostracan clade, only distantly related to the Copepoda–Thecostraca, has also been proposed by another molecular study (Abele *et al.*, 1992). The Oligostraca, in addition, share short oligomeric trunks, which are not obvious in pentastomids owing to elongation of their non-segmental posterior end (Walossek and Müller, 1994). The Ichthyostraca are characterized by reduction of antennules, unpaired ovaries, replacement of the sperm acrosome by a pseudoacrosome (Storch and Jamieson, 1992), and parasitism on vertebrates. Moreover, their monophyly is also supported by molecular data (Abele *et al.*, 1989, 1992; Spears and Abele, 1997, this volume) (Figure 9.2(b)).

## 9.5.4 THE BASAL PHYLOGENY OF THE PANARTHROPODA

The basal sequence of clades splitting off from the panarthropod stem is probably Onychophora, Pycnogonida, Euchelicerata, Myriapoda and Pancrustacea, with an uncertain position of remipedes. The Tardigrada, not included in the present study, are now generally regarded as belonging to the panarthropod clade (Nielsen, 1995; Giribet *et al.*, 1996).

Monophyly of Arachnomorpha has rarely been questioned by morphologists, but the Cormogonida hypothesis emerges from our molecular and TE analyses, in accord with the tree of Sharov (1966). The combined 18S rDNA–UBQ tree published by Wheeler *et al.* (1993) includes the Arachnomorpha, but pycnogonids may be pulled out of this clade to the arthropod base after application of SACW.

The monophyly of the Myriapoda–Euchelicerata clade suggested by several recent molecular analyses (Friedrich and Tautz, 1995; Giribet *et al.*, 1996) may be supported by a few characters concerning neuroendocrinological system and eyes (Paulus, 1979), though, in general, these characters are not well known comparatively and are of disputed homology. On the other hand, we believe that monophyly of the Mandibulata is reasonably well supported by morphological, developmental, endocrinological and neurobiological data (reviewed by Wägele, 1993), by patterns of mitochondrial DNA rearrangements (Boore *et al.*, 1995), as well as by TE analyses (Wheeler *et al.*, 1993). The third possible phylogeny including monophyletic schizoramians has been provided by our morphological trees (Figures 9.2(b) and 9.3(a)) but not by combined analysis.

All our analyses, either morphological, molecular or combined, suggest that Euarthropoda are monophyletic and sister group of the Onychophora. This viewpoint is shared by most authors, except for the 'polyphyletic' school (Anderson, 1973; Manton, 1977; Fryer, 1997, this volume). We have found no support either for the Uniramia, or for including onychophorans within euarthropods (Ballard *et al.*, 1992). The possible relationships of onychophorans and polychaetes (as suggested by Giribet *et al.*, 1996) cannot be supported by analysis of the present list of OTUs. We consider this possibility improbable when confronted with the other studies proposing close relationships of onychophorans and euarthropods (Boudreaux, 1979; Weygoldt, 1986; Ballard *et al.*, 1992; Wheeler *et al.*, 1993; Nielsen, 1995; Wills *et al.*, 1995), and/or only distant relationships of panarthropods and annelids (e.g. Field *et al.*, 1988; Eernisse *et al.*, 1992; Winnepenninckx *et al.*, 1995).

## 9.5.5 POSITION OF FOSSIL CLADES

Fossil clades have been studied cladistically by Briggs *et al.* (1992) and Wills *et al.* (1994, 1995); however, their cladograms, including both fossil and Recent species, and their methodology are incompatible with the present analysis. In general, tardigrades, lobopods (including onychophorans), and extinct dinocarids are shown as successive outgroups of the Euarthropoda. The euarthropods are subdivided to Atelocerata and Schizoramia, the latter group being split into extinct Marrellomorpha, Arachnomorpha (including euchelicerates, trilobites, and other fossils), and 'Crustaceanomorpha' (crustaceans including remipedes and mystacocarids, and many fossils).

It has been demonstrated that inclusion of fossils into cladistic analyses may be essential for a proper understanding of the phylogenetic relationships of extant groups (Novacek, 1992; Eernisse and Kluge, 1993; Wills *et al.*, 1995). Nevertheless, we are not convinced that the critical arthropod fossils are known sufficiently to be included to a combined fossil-extant data matrix. Are, for instance, the 'great appendages' of the possible arachnomorph *Yohoia* homologous to deuterocerebral or tritocerebral, or any other limbs (keeping in mind that even homology of head segments of extant arthropods is quite unclear)? For the present, we prefer to interpolate the critical fossils into a phylogeny based predominantly on living groups.

## ACKNOWLEDGEMENTS

The authors are indebted to R.H. Thomas, P. Štys, P. Švácha, and O. Nedvěd for their comments on drafts of the present paper.

## REFERENCES

Abele, L.G., Kim, W. and Felgenhauer, B.E. (1989) Molecular evidence for inclusion of the phylum Pentastomida in the Crustacea. *Molecular Biology and Evolution*, **6**, 685–91.

Abele, L.G., Spears, T., Kim, W. and Applegate, M. (1992) Phylogeny of selected maxillopodan and other crustacean taxa based on 18S ribosomal nucleotide sequences: a preliminary analysis. *Acta Zoologica*, **73**, 373–82.

Adoutte, A. and Philippe, H. (1993) The major lines of metazoan evolution: summary of traditional evidence and lessons from ribosomal RNA sequence analysis, in *Comparative Molecular Neurobiology* (ed. Y. Pichon), Birkhäuser, Basel, pp. 1–30.

Anderson, D.T. (1973) *Embryology and Phylogeny in Annelids and Arthropods*, Pergamon Press, Oxford.

Averof, M. and Akam, M. (1995a) *Hox* genes and the diversification of insect and crustacean body plans. *Nature*, **376**, 420–3.

Averof, M. and Akam, M., (1995b) Insect–crustacean relationships: insights from comparative developmental and molecular studies. *Philosophical Transactions of the Royal Society of London Series B, Biological Sciences*, **347**, 293–303.

Ballard, J.W.O., Olsen, G.J., Faith, D.P., Odgers, W.A., Rowell, D.M. and Atkinson, P.W. (1992) Evidence from 12S ribosomal RNA sequences that Onychophora are modified arthropods. *Science*, **258**, 1345–8.

Bergström, J. (1979) Morphology of fossil arthropods as a guide to phylogenetic relationships, in *Arthropod Phylogeny* (ed. A.P. Gupta), Van Nostrand Reinhold Co., New York, pp. 3–56.

Boore, J.L., Collins, T.M., Stanton, D., Daehler, L.L. and Brown, W.M. (1995) Deducing the pattern of arthropod phylogeny from mitochondrial DNA rearrangements. *Nature*, **376**, 163–5.

Boudreaux, H.B. (1979) *Arthropod Phylogeny, with Special Reference to Insects*, J. Wiley and Sons, New York.

Briggs, D.E.G., Fortey, R.A. and Wills, M.A. (1992) Morphological disparity in the Cambrian. *Science*, **256**, 1670–3.

Budd, G. (1993) A Cambrian gilled lobopod from Greenland. *Nature*, **364**, 709–11.

Cisne, J. L. (1974) Trilobites and the origin of arthropods. *Science*, **186**, 13–18.

de Queiroz, A. (1993) For consensus (sometimes). *Systematic Biology*, **42**, 368–72.

Dohle, W. (1988) *Myriapoda and the Ancestry of Insects*. C.H. Brookes Memorial Lecture (1986), Manchester Polytechnic, Manchester.

Eernisse D.J. and Kluge A.G. (1993) Taxonomic congruence versus total evidence, and amniote phylogeny inferred from fossils, molecules, and morphology. *Molecular Biology and Evolution*, **10**, 1170–95.

Eernisse, D.J., Albert, J.S. and Anderson, F.E. (1992) Annelida and Arthropoda are not sister taxa – a phylogenetic analysis of spiralian metazoan morphology. *Systematic Biology*, **41**, 305–30.

Farris, J.S. (1969) A successive approximations approach to character weighting. *Systematic Zoology*, **18**, 374–85.

Farris, J.S. (1988) Hennig86 version 1.5 manual, software and MSDOS program, Published by the author, New York.

Field, K.G., Olsen, G.J., Lane, D.J., Giovannoni, S.J., Ghiselin, M.T., Raff, E.C., Pace, N.R. and Raff, R.A. (1988) Molecular phylogeny of the animal kingdom. *Science*, **239**, 748–53.

Friedrich, M. and Tautz, D. (1995) Ribosomal DNA phylogeny of the major extant arthropod classes and the evolution of myriapods. *Nature*, **376**, 165–7.

Giribet, G., Carranza, S., Baguñá, J., Riutort, M. and Ribera, C. (1996) First molecular evidence for the existence of a Tardigrada + Arthropoda clade. *Molecular Biology and Evolution*, **13**, 76– 84.

Higgins, D. G., Bleasby, A. J. and Fuchs, R. (1992) CLUSTAL V: improved software for multiple sequence alignment. *CABIOS*, **8**, 189–91.

Huelsenbeck, J.P., Bull, J.J. and Cunningham, C.W. (1996) Combining data in phylogenetic analysis. *Trends in Ecology and Evolution*, **11**, 152–8.

Huys, R., Boxshall, G.A. and Lincoln, R.J. (1993) The tantulocaridan life cycle: the circle closed? *Journal of Crustacean Biology*, 13, 432–42.

Kraus, O. and Kraus M. (1994) Phylogenetic system of the Tracheata (Mandibulata): on 'Myriapoda'–Insecta interrelationships, phylogenetic age and primary ecological niches. *Verhandlungen des Naturwissenschaftlichen Vereins in Hamburg, NF*, **34**, 5–31.

Kristensen, N.P. (1995) Forty years' insect phylogenetic systematics: Hennig's 'Kritische Bemerkungen...' and subsequent developments. *Zoologische Beiträge, NF*, **36**, 83–124.

Kukalová-Peck, J. (1991) Fossil history and evolution of hexapod structures, in *The Insects of Australia*, vol. 1 (ed. I.D. Naumann and CSIRO), CSIRO, Melbourne University Press, Carlton, pp. 141–79.

Manton, S.M. (1977) *The Arthropoda: Habits, Functional Morphology, and Evolution*, Clarendon Press, Oxford.

Nielsen, C. (1995) *Animal Evolution: Interrelationships of the Living Phyla*, Oxford University Press, Oxford.

Novacek, M.J. (1992) Fossils as critical data for phylogeny, in *Extinction and Phylogeny* (eds M.J. Novacek and Q.D. Wheeler), Columbia University Press, New York, pp. 46–88.

Osorio, D. and Bacon, J.P. (1994) A good eye for arthropod evolution. *BioEssays*, **16**, 419–23.

Osorio, D., Averof, M. and Bacon, J.P. (1995) Arthropod evolution: great brains, beautiful bodies. *Trends in Ecology and Evolution*, **10**, 449–54.

Panganiban, G., Nagy, L. and Carroll, S.B. (1995) The role of the Distal-less gene in the development and evolution of insect limbs. *Current Biology*, **4**, 671–5.

Paulus, H.F. (1979) Eye structure and the monophyly of Arthropoda, in *Arthropod Phylogeny* (ed. A.P. Gupta), Van Nostrand Reinhold Co., New York, pp. 299–383.

Popadić, A., Rusch, D., Peterson, M., Rogers, B.T. and Kaufman, T.C. (1996) Origin of the mandibulate mandible. *Nature*, **380**, 395.

Schram, F.R. (1986) *Crustacea*, Oxford University Press, New York.

Schram, F.R. and Hoeg, J.T. (1995) New frontiers in barnacle evolution, in *New Frontiers in Barnacle Evolution* (eds F.R. Schram and J.T. Høeg), A.A. Balkema, Rotterdam, pp. 297–312.

Sharov, A.G. (1966) *Basic Arthropodan Stock with Special Reference to Insects*, Pergamon Press, Oxford.

Storch, V. and Jamieson, B.G.M. (1992) Further spermatological evidence for including the Pentastomida (tongue worms) in the Crustacea. *International Journal of Parasitology*, **22**, 95–108.

Štys, P. and Zrzavý, J. (1994) Phylogeny and classification of extant Arthropoda: review of hypotheses and nomenclature. *European Journal of Entomology*, **91**, 257–75.

Telford, M.J. and Thomas, R.H. (1995) Demise of the Atelocerata? *Nature*, **376**, 123–4.

Turbeville, J.M., Pfeifer, D.M., Field, K.G. and Raff, R.A. (1991) The phylogenetic status of arthropods, as inferred from 18S rRNA sequences. *Molecular Biology and Evolution*, **8**, 669–86.

Wägele, J.W. (1993) Rejection of the 'Uniramia' hypothesis and implications of the Mandibulata concept. *Zoologische Jahrbücher, Abteilung für Systematik*, **120**, 253–88.

Wägele, J.W. and Stanjek, G. (1995) Arthropod phylogeny inferred from partial 12S rRNA revisited – monophyly of the Tracheata depends on sequence alignment. *Journal of Zoological Systematics and Evolutionary Research*, **33**, 75–80.

Walossek, D. and Müller, K.J. (1994) Pentastomid parasites from the lower Paleozoic of Sweden. *Transactions of the Royal Society of Edinburgh – Earth Sciences*, **85**, 1–37.

Weygoldt, P. (1986) Arthropod interrelationships – the phylogenetic–systematic approach. *Zeitschrift für Zoologische, Systematik und Evolutionsforschung*, **24**, 19–35.

Wheeler, W.C. (1994) Sequence alignment, parameter sensitivity and the phylogenetic analysis of molecular data. *Systematic Biology*, 44, 321–31.

Wheeler, W.C. and Gladstein, D.S. (1992) MALIGN, program and documentation, American Museum of Natural History.

Wheeler, W.C., Cartwright, P. and Hayashi, C.Y. (1993) Arthropod phylogeny: a combined approach. *Cladistics*, **9**, 1–39.

Whitington, P.M., Meier, T. and King, P. (1991) Segmentation, neurogenesis and formation of early axonal pathways in the centipede, *Ethmostigmus rubripes* (Brandt). *Roux's Archive for Developmental Biology*, **199**, 349–63.

Wills, M.A., Briggs, D.E.G. and Fortey, R.A. (1994) Disparity as an evolutionary index: a comparison of Cambrian and Recent arthropods. *Paleobiology*, **20**, 93–130.

Wills, M.A., Briggs, D.E.G., Fortey, R.A. and Wilkinson, M. (1995) The significance of fossils in understanding arthropod evolution. *Verhandlungen der Deutschen Zoologischen Gesellschaft*, **88**, 203–15.

Winnepenninckx, B., Backeljau, T., Mackey, L.Y., Brooks, J.M., De Wachter, R., Kumar, S. and Garey, J.R. (1995) 18S rRNA data indicate that Aschelminthes are polyphyletic in origin and consist of at least three distinct clades. *Molecular Biology and Evolution*, **12**, 1132–7.

Zrzavý, J. and Štys, P. (1997) The basic body plan of arthropods: insights from evolutionary morphology and developmental biology. *Journal of Evolutionary Biology*, **10**, 353–67.

# 10 The place of tardigrades in arthropod evolution

*R.A. Dewel and W.C. Dewel*

*Department of Biology, Appalachian State University, Boone, NC 28608, USA*
*email: dewelra@appstate.edu*
*Department of Biology, Appalachian State University, Boone, NC 28608, USA*
*email: dewelwc@appstate.edu*

## 10.1 INTRODUCTION

The phylum Tardigrada is an engaging but enigmatic group composed of minute metazoans with four pairs of stubby lobopodous appendages. Although tardigrades are thought to share important derived characters with arthropods, many of the designated attributes are actually plesiomorphies and, as a consequence, the phylogenetic position of the taxon remains uncertain. Embryological data indicating, for example, an enterocoelous formation of mesoderm, are questionable (Marcus, 1929; Pollock, 1975; Nelson, 1982; Bertolani, 1989), and the fossil record from Cretaceous amber (Cooper, 1964) and Quaternary travertine (Durante and Maucci, 1972) reveals only representatives from well-established taxonomic categories. Although similarities between tardigrades and the Middle Cambrian lobopod *Aysheaia* have been recognized (Renaud-Mornant, 1982; Grimaldi de Zio *et al.*, 1987a; Simonetta and Delle Cave, 1991), the relationship of *Aysheaia* to other lobopods (Ramsköld, 1992; Hou and Bergström, 1995) and of both to arthropods remains elusive. This low level of resolution in the phylogenetic position of tardigrades has not been refined by recent molecular studies that place them as either basal to both arthropods and coelomate protostomes (Moon and Kim, 1996) or on a moderately supported clade between Priapulida and Arthropoda (Garey *et al.*, 1996; Giribet *et al.*, 1996; see also Winnepenninckx *et al.*, 1995, for the relationship of Priapulida to arthropods). Unfortunately onychophorans, whose position has been proposed to be either within the arthropods (Ballard *et al.*, 1992) or as the sister group of all arthropods (Wheeler *et al.*, 1993) have not hitherto been included in an analysis with tardigrades.

The present study assesses tardigrade characters both in terms of their suitability for determining the position of the phylum on the lobopod and arthropod clade and for their effectiveness in elucidating the phylogeny of major arthropod groups. The incorporation of lobopod and stem lineage arthropod characters viewed from the tardigrade perspective will aid in unravelling deep arthropod relationships. This study concentrates on characters viewed as relevant to arthropod phylogeny (Greven, 1980; Bertolani, 1982; Nelson, 1982, 1991; Nelson and Higgins, 1990; Dewel *et al.*, 1993 and Kinchin, 1994 for reviews). Additional discussions on tardigrade and arthropod phylogeny have been offered by Weygoldt (1986), Ax (1987), Wright and Luke (1989), Kristensen (1991), Eernisse *et al.* (1992), Backeljau *et al.* (1993), Dzik (1993), Raff *et al.* (1994), Hou *et al.* (1995), Monje-Najera (1995), Nielsen (1995), Wills *et al.* (1995), Budd (1996), Garey *et al.* (1996), Nielsen *et al.* (1996) and Waggoner (1996). Due to the microscopic size of the animals (most are no larger than 100–500 μm) some morphological characters are cellular or even subcellular and can only be analysed using electron microscopes. It is assumed that the group has undergone a significant reduction in size having arisen from ancestors of milli- to centimetric size. Progenesis is a likely evolutionary mechanism producing this miniaturization, and characters are interpreted as potentially appearing less differentiated or specialized and hence more plesiomorphic than expected.

## 10.2. MORPHOLOGY OF TARDIGRADES

### 10.2.1 CHARACTERS

**(a) General**

The tardigrade body is bilaterally symmetrical, generally flattened on the ventral surface and convex on the dorsal surface (Figures 10.1, 10.2(a,b) and 10.3). The body wall is thin, consisting of a simple epidermis and a cuticle that is periodically moulted. Internally, the body cavity is filled with nutrient storage cells and is criss-crossed by numerous somatic muscles. The digestive tract is complete and extends from an anterior or ventral mouth to either a terminal anus positioned between the last pair of legs in heterotardigrades or a ventral, subterminal, cloaca in eutardigrades. The central part of the tardigrade nervous system lies free in the body cavity and consists of a dorsal brain and a ladder-type chain of paired ventral ganglia

*Arthropod Relationships*, Systematics Association Special Volume Series 55, Edited by R.A. Fortey and R.H. Thomas. Published in 1997 by Chapman & Hall, London. ISBN 0 412 75420 7

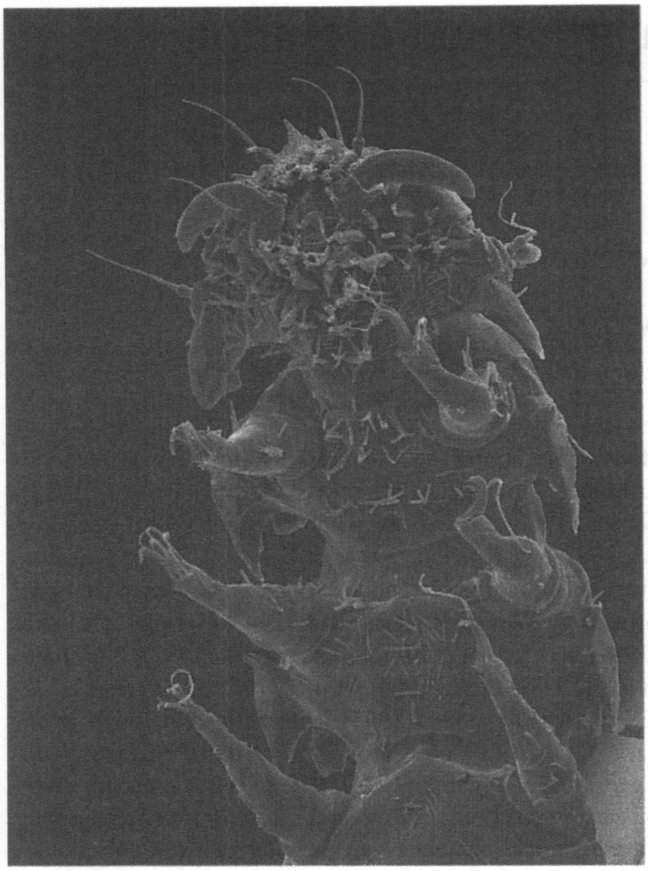

**Figure 10.1** *Stygarctus* sp. Ventral view reveals numerous bacteria decorating the cuticle, subterminal mouth, legs terminating in claws, lateral projections of the dorsal plates and prominent head appendages. Cephalic sensillae include a single median cirrus and paired internal and external cirri and secondary clavae positioned proximally and primary clavae and lateral cephalic cirri located distally on the head appendages. Magnification x4000. (Photograph by David Scharf.)

(Figure 10.3). Although marine arthrotardigrades lack excretory organs, eutardigrades have Malpighian tubules and the terrestrial echiniscids possess two serially repeated epidermal organs that appear to function in excretion and osmoregulation. A testis or ovary overlies the gut; in eutardigrades the gonoducts enter the cloaca, whereas in heterotardigrades they join a single midventral gonopore (Figure 10.3). Probably as a result of miniaturization many tissues exhibit eutely, or a constant cell number.

The phylum was divided into two major classes, the Heterotardigrada and the Eutardigrada (Table 10.1), by Marcus (1929). The Heterotardigrada exhibit greater morphological diversity and within that group the marine species occupy a larger morphospace, varying more in body shape and overall appearance than semi-terrestrial forms (Ramazzotti and Maucci, 1982). This diversity is taken to indicate that the marine arthrotardigrades are the most basal members of the phylum and hence are likely to retain the greatest number of plesiomorphic character states (Renaud-Mornant, 1982; Kristensen and Renaud-Mornant, 1983; Grimaldi de Zio *et al.*, 1987a).

**(b) Segmentation**

Tardigrades are composed of a total of seven or eight somewhat indistinct segments. Only larval or parasitic arthropods or the Cambrian lobopod *Paucipodia* (Chen *et al.*, 1995) have a comparably reduced number of segments. From studies of the nervous system it appears that the cephalon contributes three of the segments (Kristensen and Higgins, 1984, 1989; Dewel and Dewel, 1996). The trunk is formed from either four segments, one for each pair of legs with the gonads and their ducts derived from the last segment, or five segments with the reproductive structures provided by a vestigial segment that is now limbless (Figures 10.2(a,b) and 10.3). Of the two choices, the second appears to be better supported since the fourth ganglia of the ventral chain innervates the gonads whereas a fifth 'nebenganglion' innervates the last pair of legs (Marcus, 1929, Figure 27 for *Macrobiotus*).

**(c) Dorsal plates**

Segmentally arranged dorsal and ventral sclerotized plates have been proposed to be plesiomorphic features of the phylum (Figures 10.1 and 10.2(a,b)) (Kristensen and Higgins, 1984; Kristensen, 1987). Although both dorsal and ventral plates are found in the semi-terrestrial family Echiniscidae, less-derived states probably are represented by the diversity of dorsal plates found in the Stygarctidae and Renaudarctidae. Within these families, the genera *Stygarctus*, *Pseudostygarctus*, *Megastygarctus* and *Renaudarctus* possess additional dorsal intersegmental divisions with nodes (Figure 10.2(b)) (McKirdy *et al.*, 1976; Kristensen and Higgins, 1984). In addition, *Stygarctus*, *Neostygarctus*, *Parastygarctus*, *Pseudostygarctus* together with *Renaudarctus* have flanges or cuticular folds projecting mid-laterally from the dorsal plates (Figures 10.1, 10.2(a,b) and 10.3) (McKirdy *et al.*, 1976; Grimaldi de Zio *et al.*, 1982). Other genera (*Florarctus*, *Actinarctus*, *Paradoxipus* and *Halechiniscus*) have cuticular expansions (alae) that may be homologous to flanges of the stygarctids. Possible homologues of the flanged plates are discussed below.

**(d) Cuticle and epidermis**

The epidermis secretes a complex two- or three-layered cuticle that is bounded apically as in onychophorans and arthropods by a multi-layered epicuticle (Rieger and Rieger, 1977; Greven, 1983; Wright and Luke, 1989). Chitin is present in the cuticle of all three groups where it is usually localized in the innermost layer, the endocuticle (Greven and Peters, 1986). Although the type of chitin is not known, in tardigrades the ultrastructure of the fibres indicates that

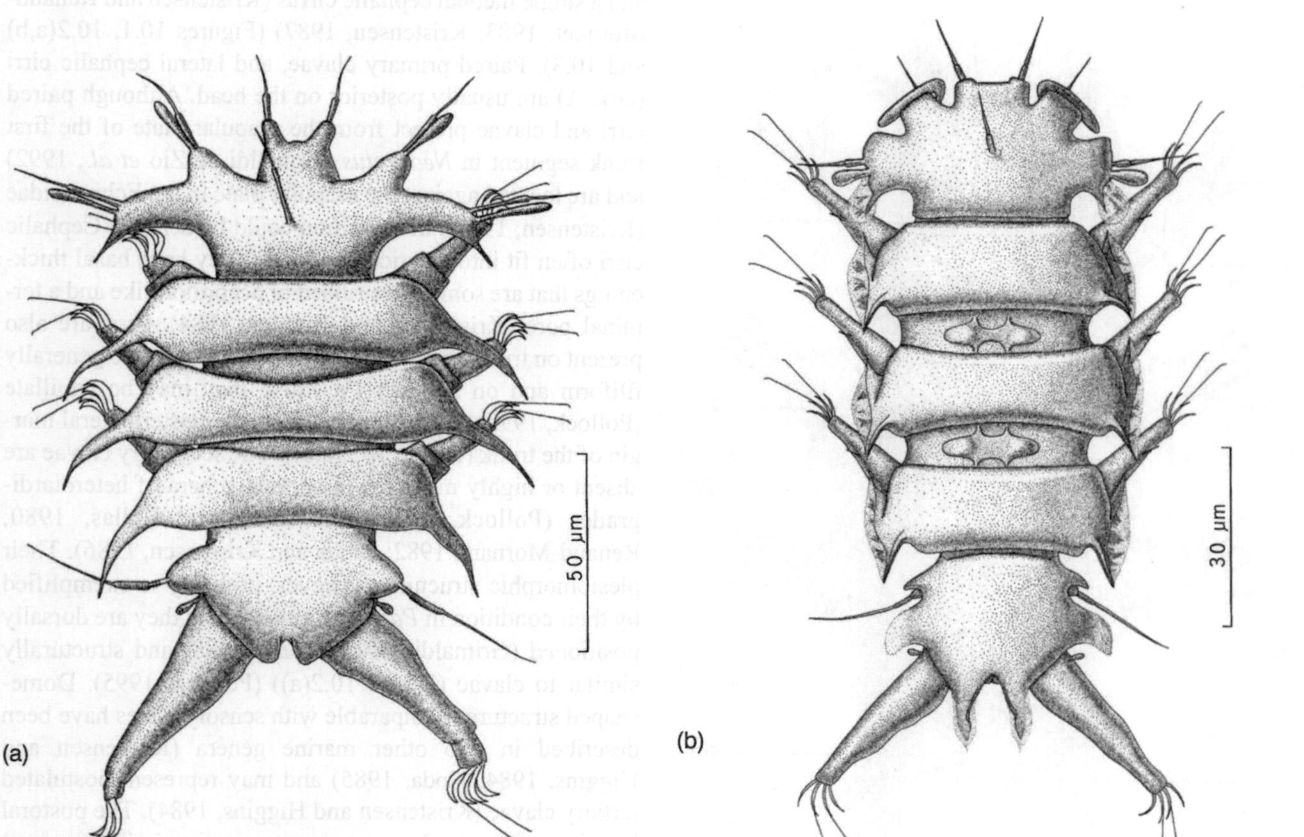


**Figure 10.2** (a) *Parastygarctus sterreri*. Note the head appendages and lateral projections or flanges of the dorsal plates. (b) *Stygarctus abornatus*. Node-like intersegmental thickenings are visible posterior to the first and second dorsal trunk plates. (c) Foot and claws from leg I (lateral view) of *Megastygarctus orbiculatus*. Claws of the illustrated stygarctids and *Renaudarctus* have accessory spines. The uncus-like swelling of the claws of *Megastygarctus* and many other tardigrades is similar to that seen on many Cambrian lobopods. (d) Bucco-pharyngeal apparatus and mouth cone of *Paradoxipus oryzeliscoides*. The paired stylets are projected through the mouth opening and punctured cell contents are ingested by pumping of the muscular pharynx. The mouth is surrounded by the mouth cone that is innervated with different types of sensillae exhibiting the pattern of biradial symmetry indicated on the right. [(a–c) from McKirdy *et al.* (1976); (d) bucco-pharyngeal apparatus from Kristensen and Higgins (1989).]

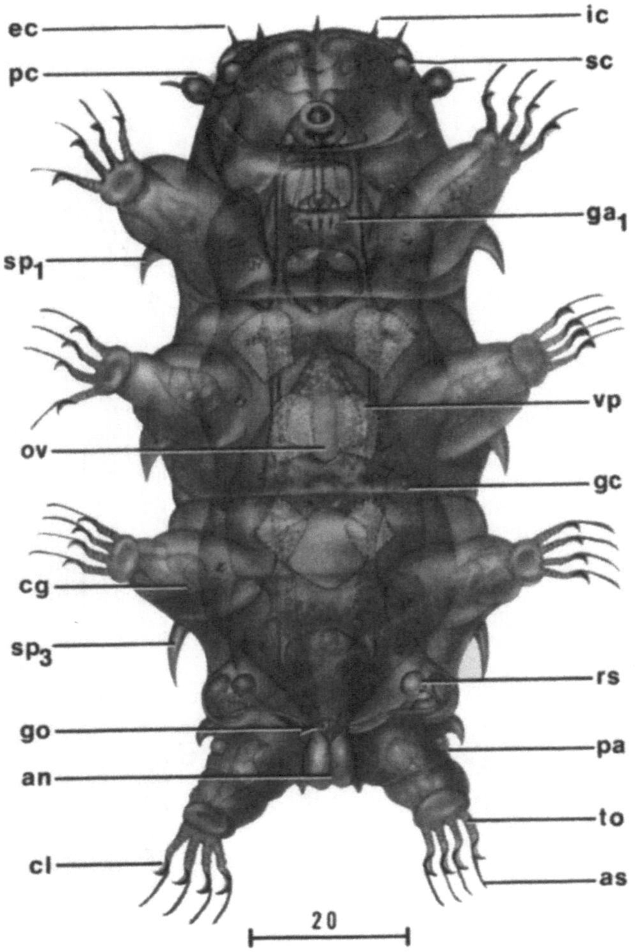


**Figure 10.3** Female *Renaudarctus psammocryptus*, ventral view. a, anus; as, accessory hook of claw; cg, claw gland; cl, claw; ec, external cirrus; ga, ventral ganglion 1; gc, gut crystal; go, gonopore; ic, internal cirrus; ov, ovary; pa, leg papilla; pc, primary clava; rs, seminal receptacle; sc, secondary clava; sp1, lateral flange (spine); sp3, lateral flange (spine), to; toes; vp ventral plate. Scale bar = 20 mm. (From Kristensen and Higgins, 1984.)

they orient randomly without the helicoids or laminae that are common in arthropod cuticles (Wright and Luke, 1989). Within the inner layer of endocuticle the chitin must exist in short chains, since in living animals this layer is in a liquid state and freely transports lipid-like droplets or excretory wastes (Dewel *et al.*, 1992). The epidermis is also responsible for a curious synapomorphy between tardigrades and arthropods, bismuth staining of Golgi complexes, a property not shared with onychophorans (Locke and Huie, 1977).

### (e) Sensilla

The primitive complement of sensory appendages apparent in the marine tardigrades consists of paired internal and external cephalic cirri, secondary clavae (cephalic papillae) and a single median cephalic cirrus (Kristensen and Renaud-Mornant, 1983; Kristensen, 1987) (Figures 10.1, 10.2(a,b) and 10.3). Paired primary clavae, and lateral cephalic cirri (cirri A) are usually posterior on the head. Although paired cirri and clavae project from the scapular plate of the first trunk segment in *Neoarctus* (Grimaldi de Zio *et al.*, 1992) and are located against the scapular plate in the Echiniscidae (Kristensen, 1987), that position could be derived. Cephalic cirri often fit into a cuticular socket. They have basal thickenings that are sometimes ribbed or accordion-like and a terminal pore (Kristensen and Higgins, 1984). Cirri are also present on trunk appendages (I–III) where they are generally filiform and on leg pair IV where they may be papillate (Pollock, 1995). Cirri E project from the posterolateral margin of the trunk (Figure 10.2(a,b)). The secondary clavae are absent or highly modified in various genera of heterotardigrades (Pollock, 1970; Kristensen and Hallas, 1980; Renaud-Mornant, 1982; Binda and Kristensen, 1986). Their plesiomorphic structure, however, probably is exemplified by their condition in *Parastygarctus* where they are dorsally positioned (Grimaldi de Zio *et al.*, 1988) and structurally similar to clavae (Figure 10.2(a)) (Pollock, 1995). Dome-shaped structures comparable with sensory plates have been described in two other marine genera (Kristensen and Higgins, 1984; Noda, 1985) and may represent postulated tertiary clavae (Kristensen and Higgins, 1984). The postoral location of these clavae and presence in similar postoral locations of plate-like thickenings in *Batillipes* (Kristensen, 1978) and *Renaudarctus* (Kristensen and Higgins, 1984), deep folds in *Diplodarctus* (Pollock, 1995) and *Hemiarctus* (Grimaldi de Zio *et al.*, 1996) and ventral cephalic plates in echiniscids (Kristensen, 1987) suggest the presence of an innervated postoral plate (see Phylogenetic analysis, p. 118, for proposed homologues in other lobopods and arthropods).

All sensory appendages in tardigrades are individual sensilla, not modified limbs such as antennae, but they are perhaps the sole externally visible remnants of such appendages. In ultrastructure the sensilla are arthropod-like (Kristensen, 1981) but the one defining synapomorphy, the presence of three or more support cells, is questionable. We observe that tardigrade sensilla appear to have two mesaxon support cells, the same number found, for example, in nematodes (Jones, 1979; Wright, 1991). If the arrangement of sensilla on the 'head appendages' of stygarctids, principally *Parastygarctus* (Figure 10.2(a)), is taken as plesiomorphic, then it is plausible that the secondary clavae, external cirri and the lateral cirri and primary clavae represent respectively proximal and distal sensory fields on frontal appendages of the head (Dewel and Dewel, 1996).

### (f) Legs

The first three pairs of legs, located ventrolaterally, are the primary means of locomotion; whereas the fourth pair, directed posteriorly, with a reversed claw curvature, as in

**Table 10.1** Tardigrade classification*

| I. Class Heterotardigrada | II. Class Eutardigrada |
|---|---|
| A. Order Arthrotardigrada | A. Order Parochela |
| 1. Family Halechiniscidae *Orzeliscus, Halechiniscus, Paradoxipus, Florarctus, Styraconyx, Tetrakentron, Tanarctus, Archechiniscus* | 1. Family Macrobiotidae *Macrobiotus, Richtersius, Pseudodiphascon, Minibiotus, Calcarobiotus, Adorybiotus* |
| 2. Family Stygarctidae *Stygarctus, Pseudostygarctus, Parastygarctus, Neoarctus, Megastygarctus, Neostygarctus* | 2. Family Eohypsibiidae *Eohypsibius* |
| 3. Family Renaudarctic *Renaudarctus* | 3. Family Calohypsibiidae *Calohypsibius* |
| 4. Family Coronarctidae *Coronarctus* | 4. Family Necopinatidae *Necopinatum* |
| 5. Family Batillipedidae *Batillipes* | 5. Family Hypsibiidae *Hypsibius, Ramazzottius, Pseudobiotus, Isohypsibius, Doryphoribius, Itaquascon, Eremobiotus* |
| B. Order Echiniscoidae | B. Order Apochela |
| 1. Family Echiniscoididae *Anisonyches, Echiniscoides* | 1. Family Milnesiidae *Milnesium, Limmenius* |
| 2. Family Oreellidae *Oreella* | **III. Class Mesotardigrada** |
| | A. Order Themzodia |
| 3. Family Echiniscidae *Echiniscus, Antechiniscus, Bryochoerus, Hypechiniscus, Parechiniscus, Novechiniscus, Pseudechiniscus* | 1. Family Thermzodiidae *Thermozodium* [no type specimen] [no type location] |

* Only selected genera listed

the lobopod *Aysheaia* and *Onychodictyon*, are used for grasping the substrate and occasionally for movement (Figure 10.2(a,b)). Telescopic legs, characteristic of the marine arthrotardigrades, are considered to be autapomorphic (Table 10.2) (Renaud-Mornant, 1982; Kristensen and Higgins, 1984, 1989; Grimaldi de Zio *et al.*, 1987a,b). As in onychophorans and arthropods, the legs are controlled by both intrinsic and extrinsic musculature. Each leg is divided into three parts; the distal portion is an extracellular cuticular thickening (probably incorrectly termed a 'tarsus') that terminates in either claws or digits (toes) with adhesion discs and/or claws (Figures 10.1, 10.2(a,b) and 10.3). Both claws and toes are secreted by epidermal thickenings, the claw or foot glands. Claws inserting directly on the legs, the condition found in stygarctids, are thought to be plesiomorphic (Figure 10.2(c)). Claws are parallel-sided, curved and pointed. Distal accessory spines or hooks on either all or the medial pair of claws are present in Stygarctidae and Renaudarctidae (Figures 10.2(a–c) and 10.3) (Kristensen and Higgins, 1984; Grimaldi de Zio *et al.*, 1987a) and a proximal uncus on each claw is found in the stygarctids and other genera lacking toes (Figures 10.1 and 10.3) (Grimaldi de Zio *et al.*, 1987b). A strong resemblance between claws of tardigrades and Cambrian lobopods has been suggested (Whittington, 1978; Dzik and Krumbiegel, 1989) and an expansion of the comparison to include details of substructure, for example the uncus-like protuberances on the claws of *Onychodictyon* (Ramsköld, 1992), would be worthwhile, although there are problems in determining homologies.

### (g) Musculature

Somatic muscles are elongated cells that insert on the body wall. They are intermediate between smooth and obliquely striated in eutardigades (Walz, 1974) and cross-striated in marine arthrotardigrades (Kristensen, 1978). Although the pattern of striation is of limited phylogenetic value because smooth, oblique and cross-striated muscles are widely distributed in metazoans, details of the Z system reveal a close similarity to onychophoran muscles (Wright and Luke, 1989; Storch and Ruhberg, 1993). The muscular system also contains visceral muscles, cross-striated stylet muscles and the muscle cells of the pharyngeal bulb. The pharyngeal bulb is strengthened apically with a cuticle containing reinforcing bars called placoids and basally with a thickened basement membrane. In cross-section, the lumen is triradiate. The cells

**Table 10.2** Character summary

| | |
|---|---|
| **Symplesiomorphies** | **Homoplasies resulting from miniaturization** |
| Mouth cone* | Reduced sensory support cell number (?) |
| Reduced sensory support cell number (?) | Haemocoelic body cavity (?) |
| Orthogonal [tetraneural] pattern of ganglia in CNS | Lack of circulatory system |
| Triradiate pharynx (?) | Lack of circular body wall muscle |
| Cerebral ocelli (?) | Lack of serial nephridia |
| Gut diverticula* | Lobopodous legs (?)* |
| Haemocoelic body cavity (?) | Cerebral ocelli (?) |
| Striated muscle | Reduced segment number |
| Muscle attachments with intermediate filaments | Eutely and low cell number |
| Lobopodous legs (?)* | **Autapomorphies** |
| Intrinsic and extrinsic leg musculature | Stylet apparatus, flexible buccal tube |
| Flexible chitinous cuticle* | Telescopic legs |
| Growth by ecdysis | |
| Postoral cephalic plate* | **Synapomorphies with arthropods** |
| Dorsal gonad | Dorsal and ventral plates and flanges (?)* |
| **Homoplasies** | Head with three segments, frontal appendage fully incorporated into cephalon* |
| Triradiate pharynx (?) | Arthropod-like sensilla |
| Malpighian tubules, eutardigrades | Unciliated epithelium |
| Cryptobiosis | Bismuth staining of Golgi apparatus |
| Rectal thickenings, eutardigrades | |

Characters found in protostome phyla other than Arthropoda *sensu stricto* are listed as plesiomorphies. * Used in phylogenetic analysis with PAUP.

comprise an ectodermally derived myoepithelium; those capping the clefts of the lumen contain intermediate filaments while the contractile cells between the clefts display myofilaments forming a single sarcomere (Eibye-Jacobsen, 1996). Of greatest phylogenetic interest is the presence of intermediate-sized filaments (Dewel and Dewel, 1979) rather than microtubules in tendon cells at muscle attachment sites. Intermediate filaments have a nearly ubiquitous distribution in eumetazoa but apparently have been replaced by microtubules in all arthropods. That replacement is regarded as strong support for the monophyly of Arthropoda (Bartnik and Weber, 1989).

### (h) Body cavity

There is a large body cavity filled with numerous cells, free-floating in eutardigades and anchored to the body wall in heterotardigrades. The ultrastructure of these cells indicate that they store lipids and polysaccharides (Weglarska, 1975) and function as phagocytes (Dewel *et al.*, 1993). They do not appear to contain oxygen-transporting proteins. The body cavity usually is considered a haemocoel or mixocoel, but its ontogeny is poorly known. Coelomic cavities other than that formed by the gonadal cavity are thought to have been lost secondarily.

### (i) Mouth

The mouth opening of tardigrades is positioned at the end of a telescoping mouth cone. It is surrounded by structures exhibiting radial or biradial symmetry (Figures 10.2(d) and 10.3). Lamellae, either grouped into lips or arranged in a continuous ring form the anterior portion of the buccal ring. Sensory structures both on the outer cuticle of the mouth cone and the buccal tube show biradial symmetry around the mouth opening (Figure 10.2(d)). The general radial pattern of structures around the mouth opening is comparable with that seen in onychophorans and certain aschelminths while the particular biradial symmetry of sensory elements is thought to be similar to that seen in the anomalocaridids and perhaps other stem lineage arthropods (Hou *et al.*, 1995; Budd, 1996; Dewel and Dewel, 1996; Waggoner, 1996).

### (j) Stylet apparatus

Lateral to the buccal tube is the stylet mechanism which consists of two protrusible stylets, stylet sheaths or flanges, stylet supports, and associated protractor and retractor muscles (Figure 10.2(d)) (Schuster *et al.*, 1980). The paired stylets, which may be calcified, enter the buccal tube through stylet sheaths on each side of the anterior end of the tube. The stylets are projected through the mouth opening to pierce plant or animal cells during feeding. Both the stylets and stylet supports are enveloped by paired salivary glands that secrete the cuticular components of the apparatus before ecdysis. In the intermoult, the salivary glands produce an unknown secretion that accumulates around the stylets and probably is released when the stylets are protruded during feeding. The thickened posterior end of each stylet forms a furca (condyle) (Figure 10.2(d)). Lateral stylet supports, which may be absent (Stygarctidae, Renaudarctidae) or poorly developed in some genera, attach the furcae to the buccal tube. Stylet protractor muscles extend from the furcae to the buccal tube; retractor muscles insert on the pharynx. Although a hypothesis that the stylets are modified claws (Nielsen, 1995) is supported by the origin of the jaw from appendage-like protrusions of the second segment in onychophorans (Walker and Campiglia, 1988), an alternative proposition that they derive from one or more posterior segments should also be considered. Only careful analysis of this character and, if possible, observation of its ontogeny will help determine its segmental identity.

**(k) Gut and excretory system**

The digestive tract consists of a foregut including the buccal tube, pharynx and oesophagus, midgut and hindgut. The foregut and hindgut are lined with cuticle that is shed during moulting. In many heterotardigrades the midgut is characterized by sac-like diverticula that even occasionally extend into the cavity of the legs. Excretion and osmoregulation is performed in eutardigrades by the rectum and three (four in *Milnesium* females) Malpighian tubules that open via a pylorus into the digestive tract where the midgut joins the hindgut (Dewel and Dewel, 1979; Greven, 1979; Weglarska, 1980, 1987a,b; Møbjerg and Dahl, 1996). The pylorus does not have a cuticular lining and is, therefore, of endodermal origin. Although excretory organs are absent in marine heterotardigrades, terrestrial echiniscids have evolved two serially repeated epidermal organs that appear to perform similar excretory and osmoregulatory functions (Dewel *et al.*, 1992). Both organs are composed of only three cells and lie between the second and third pair of legs respectively. Neither system found in tardigrades appears to be synapomorphic with excretory or osmoregulatory organs in onychophorans and arthropods (Greven, 1982; Dewel *et al.*, 1992).

**(l) Reproductive system**

In marine tardigrades the reproductive system comprises a single testis or ovary that overlies the gut (all species examined are gonochoristic with the exception of a single hermaphroditic population of *Orzeliscus* (Bertolani, 1987)). The testis is joined by paired vasa deferentia to a papillate gonopore with a crescent-shaped opening rostral to the anus. The ovary has only a single oviduct which opens rostral to the anus via a gonopore surrounded by a rosette of six or seven cells each with a cuticular plate. *Renaudarctus* which has a simple oval gonopore immediately adjacent to the anus is an exception (Figure 10.3) (Kristensen and Higgins, 1984). In some females paired seminal receptacles with often highly convoluted ducts open lateral to the gonopore (Figure 10.3). Except for the gonopores very little sexual dimorphism is evident in marine tardigrades.

**(m) Nervous system**

The nervous system is composed of a group of dorsal or lateral supraoesophageal ganglia, a circumbuccal ring of ganglia and a ventral chain of four paired trunk ganglia (Figure 10.3) (Dewel and Dewel, 1996; Wiederhöft and Greven, 1996). The arrangement of the cephalic ganglia is essentially orthogonal (tetraneural) with a series of paired dorsal centres connected to paired ventral or lateral ones and to the first ganglia of the ventral chain. Analysis of the central components suggest that they are derived from neural elements of 3½ segments (Dewel and Dewel, 1996). The pattern of innervation of sensory receptors in the mouth cone, internal and external ciri, secondary clavae and lateral cirri and the primary clavae is consistent with this proposal. Based on a model of the brain that relates neurophil centres for these receptors to hypothetical segmental ganglia, the mouth cone appears to be innervated by the putative second segment and the secondary clavae, external cirri, lateral cirri and primary clavae, and the cerebral ocelli by the third.

## 10.2.2 DISCUSSION OF CHARACTERS

Many of the above characters are summarized and categorized in Table 10.2. Possible synapomorphies of tardigrades and stem lineage arthropods are tabulated along with putative symplesiomorphies, autapomorphies and characters that would have a significant probability of arising convergently or exhibiting a deceptively plesiomorphic state as a result of miniaturization through progenesis. Only a partial listing of plesiomorphic characters is given. The latter are present in several potential outgroup phyla including Annelida, Priapulida, Mollusca, Nematoda, Kinorhyncha, Loricifera and Onychophora. The onychophorans are considered a sister group of the Arthropoda (Lauterbach, 1978; Eernisse *et al.*, 1992; Storch and Ruhberg, 1993; Wheeler *et al.*, 1993) basal to a tardigrade–arthropod clade (Wright and Luke, 1989; Monje-Najera, 1995; Nielsen, 1995; Wills *et al.*, 1995; Budd, 1996; Dewel and Dewel, 1996; Nielsen *et al.*, 1996). Some of the characters have resulted from progenesis and are indicated separately. The dearth of apomorphies that tardigrades share with arthropods is readily apparent and, of the characters listed, all are somewhat ambiguous.

Although the evolutionary history of tardigrades undoubtedly included a process of miniaturization, and progenesis was probably the operating mechanism, the degree to which it influenced character morphology is not known. If progenesis affected only later ontogenetic stages, modifications would be minimal and localized (Gould, 1977; McKinney and McNamara, 1991; Raff, 1994). For example, restriction on cytokinesis and growth but not differentiation would transform an ancestor into a well-proportioned, and therefore easily recognized dwarf descendent (Gould, 1977). However, the enigmatic status of tardigrades, due in part to the presumed absence of several different tissues and organs (coelomic linings, nephridia, circular muscle, blood vessels), suggests that at least some of the heterochronic changes disrupted tissue interactions during critical stages in differentiation with profound consequences (McKinney and McNamara, 1991; Raff, 1994). Thus, if progenesis halted differentiation or operated early in development, characters would be unrecognizable or absent. Although these paedomorphic character states are derived (homoplastic), they have the potential of being interpreted erroneously as plesiomorphies (Table 10.2) (Rieppel, 1993).

It is possible that tardigrades could have arisen from an arthropod ancestor through sexual maturation of an early larval stage. For example, in arthropod 'head larva', such as the 'Orsten' larvae (Müller and Walossek, 1986), ganglia for the larval appendages A1, A2 and the mandible are presumably uncondensed and postoral. Conceivably, an extreme case of progenesis in stem or crown lineage arthropod larvae could have resulted in a paedomorphic descendent having a nervous system with an organization similar to that found in tardigrades.

Nevertheless, it seems unlikely that progenesis would have resulted in the loss of other arthropod characters, such as microtubules at muscle attachment sites. Microtubular tendons are diagnostic of all extant arthropods (Bartnik and Weber, 1989). Tardigrades have intermediate filaments at attachment sites, but mites and other similarly reduced living arthropods do not; therefore, tardigrades, like onychophorans, appear to be excluded on the basis of this character from living arthropod groups (Dewel and Dewel, 1979; Raff *et al.*, 1994).

If it is unlikely that tardigrades arose from within crown lineage arthropods, could they have derived from stem lineage forms? More specifically, does the stem lineage larva described by Müller *et al.* (1995) share any derived features with extant tardigrades? The proposed similarities in structure of the claws, putative sensory receptors, cuticular pillars and details of the mouth are either structurally dissimilar or are plesiomorphic. Pillars are well developed in certain nematodes (Malakhov, 1994) and juxtapositioned hair or peg sensilla and chemoreceptive sensilla are found in onychophorans (Storch and Ruhberg, 1977, 1993). This however, does not entirely rule out stem tardigrade status for this animal and additional material may introduce convincing synapomorphies. The posterior hooks of the claws of the limbs seem very similar to arthropod gnathobases and under certain ecological conditions gnathobases develop spines (Butterfield, 1994) that are similar to claws. It may be therefore worth considering whether tardigrades could have derived from stem lineage arthropods with gnathobases. The basipods, having lost endo- and exopodal rami, could have become stubby legs with the gnathobases providing claw-like spines.

## 10.3 PHYLOGENETIC ANALYSIS

### 10.3.1 HOMOLOGIES

A phylogenetic analysis to clarify the relationship of tardigrades to arthropods was carried out with 34 fossil and modern taxa using a database of lobopod and arthropod characters (see http://www.sysassoc.chapmanhall.com). To obtain the most ancestral character states, taxa chosen in the analysis are almost exclusively from the Cambrian. Only those taxa preserved with sufficient detail for obtaining an adequate proportion of the character states were used. Accordingly, the Vendian biota (Waggoner, 1996) and lobopods, except for the Cambrian *Aysheaia*, *Cardiodictyon* and *Hallucigenia*, were excluded. Many of the arthropod characters were coded according to Briggs and Fortey (1992), Budd (1993, 1996), Chen *et al.*, (1994), Fortey and Theron (1994) and Wills *et al.*, (1994, 1995). Of the large set of tardigrade characters relevant to understanding lobopod and arthropod relationships, only a small subset are suitable for comparison with fossilized material. Hypotheses regarding the homology of questionable characters are noted below. These hypotheses are offered in an attempt to bridge lobopod and arthropod morphological disparities and derive from a reinterpretation of the characters from the perspective of tardigrade morphology.

#### (a) Dorsal plates

Potential homologues for dorsal plates, flanges and/or alae of tardigrades are the dorsal plates of Cambrian lobopods (Ramsköld, 1992; Hou and Bergström, 1995), dorsal protuberances and lateral flaps of *Kerygmachela* (Budd, 1993), terga and lateral flaps of *Opabinia* (Whittington, 1975) and *Anomalocaris* (Whittington and Briggs, 1985; see also Bergström, 1986, 1987; Briggs and Whittington, 1987) or the terga and pleurae of euarthropods. Although precise homologues for this character are difficult to determine (Budd, 1996), the finding that lateral flaps in anomalocaridids constitute the biramous limbs of arthropods (Hou *et al.*, 1995) appears to weaken any proposed relationship between extensions of the dorsal plates in tardigrades and lateral flaps in stem lineage arthropods. Nevertheless, the alternating pattern of plates and annulae with nodes in *Stygarctus*, *Megastygarctus* and perhaps *Renaudarctus* appear to correspond more closely to the alternating pattern of annuli and sclerotized plates found in some lobopods and *Kerygmachela*, than to the terga and pleurae of arthropods.

#### (b) Cephalic plates

Hou and Bergström (1995) describe 'a pair of large sclerites reminiscent of bivalve shells' covering the heads of *Cardiodictyon* and *Hallucigenia*. Poor preservation makes this interpretation inconclusive and the character is reinterpreted here as a single dorsal head plate that is associated with ventral postoral frontal appendages (Hou and Bergström, 1995, Figure 1). It is homologized on the basis of its position relative to the frontal appendage and on grounds of parsimony to rostral or ventral head plates in other taxa. The position in *Opabinia* (Whittington, 1975; Figures 5–7), and *Anomalocaris* (Chen *et al.*, 1994; Figures 1 and 2) is thought to have resulted from ventrocaudal rotation of the mouth opening. Although the hypostome is in a similar location and probably formed in part or whole from the ventral plate, it is provisionally considered a separate structure because of its distinctive morphology (anterior wings, etc.) in most arthropods. The character is also speculated to include the ventral postoral plate of *Aysheaia* (Whittington, 1978; Figures 77 and 80) and the innervated postoral ventral 'plates' of some marine arthrotardigrades (Sensilla above). This conjecture, however, requires an improbable macroevolutionary change in the position of the mouth relative to the plate and is offered only as part of an initial hypothesis that homologizes all cephalic plates closely connected to a frontal appendage.

**(c) Frontal appendages**

The single, grasping proboscis-like structure of *Opabinia* is considered to be derived from the fusion of the basal portion of paired frontal appendages (Bergström, 1986, 1987; Chen *et al.*, 1994; Budd, 1996; for an opposing view see Briggs and Whittington, 1987). That appendage, together with the frontal appendages of *Aysheaia*, *Parastygarctus*, *Kerygmachela* and *Anomalocaris* (Chen *et al.*, 1994; Budd, 1996), are tentatively homologized with the great appendages of *Alalcomenaeus*, *Yohoia* and *Leanchoilia*. The five pre-antennal limbs of *Sanctacaris* (Briggs and Collins, 1988) are speculated to represent supernumerary rami of a single pair of frontal appendages. On the basis of their large size and position on the head tagma ventral to the compound eyes or dorsal to the ventral plate, frontal appendages are assigned to the preantennal (ocular) segment. Although it is possible that great appendages could have arisen convergently through ectopic expression of genes of posterior segments (Jacobs, 1992), it is reasonable to consider as an initial hypothesis that all of the great appendages originated from the same segment and that segment gave rise to the compound eyes. The anterior appendages of *Hallucigenia* and *Cardiodictyon* (Hou and Bergström, 1995) are presumably frontal but they are clearly ventral and postoral.

**(d) Head segmentation**

It is proposed from a study of the nervous system (Dewel and Dewel, 1996) that the mouth cone and head appendages or cephalic sensilla posterior to the internal cirri of tardigrades are derived from segments homologous to the second (jaws) and third (oral papillae) cephalic segments of onychophorans (Walker and Campiglia, 1988) and the segments that give rise to the mouth cone and frontal appendages of other lobopods and stem lineage arthropods. The segment of the frontal appendages is tentatively homologized to the ocular segment of crown lineage arthropods. The position of the compound eyes immediately dorsal to the frontal appendages in stem forms (*Opabinia*, *Laggania*, *Alalcomenaeus*, *Acteus*, *Yohoia*) suggests that they developed on the dorsum of that segment. Moreover, the observation of a small group (5–8) of ommatidial-like units positioned just medial to the frontal appendages on the head of *Kerygmachela* (Conway Morris, 1994, Figure 2B; also Budd, 1997, this volume) supports that conclusion. If this proposal is correct, the head tagma of lobopods, tardigrades, *Kerygmachela* and *Opabinia* would correspond to preantennal regions of the euarthropod head. Accordingly, as an initial hypothesis, the remaining cephalic segments are can be designated as postocular.

Nevertheless, the segmental identity of the frontal appendages should remain provisional because of a lack of unequivocal fossil evidence and the large amount of morphological variation in the appendages coded as being frontal. The latter observation suggests that some or all of these appendages may have originated from a more posterior segment and moved subsequently into the position subjacent to the eyes. Their position may simply reflect the convergent rostral migration of various grasping, sweeping or raptorial limbs. In any event if these appendages correspond to the first antennae, then the basal organization of the euarthropod head, exemplified by a preoral commissure for A1 and a postoral commissure for A2 as found in the notostracan *Apus* (*Triops*) (Henry, 1948) and the anostracan *Artemia* (Benesch, 1969), may be inferred to have been already present in some lobopods and stem lineage arthropods with frontal appendages.

An origin for the frontal appendages more posterior than the first antenna, however, seems improbable. Fixation of the head tagma appears to have taken place relatively early in arthropod evolution, probably basal to the splitting of the arachnomorph and crustaceanomorph clades since the pattern of anterior segments appears to be invariant in all major groups of living arthropods. The similarities are evident in both embryological (Dohle, 1974; Weygoldt, 1975, 1979) and molecular genetics studies (Averof and Akam, 1995). In the latter, the analysis of neural elements in *Drosophila* embryos indicate that the optic lobes are of segmental origin and that they derive from a segment anterior to the antennal segment and posterior to the clypeolabrum (Schmidt-Ott *et al.*, 1995). Comparable results with *engrailed* labelling in crustaceans indicate that optic ganglia lie anteriorly adjacent to the second stripe in the first antennal segment (Scholtz, 1995). Although *engrailed* labelling of the labrum is absent in this study, it may simply indicate that the 'labrum' of crown lineage crustaceans is an autapomorphic outgrowth of the hypostome found in stem lineage forms (Walossek and Müller, 1990) or, perhaps, it signifies that labral labelling was indistinguishable from that associated with the stomatogastric system.

**(e) Trunk appendages**

The flap-like limbs of *Anomalocaris* are considered to be homologous to arthropod biramous appendages (Budd,

1996; Collins, 1996), contrary to the conclusions of Hou *et al.* (1995) and Wills *et al.* (1995). According to recent evidence on certain anomalocaridids from the Chengjiang fauna the elements making up lateral flaps show a close correspondence to component parts of biramous limbs (Hou *et al.*, 1995, Figure 10). Both are composed of a basipod (propod), an endopod and an exopod. In anomalocaridids, as in basal arthropods, a proximal endite or coxa is absent (Walossek and Müller, 1990; Walossek, 1995) but a basipod having multiple 'endites' or gnathobasic protuberances is present. In the anomalocaridid the exopod is flattened and lacks gill blades or lamellae. The endopod is either annulated (wrinkled) or articulated and appears to derive from the anterior margin of the basipod. The lateral flaps of the anomalocaridids are hypothesized to be homologous to flaps in *Kerygmachela* and *Opabinia.* In *Kerygmachela* the exopod (posterior) portion of the flap is finely annulated (Budd, 1993) but in *Opabinia* it exhibits gill lamellae that are free on the outer margin (Bergström, 1986; Budd, 1996). Lobopodous elements (endopods?) have been reported for both *Kerygmachela* and *Opabinia* (Budd, 1993, 1996) but because of poor preservation the exact relationship of these structures to exopods is uncertain. Thus, lobopodial appendages are thought to be, at least in part, homologous to lateral flaps and biramous limbs, but specific correspondences between component structures remain obscure.

**(f) Telson**

The definition of the telson is expanded to accommodate an apparatus (tail fan, *sensu* Whittington, 1975) composed of dorsally directed lateral flaps. These are regarded as free in *Opabinia* and *Anomalocaris* (Briggs and Whittington, 1987) but overlapping or partially fused into a caudally directed paddle in *Alacomenaeus*, *Yohoia* and *Sanctacaris.* Further modification of the same basic apparatus is found in *Leanchoilia* (spinous), *Sidneyia* (additional fused segments) and perhaps *Fuxianhuia.*

### 10.3.2 PHYLOGENETIC TREES

The data were analysed with PAUP 3.1.1. (Swofford, 1993) using the conventional criterion of Fitch parsimony (unordered characters). Characters were scored as missing (?) if their state could not be determined from the available material. The data matrix was analysed employing 10 replicates of random addition heuristic searches. When *Hallucigenia* was used as the outgroup, 10 equally parsimonious trees (length 249 steps, C.I. = 0.542, R.I. = 755) were obtained. A strict consensus tree of those 10 trees is shown in Figure 10.4. The composition of two major clades, the arachnomorphs and the crustaceanomorphs, is comparable to those groupings in other published phylogenetic analyses of fossil arthropods (Briggs and Fortey, 1992; Wills *et al.*, 1994, 1995). The inclusion of *Fuxianhuia* and *Marrella* in

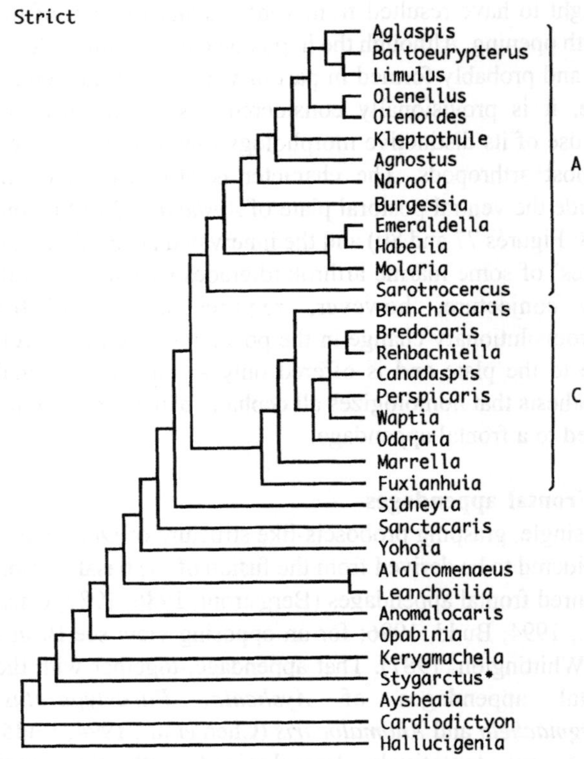


**Figure 10.4** Strict consensus tree from 10 equally parsimonious trees. *, the tardigrade, *Stygarctus*; A, arachnomorph clade; C, crustaceanomorph clade.

the crustaceanomorphs is unusual. *Fuxianhuia* is united on the basis of the structure of forehead, stalked eyes and proliferated numbers of metameres. The placement of *Marrella*, which depends on character reversal for trilobation, presence of pleurae and tergal overlap as well as diminishing trunk limb size, is, in view of considerable probability of homoplasy, less secure. Figure 10.5, derived from one of the most parsimonious trees, is used to illustrate apomorphies for the more basal taxa. All genera except the sister group of *Alalcomenaeus* and *Leanchoilia* form a series of plesions leading to the AC clade. These branch in an order largely determined by character state changes in the head plate, rostral cirri, frontal appendage, biramous limbs and telson. Although the ventral head plate and frontal appendage are also symplesiomorphies of the clade, they are recognizable in a number of taxa and their structure and position are sufficiently varied to make them phylogenetically useful.

The AC clade plus *Sidneyia* is characterized by the absence of frontal appendages, the presence of antenniform rami on the first postocular appendage and the incorporation of more rostral segments into a paddle-shaped tail. The AC clade alone is united by the absence of a ventral plate and presence of a hypostome with derived features such as anterior wings. The node may appear to be better supported than

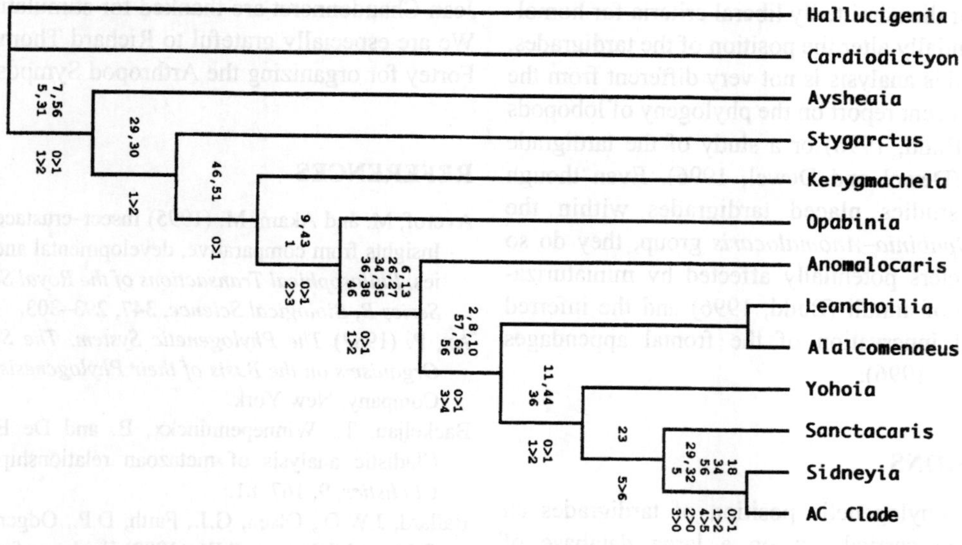


**Figure 10.5** Summary of character state changes in one of the 10 most parsimonious trees. Only synapomorphies that the indicated taxon shares with the remainder of the clade are given. *Aysheaia*: rostral cirri or spines present (7); telson or terminal segment reduced or absent (56); cephalic plate ventral, mouth anterior (5); spines on frontal appendage robust (31). *Stygarctus*: frontal appendages dorsal and/or preoral (29); frontal appendage lobopodous and flattened (30). *Kerygmachela*: trunk appendages lobopodous and flattened (46); posterior (outer) ramus unsegmented (annulated) (51). *Opabinia*: tergites present (9); postcephalic articulation with overlap (43); trunk not annulated (1). *Anomalocaris*: head tagma with postocular segments (6); doublure present (13); apparent number of cephalic segments 4–6 (26); cephalic gnathobases present (27); appendages on fourth or first postocular segment biramous (34); appendages on first, second and third postocular segments similar to trunk appendages (35,36,38); trunk appendages with articulating ramus (46). [Anomalocaridids constitute a diverse group with unsettled systematics (Hou *et al.*, 1995; Collins, 1996). For this reason, several of the character states were based not strictly on *Anomalocaris* but encompass evidence from other recently described genera within the family]. *Alalcomenaeus* and *Leanchoilia*: exoskeleton sclerotized (2); cephalic shield entire (8); pleurae unfurrowed (*Alal*), furrowed (*Lean*) (10); telson paddle shaped, articulating at base (57); anus rostral to telson (63); telson composed of caudally directed fused or overlapping lateral flaps (56). *Yohoia*: trilobation present (11); abdominal trunk tagma present (44); first postocular cephalic segment with elongated endopod (36). *Sanctacaris*: compound eyes stalked lateral to head tagma (23). *Sidneyia* and AC clade: hypostome conterminant (18); appendages on first postocular segment uniramous (34); telson composed of fused lateral flaps plus more rostral body pleurae (56); frontal appendages absent (29,32); cephalic plate ventral; mouth posterior (5).

it actually is because some character changes may not be independent. Although the plesions basal to the AC clade plus *Sidneyia* are excluded on the basis of their frontal appendages and apparent lack of a hypostome, they exhibit a number of typical arthropod features (albeit with an ambiguous distribution). However, inasmuch as the coding of the great or grasping appendages is problematic, the alternative hypothesis that they are equivalent to the first antennae should be tried. Thus recoded, the taxa may assort to one branch or the other of the AC clade.

As discussed previously, the position of *Stygarctus* is conjectural due to difficulties in coding characters in minute organisms. Despite this, the range of credible nodes for tardigrades is limited. At the base of the tree the dorsal position of the head plate clearly separates *Hallucigenia* and *Cardiodictyon* from the remainder of the clade. Although several apomorphies appear to unite tardigrades with either *Aysheaia* or *Kerygmachela*, they fail to produce a monophyletic grouping. The mouth opening is anterior to the ventral plate in *Aysheaia* and tardigrades, but its location is unknown in *Kerygmachela*. The structure and position of the frontal appendages are comparable in all three taxa, but they are more similar to those in *Kerygmachela* because of their anterior position and flattened morphology. In addition, annuli with nodes alternating with dorsal plates or tubercles are similar to those in *Kerygmachela*, but rostral cirri or spines typify all three taxa. Finally, other apomorphies, such as the rostral plate or postcephalic articulation with overlap, separate tardigrades from *Opabinia* and the rest of the clade. Thus, assuming that the characters were minimally affected by progenesis, the branch for tardigrades lies between nodes for *Aysheaia* and *Kerygmachela*.

This placement seems to define clearly the phylogenetic position of tardigrades, but because it falls between taxa of morphologically disparate organisms, lobopods and animals with lateral flaps (anomalopods; Waggoner, 1996), the resolution may be more apparent than real. A slight reinterpretation of the admittedly liberal criteria for homology could substantially alter the position of the tardigrades.

Nevertheless, this analysis is not very different from the conclusions of a recent report on the phylogeny of lobopods and arthropods (Budd, 1996) or a study of the tardigrade nervous system (Dewel and Dewel, 1996). Even though both of those studies placed tardigrades within the *Kerygmachela–Opabinia–Anomalocaris* group, they do so employing characters potentially affected by miniaturization, the absence of annuli (Budd, 1996) and the inferred supraoesophageal innervation of the frontal appendages (Dewel and Dewel, 1996).

## 10.4 CONCLUSIONS

To describe the phylogenetic position of tardigrades an analysis has been carried out on a large database of Cambrian lobopods and arthropods. Characters were coded from the perspective of tardigrade morphology, and homologies are interpreted to bridge morphological disparities between lobopods and arthropods. Few of the characters in living tardigrades are suitable for comparison with the fossilized material. Of those chosen, particular emphasis is given to morphology of the integument, pattern of head segmentation and the structure of cephalic and trunk appendages. Existing evidence suggests that the tardigrade head largely derives from three segments whose homologous metameres among arthropods would be pre-antennular and, except for the ocular segment, somewhat vestigial. The third cephalic segment is tentatively homologized with the ocular and frontal appendage segment of arthropods, and the configuration of cephalic appendages in other members of that phylum is based on that ancestral state. The phylogenetic analysis indicates that the node for tardigrades falls between the lobopod *Aysheaia* and the lateral flap animal *Kerygmachela*. At least two factors make this placement conjectural. Firstly, the coding of tardigrade characters is problematic since their miniaturization probably derives from progenesis, a potential source of homoplasy. Secondly, their placement between groups that exhibit significant morphological differences will be unstable until more apomorphies bridging the transition are found.

## ACKNOWLEDGEMENTS

The authors are grateful for illustrations provided by David McKirdy, David Scharf and Reinhardt Kristensen. Kelly P. Steele gave valuable advice concerning the Phylogenetic Analysis. Frederick Collier, Elizabeth Valiulis and Doug Erwin, National Museum of Natural History, Smithsonian Institution, Washington, DC, provided the senior author the opportunity to examine Burgess Shale fossils on two occasions. Graham Budd, Dieter Walossek, Lars Ramsköld and Jean Chaudonneret are thanked for stimulating discussions. We are especially grateful to Richard Thomas and Richard Fortey for organizing the Arthropod Symposium.

## REFERENCES

Averof, M. and Akam, M. (1995) Insect–crustacean relationships: Insights from comparative, developmental and molecular studies. *Philosophical Transactions of the Royal Society of London, Series B, Biological Science*, **347**, 293–303.

Ax, P. (1987) *The Phylogenetic System. The Systematization of Organisms on the Basis of their Phylogenesis*, John Wiley and Company, New York.

Backeljau, T., Winnepenninckx, B. and De Bruyn, L. (1993) Cladistic analysis of metazoan relationships: a reappraisal. *Cladistics*, **9**, 167–81.

Ballard, J.W.O., Olsen, G.J., Faith, D.P., Odgers, W.A., Rowell, D.M. and Atkinson, P.W. (1992) Evidence from 12S ribosomal RNA sequences that onychophorans are modified arthropods. *Science*, **258**, 1345–8.

Bartnik, E. and Weber, K. (1989) Widespread appearance of intermediate filaments in invertebrates; common principles and aspects of diversion. *European Journal of Cell Biology*, **50**, 17–33.

Benesch, R. (1969) Zur Ontogenie und Morphologie von *Artemia salina* L. *Zoologische Jahrbücher Abtielung für Anatomie und Ontogenie der Tiere*, **86**, 307–458.

Bergström, J. (1986) *Opabinia* and *Anomalocaris*, unique Cambrian 'arthropods'. *Lethaia*, **19**, 241–6.

Bergström, J. (1987) The Cambrian *Opabinia* and *Anomalocaris*. *Lethaia*, **20**, 187–8.

Bertolani, R. (1982) Tardigradi (Tardigrada). *Guide per il Riconoscimento delle Specie Animale delle Acque Interne Italiane*. Consiglio Nazionale delle Richerche, Verona, Italy.

Bertolani, R. (1987) Sexuality, reproduction and propagation in Tardigrada, in *Biology of Tardigrades. Selected Symposia and Monographs*. (ed. R. Bertolani), Collana U.Z.I. Mucchi, Modena, Italy, pp. 93–101.

Bertolani, R. (1989) Tardigrada, in *Reproductive Biology of the Invertebrates, Vol. IV, Part B: Fertilization, Development, and Maternal Care* (eds K.G. Adiyodi and R.G. Adiyodi), Oxford & IBH Publishing Co. Pvt. Ltd, New Delhi, India, pp. 49–60.

Binda, M.G. and Kristensen, R.M. (1986) Notes on the genus *Oreella* (Oreellidae) and the systematic position of *Carphania fluviatilis* Binda, 1978 (Carphanidae Fam. Nov., Heterotardigrada). *Animalia*, **13**, 9–20.

Briggs, D.E.G. and Collins, D. (1988) A Middle Cambrian chelicerate from Mount Stephen, British Columbia. *Paleontology*, **31**, 779–98.

Briggs, D.E.G. and Fortey, R.A. (1992) The Early Cambrian Radiation of Arthropods, in *Origin and Early Evolution of the Metazoa* (eds J.H. Lipps and P.W. Signor), Plenum Press, New York, pp. 335–73.

Briggs, D.E.G. and Whittington, H.B. (1987) The affinities of the Cambrian animals *Anomalocaris* and *Opabinia*. *Lethaia*, **20**, 185–6.

Budd, G.E. (1993) A Cambrian gilled lobopod from Greenland. *Nature*, **364**, 709–11.

Budd, G.E. (1996) The morphology of *Opabinia regalis* and the reconstruction of the arthropod stem group. *Lethaia*, **29**, 1–14.

Butterfield, N.J. (1994) Burgess shale-type fossils from a lower Cambrian shallow-shelf sequence in northwestern Canada. *Nature*, **369**, 477–9.

Chen, J.-Y., Ramsköld, L. and Zhou, G.-Q. (1994) Evidence for monophyly and arthropod affinity of Cambrian giant predators. *Science*, **264**, 1304–8.

Chen, J.-Y., Zhou, G.-Q. and Ramsköld, L. (1995) A new early Cambrian onychophoran-like animal, *Paucipodia* gen. nov., from the Chengjiang fauna, China. *Transactions of the Royal Society of Edinburgh*, **85**, 275–82.

Collins, D. (1996) The 'evolution' of *Anomalocaris* and its classification in the arthropod class Dinocarida (nov.) and Order Radiodonta (nov.). *Journal of Paleontology*, **70**, 280–93.

Conway Morris, S. (1994) Why molecular biology needs paleontology. *Development*, *1994* Supplement, 1–13.

Cooper, D.W. (1964) The first fossil tardigrade: *Beorn leggi* Cooper from Cretaceous amber. *Psyche*, **71**, 41–8.

Dewel, R.A. and Dewel, W.C. (1979) Studies on the tardigrades. IV. Fine structure of the hindgut of *Milnesiun tardigradum* Doyère. *Journal of Morphology*, **161**, 79–109.

Dewel, R.A. and Dewel, W.C. (1996) The brain of *Echiniscus viridissimus* (Heterotardigrada): a key to understanding the phylogenetic position of tardigrades and the evolution of the arthropod head. *Zoological Journal of the Linnean Society*, **116**, 35–49.

Dewel, R.A., Dewel, W.C. and Roush, B.G. (1992) Unusual cuticle-associated organs in the heterotardigrade *Echiniscus viridissimus*. *Journal of Morphology*, **212**, 123–40.

Dewel, R.A., Nelson, D.R. and Dewel, W.C. (1993) Tardigrada, in *Vol. 12. Onychophora, Chilopoda and Lesser Protostomata. Microscopic Anatomy of Invertebrates* (eds F.W. Harrison and M.E. Rice), Wiley-Liss, New York, pp. 143–83.

Dohle, W. (1974) The segmentation of the germ band of Diplopoda compared with other classes of arthropods. Symposium. *Zoological Society of London*, **32**, 143–61.

Durante, M.V. and Maucci, W. (1972) Descrizione di *Hybsibius* (*Isohyps.*) *basalovoi* sp. nov. e altre notizie su tardigradi del Veronese. *Memorie del Museo Civico di Storia Naturale di Verona*, **20**, 275–81.

Dzik, J. (1993) Early metazoan evolution and the meaning of its fossil record. *Evolutionary Biology*, **27**, 339–86.

Dzik, J. and Krumbiegel, G. (1989) The oldest 'Onychophroan' *Xenusion*: a link connecting phyla? *Lethaia*, **22**, 169–81.

Eernisse, D.J., Albert, J.S. and Anderson, F.E. (1992) Annelida and arthropoda are not sister taxa: a phylogenetic analysis of spirilian metazoan morphology. *Systematic Biology*, **41**, 305–30.

Eibye-Jacobsen, J. (1996) On the nature of pharyngeal muscle cells in the Tardigrada. *Zoological Journal of the Linnean Society*, **116**, 123–38.

Fortey, R.A. and Theron, J.N. (1994) A new Ordovician arthropod, *Soomaspis*, and the agnostid problem. *Palaeontology*, **37**, 841–61.

Garey, J.R., Krotec, M., Nelson, D.R. and Brooks, J. (1996) Molecular analysis supports a tardigrade–arthropod association. *Invertebrate Biology*, **115**, 79–88.

Giribet, G., Carranza, S., Baguñá, J., Riutort, M. and Ribera, C. (1996) First molecular evidence for the existence of a Tardigrada + Arthropoda clade. *Molecular Biology and Evolution*, **13**, 76–84.

Gould, S.J. (1977) *Ontogeny and Phylogeny*, Harvard University Press, Boston.

Greven, H. (1979) Notes on the structure of vasa Malpighii in the eutardigrade *Isohypsibius augusti* (Murray). *Zeszyty Naukowe Uniwersytetu Jagiellonskiego, Prace Zoologiczne*, **25**, 87–95.

Greven, H. (1980) Die Bartierchen. *Die Neue Brehm-Bucherei*, Vol. 537, Ziemsen Verlag, Wittenburg-Lutherstadt, Germany.

Greven, H. (1982) Homologues or analogues? A survey of some structural patterns in Tardigrada. in *Proceedings of the Third International Symposium on the Tardigrada*. (ed. D.R. Nelson), East Tennessee State University Press, Johnson City, TN, USA, pp 55–76.

Greven, H. (1983) Tardigrada, in *Biology of the Integument. Vol. I Invertebrates* (eds J. Bereiter-Hahn, A.G. Matolsky and K.S. Richards), Springer-Verlag, Berlin, pp. 714–27.

Greven, H. and Peters, W. (1986) Localization of chitin in the cuticle of Tardigrada using wheat germ agglutinin gold conjugate as a specific electron dense marker. *Tissue and Cell*, **18**, 297–304.

Grimaldi de Zio, S., D'Addabbo Gallo, M. and Morone De Lucia, R.M. (1982) *Neostygarctus acanthophorus*, n. gen. n. sp., nuovo tardigrado marino del Mediterraneo. *Cahiers di Biologie Marine*, **23**, 319–23.

Grimaldi de Zio, S., D'Addabbo Gallo, M. and Morone De Lucia, R.M. (1987a) Adaptive radiation and phylogenesis in marine Tardigrada and the establishment of Neostygarctidae, a new family of Heterotardigrada. *Bollettino di Zoologia*, **54**, 27–33.

Grimaldi de Zio, S., D'Addabbo Gallo, M., Morone De Lucia, R.M. and D'Addabbo, L. (1987b) Marine Arthrotardigrada and Echiniscoidea (Tardigrada, Heterotardigrada) from the Indian Ocean. *Bollettino di Zoologia*, **54**, 347–57.

Grimaldi de Zio, S., D'Addabbo Gallo, M. and Morone De Lucia, R.M. (1988) Two new Mediterranean species of the genus *Halechiniscus* (Tardigrada, Heterotardigrada). *Bollettino di Zoologia*, **55**, 205–11.

Grimaldi de Zio, S., D'Addabbo Gallo, M. and Morone De Lucia, R.M. (1992) *Neoarctus primigenius* n. g. n. sp., a new Stygarctidae of the Tyrrhenian Sea (Tardigrada, Arthrotardigrada). *Bollettino di Zoologia*, **59**, 309–13.

Grimaldi de Zio, S., D'Addabbo Gallo, M., Morone De Lucia, R.M. and Troccoli, A. (1996) *Hemitanarctus chimaera* n.g., n.sp., new Halechiniscidae from the Ionian Sea (Tardigrada, Heterotardigrada). *Zoologischer Anzeiger*, **234**, 167–74.

Henry, L.M. (1948) The nervous system and the segmentation of the head in the Annulata. *Microentomology*, **13**, 1–26.

Hou, X.-G. and Bergström, J. (1995) Cambrian lobopodans-ancestors of extant onychophorans? *Zoological Journal of the Linnean Society*, **114**, 3–19.

Hou, X.-G., Bergström, J. and Ahlberg, P. (1995) *Anomalocaris* and other large animals in the lower Cambrian Chengjiang fauna of southwest China. *Geologiska Foreningen i Stockholm Förhandlingar*, **117**, 163–83.

Jacobs, D.K. (1992) Two applications of developmental genetics to paleontology: Segmentation genes in molluscs and preoral appendages in taxa of uncertain affinity. *Paleontological Society Special Publications*, **6**, 145 (abstract)

Jones, G.M. (1979) The development of amphids and amphidial glands in adult *Syngamus trachea* (Nematoda: Syngamidae). *Journal of Morphology*, **160**, 299–322.

Kinchin, I.M. (1994) *The Biology of Tardigrades*, Portland Press, London.

Kristensen, R.M. (1978) On the structure of *Batillipes noerrevangi Kristensen* 1978. 2. The muscle attachments and the true cross-striated muscles. *Zoologische Anzeiger*, **200**, 173–84.

Kristensen, R.M. (1981) Sense organs of two marine arthrotardigrades (Heterotardigrada, Tardigrada). *Acta Zoologica*, **62**, 27–41.

Kristensen, R.M. (1987) Generic revision of the Echiniscidae (Heterotardigrada) with a discussion of the origin of the family, in *Biology of Tardigrades. Selected Symposia and Monographs* (ed. R. Bertolani), Collana U.Z.I., Mucchi, Modena, Italy, pp. 261–335.

Kristensen, R.M. (1991) Loricifera – A general and phylogenetic overview. *Verhandlungen der Deutschen Zoologischen Gesellschaft*, **84**, 231–46.

Kristensen, R.M. and Hallas, T.E. (1980) The tidal genus *Echiniscoides* and its variability, with the erection of Echiniscoididae fam. n. (Tardigrada). *Zoologica Scripta*, **9**, 113–27.

Kristensen, R.M. and Higgins, R.P. (1984) Revision of *Styraconyx* (Tardigrada: Halechiniscidae), with descriptions of two new species from Disko Bay, West Greenland. *Smithsonian Contributions in Zoology*, **391**, 1–40.

Kristensen, R.M. and Higgins, R.P. (1989) Marine Tardigrada from the southeastern United States coastal waters. I. *Paradoxipus orzeliscoides* n. gen., n. sp. (Arthrotardigrada: Halechiniscidae). *Transactions of the American Microscopical Society*, **108**, 262–82.

Kristensen, R.M. and Renaud-Mornant, J. (1983) Existence d'arthrotardigrades semi-benthiques de genres nouveaux de la sous-famille des Styraconyxinae subfam. nov. *Cahiers de Biologie Marine*, **24**, 337–53.

Lauterbach, K.E. (1978) Gedanken zur Evolution der Euarthropoden-Extremität. *Zoologische Jahrbücher Abteilung für Anatomie und Ontogenie der Tiere*, **99**, 64–92.

Locke, M. and Huie, P. (1977) Bismuth staining of Golgi complex is a characteristic arthropod feature lacking in *Peripatus*. *Nature*, **270**, 341–3.

Malakhov, V.V. (1994) *Nematodes: Structure, Development Classification, and Phylogeny*, Smithsonian Institution Press, Washington, DC.

Marcus, E. (1929) Tardigrada. in *Klassen und Ordnungen des Tierreichs*, Bd. 5, Abtlg. IV. Buch 6 (ed. H.G. Bronn), Akademische Verlagsgellschaft, Leipzig, Germany.

McKinney, M.L. and McNamara, K.J. (1991) *Heterochrony: The Evolution of Ontogeny*, Plenum Press, New York.

McKirdy, D., Schmidt, P. and McGinty-Bayly, M. (1976) Tardigrada. *Mikrofauna des Meeresboden*, **58**, 1–43.

Møbjerg, N. and Dahl C. (1996) Studies on the morphology and ultrastructure of the Malpighian tubules of *Halobiotus crispae* Kristensen, 1982 (Eutardigrada). *Zoological Journal of the Linnean Society*, **116**, 85–99.

Monje-Najera, J. (1995) Phylogeny, biogeography and reproductive trends in the Onychophora. *Zoological Journal of the Linnean Society*, **114**, 21–60.

Moon, S.Y. and Kim, W. (1996) Phylogenetic position of the Tardigrada based on the 18S ribosomal RNA gene sequences. *Zoological Journal of the Linnean Society*, **116**, 61–9.

Müller, K.J. and Walossek, D. (1986) Arthropod larvae from the Upper Cambrian of Sweden. *Transactions of the Royal Society of Edinburgh: Earth Sciences*, **77**, 157–79.

Müller, K.J., Walossek, D. and Zakharov, A. (1995) 'Orsten' type phosphatized soft-integument preservation and a new record from the Middle Cambrian Kuonamka formation in Siberia. *Neue Jahrbucher für Geologische und Palaontologische Abhandlungen*, **197**, 101–18.

Nelson, D.R. (1982) Developmental Biology of the Tardigrada, in *Developmental Biology of Freshwater Invertebrates* (eds F.W. Harrison and R.R. Cowden), A.R. Liss, New York, pp. 363–98.

Nelson, D.R. (1991) Tardigrada, in *Ecology and Classification of North American Freshwater Invertebrates* (eds J.H. Thorp and A.P. Covich), Academic Press, New York, pp. 501–22.

Nelson, D.R. and Higgins, R.P. (1990) Tardigrada, in *Soil Biology Guide* (ed. D.L. Dindal), John Wiley and Sons, New York, pp. 393–419.

Nielsen, C. (1995) *Animal Evolution: Interrelationships of the Living Phyla*, Oxford University Press, Oxford, UK.

Nielsen, C., Scharff, N. and Eibye-Jacobsen, D. (1996) Cladistic analysis of the animal kingdom. *Biological Journal of the Linnean Society*, **57**, 385–410.

Noda, H. (1985) Description of a new subspecies of *Angursa biscupis* Pollock (Heterotardigrada, Halechiniscidae) from Tanabe Bay. *Seto Marine Biological Laboratory Publications*, **30**, 269–76.

Pollock, L.W. (1970) *Batillipes dicrocercus* n. sp., *Stygarctus granulatus* n. sp. and other Tardigrada from Woods Hole, Massachusetts, USA. *Transactions of the American Microscopical Society*, **89**, 38–53.

Pollock, L.W. (1975) Tardigrada, in *Reproduction of Marine Invertebrates, Vol. II* (eds A.C. Giese and J.S. Pearse), Academic Press, Inc, New York, pp. 43–54.

Pollock, L.W. (1995) New marine tardigrades from Hawaiian beach sand and phylogeny of the family Halechiniscidae. *Invertebrate Biology*, **114**, 220–35.

Raff, R.A., Marshall, C.R. and Turbeville, J.M. (1994) Using DNA sequences to unravel the Cambrian radiation of the animal phyla. *Annual Review of Ecology and Systematics*, **25**, 351–75.

Ramazzotti, G. and Maucci, W. (1982) A history of tardigrade taxonomy, in *Proceedings of the Third International Symposium on the Tardigrada* (ed. D.R. Nelson), East Tennessee State University Press, Johnson City, TN, USA, pp. 11–30.

Ramsköld, L. (1992) Homologies in Cambrian Onychophora. *Lethaia*, **25**, 443–60.

Renaud-Mornant, J. (1982) Species diversity in marine Tardigrada, in *Proceedings of the Third International Symposium on the Tardigrada* (ed. D.R. Nelson), East Tennessee State University Press, Johnson City, TN, USA, pp. 149–77.

Rieger, G.E. and Rieger, R.M. (1977) Comparative fine structural study of the gastrotrich cuticle and aspects of cuticle evolution within the Aschelminthes. *Zeitschrift für Zoologische Systematik und Evolutionsforschung*, **15**, 81–124.

Rieppel, O. (1993) The conceptual relationship of ontogeny, phylogeny, and classification: the taxic approach. *Evolutionary Biology*, **27**, 1–32.

Schmidt-Ott, U., Gonzalez-Gaitan, M. and Technau, G.M. (1995) Analysis of neural elements in head-mutant *Drosophila* embryos suggests segmental origin of optic lobes. *Roux's Archives of Developmental Biology*, **205**, 31–44.

Scholtz, G. (1995) Head segmentation in Crustacea – an immunochemical study. *Zoology*, **98**, 104–14.

Schuster, R.O., Nelson, D.R., Grigarick, A.A. and Christenberry, D. (1980) Systematic criteria of the Eutardigrada. *Transactions of the American Microscopical Society*, **99**, 284–303.

Simonetta, A.M. and Delle Cave, L. (1991) Early Paleozoic arthropods and problems of arthropod phylogeny; with some notes on taxa of doubtful affinities, in *The Early Evolution of Metazoa and the Significance of Problematic Taxa* (eds A.M. Simonetta and S. Conway Morris), Cambridge University Press, Cambridge, UK, pp. 189–244.

Storch, V. and Ruhberg, H. (1977) Fine structure of the sensilla of *Peripatus moseleyi* (Onychophora). *Cell and Tissue Research*, **177**, 539–53.

Storch, V. and Ruhberg, H. (1993) Onychophora, in *Vol. 12. Onychophora, Chilopoda and Lesser Protostomata. Microscopic Anatomy of Invertebrates* (eds F.W. Harrison and M.E. Rice), Wiley-Liss, New York, pp. 11–56.

Swofford, D.L. (1993) *PAUP: Phylogenetic analysis using parsimony, version 3.1.1.* Illinois Natural History Survey, Champaign, Illinois, USA.

Waggoner, B.M. (1996) Phylogenetic hypotheses of the relationships of arthropods to Precambrian and Cambrian problematic fossil taxa. *Systematic Biology*, **45**, 190–222.

Walker, M.H. and Campiglia, S. (1988) Some aspects of segment formation and post-placental development in *Peripatus acacioi* Marcus and Marcus (Onychophora). *Journal of Morphology*, **195**, 123–40.

Walossek, D. (1995) The Upper Cambrian *Rehbachiella*, its larval development, morphology and significance for the phylogeny of Branchiopoda and Crustacea. *Hydrobiologia*, **298**, 1–13.

Walossek, D. and Müller, K.J. (1990) Upper Cambrian stem-lineage crustaceans and their bearing upon the monophyletic origin of Crustacea and the position of *Agnostus*. *Lethaia*, **23**, 409–27.

Walz, B. (1974) The fine structure of somatic muscles of Tardigrada. *Cell and Tissue Research*, **149**, 81–9.

Weglarska, B. (1975) Studies on the morphology of *Macrobiotus richtersi* Murray. *Memorie Istituto Italiano di Idrobiologica*, Supplement **32**, 445–64.

Weglarska, B. (1980) Light and electron microscopic studies on the excretory system of *Macrobiotus richtersi*, Murray 1911 (Eutardigrada). *Cell and Tissue Research*, **207**, 171–82.

Weglarska, B. (1987a) Studies on the excretory system of *Isohypsibius granulifer* Thulin Eutardigrada, in *Biology of Tardigrades*. Selected Symposia and monographs, I. (ed. R. Bertolani), Mucchi Editore, Collana U.Z.I., Modena, Italy, pp. 15–24.

Weglarska, B. (1987b) Morphology and ultrastructure of excretory system in *Dactylobiotus dispar* (Murray) (Eutardigrada), in *Biology of Tardigrades*. Selected Symposia and monographs, I. (ed. R. Bertolani), Mucchi Editore, Collana U.Z.I., Modena, Italy, pp. 25–33.

Weygoldt, P. (1975) Untersuchen zur embryologie und morphologie der geisselspinne *Tarantula marginemaculata* C.L. Koch (Arachnida, Amblypygi, Tarantulidae). *Zoomorphologie*, **82**, 137–99.

Weygoldt, P. (1979) Significance of later embryonic stages and head development in arthropod phylogeny, in *Arthropod Phylogeny* (ed. A.P. Gupta), Van Nostrand Reinhold, New York, pp. 107–35.

Weygoldt, P. (1986) Arthropod interrelationships – the phylogenetic systematic approach. *Zeitschrift für Systematik Evolutionsforschung*, **24**, 19–35.

Wheeler, W.C., Cartwright, P. and Hayashi, C.Y. (1993) Arthropod phylogeny: a combined approach. *Cladistics*, **9**, 1–39.

Whittington, H.B. (1975) The enigmatic animal *Opabinia regalis*, Middle Cambrian, Burgess Shale, British Columbia. *Philosophical Transactions of the Royal Society of London, Series B, Biological Science*, **271**, 1–43.

Whittington, H.B. (1978) The lobopod animal *Aysheaia pedunculata* Walcott, Middle Cambrian, Burgess Shale, British Columbia. *Philosophical Transactions of the Royal Society of London, Series B, Biological Science*, **284**, 165–97.

Whittington, H.B. and Briggs, D.E.G. (1985) The largest Cambrian animal, *Anomalocaris*, Burgess Shale, British Columbia. *Philosophical Transactions of the Royal Society of London, Series B, Biological Science*, **309**, 569–609.

Wiederhöft, H. and Greven, H. (1996) The cerebral ganglia of *Milnesium tardigradum* Doyère (Apochela, Tardigrada): Three dimensional reconstruction and notes on their ultrastructure. *Zoological Journal of the Linnean Society*, **116**, 71–84.

Wills, M.A., Briggs, D.E.G. and Fortey, R.A. (1994) Disparity as an evolutionary index: a comparison of Cambrian and Recent arthropods. *Paleobiology*, **20**, 93–130.

Wills, M.A., Briggs, D.E.G., Fortey, R.A. and Wilkinson, M. (1995) The significance of fossils in understanding arthropod evolution. *Verhandlungen Deutsche Zoologische Gesellschaft*, **88**, 203–15.

Winnepenninckx, B., Backeljau, T., Mackey, L.Y., Brooks, J.M., De Wachter, R., Kumar, S. and Garey, J.R. (1995) 18S rRNA data indicate that Aschelminthes are polyphyletic in origin and consist of at least three distinct clades. *Molecular Biology and Evolution*, **12**, 1132–7.

Wright, J.C. and Luke, B.M. (1989) Ultrastructural and histochemical investigations of *Peripatus* integument. *Tissue and Cell*, **21**, 605–25.

Wright, K.A. (1991) Nematoda, in *Vol. 4. Microscopic Anatomy of Invertebrates* (eds F.W. Harrison and E.E. Ruppert), Wiley-Liss, New York, pp. 111–95.

# 11 Stem group arthropods from the Lower Cambrian Sirius Passet fauna of North Greenland

*G.E. Budd*

*Department of Historical Geology and Palaeontology, Institute of Earth Sciences, Norbyvägen 22, S-75236 Uppsala, Sweden*
*email: Graham.Budd@pal.uu.se*

## 11.1 INTRODUCTION

Discussion of fossil evidence for the origin and early evolution of the arthropods has been dominated for many years by the evidence from the Middle Cambrian Burgess Shale from British Columbia (Whittington, 1979; Gould, 1989). What these fossils mean, however, both in terms of arthropod classification and the early evolution of the phylum is far from clear: no single opinion has won universal assent. In addition, the Burgess Shale has also yielded some celebrated problematica such as *Opabinia* (Whittington, 1975), *Anomalocaris* (Whittington and Briggs, 1985; Collins, 1996) and *Hallucigenia* (Conway Morris, 1977; Ramsköld and Hou, 1991). No one would dispute that these fossils **are** problematic, in the sense that they are difficult to understand. However, that methodological difficulty should not be confused with the possibility that these fossils have only remote affinities with all living groups. The sense that these fossils can indeed be understood has strengthened in recent years with the discovery of several more important Cambrian fossil localities. They have shed important new light on the Burgess Shale itself, and thus on one important source of information about arthropod phylogeny. One of these new biotas, the Sirius Passet fauna, will be discussed here, with particular regard to the light it sheds on the origins of the euarthropods.

## 11.2 SIRIUS PASSET FOSSILS

The Sirius Passet fauna was discovered by chance in 1984. The rock unit that yields it, the Lower Cambrian Buen Formation, crops out in the far north of Greenland, and was probably deposited on the outer continental shelf. Several collecting trips have been made to the site, and perhaps 10 000 individual fossils have been collected, rather fewer than are known from either the Burgess Shale or the Chengjiang fauna. The preservation of the fossils is not in general spectacular, and careful study of the specimens is required in order to understand their morphology. At least 15 arthropod taxa are present in the fauna, together with several other taxa whose affinities may lie with the arthropods.

### 11.2.1 SIRIUS PASSET EUARTHROPODS

Sirius Passet arthropods provide a good cross-section of Cambrian arthropod morphology, from a single trilobite genus (*Buenellus*; see Blaker, 1988) and a probable stem group trilobite *Kleptothule* (Budd, 1995), through chelicero-morphs and crustaceanomorphs. As with many of the Cambrian arthropods, the relationships of these taxa are not always clear, although progress is being made in elucidating the phylogeny. Part of the problem is that with no certain out-group identified, it is difficult to polarize characters within the group as a whole. Without this sort of anchor, resolving issues such as whether the crustaceans are paraphyletic and basal or not (Briggs and Fortey, 1989) is bound to prove difficult. The only taxa that have traditionally been placed as close sister groups of the arthropods are the onychophorans and the tardigrades, but these do not fully resolve the order of acquisition of most arthropodan features.

### 11.2.2 SIRIUS PASSET TAXA AND THE ARTHROPOD STEM GROUP

The Sirius Passet taxon *Kerygmachela* has been proposed as a lobopod, as well as a relative of *Opabinia* and *Anomalocaris* (Budd, 1993, 1996). Given the various arthropod-like characters of *Anomalocaris* recognized (Hou *et al.*, 1995) anomalocaridid-like taxa would now reside in the arthropod stem group, and thus contribute to an understanding of how the first arthropods evolved. At the base of the stem group lies the onychophorans and other lobopods (Budd, 1996). Such a model for arthropod evolution has proved to be controversial. In part this is because the presence of lobopodous limbs in *Kerygmachela* has not been demonstrated to everyone's satis-

*Arthropod Relationships*, Systematics Association Special Volume Series 55, Edited by R.A. Fortey and R.H. Thomas. Published in 1997 by Chapman & Hall, London. ISBN 0 412 75420 7

faction (Chen *et al.*, 1994; Peterson, 1995). An allied suggestion is that this genus, along with *Anomalocaris* and *Opabinia*, form a monophyletic clade within the euarthropods (Chen *et al.*, 1994). While this is possible, it requires a number of important character reversals. In particular, *Kerygmachela* would be required to have secondarily lost ancestral dorsal articulating segmentation and to have gained flexibility through an annulated, relatively soft and unsclerotized exoskeleton. Furthermore, the **detail** of the construction of the body of *Kerygmachela* is very similar to that of the extant onychophorans, especially with regard to the axial structures including a putative pericardial dorsal sinus (Figure 11.1). Unlike other Sirius Passet taxa (see *Pambdelurion* nov. below) *Kerygmachela* shows no signs of possessing discrete bundles of musculature working on a lever system. Although the documentation of trunk limbs is limited, the frontal appendages at least are well preserved, and these are clearly unsegmented (Figure 11.2). These characters are all present in lobopods, and so there an *a priori* reason for thinking they may be plesiomorphic rather than secondary modifications from a fully arthropodized state. This conclusion is supported by cladistic analysis (see below). Similarly, the 'Peytoia' mouth part of the anomalocaridids **may** be a derived feature within the arthropods, as recently suggested by Collins (1996), but given the presence of apparently rather similar structures in onychophorans, tardigrades (Dewel and Dewel, 1996, 1997, this volume) and indeed some pseudocoelomates (Hou *et al.*, 1995), it is more plausible to see it as a retained plesiomorphy within the arthropod stem-lineage. A further objection to a model placing *Anomalocaris* and its relatives within the arthropod stem group comes from opposition to the idea that phylum-level evolutionary patterns are discernible in the fossil record (Peterson, 1995), an idea based on the rapid evolution of phyla from cryptic ancestors (Davidson *et al.*, 1995). More data on the morphology of anomalocaridid-like taxa are evidently relevant to this problem: a new example, *Pambdelurion whittingtoni* nov. is described here.

## 11.3 SYSTEMATIC PALAEONTOLOGY

SUPERPHYLUM LOBOPODIA SNODGRASS 1938

*Pambdelurion whittingtoni* gen. et sp. nov.

### Etymology

*Pambdelurion*, from pambdelyrion (Greek), all-abominable or all-loathsome, in honour of the new taxon's formidable appearance.

*whittingtoni*, in honour of Professor H.B. Whittington, FRS, and to celebrate his 80th birthday. The gender is masculine.

### Holotype

MGUH 24508 from GGU 340103.3637 (Figures 11.3–11.5) (GGU, Geological Survey of Greenland, Copenhagen (now part of Geological Survey of Denmark and Greenland). MGUH, Geological Museum, Copenhagen.)

### Paratypes

MGUH 24509 from GGU: 340103.3103a (Figures 11.6–11.8); MGUH 24510 from GGU: 3403103.3103b; MGUH 24511 from GGU 340103.1573 (Figures 11.6, 11.7); MGUH 24511 from GGG 340103.3807; MGUH 24513 from GGU 340103.305.

### Other material

About 300 other specimens are known, many of which are very poorly preserved.

### Horizon and locality

Buen Formation (Atdabanian, Lower Cambrian), J.P. Koch Fjord, Peary Land, North Greenland. For discussion of locality and geological setting, see Conway Morris and Peel (1995), and Budd (1995).

### Diagnosis

Anomalocaridid-like taxon, but with annulated and flexible spinose frontal appendages. No eyes known. 'Peytoia' mouth-part poorly sclerotized. Trunk bears lobopodous limb pairs probably unconnected to lateral flaps. Tail poorly known, apparently consisting of sub-circular flap.

### Description

*Pambdelurion* is one of the more common animals in the Sirius Passet fauna, with some 300 specimens in collections. No complete specimens are known, but several specimens lack only part of the tail region; they are often very poorly preserved (compared with other Sirius Passet arthropods), especially in their lateral regions, indicating a weakly sclerotized cuticle.

A wide size range of *Pambdelurion* is known, with body lengths (excluding appendages) ranging from 45 mm to an estimated 290 mm.

The head region is small, and bears a ventral mouth, surrounded by a probably complete circle of cuneiform plates (Figures 11.4 and 11.5; the mouth part in this specimen is, however, incomplete). This is 'the 'Peytoia' mouth structure, flanked anteriorly by a pair of ventrally attached, flexible, annulated appendages (Figures 11.3–11.5), each bearing a paired series of thick, flexible spines along the anterior margin; the appendages terminate in a cluster of four much longer spines around a pointed appendage tip. The trunk shows traces of axial transverse wrinkling, but with no clear signs of segmentation. It bears a series of 11 lateral flaps which are distally longitudinally wrinkled. Additionally, two or three smaller flaps may be present in the cephalic region ('occipital flaps'), although the evidence is at best equivocal (Figure 11.4).

Complete limbs are known from several fortuitously preserved specimens that were buried at an angle to bedding, so allowing the limbs to become separated from the lateral lobes

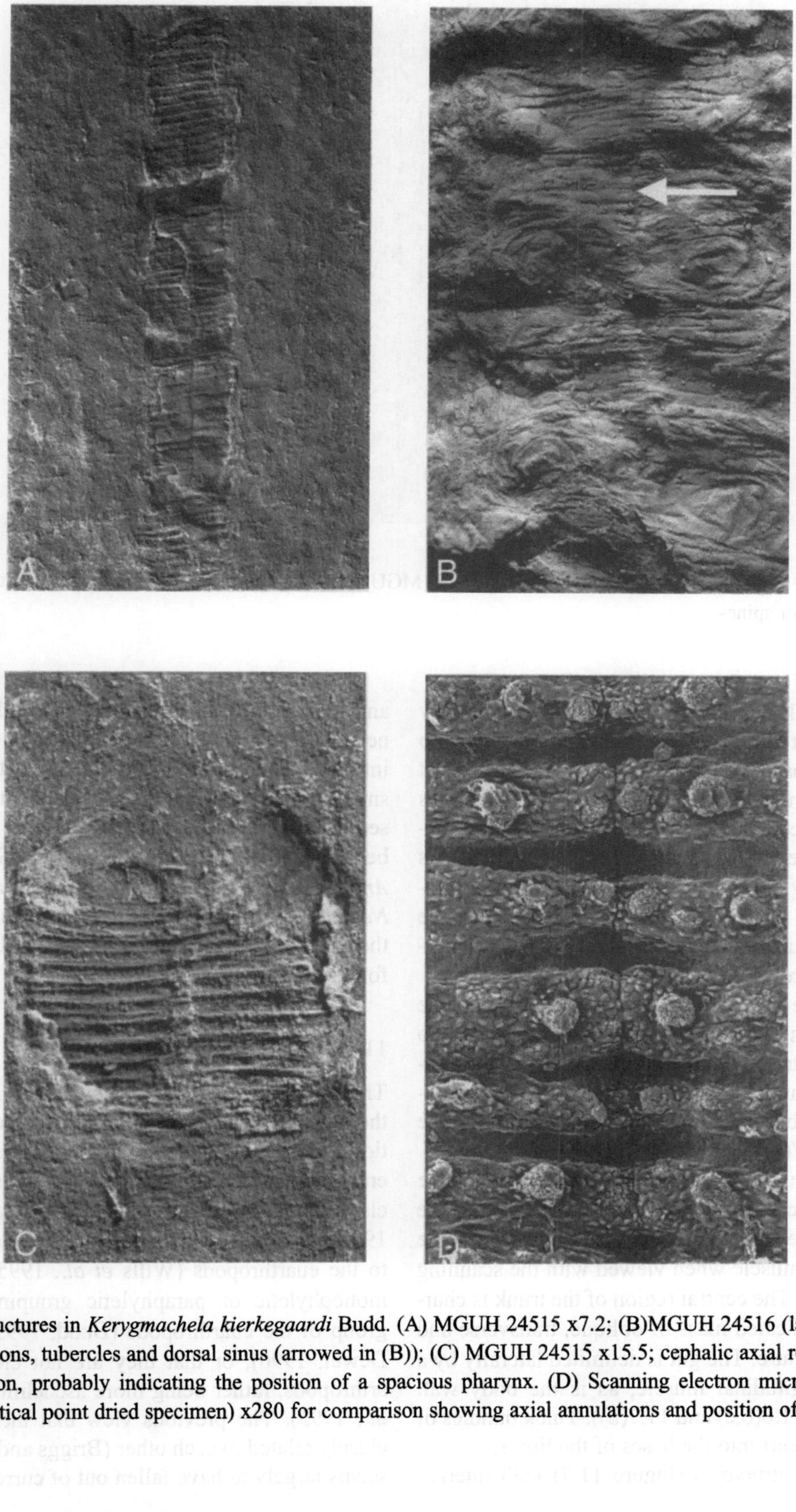


**Figure 11.1** Axial structures in *Kerygmachela kierkegaardi* Budd. (A) MGUH 24515 x7.2; (B)MGUH 24516 (latex) x9.2; dorsal surfaces showing annulations, tubercles and dorsal sinus (arrowed in (B)); (C) MGUH 24515 x15.5; cephalic axial region showing insular region of mineralization, probably indicating the position of a spacious pharynx. (D) Scanning electron micrograph of dorsum of *Euperipatoides* sp. (critical point dried specimen) x280 for comparison showing axial annulations and position of dorsal sinus.

**Figure 11.2** Cephalic region of *Kerygmachela kierkegaardi*. MGUH 22.084 (paratype) x5.0 showing annulated, frontal lateral appendages with anterior spines.

(Figures 11.6–11.8). The limbs seem to have been smoothly flexible, and bear least 50 annulations extending from the limb base to the tip. The bases of the limbs are broad, but do not appear to bear any structures such as gnathobases. The limbs are also preserved in conjunction with the lateral lobes in several specimens (Figures 11.6(b) and 11.7(b)), although there is no evidence that they are directly connected to form a biramous structure. The posterior of the animal seems to be marked by a sub-circular tail structure, although this is unfortunately not clearly preserved in any known specimens.

The dorsal surface of the animal is poorly known, and the presence of eyes has yet to be demonstrated. There is no clear sign of segmentation or annulation. However, the ventral surface of the animal shows sets of transverse annulations, which may be interrupted where the limbs are attached (Figure 11.7(a)).

A great deal of musculature has been preserved in some specimens. The muscle fibres – the oldest known from the fossil record – can be seen to possess the typical transverse structure of striated muscle when viewed with the scanning electron microscope. The central region of the trunk is characterized by a complicated mass of oblique, transverse and longitudinal musculature. The gut is delimited laterally by a thick sheath of longitudinal muscle, as is the body wall (Figures 11.3, 11.4, 11.6(b,c) and 11.7(b)). Thick bundles of muscle fibres also insert into the bases of the limbs.

Pairs of reniform structures (Figure 11.9) with internal anastomoses are situated either side of the gut, and are connected to it by short transverse tubes. In some specimens the internal structures are linear and elongate, in others they are small and circular, suggesting different sections through a set of radiating tubes. Similar reniform structures appear to be present in *Opabinia* (Budd, 1996, Figure 1(c)), *Anomalocaris*, and many Cambrian arthropods such as *Naraoia* (Whittington, 1977). Given their association with the gut, they are likely to represent mid-gut diverticulae, as found in extant chelicerates.

### 11.3.1 DISCUSSION

The problematic forms *Opabinia* and *Anomalocaris* from the Burgess Shale have been the subject of much speculation (see review in Budd, 1996). At present, there are several theories concerning their affinities: that they form a clade within the euarthropods (Chen *et al.*, 1994; Collins, 1996); that they form a monophyletic clade distantly related to the euarthropods (Wills *et al.*, 1995); that they form a monophyletic or paraphyletic grouping within the stem group of the euarthropods (Budd, 1993, 1996; Dewel and Dewel, 1996); or that they are not closely related to the arthropods, rather being more aschelminth in form (Hou *et al.*, 1995). The previous view that the two taxa were not closely related to each other (Briggs and Whittington, 1987) seems largely to have fallen out of currency.

**Figure 11.3** *Pambdelurion whittingtoni* nov. gen. et sp. MGUH 24508 x0.5 (Holotype), almost complete specimen lacking only tail region. Compare Figure 11.4.

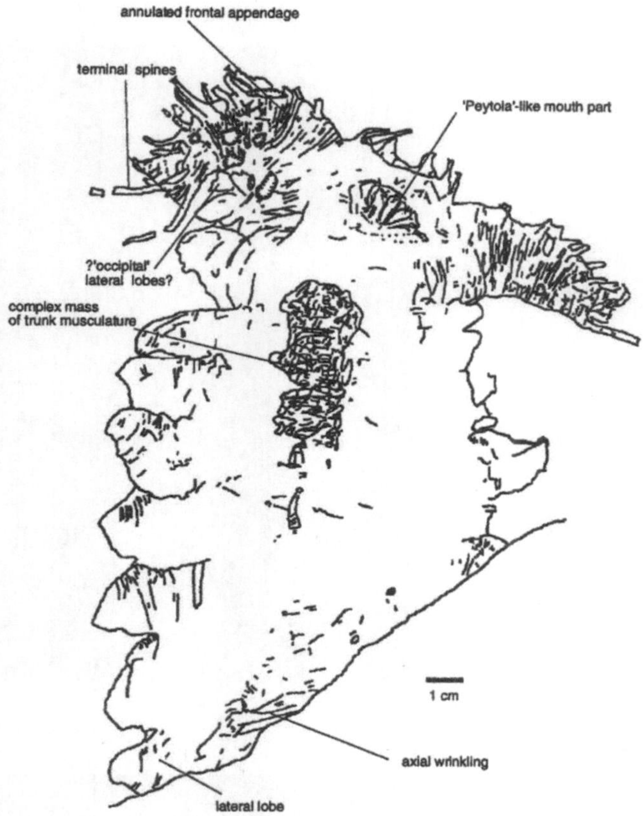


**Figure 11.4** Drawing of *Pambdelurion whittingtoni*, MGUH 24508 (Holotype), ventral view showing details of cephalic region. It is possible that the small structures just posterior to the right appendage represent small, occipital lobes which are also known from *Anomalocaris*.

The view of anomalocaridid relationships envisaged by Budd (1993, 1996) and Dewel and Dewel (1996) is supported by the discovery of *Pambdelurion*. This genus is similar to described anomalocaridids, with shared features including the presence of ventral frontal appendages surrounding a 'Peytoia'-like mouth-part, a body containing eleven segments with lateral, striated lobes and limbs, and gut diverticulae (Chen *et al.*, 1994). However, the differences in appendage morphology are profound. The frontal appendages of *Pambdelurion* are highly flexible and annulated, being able to bend backward as well as forward, unlike the precisely articulated appendages of *Anomalocaris* and its close relatives (Whittington and Briggs, 1985; Chen *et al.*, 1994; Hou *et al.*, 1995; Nedin, 1996), which could only flex in one direction. Furthermore, the trunk appendages are clearly lobopodous, and show no clear signs of possessing gnathobases or being biramous. This contrasts with the gnathobasic, biramous limbs possessed by *Parapeytoia* from the Chengjiang fauna (Hou *et al.*, 1995). The morphology of *Pambdelurion* thus combines features from both lobopodous and more arthropodous taxa.

Parsimony analysis of lobopods and arthropods together also supports the view that known lobopods are paraphyletic. A cladistic analysis of 30 characters and seven taxa produced one most parsimonious tree (Figure 11.10; Table 11.1; Appendix 11.1A). By optimizing the character state changes to favour multiple loss over multiple gain of characters, it is possible to reconstruct the character states of the internal nodes of the tree. In particular, the character states of the last common ancestor of the euarthropods and anomalocaridids are almost identical with those of the anomalocaridids themselves, suggesting that they may represent a metataxon (i.e. a

**Figure 11.5** *Pambdelurion whittingtoni* MGUH 24508 (Holotype) showing details of head and appendage structure. Right appendage illustrated: (A) x2.2; (B) x7.6. Notice appendage annulations and relatively broad, flexible spines. Compare Figure 11.4.

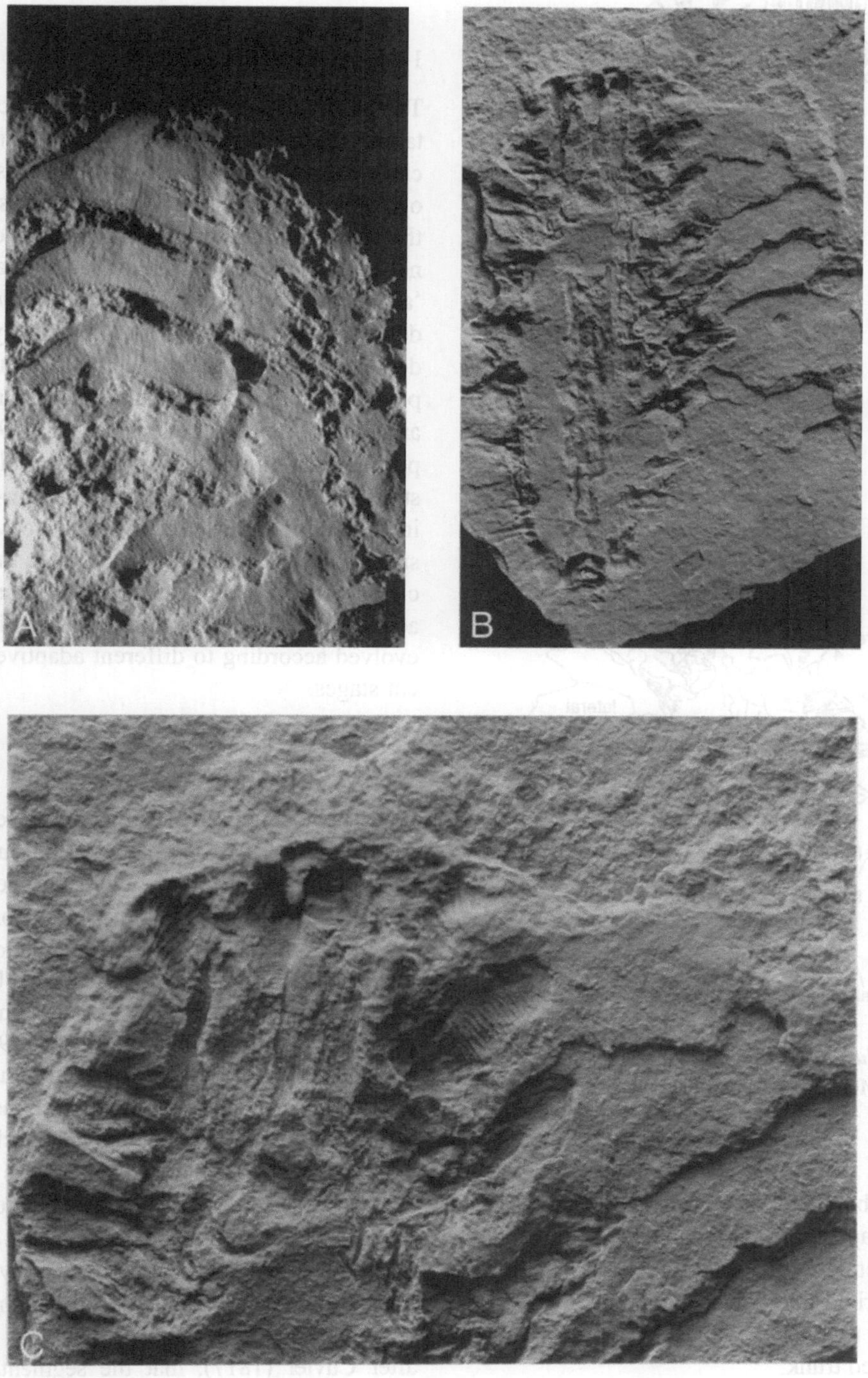


**Figure 11.6.** ***Pambdelurion whittingtoni*****. Limb-bearing specimens. (A) MGUH 24509 x0.95. Ventral view showing ventral surface and limb series of the right side. Compare Figures 11.7(A) and 11.8. (B,C) MGUH 24511. Compare Figure 11.7(B). (A) x2.1, small *Pambdelurion* specimen showing lateral lobes and musculature delineating gut and body wall. (B) x5.0, close-up of cephalic region showing bases of annulated limbs preserved as concavities in the lateral lobes.**

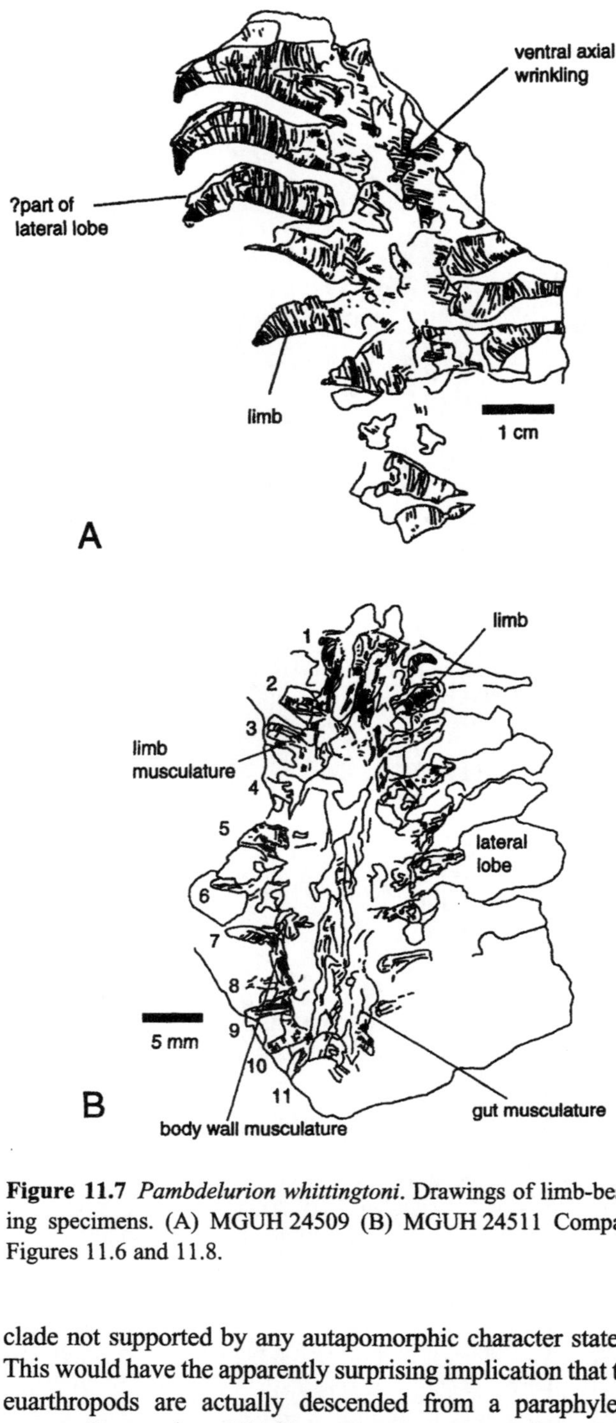


**Figure 11.7** *Pambdelurion whittingtoni*. Drawings of limb-bearing specimens. (A) MGUH 24509 (B) MGUH 24511 Compare Figures 11.6 and 11.8.

clade not supported by any autapomorphic character states). This would have the apparently surprising implication that the euarthropods are actually descended from a paraphyletic group of anomalocaridid-like animals, complete with differentiated head and tail regions, 'Peytoia'-like mouth part, large size and 11-segmented trunk.

## 11.4 CHARACTER EVOLUTION IN STEM GROUP ARTHROPODS

If the view of arthropod relationships presented here is accepted, then it is possible to make further comments on character evolution. Rather than being a set of totally interdependent characters, 'arthropodization' appears to be a complex and – phylogenetically, if not temporally – drawn-out process. This may be illustrated by two examples of character change, sclerotization and segmentation.

### 11.4.1 SCLEROTIZATION

The ability to sclerotize the cuticle by chitinoproteinaceous tanning seems to be plesiomorphic within the entire lobopod clade. However, the extent to which this occurs within various taxa varies greatly. Within the most basal taxa such as the onychophorans and *Aysheaia*, it is confined to the terminal limb claws and (in the onychophorans) the jaws. In the 'armoured lobopods' such as *Hallucigenia* there is a further development of sclerotized, presumably defensive pairs of dorsal spines. Towards the crown-group, anomalocaridids possess both sclerotized but non-articulated ventral sclerites and jointed appendages (Hou *et al.*, 1995). However, their poor preservation and flexibility (Collins, 1996) demonstrates that the cuticle overall was non-sclerotized. Finally, in the euarthropods, sclerotization covers the entire body surface. Despite the common opinion that a sclerotized cuticle was the key to the success of the arthropods, the actual acquisition of such a character seems diffuse, and may have evolved according to different adaptive pressures at different stages.

### 11.4.2 SEGMENTATION

If the phylogenetic reconstruction presented here is correct, segmentation as represented by the euarthropods does not evolve until some point within the stem group. The most primitive state seems to be a homonomous ectoderm, followed by the heteronomous annulation of the armoured lobopods and *Kerygmachela*, followed by the hemi-segmentation of the tardigrades (see discussion below) and perhaps *Opabinia* (Whittington, 1975; Budd, 1996). The state in the anomalocaridids is less clear, because trunk segmentation has never been adequately described from these taxa. However, in the euarthropods, a segmentation into sclerotized, articulating tergites is seen (Chen *et al.*, 1994). This reconstruction of the character state changes is in principle testable by examining developmental homologies in segment determination within extant lobopods and arthropods (e.g. by tracing homologous homeotic domain boundaries). It has important implications for arthropod relationships. It has been commonly assumed, after Cuvier (1817), that the segmentation present in the arthropods is directly homologous with that of the annelids. Apart from a slight shift in the segment boundaries relative to the positions of muscle insertion, arthropods are in this view are more or less sclerotized annelids (Snodgrass, 1938). Other arthropodan features such as the haemocoel, formed from the breakdown of a spacious coelom, followed on as correlated changes. However, even considering the evidence from living

**Figure 11.8** *Pambdelurion whittingtoni*. MGUH 24509. Details of annulated limbs; see also Figures 11.6 (A) and 11.7(A). Note that both limbs show a similar unannulated structure along the top surface of the limb. It is not clear what these structures represent, but they may be parts of the lateral lobes. Magnification: (A) x3.8; (B) x4.2.

**Figure 11.9** *Pambdelurion whittingtoni*. MGUH 24514 x4.7 showing detail of a pair of diverticulae situated either side of the broad gut.

taxa alone, this assertion seems doubtful. Neither the onychophorans nor the tardigrades, although serially repetitive in their internal organs and their limbs, express full annelidan segmentation. The onychophorans, for example, possess neither a segmentally arranged ventral nerve ladder (the transverse connectives are highly irregular), nor segment boundary expression in the musculature and cuticle, nor internal septae. These contradictions in the 'sclerotized annelid' model are often dismissed with a reference to the terrestrialization of the onychophorans and the necessary changes forced on them by this move. However, external segmentation is possessed neither by the Cambrian armoured lobopods nor *Kerygmachela*, and reconstruction of the character state along the euarthropod stem based on the phylogeny presented shows that arthropodan segmentation is derived. The segmental organization of the arthropods therefore cannot be homologous (in a simple way) to the segmentation present in the annelids [see Eernisse *et al.* (1992) and Weisblat *et al.* (1988) for doubts about the similarity between segmental development between arthropods and annelids]. Nevertheless, it is clear that serialization, in the sense of repetition of body units, is to a greater or lesser extent exhibited by many metazoans such as the kinorhynchs: its origin may lie very deeply within the clade. Expression of such a feature in repeated lineages may reflect both the adaptive utility of such a form of bodily organization in terms of development and growth, and conceivably the repeated deployment of an underlying, common set of genes within different lineages.

## 11.5 PROBLEMATIC TAXA

Given the general scheme presented above, several taxa remain difficult to place. These include the tardigrades, pentastomids and myriapods.

### 11.5.1 TARDIGRADES

Tardigrades have been notoriously difficult to classify (see review in Kinchin, 1994). Generally, it is assumed that as miniaturized, highly specialized bilaterians, they have shed many phylogenetically useful features, while at the same time possibly gaining characters that are convergently similar to other small metazoans such as the so-called aschelminth taxa (see Conway Morris *et al.*, 1985 for discussions of some of these taxa). A putative Cambrian tardigrade has now been reported (Müller *et al.*, 1995). Recent attempts to place them systematically using molecular techniques have proved to give contradictory results (Giribet *et al.*, 1996; Moon and Kim, 1996). Tardigrades were originally included in the cladistic analysis presented here. Subsequently they were excluded because they proved difficult to code, and rendered the whole analysis unstable. Recently, a suggestion has been made based on head segmentation that their affinities lie somewhere within the group of organisms considered as stem group arthropods here, i.e. within the group bracketed by *Kerygmachela* and *Anomalocaris* (Dewel and Dewel, 1996, 1997, this volume; see also Budd, 1996).

**Table 11.1** Character table for cladistic analysis presented in Figure 11.10: see Appendix 11A for details of characters.

| | 1 | 2 | 3 | 4 | 5 | 6 | 7 | 8 | 9 | 10 | 11 | 12 | 13 | 14 | 15 | 16 | 17 | 18 | 19 | 20 | 21 | 22 | 23 | 24 | 25 | 26 | 27 | 28 | 29 | 30 |
|---|---|---|---|---|---|---|---|---|---|---|---|---|---|---|---|---|---|---|---|---|---|---|---|---|---|---|---|---|---|---|
| *Peripatus* | 1 | 1 | 0 | ? | 2 | 1 | 0 | 0 | 0 | 0 | 0 | 0 | 0 | x | 0 | 0 | 0 | 0 | 1 | 0 | 1 | 1 | 0 | 0 | 0 | x | 1 | 1 | 0 | 0 |
| *Aysheaia* | ? | 1 | 0 | 1 | 0 | 0 | 0 | 0 | 0 | 0 | 0 | 0 | 0 | x | 0 | 0 | 0 | 0 | 1 | 0 | 1 | 0 | x | 0 | 0 | x | 1 | 1 | 0 | 0 |
| *Kerygmachela* | 1 | ? | ? | 1 | 0 | 1 | 0 | 1 | 0 | 1 | 1 | 0 | 1 | 0 | 1 | 1 | 0 | 0 | 1 | 0 | ? | ? | 0 | 0 | 0 | x | 1 | 1 | 1 | 0 |
| *Opabinia* | 0 | 1 | 0 | 1 | 1 | 1 | 0 | 0 | 0 | 0 | 1 | 0 | 1 | 1 | 1 | 1 | 0 | 1 | 1 | ? | 1 | 1 | 1 | 0 | 0 | x | 1 | 1 | 1 | 0 |
| *Pambdelurion* | 0 | 1 | 1 | 1 | 1 | 1 | 0 | 1 | ? | 1 | 1 | 0 | ? | ? | 1 | ? | 1 | 1 | 1 | 1 | 0 | 0 | x | 0 | 0 | x | 1 | 0 | 1 | 0 |
| Anomalocaridids | 0 | 1 | 1 | 1 | 1 | 1 | 1 | 0 | 1 | 1 | 1 | 1 | 1 | 1 | 1 | 0 | 1 | 1 | ? | ? | 1 | 1 | 1 | 1 | 1 | 0 | 0 | 0 | 1 | 1 |
| CCT | 0 | 0 | x | 1 | 1 | 1 | 1 | 0 | 1 | ? | 1 | 1 | 1 | 1 | 1 | ? | ? | 1 | 0 | 1 | 1 | 1 | 1 | 1 | 1 | 1 | 0 | 0 | 1 | ? |

Coding: uncertain character states: ?; inapplicable characters states: x. The anomalocaridids are coded as a composite (which may not be entirely accurate) because data from them are patchy (*Parapeytoia* is known with biramous legs, for example, but not eyes).

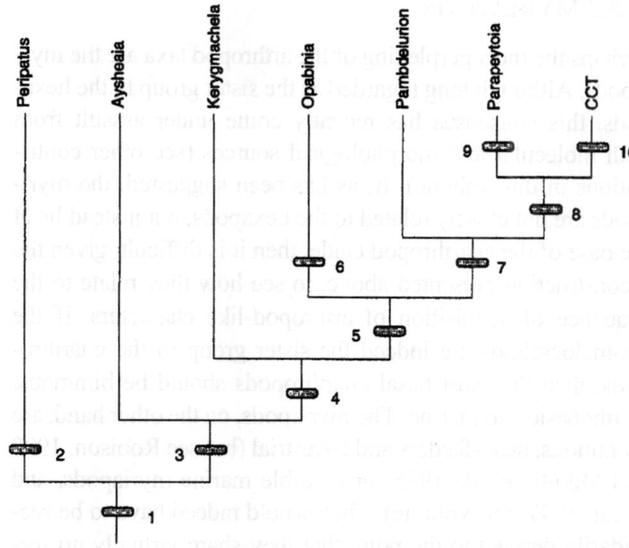


**Figure 11.10** Cladistic analysis using PAUP of arthropod stem group taxa, with seven taxa and 30 characters (not including autapomorphies). Rescaled consistency index = 0.741. Tree rooted by taking *Aysheaia* and *Peripatus* as an out group. 'CCT', Crustacean, Chelicerate, Trilobite clade of Briggs and Fortey (1989). *Parapeytoia* stands proxy for anomalocaridids. Apomorphies, optimized with accelerated transformation (favours multiple loss over multiple gain): 1. Lateral modified anterior appendages; dorsal pericardial sinus, other characters as in the plexus of arthropod characters of Budd (1993) 2. Onychophoran autapomorphies (slime glands, oblique body musculature, jaws, antennal structures). 3. Lateral branched frontal appendages. 4. Heteronomous annulation; 11-segmented trunk; lateral lobes, gill structures. 5. Loss of dorsal sinus; anterior appendages latero-ventral; gill attachment free; gut with diverticulae; lever system of musculature*[1]; compound eyes on stalks; segmentation. 6. Loss of 11-segmented body; anterior ventral 'trunk' bearing appendage at anterior tip. 7. 'Peytoia'-like mouth-parts; modified occipital segments; gigantism; loss of dorsal annulation. 8. Anterior and trunk appendages arthropodized; biramous limbs with gnathobases; loss of circular body musculature*[2]; ventral sclerites; strongly sclerotized frontal appendages*[3] 9. Loss (or transformation) of 'Peytoia'-like mouth-parts; Complete cuticular sclerotization with tergite articulation. Marked character state changes (*) are those whose position is uncertain on the cladogram because of missing data. 1: may plot at nodes 5 or 7 because state is not known in *Opabina*; 2: may plot at nodes 8 or 10 because state not known in anomalocaridids; 3: may plot at nodes 8 or 9 because it may characterize basal CCT arthropods.

### 11.5.2 PENTASTOMIDS

These unusual lobopodous parasites have sometimes been considered to have affinities with other lobopod animals, although rRNA and ontogenetic data suggests they are derived crustaceans (Abele *et al.*, 1989). A Cambrian fossil record of pentastomids is now known (Walossek and Müller, 1994) but, as is the case of the tardigrades, it sheds little phylogenetic light on the subject.

11.5.3 MYRIAPODS

Perhaps the most perplexing of the arthropod taxa are the myriapods. Although long regarded as the sister group to the hexapods, this consensus has recently come under assault from both molecular and morphological sources (see other contributions in this volume). If, as has been suggested, the myriapods are not closely related to the hexapods, but instead lie at the base of the euarthropod clade, then it is difficult, given the reconstruction presented above, to see how they relate to the sequence of acquisition of arthropod-like characters. If the anomalocaridids are indeed the sister group to the euarthropods, then the most basal euarthropods should be biramous, gnathobasic and marine. The myriapods, on the other hand, are uniramous, head-feeders and terrestrial (but see Robison, 1990 and Mikulic *et al.*, 1985 for possible marine myriapods, and Shear, 1997, this volume). They would indeed have to be secondarily derived to the point that they share virtually no features with the most basal euarthropods apart from the sclerotized cuticle. Their late appearance in the fossil record [probably in the Silurian; Jeram *et al.* (1990), although a putative Cambrian example has been reported (Robison, 1990)] is also puzzling. Given these features, the possibility of their having been separately derived from a lobopodous ancestor cannot as yet be discounted. However, this may be difficult either to prove or refute, because with the possible exception of the tardigrades, the only taxa intervening between the myriapods and the euarthropods would be extinct Cambrian ones. Thus, all analyses of Recent taxa will yield a monophyletic euarthropod clade with the myriapods at the base, even though they may have gained some characters such as the sclerotized cuticle in parallel with the biramous euarthropods.

## 11.6 CONCLUSIONS

Although the Cambrian arthropod fossil record remains controversial, a strong case can be made for interpreting supposedly problematic taxa, several of which are known from the Sirius Passet fauna, as lying within the stem group of extant euarthropods. In such a case, the early fossil record of the arthropods provides an almost unique documentation of the origin of an invertebrate phylum (see also Chen *et al.*, 1995; Conway Morris and Peel, 1995). In the reconstruction presented here, the euarthropods are seen to have evolved out of a diverse and paraphyletic assemblage of lobopodous animals, although the position of several taxa, notably the myriapods, remains problematic. The stages in this process can be elucidated to some extent: the first arthropod-like character acquired was probably the frontal, branched appendages seen in taxa like *Aysheaia*, followed by heteronomous trunk annulation, lateral lobes, compound eyes, the biramous limb and ventral sclerotization, and finally full sclerotization. In time it may also be possible to describe the adaptive significance of these features. The relative order of appearance of the essential characters of arthropods may, therefore, be traced in the fossil record, suggesting that the evolution of the phylum took place in such a way that allowed diversification of intermediate stages. This in turn brings into question, as did Fortey *et al.* (1996), the notion that the 'Cambrian Explosion' involved sudden and disjunct morphological change, perhaps involving unusual genetic mechanisms.

ACKOWLEDGEMENTS

I thank S. Conway Morris and R.A. Fortey for comments on a previous version of the manuscript, and J.S. Peel for advice and support. M.S. Reid kindly provided specimens of *Euperipatoides*. Financial support is gratefully acknowledged from the Carlsberg Foundation and National Geographic. This work was undertaken whilst in receipt of a Swedish Natural Sciences Research Council (NFR) postdoctoral fellowship.

**Appendix 11A** Character states for the cladistic analysis of Figure 11.10

| | | | |
|---|---|---|---|
| 1. Dorsal sinus. | 0: Absent | 1: Present | |
| 2. Circum-oral structures. | 0: Absent | 1: Present | |
| 3. Oral structures like '*Peytoia*'. | 0: No | 1: Yes | |
| 4. Modified frontal appendages. | 0: Absent | 1: Present | |
| 5. Anterior appendage position. | 0: Lateral | 1: Latero-ventral | 2: Dorsal |
| 6. Mouth position. | 0: Terminal | 1: Ventral | |
| 7. Anterior appendage state. | 0: Lobopodous/annulated | 1: Arthropodous | |
| 8. Extended terminal appendage spines. | 0: Absent | 1: Present | |
| 9. Modified occipital appendages. | 0: Absent | 1: Present | |
| 10. 11 segmented trunk. | 0: Absent | 1: Present | |
| 11. Lateral lobes. | 0: Absent | 1: Present | |
| 12. Biramous limbs. | 0: Absent | 1: Present | |
| 13. Gill structures. | 0: Absent | 1: Present | |
| 14. Gill attachment. | 0: Fused to lobe | 1: Free apart from anterior margin | |
| 15. Modified tail structures. | 0: Absent | 1: Present | |

**Appendix 11A** (continued)

| | | |
|---|---|---|
| 16. Tail spines. | 0: Absent | 1: Present |
| 17. Body size. | 0: 'Normal' | 1: Giant (>20 cm) |
| 18. Gut with diverticulae. | 0: Absent | 1: Present |
| 19. Circular body muscle. | 0: Absent | 1: Present |
| 20. Lever muscle system. | 0: Absent | 1: Present |
| 21. Terminal limb claws. | 0: Absent | 1: Present |
| 22. Eyes. | 0: Absent | 1: Present |
| 23. Eye state. | 0: Simple | 1: Compound and stalked |
| 24. Trunk limbs. | 0: Lobopod | 1: Arthropod |
| 25. Cuticular sclerotization. | 0: Absent | 1: Present |
| 26. Sclerotization extent. | 0: Ventral only | 1: Complete |
| 27. Trunk annulation. | 0: Absent | 1: Present |
| 28. Dorsal annulation. | 0: Absent | 1: Present |
| 29. Heteronomous annulation. | 0: Absent | 1: Present or fully segmented |
| 30 Strongly sclerotized frontal appendages. | 0: Absent | 1: Present |

## REFERENCES


Abele, L.G., Kim, W. and Felgenhauer, B.E. (1989) Molecular evidence for the inclusion of the phylum Pentastomida in the Crustacea. *Molecular Biology and Evolution*, **6**, 685–91.

Blaker, M.R. (1988) A new genus of nevadiid trilobite from the Buen Formation (Early Cambrian) of Peary Land, central North Greenland. *Rapport Grønlands geologiske Undersøgelse*, **137**, 33–41.

Briggs, D.E.G. and Fortey, R.A. (1989) The early radiation and relationships of the major arthropod groups. *Science*, **246**, 241–3.

Briggs, D.E.G. and Whittington, H.B. (1987) The affinities of the Cambrian animals *Anomalocaris* and *Opabinia*. *Lethaia*, **20**, 185–6.

Budd, G. (1993) A Cambrian gilled lobopod from Greenland. *Nature*, **364**, 709–11.

Budd, G.E. (1995) *Kleptothule rasmusseni* gen. et sp. nov.: an ?olenellinid-like trilobite from the Sirius Passet fauna (Buen Formation, Lower Cambrian, North Greenland). *Transactions of the Royal Society of Edinburgh: Earth Sciences*, **86**, 1–12.

Budd, G.E. (1996) The morphology of *Opabinia regalis* and the reconstruction of the arthropod stem group. *Lethaia*, **29**, 1–14.

Chen, J.Y., Ramsköld, L. and Zhou, G.Q. (1994) Evidence for monophyly and arthropod affinity of Cambrian giant predators. *Science*, **264**, 1304–8.

Chen, J.Y., Edgecombe, G.D., Ramsköld, L and Zhou, G.Q. (1995) Head segmentation in Early Cambrian *Fuxianhuia*: implications for arthropod evolution. *Science*, **268**, 1339–43.

Collins, D. (1996) The 'evolution' of *Anomalocaris* and its classification in the arthropod class Dinocarida (Nov.) and Order Radiodonta (Nov.). *Journal of Paleontology*, **70**, 280–93.

Conway Morris, S. (1977) A new Metazoan from the Burgess Shale of British Columbia. *Palaeontology*, **20**, 623–40.

Conway Morris, S. and Peel, S.J. (1995) Articulated halkieriids from the Lower Cambrian of North Greenland and their role in early protostome evolution. *Philosophical Transactions of the Royal Society of London B*, **347**, 305–58.

Conway Morris, S, George, J.D., Gibson, R. and Platt, H.M. (eds) (1985) *The Origins and Relationships of Lower Invertebrates*, Systematics Association Special Volume 28, Oxford University Press, Oxford.

Cuvier, G. (1817) *Le règne animal distribué d'après son organisation, pour servir de base à l'histoire naturelle des animaux et d'introduction à l'anatomie compareé.* Volume 2, Deterville, Paris.

Davidson, E.H., Peterson, K.J. and Cameron, R.A. (1995) Origin of bilaterian body plans – evolution of developmental regulatory mechanisms. *Science*, **270**, 1319–25.

Dewel, R.A. and Dewel, W.C. (1996) The brain of *Echiniscus viridissimus* Peterfi 1956 (Heterotardigrada): a key to understanding the phylogenetic position of tardigrades and the evolution of the arthropod head. *Zoological Journal of the Linnean Society of London*, **116**, 35–49.

Eernisse, D.J., Albert, J.S. and Anderson, F.E. (1992) Annelida and Arthropoda are not sister taxa: a phylogenetic analysis of spiralian metazoan morphology. *Systematic Biology*, **41**, 305–30.

Fortey, R.A., Briggs, D.E.G. and Wills, M.A. (1996) The Cambrian evolutionary 'explosion': decoupling cladogenesis from morphological disparity. *Biological Journal of the Linnean Society*, **57**, 13–33.

Giribet, G., Carranza, S., Baguna, J., Riutort, M. and Ribera, C. (1996) First molecular evidence for the existence of a Tardigrada plus Arthropoda clade. *Molecular Biology and Evolution*, **13**, 76–84.

Gould, S.J. (1989) *Wonderful Life. The Burgess Shale and the Nature of History*, W.W. Norton and Co., New York.

Hou, X.G., Bergström, J. and Ahlberg, P. (1995) *Anomalocaris* and other large animals in the Lower Cambrian Chengjiang fauna of southwest China. *Geologisk Föreningens i Stockholm Förhandlingar*, **117**, 163–83.

Jeram, A.J., Selden, P.A. and Edwards, D. (1990) Land animals in the Silurian: arachnids and myriapods from Shropshire, England. *Science*, **250**, 658–61.

Kinchlin, I.M. (1994) *The Biology of Tardigrades*, Portland Press, London.

Mikulic, D.G., Briggs, D.E.G. and Klessendorf, J. (1985) A new exceptionally preserved biota from the Lower Silurian of

Wisconsin, U.S.A. *Philosophical Transactions of the Royal Society of London, B*, **311**, 75–85.

Moon, S.Y. and Kim, W. (1996) Phylogenetic position of the Tardigrada based on the 18S ribosomal RNA gene sequences. *Zoological Journal of the Linnean Society of London*, **116**, 61–9.

Müller, K.J., Walossek, D. and Zakharov, A. (1995) 'Orsten' type phosphatized soft-integument preservation and a new record from the Middle Cambrian Kuonamka Formation in Siberia. *Neues Jahrbuch für Geologie und Paläontologie Abhanlungen*, **197**, 101–18.

Nedin, C. (1996) The Emu Bay Shale, a Lower Cambrian fossil *Lägerstätten*, Kangaroo Island, South Australia. *Memoirs of the Association of Australasian Palaeontologists*, **18**, 31–40.

Peterson, K.J. (1995) A phylogenetic test of the calcichordate scenario. *Lethaia*, **28**, 25–38.

Ramsköld, L. and Hou, X.G. (1991) New Early Cambrian animal and onychophoran affinities of enigmatic metazoans. *Nature*, **351**, 225–8.

Robison, R.A. (1990) Earliest known uniramous arthropod. *Nature*, **343**, 163–4.

Snodgrass, R.E. (1938) Evolution of the Annelida, Onychophora and Arthropoda. *Smithsonian Miscellaneous Collections*, **97**, 1–159.

Walossek, D. and Müller, K.J. (1994) Pentastomid parasites from the Lower Palaeozoic of Sweden. *Transactions of the Royal Society of Edinburgh: Earth Sciences*, **85**, 1–37.

Weisblat, D.A., Price, D.J. and Wedeen, C.J. (1988) Segmentation in leech development. *Development*, **104** Supplement, 161–8.

Whittington, H.B. (1975) The enigmatic animal *Opabinia regalis*, Middle Cambrian, Burgess Shale, British Columbia. *Philosophical Transactions of the Royal Society of London, B*, **271**, 1–43.

Whittington, H.B. (1977) The Middle Cambrian Trilobite *Naraoia*, Burgess Shale, British Columbia. *Transactions of the Royal Society of London, B*, **280**, 409–43.

Whittington, H.B. (1979) Early arthropods, their appendages and relationships, in *The Origin of Major Invertebrate Groups* (ed. M.R. House), Systematics Association Special Volume 12, Academic Press, London, pp. 253–68.

Whittington, H.B. and Briggs, D.E.G. (1985) The largest Cambrian animal, *Anomalocaris*, Burgess Shale, British Columbia. *Philosophical Transactions of the Royal Society of London, B*, **309**, 569–609.

Wills, M.A., Briggs, D.E.G., Fortey, R.A. and Wilkinson, M. (1995) The significance of fossils in understanding arthropod evolution. *Verhandlungen der Deutschen Zoologischen Gesellschaft*, **88**, 203–15.

# 12 Cambrian 'Orsten'-type arthropods and the phylogeny of Crustacea

*D. Walossek and K.J. Müller*

*Sektion für Biosystematische Dokumentation, Universität Ulm, Liststrasse 3, D-89079 Ulm, Germany email: dieter.walossek@biologie.uni-ulm.de*
*Institut für Paläontologie, Universität Bonn, Nussallee 8, D-53115 Bonn, Germany*

## 12.1 INTRODUCTION

'Orsten' is a special type of anthraconitic, organic-rich, concretionary limestone which is intercalated in the Upper Cambrian Alum Shale of southern Sweden. In most localities it contains abundant megafossils, mainly trilobites. We have extended its meaning to embrace a type of *lagerstätten* with a particular type of fossil preservation, i.e. the three-dimensional preservation of phosphatized cuticle-bearing organisms. Since their discovery by Müller (1979, 1982, 1983, 1985, 1990) on the isle of Öland and in Västergötland, such *lagerstätten* have yielded more than 100 000 microfossils, mainly arthropods. They have been isolated from the limestone with 15% acetic acid. The vast majority are bivalved arthropods called phosphatocopines. The cuticular surface of these arthropods is still present in full detail, revealing eyes and limbs, hairs and minute bristles, pores of sensilla, gland openings, and even cellular patterns and grooves of muscle attachments underneath. The secondary bristles on filtratory setae on the limbs of some of these fossils are thinner than 0.5 µm.

The maximum size of specimens recovered in this type of preservation does not exceed 2 mm, while the best-preserved animals measure between 100 and 500 µm (Figure 12.1). Accordingly, much of the material comprises small-sized, bottom-living meiofaunal animals and larvae, which, in several cases, have been found in sets of successive instars. The ontogenetic changes of morphological characters can thus also be added to phylogenetic analysis (Müller and Walossek, 1985a, 1986a,b, 1987, 1988a,b, 1991a,b; Walossek and Müller, 1990, 1991, 1994, 1997; Walossek, 1993, 1995, 1996). 'Orsten'-type preservation displays the whole external morphology of these early animals and permits reconstruction of their life habits.

Besides the discoveries in southern Sweden (Öland, Västergötland), 'Orsten'-type fossils have been recovered from several other geological ages and areas, ranging from the Lower Cambrian of Shropshire, United Kingdom (Hinz, 1987), the Middle Cambrian of Siberia, Russia (Müller *et al.*, 1995), and the Georgina Basin, Australia (Müller and Hinz, 1992; Müller and Hinz-Schallreuter, 1993; Walossek *et al.*, 1993), the Upper Cambrian of Poland (Walossek and Szaniawski, 1991), to the possibly Lower Ordovician of the Isle of Öland, Sweden (Andres, 1989) and the Upper Cambrian/Lower Ordovician of Newfoundland, Canada (Roy and Fåraeus, 1989; Walossek *et al.*, 1994). 'Orsten'-type fossilization is variable in kinds and quality. In many cases it replaced the outer layer of arthropod (or arthropod-

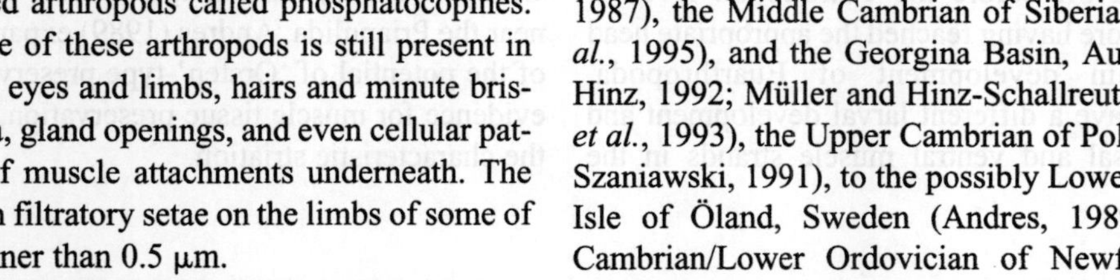


**Figure 12.1** Scanning electron micrographs of nauplius-like arthropod larvae (length approx. 100 µm). (a) Dorsal view of a complete specimen, showing three pairs of hook-like structures which may refer to the three pairs of limbs (for muscle attachment?; from Müller and Walossek, 1986b). (b) Ventral view of a second larvae of the same type having shorter tail spines (from Walossek and Müller, 1989). (c) Ventral view of a specimen from the Middle Cambrian of Australia (from Walossek *et al.*, 1993).

*Arthropod Relationships*, Systematics Association Special Volume Series 55, Edited by R.A. Fortey and R.H. Thomas. Published in 1997 by Chapman & Hall, London. ISBN 0 412 75420 7

like) cuticle, the epicuticle, by phosphatic matter, leading to the preservation of the finest details. The interior, in most cases, is empty or filled by amorphous phosphatic matter. More rarely, the specimens are massive, but in these cases the cuticle appears to be less well preserved.

Our joint project since 1983 was more or less confined to the Arthropoda. We had been able to identify non-crustacean arthropods including *Agnostus pisiformis* and its complete early larval series, and a chelicerate larva (Figure 12.2(a,b)). Early representatives of extant crustacean taxa (members of the crown group) include: *Bredocaris admirabilis* Müller, 1983, and Skaracarida Müller and Walossek, 1985, as Maxillopoda, and *Rehbachiella kinnekullensis* Müller, 1983, as a branchiopod. Several forms are recognized representatives of the stem line of Crustacea: the phosphatocopines (Figure 12.3; Müller, 1979, 1982, Walossek *et al.*, 1993, see Walossek and Müller, 1997 for discussion of their phylogenetic position), *Cambropachycope clarksoni* Walossek and Müller, 1990, *Goticaris longispinosa* Walossek and Müller, 1990, *Henningsmoenicaris scutula* (Walossek and Müller, 1990), *Martinssonia elongata* Müller and Walossek, 1986 (Walossek and Müller, 1990 for discussion of its phylogenetic position), and *Cambrocaris baltica* Walossek and Szaniawski, 1991.

In 1994 we described a set of larvae from the 'Orsten' as early representatives of the parasitic Pentastomida (Figure 12.4(a,b)), with the support of the few specialists on this group, W. Böckeler of Kiel, and J. Riley of Dundee. We consider this taxon as truly arthropodan because of its arthropodial legs with pivot joints. However, this group must have branched off before the euarthropod level of organization and before having reached the appropriate head tagmosis and brain development of Euarthropoda. Pentastomida also have a different larval development and straight, paired dorsal and ventral muscle strands in the trunk. These features can be obtained from Andres' (1989) advanced larval stages from the Lower Ordovician of the isle of Öland, Sweden (a new later larval specimen from Stora Backor quarry, Västergötland, Sweden, discovered by J.E. Repetski recently is shown in Figure 12.4(c)). Our interpretations confirm traditional assumptions on the relationships of Pentastomida but are in striking contrast to hypotheses based on sperm data (Wingstrand, 1972) and molecular data (Abele *et al.*, 1989), which postulate affinities with a particular crustacean taxon, the parasitic Branchiura. Fish lice are a group that has no fossil record as yet, but several morphological features indicate that they are representatives of the Maxillopoda. It may even prove that our evidence illustrates the limitations of molecular studies.

To what might be termed 'pre-euarthropods', especially with the recognition of Cambrian lobopods as the ancestral stock of Onychophora (Ramsköld and Hou, 1991), another group may be added: we recently reported the discovery of tardigrade fossils from the Middle Cambrian of Siberia (Müller *et al.*, 1995). The previous fossil record of this group is from the Cretaceous. These minute fossils are remarkably modern in design and of the same gross size, about 250–350 µm long, and shape as extant tardigrades or their larval stages (Figure 12.4(d,e)).

We are currently studying larvae of Priapulida (Nemathelminthes) from the Middle Cambrian of Australia (of the *Tubiluchus* type). Furthermore, the Palaeoscolecida, a variety of worm-shaped fossils described by Müller and Hinz-Schallreuter (1993) from the Middle Cambrian of the Georgina Basin, Australia, may also be placed within or near the Priapulida. Andres (1989) expanded our knowledge of the potential of 'Orsten'-type preservation by presenting evidence for muscle tissue preservation, some still showing the characteristic striation.

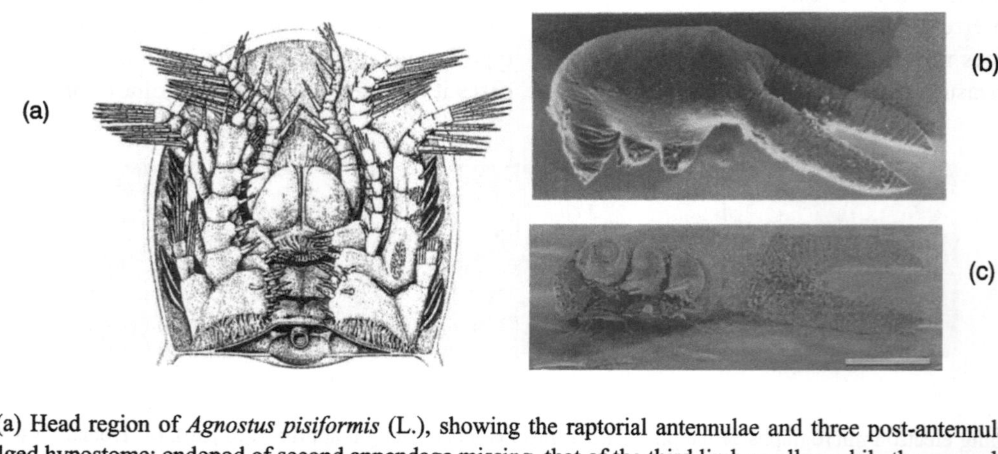


**Figure 12.2** (a) Head region of *Agnostus pisiformis* (L.), showing the raptorial antennulae and three post-antennular limbs arranged around the bulged hypostome; endopod of second appendage missing, that of the third limb smaller, while the exopods are prominent in both these anterior limbs; last head limb of the same shape as the trunk limbs and with large endopod but short exopod (from Müller and Walossek, 1987). (b,c) Scanning electron micrographs of an 'Orsten' chelicerate larva in oblique lateral and ventral views (from Müller and Walossek, 1986b; flipped head to the left).

(a) 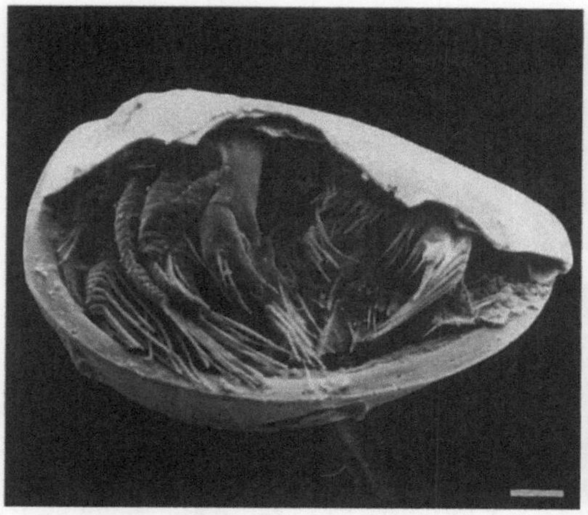 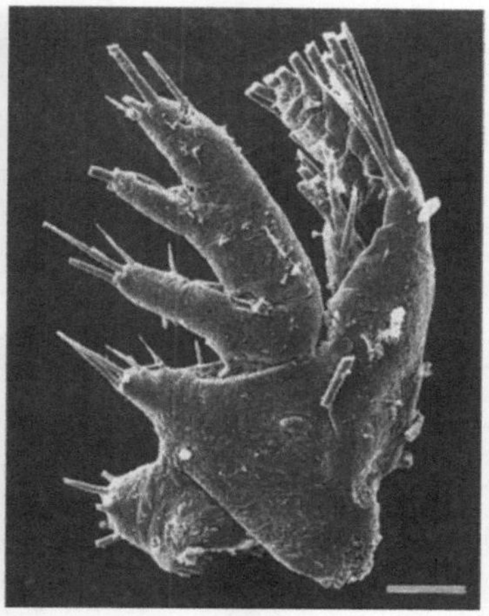 (b)

**Figure 12.3** (a) Scanning electron micrograph of *Hesslandona unisulcata* Müller, 1982 (Phosphatocopida), one of the valve-like shield sides is peeled off to view the limbs with their long setation; the prominent limb in the centre is the mandible (from Müller, 1982; flipped horizontally to have the anterior to the left). (b) Isolated postmandibular limb of a phosphatocopine, with the basipod(ite) bearing the proximal endite at its inner proximal edge and carrying a three-segmented endopod and multiannulate exopod (from Walossek *et al.*, 1993, Figure 4d; see also this reference for Middle Cambrian relatives lacking the exopods on their limbs).

## 12.2 REMARKS ON CURRENT OPINIONS OF ARTHROPOD CHARACTERS

For tracing arthropod phylogeny one has critically to evaluate traditional hypotheses and terminologies. By investigating real, three-dimensional-preserved ancient fossils, rather than just constructing models, it becomes apparent that several current opinions are based on unsustainable assumptions. Three pertinent examples are given below, which concern the tagmosis of the head, the development of the brain, and the morphology and terminology of arthropod limbs.

### 12.2.1 ARTHROPOD BRAIN

In entomostracan crustaceans – which we consider as a valid monophylum and sister taxon of the Malacostraca (Walossek and Müller, 1997) – the tritocerebrum, as the ganglion of the antennal segment (a2), is not included in a pre-oesophageal 'brain'. Calman (1909, p. 51) also remarked on branchiopod morphology, particularly tagmosis, trunk limbs and the nervous system:

> ... Their primitive character is further shown, as has been pointed out, by the general uniformity of the trunk-somites and their appendages, by the presence of gnathobases [remarks: our proximal endite, see below] on all the trunk limbs, by the 'ladder-like' form of the ventral nerve-chain and the post-oral position of the antennal ganglia, and by the tubular heart and its segmentally arranged ostia....

The same is true for isopod Eumalacostraca and possibly various other arthropods of different groups. This implies that the a2-ganglion was not part of a supraoesophageal brain at the base of euarthropod divergence (ground pattern of Euarthropoda), nor can a brain including this ganglion be regarded as part of the ground pattern of Crustacea s. str. (Figure 12.5). Considering separation as the plesiomorphic character state, its inclusion in a number of taxa must be considered as an apomorphy achieved by convergent development. This is important particularly for discussion of the relationships of tracheates and crustaceans.

### 12.2.2 HEAD SEGMENTS

Another widespread, but long-falsified opinion concerns the segment array of the original euarthropod head as including the first antennae or antennulae (a1) and four pairs of limb-bearing body somites. This hypothesis, which dates back at least to Størmer's interpretations of fossil arthropods from the first half of this century, has been falsified by Cisne (1975) in his restudy of the trilobite *Triarthrus eatoni*: discovering that this species had only three pairs of post-antennular appendages behind the antennulae in its head. Few recent textbooks, such as those by Brusca and Brusca (1990) show this new reconstruction. Subsequent studies confirmed this segmentation pattern for a variety of other trilobites and related forms (see list of references in Müller and Walossek, 1987; Figure 12.2(a) for *Agnostus*), and also for aglaspidids (Briggs *et al.*, 1979), a group which traditionally has been placed at the base of Chelicerata, a view that cannot be upheld any longer. Moreover, some of the representatives of the early stem group of Crustacea (Walossek and Müller, 1990; Figure 8.1) reveal a similarly 'short' head, i.e. they

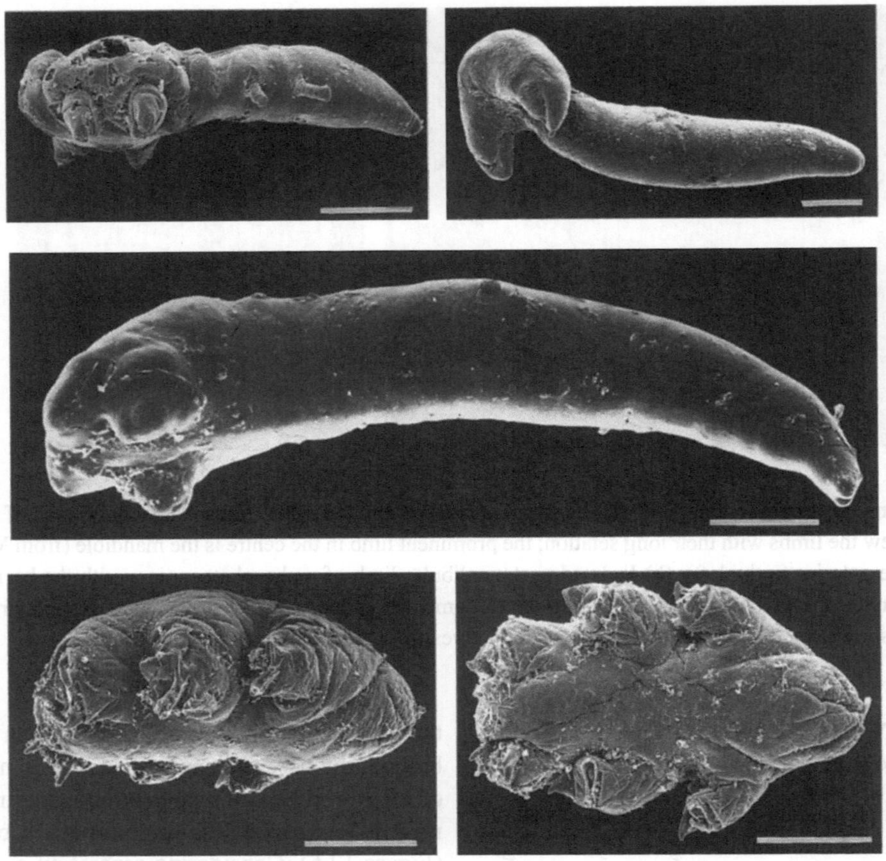


**Figure 12.4** Scanning electron micrographs of early larvae of the Upper Cambrian stem-line pentastomids (a,b). A new, yet undescribed later pentastomid larva (c) from Stora Backor quarry, Västergötland, Sweden (provided by J.E. Repetski, Reston), and a most likely larval tardigrade from the Middle Cambrian of Siberia (d,e). (a) *Boeckelericambria pelturae* Walossek and Müller, 1994. (b) *Heymonsicambria repetskii* Walossek and Müller, 1994 (from Walossek and Müller, 1994). (c) Ventro-lateral view. (d) Lateral view. (e) Ventral view (from Müller *et al.*, 1995; size about 250–350 μm).

also had only antennae and three pairs of biramous head appendages.

Euarthropod level of organization had, hence, not achieved a segmental state as has been assumed traditionally, but the different taxa diverged from a lower level of organization. Accepting a shorter head tagma at the beginning of the euarthropod line, makes interpretations of further changes within arthropod taxa not only easier but also more parsimonious.

The chelicerate line, for example, incorporated three more – and not two – segments progressively into their 'head' to form the characteristic prosoma (formula: 1 + 6, later 0 + chelicera + 5, and, later 0 + chelicera + pedipalp + 4). The earliest representatives of the crustacean line also still had only three post-antennular limbs, as seen in *Goticaris* and *Cambropachycope* (Figure 12.6(a,e)). The fourth one is only partially added to the head in *Martinssonia* (Figure 12.7). The so-called maxillary segment as the fourth post-antennular segment was included later during crustacean evolution, characterizing at least all representatives of the crown group of Crustacea. It must be assumed that offspring of this early line will be discovered, which had at least a visible chelicerate character but still between three and six post-antennular limb-bearing head segments, and thus the same number as crown-group crustaceans and tracheates.

### 12.2.3 LARVAE

With regard to a possibly shorter 'early' head at the base of euarthropod evolution, the larvae are of particular interest. Trilobite protaspids are assumed to have had three pairs of post-antennal limbs, but there is no strong evidence available as yet. The earliest known larvae of representatives of the crustacean stem line *Goticaris* (Walossek and Müller, 1990), *Cambropachycope* (D. Walossek and K.J. Müller, manuscript in preparation), *Martinssonia* (Müller and Walossek, 1986a) (Figure 12.7) and *Henningsmoenicaris*

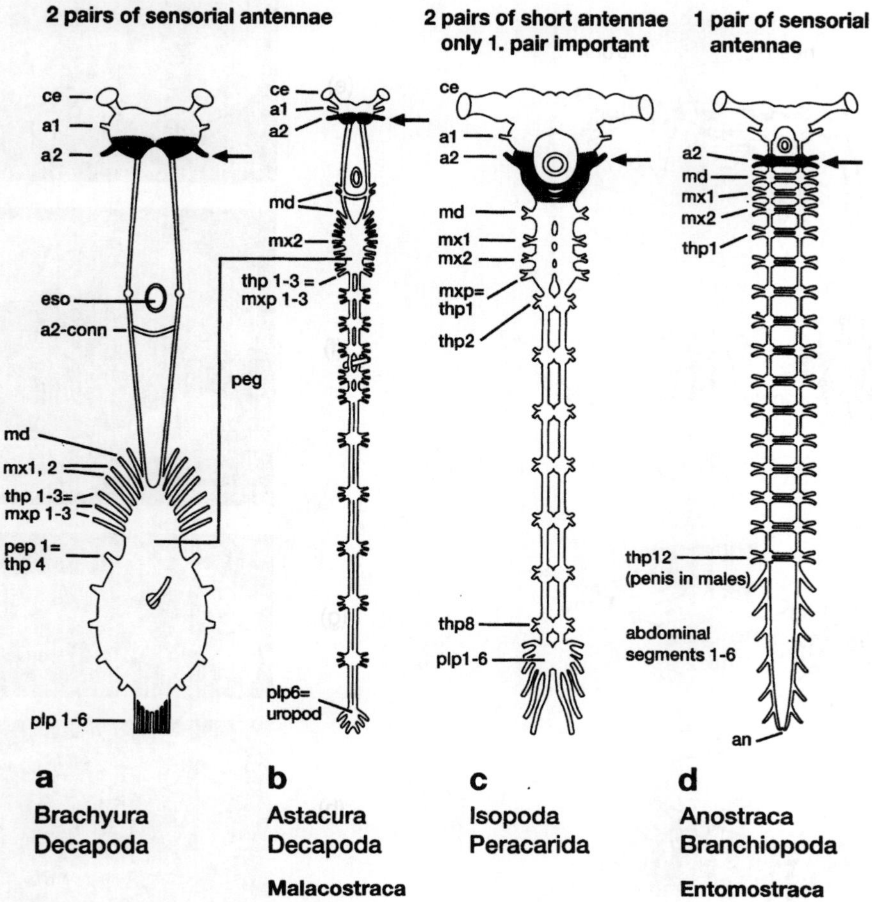


**Figure 12.5** Examples of nervous systems of crustacean taxa. Antennal ganglia emphasized by arrows. (Modified from various sources, e.g. Gruner, 1993; Forest, 1994.) For abbreviations, see page 152.

(Walossek and Müller, 1990, 1991) all had three functional pairs of post-antennular limbs. They have been termed 'head larvae' (Walossek and Müller, 1990) to enhance the difference from the crustacean nauplius as a 'part-head larva', carrying antennulae and only two pairs of functional limbs, the antennae and mandibulae. Notably, the earliest pantopod larva, the protonymph, and an Upper Cambrian larva quite similar to it (Müller and Walossek, 1986b, 1988a) (Figure 12.2(b)) also have three pairs of post-antennal limbs. Accepting that the antennulae have been lost in the early evolution of the Chelicerata, they are homologized with the chelicerae and two pairs of limbs.

A tentative model in Figure 12.8 shows how larvae and head tagmosis of non-trilobite Cambrian arthropods fit into a phylogenetic framework. It implies that the early larvae were composed of the complete ground-pattern head (1+3) and a trunk bud, which grew out by budding off somites sequentially, accompanied by progressive development of limb anlagen. Originally, the limbs of early euarthropods were principally of the same design (see below) and aided in locomotion and food intake. Hence we regard the nauplius of the Crustacea s. str. as a novel larval type, which has a set of features unknown from the larvae of the stem group crustaceans. The nauplius, regarded as an apomorphy of the monophylum Crustacea s. str. (crown group of Crustacea), has:

1. Fewer but highly specialized limbs (antennulae, antennae and mandibulae), a mouth internalized into an atrium oris and covered by the labrum with lateral bristles (this is not the hypostome which is still present in the stem group representatives; see Figure 12.6(a,b), and (also Walossek and Müller, 1990, their Figure 5).
2. A post-oral sternum (= the fused sternites of antennal and mandibular sternites) with characteristic sets of bristles.
3. Paragnath humps representing protrusions on the mandibular sternite (see Walossek, 1993 for morphogenesis of this structure).

The nauplius larva ('orthonauplius') acts as a kind of 'locomotive' for the developing trunk which takes over walking and feeding functions much later during post-embryonic growth (e.g. in *Artemia salina* late in its post-larval differentiation phase; see Walossek, 1993). The earliest larva known

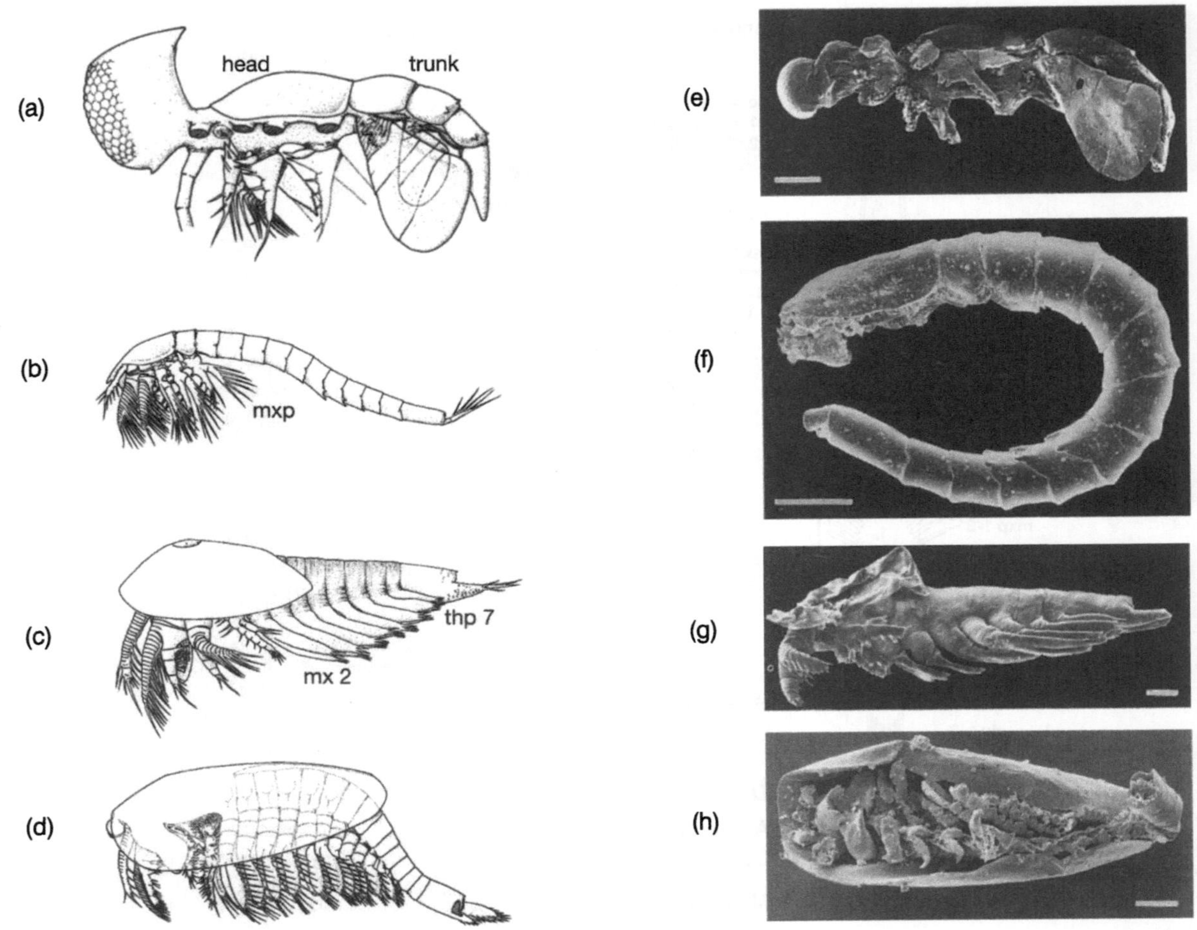


**Figure 12.6** Selection of reconstructions of Upper Cambrian 'Orsten' Crustacea s. l. and s. str.. (a–d) and scanning electron micrographs of these (e–h). (a) Stem-group crustacean *Cambropachycope clarksoni* Walossek and Müller, 1990. (b) Maxillopod *Skara annulata* Müller, 1983, assumed to be more closely related to Mystacocarida and Copepoda (reconstructions from Müller and Walossek, 1985b). (c) *Bredocaris admirabilis* Müller, 1983, assumed to represent the sister taxon of the thecostracan Maxillopoda (Ascothoracida, Facetoteca, Acrothoracida, Cirripedia, possibly also Tantulocarida. (d) *Rehbachiella kinnekullensis* Müller, 1983, as a representative of the Branchiopoda, possibly in the stem line of Anostraca (reconstructions from Walossek and Müller, 1992). (e–h) Corresponding photographs [(e) and (h) flipped head to the left.]

from phosphatocopines has a labrum, sternum and paragnath humps, but also one more pair of limbs, accordingly representing a 'head-larva'. Again, phosphatocopines have post-mandibular limbs all of the same shape, and no special first maxilla or maxillula is present, as in the representatives of the crown group (see below).This confirms a position for this group between the known representatives of the stem line of Crustacea and the crown group (i.e. representing the sister group of Crustacea s. str. in cladistic terms).

## 12.3 LIMB EVOLUTION IN ARTHROPODS

The limb pattern and general design has been debated since the early days of discussion about arthropod phylogeny. Cisne (1975) showed that in the trilobite *Triarthrus* the limb consisted only of an undivided basal portion and two rami, the seven-segmented endopod (as the distal elongation of the stem) and the leaf-shaped exopod arising from the slanting outer rim of the stem (thus being completely different from the endopod and most likely with a different genesis). Subsequent studies confirmed the results of Cisne, and Chen *et al.* (1991) added the impressive picture of a *Naraoia* limb from the Lower Cambrian of China (Figure 12.9(a)). In our description of *Agnostus pisiformis* we presented a three-dimensional view of such an early limb type (Figure 12.10(a)).

Based on such a design, we recognized similar limbs in our representatives of the crustacean stem lineage, but discovered that they bore one more element at the inner proximal edge of their limb stem which we termed the 'proximal endite' (Walossek and Müller, 1990; Figure 12.10(c)). Since

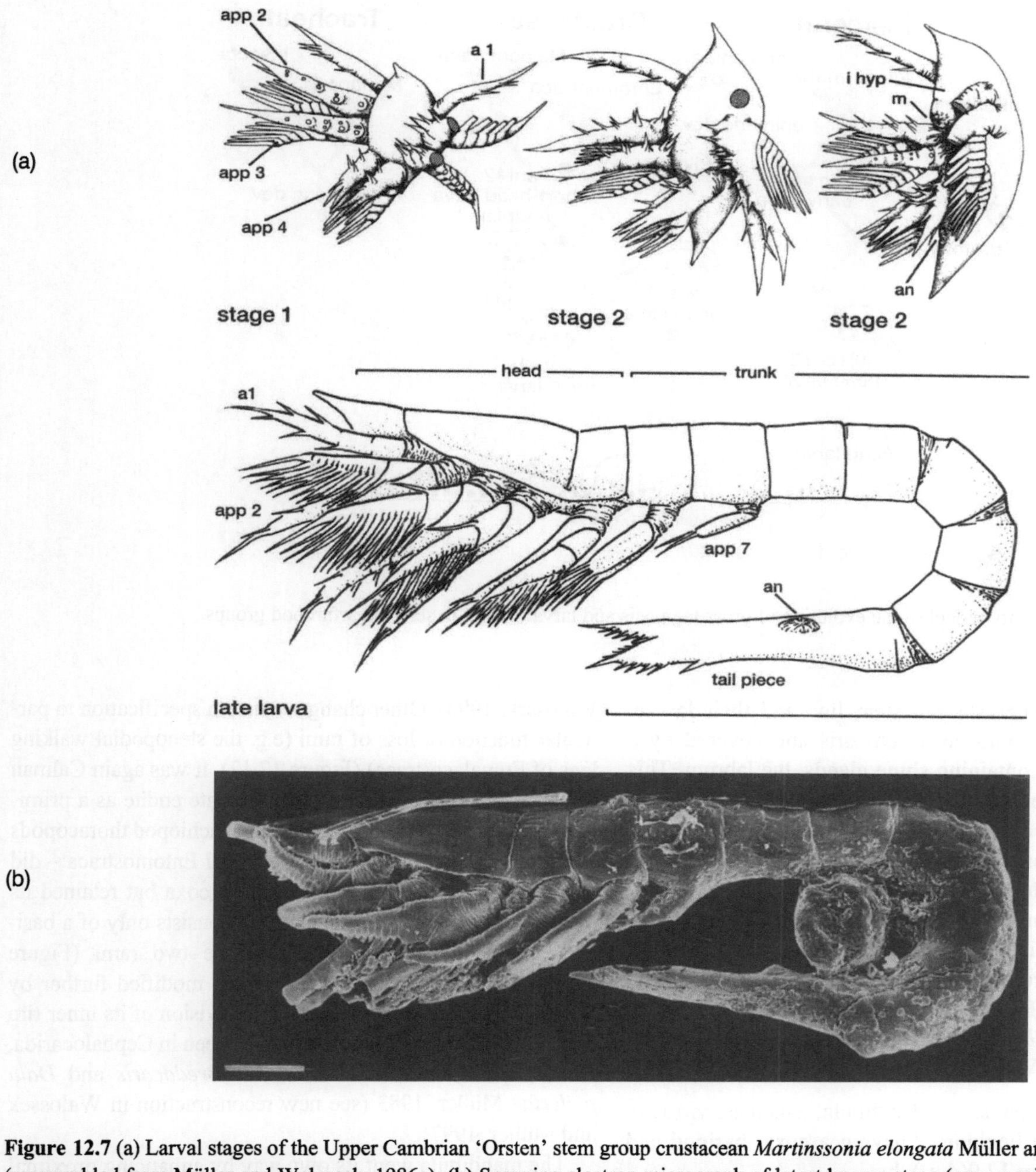


**Figure 12.7** (a) Larval stages of the Upper Cambrian 'Orsten' stem group crustacean *Martinssonia elongata* Müller and Walossek, 1986 (reassembled from Müller and Walossek, 1986a). (b) Scanning electron micrograph of latest known stage (new specimen, illustrated in Walossek and Müller, 1990).

the portion carrying the rami has long been called basipod(ite) in crustaceans, we changed the terminology of the euarthropod limb stem and called it basipod and no longer coxa. Notably, the basipod has the same basic setation pattern in, for example, *Martinssonia* and in the naupliar antenna of a living barnacle (Figure 12.9(b,c)).

We assume that the crustacean line developed different limb types in parallel with changes in locomotory and feeding strategies: the antennulae became more stalked (fewer but longer articles) to help with food intake, while at least some of the exopods became multi-annulate and received setation only along their inner margin, pointing toward the endopod (five-segmented originally) (Figures 12.5, 12.9(c), 12.10(c) and 12.11) which changed both the mode of swimming and fluid currents around the body. The proximal endite may have facilitated food intake near the body surface, which helped to separate the proximal feeding function from the distal locomotory activities of the limbs. The mouth, originally placed at the rear of the bulging hypostome, as seen in *Agnostus* and in some of the 'Orsten' rep-

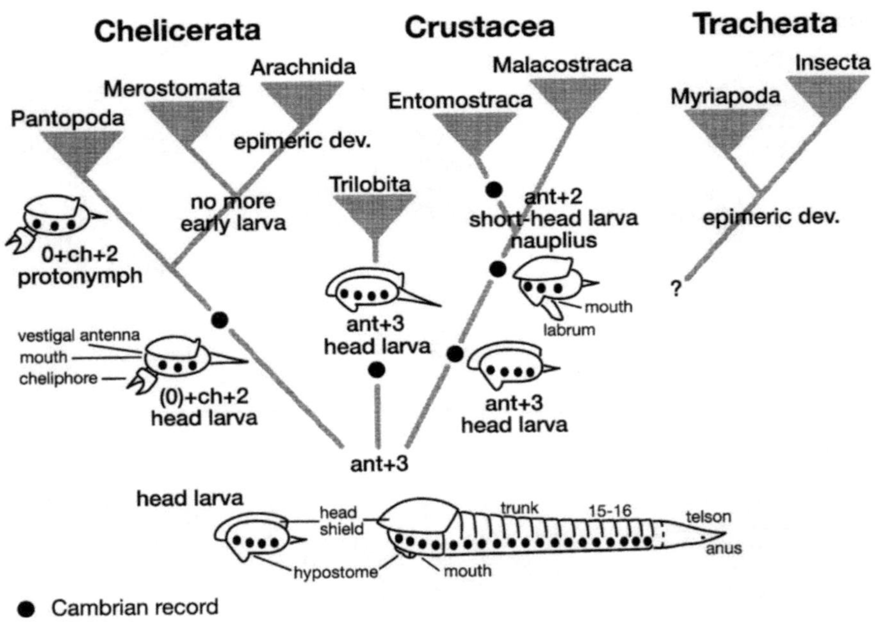


**Figure 12.8** Preliminary model of the evolution of gross tagmosis and larval design in selected arthropod groups.

resentatives of the crustacean stem line and their larvae, became withdrawn into an atrium oris and covered by a bulging structure containing slime glands, the labrum. This organ must not be confused with the hypostome, which is still to some degree retained in extant crustaceans.

This recognition of the 'proximal endite' led us to reinterpret the traditional models of limb design and limb evolution in crustaceans. The endite, present on naupliar limbs of feeding nauplii, underwent progressive transformation into the stem or protopodal portion 'coxa' in Crustacea (see Walossek, 1993, Figure 17 for the morphogenesis of the mandibula of *Rehbachiella*). However this holds true only for certain limbs. Well-developed coxal structures occur:

- in the second antenna and mandibula, which are virtually identical in the beginning, having a coxa, basipod and two rami (Figure 12.9(d,g)), but
- not in the first and second maxillae of all entomostracan taxa (Figure 12.11 upper row),
- not in thoracopods of any entomostracan taxon, at least basically (see Walossek, 1993, Figure 27 for the trunk limb of *Rehbachiella*),
- possibly not in the second maxilla of Malacostraca, but in their first maxilla (Figure 12.11, lower row) and in all of their anterior eight thoracopods (Figure 12.12; status of pleopods uncertain).

Further modifications connected with subdividing the basipod may occur as a result of functional requirements to develop efficient and flexible limb stems (e.g. in the anterior trunk limbs of Calmanostraca = Notostraca and Kazacharthra; see Walossek, 1993). Other changes concern specification to particular function or loss of rami (e.g. the stenopodial walking legs of Eumalacostraca) (Figure 12.12). It was again Calman (1909, p. 51) who pointed to this separate endite as a primitive trait of branchiopod Crustacea. Branchiopod thoracopods – and this can be expanded to almost all Entomostraca – did not develop the proximal endite into a coxa but retained its plesiomorphic shape, i.e. such a limb consists only of a basipod and its proximal endite plus the two rami (Figure 12.10(h,i)). Yet this limb type became modified further by elongation of the basipod and the subdivision of its inner rim into small setiferous endites, as can be seen in Cephalocarida, Branchiopoda, and the Maxillopoda *Bredocaris* and *Dala peilertae* Müller, 1983 (see new reconstruction in Walossek and Müller, 1997).

The mandibula went its own way by enhancing proximal endite to a the coxa as the major organ of food destruction, but by reducing the 'palp' comprising the basipod and the rami (Figure 12.10(e)). In the Entomostraca, this process is completed at the adult state, with the exception of the most possibly paedomorphic Maxillopoda, while in the Malacostraca during later larval development (if present) the 'palp' reappears as a basically three-segmented, setiferous ramus, possibly adapted for cleaning.

Traditionally, the arthropod limb stem should have comprised three portions: the pre-coxa, the coxa and the basipod. In our view, the pre-coxa has several origins:

1. Størmer added a portion proximal to his 'coxa' – in fact our basis – and by placing the outer ramus – our exopod

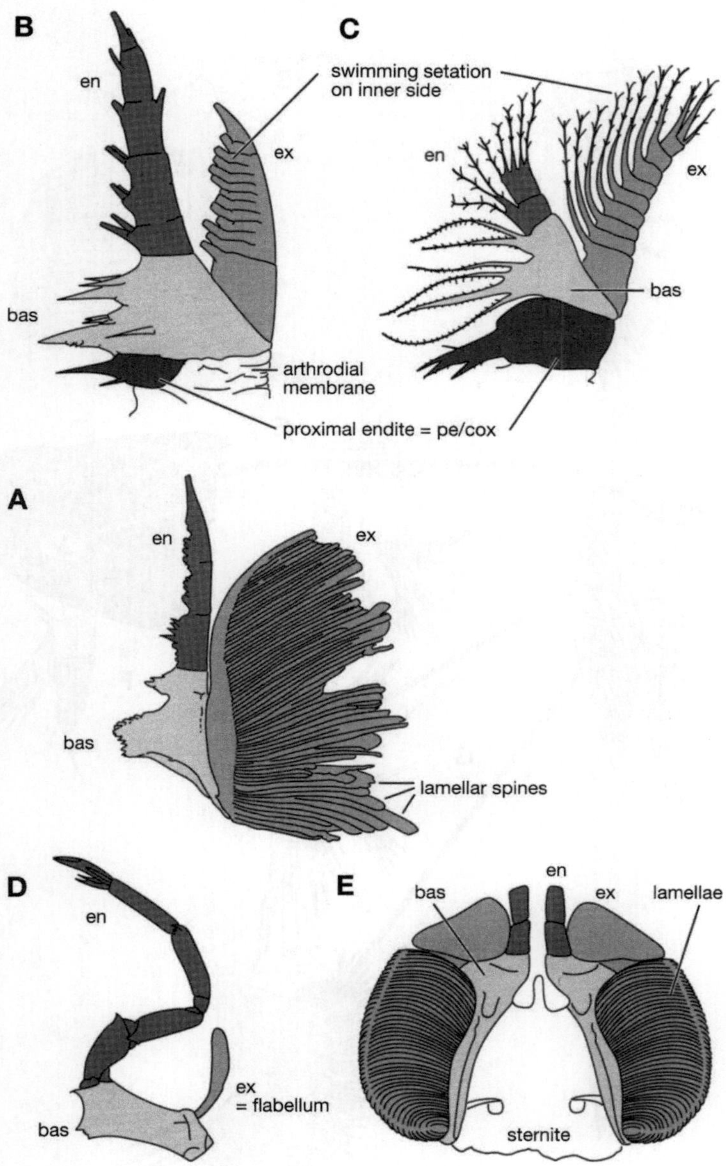


**Figure 12.9** Drawings of different arthropod limbs to show the homology of their components. (A) *Naraoia* limb (redrawn after Chen *et al.*, 1991). (B) *Martinssonia* limb (redrawn from Müller and Walossek, 1986a). (C) Second antenna of a naupliar barnacle with large gnathobasic coxa (!). Arrows connect identical spines and setae of *Martinssonia* as a representative of the crustacean stem line and the naupliar limb (redrawn from Costlov and Bookhout, 1957). (D) Last prosomal leg of *Limulus* sp. with its exopodal rudiment, the so-called flabellum. (E) One of the flat opisthosomal gill limbs of *Limulus*. (Illustrations (D,E) from Siewing, 1985, Figure 838.) Shading, see Figure 12.10.

– on his 'pre-coxa', the trilobite outer ramus turned into a pre-epipod (gill). We note that an exopod cannot arise from a coxa or pre-coxa, but the epipodial gill, as seen in crustaceans.

2. The pre-coxa of other authors is only the arthrodial membrane between limb and body.
3. Kaestner (1970, Figure 661) termed the arthrodial membrane between basipod and coxa of a copepod mandibula the 'coxa', and, in consequence, called the true coxa a 'pre-coxa'.

Our hypothesis for the evolution of arthropod limbs abandons the idea of a pre-coxa in any arthropod limbs. The homologization of the different limb portions and their modifications in the various crustacean taxa is illustrated in Figures 12.9 and 12.12. To simplify: if there is no coxa

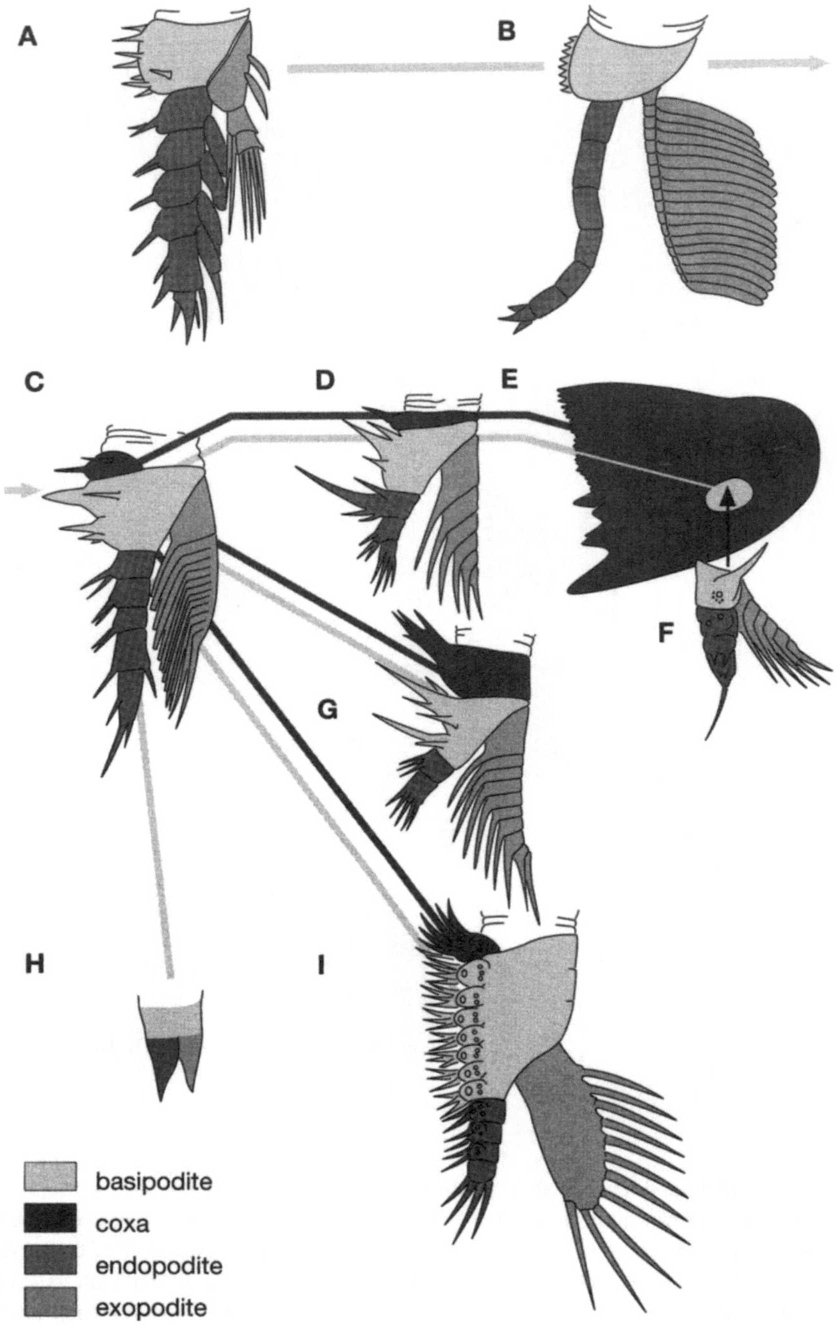


**Figure 12.10** Hypothesis of the evolutionary path toward the different limb types of Crustacea s. str. from a type known from Early Palaeozoic arthropods. (A,B,C,D) Trilobites (*Naraoia*) and *Agnostus*, etc. through a stage seen in stem group crustaceans such as the 'Orsten' fossil *Martinssonia elongata*. (E,F) Second antenna and mandibula of naupliar cirripeds. (E,F) A late larval mandibula of *Rehbachiella* still with the basipod and rami present but reduced. (G,H) Trunk-limb bud of *Rehbachiella*. (I) Late larval trunk limb of *Rehbachiella*. Shadings for limb patterns shown in boxes also apply to figures 12.9 and 12.12. (For further details, see text; modified from Walossek, 1993.)

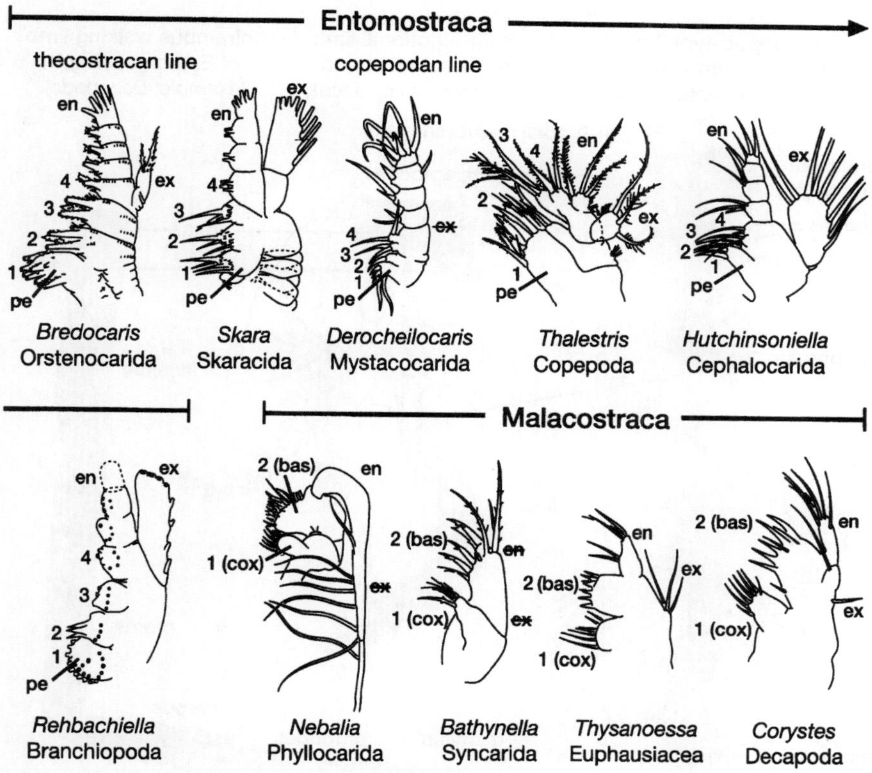


**Figure 12.11** Shapes of maxillulae (first maxillae, mx1) in different crustacean taxa. (Illustrations from different sources.)

basally, then neither can there be a pre-coxa. This model can be readily expanded to Trilobita and Chelicerata: their basal portion is the basipod, not the coxa. Accordingly the rami are the endopod and the exopod, as in Crustacea. Chelicerate evolution includes loss of the exopod to form uniramous walking legs. Merostomata (*Limulus*) retain the exopod in the last prosomal leg (Figure 12.10(d)) and in the flat opisthosomal legs (Figure 12.10(e)). Their numerous flat gill lamellae correspond well with the lamellar spines of the *Naraoia* leg (Figure 12.10(a)), which makes clear the evolutionary history of chelicerate gills and, later, book lungs.

Difficulties arise from the fact that some authors still omit proximal limb portions from descriptions, although this region is crucial for the understanding of limb homologies. It is also a consequence of our argument that the limbs at the head–trunk border, that is, the maxillulae and maxillae, are of importance for evaluating the divergence in the major lines of the Crustacea. The maxillulae (mx1) are modified at the crown-group level (Figure 12.11), while the maxillae (mx2) retained their trunk limb design for much longer, and became modified within the different lineages. A trunk limb-like maxilla is present at least in Cephalocarida, the fossil maxillopod *Bredocaris*, the fossil branchiopod *Rehbachiella*, and possibly also in *Lepidocaris* from the Devonian (see Walossek, 1993; assumed to be a representative of the anostracan line). Study of the history of these two pairs of post-mandibular limbs would be phylogenetically interestingly. In some taxa, more segments and/or their limbs are added to form a larger head tagma or cephalic feeding system, the 'cephalothorax' – which invites comparison with the Chelicerata. Potentially each limb provides its own set of features in the evolution of Crustacea (Figure 12.12; also Boxshall, 1997, this volume).

## 12.4 BODY TAGMOSIS

Associated with limb development is the clustering of limbs into functional compartments, often closely linked with general body tagmosis. Since there is considerable inconsistency in the use of the term 'abdomen', particularly in Crustacea, which may either bear or lack limbs, we suggest a tagmosis model in Figure 12.13. Crustaceans show a high plasticity of segment patterns, ranging from forms with up to 71 somites and more than 30 pairs of limbs behind the head, as in triopsid Branchiopoda (apparently developed by poly-metamerism and poly-pody), to forms which show a high degree of segment reduction, as in Cladocera and advanced types of Ostracoda.

Walossek (1993) showed that the ontogeny of Entomostraca in general is a two-step process, first develop-

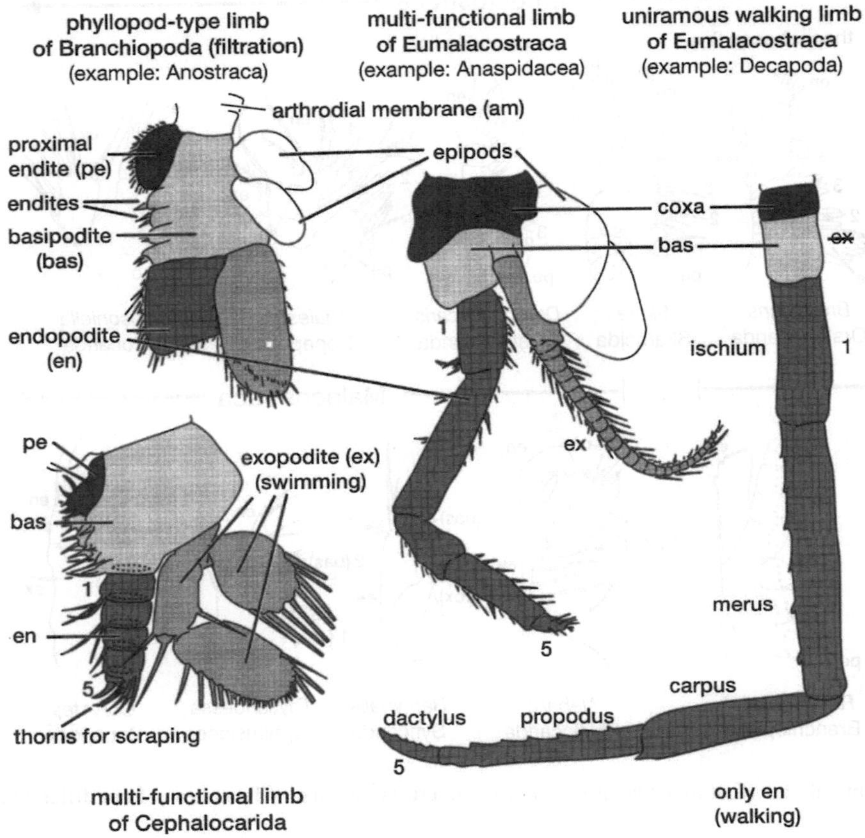


**Figure 12.12** Selection of trunk limb types in Crustacea s. str. (shadings as shown in Figure 12.10). (Modified from different sources.)

ing the thorax and its limbs, and, later on, the abdominal segments. The largest block of trunk segments is developed in the Anostraca with 13 (Benesch, 1969), 12 of which bear limbs (the last modified to genital structures). Cephalocarida have nine limb-bearing thoracomeres (last limb modified to egg carriers; Sanders, 1963) and 10 abdominal segments, while Maxillopoda have basically seven limb-bearing thoracomeres and four abdominal segments. Malacostraca have a block of 14 limb-bearing trunk segments, divided into two sets of eight and six, and just one apodous segment only in the Phyllocarida. No phase of development of abdominal segments occurs in this group.

It becomes apparent that the malacostracan pleon is not an abdomen in the sense used above, but should be regarded as the second set of thoracic segments. On the other hand, the abdomen of *Artemia* is no pleon either. It is evidently critical to use the term abdomen for the posterior part of insects, since limbs were only partially lost there, and indeed are still present at the rear in form of genital limbs and cerci. The same is true for Chelicerata, which have an opisthoma behind their prosoma which bore legs originally.

It is still difficult to estimate which one is the plesiomorphic state – whether with, or without an abdomen of the type developed in Entomostraca. However, the model can serve to characterize both groups of Crustacea and adds further evidence for a sister group relationship of Entomostraca and Malacostraca.

## 12.5 GROUND PATTERNS AND EVOLUTION

Our review of arthropod and crustacean phylogeny is based primarily on 'Orsten' and other fossil material. It can include only a fraction of the morphological characters available, but does show the potential available for future work. Our main intention is to demonstrate that fossils of the quality of 'Orsten' material provide a huge 'toolbox' to be used for reconstructing arthropod, and particularly crustacean, phylogeny. This large data set in our opinion is more reliable for constructing trees than data recycled from published character matrices.

Terminology must be as consistent as possible. We believe that ground patterns (groundplans of other authors in this book) are essential in phylogeny reconstruction (Walossek and Müller, 1997; Kukalová-Peck, 1997, this volume). The ground

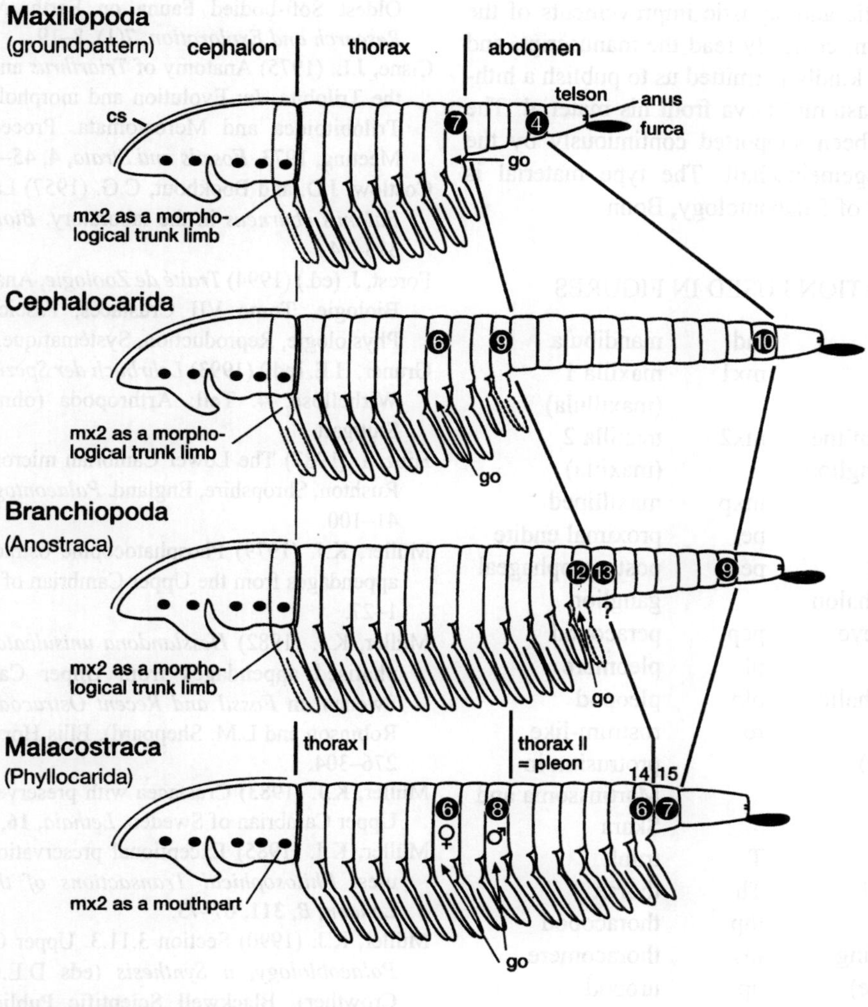


**Figure 12.13** Hypothesis of body tagmosis in selected major crustacean taxa.

pattern, however, is not a kind of 'paper model' which never existed in real animals, but rather includes the whole set of features of an organism at a basal evolutionary level, and combines plesiomorphies as well as apomorphies. It carries the genetic potential from which a derived arthropod could draw for adaptation to a specific ecological niche, associated with the development of new structures. As an example, the apparently bizarre *Goticaris* is not just a multi-segmented early crustacean but has its own specialities: an unpaired compound eye, an uncovered mouth on its head surface, and a sack-like body with four pairs of uniramous, paddle-shaped limbs, finally terminating in a long spine. The related *Cambropachycope* had a feeble head shield and faint trunk segmentation, developed a huge paddle instead of a normal first trunk limb and only one more smaller paddle limb behind. Yet, both can be recognized as representatives of the stem line of Crustacea because of apomorphic characters which are not found in the appropriate outgroups.

Reconstruction of ground patterns is also valuable because all characters are subject to evolutionary changes; rarely does a character remain unmodified into extant taxa. The evolutionary path of a taxon is inaccessible without knowledge of the selective forces at the beginning of a group and the order of appearance of their characters. The construction and homologization of arthropod limb types can only be understood from evidence received from the earliest known stages of euarthropod evolution. Again, many anatomical features are functionally controlled, and it is always necessary to consider the original design of a structure, i.e. reconstructing the plesiomorphic state of a feature to understand its particular changes. That caution can be necessary has been shown by the nervous system of arthropods – for which a character state developed only in certain taxa has been taken as a ground-pattern feature of Arthropoda.

## ACKNOWLEDGEMENTS

D.W. wishes to express his gratitude to the organizers of this conference, R.H. Thomas and R.A. Fortey, London, for the invitation and linguistic and stylistic improvements of the text. W. Ahlrichs, Ulm, critically read the manuscript, and J.E. Repetski, Reston, kindly permitted us to publish a hitherto undescribed pentastomid larva from his material. The 'Orsten' project has been supported continuously by the Deutsche Forschungsgemeinschaft. The type material is housed at the Institute of Palaeontology, Bonn.

## LIST OF ABBREVIATIONS USED IN FIGURES

| | | | |
|---|---|---|---|
| a1 | antenna 1, antennula | md | mandibula |
| | | mx1 | maxilla 1 (maxillula) |
| a2 | antenna 2 | | |
| a2-conn | connective of the antennal ganglion | mx2 | maxilla 2 (maxilla) |
| Abd | abdomen | mxp | maxilliped |
| app | appendage | pe | proximal endite |
| bas | basipod(ite) | peg | post-oesophageal ganglion |
| C | head or cephalon | | |
| ce | compound eye (faceted) | pep | peraeopod |
| | | pl | pleomere |
| cs | head or cephalic shield | plp | pleopod |
| | | ro | rostrum-like protrusion in Martinssonia and Skara |
| en | endopod(ite) | | |
| epi | epipod(ite) | | |
| eso | oesophagus | | |
| ex | exopod(ite) | T | trunk |
| fu | furca (furcal rami) | Th | thorax |
| | | thp | thoracopod |
| go | gonad opening (genital pore) | ths | thoracomere |
| | | up | uropod |
| m | mouth | tel | telson |

## REFERENCES


Abele, L.G, Kim, W. and Felgenhauer, B.E. (1989) Molecular evidence for inclusion of the Phylum Pentastomida in the Crustacea. *Molecular Biology and Evolution,* **6**(6), 685–91.

Andres, D. (1989) Phosphatisierte Fossilien aus dem unteren Ordoviz von Südschweden. *Berliner geowissenschaftliche Abhandlungen (A)*, **106**, 9–19.

Benesch, R. (1969) Zur Ontogenie und Morphologie von Artemia salina L. *Zoologische Jahrbücher der Anatomie und Ontogenie*, **86**, 307–458.

Briggs, D.E.G., Bruton, D.L. and Whittington, H.B. (1979) Appendages of the arthropod *Aglaspis spinifer* (Upper Cambrian, Wisconsin) and their significance. *Palaeontology*, **22**(1), 167–80.

Brusca, R.C. and Brusca, G.J. (1990) *Invertebrates*, Sinauer Associates, Inc., Sunderland, Massachusetts.

Calman, C. (1909) Part VII. Appendiculata. 3rd vol. Crustacea, in *A Treatise on Zoology* (ed. R. Lankester), Adam and Charles Black, London.

Chen Jun-Yuan, Bergström, J., Lindström, M. and Hou Xianguang (1991) Fossilized Soft-bodied Fauna. The Chengjiang Fauna – Oldest Soft-bodied Fauna on Earth. *National Geographic Research and Exploration*, **7**(1), 8–19.

Cisne, J.L. (1975) Anatomy of *Triarthrus* and the relationships of the Trilobita. *In*: Evolution and morphology of the Trilobita, Trilobitoidea and Merostomata. Proceedings of the Oslo Meeting, 1973, *Fossils and Strata*, **4**, 45–63.

Costlow, J.D. and Bookhout, C.G. (1957) Larval development of *Balanus eburneus* in the laboratory. *Biological Bulletin*, **112**, 313–24.

Forest, J. (ed.) (1994) *Traité de Zoologie*, Anatomie, Systématique, Biologie, Tome VII Crustacés, Fascicule 1 Morphologie, Physiologie, Reproduction, Systématique, Masson, Paris.

Gruner, H.E. (ed.) (1993) *Lehrbuch der Speziellen Zoologie*, Bd. 1, Wirbellose, 4. Teil: Arthropoda (ohne Insecta), Fischer, Stuttgart.

Hinz, I. (1987) The Lower Cambrian microfauna of Comley and Rushton, Shropshire, England. *Palaeontographica A*, **198**(1–3), 41–100.

Müller, K.J. (1979) Phosphatocopine ostracodes with preserved appendages from the Upper Cambrian of Sweden. *Lethaia*, **12**, 1–27.

Müller, K.J. (1982) *Hesslandona unisulcata* sp. nov. with phosphatised appendages from Upper Cambrian 'Orsten' of Sweden, in *Fossil and Recent Ostracods* (eds R.H. Bate, E. Robinson and L.M. Sheppard), Ellis Horwood, Chichester, pp. 276–304.

Müller, K.J. (1983) Crustacea with preserved soft parts from the Upper Cambrian of Sweden. *Lethaia*, **16**, 93–109.

Müller, K.J. (1985) Exceptional preservation in calcareous nodules. *Philosophical Transactions of the Royal Society of London, B*, **311**, 67–73.

Müller, K.J. (1990) Section 3.11.3. Upper Cambrian 'Orsten', in *Palaeobiology, a Synthesis* (eds D.E.G. Briggs and P.R. Crowther), Blackwell Scientific Publications, Oxford, pp. 274–7.

Müller, K.J. and Hinz, I. (1992) Cambrogeorginidae fam. nov., soft-integumented Problematica from the Middle Cambrian of Australia, *Alcheringa*, **16**, 333–53.

Müller, K.J. and Hinz-Schallreuter, I. (1993) Palaeoscolecid worms from the Middle Cambrian of Australia, *Palaeontology*, **36**(3), 549–92.

Müller, K.J. and Walossek, D. (1985a) A remarkable arthropod fauna from the Upper Cambrian 'Orsten' of Sweden. *Transactions of the Royal Society of Edinburgh: Earth Sciences*, **76**, 161–72.

Müller, K.J. and Walossek, D. (1985b) Skaracarida, a new order of Crustacea from the Upper Cambrian of Västergötland, Sweden, *Fossils and Strata*, **17**, 1–65; plates 1–17.

Müller, K.J. and Walossek, D. (1986a) *Martinssonia elongata* gen. et sp. n., a crustacean-like euarthropod from the Upper Cambrian 'Orsten' of Sweden. *Zoologica Scripta*, **15**(1), 73–92.

Müller, K.J. and Walossek, D. (1986b) Arthropod larvae from the Upper Cambrian of Sweden. *Transactions of the Royal Society of Edinburgh: Earth Sciences*, **77**, 157–79.

Müller, K.J. and Walossek, D. (1987) Morphology, ontogeny and life habit of *Agnostus pisiformis* from the Upper Cambrian of Sweden. *Fossils and Strata*, **19**, 1–124.

Müller, K.J. and Walossek, D. (1988a) Eine parasitische Cheliceraten-Larve aus dem Kambrium. *Fossilien*, **1**, 40–2.

Müller, K.J. and Walossek, D. (1988b) External morphology and larval development of the Upper Cambrian maxillopod *Bredocaris admirabilis*. *Fossils and Strata*, **23**, 1–70, 16 plates.

Müller, K.J. and Walossek, D. (1991a) 'Orsten' arthropods – small in size but of great impact on biological and phylogenetic interpretations. *Geologiska Föreningens i Stockholm Förhandlingar* Meeting Proceedings, **113**(1), 88–90.

Müller, K.J. and Walossek, D. (1991b) Ein Blick durch das 'Orsten'-Fenster in die Arthropodenwelt vor 500 Millionen Jahren. *Verhandlungen der Deutschen Zoologischen Gesellschaft*, **84**, 281–94.

Müller, K.J., Walossek, D. and Zakharov, A. (1995) 'Orsten' type phosphatized soft-integument preservation and a new record from the Middle Cambrian Kuonamka Formation in Siberia. *Neues Jahrbuch der Geologie und Palaeontologie, Abhandlungen*, **191**(1), 101–18.

Ramsköld, L. and Hou X. (1991) New early Cambrian animals and onychophoran affinities of enigmatic metazoans. *Nature*, **351**, 225–8.

Roy, K. and Fåhræus, L.E. (1989) Tremadocian (Early Ordovician) nauplius-like larvae from the Middle Arm Point Formation, Bay of Islands, western Newfoundland. *Canadian Journal of Earth Sciences*, **26**, 1802–6.

Sanders, H.L. (1963) The Cephalocarida. Functional Morphology, Larval Development, Comparative External Anatomy. *Memoirs of the Connecticut Academy of Arts and Sciences*, **15**, 1–80.

Siewing, R. (1985) *Lehrbuch der Zoologie*, Band 2, Systematik, Fischer, Stuttgart.

Walossek, D. (1993) The Upper Cambrian *Rehbachiella kinnekullensis* and the phylogeny of Branchiopoda and Crustacea. *Fossils and Strata*, **32**, 1–202.

Walossek, D. (1995) The Upper Cambrian *Rehbachiella*, its larval development, morphology and significance for the phylogeny of Branchiopoda and Crustacea. *Hydrobiologia*, **298**, 1–13.

Walossek, D. (1996) *Rehbachiella*, der bisher älteste Branchiopode. *Stapfia*, **42**, also *Kataloge des O.Ö. Landesmuseum, NF*, **100**, 21–8.

Walossek, D. and Müller, K.J. (1989) A second type A-nauplius from the Upper Cambrian of Sweden. *Lethaia*, **22**, 301–6.

Walossek, D. and Müller, K.J. (1990) Stem-lineage crustaceans from the Upper Cambrian of Sweden and their bearing upon the position of *Agnostus*. *Lethaia*, **23**, 409–27.

Walossek, D. and Müller, K.J. (1991) Lethaia Forum: *Henningsmoenicaris* n. gen. for *Henningsmoenia* Walossek and Müller, 1990 – correction of name. *Lethaia*, **24**, 138.

Walossek, D. and Müller, K.J. (1992) The 'Alum Shale Window' Ą Contribution of 'Orsten' Arthropods to the Phylogeny of Crustacea. *Acta Zoologica*, **73**, 305–12.

Walossek, D. and Müller, K.J. (1994) Pentastomid parasites from the Lower Palaeozoic of Sweden, *Transactions of the Royal Society of Edinburgh: Earth Sciences*, **85**, 1–37.

Walossek, D. and Müller, K.J. (1997) Early Arthropod Phylogeny in the Light of the Cambrian 'Orsten' Fossils, in *Arthropod Fossils and Phylogeny* (ed. G. Edgecombe), Columbia University Press, New York (in press)

Walossek, D. and Szaniawski, H. (1991) *Cambrocaris baltica* n. gen. n. sp., a possible stem-lineage crustacean from the Upper Cambrian of Poland. *Lethaia*, **24**, 363–78.

Walossek, D., Hinz-Schallreuter, I., Shergold, J.H. and Müller, K.J. (1993) Three-dimensional preservation of arthropod integument from the Middle Cambrian of Australia. *Lethaia*, **26**, 7–15.

Walossek, D., Repetski, J.E. and Müller, K.J. (1994) An exceptionally preserved parasitic arthropod, *Heymonsicambria taylori* n. sp. (Arthropoda incertae sedis: Pentastomida), from Cambrian–Ordovician boundary beds of Newfoundland, Canada. *Canadian Journal of Earth Sciences*, **31**, 1664–71.

Wingstrand, K.G. (1972) Comparative spermatology of a pentastomid *Raeillietiella hemidactyli* and a branchiuran crustacean *Argulus foliaceus* with a discussion of pentastomid relationships. *Biologiske Skrifter*, **19**, 1–72.

# 13 Comparative limb morphology in major crustacean groups: the coxa–basis joint in postmandibular limbs

*G. Boxshall*

*Department of Zoology, The Natural History Museum, Cromwell Road, London, SW7 5BD, UK*
*email: G.Boxshall@nhm.ac.uk*

## 13.1 INTRODUCTION

Although it may have been plesiomorphic to a wider group of arthropods, one characteristic of the superclass Crustacea is the possession of a nauplius larva bearing three pairs of functional appendages, antennules, antennae and mandibles. The nauplius larva exhibits basically the same organization across the range of crustacean classes in which it is retained, and it is in the immediate postnaupliar phase, during the formation of the cephalon with its complement of five appendages, that the great diversity of crustacean form first becomes apparent. Cephalization has been a key engine driving crustacean evolution and diversification, and the specialization of the postmandibular limbs to form part of the cephalic feeding apparatus has been a major part of the process (Walossek, 1993). Classical comparative anatomy has revealed a wealth of new data concerning the musculature of the maxillules, maxillae and maxillipeds (= first thoracopods) in the Crustacea (Boxshall and Huys, 1992). These new data are used here to identify the basic limb segmentation patterns across the Crustacea, to consider whether these are robust crustacean apomorphies or simply symplesiomorphies shared with a wider group of arthropods, and to determine whether certain problematic fossils from the Cambrian should be classified as crustaceans.

This brief account focuses primarily on the maxillules in order to demonstrate the value of classical comparative anatomy in formulating and testing hypotheses concerning the homology of appendage segments. The maxillule was chosen since it is the first of the postnaupliar appendages to be added during development. Some observations are also included on maxillae and maxillipeds to illustrate its potential application, for example, in solving long-standing confusion surrounding the identity of the appendage segments of ostracods as well as to demonstrate that the approach is appropriate for a range of appendages and is not restricted to maxillules.

Walossek and Müller (1997, this volume) also consider limb structure in crustaceans but employ a different terminology for the limb segments from that adopted here. The following comparison of terms should facilitate cross-reference between the two systems. When undivided, the proximal segment (articulating proximally with the body somite and bearing the rami distally) of the postmandibular limbs of crustaceans is referred to here as the protopod. Walossek and Müller refer to this as the basipodite. The protopod is typically subdivided by a transverse articulation, the coxa–basis articulation (see below). The segment distal to this articulation is referred to here as the basis. This segment is also referred to as the basipodite by Walossek and Müller. The part of the divided protopod lying proximal to the coxa–basis articulation is referred to here as a syncoxa or, if further subdivided into two segments by another transverse articulation, as the precoxa and coxa. The entire part of the limb proximal to the coxa–basis articulation is referred to as the coxa by Walossek and Müller.

## 13.2 THE COXA–BASIS ARTICULATION

It is difficult to interpret the patterns of segmentation of arthropod limbs by external examination alone because it is easy to confuse fold lines and sutures with articulations between segments. Study of the musculature of an appendage can help to determine segmental identity because the arrangement of muscle origins, insertions and intermediate, tendinous attachment points typically reflects the ancestral pattern of articulations between segments. Studies of the musculature of a range of copepod orders (Boxshall, 1982, 1985, 1990) revealed a particular muscle signature pattern which unequivocally identified one articulation, that between coxa and basis, within the protopodal part of the postmandibular limbs. A good illustration of the muscle signature is provided by the thoracic swimming legs of copepods. These are biramous and each ramus is typically three-segmented. The protopodal part of the leg is also three-segmented, comprising a minute, crescentic precoxa (described as a pouch of integument on which thoracopod

*Arthropod Relationships*, Systematics Association Special Volume Series 55, Edited by R.A. Fortey and R.H. Thomas. Published in 1997 by Chapman & Hall, London. ISBN 0 412 75420 7

promotor muscle 3 inserts by Boxshall, 1985, pp. 363, 365), a large coxa which is permanently fused to the opposite member of the pair via a rigid intercoxal sclerite, and a basis which carries the rami. In all copepods studied, all the intrinsic muscles inserting on either ramus originate in the basis and all of the extrinsic muscles insert either in the precoxa and coxa, or inside the basis close to the coxa–basis plane (Boxshall, 1985, Figure 78). No muscles originating on the body wall (extrinsic muscles) nor any intrinsic muscles originating in the precoxa or coxa pass through the coxa–basis plane to insert on either ramus. The hiatus in musculature between coxa and basis in copepod swimming legs serves as a signature that can be used to identify the coxa–basis plane as a point of reference, even though this hiatus may not be complete in some groups.

This muscle signature is also recognisable in the copepod maxillule, maxilla and maxilliped. In the maxilliped of *Euaugaptilus placitus* Scott, for example, a total hiatus is found at the coxa–basis joint. No muscles pass through from the syncoxa (fused precoxa and coxa) to insert on the ramus and all the muscles to the ramal segments originate either within the basis or inside the ramus itself (Boxshall, 1985, Figure 23). The maxilla of this calanoid is similar (Boxshall, 1985, Figure 22) with nine muscles inserting or originating around the coxa–basis plane, but a single muscle (mx.r.rem.1) originates in the precoxa and passes through the coxa plane to insert on the ramus. The maxilla of *Mormonilla phasma* Giesbrecht exhibits a total hiatus at the coxa–basis plane as does the maxillule of the same species (Boxshall, 1985, Figures 29 and 30) but a single muscle passes through the plane in the maxilliped (Boxshall, 1985, Figure 31).

The coxa–basis hiatus in musculature is not absolute. Muscles do pass through the coxa–basis plane and into the ramus without inserting but, for the copepods, these are exceptions and do not prevent recognition of the coxa–basis plane. This muscular signature of the coxa–basis plane will be used below to help in interpreting segmentation patterns of postmandibular appendages of other major crustacean taxa.

## 13.3 THE CRUSTACEAN MAXILLULE

### 13.3.1 COPEPODA

The first postmandibular limb in crustaceans is the maxillule. The maxillule of copepods can be used to illustrate the basic segmentation pattern in the Crustacea. The copepod maxillule has three protopodal segments (precoxa, coxa and basis) and is biramous with a three-segmented endopod and an unsegmented exopod (Huys and Boxshall, 1991). In greater detail, the precoxa has a single setiferous endite, the coxa has a single setiferous endite and a lobate setiferous outer lobe (the epipodite), and the basis has two defined endites plus a lobate setiferous exite. The endopodal segments carry setae along their inner margins and on the apex of the ramus. The exopod bears an array of setae along its distal and outer margins. The exopod and epipodite are adapted to present a large interface with the water and they function by moving water and generating flow fields around the copepod.

### 13.3.2 OSTRACODA

The maxillule of myodocopid ostracods is similar to that of copepods in basic organization. The most complex maxillule found in the Myodocopida is that of *Azygocypridina* Sylvester-Bradley (Figures 13.1 and 13.2(b)). In *A. imperalis* (Stebbing) the protopod is three-segmented, comprising a large precoxa with a single well developed endite, a coxa bearing an endite and a single outer seta possibly representing the epipodite, and a basis with two endites. The exopod of *A. imperalis* is unsegmented and the endopod is two-segmented. However, a three-ségmented endopod is found within the Myodocopida and represents the plesiomorphic condition of this ramus in the group. The coxa–basis plane is readily identified by its muscular signature, a complete hiatus with no muscles passing through from proximal to the coxa–basis plane into the rami. The two segments proximal to the coxa–basis plane are the precoxa and coxa.

The entire appendage is orientated differently from the maxillule of a copepod. It is a more three-dimensional appendage in ostracods, presumably to function in the confined space within the paired valves of the ostracod shell. The flattened endopod is rotated with respect to the protopodal part and two views of the appendage (Figures 13.1 and 13.2(b)) are necessary to show the relative positions of the segments and their muscles. The exopod and reduced epipodite are not adapted to generate flow fields.

### 13.3.3 MYSTACOCARIDA

The maxillule of mystacocarids is uniramous, lacking an exopod. There is a well defined hiatus in the musculature indicating the coxa–basis plane (Figure 13.3(a)). Proximal to this plane are two incompletely separated segments, each bearing a setose endite. These two segments are here identified as the precoxa and coxa. The next segment is the basis and carries two endites. The remaining four segments represent the endopod.

The maxillulary protopod of mystacocarids shares the same basic pattern as that of copepods and ostracods: consisting of precoxa and coxa, each with one endite, and basis with two endites. The endopod has one more segment than in copepods and ostracods. Mystacocarids feed by collecting microorganisms and organic matter from the surface of sediment particles (Baccari and Renaud-Mornant, 1974). They do not generate flow fields and lack any vestige of the exopod.

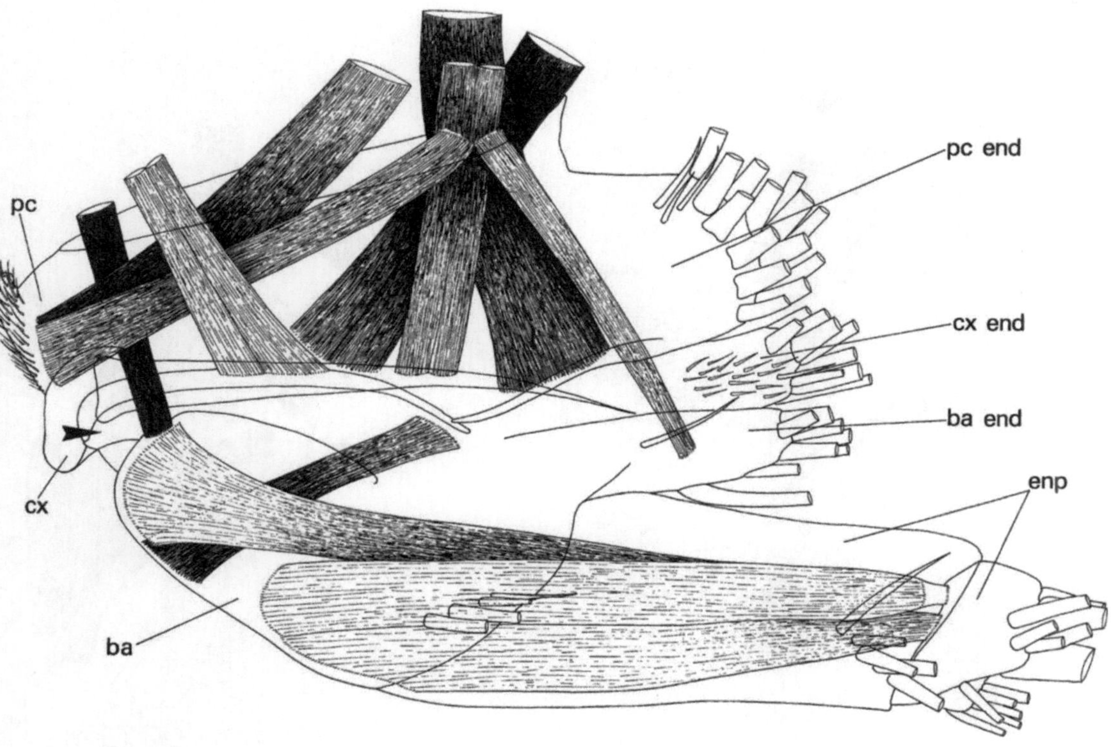


**Figure 13.1** Maxillule of the myodocopid ostracod *Azygocypridina imperalis*, anteromedial view showing musculature. Only basal parts of long setation elements on endites shown. Abbreviations: cx, coxa; ba, basis; enp, endopod; exp, exopod; end, endite; epi, epipodite; pc, precoxa; sc, syncoxa. The coxa–basis plane is arrowed.

### 13.3.4 REMIPEDIA

The maxillule of remipedes is uniramous and also comprises seven segments, just as in mystacocarids. The coxa–basis plane is well defined in the musculature and there are no muscles passing through this plane into the ramus (Figure 13.3(b)). Proximal to the coxa–basis plane are two segments, the precoxa and coxa, each with a single endite. Distal to the plane lies the basis, which carries only a single endite. The endopod is four-segmented, with the fourth segment bearing an apical claw. A large venom gland lies in the head that secretes venom into the sac-like reservoir in the endopod of the maxillule. The venom is injected into prey organisms via the apical claw (Schram and Lewis, 1989). The exopod is absent.

The remipede maxillule is a robust, highly muscular, subchelate limb. The primary articulation lies between the first and second endopodal segments and adducts the distal subchela. The muscles within the second endopodal segment have a pinnate arrangement, inserting on a long apodeme-like extension from the apical segment. A pinnate arrangement of fibres allows muscles to contract without increasing in volume (Baskin and Paolini, 1966), and is commonly found in positions where space is limited, such as within the chelipeds of a crab (Alexander, 1968).

Apart from the presence of only one endite on the basis rather than two, the segmentation pattern and numbers of segmental endites of remipedes are the same as those of mystacocarids.

### 13.3.5 BRANCHIURA

The maxillules of adult *Argulus* are modified as suckers (Figure 13.4(a)). The musculature is profoundly modified to move the entire sucker and to generate the necessary suction (Gresty *et al.*, 1993). In the branchiuran genus *Dolops*, and in the early larval stages of *Argulus*, the maxillule is a uniramous clawed appendage. The adult sucker develops in the syncoxal part of the protopod (Figure 13.4(b)) and functional continuity is maintained as the change from attachment via the barbed apical claws to attachment via the proximal sucker takes place. The fifth larval stage appears to represent the transition phase between the two attachment mechanisms in *Argulus foliaceus* (Rushton-Mellor and Boxshall, 1994). The musculature of the maxillule in the first larval stage shows a hiatus at the first articulation (Figure 13.3(c)). Although there are three extrinsic muscles passing through this articulation towards insertions in the ramus, nine muscles originate or

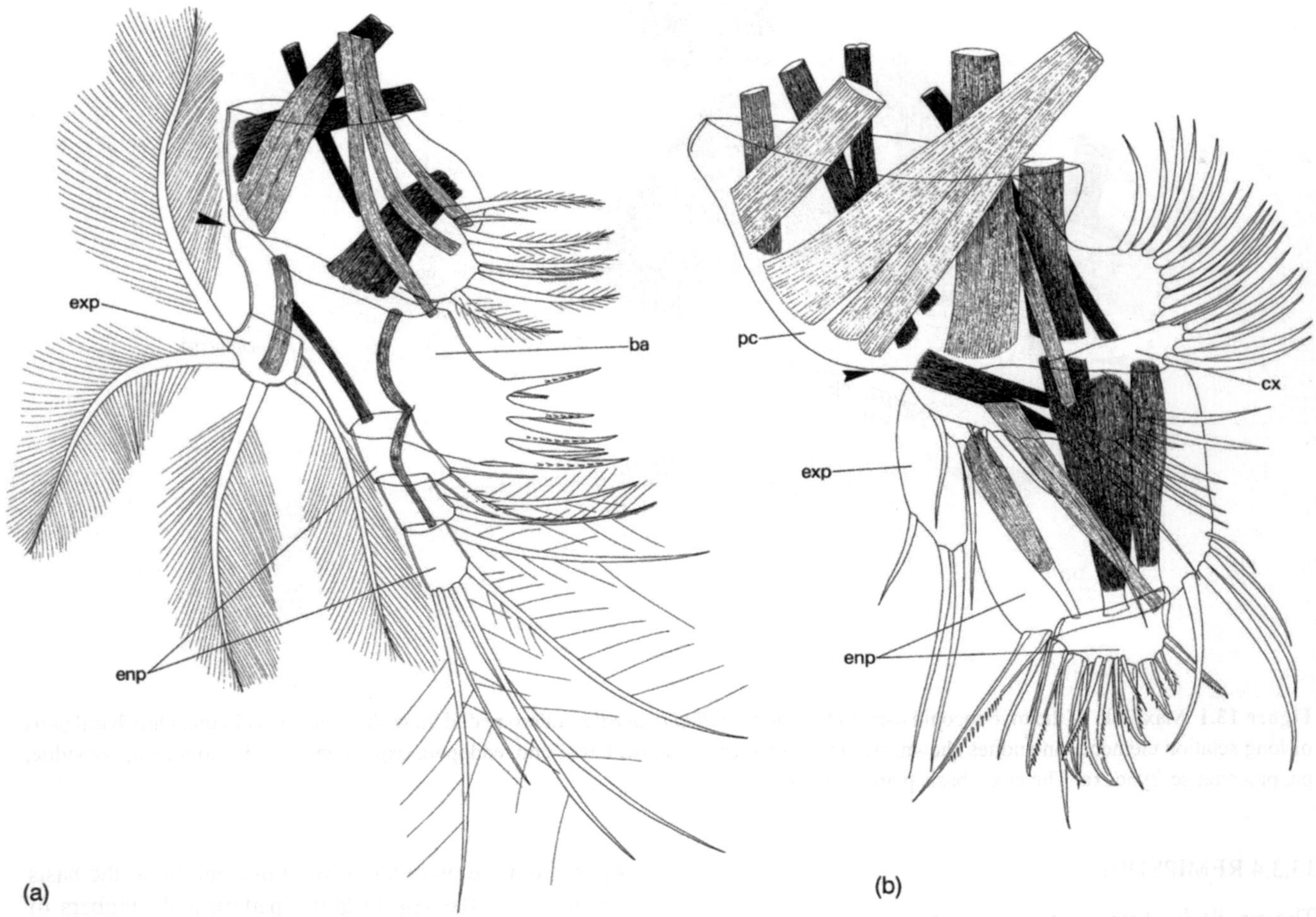


**Figure 13.2** (a) Maxillule of the first protozoeal stage of the penaeidean malacostracan *Pleoticus muelleri*, anterior view showing musculature. (b) Maxillule of the myodocopid ostracod *Azygocypridina imperalis*, postero-lateral view. Abbreviations as for Figure 13.1.

insert around this articulation and it is here identified as the coxa–basis plane.

There is only a single segment proximal to the coxa–basis plane and it lacks endites. It is tentatively identified as a syncoxa. The lack of protopodal endites in the branchiuran maxillule can be correlated with its specialization as an attachment organ from the first larval stage. Distal to the coxa–basis plane lies the basis and three distinct endopodal segments. The apical claws are here interpreted as derived from a fourth endopodal segment.

### 13.3.6 BRANCHIOPODA

Extant branchiopods are characterized by their reduced maxillules (Calman, 1909). They provide no information relevant to this investigation of musculature patterns. Fossil branchiopods will be discussed below.

### 13.3.7 THECOSTRACA AND TANTULOCARIDA

The maxillule of thecostracans is reduced and does not provide information useful in this analysis. Tantulocarids lack maxillules at all stages in their life cycle.

### 13.3.8 MALACOSTRACA

Penaeidean shrimps retain the most plesiomorphic developmental pattern of any malacostracan. The maxillule of a first protozoeal stages of a penaeidean, *Pleoticus muelleri* (Bate), has been used as a representative of the Malacostraca. As in all malacostracans, the protopodal part consists of only two segments, each bearing a single endite (Figure 13.2(a)). These two segments are separated by a complete hiatus in the musculature that is readily identifiable as the coxa–basis articulation. The single segment proximal to the coxa–basis

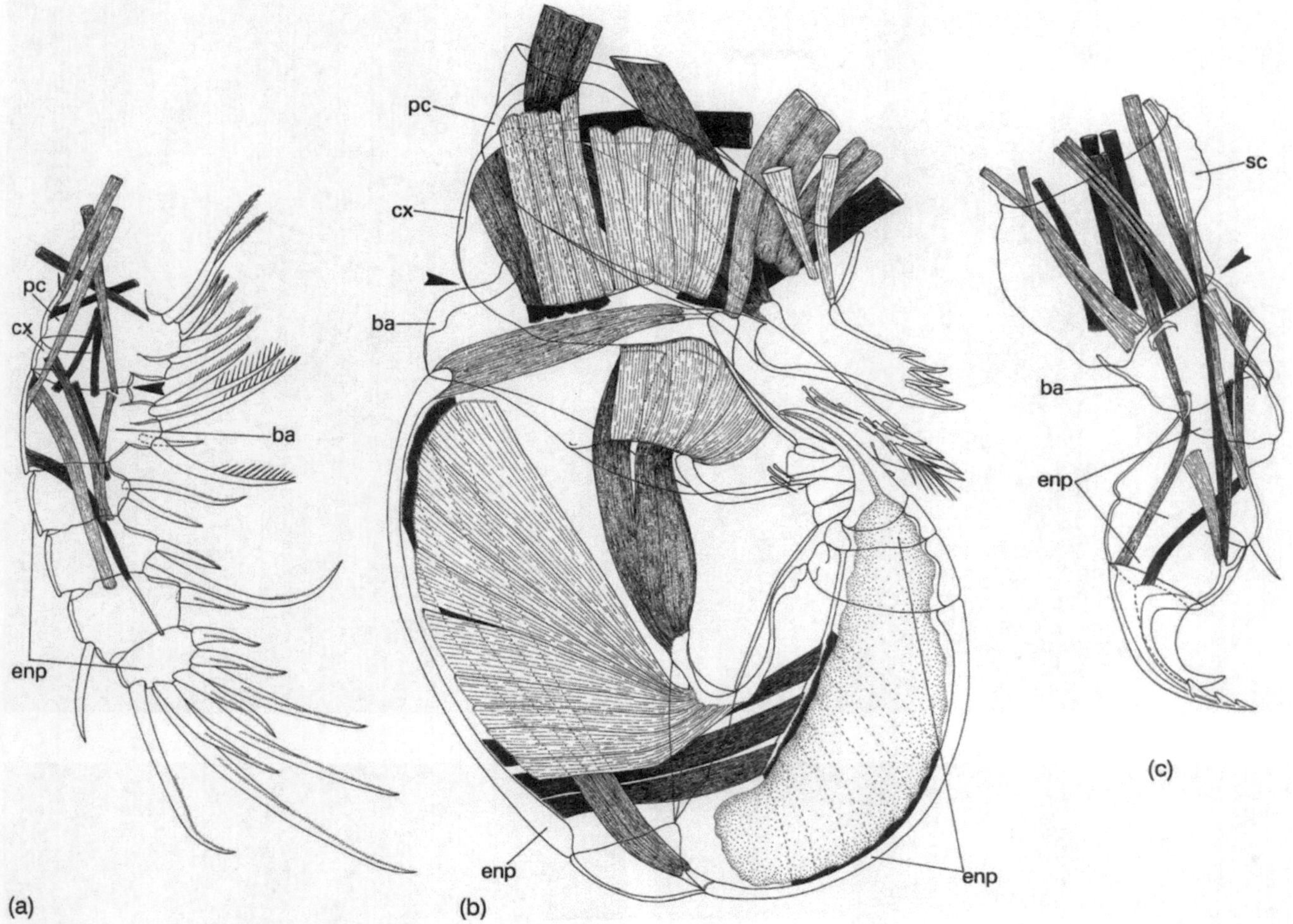


**Figure 13.3** (a) Maxillule of the mystacocarid *Derocheilocaris remanei*, anterior view showing musculature. (b) Maxillule of the remipede *Speleonectes tulamensis*, anterior view showing musculature. (c) Maxillule of first-stage larva of the branchiuran *Argulus foliaceus*, anterior view showing musculature. Abbreviations as for Figure 13.1.

plane may represent either a syncoxa (homologous with both precoxa and coxa) or a coxa, with the precoxa being lost. It is not possible to distinguish between these alternatives on the currently available data. The distal protopodal segment is the basis and carries the three-segmented endopod and unsegmented exopod. There is no coxal epipodite in malacostracans.

### 13.3.9 CEPHALOCARIDA

The work of Hessler (1964) on the musculature of *Hutchinsoniella macracantha* Sanders revealed a rather specialized maxillule in adult cephalocarids, with a single elongate endite present on the proximal segment of a two-segmented protopod. The two protopodal segments were identified by Hessler as coxa (coxopod) and basis (basopod). The basis carries a four-segmented endopod and an unsegmented exopod which is setiferous along its outer and distal margins (Hessler, 1964, Figure 11). The coxa–basis plane is well defined by its muscular signature.

In the larval stage of *H. macracantha* (Hessler, 1964, Figure 18) however, the inner margin of the protopod carries a total of four equally developed endites but the protopod is not clearly subdivided into coxa and basis. The muscle signature of coxa–basis plane is poorly defined but recognizable. There is no coxal epipodite in either larvae or adult cephalocarids.

### 13.3.10 HYPOTHETICAL ANCESTRAL STATE

The protopodal part of the maxillule is differentiated into a maximum of three distinct segments within the Crustacea. All three segments, precoxa, coxa and basis, are expressed in the Copepoda, Ostracoda, Mystacocarida and Remipedia.

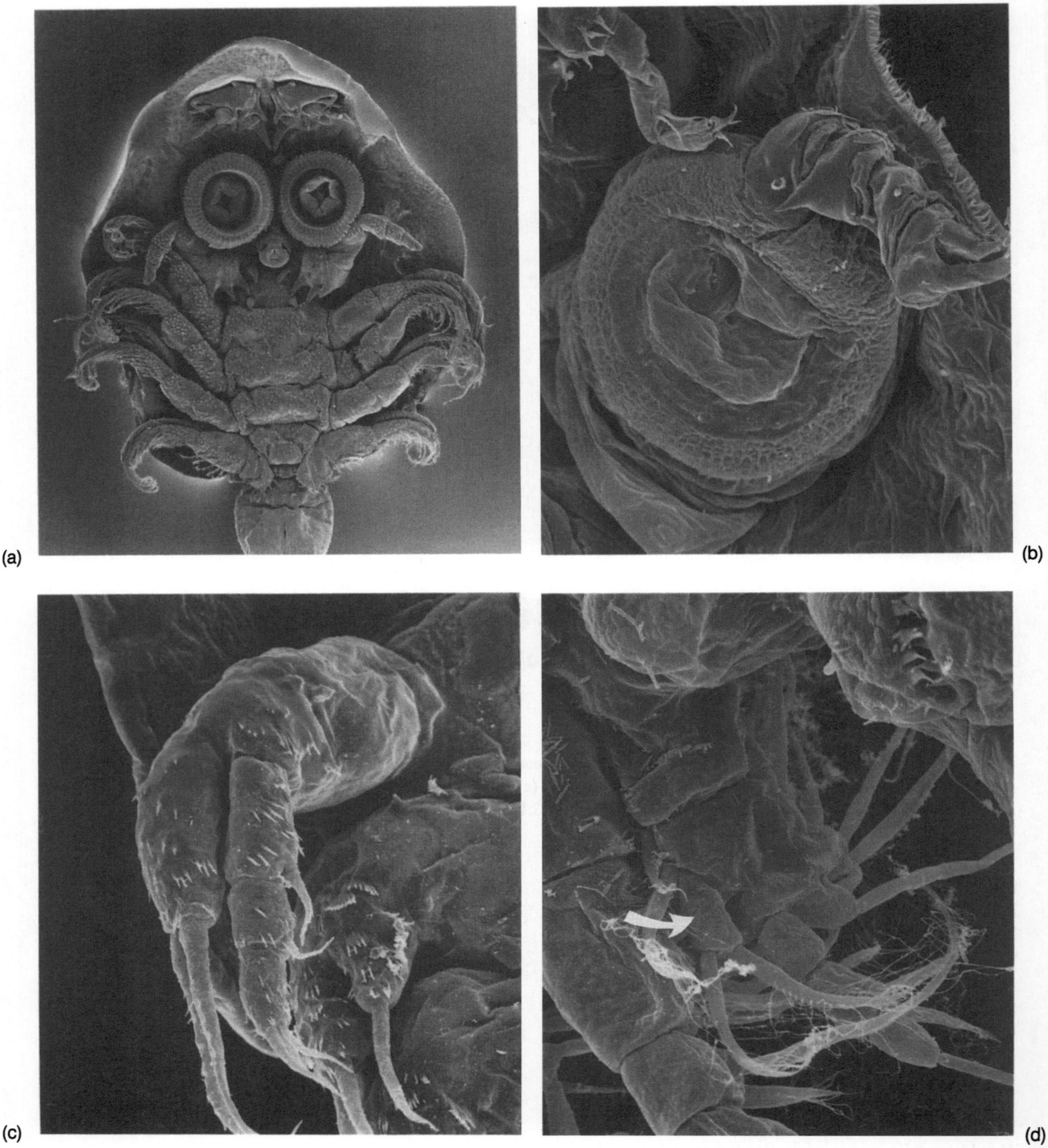

**Figure 13.4** Scanning electron micrographs. (a) Adult *Argulus foliaceus*, ventral view showing sucker-like form of maxillules. (b) Maxillule of fifth-stage larva of *A. foliaceus* showing barbed claws distally and developing sucker within syncoxa proximally (from Rushton-Mellor and Boxshall, 1994). (c) First thoracopod of first-larval stage of *A. foliaceus*, showing three-segmented endopod and unsegmented exopod. (d) First thoracopod (= maxilliped) of mystacocarid *Derocheilocaris remanei*, showing three-segmented endopod and unsegmented exopod (arrowed).

In the Cephalocarida, Branchiura and Malacostraca the protopodal part is differentiated into only two segments, by an articulation which is identifiable by its muscular signature as the coxa–basis joint. The segment distal to the articulation is the basis. The proximal protopodal segment of malacostracans could be homologized with either the precoxa, the coxa, or with the precoxal and coxal segments combined of other crustaceans. In the Branchiura the part proximal to the coxa–basis joint is not subdivided and lacks endites. This is here interpreted as a secondary, autapomorphic condition correlated with the profound modification of the maxillule for attachment to the host. In larval cephalocarids the only indication of subdivision of the proximal segment is the possibility that this part carries two of the four protopodal endites. The precise position of the larval endites relative to the coxa–basis articulation expressed in the adult cannot be discerned from data in the work of Hessler (1964).

Four endites are present along the margin of the undivided protopod of the larval maxillule in cephalocarids. Four endites (one precoxal, one coxal and two basal) are present in copepods, ostracods and mystacocarids. One of the basal endites is lost in remipedes but they retain the precoxal and coxal endites. In the Malacostraca only two endites remain, one (distal) derived from the basis and the other (proximal) of uncertain derivation (precoxal, coxal or syncoxal).

A total of four segments is expressed in the maxillulary endopod in Crustaceans. Four segments are retained in cephalocarids, remipedes, mystacocarids and branchiurans but a maximum of only three is retained in copepods, ostracods and malacostracans.

The exopod is lost in some groups but in all those crustacean taxa in which an exopod is present this ramus comprises a single segment, defined at its base and bearing setae along the outer and distal margins. It is referred to here as unsegmented.

A coxal epipodite is present as a well-developed setose lobe only in the Copepoda. It may be represented by the single seta in ostracods. A maxillulary epipodite is absent in all other crustacean groups. Its presence is here interpreted as an apomorphy of the Copepoda and Ostracoda.

By combining these states of maximum expression, it is postulated here that the ancestral crustacean maxillule (Figure 13.5) comprised a protopodal part divided into two segments by the coxa–basis articulation. The medial margin of the protopod was differentiated into four endites, two proximal to the coxa–basis articulation and two distal to the articulation, i.e. on the basis. The basis carried a four-segmented endopod and an unsegmented exopod. The proximal protopodal segment was not primitively differentiated into precoxa and coxa. Differentiation into a proximal precoxa bearing one endite and a distal coxa also bearing one endite occurred within the Crustacea, on the lineage leading to the maxillopodan assemblage and the Remipedia.

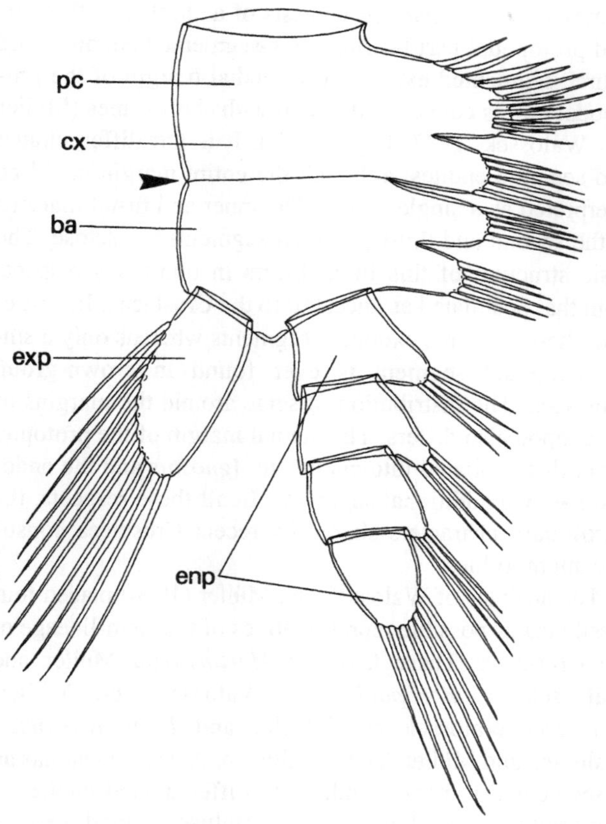


**Figure 13.5** Schematic reconstruction of hypothetical ancestral crustacean maxillule.

### 13.3.11 OUTGROUP COMPARISONS

Walossek and Müller (1990) identified a number of Upper Cambrian fossils that they placed on the stem-lineage of the Crustacea and discussed the monophyly of the Crustacea with respect to these taxa. Recent analyses have questioned the monophyletic status of the Crustacea (Averhof and Akam, 1995) or have included within the Crustacea numerous fossil taxa of dubious affinity (Wills *et al.*, 1995). These considerations indicate that the superbly preserved Cambrian fossils from the Orsten formation (Müller, 1982; Müller and Walossek, 1985, 1986, 1987, 1988; Walossek and Müller, 1990; Walossek, 1993) and those from the more famous but less well-preserved Burgess Shale formation (Briggs, 1976, 1978) might provide a suitable outgroup for the crown-group Crustacea.

The analysis of Wills *et al.* (1995) suggests that the Trilobita might be an appropriate outgroup. The best available description of trilobite appendages is that of *Agnostus pisiformis* (Linnaeus) by Müller and Walossek (1987), who did not consider *Agnostus* to form part of the main trilobite clade; however, the comprehensive analysis of Fortey and Theron (1994) which suggested that agnostids are derived, neotenic trilobites, is followed here. The fourth appendage

(= maxillule) of *Agnostus* consists of a short, undifferentiated protopodal part bearing a six-segmented endopod and a three-segmented exopod. The medial margin of the protopodal part is convex and armed with short spines (Müller and Walossek, 1987, Figure 6(d)). It is not differentiated into separate endites, although the entire margin could be interpreted as a single endite. The inner and distal margins of the second and third exopodal segments are setose. The basic structure of this limb differs in numerous respects from that postulated as ancestral to the Crustacea. In particular, there are three exopodal segments whereas only a single exopodal segment is ever found in crown-group Crustacea. The distribution of setae around the margins of the exopod also differs. The medial margin of the protopod is not differentiated into endites in *Agnostus* and the endopod has two additional segments. In all these respects, the maxillulary characters shared by recent Crustacea constitute apomorphies.

The analysis of Walossek and Müller (1990) placed four fossil taxa as possible representatives of the stem-lineage of the Crustacea. These taxa are *Martinssonia* Müller and Walossek, *Cambropachycope* Walossek and Müller, *Goticaris* Walossek and Müller and *Henningsmoenia* Walossek and Müller. The maxillule of *Martinssonia* has an unsegmented protopod with three differentiated endites, a five-segmented endopod and a multisegmented exopod bearing setae along the inner margin (Müller and Walossek, 1986, Figure 4(c)). The maxillule belongs to a homonomous series of similar appendages including the antennae, mandibles, maxillules and maxillae. There is no significant difference in basic organization of the naupliar and the postmandibular limbs, as found in crown-group Crustacea (Walossek, 1993). The maxillule of *Martinssonia* does not conform to that postulated as synapomorphic for the crustaceans, the most profound difference being the multisegmented state of the exopod. The maxillules of *Goticaris* and *Cambropachycope* both exhibit the same organization as that of *Martinssonia* (Walossek and Müller, 1990). Selection of any of these taxa as the outgroup would indicate that the postulated form of crustacean maxillule is apomorphic and serves as a diagnostic character for the Crustacea.

The maxillule of *Henningsmoenia* is interesting. It is only partially known but according to the illustrations of Walossek and Müller (1990, Figure 5D6) the exopod is unsegmented with setae along outer and distal margins, as in the crustacean type of maxillule. It possesses only two differentiated endites on the protopodal margin. Further information on this species is required before the form of its maxillule and its bearing on the value of the maxillule as a diagnostic crown-group crustacean character can be assessed.

Treating either the trilobite *Agnostus* or the stem-lineage crustacean forms recognized by Walossek and Müller (1990) as the outgroup of crown-group Crustacea confirms that the postulated ancestral crustacean maxillule provides a robust and complex synapomorphy of the Crustacea. Considering the appendage in isolation, I conclude that maxillulary structure provides a reliable diagnostic character for true crustaceans.

### 13.3.12 FOSSILS ATTRIBUTED TO THE CRUSTACEA

The maxillules of several Cambrian fossils previously attributed to the Crustacea are sufficiently well preserved to allow comparison with the hypothetical ancestral crustacean maxillulary pattern postulated above. Considerable caution is necessary here in the interpretation of segmental boundaries in fossils since the musculature is usually not available as a reference and cannot be used for unequivocal identification of the coxa–basis plane. The regrettable tendency to identify every fold in the surface of fossil arthropod appendages as an articulation has led to some of the more fanciful interpretations of segmentation patterns. For example, the claims of Kukalová-Peck (1991, 1992) that the ancestral arthropodan leg contained 11 segments in its groundplan are, as Fryer (1996) has pointed out, unsupported by any reliable evidence and are contradictory to a substantial body of embryological data. Further, the uncritical acceptance of such unsupported characters (Shear, 1992) reduces the rigour of subsequent analyses.

#### (a) *Rehbachiella* Müller

*Rehbachiella* was placed in the Branchiopoda by Walossek (1993). The superbly detailed account of the development of *Rehbachiella kinnekullensis* Müller by Walossek shows that the maxillule conforms to the postulated pattern in all important features. The medial margin of the protopodal part is differentiated into four endites but, as indicated by Walossek (1993), is not clearly subdivided into segments. The endopod is four-segmented and the exopod is unsegmented with setae along its outer and distal margins. This structure is shared with the Crustacea and strongly supports the classification of *Rehbachiella* in the Crustacea. Its branchiopodan affinities are indicated by the presence of the typical branchiopodan dorsal organ on the dorsal surface of the cephalic shield.

#### (b) Phosphatocopida

The phosphatocopines have previously been considered as members of the Crustacea Ostracoda (Müller, 1982). *Hesslandona sulcata* Müller possesses a series of homonomous appendages from the antenna and mandible through to the next three pairs of postmandibular limbs (Müller, 1982), just as in *Martinssonia*. The postantennular naupliar and the postmandibular limbs all have multisegmented exopods. Phosphatocopines cannot be placed either in the Ostracoda or in the crown-group Crustacea. The detailed similarities between their appendages and those of

*Martinssonia* suggest that they may also belong on the stem-lineage of the Crustacea.

**(c) *Bredocaris* Müller**

*Bredocaris* was placed in a new order, the Orstenocarida, in the maxillopodan lineage within the Crustacea by Müller and Walossek (1988). The maxillule of *Bredocaris admirabilis* Müller is identical in its basic organization to that of *Rehbachiella*. This supports its classification within the Crustacea. The close resemblance between the appendages of *Bredocaris* and those of the larval stages of *Rehbachiella* suggests a close affinity between these taxa. The well-developed branchiopodan-like dorsal organ on the cephalic shield of *Bredocaris* is here interpreted as further evidence of the branchiopodan affinities of the Orstenocarida. It is proposed here to transfer the Orstenocarida to the Branchiopoda.

**(d) *Skara* Müller**

*Skara* was placed in a new order, the Skaracarida, by Müller and Walossek (1985) and this order was identified as showing affinities to the maxillopodans within the Crustacea. The maxillule of both species of *Skara* conforms to the basic plan as exhibited by the recent Crustacea, by *Rehbachiella* and by *Bredocaris*. It differs only in the reduction of the endopod to three segments, rather than four. Its classification in the Crustacea is supported here.

**(e) *Canadaspis* Novozhilov**

*Canadaspis* was classified as a member of the Crustacea Malacostraca by Briggs (1978) and was placed in the malacostracan subclass Phyllocarida. The basis for this classification was primarily body tagmosis. Briggs (1978) showed that the postcephalic trunk of *Canadaspis perfecta* (Walcott) is divided into eight thoracic and seven abdominal somites, plus the telson. This corresponds to the trunk tagmosis exhibited by recent phyllocarids. However, the maxillule of *Canadaspis* differs markedly from that described above as typical for crustaceans. The maxillule of *Canadaspis* has about 14 segments in the inner ramus (endopod?) and a lamellate outer ramus divided into rays arising from a thickened proximal lobe (Briggs, 1978). The structure of the protopodal part could not be determined in detail but it lacks lobate endites.

The maxillule of *Canadaspis* does not present the diagnostic apomorphies of the Crustacea. On the basis of this limb alone *Canadaspis* should be excluded from the crown-group Crustacea. It does not even belong on the stem-lineage of the Crustacea.

## 13.4 THE CRUSTACEAN MAXILLA

The basic structure of the maxilla in the Crustacea is well illustrated by the cephalocarids, in which the maxilla forms the first in a series of virtually identical postcephalic trunk limbs (Sanders, 1963; Hessler, 1964). The limb has an undivided protopodal part, a six-segmented endopod, a two-segmented exopod and an unsegmented exite located distally (referred to as the pseudepipodite). The medial margin of the protopod is differentiated into five or six spinose endites. As indicated by its musculature and by the presence of transverse folds on the surface of the limb, the protopod shows some signs of subdivision into proximal and distal parts. Hessler (1964) identified a transverse feature, the 'posterior oblique furrow', around which numerous muscles inserted. This feature could be interpreted as the incipient coxa–basis articulation on the basis of this muscle signature. This interpretation would suggest that the exite originates from the basis and is, therefore, not homologous with a coxal epipodite.

The maxillary exopod of cephalocarids is unique within the Crustacea in being two-segmented. All other crustacean classes have an unsegmented exopod. In the Malacostraca the exopod may be extremely well developed, forming the large setose scaphognathite, but it is primitively unsegmented. In recent Branchiopoda, the entire maxilla is reduced but in the fossil branchiopods *Rehbachiella* and *Lepidocaris* Scourfield (Scourfield, 1926; Walossek, 1993) the maxilla is well developed and has an unsegmented exopod. The maxilla is uniramous, lacking the exopod, in copepods, remipedes, mystacocarids and branchiurans. In thecostracans the entire limb is very reduced and in tantulocaridans it is absent.

Remarkably, the maxilla of myodocopidan ostracods has traditionally been described as having a three-segmented exopod. For example, Kornicker and Yager (1996) recently described a new species of *Spelaeoecia* Angel and Iliffe and interpreted both the maxilla (= fifth limb) and the maxilliped (= sixth limb) as having three-segmented exopods. The former would be a unique state for the Crustacea since a maximum of two segments (as in cephalocarids) is otherwise known for the maxillary exopod. Examination of material of *Spelaeoecia bermudensis* Angel and Iliffe revealed that the musculature of the maxilla and maxilliped is similar (Figure 13.6(b,c)). The coxa–basis plane is identifiable, although not strongly expressed, in both limbs and facilitates comparison. The segment distal to the coxa–basis plane is identified as the basis which, in the maxilla, Kornicker and Yager interpreted as the first exopodal segment. On the outer margin is a group of three setae. Inserting on the outer surface of the basis about in this position are two muscles (labelled m1 and m2 in Figure 13.6(b)) which pass through the coxa–basis plane from origins near the epipodite. Serial homologues (Figure 13.6(c)) of these two muscles can be identified in the maxilliped, passing through the coxa–basis plane from origins near the epipodite. These insert on the unsegmented ramus interpreted as the endopod by Kornicker and Yager (1996). This

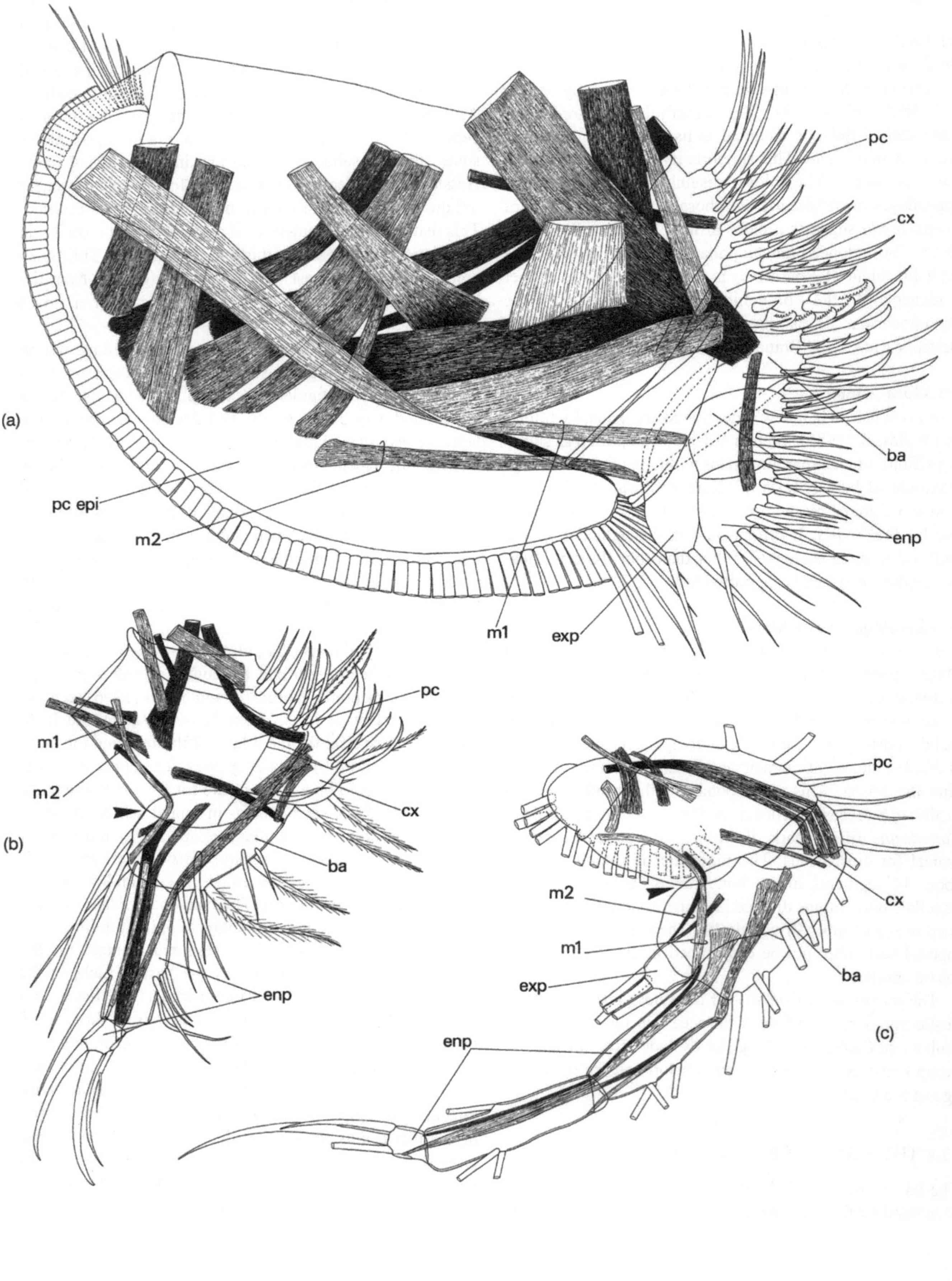
(a)
pc
cx
ba
enp
exp
m1
m2
pc epi
(b)
pc
m1
m2
cx
ba
enp
(c)
pc
cx
m2
m1
exp
ba
enp

ramus is located externally on the basis and is here reinterpreted as the unsegmented exopod. Based on the new evidence from the musculature, reinterpretation of the maxilla suggests that the exopod has been incorporated into the basis and is represented by the group of three setae on its outer surface. The long ramus, reinterpreted here as the endopod, is two-segmented in the maxilla and three-segmented in the maxilliped of *Spelaeoecia*.

Consideration of the maxilla of *Azygocypridina*, another myodocopid ostracod, supports the reinterpretation. The maxilla of *Azygocypridina* (Figure 13.6(a)) is profoundly modified by extreme development of the epipodite, which superficially resembles the enormous scaphognathite (derived from the exopod) of some adult malacostracans. This ostracod epipodite is precoxal in origin. A small coxa and basis can be recognized. The endopod is three-segmented, although the first and second segments are partly fused and the entire ramus is partly fused to the basis. The exopod is not separated proximally from the basis. It is lobate and carries three setae. This is the most fully expressed maxillary exopod in the myodocopid Ostracoda.

The number of maxillary endopodal segments varies widely from class to class within the Crustacea. Six is the maximum. The crustacean maxilla typically carries an unsegmented exopod, although two segments occur in cephalocarids. The exopod is lost in most crustaceans including the copepods. The alternative interpretation proposed recently by Ferrari and Steinberg (1993) on the basis of setation development patterns, that the four-segmented ramus remaining in the copepod maxilla is the exopod, is not supported by comparative studies of other crustacean groups. Ferrari's (1995) attempt to support this interpretation based on the application of Schram and Emerson's (1991) arthropod pattern theory to copepods was unlikely to succeed since this theory is itself constructed around a weakly supported interpretation of the fossil *Tesnusocaris* Brooks. It is also contradicted by evidence from developmental studies (Panganiban *et al.*, 1995) which shows that the exopods and endopods of biramous crustacean limbs differentiate on a dorsoventral axis and not along an anterior–posterior axis, as would be implied by Ferrari's hypothesis.

## 13.5 THE TRUNK LIMBS

The first postcephalic trunk limb of crustaceans is often modified for a role in raptorial feeding, in which case it is usually referred to as a maxilliped. The second and third limbs are also sometimes involved in feeding and are, then, also referred to as maxillipeds. This functionally based nomenclature has its anomalies since in the mystacocarid *Derocheilocaris* Pennak and Zinn the first trunk limb has been referred to as a maxilliped although it is not markedly modified for a raptorial function.

The first trunk limbs in copepods and remipedes are uniramous due to the loss of the exopod and may be termed maxillipeds. The endopod may have a maximum of six segments, as found in copepods (Huys and Boxshall, 1991). The first trunk limb of ostracods has a three-segmented endopod and an unsegmented exopod (Figure 13.6(c)). The same pattern of ramal segmentation is exhibited in the first postcephalic trunk limb of the free-swimming larval stage of branchiuran fish lice (Figure 13.4(c)) and also in the Mystacocarida (Figure 13.4(d)). In the Branchiopoda, the first trunk limb of females of the onchyopodan *Polyphemus pediculus* (Linn.) also has a one-segmented exopod and three-segmented endopod (Lilljeborg, 1900, Plate LXXX, Figure 1) and other branchiopods have unsegmented exopods. Several groups of crustaceans are characterized by a three-segmented endopod and an unsegmented exopod on the first trunk limb. In penaeid malacostracans, the first trunk limb has a four-segmented endopod and an unsegmented exopod in the first protozoeal stage (Dobkin, 1961).

The number of endopodal segments is often reduced from its maximum of six and these reductions appear to have occurred independently on several occasions within crustacean groups, so levels of homoplasy are probably high. The actual number of endopodal segments is therefore not regarded as especially informative about phylogenetic relationships. The number of exopodal segments may, however, be more informative. It is suggested here that the unsegmented exopod on the postmaxillipedal trunk appendages of crustaceans is plesiomorphic. This inference is based partly on ontogenetic data since the exopod of the first trunk limb on first appearance is typically unsegmented whereas the endopod typically comprises three or more segments (Dobkin, 1961; Rushton-Mellor and Boxshall, 1994). The Cambrian fossil crustaceans such as *Rehbachiella* and *Bredocaris*, with their unsegmented exopods, are in accord with this interpretation. If the unsegmented exopod is the plesiomorphic condition for the Crustacea then the two-segmented state of the exopod in cephalocaridans would be derived as would the three-segmented state in other crustacean taxa, such as copepods, remipedes and thecostracans.

**Figure 13.6** (a) Maxilla of the myodocopid ostracod *Azygocypridina imperalis*, anterolateral view showing musculature. Setal array along margin of precoxal exite with only bases shown. (b) Maxilla of the myodocopid ostracod *Spelaeoecia bermudensis*, anterolateral view showing muscles (m1 and m2) to incorporated exopod represented by three setae on surface of segment. (c) Maxilliped of the myodocopid ostracod *Spelaeoecia bermudensis*, anterolateral view showing musculature to unsegmented exopod as in Figure 13.6(b). Abbreviations as for Figure 13.1.

## 13.6 DISCUSSION

This paper draws attention to the value of anatomical studies. The musculature of crustacean appendages provides a wealth of data that facilitates the identification of segmental homologies and allows the recognition of characters of value in phylogenetic studies. I have largely refrained from making phylogenetic inferences because comprehensive data are presented here on only a single appendage, the maxillule, and it is necessary to consider the entire body of evidence before making inferences. However, I would present one inference concerning the Remipedia. In most recent analyses the Remipedia is typically placed basally in the Crustacean clade or even outside the crown group Crustacea (Wills *et al.*, 1995). This position can usually be traced back to pivotal character states based on lack of differentiation of the postcephalic trunk into thorax and abdomen, and to the large number of trunk somites. Consideration of appendage characters, after comparison with other extant crustaceans, suggests a different placement, namely that the Remipedia belongs in the maxillopodan lineage. The presence of a defined precoxa in the maxillule, the loss of the maxillary exopod and the three-segmented state of the exopods of the trunk limbs are all interpreted here as derived states and are shared only with taxa traditionally placed within the Maxillopoda. These and other limb characters indicate that the Remipedia are derived within the Crustacea and that their basal position in cladograms is a result of overemphasis on trunk tagmosis and somite numbers.

Except for the Cambrian Orsten fossils described in a series of papers by Müller and Walossek, and for the Devonian *Lepidocaris* Scourfield, virtually all palaeozoic arthropods are too poorly preserved to permit accurate analysis of the segmental homologies of their appendages, especially of the protopodal parts. This unfortunate situation creates voids in character matrices and there appears to be a real danger that important appendage characters are being omitted or defined without due rigour, resulting in an undue bias towards characters, such as body tagmosis, carapace and telson, that are well preserved in fossils (Briggs and Fortey, 1989). Simply applying parsimony-based algorithms to biased character matrices neither generates insight nor furthers our understanding of arthropod phylogeny.

Even with the best-preserved Cambrian fossils it is still difficult to homologize the components of the protopods of their limbs with those of extant crustaceans. Walossek and Müller (1990) correctly recognized the possession of a discrete 'proximal endite' located on the limb base of stem-lineage crustaceans as an important character, lacking in 'trilobites, other arachnatans and tracheates/uniramians'. They homologized this proximal endite with the coxal portion of the protopods of the maxillule, and of the antenna and mandible. Considering only the maxillule, it is not possible to confirm whether the proximal endite in these forms is homologous with either or both of the two endites (precoxal and coxal) found in copepods and remipedes for example.

Manton's work on mandibles in arthropods drew attention to differences in the origin of the biting surface in different classes (see Manton, 1973 for brief review). According to Manton, in the Crustacea the mandibular biting surface was derived from the proximal gnathobase, whereas in the Uniramia the biting surface was derived from the distal tip of the appendage. Manton's support of arthropod polyphyly was based partly on the implausibility of any functional intermediate between these two types. The example of the branchiuran maxillule is instructive in this context. The maxillule of *Argulus* is the primary attachment organ throughout its life cycle. The first five larval stages attach by means of the barbed distal claw, whereas the later larval stages and adults attach by means of the proximal sucker. The claw eventually degenerates to a vestigial setiferous lobe concealed beneath the rim of the sucker in the adult (Rushton-Mellor and Boxshall, 1994). The maxillule of *Argulus* provides the clearest example of how an arthropod limb can maintain functional continuity during a remarkable ontogenetic change from a distally located to a proximally located structure. It serves to remind us that study of the adaptive radiation in form and function of the jointed limbs of arthropods is the key to understanding the long and complex evolutionary history of the phylum Arthropoda.

ACKNOWLEDGEMENTS

I am grateful to Bob Hessler (Scripps Institution of Oceanography) for his valuable criticism of this paper.

## REFERENCES

Alexander, R. McN. (1968) *Animal Mechanics*, Sidgwick and Jackson Ltd., London.

Averhof, M. and Akam, M. (1995) Insect–crustacean relationships: insights from comparative developmental and molecular studies. *Philosophical Transactions of the Royal Society, B*, **347**, 293–303.

Baccari, S. and Renaud-Mornant, J. (1974) Anatomie du tube digestif de *Derocheilocaris remanei* Delamare et Chappuis, 1951 (Crustacea: Mystacocarida). *Archives de Zoologie Expérimentale et Générale*, **115**, 607–620.

Baskin, R.J. and Paolini, P.J. (1966) Muscle volume changes. *Journal of General Physiology*, **49**, 387–404.

Boxshall, G.A. (1982) On the anatomy of the misophrioid copepods, with special reference to *Benthomisophria palliata* Sars. *Philosophical Transactions of the Royal Society, B*, **297**, 125–81.

Boxshall, G.A. (1985) The comparative anatomy of two copepods, a predatory calanoid and a particle-feeding mormonilloid. *Philosophical Transactions of the Royal Society, B*, **311**, 303–77.

Boxshall, G.A. (1990) The skeletomusculature of siphonostomatoid copepods, with an analysis of adaptive radiation in structure of the oral cone. *Philosophical Transactions of the Royal Society, B*, **328**, 167–212.

Boxshall, G.A. and Huys, R. (1992) A homage to homology: patterns of copepod evolution. *Acta Zoologica*, **73**, 327–34.

Briggs, D.G. (1976) The arthropod *Branchiocaris* n. gen., Middle Cambrian, Burgess Shale, British Columbia. *Bulletin of the Geological Survey of Canada*, **264**, 1–29.

Briggs, D.G. (1978) The morphology, mode of life, and affinities of *Canadaspis perfecta* (Crustacea: Phyllocarida), Middle Cambrian, Burgess Shale, British Columbia. *Philosophical Transactions of the Royal Society, B*, **281**, 439–87.

Briggs, D.G. and Fortey, R.A. (1989) The early radiation and relationships of major arthropod groups. *Science*, **246**, 241–3.

Calman, W.T. (1909) Crustacea, in *A Treatise on Zoology* (ed. R. Lankester) part 7, fasc. 3. Adam and Charles Black, London.

Dobkin, S. (1961) Early developmental stages of pink shrimp, *Penaeus duorarum* from Florida waters. *Fishery Bulletin*, **190**, 321–49.

Ferrari, F.D. (1995) Identity of the distal segments of the maxilla 2 and maxilliped in copepods: new teeth for Carl Claus' old saw. *Crustaceana*, **68**, 103–110.

Ferrrari, F.D. and Steinberg, D. (1993) *Scopalatum vorax* (Esterly, 1911) and *Scolecithricella lobophora* Park, 1970 calanoid copepods (Scolecitrichidae) associated with a pelagic tunicate in Monterey Bay. *Proceedings of the Biological Society of Washington*, **106**, 467–89.

Fortey, R.A. and Theron, J.N. (1994) A new Ordovician arthropod, *Soomaspis*, and the agnostid problem. *Palaeontology*, **37**, 841–61.

Fryer, G. (1996) Reflections on arthropod evolution. *Biological Journal of the Linnean Society*, **58**, 1–55.

Gresty, K.J., Boxshall, G.A. and Nagasawa, K. (1993) The fine structure and function of the cephalic appendages of the branchiuran parasite, *Argulus japonicus* Thiele. *Philosophical Transactions of the Royal Society, B*, **339**, 119–35.

Hessler, R.R. (1964) The Cephalocarida. Comparative skeletomusculature. *Memoirs of the Connecticut Academy of Arts and Sciences*, **16**, 1–97.

Huys, R. and Boxshall, G.A. (1991) *Copepod Evolution*, The Ray Society, London.

Kornicker, L.S. and Yager, J. (1996) The troglobitic Halocyprid Ostracoda of anchialine caves in Cuba. *Smithsonian Contributions to Zoology*, **580**, 1–16.

Kukalová-Peck, J. (1991) Fossil history and evolution of hexapod structures, in *The Insects of Australia*, 2nd edn (ed. I.D. Naumann), Melbourne University Press, Melbourne, pp. 141–79.

Kukalová-Peck, J. (1992) The 'Uniramia' do not exist: the groundplan of the Pterygota as revealed by Permian Diaphanopterodea from Russia (Insecta: Palaeodichtyopteroidea). *Canadian Journal of Zoology*, **70**, 236–55.

Lilljeborg, W. (1900) *Cladocera Sueciae*, Vol. III, Akademischen Buchdruckerei, Uppsala.

Manton, S.M. (1973) Arthropod phylogeny – a modern synthesis. *Journal of Zoology*, **171**, 111–30.

Müller, K.J. (1982) *Hesslandona unisulcata* sp. nov. (Ostracoda) with phosphatized appendages from Upper Cambrian 'Orsten' of Sweden, in *A Research Manual of Fossil and Recent Ostracodes* (eds R.H. Bate, E. Robinson and L. Shepard), Ellis Horwood, Chichester, pp. 276–307.

Müller, K.J. and Walossek, D. (1985) Skaracarida, a new order of Crustacea from the Upper Cambrian of Västergötland, Sweden. *Fossils and Strata*, **17**, 1–65.

Müller, K.J. and Walossek, D. (1986) *Martinssonia elongata* gen. et sp. n., a crustacean-like euarthropod from the Upper Cambrian 'Orsten' of Sweden. *Zoologica Scripta*, **15**, 73–92.

Müller, K.J. and Walossek, D. (1987) Morphology, ontogeny and life habit of *Agnostus pisiformis* from the Upper Cambrian of Sweden. *Fossils and Strata*, **19**, 1–124.

Müller, K.J. and Walossek, D. (1988) External morphology and larval development of the Upper Cambrian maxillopod *Bredocaris admirabilis*. *Fossils and Strata*, **23**, 1–70.

Panganiban, G., Sebring, A., Nagy, L. and Carroll, S. (1995) The development of crustacean limbs and the evolution of Arthropods. *Science*, **270**, 1363–6.

Rushton Mellor, S.K. and Boxshall, G.A. 1994 The developmental sequence of *Argulus foliaceus* (Crustacea: Branchiura). *Journal of Natural History*, **28**, 763–85.

Sanders, H.L. (1963) The Cephalocarida. Functional morphology, larval development, comparative external anatomy. *Memoirs of the Connecticut Academy of Arts and Sciences*, **15**, 1–80.

Schram, F.R. and Emerson, M.J. (1991) Arthropod pattern theory: a new approach to arthropod phylogeny. *Memoirs of the Queensland Museum*, **31**, 1–18.

Schram, F.R. and Lewis, C.A. (1989) Functional morphology of feeding in the Nectiopoda, in *Crustacean Issues 6, Functional Morphology of Feeding and Grooming in Crustacea* (eds B.E. Felgenhauer, L. Watling and A.B. Thistle), Balkema, Rotterdam and Brookfield, pp. 115–22.

Scourfield, D.J. (1926) On a new type of crustacean from the Old red Sandstone (Rhynie Chert Bed), Aberdeenshire) – *Lepidocaris rhyniensis*. *Proceedings of the Linnean Society London*, **152**, 290–8.

Shear, W.A. (1992) End of the 'Uniramia' taxon. *Nature*, **359**, 477–8.

Walossek, D. (1993) The Upper Cambrian *Rehbachiella* and the phylogeny of Branchiopoda and Crustacea. *Fossils and Strata*, **32**, 1–202.

Walossek, D. and Müller, K.J. (1990) Upper Cambrian stem-lineage crustaceans and their bearing upon the monophyletic origin of the Crustacea and the position of *Agnostus*. *Lethaia*, **23**, 409–27.

Wills, M.A., Briggs, D.G., Fortey, R.A. and Wilkinson, M. (1995) The significance of fossils in understanding arthropod evolution. *Verhandlungen der Deutschen Zoologischen Gesellschaft*, **88**, 203–15.

# 14 Crustacean phylogeny inferred from 18S rDNA

*T. Spears and L.G. Abele*

*Department of Biological Science, Florida State University, Tallahassee, FL 32306-2043, USA, email: spears@bio.fsu.edu*
*Office of the Provost, Florida State University, Tallahassee, FL 32306-3020, USA, email: labele@aa.fsu.edu*

## 14.1 INTRODUCTION

Resolution concerning issues of higher-order crustacean phylogeny remains elusive even after years of thorough morphological and palaeontological scrutiny. Surprisingly, there is as yet no consensus regarding even the number of constituent crustacean classes. One view (Schram, 1986) based on a cladistic analysis of morphological characters suggests there are four classes: Remipedia, Phyllocarida, Maxillopoda and Malacostraca (Figure 14.1(a)); an alternative cladistic analysis (Brusca and Brusca, 1990) suggests there are five: Remipedia, Branchiopoda, Cephalocarida, Maxillopoda and Malacostraca (Figure 14.1(b)). Still another view (Bowman and Abele, 1982), presented as a classification rather than a phylogeny, divides crustaceans into six classes: the five aforementioned groups and the Ostracoda (Figure 14.1(c)). Finally, there is the classification by Starobogatov (1986) (English translation by Grygier, 1988), wherein the Crustacea are divided into four classes (Branchipodioides, Carcinioides, Ascothoracioides and Halicynioides) based on the principle of primary heteronomy (Ivanov, 1944) and the relationship between larval prototagmata and adult tagmosis (Figure 14.1(d)). As seen in Figure 14.1, these views also present conflicting ideas concerning the relationships among classes and hence, the course of crustacean evolution.

There are several reasons for these differences. An extensive fossil record suggests that most of the major lineages of crustaceans arose and diversified very early (most likely in Precambrian times) and have subsequently undergone a long period of independent evolution (Figure 14.2). This has led to extreme diversity both within and among groups that is manifested in many ways. For instance, some taxa have become highly modified parasites and lack typical diagnostic features. Crustaceans are also genetically and developmentally very labile, making paedomorphosis and character convergence and reversal relatively common phenomena. Hence, it is problematic to select and interpret useful phylogenetic characters.

We present the results of a molecular approach to higher-order crustacean phylogeny using small subunit (SSU) 18S rDNA sequence data. Studies to date have shown evidence of this sort to be effective in constructing a phylogenetic framework independent of morphology against which various evolutionary hypotheses can be tested. SSU genes occur in the genomes of all living organisms and evolve at a relatively conserved rate owing to the functionally important role they play in ribosome formation and protein synthesis. These features permit phylogenetic analyses to be performed on diverse and distant taxa. As such, 18S rDNA sequence data are perceived to be well-suited to recover ancient patterns of lineage branching such as those that gave rise to crustaceans and their arthropod relatives.

We specifically address these questions:

1. Are phyllocarid crustaceans members of the Phyllopoda or members of the Malacostraca?
2. Is there evidence for a monophyletic Maxillopoda, and what are its component taxa?
3. What are the phylogenetic positions of the Remipedia and Cephalocarida?
4. How many crustacean 'classes' are there and what are the relationships among crustacean lineages?
5. Which arthropod taxon is the sister group of crustaceans?

We describe the results of parsimony analyses based on new and previously published sequence data. More detailed investigations will be reported elsewhere.

## 14.2 MATERIALS AND METHODS

### 14.2.1 STUDY TAXA

The species used in the present study, their 18S rDNA sequence GenBank accession numbers, and their higher taxon designations used in the accompanying cladograms are listed in Table 14.1. Crustacean taxon designations are

*Arthropod Relationships*, Systematics Association Special Volume Series 55, Edited by R.A. Fortey and R.H. Thomas. Published in 1997 by Chapman & Hall, London. ISBN 0 412 75420 7

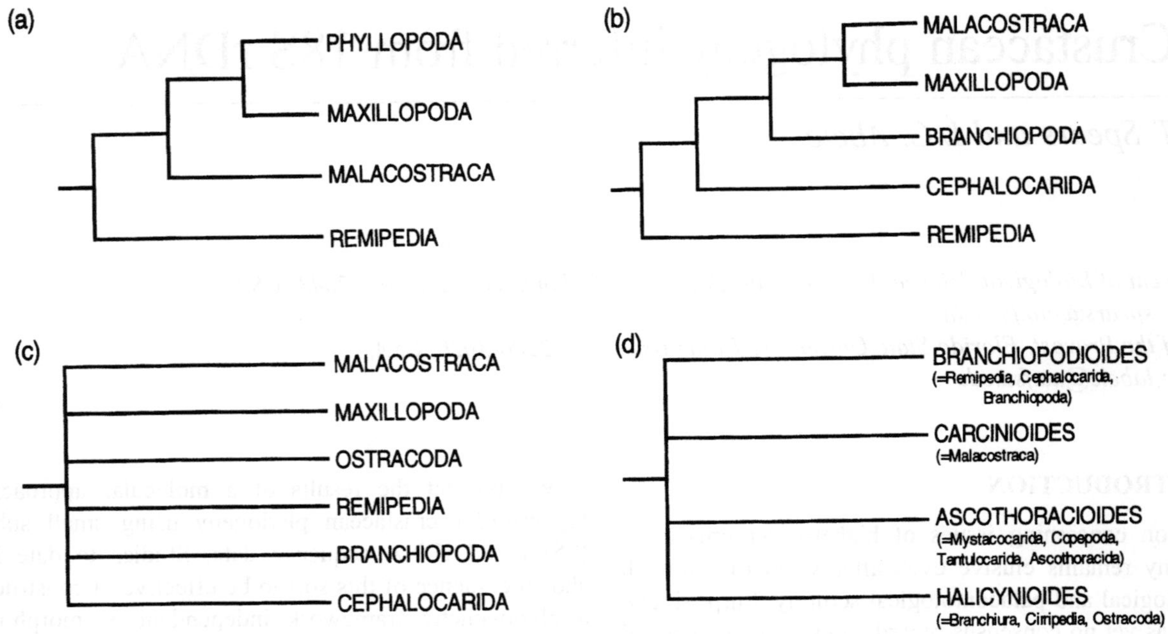


**Figure 14.1** Four alternative views of crustacean phylogeny assigning crustaceans to either four, five or six classes. (a) Four classes of crustaceans, with remipedes among the most primitive; based on Schram (1986). (b) Five classes of crustaceans, also with remipedes among the most primitive, but with the Malacostraca in a more derived position than in phylogeny (a); in this scheme, branchiopods and cephalocarids are accorded class rank, having been removed from the Phyllopoda as in phylogeny (a); based on Brusca and Brusca (1990). (c) Six classes of crustaceans based on the classification presented in Bowman and Abele (1982) and intended to reflect phylogenetic uncertainty among the classes; in this scheme, ostracods are removed from the Maxillopoda and are accorded class rank. (d) Four classes based on Starobogatov (1986); because of the novel nomenclature used in this classification, the names of constituent taxa are presented to facilitate comparisons with other schemes.

according to Bowman and Abele's (1982) classification; taxon labels for non-crustacean arthropod species are taken from their GenBank entries.

### 14.2.2 DNA EXTRACTION, AMPLIFICATION AND SEQUENCING

Previously unpublished crustacean sequences obtained by Abele, Spears and co-workers and used here were obtained by extracting total genomic DNA from fresh or ethanol-preserved specimens, amplification of the 18S rDNA gene by the polymerase chain reaction (PCR) and DNA sequencing using standard protocols outlined in Spears *et al.* (1994). Some DNA extractions were carried out using the G-nome DNA Kit (BIO 101; Vista, CA). PCR products resulting in a single band were purified using QIAquick-spin columns (Qiagen, Inc.; Chatsworth, CA). When non-specific PCR products were also evident, either agarose or acrylamide gel purification was performed to obtain the targeted PCR product. On occasion, low-yield PCR products were cloned (TA Cloning™ System; Invitrogen Corp., San Diego, CA). Double-stranded cycle sequencing was performed using fluorescent-dye terminators (Perkin-Elmer; Foster City, CA) followed by electrophoresis on an Applied Biosystems, Inc. Model 373A Automated Sequencing System (Foster City, CA).

### 14.2.3 SEQUENCE ALIGNMENT

Sequences were aligned in CLUSTAL W (Thompson *et al.*, 1994) using default parameters (slow pairwise option: gap opening penalty = 10, gap extension penalty = 0.10; multiple alignment option: gap opening penalty = 10, gap extension penalty = 0.05, sequences with > 40% divergence delayed, gaps reset between alignments); all substitutions were equally weighted. Final adjustments were made with reference to the secondary structure model for the eukaryotic SSU molecule (De Rijk *et al.*, 1992). Regions of ambiguous alignment (i.e. where nucleotide homology is questionable, as in length-variable regions) were delimited by identifying the first parsimony-uninformative nucleotide on each side of an ambiguous region and were excluded from subsequent phylogenetic analyses. Separate alignments were constructed in this manner with different sets of taxa for the various phylogenetic questions under investigation so as to minimize alignment ambiguities.

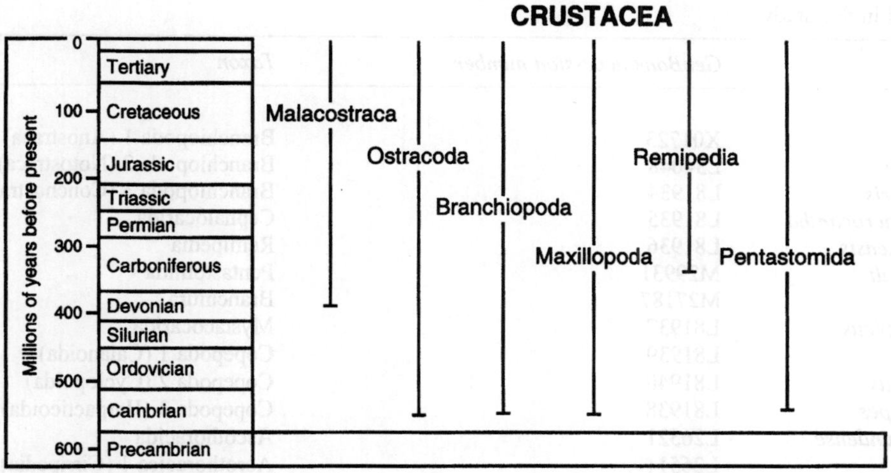


**Figure 14.2** Stratigraphic range of the various higher-order extant crustacean taxa (cephalocarids are unknown as fossils). [Based on Briggs *et al.* (1993), Whatley *et al.* (1993), Walossek (1993), and Walossek and Müller (1994).]

### 14.2.4 PHYLOGENETIC INFERENCE

Aligned data sets were assessed for phylogenetic signal (i.e. non-randomness) based on the $g_1$ statistic (Hillis and Huelsenbeck, 1992) for 10 000 random trees using the random trees option in PAUP 3.1.1 (Swofford, 1993). Parsimony analyses were performed (also in PAUP) on parsimony-informative characters using delayed character transformation and the branch-and-bound search procedure on 100 replicates of a random addition sequence for the malacostracan and remipede-cephalocarid data sets; for the larger maxillopodan and arthropodan data sets, a heuristic search procedure was used with tree bisection and reconnection (TBR) branch-swapping and the MULPARS option. The delayed transformation option was used for character-state optimization. The initial upper bound for branch-and-bound searches was determined from a preliminary heuristic search. In all analyses, character states were unordered and weighted equally, and alignment gaps were coded as missing data. Consistency indices (CI) (Kluge and Farris, 1969) and retention indices (RI) (Farris, 1989) were also reported by PAUP.

Support for relationships revealed by the most parsimonious (MP) topology was evaluated by bootstrap resampling (Felsenstein, 1985a) for 500 replicates using the heuristic search option in PAUP with a simple addition sequence. Another indicator of support, the decay index (Bremer, 1988, 1994), was determined for the malacostracan, maxillopodan and remipede–cephalocarid data sets by comparing the topology of strict consensus trees successively longer than the MP tree in one-step increments and noting the number of extra steps required to collapse a particular node.

Alternative phylogenetic hypotheses were investigated by enforcing topological constraints in PAUP. A modified version (Felsenstein, 1985b) of Templeton's (1983) test, available under the DNAPARS program in PHYLIP 3.5c (Felsenstein, 1993), was used to determine if an alternative topology was significantly worse than the MP topology.

## 14.3 PHYLLOCARIDA AND THE MALACOSTRACA

Phyllocarids represent one of several taxonomic enigmas among the Crustacea. They comprise a single order (Martin *et al.*, 1996, for classification), the members of which possess relatively derived features (tagmosis, an optic chiasma, a bivalved carapace, and a well-developed abdomen with pleopods), as well as more primitive features (caudal furca and polyramous, leaf-like, phyllopodous, thoracic limbs used in filter-feeding). For this reason, they have been classified as a subclass and early lineage of the crustacean class Malacostraca (Claus, 1888; Manton, 1934; Dahl, 1987, 1992) (Figures 14.3(a,b)), or as a subclass and early lineage within an entirely different class, the Phyllopoda, along with branchiopods and cephalocarids (Milne Edwards, 1834; Schram, 1986) (Figure 14.3(c)). Jamieson's (1991) interpretation of spermatological evidence leads to a phylogenetic view similar to that shown in Figure 14.3(c), although he considers phyllocarids, not cephalocarids, to be the sister group of branchiopods. Alternatively, Martin *et al.* (1996) suggest that the thoracic limbs of nebaliacean phyllocarids and branchiopods are distinct enough to justify placing phyllocarids into a different class.

This latter view is strongly supported in our molecular-based phylogeny (Figure 14.4) inferred from a maximum parsimony analysis of 18S rDNA sequences for a cephalocarid, three branchiopods, a phyllocarid and four malacostracans (two hoplocarids, a syncarid and a eucarid). Two copepods and two chelicerate sequences (the latter as outgroup taxa)

**Table 14.1** Taxa used in this study

| *Species* | *GenBank accession number* | *Taxon* |
|---|---|---|
| Crustacea | | |
| *Artemia salina* | X01723 | Branchiopoda 1 (Anostraca) |
| *Lepidurus packardi* | L34048 | Branchiopoda 2 (Notostraca) |
| *Limnadia lenticularis* | L81934 | Branchiopoda 3 (Conchostraca) |
| *Hutchinsoniella macrocantha* | L81935 | Cephalocarida |
| *Speleonectes tulumensis* | L81936 | Remipedia |
| *Porocephalus crotali* | M29931 | Pentastomida |
| *Argulus nobilis* | M27187 | Branchiura |
| *Derocheilocaris typicus* | L81937 | Mystacocarida |
| *Calanus pacificus* | L81939 | Copepoda 1 (Calanoida) |
| *Encyclops serrulatus* | L81940 | Copepoda 2 (Cyclopoida) |
| *Cancrincola plumipes* | L81938 | Copepoda 3 (Harpacticoida) |
| *Ulophysema oeresundense* | L26521 | Ascothoracida |
| *Berndtia purpurea* | L26511 | Acrothoracica 1 (Cirripedia) |
| *Trypetesa lampas* | L26520 | Acrothoracica 2 (Cirripedia) |
| *Euphilomedes cacharodonta* | L81941 | Ostracoda 1 (Myodocopida) |
| *Rutiderma* sp. | L81942 | Ostracoda 2 (Myodocopida) |
| *Bairdia* sp. | L81943 | Ostracoda 3 (Podocopida) |
| *Heterocypris* sp. | L81944 | Ostracoda 4 (Podocopida) |
| *Nebalia bipes* | L81945 | Phyllocarida (Malacostraca) |
| *Squilla empusa* | L81946 | Hoplocarida 1 (Malacostraca) |
| *Gonodactylus* sp. | L81947 | Hoplocarida 2 (Malacostraca) |
| *Anaspides tasmaniae* | L81948 | Syncarida (Malacostraca) |
| *Procambarus leonensis* | M34363 | Eucarida (Malacostraca) |
| Myriapoda | | |
| *Megaphyllum* sp. | X90658 | Myriapoda 1 (Diplopoda) |
| *Lithobius forficatus* | X90654 | Myriapoda 2 (Chilopoda) |
| *Scolopendra cingulata* | U29493 | Myriapoda 3 (Chilopoda) |
| Chelicerata | | |
| *Limulus polyphemus* | L81949 | Chelicerata 1 (Merostomata) |
| *Eurypelma californica* | X13457 | Chelicerata 2 (Arachnida) |
| Insecta | | |
| *Crossodonthina koreana* | Z36893 | Collembola 1 |
| *Hypogastrura dolsana* | Z26765 | Collembola 2 |
| *Petrobius brevistylis* | X89808 | Archaeognatha |
| *Aeschna cyanea* | X89481 | Odonata |
| *Carausius morosus* | X89488 | Phasmatoptera |
| *Lygus hesperus* | U06476 | Hemiptera |
| *Meloe proscarabaeus* | X77786 | Holometabola |
| Annelida | | |
| *Eisenia fetida* | X79872 | Annelida 1 |
| *Lanice conchilega* | X79873 | Annelida 2 |

were included to investigate the relative branching order of the taxa of interest (Table 14.1). The $g_1$ statistic (–0.9893) based on 10 000 random trees indicates the presence of phylogenetic signal within the aligned data set of 232 parsimony-informative characters.

The result shown in Figure 14.4 does not support a monophyletic Phyllopoda comprised of the [Phyllocarida + (Branchiopoda + Cephalocarida)]. Instead, the first branch leads to a branchiopod clade (bootstrap proportion, BP = 100%), then a (cephalocarid + copepod) clade (BP = 90%), and lastly a strongly supported malacostracan clade (BP = 98%) with the Phyllocarida as the basal lineage. The (phyllocarid + malacostracan) relationship persists in all trees up to 12 steps longer than the two MP trees. The alternative phyllopodan hypothesis (*sensu* Schram, 1986) (Figure 14.3(c)) requires 22 additional steps, has a lower consistency index (0.622 versus 0.648 for the MP trees), and is judged to be significantly worse by the modified version of Templeton's (1983) test as performed by DNAPARS in PHYLIP 3.5c. This result suggests that the presence of foliaceous limbs and the associated mode of filter-feeding observed among extant phyllocarids, cephalocarids and

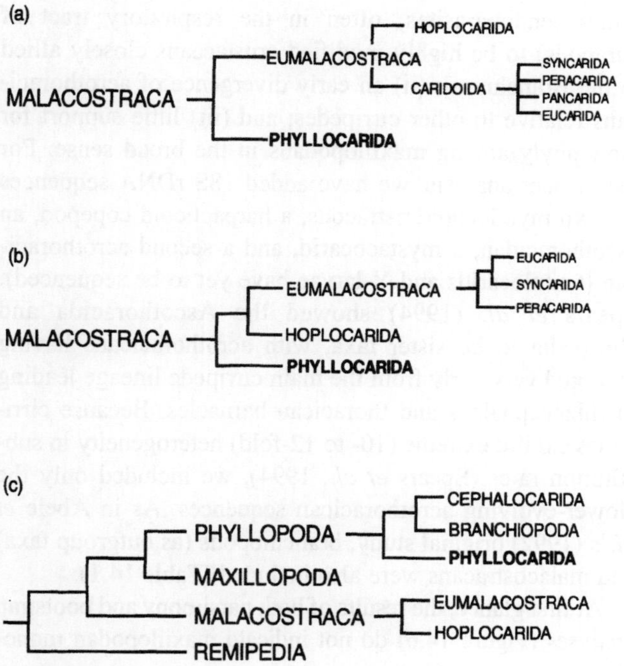


**Figure 14.3** Three alternative views of extant malacostracan phylogeny indicating the variable phylogenetic position of the Phyllocarida (shown in bold letters) within the Crustacea. (a) Phyllocarids as members of the crustacean class Malacostraca, and hoplocarids as eumalacostracans; based on Hessler (1983). (b) Phyllocarids as malacostracans, and hoplocarids as an independent malacostracan lineage; based on Dahl (1983). (c) Phyllocarids as members of the crustacean class Phyllopoda, and hoplocarids as malacostracans; based on Schram (1986).

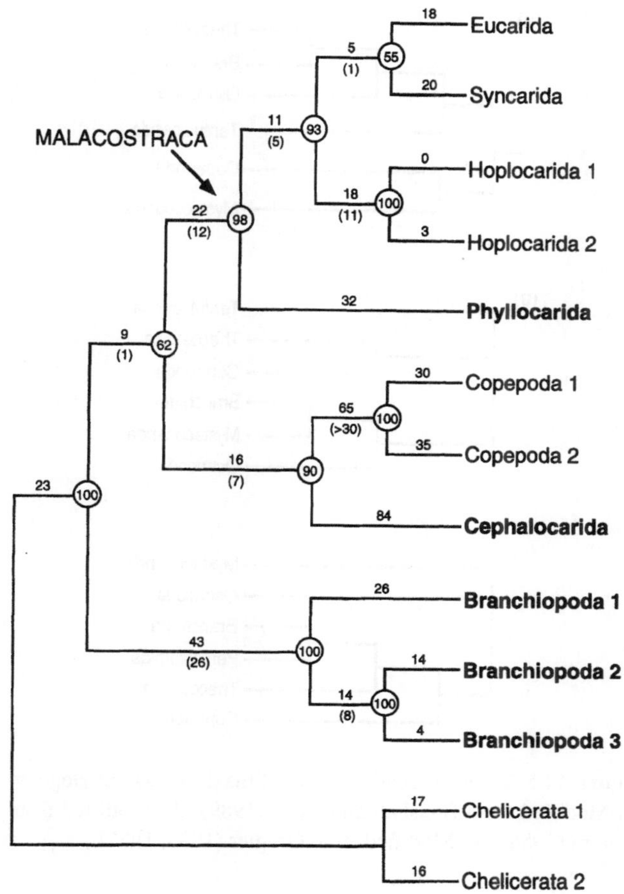


**Figure 14.4** One of two most-parsimonious (MP) trees resulting from a maximum parsimony analysis (branch-and-bound) of 232 parsimony-informative characters based on 18S rDNA sequences. The alternative topology placed the Syncarida as the sister taxon to the Hoplocarida. Uncircled numbers indicate branch lengths (i.e. the number of character changes along a branch). Total tree length = 525 steps; C.I. = 0.648; R.I. = 0.657. Numbers in parentheses are decay indices. Circled numbers at nodes indicate bootstrap proportions ⩾ 50% for relationships shown based on bootstrap resampling (500 replicates). Arrow indicates the node for the crustacean class Malacostraca, which, in this view, includes the Phyllocarida. An alternative view (Schram, 1986) for a monophyletic Phyllopoda (comprised of taxa written in bold letters) is not supported.

branchiopods either arose independently or persist as symplesiomorphies.

As an aside, the phylogenetic position of the Hoplocarida relative to the Eumalacostraca is equivocal as indicated by the low bootstrap value (55%) for the (Syncarida + Eucarida) node in Figure 14.4. Hence, we are as yet unable to determine whether hoplocarids represent a separate, independent malacostracan lineage with taxonomic rank (subclass) equivalent to that of phyllocarids and eumalacostracans (*sensu* Dahl, 1983) (Figure 14.3(b)), or whether hoplocarids represent a separate lineage within the Eumalacostraca (*sensu* Hessler, 1983) (Figure 14.3(a)). Further investigation into this question will include representatives of several additional eumalacostracan taxa.

## 14.4 MAXILLOPODA

There are two interrelated issues relevant to maxillopodan classification and phylogeny. The first is whether the Maxillopoda should be considered a monophyletic taxon and, if so, what are the constituent taxa; the second concerns phylogenetic relationships among the various lineages. First proposed by Dahl (1956) to include the subclasses Ascothoracida, Branchiura, Cirripedia (= Thoracica + Acrothoracica + Rhizocephala), Copepoda, and Mystacocarida, it has since been proposed that the Maxillopoda also include the Ostracoda (Siewing, 1960), Pentastomida (Grygier, 1983), Facetotecta (= Y-larvae; Grygier, 1983), and Tantulocarida (as members of the Thecostraca, *sensu* Grygier, 1987; see also Boxshall and Lincoln, 1983). As seen in Figure 14.5, opinions also differ

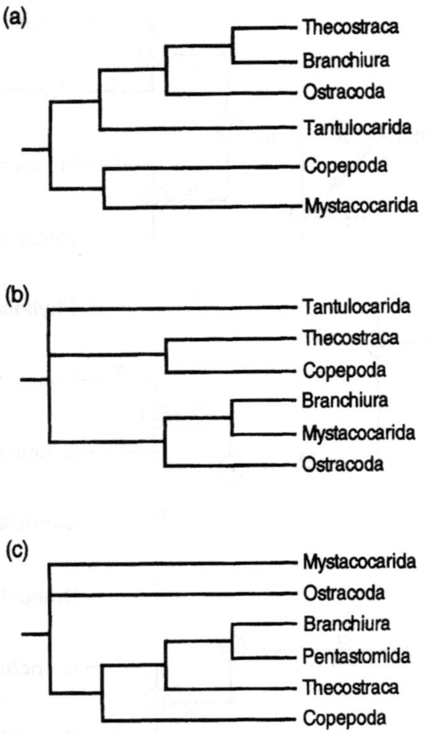


**Figure 14.5** Three alternative views of maxillopodan phylogeny. (a) Modified from Boxshall and Huys (1989). (b) Modified from Schram (1986). (c) Modified from Grygier (1983, 1987).

on the relationships among these taxa because among these diverse groups are several highly modified free-living and parasitic forms for which it is difficult to identify morphological homologies. In general, maxillopods are characterized by a basic 5-6-5 Bauplän, well-developed mandibular palps (in adults), and maxillules and maxillae used for filter-feeding. However, several maxillopodan lineages are believed to have arisen via paedomorphic processes which can confound comparative interpretations of segment number and appendage characteristics across taxa. Relatively recent discovery of new forms (Tantulocarida and Facetotecta, and the fossil Skaracarida and Orstenocarida), have led to further reassessment of maxillopodan unity (Boxshall, 1983; Grygier, 1983; Boxshall and Huys, 1989; Newman, 1992). Alternatives to adult features, such as larval or spermatozoan characters, have uncovered relationships of modified forms such as pentastomids (Wingstrand, 1972; Grygier, 1983), Hansen's Y-larvae (Grygier, 1987), and ascothoracidans (Grygier, 1982) (see also Jamieson, 1991), but much phylogenetic uncertainty persists.

Prompted by this uncertainty, Abele *et al.* (1992) undertook an investigation of maxillopodan relationships using parsimony analyses of 18S rRNA and rDNA sequences. Their results showed: (i) a (branchiuran + pentastome + ostracod) relationship [Abele *et al.*, 1989, showed pentastomes (endoparasites, often in the respiratory tract of tetrapods) to be highly modified crustaceans closely allied to the Branchiura]; (ii) an early divergence of acrothoracicans relative to other cirripedes; and (iii) little support for monophyly among maxillopodans in the broad sense. For the present analysis, we have added 18S rDNA sequences for two myodocopid ostracods, a harpacticoid copepod, an ascothoracidan, a mystacocarid, and a second acrothoracican (tantulocarids and Y-larvae have yet to be sequenced). Spears *et al.* (1994) showed the Ascothoracida and Cirripedia to be sister taxa, with acrothoracicans having diverged very early from the main cirripede lineage leading to rhizocephalans and thoracican barnacles. Because cirripedes exhibit extreme (10- to 12-fold) heterogeneity in substitution rates (Spears *et al.*, 1994), we included only the slower-evolving acrothoracican sequences. As in Abele *et al.*'s (1992) original study, branchiopods (as outgroup taxa) and malacostracans were also included (Table 14.1).

At first glance, the results of both parsimony and bootstrap analyses (Figure 14.6) do not indicate maxillopodan mono-

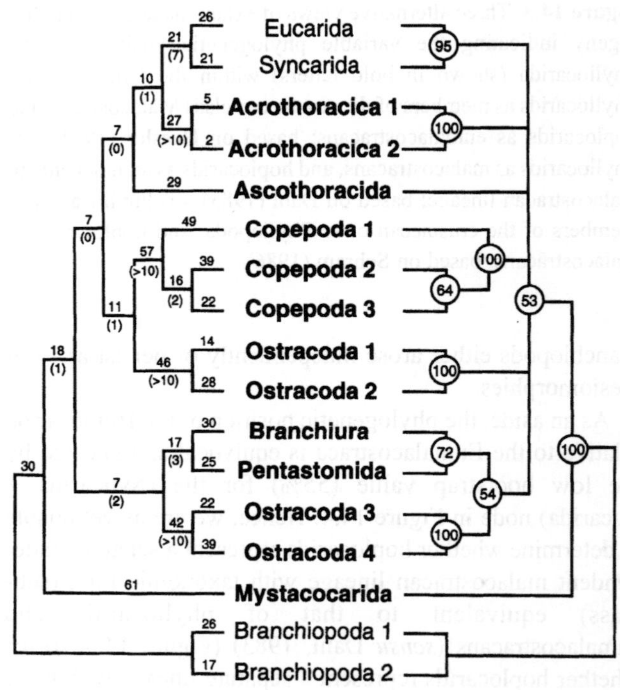


**Figure 14.6** Left: one of two most-parsimonious (MP) trees resulting from a maximum parsimony analysis (heuristic search) of 297 parsimony-informative characters based on 18S rDNA sequences. Putative maxillopodan taxa are shown in bold. The alternative topology places the Ascothoracida as the sister taxon to a {(Copepoda + Ostracoda) + [Ostracoda + (Branchiura + Pentastomida)]} clade. Uncircled numbers indicate branch lengths. Total tree length = 771 steps; C.I. = 0.563; R.I. = 0.569. Numbers in parentheses are decay indices. Right: circled numbers at nodes indicate bootstrap proportions ⩾ 50% for relationships shown based on bootstrap resampling (500 replicates).

phyly. The $g_1$ statistic (–1.3795) suggests the presence of hierarchical structure among the 297 parsimony-informative characters, and yet the number of unresolved polytomies in the bootstrap consensus tree (Figure 14.6, right) indicates little support (BP ≤ 50%) for particular relationships among these lineages. This is most likely due to the early origin and rapid radiation of these groups (Figure 14.2), resulting in highly divergent and homoplasious sequences. Hence, the presence of phylogenetic structure indicated by the $g_1$ statistic can likely be attributed to the well-supported lineages (copepods, malacostracans, acrothoracicans, and the myodocopid and podocopid ostracods).

The long branch leading to the first lineage, the Mystacocarida, indicates extensive divergence relative to other crustaceans (Figure 14.6, left). Pennak and Zinn (1943) classified mystacocarids as a subclass within the Maxillopoda, whereas Armstrong (1949) considered them as an order within the Copepoda. Although Dahl (1952) concurred with Pennak and Zinn (1943), he noted that certain aspects of the internal anatomy of mystacocarids are suggestive of a very primitive crustacean and a 'kinship with ancient Arthropoda' (Dahl, 1952, p. 36). There is no suggestion of a (mystacocarid + copepod) relationship in Figure 14.6; however, if maxillopods (*sensu lato*) are constrained to be monophyletic, then mystacocarids emerge as the sister taxon to a copepod clade. Although this topology (not shown) is 14 steps longer than the MP tree in Figure 14.6, it is not significantly worse by the modified version of Templeton's (1983) test as performed with DNAPARS in PHYLIP 3.5c. As already noted, the 18S rDNA sequence for the mystacocarid *Derocheilocaris typica* used herein is very different from those of other crustaceans. This might be viewed as support for Dahl's (1952) suggestion regarding their ancient and primitive status which it is important to confirm, despite the difficulties of obtaining them in sufficient quantity for 18S rDNA amplification and sequencing.

The second notable feature in Figure 14.6 is the polyphyletic nature of the Ostracoda. Podocopid ostracods (ostracods 3 and 4 in Figure 14.6) appear as the sister group of a (branchiuran + pentastomid) clade, and not the sister group of the Myodocopida (ostracods 1 and 2 in Figure 14.6). When ostracods are constrained to be monophyletic, however, the resulting tree is only three steps longer than the unconstrained MP tree shown in Figure 14.6 (left), the consistency index is lowered by only 0.002, and the (branchiuran + pentastomid) clade still appears as the sister group to the ostracod clade in the constrained tree. No significant difference is detected between these two alternative topologies by the modified version of Templeton's (1983) test using DNAPARS in PHYLIP 3.5c.

The affinity between pentastomes and branchiurans noted above has been suggested on the basis of spermatozoan morphology (Wingstrand, 1972; Storch and Jamieson, 1992), spermatogenesis (Riley *et al.*, 1978), and previous 18S rDNA analysis (Abele *et al.*, 1989). This relationship persists with moderate support (BP = 72%) in the present analysis (Figure 14.6) despite the inclusion of sequences for several additional maxillopodan taxa. However, Walossek and Müller (1994) reject this relationship based on what they claim to be late Cambrian larval stem group pentastomes for which they are unable to detect specific arthropod, let alone crustacean, affinities.

The present analysis (Figure 14.6) does not indicate a sister group relationship between the thecostracan taxa Ascothoracida and Acrothoracica as seen in Spears *et al.* (1994). Malacostracan taxa were not included in that earlier study as they are herein (Eucarida and Syncarida; Figure 14.6), and when these taxa are deleted from the present data set, a single MP tree is obtained that recovers the (ascothoracidan + acrothoracican) relationship. Constraining this relationship with all taxa included lengthens the original MP tree by three steps and lowers the consistency index by 0.002. Identical to the MP topology in Figure 14.6 save for the constrained thecostracan clade, the constrained tree is not significantly worse according to the modified version of Templeton's (1983) test using DNAPARS in PHYLIP 3.5c. It therefore seems reasonable to accept thecostracan monophyly, pending sequence data for taxa not represented (Facetotecta and Tantulocarida).

In summary, it does not appear from the cladogram in Figure 14.6 that the putative maxillopodan taxa share a common origin, and similarities in body plan and sperm morphology may best be explained as a combination of symplesiomorphic characters along with convergences that have accrued over these lineages' long history.

## 14.5 CRUSTACEA 'PROBLEMATICA': REMIPEDIA, CEPHALOCARIDA AND THE BASAL CRUSTACEAN LINEAGE

There is considerable interest in the recently discovered (Yager, 1981; Schram *et al.*, 1986) and enigmatic crustacean class Remipedia which has played a central role in views on crustacean and arthropod evolution. Their phylogenetic position remains uncertain (Figure 14.1) largely because remipedes are characterized by primitive features (a cephalic shield, and a long, untagmatized trunk with serially repeated biramous swimming appendages on every somite) in addition to clearly derived characters (fusion of the first trunk segment to the cephalon, uniramous mouth parts, and the presence of a pair of maxillipeds). Schram (1986) and Brusca and Brusca's (1990) morphology-based cladistic analyses place them at the base of the crustacean tree (Figure 14.1(a,b)). In contrast, Itô (1989) analysed limb form and musculature and by implication placed remipedes close to copepods.

Based on their duplosegment theory Emerson and Schram (1990, this volume) and Schram and Emerson (1991) inferred that remipedes exhibit the most primitive of crustacean body plans and represent the sister group to all other crustaceans (Figure 14.1(a)). Recent evidence from developmental gene expression studies and sequence-based phylogenies does not support this view of arthropod limb evolution, however, suggesting instead a closer relationship between uniramous insects and crustaceans (Zrzavý and Štys, 1994, 1997, this volume).

An alternative view considers cephalocarids as the most primitive of crustaceans derived from a biramous trilobitomorph ancestor, and not a uniramian (Sanders, 1957, 1963; Hessler, 1964; Hessler and Newman, 1975). This view maintains that the limb pattern of cephalocarids (a simple second maxilla little differentiated from the thoracopods in the adult, postoral cephalic appendages in the larvae also similar to thoracic limbs, and thoracic limbs with pseudepipods and foliaceous exopods) closely resembles the ancestral crustacean condition, with subsequent crustacean lineages derived via simplification and reduction of this limb pattern. Itô (1989) suggests a cephalocarid → remipede → copepod transition series for the derivation of copepod limb morphology. Another possible synapomorphy of remipedes and copepods is the fusion of the first thoracic segment with the cephalon in both. Similarity in limb morphology (Hessler, 1992) between cephalocarids and the Upper Cambrian Orsten fauna also confirms their basal phylogenetic position.

A third view (Fryer, 1992, 1996) differs greatly from the previous two in its assumption that the ur-crustacean lacked jointed appendages and was therefore derived from a worm-like, pre-arthropod lineage. Among extant crustaceans, branchiopod anostracans (fairy and brine shrimp) most resemble this ancestral condition, based in part on their primitive, anamorphic development and their possession of biramous, foliaceous and unsegmented appendages.

We test these competing ideas using sequence data from 18S rDNA genes. Analyses were performed on sequences from two malacostracan, three copepod, one remipede, one cephalocarid, and three branchiopod taxa. In the first analysis, sequences from two myriapods were included to test the hypothesis that Recent remipedes represent primitive crustaceans possessing biramous limbs derived from the duplosegment condition via myriapods. In the second analysis, two chelicerate sequences were used to investigate the alternative hypothesis that crustaceans and chelicerates are sister taxa derived from a common trilobitomorph ancestor. In this view, cephalocarids represent the primitive crustacean condition. Finally, the inclusion of representatives from three different orders of branchiopods in both of these analyses allows us to test the third hypothesis that this group best represents the basal crustacean lineage.

In the first analysis (Figure 14.7, left) with a myriapod (centipede) as an outgroup, the first branch leads to a second myriapod, a diplopodous millipede. The first crustacean lineage is a branchiopod clade, followed by a (remipede + cephalocarid) clade that is the sister group to a copepod clade. This lineage in turn comprises the sister group to the malacostracans. These relationships are not supported by bootstrap analysis (Figure 14.7, right), although individual support for the branchiopod, copepod (cephalocarid + remipede), and malacostracan clades remains high (BP ≥95%). These clades presumably account for the strong hierarchical structure in these data indicated by the $g_1$ statistic (–1.001). If Remipedia are constrained to be the sister group to all other crustaceans, two alternative topologies result that are 41 steps longer than the MP tree and significantly worse as determined by the modified version of Templeton's (1983) test using DNAPARS in PHYLIP.

Branchiopods also emerge as the sister group to all other crustaceans in a second analysis of the same set of crustacean sequences with a chelicerate (*Limulus polyphemus*) as an outgroup. In this case, the single MP tree is 11 steps shorter than the one in Figure 14.7, with identical relationships among the crustacean taxa. Constraining cephalocarids to be the basal crustacean lineage results in a significantly worse topology that is 46 steps longer than the MP outcome.

We conclude from the results of both of these analyses that the most parsimonious view of crustacean evolution encompasses an anostracan-like ur-crustacean from which first the Branchiopoda and then subsequent lineages are derived, similar to the scenario outlined by Fryer (1992, 1996). However, a more definitive answer requires that other crustaceans with a long fossil record (ostracods) also be considered. Finally, the evidence presented here clearly supports Itô's (1989) suggestion of a (cephalocarid + remipede + copepod) relationship. The remipede and cephalocarid branch lengths in Figure 14.7 are relatively long, however, and the resulting (remipede + cephalocarid) clade as well as the sister group relationship of this clade to the copepod clade might be construed as a case of 'long-branch attraction' (Felsenstein, 1978), even though neither remipedes nor cephalocarids are attracted to a similarly long branchiopod branch. However, these results are sensitive to the addition of other crustacean and arthropod taxa.

It is clearly desirable to sequence more species of these 'problematic' taxa. This is especially true for Remipedia, in light of evidence we have that several 18S rDNA pseudogenes are present within the genome of the species represented here, *Speleonectes tulumensis*. It is therefore critical that the highly divergent nature of the sequence we have obtained be confirmed for other remipede species. Interestingly, the presence of 18S rDNA pseudogenes have also been reported for another 'fast-clock' arthropod, *Drosophila melanogaster* (Benevolenskaya *et al.*, 1994).

## 14.6 CRUSTACEAN PHYLOGENY AND CRUSTACEAN–ARTHROPOD AFFINITIES

We now examine crustacean monophyly, relationships among crustaceans when all the major lineages are consid-

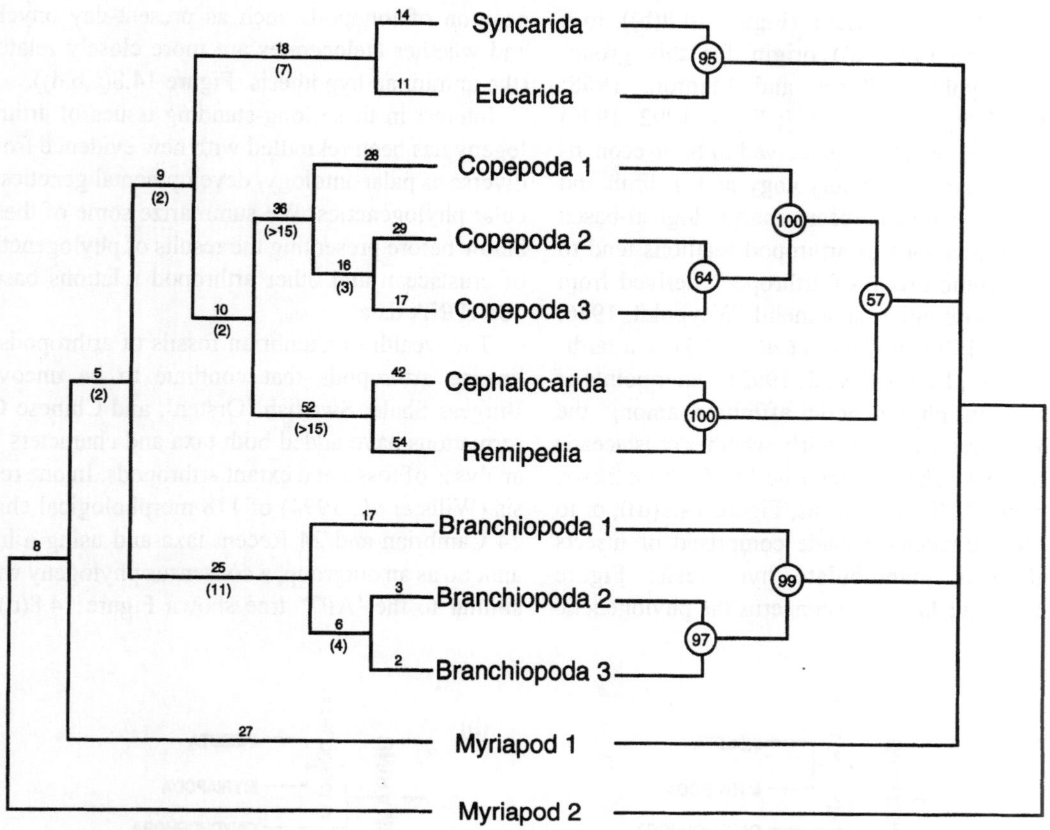


**Figure 14.7** Left: single most-parsimonious tree (branch-and-bound search) with a centipede (Myriapod 2) as an outgroup based on 185 parsimony-informative characters. Uncircled numbers indicate branch lengths. Total tree length = 429 steps; C.I. = 0.643; R.I. = 0.627. Numbers in parentheses are decay indices. Right: circled numbers at nodes indicate bootstrap proportions ⩾ 50% for relationships shown based on bootstrap resampling (500 replicates).

ered jointly, and the relationship of crustaceans to other arthropod groups. We first briefly review the more commonly held views and discuss results we have obtained based on parsimony analyses of a comprehensive data set representing a diversity of arthropods.

Crustacean monophyly is primarily based on the majority of crustaceans possessing a characteristic assortment of head appendages (two pairs of antennae, a pair of mandibles, and two pairs of maxillae) and a unique nauplius larva. However, there is little consensus regarding the pattern of evolution both within and among the diverse lineages of this group (Figures 14.1, 14.3 and 14.5). Wilson's (1992) cladistic analysis of higher-order crustacean relationships based on 59 morphological characters compiled by participants at a workshop on crustacean evolution held in 1990 at the Kristineberg Marine Biological Station, Sweden, detected little support for any of the hierarchical arrangements shown in Figure 14.1. Maxillopods do not appear monophyletic, and presumably primitive taxa (cephalocarids, branchiopods, remipedes) are not the first to diverge in Wilson's (1992) MP tree when rooted with a hypothetical ur-crustacean. There appear instead two branches: one comprising [ostracods + (branchiopods + malacostracans)], the other comprised of fossil forms followed by [tantulocarids + (mystacocarids + copepods)] and a clade comprising {cephalocarids + [remipedes + (branchiurans + thecostracans)]}. A similar topology results when the ur-crustacean is omitted, however ostracods move to the (mystacocarid + copepod) clade. Some of the 18S rDNA-based results discussed previously in this chapter are consistent with these findings (the indication that cephalocarids and remipedes are not necessarily primitive taxa, and uncertainty regarding maxillopodan monophyly); in other cases, the results differ (the relationship between ostracods and branchiurans suggested in the molecular phylogeny is absent in the morphological one).

## 14.7 MAJOR ARTHROPOD PHYLOGENY

Consensus on arthropod phylogeny is equally elusive (Schram, 1986; Wägele, 1993; Wheeler *et al.*, 1993; Raff *et al.*, 1994; Fryer, 1996), with opinions ranging from a polyphyletic (Figure 14.8(a)), diphyletic (Figure 14.8(b)), to a monophyletic (Figure 14.8(c,d)) origin for this group. Proponents of polyphyly (Tiegs and Manton, 1958; Anderson, 1973, 1979; Manton, 1977; Fryer, 1992, 1996) base their argument on what are perceived to be irreconcilable differences primarily in embryology and in limb and jaw morphology. In contrast, recent morphological-based investigations into questions of arthropod relations tend to support a monophyletic origin of arthropods derived from an ancestor shared with either an annelid (Weygoldt, 1986; Brusca and Brusca, 1990; Wheeler *et al.*, 1993) or a tardigrade (Ghiselin, 1988; Eernisse *et al.*, 1992). Other points of contention focus on phylogenetic affinities among the diverse arthropod lineages, particularly whether crustaceans are the sister taxon to chelicerates (the 'TCC' hypothesis, Figure 14.8(b); the 'APT' hypothesis, Figure 14.8(d)), or to an atelocerate (or 'tracheate') clade comprised of insects and myriapods (the mandibulate hypothesis; Figure 14.8(c)). Central to this last issue concerns the phylogenetic position of lobopods such as present-day onychophorans, and whether atelocerates are more closely related to these (the uniramian hypothesis; Figure 14.8(a,b,d)).

Interest in these long-standing issues of arthropod phylogeny has been rekindled with new evidence from fields as diverse as palaeontology, developmental genetics and molecular phylogenetics. We summarize some of these findings below before presenting the results of phylogenetic analyses of crustacean and other arthropod relations based on new 18S rDNA data.

The wealth of Cambrian fossils of arthropods and stem-lineage arthropods that continue to be uncovered from Burgess Shale, Swedish 'Orsten', and Chinese Chengjiang formations have added both taxa and characters to cladistic analyses of fossil and extant arthropods. In one recent analysis (Wills *et al.*, 1994) of 118 morphological characters for 24 Cambrian and 24 Recent taxa and using a hypothetical annelid as an outgroup, a consensus phylogeny was obtained similar to the 'APT' tree shown Figure 14.8(d): lobopods

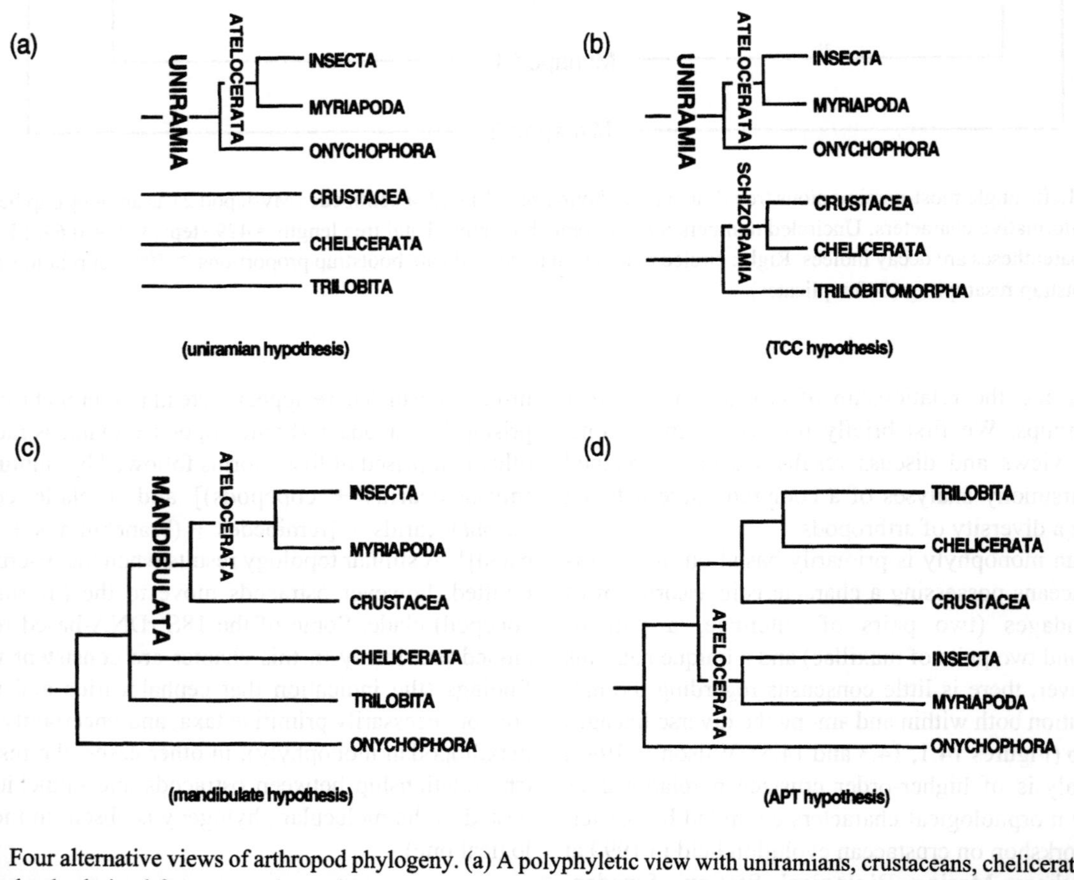


**Figure 14.8** Four alternative views of arthropod phylogeny. (a) A polyphyletic view with uniramians, crustaceans, chelicerates and trilobites independently derived from pre-arthropod ancestors (Anderson, 1973, 1979; Manton, 1977; Fryer, 1992, 1996). (b) A diphyletic view with separate 'TCC' [trilobite (T) + chelicerate (C) + crustacean (C)] and uniramian (insect + myriapod + onycophoran) clades independently derived from pre-arthropod ancestors (Tiegs and Manton, 1958; Cisne, 1974; Hessler and Newman, 1975; Bergström, 1980, 1992; Hessler, 1992). (c) A monophyletic view with a 'mandibulate' (crustacean + atelocerate) clade excluding chelicerates (Paulus, 1979; Weygoldt, 1979, 1986; Brusca and Brusca, 1990; Wägele, 1993). (d) Schram and Emerson's (1991) monophyletic view based on arthropod pattern theory (APT) that places crustaceans as the sister group to a (chelicerate + trilobite) clade.

are the sister group to arthropods; arthropods in turn are composed of the Atelocerata (myriapods + insects), which is the sister group to a larger clade comprising the (Crustacea + Crustaceanomorpha) and the Arachnomorpha (chelicerates + cheliceriformes + trilobites). It is interesting that in this analysis, crustaceans appear polyphyletic with respect to several crustaceanomorphs; among only extant crustaceans, however, remipedes diverge first, followed by branchiopods, a (copepod + mystacocarid) clade, branchiurans, thoracicans and phyllocarids. Waggoner's (1996) cladistic analysis of primarily fossil forms (extant crustaceans were not considered) arrived at a similar view of arthropod phylogeny: uniramous atelocerates (possibly polyphyletic in this analysis) appear less derived than biramous arthropods, the latter composed of monophyletic crustaceanomorph and arachnomorph clades. Support for the Mandibulata (myriapods + insects + crustaceans) is lacking in both of these analyses.

Promising evidence for understanding arthropod relations is emerging from comparative gene expression studies (Patel, 1994; Zrzavý and Štys, 1995). For example, Panganiban *et al.* (1995) report that the same gene (*Distal-less*, or *Dll*) is responsible for the initiation of development of uniramous limbs in the insect *Drosophila* and biramous limbs in two crustaceans, *Artemia* and *Mysidopsis*, and cite this as evidence for a common origin for insect and crustacean limbs in particular, and for arthropods in general (pending additional taxonomic sampling). Differences in limb branching and the spatial pattern of limbs along the body of an organism have been attributed to the control of homeotic genes such as the *Hox* gene clusters regulating *Dll* expression in different ways during development (Akam *et al.*, 1994), and Panganiban *et al.* (1995) advocate caution when using limb morphology as a phylogenetic marker owing to this potential for plasticity in *Dll* gene expression, even within an individual. Additional support for an (insect + crustacean) relationship comes from gene expression evidence for the *engrailed* (*en*) segmentation gene that has identified positional homology of limbs and segments in representatives of both of these taxa (Dohle and Scholtz, 1988; Patel *et al.*, 1989; Scholtz *et al.*, 1994). However, a clearer understanding of relationships based on the varied gene expression evidence cited above awaits demonstration that the homologies noted are not merely symplesiomorphic (Fryer, 1996, 1997, this volume).

Ribosomal RNA sequence analysis appeared promising as a tool for inferring ancient evolutionary patterns among crustaceans and other arthropods owing to the conservative rate of nucleotide substitution in ribosomal genes. As evidence accrued, however, it became apparent that lineages can exhibit extremely variable substitution rates, resulting in high levels of homoplasy and adversely affecting the reliability of virtually all tree-building methods. Differences in alignment procedures, taxonomic sampling and phylogenetic inference methodologies may also make it difficult to reconcile the results of ribosomal RNA studies of higher-order arthropod relationships. For instance, Field *et al.*'s (1988) distance analysis of partial 18S rRNA data supports arthropod monophyly, although it is uncertain from which metazoan lineage they arose. Myriapods diverge first, followed by chelicerates, crustaceans, and then insects; hence, neither atelocerate (myriapod + insect) nor mandibulate (atelocerate + crustacean) monophyly is supported. In contrast, arthropods appear paraphyletic with respect to molluscs and annelids in Lake's (1990) reanalysis of the same data set using evolutionary parsimony. Myriapods and insects are the first arthropods to diverge in Lake's (1990) analysis, but atelocerate monophyly is uncertain and a mandibulate grouping is not supported. Inconsistent results with respect to metazoan relationships also appear in Patterson's (1989) maximum parsimony analysis of the Field *et al.* (1988) data, although arthropod monophyly was similarly indicated. Turbeville *et al.* (1991) sequenced additional taxa and found moderate to weak bootstrap support for arthropod monophyly under both distance and parsimony-based methods of inference (82% and 55%, respectively). Their result shows weak bootstrap support (≤ 58%) for (annelids + molluscs) as the sister group of arthropods. Among arthropods, chelicerates are the sister group to a [myriapod + (crustacean + insect)] clade by distance analysis; under parsimony, however, there is weak bootstrap support (40%) for a (myriapod + chelicerate) relationship. Both methods support a (crustacean + insect) clade [BP = 85% (distance) and BP = 96% (parsimony)], providing strong evidence against the Atelocerata. Evolutionary parsimony found varying support for arthropod monophyly depending on the subset of taxa analysed, suggesting that results based on this method of inference can also be biased by long-branched taxa and sampling effects.

Friedrich and Tautz (1995) used partial 28S rDNA data in addition to 18S rDNA sequences to infer arthropod relationships using neighbour-joining, parsimony and maximum likelihood methods. Arthropod monophyly received strong support (BP = 96%) under maximum likelihood, with arthropods divided into two clades: a (myriapod + chelicerate) clade was the sister group to (crustaceans + insects). The (crustacean + insect) relationship received strong bootstrap support (BP = 100%), a result similar to that obtained by Turbeville *et al.* (1991). Surprisingly, there was little support (BP = 65%) of monophyly between the two crustacean taxa represented in the study (a crayfish, *Procambarus*, and a branchiopod brine shrimp, *Artemia*). Adoutte and Philippe (1993) actually found some evidence for crustacean paraphyly with respect to insects based on a parsimony analysis of 18S rDNA that included sequences for a branchiuran, a branchiopod (both crustaceans), and a beetle, although bootstrap support was low (BP ≤ 50%). Finally, Ballard *et al.*'s (1992) parsimony analysis of mitochondrial 12S rRNA data indicates an (annelid + molluscan) clade as the sister group to a monophyletic Arthropoda that includes onychophorans. Neither uniramian nor atelocerate monophyly are supported by parsimony analysis, however, as myriapods diverge first, followed by onychophorans, chelicerates, and a monophyletic group comprised of crustaceans and insects as sister taxa.

Boore *et al.* (1995) studied mitochondrial gene arrangements as molecular characters to infer phylogeny, supporting arthropod and mandibulate (atelocerate + crustacean) monophyly. Resolution was lacking among the mandibulate taxa and between arthropods and related taxa.

Alternatively, multiple data sets have been analysed either jointly or separately to look for congruence among the results (Huelsenbeck *et al.*, 1996; see also Eernisse, 1997, Wheeler, 1997, and Zrzavý *et al.*, 1997, all this volume). Wheeler *et al.* (1993) considered both molecular (18S rDNA and ubiquitin sequences) and morphological characters. The combined and separate analyses support arthropod, 'arachnate' (trilobite and chelicerate), mandibulate and atelocerate monophyly, and reject a uniramian phylogeny. The two molecular data sets differ markedly, and although both lack resolution, the 18S rDNA evidence is more congruent with the morphological evidence (although insects are paraphyletic with respect to crustaceans). Wägele's (1993) interpretation of morphological, physiological, and biochemical evidence is similar to Wheeler *et al.*'s (1993) overall result, favouring mandibulate over uniramian monophyly with crustaceans rather than onychophorans as the sister group of the Atelocerata.

Here, we extensively sample crustaceans and use sequences for the entire 18S rDNA gene (except in the case of the pentastomes). As suggested by Wheeler *et al.* (1993) and Wheeler (1995), the observed lack of phylogenetic stability within and among several of the molecular studies can be attributed to small and disparate sampling regimes. We consider 18 sequences representing all of the major crustacean lineages in a multiple alignment that also contains sequences from GenBank for two annelids, two myriapods, two chelicerates, and seven insects (Table 14.1). Most parsimonious tree(s) are compared with alternative topologies with all taxa included, and again with 'problematic' long-branch crustacean sequences excluded, in order to assess the effect of these rapidly evolving sequences on the inferred topology. Finally, we look at the stability of crustacean relationships by using different arthropod outgroups.

In the first analysis (all taxa included), neither crustaceans nor insects are monophyletic in any of the four equally parsimonious trees (Figure 14.9(a–d)), for which the 50% consensus is shown in Figure 14.10 (left). We present all four topologies in Figure 14.9 to illustrate both branch lengths and specific variation among relationships. The clade, myriapods + chelicerates (Figure 14.9(a–d)), 'attracts' another clade comprised of three fast-evolving crustaceans (Remipedia, Cephalocarida and Mystacocarida) in three of the trees. From the 50% majority-rule consensus topology (Figure 14.10, left), it can be seen that primitive collembolid insects are the sister group to branchiopod crustaceans in all four MP trees, so in some respects these data corroborate the close relationship between insects and crustaceans inferred by other rDNA-based studies. However, the unlikely (collembolid + branchiopod) relationship is probably an artefact of long-branch attraction: support for this interpretation is seen when crustaceans are constrained to be monophyletic (crustaceans appear polyphyletic in some respect in all four MP trees). The constrained topology (not shown) requires the same number of steps as the unconstrained MP trees and has a (myriapod + chelicerate) clade coming off first, followed by a monophyletic insect clade that is the sister group to crustaceans. This result also indicates that the original heuristic search procedure, used because of the large number of taxa, failed to identify all equally most-parsimonious trees. Among crustaceans constrained in this manner, a [pentastome + (branchiuran + podocopid ostracod 3 and 4)] clade emerges first, followed by branchiopods as the sister taxon to a [mystacocarid + (remipede + cephalocarid)] clade, then a (copepod + myodocopid ostracod) clade, with the remaining crustacean taxa as in Figure 14.9(a–d). Branchiopods attract the fast-evolving (remipede + cephalocarid + mystacocarid) clade rather than the collembolid insects. When the topology is constrained to favour either mandibulate, TCC, or APT hypotheses, cladograms are found that are, respectively, seven, zero, or eight steps longer than the unconstrained MP trees, and none is judged to be significantly worse than the original MP topology by the modified version of Templeton's test as performed by DNAPARS in PHYLIP. So although the evidence does not support any of these alternative hypotheses, neither does it unambiguously refute them.

The bootstrap consensus tree (Figure 14.10, right) reflects this lack of resolution. Neither crustacean nor insect monophyly is indicated, and there is no discernible relationship among any of the arthropod lineages. In contrast, individual clades within lineages (malacostracans, cirripedes, copepods, myodocopid ostracods, podocopid ostracods, branchiopods, collembolid insects and chelicerates) are

**Figure 14.9** Arthropod phylogeny inferred by parsimony analysis of 18S rDNA. Shown are the four (a–d) equally most-parsimonious trees with an annelid as an outgroup based on 314 parsimony-informative characters using a heuristic search with a simple addition sequence. Uncircled numbers indicate branch lengths. Total tree length = 1078 steps; C.I. = 0.468; R.I. = 0.482. Crustacean lineages are represented by bold branches; insect taxa are designated by bold font.

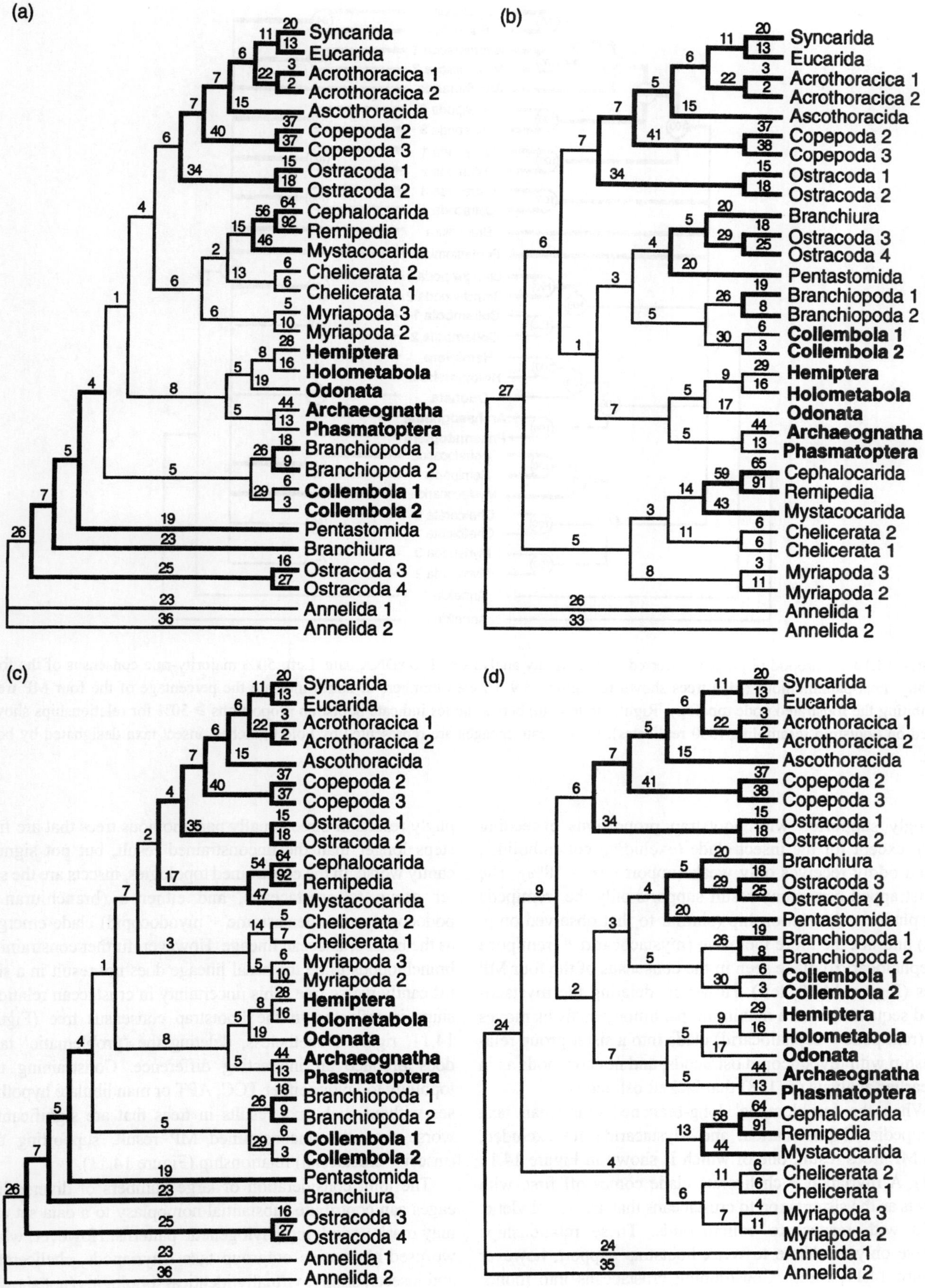
(a)
Syncarida
Eucarida
Acrothoracica 1
Acrothoracica 2
Ascothoracida
Copepoda 2
Copepoda 3
Ostracoda 1
Ostracoda 2
Cephalocarida
Remipedia
Mystacocarida
Chelicerata 2
Chelicerata 1
Myriapoda 3
Myriapoda 2
Hemiptera
Holometabola
Odonata
Archaeognatha
Phasmatoptera
Branchiopoda 1
Branchiopoda 2
Collembola 1
Collembola 2
Pentastomida
Branchiura
Ostracoda 3
Ostracoda 4
Annelida 1
Annelida 2
(b)
Syncarida
Eucarida
Acrothoracica 1
Acrothoracica 2
Ascothoracida
Copepoda 2
Copepoda 3
Ostracoda 1
Ostracoda 2
Branchiura
Ostracoda 3
Ostracoda 4
Pentastomida
Branchiopoda 1
Branchiopoda 2
Collembola 1
Collembola 2
Hemiptera
Holometabola
Odonata
Archaeognatha
Phasmatoptera
Cephalocarida
Remipedia
Mystacocarida
Chelicerata 2
Chelicerata 1
Myriapoda 3
Myriapoda 2
Annelida 1
Annelida 2
(c)
Syncarida
Eucarida
Acrothoracica 1
Acrothoracica 2
Ascothoracida
Copepoda 2
Copepoda 3
Ostracoda 1
Ostracoda 2
Cephalocarida
Remipedia
Mystacocarida
Chelicerata 2
Chelicerata 1
Myriapoda 3
Myriapoda 2
Hemiptera
Holometabola
Odonata
Archaeognatha
Phasmatoptera
Branchiopoda 1
Branchiopoda 2
Collembola 1
Collembola 2
Pentastomida
Branchiura
Ostracoda 3
Ostracoda 4
Annelida 1
Annelida 2
(d)
Syncarida
Eucarida
Acrothoracica 1
Acrothoracica 2
Ascothoracida
Copepoda 2
Copepoda 3
Ostracoda 1
Ostracoda 2
Branchiura
Ostracoda 3
Ostracoda 4
Pentastomida
Branchiopoda 1
Branchiopoda 2
Collembola 1
Collembola 2
Hemiptera
Holometabola
Odonata
Archaeognatha
Phasmatoptera
Cephalocarida
Remipedia
Mystacocarida
Chelicerata 2
Chelicerata 1
Myriapoda 3
Myriapoda 2
Annelida 1
Annelida 2

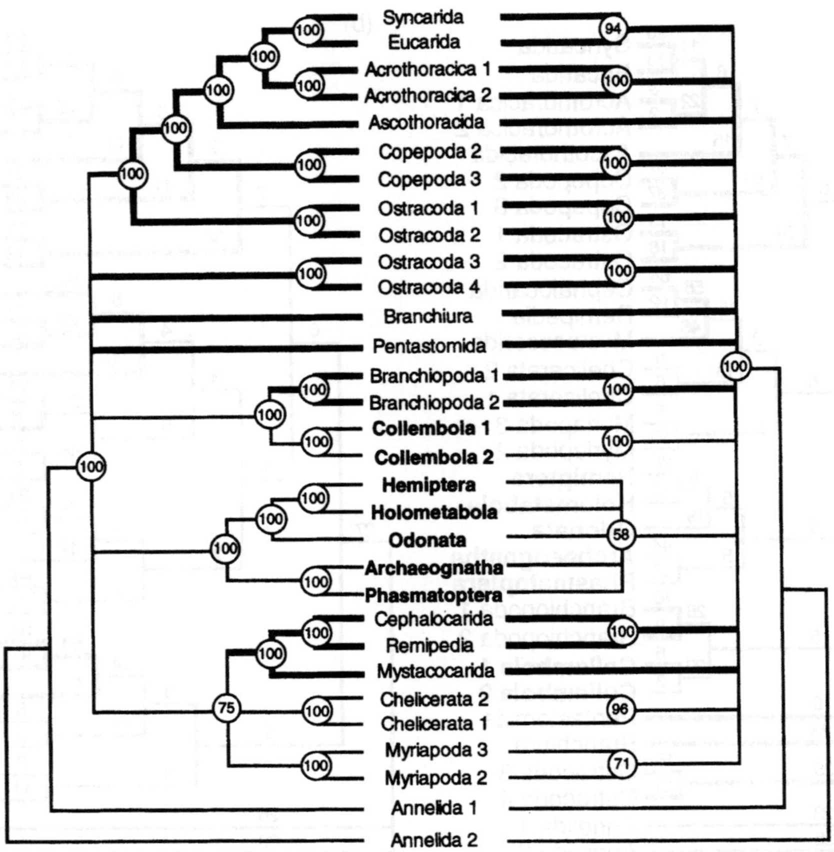


**Figure 14.10** Arthropod phylogeny inferred by parsimony analysis of 18S rDNA data. Left: 50% majority-rule consensus of the four equally most-parsimonious (MP) trees shown in Figure 14.9. Circled numbers at nodes indicate the percentage of the four MP trees exhibiting the illustrated node topology. Right: circled numbers at nodes indicate bootstrap proportions ⩾ 50% for relationships shown based on bootstrap resampling (500 replicates). Crustacean lineages are represented by bold branches; insect taxa designated by bold font.

strongly supported with bootstrap proportions exceeding 96%, except for the insect clade (excluding collembolids), which oddly receives only weak support (BP = 58%). The bootstrapped parsimony result supports only the (remipede + cephalocarid) relationship (similar to that observed on p. 182) and not the more inclusive [mystacocarid + (remipede + cephalocarid)] clade seen in the consensus of the four MP trees (Figure 14.10, left). However, deleting the mystacocarid sequence from a maximum parsimony analysis moves the (remipede + cephalocarid) clade into a sister group relationship with myodocopid ostracods, and not copepods as in an earlier analysis (p. 178) that lacked ostracods.

When the 'problematic' long-branched crustacean taxa Remipedia, Cephalocarida, and Mystacarida are excluded, two MP trees result, one of which is shown in Figure 14.11 (left). A myriapod + chelicerate clade comes off first, with insects as the sister taxon to crustaceans that are paraphyletic, again with respect to collembolids. These relationships receive only moderate to weak bootstrap support, however (Figure 14.11, right). Constraining crustaceans into monophyly results in three equally parsimonious trees that are five steps longer than the unconstrained result, but not significantly worse. In the constrained topologies, insects are the sister group to crustaceans, and either a (branchiuran + podocopid) or a (pentastome + myodocopid) clade emerges as the basal crustacean lineage. However, further constraining branchiopods to be the basal lineage does not result in a significantly worse tree. This uncertainty in crustacean relationships is reflected in the bootstrap consensus tree (Figure 14.11, right). Nonetheless, deleting the 'problematic' taxa does produce an interesting difference. Constraining the topology to favour either TCC, APT or mandibulate hypotheses without such taxa results in trees that are significantly worse than the unconstrained MP result, supporting the (insect + crustacean) relationship (Figure 14.11).

The joint consideration of large numbers of diverse lineages can contribute substantial homoplasy to a data set and may mask underlying phylogenetic patterns. However, when we used alternative outgroup taxa (myriapods, chelicerates and insects) singly, virtually identical topologies (not shown)

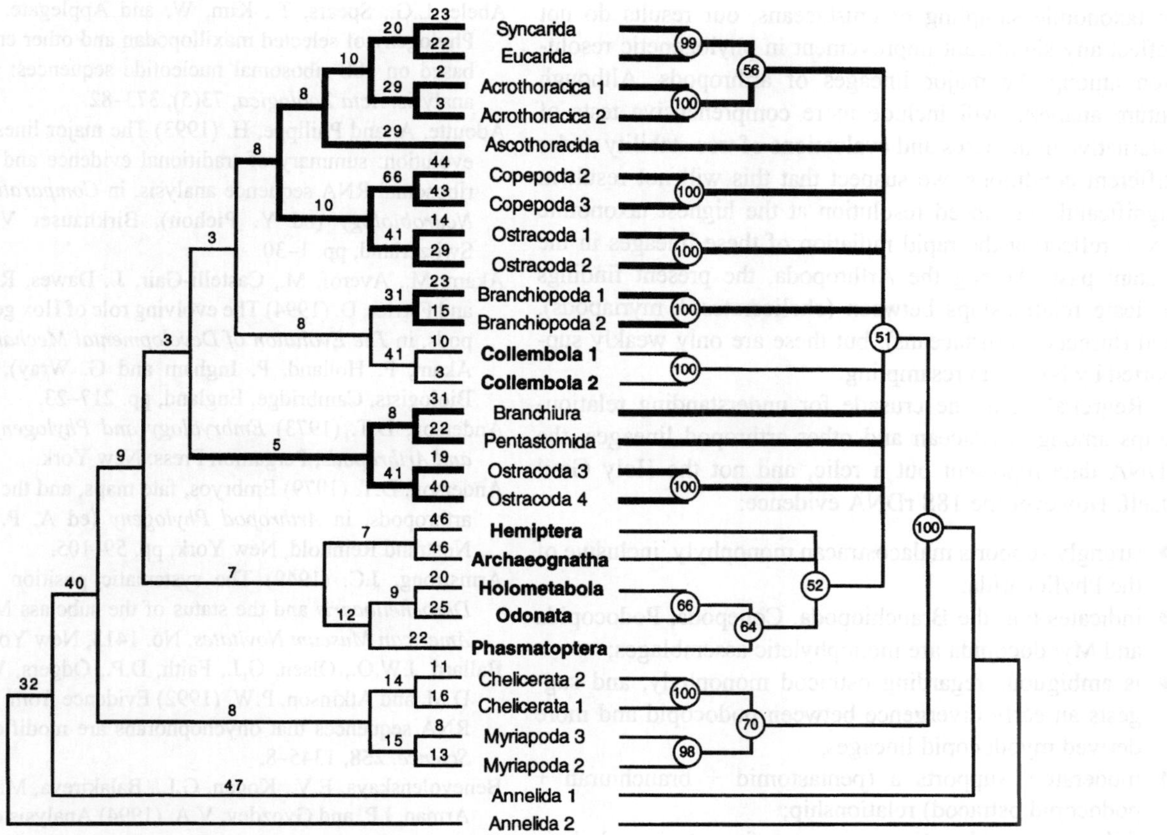


**Figure 14.11** Arthropod phylogeny inferred by parsimony analysis (heuristic search) of 18S rDNA data as in Figure 14.9 but with 'problematic' long-branched taxa omitted (Remipedia, Cephalocarida and Mystacocarida). Left: one of two equally most-parsimonious (MP) trees with an annelid as an outgroup based on 353 parsimony-informative characters. Total tree length = 1111 steps; C.I. = 0.474; R.I. = 0.510. Uncircled numbers indicate branch lengths. The alternative MP topology differs only in the relative position of the Pentastomida within the (Pentastomida + Branchiura + Ostracoda 3 and 4) clade. Right: circled numbers at nodes indicate bootstrap proportions ňA 50% for relationships shown based on bootstrap resampling (500 replicates). Crustacean lineages are represented by bold branches; insect taxa designated by bold font.

were found regardless of outgroup choice: the first crustacean branch consisted of a 'fast-evolving' [mystacocarid + (remipede + cephalocarid)] clade, sister group to a branchiopod clade; the next branch led to a group comprising the [Pentastomida + (Branchiura + podocopid ostracods 3 and 4)], although in one case this group appeared as a bifurcation of the first branch; myodocopid ostracods 1 and 2 came off next, and in one case this clade appeared as a bifurcation with the next branch, the copepods; the final three lineages were, successively, ascothoracidans, acrothoracicans and malacostracans. When the three fast-evolving crustacean taxa were omitted better supported relationships were obtained, with either a branchiopod clade, or a [pentastome + (branchiuran + podocopid ostracod)] clade, or a clade comprised of both of these groups, coming off first. These results, taken together with those shown in Figures 14.9–14.11, indicate that we cannot identify which crustacean lineage is the most basal; branchiopods, pentastomes, branchiurans, and ostracods all diverged from the main crustacean lineage in relatively rapid succession. Fryer (1992, 1996) identified only the Branchiopoda as an ancestral crustacean, which we support. The phylogenetic positions of the aberrant Remipedia, Mystacocarida and Cephalocarida depend on sequence data from additional species.

## 14.8 CONCLUSIONS

Studies of arthropod phylogeny based on rDNA sequence data typically conclude with suggestions that improved resolution might result from either more taxa, or more data per taxon, or upon further methodological investigations of phylogeny. For example, it is common knowledge that sequence-based phylogenies (particularly those using rDNA evidence) can be sensitive to several parameters such as alignment, transition/transversion ratio, variable substitution rate, and variable base composition. Despite an improvement in taxonomic sampling of crustaceans, our results do not reflect any significant improvement in phylogenetic resolution among the major lineages of arthropods. Although future analyses will include more comprehensive tests of alternative hypotheses and evaluations of tree stability under different conditions, we suspect that this will not result in significantly enhanced resolution at the highest taxonomic level, reflecting the rapid radiation of these lineages in the distant past. Among the Arthropoda, the present findings indicate relationships between (chelicerates + myriapods), and (insects + crustaceans), but these are only weakly supported by bootstrap resampling.

Regrettably, in the crusade for understanding relationships among crustacean and other arthropod lineages, the rDNA data represent but a relic, and not the Holy Grail itself. However the 18S rDNA evidence:

- strongly supports malacostracan monophyly, inclusive of the Phyllocarida;
- indicates that the Branchiopoda, Copepoda, Podocopida and Myodocopida are monophyletic assemblages;
- is ambiguous regarding ostracod monophyly, and suggests an early divergence between podocopid and more derived myodocopid lineages;
- moderately supports a (pentastomid + branchiuran + podocopid ostracod) relationship;
- fails to provide strong support for a monophyletic Maxillopoda (*sensu lato*); instead, Ascothoracida and Acrothoracica lineages are more derived than copepod, branchiuran, and ostracod lineages;
- indicates a high substitution rate within mystacocarid, remipede, and cephalocarid sequences (and also within the thoracican lineage; Spears *et al.*, 1994); pending confirmation from other species, the relative phylogenetic position of these lineages remains uncertain; and
- indicates that several 18S rDNA pseudogenes are present in the remipede *Speleonectes tulumensis*, a feature that has been reported for *Drosophila melanogaster* (Benevolenskaya *et al.*, 1994), but not for any other arthropod.

## ACKNOWLEDGEMENTS

Ellyn Whitehouse and Chris Bacot of the FSU sequencing facility, Rani Dhanarajan of the FSU cloning lab, and Margaret Seavy of the FSU analytical lab provided able assistance with various aspects of this project, as did Margaret Ptacek and Travis Boline. We thank numerous colleagues for providing and/or identifying specimens.

## REFERENCES

Abele, L.G., Kim, W. and Felgenhauer, B.E. (1989) Molecular evidence for inclusion of the phylum Pentastomida in the Crustacea. *Molecular Biology and Evolution*, **6**, 685–91.

Abele, L.G., Spears, T., Kim, W. and Applegate, M.A. (1992) Phylogeny of selected maxillopodan and other crustacean taxa based on 18S ribosomal nucleotide sequences: a preliminary analysis. *Acta Zoologica*, **73**(5), 373–82.

Adoutte, A. and Philippe, H. (1993) The major lines of metazoan evolution: summary of traditional evidence and lessons from ribosomal RNA sequence analysis, in *Comparative Molecular Neurobiology* (ed Y. Pichon), Birkhauser Verlag, Basel, Switzerland, pp. 1–30.

Akam, M., Averof, M., Castelli-Gair, J., Dawes, R., Falciani, F. and Ferrier, D. (1994) The evolving role of Hox genes in arthropods, in *The Evolution of Developmental Mechanisms* (eds M. Akam, P. Holland, P. Ingham and G. Wray), Company of Biologists, Cambridge, England, pp. 217–23.

Anderson, D.T. (1973) *Embryology and Phylogeny in Annelids and Arthropods*, Pergamon Press, New York.

Anderson, D.T. (1979) Embryos, fate maps, and the phylogeny of arthropods, in *Arthropod Phylogeny* (ed A. P. Gupta), Van Nostrand Reinhold, New York, pp. 59–105.

Armstrong, J.C. (1949) The systematic position of the genus *Derocheilocaris* and the status of the subclass Mystacocarida. *American Museum Novitates*, No. 1413, New York.

Ballard, J.W.O., Olsen, G.J., Faith, D.P., Odgers, W.A., Rowell, D.M. and Atkinson, P.W. (1992) Evidence from 12S ribosomal RNA sequences that onychophorans are modified arthropods. *Science*, **258**, 1345–8.

Benevolenskaya, E.V., Kogan, G.L., Balakireva, M.D., Filipp, D., Arman, I.P. and Gvozdev, V.A. (1994) Analysis of pseudogene nucleotide sequence reveals variability of rDNA genes in *Drosophila melanogaster*. *Russian Journal of Genetics*, **30**(3), 280–6.

Bergström, J. (1980) Morphology and systematics of early arthropods. *Abhandlungen des Naturwissenschaftichen Vereins in Hamburg, NF*, **23**, 7–42.

Bergström, J. (1992) The oldest arthropods and the origin of the Crustacea. *Acta Zoologica*, **73**(5), 287–91.

Boore, J.L., Collins, T.M., Stanton, D., Daehler, L.L. and Brown, W.M. (1995) Deducing the pattern of arthropod phylogeny from mitochondrial DNA arrangements. *Nature*, **376**, 163–5.

Bowman, T.E. and Abele, L.G. (1982) Classification of the Recent Crustacea, in *Systematics, the Fossil Record, and Biogeography* (ed. L.G. Abele), *The Biology of Crustacea*, Vol. 1 (ed. D.E. Bliss), Academic Press, New York, pp. 1–27.

Boxshall, G.A. (1983) A comparative functional analysis of the major maxillopodan groups, in *Crustacean Issues 1: Crustacean Phylogeny* (ed. F.R. Schram), A.A. Balkema, Rotterdam, pp. 121–43.

Boxshall, G.A. and Huys, R. (1989) New tantulocarid, *Stygotantulus stocki*, parasite on harpacticoid copepods, with an analysis of the phylogenetic relationships within the Maxillopoda. *Journal of Crustacean Biology*, **9**, 126–40.

Boxshall, G.A. and Lincoln, R.J. (1983) Tantulocarida, a new class of Crustacea ectoparasitic on other crustaceans. *Journal of Crustacean Biology*, **3**, 1–16.

Bremer, K. (1988) The limits of amino acid sequence data in angiosperm phylogenetic reconstruction. *Evolution*, **42**(4), 795–803.

Bremer, K. (1994) Branch support and tree stability. *Cladistics*, **10**, 295–304.

Briggs, D.E.G., Weedon, M.J. and Whyte, M.A. (1993) Arthropoda (Crustacea excluding Ostracoda), in *The Fossil Record 2* (ed. M. J. Benton), Chapman & Hall, London, Chapter 18.

Brusca, R.C. and Brusca, G.J. (1990) *Invertebrates*, Sinauer, Massachusetts.

Cisne, J.L. (1974) Trilobites and the origin of arthropods. *Science*, **186**, 13–18.

Claus, C. (1888) Ueber den Oganismus der Nebaliden und die systematische Stellung der Leptostraken. *Arbeiten aus dem zoologischen Institute der Universität Wien und der zoologischen Station in Triest*, **8**, 1–148.

Dahl, E. (1952) Reports of the Lund University Chile Expedition 1948–1949. 7. Mystacocarida. *Lunds Universitets Årsskrift. N. F. Avd.* 2, **48**(6), 1–40.

Dahl, E. (1956) Some crustacean relationships, in *Bertil Hanström, Zoological Papers in Honour of his Sixty-fifth Birthday* (ed. K. G. Wingstrand), Zoological Institute, Lund, Sweden, pp. 138–47.

Dahl, E. (1983) Malacostracan phylogeny and evolution, in *Crustacean Issues 1*: *Crustacean Phylogeny* (ed. F. R. Schram), A.A. Balkema, Rotterdam, pp. 189–212.

Dahl, E. (1987) Malacostraca maltreated – the case of the Phyllocarida. *Journal of Crustacean Biology*, **7**(4), 721–6.

Dahl, E. (1992) Aspects of malacostracan evolution. *Acta Zoologica*, **73**(5), 339–46.

De Rijk, P., Neefs, J.-M., Van de Peer, Y. and De Wachter, R. (1992) Compilation of small ribosomal subunit RNA sequences. *Nucleic Acids Research*, **20** (supplement), 2075–89.

Dohle, W. and Scholtz, G. (1988) Clonal analysis of the crustacean segment: discordance between genealogical and segmental borders. *Development* (supplement), **104**, 147–60.

Eernisse, D.J., Albert, J.S. and Anderson, F.E. (1992) Annelida and Arthropoda are not sister taxa: a phylogenetic analysis of spiralian metazoan morphology. *Systematic Biology*, **41**(3), 305–30.

Emerson, M.J. and Schram, F.R. (1990) The origin of crustacean biramous appendages and the evolution of the Arthropoda. *Science*, **250**, 667–9.

Farris, J.S. (1989) The retention index and the rescaled consistency index. *Cladistics*, **5**, 417–19.

Felsenstein, J. (1978) Cases in which parsimony or compatibility methods will be positively misleading. *Systematic Zoology*, **40**, 366–75.

Felsenstein, J. (1985a) Confidence limits on phylogenies: an approach using the bootstrap. *Evolution*, **39**, 783–91.

Felsenstein, J. (1985b) Confidence limits on phylogenies with a molecular clock. *Systematic Zoology*, **34**(2), 152–61.

Felsenstein, J. (1993) PHYLIP: Phylogeny inference package, version 3.5c. Department of Genetics, University of Washington, Seattle.

Field, K.G., Olsen, G.J., Lane, D.J., Giovanni, S.J., Ghiselin, M.T., Raff, E.C., Pace, N.R. and Raff, R.A. (1988) Molecular phylogeny of the animal kingdom. *Science*, **239**, 748–53.

Friedrich, M. and Tautz, D. (1995) Ribosomal DNA phylogeny of the major extant arthropod classes and the evolution of myriapods. *Nature*, **376**, 165–7.

Fryer, G. (1992) The origin of the Crustacea. *Acta Zoologica*, **73**, 273–86.

Fryer, G. (1996) Reflections on arthropod evolution. *Biological Journal of the Linnean Society*, **58**, 1–55.

Ghiselin, M.T. (1988) The origin of molluscs in light of molecular evidence. *Oxford Surveys in Evolutionary Biology*, **5**, 66–95.

Grygier, M.J. (1982) Sperm morphology in Ascothoracida (Crustacea: Maxillopoda): confirmation of generalized nature and phylogenetic importance. *International Journal of Invertebrate Reproduction*, **4**, 323–32.

Grygier, M.J. (1983) Ascothoracida and the unity of the Maxillopoda, in *Crustacean Issues 1*: *Crustacean Phylogeny* (ed. F. R. Schram), A.A. Balkema, Rotterdam, pp. 73–104.

Grygier, M. J. (1987) New records, external and internal anatomy, and systematic position of Hansen's Y-larvae (Crustacea: Maxillopoda: Facetotecta). *Sarsia*, **72**, 261–78.

Hessler, R.R. (1964) The Cephalocarida: comparative skeletomusculature. *Memoirs of the Connecticut Academy of Arts and Sciences*, **16**, 1–97.

Hessler, R.R. (1983) A defense of the caridoid facies; wherein the early evolution of the Eumalacostraca is discussed, in *Crustacean Issues 1*: *Crustacean Phylogeny* (ed. F. R. Schram), A.A. Balkema, Rotterdam, pp. 145–64.

Hessler, R.R. (1992) Reflections on the phylogenetic position of the Cephalocarida. *Acta Zoologica*, **73**(5), 315–16.

Hessler, R.R. and Newman, W.A. (1975) A trilobitomorph origin for the Crustacea. *Fossils and Strata*, **4**, 437–59.

Hillis, D.M. and Huelsenbeck, J.P. (1992) Signal, noise, and reliability in molecular phylogenetic analyses. *Journal of Heredity*, **83**, 189–95.

Huelsenbeck, J.P., Bull, J.J. and Cunningham, C.W. (1996) Combining data in phylogenetic analysis. *Trends in Ecology and Evolution*, **11**(4), 152–8.

Itô, T. (1989) Origin of the basis in copepod limbs, with reference to remipedian and cephalocarid limbs. *Journal of Crustacean Biology*, **9**, 85–103.

Ivanov, P.P. (1944) The primary and secondary metamery of the body. *Zhurnal Obshchey Biologii*, **5**, 61–95.

Jamieson, B.J.M. (1991) Ultrastructure and phylogeny of crustacean spermatozoa. *Memoirs of the Queensland Museum*, **31**, 109–42.

Kluge, A.G. and Farris, J.S. (1969) Quantitative phyletics and the evolution of anurans. *Systematic Zoology*, **18**, 1–32.

Lake, J.A. (1990) Origin of the Metazoa. *Proceedings of the National Academy of Sciences, USA*, **87**, 763–6.

Manton, S.M. (1934) On the embryology of the crustacean, *Nebalia bipes*. *Philosophical Transactions of the Royal Society of London, B*, **223**, 163–238.

Manton, S.M. (1977) *The Arthropoda*: *Habits, Functional Morphology, and Evolution*, Clarendon Press, Oxford.

Martin, J.W., Vetter, E.W. and Cash-Clark, C.E. (1996) Description, external morphology, and natural history observations of *Nebalia hessleri*, new species (Phyllocarida: Leptostraca), from southern California, with a key to the extant families and genera of the Leptostraca. *Journal of Crustacean Biology*, **16**(2), 347–72.

Milne Edwards, H. (1834) Histoire naturelle des Crustacés comprenant l'anatomie, la physiologie et la classification de ces animaux, in *Librairie Encyclopédique de Roret*, Tome Premier, Paris.

Newman, W.A. (1992) Origin of Maxillopoda. *Acta Zoologica*, **73**, 319–22.

Panganiban, G., Sebring, A., Nagy, L. and Carroll, S. (1995) The development of crustacean limbs and the evolution of arthropods. *Science*, **270**, 1363–6.

Patel, N.H. (1994) The evolution of arthropod pattern formation: insights from comparisons of gene expression patterns, in *The Evolution of Developmental Mechanisms* (eds M. Akam, P. Holland, P. Ingham and G. Wray), Company of Biologists, Cambridge, England, pp. 217–23.

Patel, N.H., Martin-Blanco, E., Coleman, K.G., Poole, S.J., Ellis, M.C., Kornberg, T.B. and Goodman, C.S. (1989) Expression of *engrailed* proteins in arthropods, annelids, and chordates. *Cell*, **58**, 955–68.

Patterson, C. (1989) Phylogenetic relations of major groups: conclusions and prospects, in *The Hierarchy of Life* (eds B. Fernholm, K. Bremer and H. Jörnvall), Elsevier, Amsterdam, pp. 471–88

Paulus, H.F. (1979) Eye structure and the monophyly of the Arthropoda, in *Arthropod Phylogeny* (ed A. P. Gupta), Van Nostrand Reinhold, New York, pp. 299–383.

Pennak, R.W. and Zinn, D.J. (1943) Mystacocarida, a new order of Crustacea from intertidal beaches in Massachusetts and Connecticut. *Smithsonian Miscellaneous Collections*, **103**(9), 1–11.

Raff, R.A., Marshall, C.R. and Turbeville, J.M. (1994) Using DNA sequences to unravel the Cambrian radiation of the animal phyla. *Annual Review of Ecology and Systematics*, **25**, 351–75.

Riley, J., Banaja, A.A. and James, J.L. (1978) The phylogenetic relationships of the Pentastomida: the case for their inclusion within the Crustacea. *International Journal for Parasitology*, **8**, 245–54.

Sanders, H.L. (1957) The Cephalocarida and crustacean phylogeny. *Systematic Zoology*, **6**(3), 112–28.

Sanders, H.L. (1963) The Cephalocarida: functional morphology, larval development, and comparative external anatomy. *Memoirs of the Connecticut Academy of Arts and Sciences*, **15**, 1–80.

Scholtz, G., Patel, N.H. and Dohle, W. (1994) Serially homologous *engrailed* stripes are generated via different cell lineages in the germ band of amphipod crustaceans (Malacostraca, Peracarida). *International Journal of Developmental Biology*, **38**, 471–8.

Schram, F.R. (1986) *Crustacea*, Oxford University Press, New York.

Schram, F.R. and Emerson, M.J. (1991) Arthropod pattern theory: a new approach to arthropod phylogeny. *Memoirs of the Queensland Museum*, **31**, 1–18.

Schram, F.R., Yager, J. and Emerson, M.J. (1986) Remipedia Part 1. Systematics. *San Diego Society of Natural History Memoir*, **15**, 1–60.

Siewing, R. (1960) Neuere Ergebnisse der Verwandschaftsforschung bei den Crustaceen. *Wissenschaftliche Zeitschrift der Universitat Rostock, Mathematische-Naturwissenschaftliche Reihe*, **3**, 343–58.

Spears, T., Abele, L.G. and Applegate, M.A. (1994) Phylogenetic study of cirripedes and selected relatives (Thecostraca) based on 18S rDNA sequence analysis. *Journal of Crustacean Biology*, **14**(4), 641–56.

Starobogatov, Ya.I. (1986) Systematics of Crustacea. *Zoologicheskiy Zhurnal*, **65**, 1769–81.

Starobogatov, Ya.I. (1988) Systematics of Crustacea. *Journal of Crustacean Biology*, **8**, 300–11. [English translation of Starobogatov (1986) by M. J. Grygier, including additional notes by Starobogatov.]

Storch, V. and Jamieson, B.G.M. (1992) Further spermatological evidence for including the Pentastomida (tongue worms) in the Crustacea. *International Journal for Parasitology*, **22**(1), 95–108.

Swofford, D.L. (1993) PAUP: Phylogenetic analysis using parsimony, version 3.1.1. Illinois Natural History Survey, Champaign, Illinois, USA.

Templeton, A. (1983) Phylogenetic inference from restriction endonuclease cleavage site maps with particular reference to the evolution of humans and apes. *Evolution*, **37**, 221–44.

Thompson, J.D., Higgins, D.G. and Gibson, T.J. (1994) CLUSTAL W: improving the sensitivity of progressive multiple sequence alignment through sequence weighting, position specific gap penalties and weight matrix choice. *Nucleic Acids Research*, **22**(22), 4673–80.

Tiegs, O.W. and Manton, S.M. (1958) The evolution of the Arthropoda. *Biological Review*, **33**, 255–337.

Turbeville, J.M., Pfeifer, D.M., Field, K.G. and Raff, R.A. (1991) The phylogenetic status of arthropods, as inferred from 18S rRNA sequences. *Molecular Biology and Evolution*, **8**, 669–86.

Wägele, J.W. (1993) Rejection of the 'Uniramia' hypothesis and implication of the Mandibulata concept. *Zoologische Jahrbucher (Systematik)*, **120**, 253–88.

Waggoner, B.M. (1996) Phylogenetic hypotheses of the relationships of arthropods to Precambrian and Cambrian problematic fossil taxa. *Systematic Biology*, **45**(2), 190–222.

Walossek, D. (1993) The Upper Cambrian *Rehbachiella* and the phylogeny of the Branchiopoda and Crustacea. *Fossils and Strata*, **32**, 1–202.

Walossek, D. and Müller, K.J. (1994) Pentastomid parasites from the Lower Palaeozoic of Sweden. *Transactions of the Royal Society of Edinburgh: Earth Sciences*, **85**, 1–37.

Weygoldt, P. (1979) Significance of later embryonic stages and head development in arthropod phylogeny, in *Arthropod Phylogeny* (ed A.P. Gupta), Van Nostrand Reinhold, New York, pp. 107–35.

Weygoldt, P. (1986) Arthropod interrelationships – the phylogenetic–systematic approach. *Zeitschrift fur Zoologische Systematik und Evolutionsforschung*, **24**, 19–35.

Whatley, R.C., Siveter, D.J. and Boomer, I.D. (1993) Arthropoda (Crustacea: Ostracoda), in *The Fossil Record 2* (ed. M. J. Benton), Chapman & Hall, London, Chapter 19.

Wheeler, W.C. (1995) Sequence alignment, parameter sensitivity, and the phylogenetic analysis of molecular data. *Systematic Biology*, **44**(3), 321–31.

Wheeler, W.C., Cartwright, P. and Hayashi, C.Y. (1993) Arthropod phylogeny: a combined approach. *Cladistics*, **9**, 1–39.

Wills, M.A., Briggs, D.E. and Fortey, R.A. (1994) Disparity as an evolutionary index: a comparison of Cambrian and Recent arthropods. *Paleobiology*, **20**(2), 93–130.

Wilson, G.D.F. (1992) Computerized analysis of crustacean relationships. *Acta Zoologica*, **73**(5), 383–9.

Wingstrand, K.G. (1972) Comparative spermatology of a pentastomid, *Raillietiella hemidactyli*, and a branchiuran crustacean,

*Argulus foliaceous*, with a discussion of pentastomid relationships. *Kongelige Danske Videnskabernes Selskab Biologiske Skrifter*, **19**, 1–72.

Yager, J. (1981) Remipedia, a new class of Crustacea from a marine cave in the Bahamas. *Journal of Crustacean Biology*, **1**, 328–33.

Zrzavý, J. and Štys, P. (1994) Origin of the crustacean schizoramous limb: a re–analysis of the duplosegmentation hypothesis. *Journal of Evolutionary Biology*, **7**, 743–56.

Zrzavý, J. and Štys, P. (1995) Evolution of metamerism in Arthropoda: developmental and morphological perspectives. *The Quarterly Review of Biology*, **70**(3), 279–95.

# 15 A phylogeny of recent and fossil Crustacea derived from morphological characters

*M.A. Wills*

*Department of Geology, Wills Memorial Building, University of Bristol, Queen's Road, Bristol BS8 1RJ, UK*
*email: m.a.wills@bristol.ac.uk*

## 15.1 INTRODUCTION

Crustaceans exhibit enormous diversity and morphological plasticity, having evolved many distinctive morphotypes with many heteronomous character combinations since their origination in the Cambrian (Schram, 1982; Briggs, 1992). The high level of parallelism, convergence and reversals amongst the crustaceans reflects that among the arthropods in general (Schram, 1978). Their morphological flexibility is largely due to the segmented nature of the body and serially homologous appendages, giving enormous opportunity for evolutionary specialization. Nevertheless, there are certain constraints within the Crustacea on what can be a viable, functioning arrangement of parts, and there is often a gross similarity of form and function in similar habitat types (e.g. mystacocarids, bathynellaceans and ingolfiellan amphipods in interstitial habitats).

The extant Crustacea are not clearly defined or diagnosed, although Schram (1986) identified several features that are **typical** of most taxa. The cephalon of the adult is composed of five post-acronal segments. The first two (usually preoral) segments bear first and second antennae (often sensory), while the postoral segments bear gnathobasic mandibles, followed by first and typically second maxillae. There is a tendency for the cephalic segments to fuse, forming a cephalic shield. There is also a tendency for somitic tagmatization, and a concomitant functional specialization of the appendages. Ontogeny always includes either a nauplius larva or an egg–nauplius stage. Uniformity of the above characters, along with similarities of skeletomusculature, limb structure and feeding mechanisms among several of the supposedly more primitive groups (Sanders, 1963b; Hessler, 1964), has widely been used to infer that the crustaceans constitute a clade. Recent evidence suggests that some characteristics long regarded as conservative homologues may actually be genetically derived through a number of pathways (Ferrari, 1988). Nonetheless, the group has consistently emerged from recent treatments of the morphological data (Wheeler *et al.*, 1993; Wills *et al.*, 1994, 1995), although molecular work (Adoutte and Philippe, 1993; Averof and Akam, 1995) raises the possibility of a paraphyletic Crustacea, with the Hexapoda derived from their midst (see also Carpenter, 1906; Sharov, 1966).

## 15.2 PRINCIPAL ISSUES IN CRUSTACEAN PHYLOGENY

### 15.2.1 GROSS TOPOLOGY

Schram (1986) produced the first parsimony-based cladistic analysis of crustacean subclasses and orders (summarized as Figure 15.1(b)). This resolved the non-phyllocarid Malacostraca as sister group to a Maxillopoda/Phyllopoda dichotomy, with the remipedes basal to all other crustaceans. This placement for the remipedes has received considerable support (Brusca and Brusca, 1990; Briggs *et al.*, 1992; Wills *et al.*, 1994, 1995, 1997), but Schram's grouping of the thoracic-feeding brachypodans, branchiopods and phyllocarid malacostracans as a derived lineage (Phyllopoda) has been questioned (Brusca and Brusca, 1990).

Brusca and Brusca (1990) (Figure 15.1(a)) placed the Malacostraca (*sans* phyllocarids) as sister group to the Maxillopoda. This lineage opposed the Branchiopoda, followed by the Cephalocarida and Remipedia in pectinate succession. This arrangement was based on the assumption that phyllopodous feeding limbs (e.g. those found in cephalocarids, many branchiopods, leptostracans) represent the primitive crustacean type. Given the small size of the data set, however, Brusca and Brusca saw no reason to prefer their cladogram over that of Schram (1986). The distribution of character states across taxa is such that any conceivable cladogram will involve much homoplasy.

### 15.2.2 INTEGRITY OF THE MAXILLOPODA

The Mystacocarida, Copepoda, Branchiura and Cirripedia were grouped into the subclass 'Copepodoidea' by

*Arthropod Relationships*, Systematics Association Special Volume Series 55, Edited by R.A. Fortey and R.H. Thomas. Published in 1997 by Chapman & Hall, London. ISBN 0 412 75420 7

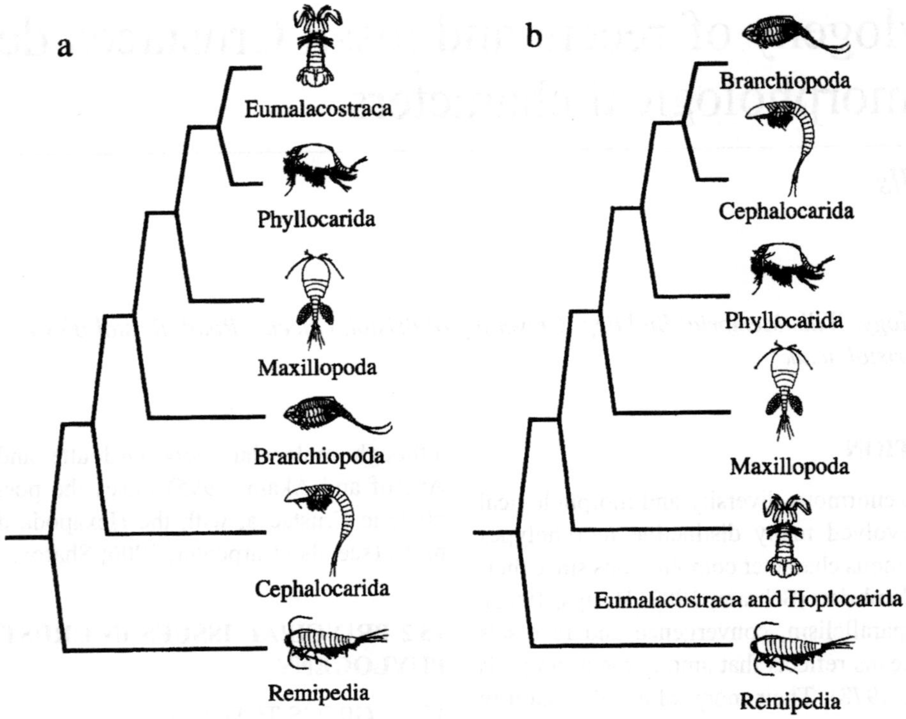


**Figure 15.1** (a). 'Traditional' model of relationships between major crustacean groups (Brusca and Brusca 1990). (b) Alternative phylogeny proposed by Schram (1986).

Beklemischev (1952, 1969), and more recently the 'Maxillopoda' by Dahl (1956, 1963) (see also Newman *et al.*, 1969; Newman, 1983). The Ostracoda are usually considered as members of this group (Siewing, 1960; Schram, 1986), as are the more recently-discovered Tantulocarida (Grygier, 1987). However construed, the subclass contains forms with a rather varied range of morphologies.

Much recent work supports a monophyletic Maxillopoda (Bowman and Abele, 1982; Grygier, 1983; Müller and Walossek, 1986, 1988; Boxshall and Huys, 1989). Boxshall and Huys (1989) recognized a core group for the taxon, centred around the shield-bearing Thecostraca (including the Ascothoracida, Cirripedia, Facetotecta and Tantulocarida). The Tantulocarida were thought to be the sister group to Recent members of the Thecostraca (*sensu* Grygier, 1985) and the Branchiura (Boxshall and Lincoln, 1987). A number of synapomorphies supposedly link the core, including the possession of seven thoracomeres (as in *Bredocaris*), and the presence (in the male at least) of gonopores on the seventh trunk segment. Grygier (1987) cited an eight-segmented antenna as a synapomorphy for the Maxillopoda, while he characterized the Thecostraca on the basis of their possession of attachment devices on the first antenna, and their lack of naupliar limb buds. However, relationships within the Maxillopoda, even in the core group, are far from satisfactorily resolved.

Analysis of 18S rRNA data led Abele *et al.* (1992) to conclude that the Maxillopoda were not monophyletic, but rather divided into two lineages: one of ostracods, branchiurans and pentastomes, the other of copepods, acrothoracicans and thoracicans. Wilson (1992) also failed to resolve the maxillopods as a clade.

### 15.2.3 STATUS OF THE PHYLLOCARIDA AND 'PHYLLOPODA'

The phyllocarids have traditionally been classified with the Eumalacostraca (Claus, 1888; Metschnikov, 1968), although Schram (1986) preferred to ally them with the Cephalocarida and Branchiopoda in the class Phyllopoda. There are few characters common to all phyllocarids, particularly if one includes the canadaspidids (Briggs, 1978, 1992). The function and morphology of the thoracopods, and the orientation of the mouth are close to those of the cephalocarids and branchiopods (Sars, 1887), possibly reflecting a primitive crustacean condition (Hessler and Newman, 1975). Other primitive features are evident in their embryology (Manton, 1934) and in the long, tubular nature of the heart. The loss of ambulatory function for the thoracic limbs, which now serve only for respiration, feeding and brood protection (Dahl, 1976), may be apomorphic.

### 15.2.4 RELATIONSHIPS WITHIN THE EUMALACOSTRACA

Calman's (1904) original classification of the Eumalacostraca assumed the caridoid facies to be the underlying groundplan for all groups (Calman, 1904, 1909). This comprised a cephalon of five fused somites, and a thorax of eight, all covered by a cephalothoracic carapace (see also Hessler, 1982, 1983). Working from the premise that a similar array of characters exemplified the primitive eumalacostracan condition, Calman (following Hansen, 1893), divided the then-known eumalacostracan orders into two higher taxa: one containing the Decapoda and Euphausiacea (Eucarida), the other comprising the Mysidacea, Cumacea, Tanaidacea, Isopoda and Amphipoda (Peracarida). Calman (1904) later erected the Syncarida to accommodate the carapace-less *Anaspides* (Thomson, 1894) (assumed to have derived from carapace-bearing ancestors) and a number of similar fossil forms.

The facies concept, particularly the assumption that the hypothetical ur-eumalacostracan possessed a well-developed cephalothoracic carapace, has needlessly constrained thinking about interrelationships within the Eumalacostraca, and their connection with the rest of the crustacean tree (see Figure 15.2) (Tiegs and Manton, 1958; Dahl, 1963, 1976; Schram, 1984). Moreover, it has recently been recognized that a carapace *sensu* Calman (1909) (deriving from the posterior margin of the cephalon) does not exist in any malacostracan or phyllopod (Dahl, 1991).

A limited number of authors have made substantial contributions since the work of Calman (1904). Schram (1981), for example, based a cladogram largely on the development and fusion of the carapace. This scheme represented a significant departure from the traditional model in two important (though not entirely novel) respects.

1. The Isopoda and Amphipoda (Acaridea) were grouped, and these taxa united with the Syncarida on the basis of the absence of the carapace (Arthrostraca) (Haeckel, 1896; Giesbrecht, 1913; Grobben, 1919).
2. The Mysidacea, Belotelsonidea and Waterstonellidea formed a clade (Mysoida), elevated to a position opposing the Eucarida.

A later cladistic analysis derived from a larger database of some 31 characters (Schram, 1984) was consistent with the above in the preservation of the Hemicaridea (Cumacea, Spelaeogriphacea and Tanaidacea), Brachycarida (Hemicaridea and Thermosbaenacea), Eucarida and Acaridea. However, neither the Arthrostraca nor the Mysoida could be reconciled as clades. The Mysidacea (Mysida, Lophogastrida and Pygocephalomorpha) were paraphyletic, regardless of whether the possession of a carapace was considered a primitive or a derived character. Schram (1986) (Figure 15.3(a)) crystallized these ideas, recognizing four major lines of eumalacostracan evolution: eucarids, syncarids, belotelsonids and waterstonellids–peracarids–pancarids. Dahl (1992) criticized the hypothesis that the Syncarida may have arisen from a common eumalacostracan stock mediated by the loss of the carapace, while the belotelsonid line was considered just one of many plesions. Dahl's (1992) greatest reservations concerned the validity of the putative waterstonellid–peracarid–pancarid lineage.

Watling's (1983) analysis of all Eumalacostraca (minus the Decapoda and Amphionidacea) focused on the morphology of the carapace, compound eyes, mandible, maxillipede, blood vascular system, and developmental patterns. This concluded the existence of three major eumalacostracan evolutionary lines. One comprised the Syncarida, Eucarida and Mysidacea. These last had previously been regarded as peracarids, and their grouping with the eucarids was a radical reinterpretation. Another lineage led from the Isopoda to the Cumacea, Spelaeogriphacea and Tanaidacea (Hemicaridea of Schram, 1986) plus the Thermosbaenacea. The third lineage contained solely the Amphipoda.

The stomatopods have long been recognized as a distinct group of malacostracans. Calman (1904) placed them in a separate superorder (Hoplocarida), assumed to have been an early branch within the Eumalacostraca. With the recognition of the related Palaeozoic orders Palaeostomatopoda and Aeschronectida (Schram, 1969a,b) came the hypothesis that the hoplocarids constituted a sister group to the Eumalacostraca (Figure 15.3(a)).

### 15.2.5 THE SIGNIFICANCE OF PROBLEMATIC FOSSIL TAXA

#### (a) Canadaspidida

*Canadaspis* contains an enigmatic mix of supposedly primitive and derived characters (Briggs, 1978), although its morphological similarities to the Malacostraca were soon noted. All authors initially agreed on an assignment to the Phyllocarida (Novozhilov, 1960), and parallels were frequently drawn with the living nebaliids (Walcott, 1912; Raymond, 1920; Henriksen, 1928). (For alternative perspectives on the affinities of canadaspidids, see Raymond, 1935; Størmer, 1944, 1959; Rolfe, 1969; Briggs, 1978; Dahl, 1983, 1984; Bergström, 1992.)

#### (b) Odaraia

Walcott (1912) considered this genus to be an unequivocal crustacean, assigning it to the order Hymenocarina. Henriksen (1928) supported this classification, but cautioned that a consideration of characters other than solely those of the carapace were necessary for such an assignment. Størmer (1944) removed the Hymenocarina to a new class (the Pseudocrustacea) within the Trilobitomorpha. Unease with this assignment eventually resulted in Størmer's (1959) removal of *Odaraia* from the

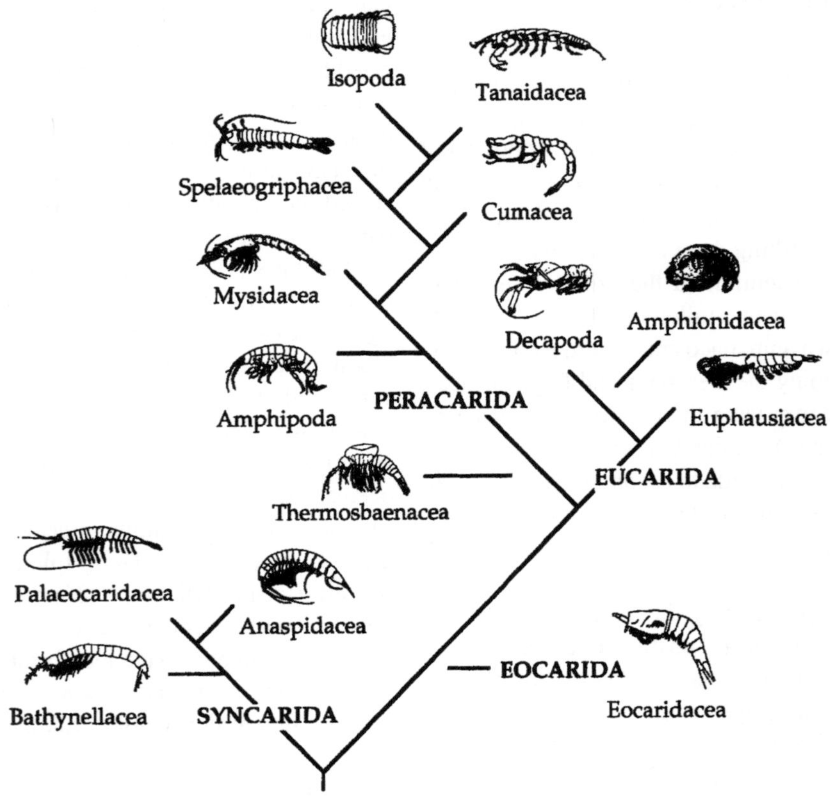


**Figure 15.2** 'Traditional' model of eumalacostracan evolution. (After Schram, 1981.)

Trilobitomorpha in compiling the *Treatise on Invertebrate Paleontology*, along with several other bivalved forms. Rolfe (in Moore, 1959) was to reinstate *Odaraia* within the Crustacea, treating it as a phyllocarid.

The debate regarding the affinities of *Odaraia* has more recently centred on the morphology of the cephalon. This is not particularly well preserved, and the presence or absence of appendages therefore provides most of the information. Briggs (1981) regarded *Odaraia* to be most comparable with the modern Branchiopoda, though he conceded that there was no compelling evidence of relationship. A second possibility is that *Odaraia* shows greatest affinity with the Phyllocarida (Wills *et al.*, 1994). None of the differences in details of appendage morphology is sufficient to exclude *Odaraia* from Briggs's (1978) emended diagnosis of the Phyllocarida, but neither are the similarities compelling. The primary objection to such an assignment remains the large number of trunk somites and the lack of tagmosis.

### (c) Lepidocaris

Most earlier authors allied lipostracans with the anostracan branchiopods (Scourfield, 1926; Sanders, 1963b), principally based on similarities of the cephalic appendages (see also Schram, 1986). Walossek (1993) and Walossek and Müller (1997, this volume) united the Recent Anostraca (Euanostraca) and *Lepidocaris* in the Sarsostraca, while the Sarsostraca were grouped with the Orsten *Rehbachiella* in an emended Anostraca. Schram (1986), by contrast, concluded that *Lepidocaris* shows greater affinity with the Brachypoda than the Branchiopoda, although he considered that the number of features in common with the branchiopods was sufficient to suggest some relatively close relationship between all three taxa.

### (d) Orsten genera

*Bredocaris*

Many features of *Bredocaris* appear similar to character-states occurring in the Maxillopoda, the group with which Müller and Walossek (1988) allied it. The structure of the fourth appendage or maxilla of *Bredocaris*, with its rudimentary exopod, shares marked similarities with the postmandibular limbs of the Mystacocarida, even in details of the arrangement of setae. The heavily enditic antennal protopods of *Bredocaris* may represent a primordial state of development (Costlow and Bookhout, 1957, 1958; Walley, 1969; Stubbings, 1975). In all other maxillopodan orders, including the Mystacocarida, the antennae have either lost or greatly reduced their proximal enditic armature (Boxshall and Lincoln, 1983, 1987; Grygier, 1985). Müller and Walossek (1988) concluded that *Bredocaris* shows greatest

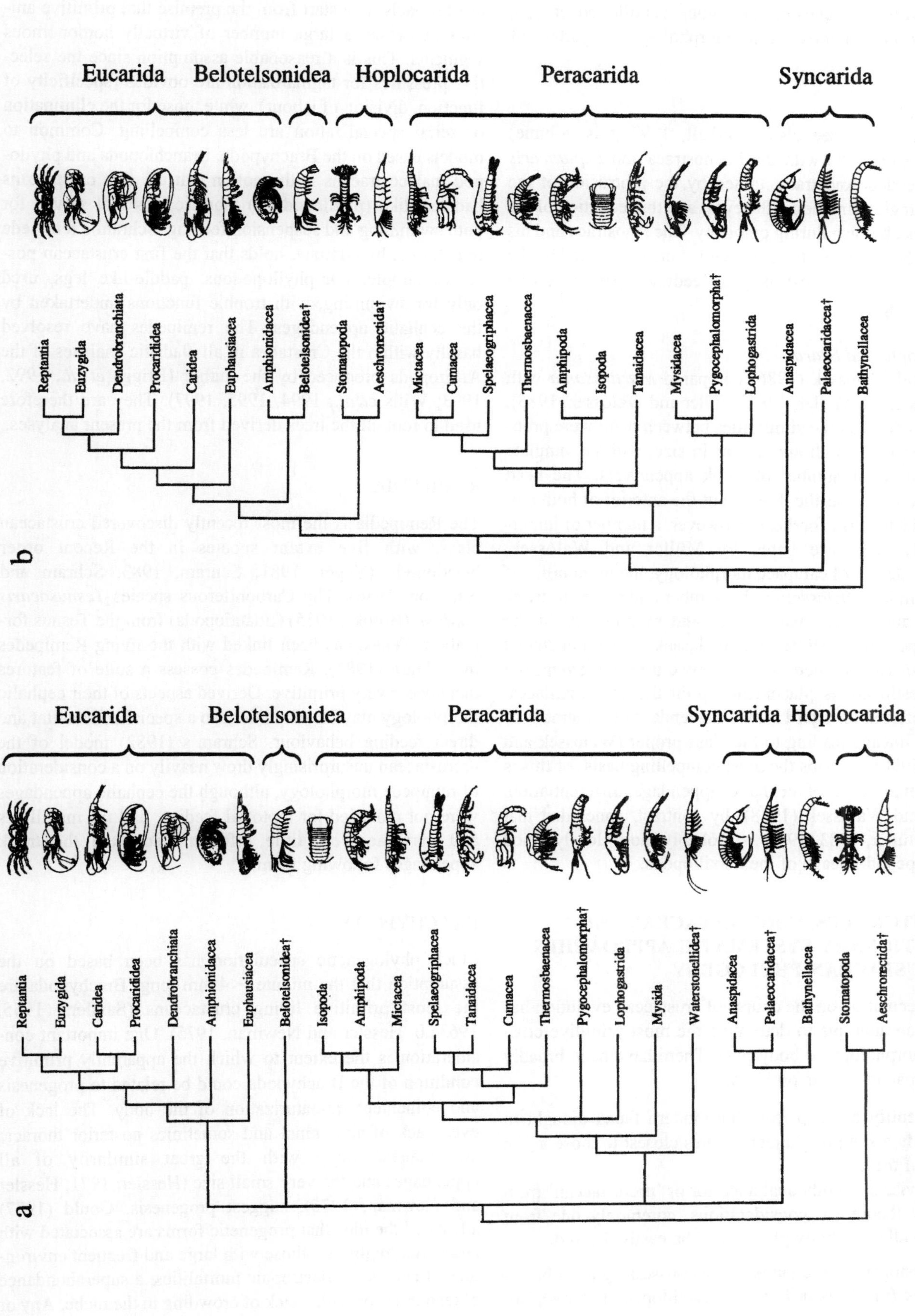


**Figure 15.3** Hypotheses of relationships between major eumalacostracan and hoplocarid taxa. (a) Model proposed by Schram (1986). (b) Model proposed herein from preferred ordering and weighting of the data (see also Figure 15.4).

affinity with the thecostracans among maxillopod groups, shown by their tagmosis, limb morphology and pattern of ontogeny.

*Rehbachiella*

Walossek (1993; see also Boxshall, 1997, this volume) joined *Rehbachiella* with the Euanostraca and *Lepidocaris* in an emended Anostraca, united by the protrusion of the forehead region (including the eyes) and the reduction of the naupliar neck organ during ontogeny. The Anostraca and his Phyllopoda (Calmanostraca and Onychura) are united by the specialization of the post-naupliar feeding apparatus to true filter-feeding.

*Martinssonia and Skara*

Müller and Walossek (1986) compared *Martinssonia* with the Skaracarida (Müller, 1983; Müller and Walossek, 1985), but concluded that the similarities between them were probably superficial. Both are similar in size, with an annulate trunk and small number of trunk appendages. The head shields do not cover the thorax, but the anterior of both animals projects into a forehead. However, a number of important differences are listed by Müller and Walossek, including details of carapace morphology, the immobility of the labrum in *Martinssonia*, the number and nature of trunk somites, and the precise number and morphology of the trunk appendages. Müller and Walossek (1986) concluded that *Martinssonia* failed to fit into any crustacean group, and even questioned its placement within the true Crustacea. They preferred to regard it as a descendant of a member of the stem lineage leading to the class proper (Walossek and Müller, 1990). Perhaps the most compelling basis for this is the minimal level of cephalic appendage differentiation. Müller and Walossek (1988), by contrast, concluded that *Skara* (Müller, 1983, 1985) is probably most closely allied to the copepod lineage of the maxillopods.

## 15.3 OUTGROUPS, 'URCRUSTACEA', AND EVOLUTIONARY SYSTEMATIC APPROACHES TO CRUSTACEAN PHYLOGENY

Much speculation on the course of crustacean evolution has focused on attempts to determine the most primitive crustacean morphology or body plan. There have been broadly two approaches to this problem:

1. To identify a group from the modern fauna thought to embody a suite of character states closest to those at the base of the tree.
2. To devise a hypothetical ancestor or 'urcrustacean' from purely theoretical considerations, commonly one from which all other body plans could be easily derived.

The ancestor was once envisaged as possessing an abbreviated trunk (much as in larval or maxillopodan forms), but most models now start from the premise that primitive animals possessed a large number of virtually homonomous segments. This is a reasonable assumption since the selective pressures for tagmatization are obvious (specificity of function, division of labour), while those for the elimination of serial specialization are less compelling. Common to models based on the Brachypoda, Branchiopoda and phyllopod malacostracans is the notion that the first crustaceans had leaf-like (phyllopodous) thoracic legs that served for both swimming and suspension feeding. Schram's remipede hypothesis, by contrast, holds that the first crustacean possessed simple, non-phyllopodous, paddle-like legs, used only for swimming, with trophic functions undertaken by the cephalic appendages. The remipedes have resolved basally within the Crustacea in all cladistic analyses of the Arthropoda produced by the author (Briggs *et al.*, 1992, 1993; Wills *et al.*, 1994, 1995, 1997). They are therefore used to root all the trees derived from the present analyses.

### REMIPEDIA

The Remipedia is the most recently discovered crustacean class, with five extant species in the Recent order Nectiopoda (Yager, 1981; Schram, 1983; Schram and Emerson, 1986). The Carboniferous species *Tesnusocaris goldichi* (Brooks, 1955) (Enantiopoda) from the Tesnus formation, Texas, has been linked with the living Remipedes by Schram (1983). Remipedes possess a suite of features that appear very primitive. Derived aspects of their cephalic morphology may be adaptations to a specialized habitat and direct feeding behaviour. Schram's (1982) model of the urcrustacean unsurprisingly drew heavily on a consideration of remipede morphology, although the cephalic appendages were not modified for raptorial feeding, and the maxillules and maxillae were little differentiated from the trunk appendages following them.

### BRACHYPODA

Much phylogenetic speculation has been based on the assumption that the minute (<4 mm long) Brachypoda are the most primitive living crustaceans (Sanders, 1955, 1963a,b; Hessler and Newman, 1975). One important consideration is the extent to which the apparently primitive condition of the Brachypoda could be related to progenesis and coincident miniaturization of the body. The lack of eyes, lack of abdominal and sometimes posterior thoracic appendages, along with the great similarity of all appendages and the very small size (Hessler, 1971; Hessler and Newman, 1975), suggest progenesis. Gould (1977) advanced the idea that progenetic forms are associated with r-selection regimes – those with large and frequent environmental changes, catastrophic mortalities, a superabundance of resources, or with a lack of crowding in the niche. Any or several of these could apply to the flocculant zones inhabited by the Brachypoda. However, fecundity is very low and reproduction highly seasonal, features atypical of r-selection strategists. This fact, together with the complete development of the trunk, led Cisne (1982) to conclude that the Brachypoda probably preserve the essential ontogenetic features of the ancestral crustacean. Schram (1982) by contrast, considered the Brachypoda too derived to be placed basally in arthropod phylogeny. No fossil record is known for the group, although Schram (1986) considered the Devonian *Lepidocaris* (Lipostraca) to constitute a sister taxon to the modern brachypods, together constituting the Cephalocarida.

The morphology of Hessler and Newman's (1975) urcrustacean was heavily influenced by their belief that the brachypods represent the most primitive living group [something that the discovery of the remipedes (Yager, 1981) was not to change (Hessler, 1992)]. Newman reconstructed the animal with a carapace (unlike most other models), while Hessler thought it likely that the ancestor possessed only a minimally-developed head shield (Hessler and Newman, 1975).

BRANCHIOPODA

Numerous authors have proposed a basal status for the branchiopods because of their large but variable number of trunk segments and long row of serially similar, 'primitive' limbs (Claus 1876; Grobben 1893; Calman, 1909; Zimmer, 1926–1927; Tiegs and Manton, 1958; Fryer, 1992). Fryer's (1992) urcrustacean model drew heavily on a comparison with the morphology of anostracans. Much of Fryer's discussion was underpinned by the assumption that while the Crustacea are monophyletic, the arthropods themselves are polyphyletic. He therefore derived the ancestral crustacean from a minimally developed, apodous, segmented worm-like animal, with a very weakly sclerotized cuticle.

## 15.3.1 METHODS

### (a) Selection of taxa

This study investigated the phylogeny of crustacean orders and suborders derived from a consideration of morphological characters. The classification scheme of Schram (1986) provided a framework for the selection of the Recent and most of the fossil taxa. The vast majority of Schram's suborders are recognized as groups by most other workers, even if not afforded the same taxonomic status. The utility of his classification here is in ensuring adequate coverage of the range of crustacean body plans. One exemplar (occasionally two) from each order or suborder was coded (see Appendix II). Where the relationships of taxa within these groupings was unambiguous, a primitive or basal representative was selected, and where no such consensus existed, genera with a 'representative' or less specialized morphology were used. Highly modified parasitic representatives were not coded.

An alternative approach would be to code the higher taxa themselves. However, in several cases, this would require the use of 'polymorphic' or 'uncertain' multistate codings to encompass the range of morphology exemplified by all the genera in an order. No methods currently exist to appropriately limit the manner in which such codes are interpreted so as to avoid nonsensical reconstructions (Nixon and Davis, 1991; Platnick *et al.*, 1991), and this approach is therefore problematic.

### (b) Characters

Establishing detailed homologies for a group as large and plastic as the Crustacea is a vexing problem (Schram, 1986). Most discussions of crustacean relationships draw heavily on considerations of tagmosis patterns, numbers of appendages, the numbers of podomeres in the rami of these appendages, and the morphological specialization of limbs (Hessler and Newman, 1975; Cisne, 1982; Hessler *et al.*, 1982; Schram, 1982; Itô, 1989; Wilson, 1992). Detailed comparisons of appendage morphology are also central in attempts to determine the affinities of enigmatic fossil taxa (Scourfield, 1926; Briggs, 1992; Walossek, 1993). Particular emphasis has been placed on the condition of the 'cephalic' (first five) limbs, deviations from the norm within this character set being sufficient for some authors to exclude taxa from the crustacean crown group (Müller and Walossek, 1986). The present database therefore codes the morphology of all five anterior appendages independently (cf. Wills *et al.*, 1994, 1995, 1997 for all arthropods). The first four post-maxillary appendages have also been coded; a number intended to capture most of the variation observed in anterior trunk appendage morphology throughout the crustaceans. While some taxa show significant differentiation of the anterior trunk appendages (e.g. reptantians) (in some cases beyond the fourth post-maxillary) others are remarkably conservative in this region (e.g. Canadaspidida). Hence, while post-maxillary appendages code differently in some groups, in others they represent what is essentially a serial homology. The resulting database is composed of 135 characters coded for 62 taxa (see http://www.systassoc.chapmanhall.com).

### (c) Parsimony models and weighting

There are several plausible ways to treat characters conveying information on numbers of somites and podomeres. Previous analyses of such data (Wilson, 1992) have banded numbers, recognizing ranges as individual character states. Such an approach requires some basis for the positioning of the bands [e.g. discontinuities in the distribution or APT zones for numbers of somites and tagma boundaries (Emerson and Schram, 1990, 1997, this volume; Schram and Emerson, 1991)]. An alternative approach, and the one adopted here, is to recognize separate states for all exemplified numbers. Both stances have very particular implications depending on the parsimony method used (Pimentel and Riggins, 1987; Hauser and Presch, 1991).

If characters are unordered (Fitch parsimony), then banding provides one method of grouping numbers which may be very similar and (in the strictest sense) phenetically rather proximate. Nonetheless, the locations of boundaries between states may still be critical. Not banding negates any numerical similarity between states, such that transitions between any number and any other are made with equal cost on the cladogram. This may be undesirable, and arguably results in a loss of useful phylogenetic information. Precise matches often do not occur within orders, and certainly not subclasses, but similar numbers typically do. Ordering provides a mechanism for treating such characters as nested sets of synapomorphies, allowing increases and decreases in podomere number to support branches. However, this practice ascribes as much weight to a transition between adjacent states in the ordered sequence as is given to a transition between the numerical extremes in an unordered analysis. Similarly, transitions between extremes in an ordered sequence with *n* states are ascribed *n*–1 times the weight of a binary character. Given the frequent variation in numbers of podomeres in specific rami within orders, such characters probably contain a weaker phylogenetic signal than most others. The practice of down-weighting some ordered characters so as to render transitions between extremes equivalent in weight to that between alternative states of binary characters (ranging) provides a viable compromise.

The present database was prepared by breaking down numerical sequences into a larger number of states in most cases. However, states were not recognized merely as numerical intermediates. Only states represented by one or more taxa were coded. This restriction is only pertinent where transition sequences are ordered.

As a consequence of the frequent homonomy of anterior trunk appendages in non-malacostracan taxa, flat-weighting of the data will effectively (by default) ascribe excessive weight to the morphology of the trunk appendages in the analysis as a whole. The number of appendages scored is also arbitrary in any case, and the down-weighting of characters relating to appendages 7, 8 and 9 is therefore no more *ad hoc* than their equal weighting.

Since a large proportion of the characters in the present databases (18%) relate to numbers of podomeres, and most of these have large numbers of states, the drawbacks discussed above for flat-weighted runs are appreciable.

#### (d) Alternative runs of the data

Two alternative ordering regimes were investigated in conjunction with the weighting schemes discussed below:

1. Those characters relating to numbers of podomeres, somites and other organs ordered.
2. Those characters relating to numbers of podomeres and other organs ordered (but those coding for somites and tagmosis patterns unordered).

In addition, a small number of runs were made with all characters unordered.

The effects of six alternative weighting regimes were investigated (clearly it is impractical to explore all possible combinations).

1. All characters equally weighted (nonetheless arbitrary for reasons discussed above).
2. Characters relating to appendages 7–9 (essentially homonomous in most taxa) weighted $^1/_3$ relative to other characters.
3. Characters relating to numbers of somites (tagmosis characters) and numbers of podomeres in appendages weighted down in proportion to the number of states recognized.
4. Characters relating to numbers of podomeres in appendages, and numbers of other organs weighted down in proportion to the number of states recognized.
5. As in '4', with the additional downweighting of characters relating to appendages 7–9 (as in '2').
6. As in '5', with the additional down-weighting of characters relating to numbers of somites (tagmosis characters) in proportion to the number of states recognized.

Two ordering schemes and six weighting regimes yield 12 possible combinations. Each of these were run with and without non-ostracodan Orsten taxa (*Bredocaris*, *Dala*, *Martinssonia*, *Rehbachiella* and *Skara*) (24 possible runs). There are compelling theoretical reasons for down-weighting the characters relating to the typically homonomous second to fourth thoracopods. As discussed above, the ranging of numbers of podomeres and numbers of other organs is also preferred, with a corollary that such characters should then be ordered. The preferred run therefore combined ordering regime 2 with weighting regime 5, and the discussion focuses principally on the topology of cladograms resulting from it. Other results are discussed for comparison and completeness.

Trees were constructed using PAUP 3.0 (Swofford, 1990), using a minimum of 30 random additions of taxa followed by TBR branch swapping. The remipedes were transferred to the outgroup in all analyses.

## 15.4 RESULTS AND DISCUSSION

### 15.4.1 GROSS TOPOLOGY

Eight trees resulted from the preferred 'weighting' and ordering of the data (RI = 0.715; RC = 0.199). A strict consensus (Figure 15.4) conforms neither to the 'traditional' tree (Brusca and Brusca, 1990) nor to that of Schram (1986), but rather incorporates some elements of both. The

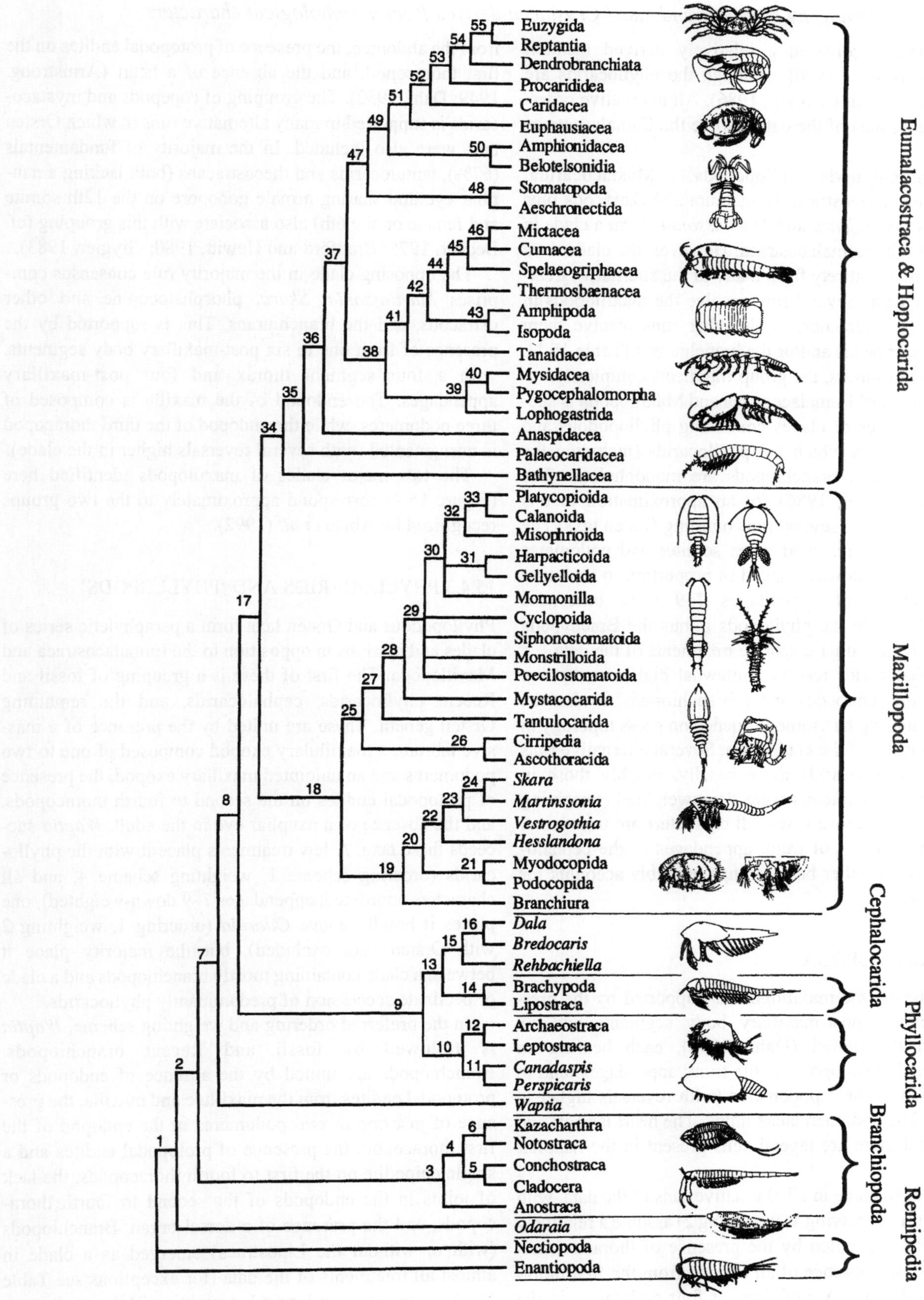


**Figure 15.4** Strict consensus of eight trees resulting from parsimony analysis of Recent and fossil crustacean taxa. Preferred ordering (1) and weighting (5) of the data. Synapomorphies listed at http://www.systassoc.chapmanhall.com.

Eumalacostraca resolve in a relatively derived location (Brusca and Brusca, 1990), although the phyllocarids are not united with them (Schram, 1986). All alternative orderings and weightings of the data resolve the Eumalacostraca as a clade.

The maxillopods (Copepoda, Mystacocarida, Tantulocarida, Thecostraca, Branchiura and Ostracoda plus the Cambrian taxa *Skara* and *Martinssonia*) form a clade in opposition to the Eumalacostraca. However, the clade does not emerge in its entirety from many of the alternative treatments of the data. Several runs resolve the maxillopods in the absence of ostracods, while other runs resolve them without the cirripedes and/or the branchiurans (Table 15.1). In other circumstances, the group fragments completely.

The grouping of Eumalacostraca and Maxillopoda is succeeded by a series of clades containing phyllopodous and Orsten taxa, within which the phyllocarids (including the Canadaspidida) and branchiopods are monophyletic. The Phyllopoda (Schram, 1986) (or an approximation to this clade) emerged from several runs omitting Orsten taxa. The phyllopods proper resolved when somites and podomeres were ordered and downweighted in proportion to the numbers of states (but appendages 7–9 were otherwise unweighted), while the phyllopods minus the Brachypoda emerged in some rather disparate treatments of the data.

The cephalocarids resolve somewhat higher in the tree than either the remipedes or the branchiopods, so that the choice of outgroup has some influence on gross topology in the lower portion of the cladogram. Several alternative runs placed the cephalocarids more basally, notably those in which tagmosis characters were downweighted (weighting runs 3 and 6) and those where all characters are unordered. The reduced number of trunk appendages in the cephalocarids relative to other basal groups probably accounts for this mobility.

### 15.4.2 MAXILLOPODA

The node below the maxillopods is supported by the presence of eleven post-maxillary body segments and six somites in the thorax (Dahl, 1956), each bearing an appendage. The endopods of the trunk appendages tend to be composed of three podomeres (with reversals higher in the clade). The abdomen lacks limbs. The heart is short and bulbous, and there are tapetal cells present in the naupliar eye.

Copepods (a clade in all alternative runs of the data, with the exception of ordering 1, weighting 2) assume a relatively derived location, united by the presence of thoracic intercoxal plates, the absence of an exopod from the first thoracopod, and the presence of three or four podomeres in the exopods of the subsequent thoracopods. The mystacocarids share with them the presence of five or more somites in the thorax (each bearing a limb), the absence of all appendages from the abdomen, the presence of protopodal endites on the first thoracopod, and the absence of a heart (Armstrong, 1949; Dahl, 1952). The grouping of copepods and mystacocarids is supported in many alternative runs in which Orsten taxa were also included. In the majority of fundamentals (67%), tantulocarids and thecostracans (both lacking a naupliar eye and sharing a male gonopore on the 12th somite and female on the 6th) also associate with this grouping (cf. Becker, 1975; Bradford and Hewitt, 1980; Grygier, 1983).

The opposing clade in the majority rule consensus comprises *Martinssonia*, *Skara*, phosphatocopine and other ostracods, and the branchiurans. This is supported by the presence of from one to six post-maxillary body segments, with a four segment thorax and four post-maxillary appendages. The endopod of the maxilla is composed of three podomeres, while the endopod of the third thoracopod is unsegmented (with several reversals higher in the clade).

The two major clades of maxillopods identified here (Figure 15.4) correspond approximately to the two groups recognized by Abele *et al.* (1992).

### 15.4.3 PHYLLOCARIDS AND 'PHYLLOPODS'

Phyllopodous and Orsten taxa form a paraphyletic series of clades and plesions in opposition to the Eumalacostraca and Maxillopoda. The first of these is a grouping of fossil and Recent phyllocarids, cephalocarids, and the remaining Orsten genera. These are united by the presence of a massive labrum, a maxillulary exopod composed of one to two podomeres and an unjointed maxillary exopod, the presence of protopodal endites on the second to fourth thoracopods, and the absence of a naupliar eye in the adult. *Waptia* succeeds these taxa. A few treatments place it with the phyllocarids (ordering scheme 1, weighting scheme 4, and all characters unordered, appendages 7–9 down-weighted), one places it basally above *Odaraia* (ordering 1, weighting 2 with Orsten taxa excluded), but the majority place it between a clade containing mostly branchiopods and a clade or pectinate succession of predominantly phyllocarids.

In the preferred ordering and weighting scheme, *Waptia* is followed by fossil and Recent branchiopods. Branchiopods are united by the absence of endopods or protopodal endites from the maxillule and maxilla, the presence of just one or two podomeres in the endopod of the first thoracopod, the presence of protopodal endites and a single epipodite on the first to fourth thoracopods, the lack of joints in the endopods of the second to fourth thoracopods, and the presence of a dorsal organ. Branchiopods (with or without the Lipostraca) emerged as a clade in almost all treatments of the data (for exceptions see Table 15.1). In one case (ordering 1, weighting 2) the cladoceran resolved at the base of a clade containing the maxillopods and Orsten genera, followed immediately by the remaining branchiopods.

**Table 15.1** Synopsis of results from alternative orderings and weightings of the data

| Ord. | 1 | 1 | 2 | 2 |
|---|---|---|---|---|
| Wt. | With Ø | Without Ø | With Ø | Without Ø |
| 1 | (M+Ø+C/B-C) [Od] | (E-Ba/M+Ba), B [Od] | (M+Ø/B), E [Od] | (E/M?)/B [Od] |
| 2 | (M+Ø+C/B-C), E [Od] | (E-Ba/M). B [Od] | (E+Bu/M+Ø-Bu), B [Od] | (E+Bu/M+Ø-Bu), B [Od] |
| 3 | (E/M-T-O), P, B, Ø-R [Bp] | (E/(M-T/Pp+T)), B, P | (E/M-T-RO), B, P [Bp] | (E/M-T-O, Pp+O, B) |
| 4 | ((E/M)/Pp?+Ø-O), P, B [Od] | (E/M), P, B [Od] | (E/M-T-RO), B, P, Ø-R [Bp] | ((E/M-T)/Pp+O)), P, B [Bp] |
| 5 | (E/M), P, B [Od] | (E/M), P, B [Od] | E, B, P, Ø-R [Bp] | E, Pp+O-Bp, P, B? [Bp] |
| 6 | E, P, B [Bp] | E, Pp-Bp, B, P [Bp] | E, B, P [Bp] | E, Pp+O-Bp, P, B [Bp] |

**Major clades emerging from alternative runs of the data set**
Parentheses indicate sister clades [e.g. (E/M) = Eumalacostraca versus Maxillopoda].
Square brackets indicate the basalmost taxon (excepting the remipedes).
The location of Orsten taxa is not considered important to the integrity of the maxillopods.
Other monophyletic groups listed, separated by commas.

**Alternative orderings**
1. Those characters relating to numbers of podomeres, somites and other organs ordered.
2. Those characters relating to numbers of podomeres and other organs ordered (but those coding for somites and tagmosis patterns unordered).

**Alternative weightings**
1. All characters equally-weighted.
2. Characters relating to appendages 7-9 (essentially homonomous in most taxa) weighted 1/3 relative to other characters.
3. Characters relating to numbers of somites (tagmosis characters) and numbers of podomeres in appendages weighted down in proportion to the number of states recognized.
4. Characters relating to numbers of podomeres in appendages, and numbers of other organs weighted down in proportion to the number of states recognized.
5. As in "4", with the additional downweighting of characters relating to appendages 7-9 [as in "2"].
6. As in "5", with the additional downweighting of characters relating to numbers of somites (tagmosis characters) in proportion to the number of states recognized.

**Key to taxa**
B = Branchiopoda
Ba = Bathynellacea
Bp = Brachypoda
Bu = Branchiura
C = Cladocera
E = Eumalacostraca and Hoplocarida
M = Maxillopoda (including the Ostracoda and Tantulocarida unless otherwise indicated)
Ø = all Orsten taxa
O = all Ostracoda
Od = *Odaraia*
P = Phyllocarida
Pp = Phyllopoda
R = *Rehbachiella*
RO = Recent Ostracoda
T = Thecostraca
? - in some fundamentals

## 15.4.4 MALACOSTRACA

The Eumalacostraca and Hoplocarida are supported by a relatively large number of character-state changes (although reversals occur higher up in the clade). The head shield tends to incorporate a cervical groove. The antennules are biramous, and the antennal exopod is usually modified as a scaphocerite. The antennal endopod typically has five or more podomeres, while that of the mandible usually has three or four. The mandible itself typically has a marked incisor process, and is bounded by paragnaths. The appendages of the terminal trunk segment are modified as uropods. The gut often distributes diverticulae into the trunk, while the abdomen has distinctive caridoid musculature (Calman, 1909). Development is typically metamorphic, and a dorsal frontal organ is usually present.

Relationships within the group (Figure 15.3(b)) conform precisely to no previously proposed schemes (Figures 15.2

and 15.3(a)). Most radically, the hoplocarids do not constitute a sister group to the eumalacostracans (*contra* Schram, 1986), but rather fall high within the clade. Two alternative treatments of the data (ordering 1, with equal weighting but excluding Orsten taxa; ordering 2, equal weighting, including Orsten genera) resolved the hoplocarids in opposition to the eumalacostracans minus the bathynellaceans. Importantly, the cladograms suggest that several of the caridoid facies (particularly the development of a carapace covering the cephalothorax) are only derived much higher within the clade. The ancestral eumalacostracan probably also had minimally specialized cephalic appendages, and certainly no maxillipedes. Caridoid trunk musculature may have been minimally developed, and uropods were probably not incorporated into a prominent tail fan (although caudal rami may have been retained). The overall distribution of characters is inconsistent with a phylogenetic classification based primarily on the extent of carapace development (Schram, 1981). However, several elements of gross topology are broadly in accord with the 'traditional' tree (Figure 15.2), and that produced (though not preferred) by Schram (1984) assuming the presence of a carapace to be a derived feature.

**(a) Eucarida**

Eucarids (plus the Belotelsonidea) fall in a derived location, and are united by a cephalothoracic shield in which all thoracomeres give rise to a branchiostegal fold (Dahl, 1991) and (typically) by the presence of a zoea larva. The resolution of the hoplocarids in opposition to this grouping is particularly innovative. Hoplocarids have a cephalic kinesis in the carapace. The antennule is triramous, but the antennal exopod is not modified as a scaphocerite. The second to fourth thoracopods have four podomeres in their endopods, and there is a total of 13 trunk limbs in a continuous run from the anterior.

Within the eucarids, the reptants and euzygids appear to be most derived, followed by the dendrobranchiates (with which they share arthrobranch and podobranch gills on the anterior thoracopods) rather than the eukyphids (as with Schram, 1986). This relationship was also true of most other runs in which thoracopods 7–9 were down-weighted. Most runs with a flat-weighting of these appendages reversed the positions of the euzygids and dendrobranchiates. The procarideans and carideans follow (paraphyletic eukyphids), (and in almost all alternative schemes) the decapods being supported by the presence of a scaphognathite on the maxilla and the development of two maxillipedes. In addition, both maxillipedes have lamellate protopods and protopodal endites directed mediad. The first has three podomeres in the endopod. The more posterior thoracopods bear pleurobranch gills, while the pleopods are modified for brooding eggs. The heart has three pairs of ostia. The euphausiaceans are opposed to the decapods (robust in almost all other weightings and orderings), and are united to them by the presence of an antennal gland and single-segment maxillary exopod. Maxillary glands are absent. The first thoracopod has protopodal endites, and the second a single epipodite. The heart is short and bulbous.

**(b) Peracarida**

The Peracarida (including the Thermosbaenacea = Pancarida) form a clade (cf. Watling, 1983) as a sister group to the eucarids, eocarids and hoplocarids. Peracarids are supported by a relatively large number of character state changes, including the presence of a mandibular spine row and lacinia mobilis, a single maxilliped with a lamellate protopod bearing endites directed distad, the development of oöstegites on the third and fourth trunk appendages, and jointed outer uropodal rami. The node is also supported by the absence of a naupliar eye or a dorsal frontal organ in the adult, and the lack of epipodites on the later trunk appendages. The peracarids are united with all the other Eumalacostraca except the syncarids by the univalved carapace (with two or more trunk somites giving rise to branchiostegal folds), the presence of a cephalic doublure, and a heart with two pairs of ostia. Peracarids resolve as a clade in all the alternative treatments of the data considered here, opposing all other eumalacostracans with the exception (in most cases) of one or more of the syncarids.

Within the peracarids, isopods and amphipods group together (Order Edriophthalma; unlike the 'traditional' model) united by the simple head-shield, uniramous antennae and thoracopods, and thoracic coxal plates. These stand in opposition to the Mictacea, Cumacea, Spelaeogriphacea and Thermosbaenacea, all these taxa being followed by the Tanaidacea (tergites articulating with no overlap, cervical groove absent from the carapace, compound eyes sessile or on the dorsal cuticle, maxilla lacking endopod). Schram (1984, 1986) resolved the Edriophthalma plus the Mictacea versus the Thermosbaenacea, Cumacea, Tanaidacea and Spelaeogriphacea (Brachycarida, Schram, 1981). Mictaceans resolved with the edriophthalmids in several alternative weightings (notably those where appendages 7–9 were not down-weighted). The Thermosbaenacea (= Pancarida) are relatively derived within the Peracarida, and there is no compelling evidence to treat them as a separate superorder (cf. Siewing, 1958). The Mysidacea here form a clade in opposition to all other peracarids (resolving basal but paraphyletic in the phylogenies of Schram, 1984, 1986). The Mysidacea did not resolve with the eucarids with any alternative run of the data (cf. Watling, 1983), but in some senses forge a link between the eucarids and the more derived peracarid lines (Siewing, 1963; Hessler, 1982).

**(c) Syncarida**

The Syncarida do not constitute a clade, but rather fall in pectinate succession (Anaspidacea, Palaeocaridacea, Bathynellacea) at the base of the Eumalacostraca. The

Syncarida is a wastebasket taxon, united by symplesiomorphies (e.g. the absence of a well-developed carapace), although their primitive status is in accord with the traditional tree, and also one scheme within Schram (1984) where the carapace was assumed to be a derived feature. Paedomorphosis in the bathynellaceans (aborting the development of the pleopods and posterior thoracopods) has probably contributed to uncertainty over their affinities. Serban (1972) removed them from the syncarids to a taxon of equivalent status. Some alternative treatments of the data caused the syncarids to dissociate further, but none resolved them as a clade. In the presence of Orsten taxa, schemes in which characters relating to appendages 7–9 were not downweighted resolved the Bathynellacea basally, but resolved the Anaspidacea and Palaeocaridaea as a small clade in opposition to the eucarids (plus the Belotelsonidea). The combination of ordering scheme 1 with simply downweighting the anterior thoracopods resolved a paraphyletic Syncarida in opposition to the eucarid/belotelsonid clade (rather than all other eumalacostracans). The effects on these relationships of removing Orsten taxa were various. Runs in which the syncarids initially constituted a pectinate series at the base of the Eumalacostaca suffered few if any changes within the whole clade upon the deletion of Orsten genera.

The node above the syncarids (all other eumalacostracans) in the preferred weighting and ordering scheme is supported by the presence of a univalved carapace incorporating two branchiostegal folds, the presence of a cephalic doublure, and a heart incorporating two pairs of ostia.

## 15.4.5 PROBLEMATIC FOSSIL TAXA

### (a) Canadaspidida

*Perspicaris* and *Canadaspis* are sister genera (Schram, 1986), sharing eight post-maxillary appendages in a continuous run from the anterior, a uniramous antenna (endopod only), and a maxilla undifferentiated from the following trunk appendages. Moreover, the assignment of the canadaspidids to the phyllocarids (as a sister taxon to the Recent Leptostraca and other fossil groups (Rolfe, 1969; Briggs, 1978, 1992; Schram, 1986; Wills *et al.*, 1994) is also supported (cf. Dahl, 1983, 1984). Phyllocarids are united here by the all-enveloping carapace (typically with a cephalic kinesis), an antenna with nine or more podomeres in its endopod, and a mandible with a marked incisor process. The presence of a thorax composed of eight somites optimizes as a synapomorphy much lower down in the cladogram (rather than supporting just the phyllocarids) but reverses on several occasions higher in the tree (e.g. within the maxillopods).

An assignment of *Canadaspis* and *Perspicaris* to the base of the crustacean clade is highly unparsimonious (*contra* Bergström, 1992). The large number of podomeres in the endopods of the maxillae and trunk appendages of *Canadaspis* (long cited as indicative of its primitive status) emerge as uniquely derived (Briggs, 1978, 1992). An immediate relationship between phyllocarid groups and the Anostraca (Sars, 1887) is not supported.

### (b) *Odaraia*

Contrary to all previous speculation, *Odaraia* resolves neither as a close relation of the phyllocarids (Wills *et al.*, 1994), nor the branchiopods (Briggs, 1981), but rather in opposition to all other crustaceans (with the exception of the remipedes). Derived crustacean characters appearing at the node below *Odaraia* include appendages that are pendant beneath the trunk (rather than laterally displaced; Bergström, 1992) the presence of carapace adductor muscles, and the retention of a naupliar eye in the adult. Nonetheless, *Odaraia* is clearly very plesiomorphic in many respects (e.g. numerous trunk somites bearing homonomous limbs, limited specialization of the cephalic appendages). Briggs (1981) and Simonetta and Delle Cave (1975) stressed the difficulty of placing *Odaraia* with any degree of confidence. Certainly, supposed synapomorphies with other groups of crustaceans are far from compelling, and its basal resolution in the present cladogram is unsurprising in this respect. Several features, such as the 'tubular' carapace and three-fluked 'tail', have clearly evolved in isolation. Others (e.g. the bivalved carapace, large stalked eyes, and the regular decrease in the length of appendages passing down the trunk) may actually be symplesiomorphies for a large part of the crustacean or wider arthropod clade. Several such gross anatomical characters are shared by taxa with otherwise radically different tagmosis patterns and limb morphologies (e.g. *Canadaspis*, *Waptia*, *Branchiocaris*, *Marrella*, *Vachonisia*).

*Odaraia* resolves basally in most alternative treatments of the data (ordering scheme 1, weighting schemes 1, 2, 4 and 5; ordering scheme 2, weighting schemes 1 and 2) (see also Table 15.1). In a few other cases it resolves in opposition to the Eumalacostraca and a paraphyletic Maxillopoda, or groups with the branchiopods where all characters were unordered and Orsten taxa omitted (cf. Briggs, 1981). A better knowledge of the cephalic morphology of *Odaraia* might enable its placement with confidence higher up the crustacean tree, but those characters that can be scored unequivocally in the present data base indicate a very primitive status.

### (c) *Lepidocaris*

*Lepidocaris* groups with *Lightiella* (Schram, 1986), united by the similar number of post-maxillary somites, simple head-shield, thoracic egg-brooding appendages in the female, and the absence of compound eyes. The wider affinities of the cephalocarids appear to lie with *Dala*,

*Bredocaris* and *Rehbachiella*, with which they share an attached labrum, an antennule with five or more podomeres in the antennulary exopod and a maxillule with three or more podomeres in the endopod. All have a maxilla scarcely differentiated from the anterior trunk appendages.

Some treatments of the data, by contrast, placed *Lepidocaris* in opposition to the branchiopods (ordering scheme 1, weightings 1 and 2 without Orsten taxa; ordering 2, weighting 1 with or without Orsten taxa) (Scourfield, 1926; Sanders, 1963a; Walossek, 1993), while most runs in which tagmosis characters were unordered (regime 2) resolved the Lipostraca with neither the Brachypoda nor the Branchiopoda.

**(d) Orsten genera**

In the preferred ordering and weighting of the data, Orsten genera group into two clades, one (*Skara* and *Martinssonia*) in association with the ostracods (including the phosphatocopines, another (*Rehbachiella*, *Bredocaris* and *Dala*) grouping with the cephalocarids (see also Table 15.1). However, these precise associations are observed in none of the other runs. All the Orsten taxa (including the phosphatocopines but excluding *Rehbachiella*) frequently group basally above the Brachypoda (ordering 1, weightings 3 and 6; ordering 2, weightings 3–6), and such a location would be consistent with the coding of larval morphologies. In a number of other runs (weightings 1 and 2), they form a clade within the maxillopods, most commonly in immediate opposition to the mystacocarids (a group thought to be paedomorphic).

The similarity between *Skara* and *Martinssonia* is more than a superficial one. They are sister genera or paraphyletic and separated by a single node in all runs of the data (cf. Müller and Walossek, 1986). In the preferred run they are united by the simple head-shield, the attachment of the labrum, the similar number of post-maxillary body segments, the presence of three podomeres in the maxillulary endopod, and the absence of a marked molar process on the mandible. The characters shared with the phosphatocopines (considerable similarity of the cephalic appendages and the absence of most post-cephalic ones) are probably instrumental in their grouping with the ostracods as a whole (this entire clade being supported by the all-enveloping carapace, cephalic doublure, natatory antennae and biramous mandibles). This assignment for *Skara* is broadly comparable with that of Müller and Walossek (1988), although, these authors predicted greatest affinity with the copepods rather than the ostracods. By contrast, Müller and Walossek (1986) believed *Martinssonia* to be a 'stem lineage' form, deriving as a plesion from the group which was to give rise to the true crustaceans.

*Rehbachiella*, *Bredocaris* and *Dala* are united by their univalved carapaces, the sessile nature of the compound eyes (where present), the ventral direction of the labrum, and a prominent dorsal or nuchal organ. They also share ten or more podomeres in the inner ramus of the antennule, a natatory antenna bearing protopodal endites and seventeen podomeres in its exopod, and a maxilla with four podomeres in the endopod. The present analysis provides no basis for considering *Bredocaris* to be a basal maxillopod (as did Müller and Walossek, 1988), nor does *Rehbachiella* resolve with the branchiopods (see Walossek, 1993).

Schram (1986) preferred to regard most Orsten taxa as larval forms. It would be totally inappropriate to score juveniles as though they were adults, since this would allow ontogenetic change to be confounded with evolutionary change on cladograms. Such a procedure might also tend to group larvae on very general, plesiomorphic features, as they would probably possess a suite of 'juvenile' characters not scored in any of the adults. It may be significant, therefore, that the 'Orsten' taxa here fall in just two clades – one with the ostracods and the other with the cephalocarids. It is possible that a genuine affinity of, for example, *Skara* with the ostracods and *Rehbachiella* with the cephalocarids might cause the other Orsten genera to resolve in these locations also. Hessler (1971) and Hessler and Newman (1975) speculated that the morphology of cephalocarids might be the result of progenetic paedomorphosis, which might also account for the grouping with Orsten taxa. The distortions produced by erroneously comparing larvae with adults will be more misleading than those introduced by omitting the data from a small number of fossils. On balance, therefore, the phylogeny of the crustaceans as a whole is best considered in the absence of Orsten genera. A meaningful appraisal of the affinities of these forms might therefore require analysis alongside the larvae of other groups. Nonetheless, the trees produced if the five non-ostracod Orsten taxa are omitted from the analysis are identical to those derived by pruning them from the results of the analysis of all taxa. However, this is not true where alternative orderings and weightings of the data are used (ordering 2 and weighting 2 shows minimal differences). The principal effect of their inclusion is to break up a clade or clades of phyllopodous taxa into a more pectinate series at the base of the tree.

## 15.5 CONCLUSIONS

1. The results presented here serve to illustrate the problems inherent in coding and analysing crustacean morphological characters, and represent a new hypothesis of relationships within the subphylum.
2. The question of weighting arises where characters that are typically serially homologous have been coded for all taxa in order to capture the variation present in a more limited number of groups. The ordering of some characters may preserve useful information, since close numerical similarities in numbers of somites,

appendages and podomeres may not otherwise be registered in the analysis. The recognition of numerous, ordered states may also recommend the ranging of certain types of characters. A consideration of these factors has prompted the selection of a preferred weighting and ordering scheme. Alternative treatments of the data may yield topological differences at all levels.

3. The Nectiopoda represent the most primitive living group of crustaceans, and almost certainly represent the most suitable taxon for rooting parsimony networks of crustaceans (Schram, 1982, 1986; Briggs *et al.*, 1992, 1993; Wills *et al.*, 1994, 1995, 1996).
4. The preferred tree incorporates elements of both the traditional phylogeny, and that of Schram (1986). Malcostracans (*sans* phyllocarids) constitute a sister group to maxillopods. Phyllopod taxa fall in a paraphyletic series in opposition to this clade (interspersed with Orsten genera where these are included). Branchiopods (a group that emerges from almost all alternative runs) fall low in the preferred tree, cephalocarids rather higher. Phyllopods resolve as a clade with some alternative treatments, but never in conjunction with a clade of maxillopods.
5. The Eumalacostraca plus the Hoplocarida always resolve as a highly derived clade, although never in opposition to the phyllocarids. The hoplocarids do not constitute a sister group to the eumalacostracans, but rather resolve in association with the eucarids. The peracarids form a clade, but the syncarids are a wastebasket taxon, united by plesiomorphic features.
6. The Maxillopoda emerges in its entirety from the preferred treatment of the data, (including the ostracods and tantulocarids but excluding all non-ostracodan Orsten genera). Most alternative treatments yield a large clade of maxillopods, some lacking the branchiurans, others the thecostracans.
7. Depending upon the scheme of weighting and ordering implemented, the inclusion of Orsten genera may have no effect on the relationships of the remaining taxa, or alternatively may exert a marked influence. There appears to be some affinity between the cephalocarids, *Dala*, *Bredocaris* and *Rehbachiella* on one hand, and *Skara*, *Martinssonia* and the ostracods on the other. Several other treatments of the data place all the Orsten genera with the maxillopods. Orsten taxa generally appear to associate closely, and it is possible that this cohesion is the result of scoring juvenile morphologies as though they were adult.
8. The Devonian *Lepidocaris* is probably more closely related to the branchiopods than the brachypodans, but all these taxa often associate closely even if not forming a clade. The Brachypoda often fall basally in trees rooted with the Nectiopoda.
9. The Burgess Shale genus *Odaraia* emerges basally from the preferred and many other runs of the data. A more rigorous assignment for this taxon would require detailed information on the morphology of the cephalic appendages, but the features preserved suggest a very primitive status.
10. The brachypodans resolve basally (adjacent to the remipedes) in many alternative runs, notably those in which tagmosis characters were down-weighted and those where all characters are unordered. Brachypodans have a reduced number of trunk appendages relative to that in purportedly basal groups, and this probably accounts for their mobility into other regions of the tree.
11. A satisfactory resolution of crustacean relationships may ultimately require the addition of sequence data to the morphological database. Molecular studies have tended to concentrate on smaller clades (Abele *et al.*, 1992; Spears *et al.*, 1994), but the unity of these groupings is uncertain from the results of this and other analyses. An holistic or 'total evidence' approach to the problem may therefore be necessary.

## ACKNOWLEDGEMENTS

My thanks to Prof. D.E.G. Briggs, Prof. F.R. Schram, Dr R.A. Fortey, Dr M.J. Benton, Dr G.A. Boxshall, Prof. D.H. Erwin and Dr C.C. Labandeira for constructive criticism of this work. My research was supported by the Leverhulme Trust (Grant F/182/AK), and a Smithsonian Postdoctoral Fellowship.

**Appendix 15.1** Character codes

## Cephalic shields and tergites

1. Carapace: Carapace and head-shield absent (0). Simple head-shield developed (1). Univalved carapace (2). Bivalved carapace (3).
2. Bivalved carapace: With a dorsal hinge (0). Lacking a distinct hinge (1).
3. Carapace: Normal (0). All-enveloping (1).
4. Number of trunk somites giving rise to branchiostegal folds: None (0). One (1). Two (2). Three (3). Four or five (4). Eight (5). (Watling 1983; Dahl 1991).
5. Articulation of tergites: With no overlap (0). With overlapping pleurae (1).
6. Anterior of carapace: Rounded (0). Produced into a fixed rostral spine (1). Articulating rostral spine (2).
7. Cervical groove: Absent (0). Present (1).
8. Cephalic kinesis / procephalon: Absent (0). Present (1).
9. Cephalic doublure: Absent (0). Present (1).
10. Carapace divided into plates: Not divided (0). Divided (1)
11. Carapace adductor muscles: Present (0). absent (1).

## Eyes and frontal organs

12. Compound eyes: Absent (0). Present (1).
13. Compound eye support: Stalked (0). Sessile or on dorsal cuticle (1).
14. Compound eye structure: Malacostracan (0). Entomostracan (1). (Watling 1983).
15. Naupliar eye of adult: Absent (0). Present (1)
16. Number of cups comprising naupliar eye in adult: Three (0). Four (1).
17. Tapetal cells in naupliar eyes: Absent (0). Present (1).
18. Dorsal frontal organ: Absent (0). Present (1)
19. Ventral frontal organ: Absent (0). Present (1).
20. Dorsal, nuchal or neck organ: Absent (0). Present (1).

## Appendage 1 (Antennule)

21. Number of podomeres in outer ramus (exopod): 10 or more (0). 9 to 4 (1). 3 to 1 (2) Outer ramus absent (3).
22. Number of podomeres in inner ramus (endopod): 10 or more (0). 9 to 1 (1).
23. Third flagellum: Antennule uniramous or biramous (0). Antennule triflagellate (the condition in Hoplocarida) (1).

## Appendage 2 (Antenna)

24. Protopodal endites: Present (0). Absent (1).
25. Number of podomeres in outer ramus (exopod): None (a). 1 (b). 2-4 (c). 5 (d). 6-9 (e). 13 (f). 14 (g). 17 (h). 18 (i). 20 plus (j).
26. Antennal exopod modified as scaphocerite: Not modified (0). Modified (1).
27. Number of podomeres in inner ramus (endopod): None (a). 1-2 (b). 3 (c). 4 (d). 5 (e). 9 plus (f).
28. Form of appendage: Undifferentiated (0). Antenniform (i.e. strictly differentiated from those which follow and fulfilling a sensory function) (1).
29. Appendage natatory: Not natatory (0). Natatory (1).
30. Antennal gland: Absent (0). Present (1).

## Appendage 3 (Mandible)

31. Number of podomeres in exopod: None (exopod absent) (a). 1 (b). 3-5 (c). 6 (d). 11 (e). 13 (f). 14-25 (g).
32. Form of exopod: Flagelliform, or otherwise developed as an elongate process (0). Developed as a broad, lamelliform paddle or blade, or otherwise non-flagelliform (1).
33. Number of podomeres in endopod: None (endopod absent) (a). 1-2 (b). 3-4 (c). 5 (d).
34. Function of appendage: Undifferentiated (0). Mandible (differentiated from those which follow by the degree of development of the protopodal endites or gnathobases) (1).
35. Mandible with marked incisor process: Absent (0). Present (1).
36. Mandible with marked molar process: Absent (0). Present (1).
37. Lacinia mobilis on one or both mandibles: Absent (0). Present (1).
38. Mandibular spine row: Absent (0). Present (1).
39. Paragnaths: Absent (0). Present (1).
40. Condition of labrum attachment: Attached (0). Detached (1).
41. Size of labrum: Small to medium (0). Massive (1).
42. Direction of labrum: Posterior (0). Ventral (1).
43. Oral cone or pyramid formed from labrum and mouthparts: Absent (0). Present (1).
44. Epistome: Absent or vestigial (0). Well-developed (1).

## Appendage 4 (Maxillule)

45. Protopodal endites: Present (0). Absent (1).
46. Number of podomeres in exopod: None (exopod absent) (a). 1-2 (b). 6-13 (c).
47. Form of exopod: Flagelliform, or otherwise developed as an elongate process (0). Developed as a broad, lamelliform paddle or blade, or otherwise non-flagelliform (1).
48. Number of podomeres in endopod: None (endopod absent) (a). 1-2 (b). 3 (c). 4 (d). 5 (e). 14 (f). 15 (g).

## Appendage 5 (Maxilla)

49. Protopodal endites: Present (0). Absent (1).
50. Number of podomeres in exopod: None (exopod absent) (a). 1 (b). 2 (c). 7-18 (d).
51. Form of exopod: Flagelliform, or otherwise developed as an elongate process (0). Developed as a broad, lamelliform paddle or blade, or otherwise non-flagelliform (1).
52. Exopod modified as a scaphognathite: Not modified (0). Modified (1).
53. Number of podomeres in endopod: None (endopod absent) (a). 1-2 (b). 3 (c). 4 (d). 5 (e). 6 (f). 14 (g). 15 (h).
54. Epipodite: Absent (0). Present (1).
55. Maxillary glands: Absent (0). Present (1).
56. Differentiation of maxilla from first trunk appendage: Undifferentiated (0). Differentiated (1).

## Appendage 6 (First thoracopod or maxillipede)

57. Protopodal endites: Present (0). Absent (1).
58. Number of podomeres in exopod: None (exopod absent) (a). 1 (b). 2-4 (c). Very numerous (d).
59. Form of exopod: Flagelliform, or otherwise developed as an elongate process (0). Developed as a broad, lamelliform paddle or blade, or otherwise non-flagelliform (1).
60. Caridean lobe on exopod: Absent (0). Present (1).
61. Number of podomeres in endopod: None (endopod absent) (a). 1-2 (b). 3 (c). 4 (d). 5 (e). 6 (f). 7 (g). 14 (h). Very numerous (i).
62. Number of epipodites: None (0). One (1). Two (2).

63. Form of epipodite: Not modified (0). Modified as a cup or spoon-shaped respiratory structure (1).
64. Maxillipedes with lamellate protopod and protopodal endites directed: Mediad (1). Distad (2). Otherwise (0).
65. Oöstegites: Absent (0). Present (1).
66. Podobranch gills: Absent (0). Present (1).
67. Arthrobranch gills: Absent (0). Present (1).

### Appendages 7, 8 and 9

68/77/86. Protopodal endites: Present (0). Absent (1).
69/78/87. Number of podomeres in exopod: None (exopod absent) (a). 1 (b). 2 (c). 3-4 (d). Very numerous (e).
70/79/88. Form of exopod: Flagelliform, or otherwise developed as an elongate process (0). Developed as a broad, lamelliform paddle or blade, or otherwise non-flagelliform (1).
71/80/89. Number of podomeres in endopod: None (endopod absent) (a). 1 (b). 2-3 (c). 4 (d). 5 (e). 6 (f). 7 (g). 14 (h). 16 (i).
72/81/90. Number of epipodites: None (0). One (1). Two (2).
73/82/91. Oöstegites: Absent (0). Present (1).
74/83/92. Podobranch gills: Absent (0). Present (1).
75/84/93. Arthrobranch gills: Absent (0). Present (1).
76/85/94. Pleurobranch gills: Absent (0). Present (1).

### General morphology of thoracic appendages

95. Attitude of trunk appendages relative to body: Pendant (0). Laterally displaced (1).
96. Relative length of trunk limbs: Approximately constant (0). Regularly decreasing down the length of the trunk (1).
97. Correlation of trunk somites and appendages: Somites and appendages correspond (0). No correspondence between somites and appendages (1) .
98. Long, thin bar like extensions deriving from most exopodal podomeres (Orsten-type appendages): Bars absent (0). Bars present (Orsten-type) (1).
99. Female egg-brooding appendages on 12th thoracic somite: Absent (0). Present (1).
100. Thoracic coxal plates: Absent (0). Present (1).
101. Thoracic intercoxal plates: Absent (0). Present (1).
102. Cirri: Absent (0). Present (1).

### Abdominal appendages

103. Number of podomeres in exopod of second abdominal appendage / pleopod: None (0). One or vestigial (1). Two (2). Three (3). Annulate (4).
104. Number of podomeres in endopod of second abdominal appendage / pleopod: None (0). One or vestigial (1). Two (2). Four (3). Five (4). Six (5). Annulate (6).
105. Epipodites of second abdominal appendage: None (0). One (1). Two (2).
106. Abdominal appendages (pleopods) modified for brooding eggs: Unmodified (0). Modified (1).
107. Ventral abdominal combs: Absent (0). Present (1).

### Posteriormost appendages

108. Appendages on terminal trunk somite: Absent or undifferentiaed (0). Modified as uropods incorporated into a fan with the telson. Lateral margins of the uropods need not necessarily meet with the lateral margins of the telson (1).
109. Number of podomeres in inner rami of uropods: None (0). One (1). Two or more (2).
110. Number of podomeres in outer rami of uropods: None (0). One (1). Two or more (2).

### Furca

111. Telson appendages: Absent (0). Present (1).
112. Articulation of telson appendages: Immobile (0). Articulated at the base (1).
113. Telson appendages with internal joints: Joints absent (0). Joints present (1).
114. Form of telson appendages: Unmodified (0). Cerci (1). Abreptors (2).

### Tagmosis

115. Number of maxillipedes: None (0). One (1). Two (2). Three or more (3).
116. Number of post-maxillary body segments, *including* the telson or anal somite: 1-6 (a). 8 (b). 11 (c). 12 (d). 14-15 (e). 16 (f). 19 (g). 20-21 (h). 28 or more (i).
117. Number of post maxillary appendages in a continuous run from the anterior. Gaps of a single appendage are discounted. This excludes the caudal rami where present: 0-1 (a). 2 (b). 4 (c). 5 (d). 6 (e). 7 (f). 8 (g). 9 (h). 10-12 (i). 13 (j). 14 (k). Very numerous (l).
118. Number of somites in the thorax: 1 (a). 2 (b). 4 (c). 5 (d). 6 (e). 7 (f). 8 (g). 10 (h). 11 (i). Very numerous (j).

### Internal organs

119. Trunk gut diverticulae: Absent (0). Present (1).
120. Cephalic gut diverticulae: Absent (0). Present (1).
121. Foregut with pyloric function: Absent (0). Present (1).
122. Position of the anus: Terminal (0). Ventral (1).
123. Heart: Absent (0). Present (1).
124. Gross morphology of the heart: Elongate (0). Short and bulbous (1). (Watling 1983).
125. Number of pairs of ostia in heart: None (0). One pair (or one pair plus one in some copepods) (1). Two pairs (2). Three pairs (3). Four or more pairs (4).
126. Caridoid musculature: Absent (0). Present (1).
127. CNS: Ventral nerve cord with unfused, paired ganglia and double ventral commisures (0). Ventral nerve cord with fused ganglia (1).

### Reproduction and Development

128. Sex: Monoecious (0). Dioecious (1).
129. Location of gonopores: Both on 8th somite (a). Both on 9th somite (b). Both on 11th somite (c). Male on 12th somite, female on 6th somite (d). Both on 12th somite (e). Male on 12th somite, female on 11th somite (f). Both on 16th somite (g). Both on 17th or 20th somite (h). Both on 20th somite (i).
130. Male genital cones: Absent (0). Present (1).
131. Petasma: Absent (0). Present (1).
132. Appendices internae / appendices masculinae: Absent (0). Present (1).
133. Development: Anamorphic (0). Metamorphic (1). Epimorphic or direct (2).
134. Number of naupliar limbs: Three (0). Four (1).
135. Larva: Zoea (1). Cypris (2). Nauplius, metanauplius or otherwise (0).

PAUP ASSUMPTION SETS

TYPESET 1 = 1-3 5-20 22-24 26 28-30 32 34-45 47 49 51-52 54-57 59-60 63-68 70 73-77 79 82-86 88 91-102 106-108 111-114 119-135, ord: 4 21 25 27 31 33 46 48 50 53 58 61-62 69 71-72 78 80-81 87 89-90 103-105 109-110 115-118;
TYPESET 2 = 1-20 22-24 26 28-30 32 34-45 47 49 51-52 54-57 59-60 63-68 70 73-77 79 82-86 88 91-102 106-108 111-135, ord: 21 25 27 31 33 46 48 50 53 58 61-62 69 71-72 78 80-81 87 89-90 103-105 109-110;
WTSET 1 = 1000, 1-135;
WTSET 2 = 1000: 1-67 95-135, 333: 68-94;
WTSET 3 = 1000: 1-3 5-20 22-24 26 28-30 32 34-45 47 49 51-52 54-57 59-60 63-68 70 73-77 79 82-86 88 91-102 106-108 111-114 119-135, 200: 4 27, 333: 21 33 50 58 115, 111: 25 118, 167: 31 48 104, 500: 46 62 72 81 90 105 109-110, 143: 53, 125: 61 71 80 89 116, 250: 69 78 87 103, 91: 117;
WTSET 4 = 1000: 1-20 22-24 26 28-30 32 34-45 47 49 51-52 54-57 59-60 63-68 70 73-77 79 82-86 88 91-102 106-108 111-135, 333: 21 33 50 58, 111: 25, 200: 27, 167: 31 48 104, 500: 46 62 72 81 90 105 109-110, 143: 53, 125: 61 71 80 89, 250: 69 78 87 103;
WTSET 5 = 1000: 1-20 22-24 26 28-30 32 34-45 47 49 51-52 54-57 59-60 63-67 95-102 106-108 111-135, 333: 21 33 50 58 68 70 73-77 79 82-86 88 91-94, 111: 25, 200: 27, 167: 31 48 72 81 90 104, 500: 46 62 105 109-110, 143: 53, 125: 61, 83: 69 78 87, 42: 71 80 89, 250: 103;
WTSET 6 = 1000: 1-3 5-20 22-24 26 28-30 32 34-45 47 49 51-52 54-57 59-60 63-67 95-102 106-108 111-114 119-135, 200: 4 27, 333: 21 33 50 58 68 70 73-77 79 82-86 88 91-94 115, 111: 25 118, 167: 31 48 72 81 90 104, 500: 46 62 105 109-110, 143: 53, 125: 61 116, 83: 69 78 87, 42: 71 80 89, 250: 103, 91: 117;

## REFERENCES


Abele, L.G., Spears, T., Kim, W. and Applegate, M. (1992) Phylogeny of selected maxillopodan and other crustacean taxa based on 18S ribosomal nucleotide sequences: a preliminary analysis. *Acta Zoologica*, **73**, 373–82.

Adouette, A. and Philippe, H. (1993) The major lines of metazoan evolution, summary of traditional evidence and lessons from ribosomal RNA analysis, in *Comparative Molecular Neurobiology* (ed. Y. Pichon), Birkhäuser Verlag, Basel, pp. 1–30.

Armstrong, J.C. (1949) The systematic position of the crustacean genus *Derocheilocaris* and the status of the subclass Mystacocarida. *American Museum Novitates*, **1413**, 1–6.

Averof, M. and Akam, M. (1995) Insect–crustacean relationships: insights from comparative developmental and molecular studies. *Philosophical Transactions of the Royal Society London, B*, **347**, 293–303.

Becker, K.-H. (1975) *Basipodella harpacticola* n. gen., n. sp. *Helg. wiss. Meers*, **27**, 96–100.

Beklemischev, V. N. (1952) *Osnovy sraviteljnoi anatomii bespozvonocnych.* 2nd edn, Nauka, Moscow.

Beklemischev, V. N. (1969) *Principles of Comparative Anatomy of Invertebrates*, Vol. 1., Oliver & Boyd, Edinburgh. (translation of 3rd (1964) edn by J. M. MacLennan).

Bergström, J. (1992) The oldest arthropods and the origin of the Crustacea. *Acta Zoologica*, 73, 287–91.

Bowman, T.E. and Abele, L.G. (1982) Classification of the Recent Crustacea, in *The Biology of Crustacea, Vol. 1.* (ed. L.G. Abele), Academic Press, New York, pp. 1–27.

Boxshall, G.A. and Huys, R. (1989) New tantulocarid, *Stygotantulus stocki*, parasitic on harpacticoid copepods, with an analysis of the phylogenetic relationships within the Maxillopoda. *Journal of Crustacean Biology*, **9**, 126–40.

Boxshall, G.A. and Lincoln, R.J. (1983) Tantulocarida, a new class of Crustacea ectoparasitic on other crustaceans. *Journal of Crustacean Biology*, **3**, 1–16.

Boxshall, G.A. and Lincoln, R.J. (1987) The life cycle of the Tantulocarida (Crustacea). *Philosophical Transactions of the Royal Society of London, B*, **315**, 267–303.

Bradford, J.M. and Hewitt, G.C. (1980) A new maxillopodan crustacean, parasitic on a myodocopid ostracod. *Crustaceana*, **38**, 67–72.

Briggs, D.E.G. (1978) The morphology, mode of life, and affinities of *Canadaspis perfecta* (Crustacea: Phyllocarida), Middle Cambrian, Burgess Shale, British Columbia. *Philosophical Transactions of the Royal Society of London, B*, **281**, 439–87.

Briggs, D.E.G. (1981) The arthropod *Odaraia alata* Walcott, Middle Cambrian, Burgess Shale, British Columbia. *Transactions of the Royal Society of London, B*, **291**, 541–85.

Briggs, D.E.G. (1992) Phylogenetic significance of the Burgess Shale crustacean *Canadaspis*. *Acta Zoologica*, **73**, 293–300.

Briggs, D.E.G., Fortey, R.A. and Wills, M.A. (1992) Morphological disparity in the Cambrian. *Science*, **256**, 1670–3.

Briggs, D.E.G., Fortey, R.A. and Wills, M.A. (1993) How big was the Cambrian explosion? A taxonomic and morphologic comparison of Cambrian and Recent arthropods, in *Evolutionary Patterns and Processes* (eds D.R. Lees and D. Edwards), Linnean Society Symposium Series, Linnean Society of London, pp. 33–44.

Brooks, H.K. (1955) A crustacean from the Tesnus Formation of Texas. *Journal of Paleontology*, **29**, 852–6.

Brusca, R.C and Brusca, G.J. 1990. *Invertebrates*, Sinauer, Massachusetts.

Calman, W.T. (1904) On the classification of the Crustacea Malacostraca. *Annals and Magazine of Natural History*, **7**, 144–58.

Calman, W.T. (1909) Crustacea, in *Treatise on Zoology VII, Appendiculata, Fascicule 3* (ed. R. Lankester), Adam & Black, London.

Carpenter, G.H. (1906) Notes on the segmentation and phylogeny of the Arthropoda, with an account of the maxillae in *Polyxenus lagurus. Quarterly Journal of the Microscopical Society*, **49**, 469–91.

Cisne, J.L. (1982) Origin of the Crustacea, in *The Biology of the Crustacea, Vol. 1. Systematics, the Fossil Record and Biogeography* (ed. L.G. Abele), Academic Press, New York, pp. 65–92.

Claus, C. (1876) *Untersuchungen zur Erforschung der genealogischen Grundlage des Crustaceensystems*, Carl Gerold's Sohn, Wien.

Claus, C. (1888) Uber den Organismus der Nebaliden und die systematische Stellung der Leptostraken. *Arbeiteten aus dem Zoologischen Institut der Universität Wien*, **6**, 1–108.

Costlow, J.D. Jr and Bookhout, C.G. (1957) Larval development of *Balanus eburneus* in the laboratory. *Biological Bulletin*, **112**, 313–24.

Costlow, J.D. Jr and Bookhout, C.G. (1958) Larval development of *Balanus amphitrite* var. *denticulata* Broch reared in the laboratory. *Biological Bulletin*, **114**, 284–95.

Dahl, E. (1952) Mystacocarida. Reports of the Lund University Chile Expedition 1948–49. *Lund University Arssk*, **48**, 1–41.

Dahl, E. (1956) Some crustacean relationships, in *Bertil Hanstrom, Zoological Papers in Honour of His Sixty-Fifth Birthday, November 20th, 1956* (ed. K. Wingstrand) Lund Zoological Institute, Lund, Sweden, pp. 138–47.

Dahl, E. (1963) Main evolutionary lines among recent Crustacea, in *Phylogeny and Evolution of Crustacea* (eds H.B. Whittington and W.D.I. Rolfe), Museum of Comparative Zoology, Cambridge, Massachusetts, pp. 1–15.

Dahl, E. (1976) Structural plans as functional models exemplified by the Crustacea Malacostraca. *Zoologica Scripta*, **5**, 163–6.

Dahl, E. (1983) Phylogenetic systematics and the Crustacea Malacostraca – a problem of prerequisites. *Verhandlungen des Naturwissenschaftlichen Vereins in Hamburg*, **26**, 355–71.

Dahl, E. (1984) The subclass Phyllocarida (Crustacea) and the status of some early fossils; a neontologist's view. *Videnskabelige Meddeleser fra Dansk naturhistorisk Forening*, **145**, 61–76.

Dahl, E. (1991) Crustacea Phyllopoda and Malacostraca, a reappraisal of cephalic dorsal shield and fold systems. *Philosophical Transactions of the Royal Society, B*, **334**, 1–26.

Dahl, E. (1992) Aspects of malacostracan evolution. *Acta Zoologica*, **73**, 339–46.

Emerson, M.J. and Schram, F.R. (1990) The origin of crustacean biramous appendages and the evolution of Arthropoda. *Science*, **250**, 667–9.

Ferrari, F.D. (1988) Developmental patterns in numbers of ramal segments of copepod post-maxillipedal legs. *Crustaceana*, **54**, 256–93.

Fryer, G. (1992) The origin of the Crustacea. *Acta Zoologica*, **73**, 273–86.

Giesebrecht, W. (1913) Crustacea, in *Handbuch der Morphologie der Wierbellosen Tiere, Arthropoden 4* (ed. A. Lang), Gustav Fischer, Jena, pp. 9–252.

Gould, S.J. (1977) *Ontogeny and Phylogeny*, Belknap Press, Cambridge, Massachusetts.

Grobben, K. (1893) A contribution to the knowledge of the genealogy and classification of the Crustacea. *Annual Magazine of Natural History*, **11**, 440–73.

Grobben, K. (1919) Uber die muskulatur des vorderkopfes der stomatopoden und die systematische stellung dieser malakostrakengruppe. *Sitzungsberichte der Kaiserlichten Academie der Wissenschaften in Wien*, **128**, 185–214.

Grygier, M.J. (1983) Ascothoracida and the unity of the Maxillopoda. *Crustacean Issues*, **1**, 73–104.

Grygier, M.J. (1985) Comparative morphology and ontogeny of the Ascothoracida, a step towards phylogeny of the Maxillopoda. *Dissertation Abstracts International*, **45**, 2466B–7B.

Grygier, M.J. (1987) Nauplii, antennular ontogeny, and the position of the Ascothoracida within the Maxillopoda. *Journal of Crustacean Biology*, **7**, 87–104.

Haeckel, E. (1896) *Systematische Phylogenie der Wirbellose Tiere 2*, George Reimer, Berlin.

Hansen, H. J. (1893) Zur morphologie der Gliedmassen und Mundtheile der Crustaceen und Insecten. *Zoologischer Anzeiger*, **16**, 193–8.

Hauser, D.L. and Presch, W. (1991) The effect of ordered characters on phylogenetic reconstruction. *Cladistics*, **7**, 243–65.

Henriksen, K.L. (1928) Critical notes on some Cambrian arthropods described by Charles D. Walcott. *Videnskabelige Meddeleser fra Dansk naturhistorisk Forening*, **86**, 1–20.

Hessler, R.R. (1964) The Cephalocarida comparative skeletomusculature. *Memoirs of the Connecticut Academy of Arts and Sciences*, **16**, 1–97.

Hessler, R.R. (1971) Biology of the Mystacocarida, A prospectus. *Smithsonian Contributions in Zoology*, **76**, 87–90.

Hessler, R.R. (1982) The structural morphology of walking mechanisms in eumalacostracan crustaceans. *Philosophical Transactions of the Royal Society of London, B*, **296**, 245–98.

Hessler, R.R. (1983) A defense of the caridoid facies, wherein the early evolution of the Malacostraca is discussed, in *Crustacean Issues, Vol. 1, Crustacean Phylogeny* (ed. F.R. Schram), Balkema, Rotterdam, pp. 145–64.

Hessler, R.R. (1992) Reflections on the phylogenetic position of the Cephalocarida. *Acta Zoologica*, **73**, 315–16.

Hessler, R.R. and Newman, W.A. (1975) A trilobitomorph origin for the Crustacea. *Fossils and Strata*, **4**, 437–59.

Hessler, R.R., Marcotte, B.M., Newman, W.A. and Maddocks, R.F. (1982) Evolution within the Crustacea, in *Biology of Crustacea* (ed. L.G. Abele), Academic Press, New York, pp. 150–234.

Itô, T. (1989) Origin of the limb basis in copepod limbs, with reference to remipedian and cephalocarid limbs. *Journal of Crustacean Biology*, **9**, 85–103.

Manton, S.M. (1934) On the embryology of *Nebalia bipes*. *Philosophical Transactions of the Royal Society London, B*, **223**, 168–238.

Metschnikov, E. (1868) The history of the development of *Nebalia*. *Zapiski Imperatorskoi Akademii Nauk, St. Petersburg*, **13**, 1–48.

Moore, R.C. (1959) (ed.). *Treatise on Invertebrate Paleontology, Part O, Arthropoda 1*, Geological Society of America and the University of Kansas, Lawrence, Kansas.

Müller, K.J. (1983) Crustacea with preserved soft parts from the Upper Cambrian of Sweden. *Lethaia*, **16**, 93–109.

Müller, K.J. (1985) Skaracarida, a new order of Crustacea from the Upper Cambrian of Västergötland. *Fossils and Strata*, **17**, 1–65.

Müller, K.J. and Walossek, D. (1986) *Martinssonia elongata*. gen. et sp.n., a crustacean-like euarthropod from the Upper Cambrian 'Orsten' of Sweden. *Zoologica Scripta*, **15**, 73–92.

Müller, K.J. and Walossek, D. (1988) External morphology and larval development of the Upper Cambrian maxillopod *Bredocaris admirabilis*. *Fossils and Strata*, **23**, 1–69.

Newman, W.A. (1983) Origin of the Maxillopoda; urmalacostracan ontogeny and progenesis, in *Crustacean Phylogeny* (ed. F.R. Schram), Balkema, Rotterdam, pp. 105–19.

Newman, W.A., Zullo, V.A. and Withers, T.H. (1969) Cirripedia, in *Treatise on Invertebrate Paleontology Part R, Arthropoda, Vol. 4* (ed. R.C. Moore), Geological Society of America and the University of Kansas, Lawrence, Kansas, pp. R206–R295.

Nixon, K.C. and Davis, J.I. (1991) Polymorphic taxa, missing entries and cladistic analysis. *Cladistics*, **7**, 233–41.

Novozhilov, N.I. (1960) Podklass Pseudocrustacea, in *Osnovy Paleontologii, Arthropoda. Trilobitomorpha and Crustacea* (ed. Yu.A. Orlov), Nedra, Moscow, p.199.

Pimentel, R.A. and Riggins, R. (1987) The nature of cladistic data. *Cladistics*, **3**, 201–9.

Platnick, N.I., Griswold, C.E. and Coddington, J.A. (1991) On missing entries in cladistic analysis. *Cladistics*, **7**, 337–43.

Raymond, P.E. (1920) The appendages, anatomy and relationships of Trilobites. *Memoirs of the Connecticut Academy of Arts and Sciences*, **7**, 1–169.

Raymond, P.E. (1935) Leanchoilia and other Mid-Cambrian Arthropoda. *Bulletin of the Museum of Comparative Zoology, Harvard*, **76**, 205–30.

Rolfe, W.D.I. (1969) Phyllocarida, in *Treatise on Invertebrate Paleontology Part R, Arthropoda, Vol. 4* (ed. R.C. Moore), Geological Society of America and the University of Kansas, Lawrence, Kansas, pp. R296–331.

Sanders, H.L. (1955) The Cephalocarida, a new subclass of Crustacea from Long Island Sound. *Proceedings of the National Academy of Science, USA*, **41**, 61–6.

Sanders, H.L. (1963a) The Cephalocarida, functional morphology, larval development, comparative external anatomy. *Memoirs of the Connecticut Academy of Arts and Sciences*, **15**, 1–80.

Sanders, H.L. (1963b) Significance of the Cephalocarida, in *Phylogeny and Evolution of Crustacea* (eds H.B. Whittington and W.D.I. Rolfe), Museum of Comparative Zoology, Cambridge, Massachusetts, pp. 163–75.

Sars, G.O. (1887) Report of the Phyllocarida collected by H.M.S. Challenger during the years 1873–76. *Challenger Report, Zoology*, **19**, 1–38.

Schram, F.R. (1969a) Polyphyly in the Eumalacostraca? *Crustaceana*, **16**, 243–50.

Schram, F.R. (1969b) Some Middle Pennsylvanian Hoplocarida and their phylogenetic significance. *Fieldiana Geology*, **12**, 235–89.

Schram, F.R. (1978) Arthropods, a convergent phenomenon. *Fieldiana Geology*, **39**, 61–108.

Schram, F.R. (1981) On the classification of Eumalacostraca. *Journal of Paleontology*, **55**, 126–37.

Schram, F.R. (1982) The fossil record and evolution of the Crustacea, in *Biology of Crustacea* (ed. L.G. Abele), Academic Press, New York, pp. 93–147.

Schram, F.R. (1983) Remipedia and crustacean phylogeny, in *Crustacean Phylogeny* (ed. F.R. Schram), Balkema, Rotterdam, pp. 23–8.

Schram, F.R. (1984) Relationships within the eumalacostracan Crustacea. *Transactions of the San Diego Society of Natural History*, **20**, 301–12.

Schram, F.R. (1986) *Crustacea*, Oxford University Press, Oxford.

Schram, F.R. and Emerson, M.J. (1986) The great Tesnus fossil expedition of 1985. *Environment Southwest*, **515**, 16–21.

Schram, F.R. and Emerson, M.J. (1991) Arthropod Pattern Theory, A new approach to arthropod phylogeny. *Memoirs of the Queensland Museum*, **31**, 1–18.

Scourfield, D.J. (1926) On a new type of crustacean from the Old Red Sandstone – *Lepidocaris rhyniensis*. *Philosophical Transactions of the Royal Society London, B*, **214**, 153–87.

Serban, E. (1972) Bathynella (Podophallocarida Bathynellacea). *Travaux de l'Institut de Spéologie 'Emile Racovitza'*, **11**, 11–224.

Sharov, A.G. (1966) *Basic Arthropodan Stock*, Pergamon, New York.

Siewing, R. (1958) Untersuchungen zur Morphologie der Malacostraca. *Zoologische Jahrbücher, Abteilung für Anatomie*, **75**, 39–176.

Siewing, R. (1960) Neuere Ergebnisse der Verwandtschaftsforschung bei den Crustaceen. *Wissenschaftliche Zeitschrift der Universität Rostock, Reihe. Mathematik-Naturwissenschaften*, **9**, 343–58.

Siewing, R. (1963) Studies in Malacostracan morphology, Results and problems, in *Phylogeny and Evolution of Crustacea* (eds H.B. Whittington and W.D.I. Rolfe), Museum of Comparative Zoology, Cambridge, Massachusetts, pp. 85–104.

Simonetta, A.M. and Delle Cave, L. (1975) The Cambrian non trilobite arthropods from the Burgess Shale of British Columbia. A study of their comparative morphology, taxonomy and evolutionary significance. *Palaeontographica Italica*, **69**, 1–37.

Spears, T., Abele, L.G. and Applegate, M.A. (1994) Phylogenetic study of cirripedes and selected relatives (thecostraca) based on 18S rDNA sequence-analysis. *Journal of Crustacean Biology*, **14**, 641–56.

Størmer, L. (1944) On the relationships and phylogeny of fossil and Recent Arachnomorpha. *Skrifter utgitt av det Norske Vidensk Academi i Oslo*, **5**, 1–158.

Størmer, L. (1959) Chelicerata, Merostomata, in *Treatise on Invertebrate Paleontology, Part P, Arthropoda 2* (ed. R.C. Moore), Geological Society of America and the University of Kansas, Lawrence, Kansas, pp. 1–41.

Stubbings, H.G. (1975) *Balanus balanoides*. L.M.B.C. *Memoirs on Typical British Marine Plants and Animals*, **37**, 70–171.

Swofford, D.L. (1990) *PAUP, Phylogenetic Analysis Using Parsimony, Version 3.0.* Computer program distributed by the Illinois Natural History Survey, Champaign, Illinois.

Thomson, G.M. (1894) On a freshwater schizopod from Tasmania. *Transactions of the Linnean Society of London, Zoology*, **6**, 285–303.

Tiegs, O.W. and Manton, S.M. (1958) The evolution of the Arthropoda. *Biological Reviews of the Cambridge Philosophical Society*, **33**, 255–337.

Walcott, C.D. (1912) Middle Cambrian Branchiopoda, Malacostraca, Trilobita and Merostomata. Cambrian geology and paleontology, II. *Smithsonian Miscellaneous Collections*, **57**, 109–44.

Walley, L.J. (1969) Studies on the larval structure and metamorphosis of *Balanus balanoides*. *Philosophical Transactions of the Royal Society of London, B*, **256**, 237–80.

Walossek, D. (1993) The Upper Cambrian *Rehbachiella* and the phylogeny of the Branchiopoda and Crustacea. *Fossils and Strata*, **32**, 1–202.

Walossek, D. and Müller, K. J. (1990) Upper Cambrian stem lineage crustaceans and their bearing upon the monophyly of Crustacea and the position of Agnostus. *Lethaia*, **23**, 409–27.

Watling, L. (1983) Peracaridan disunity and its bearing on eumalacostracan phylogeny with a redefinition of eumalacostracan superorders, in *Crustacean Issues, Vol. 1, Crustacean Phylogeny* (ed. F.R. Schram), Balkema, Rotterdam, pp. 213–28.

Wheeler, W., Cartwright, P. and Hayashi, C. Y. (1993) Arthropod phylogeny, a combined approach. *Cladistics*, **9**, 1–39.

Wills, M.A., Briggs, D.E.G. and Fortey, R.A. (1994) Disparity as an evolutionary index. A comparison of Cambrian and Recent arthropods. *Paleobiology*, **20**, 93–130.

Wills, M.A., Briggs, D.E.G., Fortey, R.A. and Wilkinson, M. (1995) The significance of fossils in understanding arthropod evolution. *Verhandlungen der Deutschen Zoologischen Gesellschaft*, **57**,13–33.

Wills, M.A., Briggs, D.E.G., Fortey, R.A., Wilkinson, M. and Sneath, P.H.A. (1997) An arthropod phylogeny based on fossil and Recent taxa, in *Arthropod Fossils and Phylogeny* (ed. G.E. Edgecombe), Columbia University Press, New York (in press).

Wilson, G.D.F. (1992) Computerized analysis of crustacean relationships. *Acta Zoologica*, 73, 383–9.

Yager, J. (1981) Remipedia. A new class of Crustacea from a marine cave in the Bahamas. *Journal of Crustacean Biology*, **1**, 328–33.

Zimmer, C. (1926–27) Crustacea = Krebse Allgemeine Einleitung in die Naturgeschichte der Crustacea, in *Handbuch der Zoologie, Vol. III, No. 1* (eds W. Kukenthal and T. Krumbach), de Gruyter, Berlin.

# 16 The fossil record and evolution of the Myriapoda

*W.A. Shear*

*Department of Biology, Hampden-Sydney College, Hampden-Sydney, VA 23943, USA*
*email: BillS@tiger.hsc.edu*

## 16.1 INTRODUCTION

'Myriapoda' comprises four living (Chilopoda, Diplopoda, Symphyla and Pauropoda) and two or more extinct classes of terrestrial arthropods with a typical atelocerate head and an untagmatized trunk. The Chilopoda, centipeds, are carnivorous and predominantly soft-bodied (see Lewis, 1981, for a summary of their biology), while the other large class, Diplopoda, millipeds, usually have a mineralized cuticle and feed mostly on decaying plant matter (biology summarized by Hopkin and Read, 1992). Pauropoda and Symphyla are two small classes of poorly known minute animals, also subsisting on detritus; superficially symphylans resemble centipeds and pauropods look like tiny millipeds. Because of several morphological synapomorphies, these four classes have been considered phylogenetically close to insects, a view exemplified by the proposal of such names as Atelocerata and Tracheata, interchangeable epithets for a taxon including Chilopoda, Hexapoda, Diplopoda, Pauropoda and Symphyla.

The traditional view has been to regard Myriapoda as monophyletic and hence a taxon. While the entire history of this standpoint cannot be rehearsed here (see Kraus and Kraus, 1994, for a brief review), it is exemplified by the systematic treatment of Boudreaux (1979). Under a Superclass Atelocerata, Myriapoda is given status as a class, including two subclasses, Collifera (Infraclasses Pauropoda and Diplopoda) and Atelopoda (Infraclasses Chilopoda and Symphyla). The classification of Štys and Zrzavý (1994) is similar in principle, but the grouping of Symphyla and Chilopoda is rejected; Myriapoda and Hexapoda are ranked as subphyla.

The conclusions of Manton (1977) are well known: 'The trunk morphology of the myriapod classes is based upon a common plan, but the class differences are so great as to suggest parallel evolution of these classes from soft-bodied ancestral Uniramia with a common head type' (Manton, 1977, p. 396).

In contrast to these viewpoints, Dohle (1965, 1974, 1980, 1988, 1997, this volume), followed by Kraus and Kraus (1994; also Kraus, 1997, this volume), performed cladistic analyses which resulted in a paraphyletic 'Myriapoda'. Chilopoda is considered the sister taxon of all other atelocerates, and Hexapoda is the sister taxon of Progoneata, which includes Symphyla as the sister taxon of Diplopoda + Pauropoda [Dignatha; the name of this taxon may be inappropriate, since Hilken and Kraus (1994) have argued convincingly that the first and second maxillae are both present in the gnathochilarium]. Dohle's arrangement has found some acceptance, though most textbooks still adhere to a monophyletic Myriapoda. However, of the five supposed synapomorphies for Myriapoda proposed by Brusca and Brusca in their 1990 textbook, four are incorrect, and both synapomorphies uniting Symphyla and Chilopoda are wrong.

These are as follows, for 'Myriapoda':

5. Presence of Organs of Tömösvary: homologous structures occur in primitive hexapods.
6. Presence of myriapod repugnatorial glands: the glands of Diplopoda and Chilopoda are certainly not homologous, glands are unknown in Pauropoda and Symphyla.
8. Loss of the compound eyes: compound eyes have not been lost but have been transformed in various ways in the different classes.
9. Loss of the palps on the first and second maxillae: these are present in Chilopoda and in pselaphognath diplopods, probably also in Symphyla.

For Symphyla + Chilopoda:

18. Medial coalescence of both pairs of maxillae: generally not so in Chilopoda.
19. Modification of the first pair of trunk appendages into raptorial poison fangs: true only of Chilopoda, in Symphyla these appendages are reduced or absent.

These erroneous data were evidently also used by Wills *et al.* (1995) to conclude that myriapods were monophyletic.

While morphology strongly supports a paraphyletic 'Myriapoda' with hexapods nested between chilopods and progoneates, the conclusions of molecular studies have been inconsistent. Field *et al.* (1988), using 18S rRNA, placed a milliped (*'Spirobolus'*) at the very base of the arthropod

*Arthropod Relationships*, Systematics Association Special Volume Series 55, Edited by R.A. Fortey and R.H. Thomas. Published in 1997 by Chapman & Hall, London. ISBN 0 412 75420 7

branch of their distance tree; a crustacean grouped with an insect. Turbeville *et al.* (1991) found different trees depending on the algorithm used to analyse data from 18S rRNA: with distance analysis, they found (((Chelicerata ((Diplopoda (Crustacea (Insecta))), but with maximum parsimony it was ((Crustacea (Insecta)) (Myriapoda (Chelicerata)). A majority consensus rule tree was unable to resolve a trichotomy between Diplopoda, Insecta + Crustacea, and Chelicerata. In all three cases Crustacea was closer to Hexapoda (Insecta) than was Diplopoda. The only myriapod used was the milliped *Spirobolus marginatus* (data from Field *et al.*, 1988); Turbeville *et al.* (1991) suggested that large numbers of substitutions, said to be present in millipeds, can force a clade deeper into the tree than its real position. Boore *et al.* (1995) strongly supported a monophyletic Mandibulata on the basis of mitochondrial gene arrangements, but could not resolve relationships within the taxon.

Using nuclear ribosomal DNA and three tree-building methods (neighbour-joining, weighted maximum parsimony and maximum likelihood), Friedrich and Tautz (1995) concluded that Crustacea, not 'Myriapoda' is the sister-taxon of Hexapoda, and supported a monophyletic Myriapoda, based on data from *Polyxenus* and *Megaphyllum* (both millipeds) and *Lithobius* (a centiped).

Wheeler *et al.* (1993) sought to combine morphological and molecular data. Using 100 characters from the literature, they were unable to resolve the trichotomy Diplopoda–Chilopoda–Hexapoda, but Crustacea appeared as the sister taxon to these atelocerates. Wheeler *et al.* (1993) rightly questioned two of the synapomorphies used by Brusca and Brusca (1990), but accepted the incorrect statement that the maxillae of diplopods and chilopods lack palpi. None of the purely molecular trees they found, however, placed myriapods close to insects. Their consensus cladogram from 18S rDNA was unable to resolve the relationships of the major arthropod groups. Ubiquitin sequence data produced a tree best described as nonsensical, showing insects paraphyletic with respect to crustaceans and chelicerates (a crab and a barnacle grouped with dragonflies; a cicada with a spider). A similar problem, perhaps as a result of the inclusion of ubiquitin data, plagued a consensus cladogram for all molecular data: Odonata and Crustacea as sister taxa. However, centipeds and millipeds appear as the sister taxon of insects with the strangely placed crustacean examples. Combining molecular and morphological data led to a very traditional tree with (Crustacea ((Myriapoda (Hexapoda)). Sensible cautions when using molecular data were voiced by Wheeler *et al.* (1993): lack of support for morphological groups by molecular data may be due to polarization problems with non-additive multi-state characters; use of one molecule and single examples of large taxa seems likely to lead to an unstable situation [note also that Turbeville *et al.* (1991) derived strikingly different cladograms depending on how their data were analysed].

Thus, at present, disagreement between molecular trees is as striking as the disagreement between molecular and morphological trees. An attempt to combine data from both sources resulted in a highly traditional tree, essentially the same as that proposed by Snodgrass in 1938. Given that some morphological trees that imply or include a monophyletic Myriapoda are based on erroneous data (Brusca and Brusca, 1990), and that molecular phylogenies are clearly still unstable, I accept the well-supported arrangement of Dohle (1980) and Kraus and Kraus (1994). Myriapoda is not monophyletic as such and is not a taxon (Shear, 1994).

Within the myriapod classes, cladistic analyses have been carried out only for Chilopoda and Diplopoda; Symphyla and Pauropoda are small classes with relatively low diversity which have not been extensively studied. For chilopods, the most strongly supported hypothesis is that of Dohle (1985), which Shear and Bonamo (1988) extended to include an extinct class, Devonobiomorpha. Enghoff (1984) established the Diplopoda as monophyletic and presented a tree of the major subtaxa. Enghoff (1990) also worked out a hypothetical groundplan for the chilognath millipeds that was close to the morphology of the primitive Order Glomeridesmida.

## 16.2 THE FOSSIL RECORD

It would be reassuring to assert that at least some of the phylogenetic controversies mentioned in the preceding section can be resolved by an examination of the fossil record. Unfortunately, while the fossil record of myriapods, particularly diplopods, is quite rich, insufficient study has left us with a mass of unanalysed and uncorrelated data, much of it archaic and of dubious accuracy. One clear conclusion, however, is that the diversity of myriapods at a high taxonomic level was significantly greater in the Palaeozoic than it is today, with at least two extinct classes present among the fossils.

### 16.2.1 CAMBRIAN

The earliest known fossil attributed to Atelocerata (as 'Uniramia') is *Cambropodus gracilis* Robison, from the Middle Cambrian Wheeler Formation of Utah (Robison, 1990). The fossil is from a marine deposit, not well preserved, and seems to consist of an anterior, subglobular head with perhaps three pairs of long, filiform appendages, followed by about nine trunk units, each with a pair of similar appendages, either with swollen bases or superimposed on some other part of the body basal to the legs. Despite the original attribution by Robison (1990), there is no evidence that *Cambropodus* is really even an arthropod, as all of its features are also compatible with an identity as a polychaete annelid. Jointed appendages are claimed on the basis of angular bends in some of them, but of the visible ones, as many have curved flexions as have distinctly angular ones, and the angular bends are not consistently distributed on the legs as one would expect if they marked the joints between podomeres; these could be elongated, tentacle-like parapodia rather than legs. The misleading reconstruction of *Cambropodus* by Simonetta and Delle Cave (1991) includes many features not in evidence on the fossil, such as leg podomeres, multisegmented antennae, and gnathobases on the anterior appendages. However, if *Cambropodus* is indeed an arthropod, it shares some features, such as the swollen leg bases, with an enigmatic marine arthropod reported from the Lower Silurian of Waukesha, Wisconsin (USA) by Mikulic *et al.* (1985); unfortunately no detailed description of this animal has yet appeared. In turn, the Waukesha animal has some characters in common with kampecarids (see below) from the Late Silurian and Early Devonian, such as an apparently apodous tagma at the posterior end.

### 16.2.2 ORDOVICIAN

No fossils of possible myriapodous arthropods are known from the Ordovician, but Johnson *et al.* (1994) have reported trace fossils, probably made in mud exposed to air, that are consistent with myriapod locomotion. These were found in the (probably) Caradoc age Borrowdale Volcanic Group of Britain.

### 16.2.3 SILURIAN AND LATER

In the Late Silurian and Early Devonian, a considerable variety of myriapod fossils has been found; this material has been ably reviewed by Almond (1985). Since Almond's review, myriapod remains have been discovered by Jeram *et al.* (1990) as part of the earliest known terrestrial fauna, from the Pridoí of Shropshire, England.

Centipeds, attributable to the extant Order Scutigerimorpha, are a part of this early fauna, represented by distinctive podomeres (Shear, *et al.*, in press). Very similar podomeres also occur in the Middle Devonian Gilboa (New York, USA) fauna (Shear *et al.*, 1984). Shear and Bonamo (1988) described an extinct order of centipeds, Devonobiomorpha, from Gilboa. The characters justifying the new order are of the same level as those used to diagnose the living orders. *Devonobius delta*, the only known species, appears to be somewhat intermediate between the Order Lithobiomorpha and the epimorphic orders Scolopendromorpha and Geophilomorpha; it probably represents a stem group of the Epimorpha. Well-preserved whole scutigerimorphs appear in siderite nodules from Mazon Creek (Illinois, USA), of Late Carboniferous Age (Mundel, 1979). Other centiped orders are also represented in the Upper Carboniferous; *Mazoscolopendra* is an evident scolopendromorph, and *Eileticus* may be a geophilomorph (Mundel, 1979). Both the scutigerimorph and scolopendromorph of the Upper Carboniferous conform in every visible respect to living members of the same orders. A few other supposed centiped fossils have been named from the Upper Carboniferous, but all require restudy. There appears to be no Mesozoic record for Chilopoda, but extant genera are fairly well represented in Tertiary ambers (Hoffman, 1969).

*Archidesmus macnicoli*, described by Peach in 1882, is a good candidate for the earliest milliped, of Gedinnian (Early Devonian) age, from the Lower Old Red Sandstone lacustrine fish beds of Scotland. About 30 diplosegments are visible in the specimen, each with a broad, lateral set of paranota, which cover the legs (Almond, 1985). The few Devonian milliped fossils have been comprehensively studied by Almond (1986) in an unpublished thesis, and a number of new taxa are diagnosed therein. Heads are rarely preserved in these Devonian fossils, but where they appear, large mandibles and a single pair of antennae can sometimes be seen, and this, coupled with the presence of diplosegments divided into anterior prozonites and posterior metazonites, mark them as millipeds. At least three or four 'orders' might be subjectively recognized among this array, none of which seems very closely related to any living millipeds (Almond, 1986). One distinguishing feature of a number of these fossils is the presence of a pair of large sternal pores located between the leg bases. Similar pores also occur in euphoberiid millipeds from the Late Carboniferous, suggesting a possible evolutionary continuity with the Devonian fossils. Such pores are unknown in living millipeds.

Evident diplotergites of an unusual form have been described by Tesakov and Alekseev (1992) from the Devonian of Kazakhstan, but more evidence is needed to determine their taxonomic identity. The name given to these specimens, *Lophodesmus*, is preoccupied in living diplopods.

A few, poorly known milliped fossils are from the Early Carboniferous. Typical is an unnamed species from the Viséan of Scotland (Figure 16.1); the head was typically diplopodan, the diplotergites bore short spines as well as the pores of repugnatorial glands (the earliest evidence of these), the sternites show prominent spiracles, and sternal pores were lacking (Shear, 1994). The greatest wealth of milliped fossils comes from the Late Carboniferous of North America and Europe. Unfortunately, the available data consist almost entirely of isolated descriptions of new finds and new taxa. Hoffman (1969) presented a partial review of the overall milliped fossil record, but was not able to comment on numerous Palaeozoic forms, and Kraus (1974) briefly examined the morphology of Palaeozoic diplopods.

Pentazonians were in existence by the Late Carboniferous, including members of the extinct Order

**Figure 16.1** An unnamed fossil milliped from the Early Carboniferous (Viséan) of East Kirkton, Scotland.

Amynilyspedida. These are short-bodied millipeds capable of complete conglobulation, and in the case of *Amynilyspes wortheni* Scudder (Figure 16.2), this defensive behaviour was enhanced by the presence of long, paired spines on the tergites, a feature not seen in any living forms. The spines would certainly make *Amynilyspes* a painful mouthful for a large amphibian (Hannibal and Feldmann, 1981). Members of the genus *Pleurojulus* Fritsch were cylindrical in form and resembled larger species of the living Orders Spirobolida and Spirostreptida, except that they had free diplopleura (Kraus, 1974). Other genera, such as *Xyloiulus* Cook and *Plagioscetus* Hoffman, were very similar to modern Spirobolida, if not members of that order, with diplosegments in the form of complete rings (Hoffman, 1963).

**Figure 16.2** Original reconstruction of *Amynilyspes wortheni* Scudder, a pentazonian milliped from the Late Carboniferous of Mazon Creek, Illinois, USA, based upon Hannibal and Feldman, 1981. The head is conjectural, drawn to resemble that of the related living glomerid millipeds. The animal probably would have been about 30 mm long and 8–9 mm wide, but similar forms up to 60 mm long are also known.

The dominant diplopods of the Late Carboniferous belonged to the Order Euphoberiida. Following previous workers, Hoffman (1969) considered the euphoberiids to be members of a distinct class, Archipolypoda. There appears to be no good reason to consider them as such, and the order should be included in Diplopoda (Burke, 1979; Hannibal and Feldmann, 1988; Hannibal, 1995). Euphoberiids are typified by species of the genera *Euphoberia* Meek and Worthen, *Acantherpestes* Meek and Worthen, and *Myriacantherpestes* Burke. The most detailed and accurate descriptions of these animals are to be found in Burke (1973, 1979). Euphoberiids were large animals, but not extraordinarily so by modern standards, being a little larger than the largest living millipeds. Like amynilyspedids, they bore metatergal spines, in some cases very long ones (Figure 16.3). It is unclear if these spines were directed dorsally

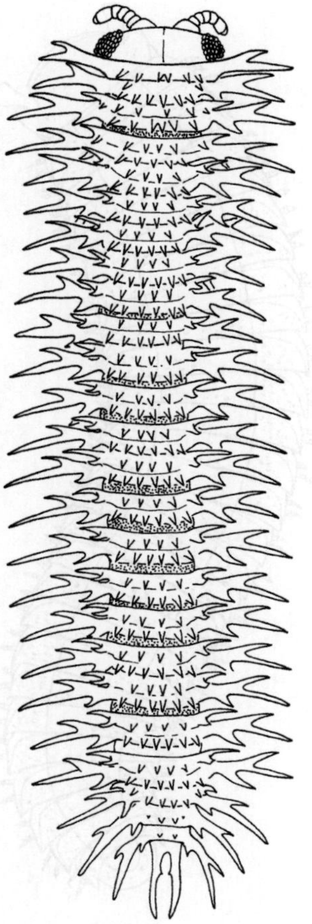

**Figure 16.3** Original reconstruction of an undescribed genus and species of euphoberiid milliped from the Late Carboniferous of Hamilton, Kansas, USA, based on specimen HM A2172, Hunterian Museum, Glasgow. The head is based on the reconstruction of *Myriacantherpestes ferox* by Burke (1979). This animal probably would have been about 130 mm long, but Hannibal and Feldmann (1988) have reported specimens of this species up to 270 mm long. The low segment count (26) relative to other euphoberiid specimens suggest that it may have been a juvenile, but the fossil shows traces of modified legs associated with segments 14–15, suspected by Hannibal (1995) of being indicative of sexual dimorphism.

from a cylindrical body, as reconstructed by Hoffman (1969) for *Euphoberia*, or were laterally directed as Burke (1979) shows for *Myriacantherpestes*. In any case, they seem clearly defensive and appear to have been broken off in life in some of the fossils. While a few have argued that the euphoberiids were weakly sclerotized (Kraus, 1974), an examination of well-preserved examples (Hannibal and Feldmann, 1988) shows a thick cuticle, deeply pitted and with brittle fractures; the tergal spines seem to have been hollow, based on the pattern of crushing. It is hard to imagine the long spines as defensive if they were not also well hardened. The defence may not have always worked: Hannibal and Feldmann (1988) found euphoberiid tergites in amphibian coprolites. Most authors have made reference to the large, faceted eyes not only of euphoberiids but also of other fossil diplopods, eyes which contrast with the loose groups of pseudocellar units or complete blindness of all living forms. Kraus (1974) estimated the number of facets in one fossil diplopod's eye as over 1000. Of course it is impossible to know if these were true compound eyes or unusually large pseudofaceted eyes, as described for modern diplopods by Paulus (1979). The tergites of euphoberiids are not divided in the midline, and there are no free pleurites. The sterna, two per segment, are also not divided but differ from living forms, and resemble their possible Devonian ancestors, in each bearing a pair of large pores, which possibly accommodated eversible sacs. Such sacs are found today on the coxae, not sterna, of some millipeds. The presence of these pores and a study of the legs led Verhoeff (1926) to propose that euphoberiid sterna resulted from the fusion of sternites, pleurites and leg coxae, moving the eversible sacs of the leg coxae to the compound coxosterna. However, careful examination of the legs of both euphoberiids and Devonian millipeds show that they have the typical seven podomeres, including a free coxa. Thus the sternal pores of these animals represent a structure not known from other millipeds, and a possible synapomorphy. Above the leg sockets are presumed tracheal stigmata, located similarly in living millipeds.

The presence of stigmata on the sterna, not the pleura, of millipeds has not been taken into account in work on the origin of the pleural sclerites in Atelocerata. The euphoberiids clearly do not have free pleurites. The tracheal stigmata are in the sterna, not pleura. This is also the case for millipeds in which the pleura are free, unfused with the terga. The articulation of diplopod legs with the sterna, not pleura, and the presence of the stigmata on the sterna, suggest that the pleura of millipeds are not homologous to the pleura of insects nor the pleural sclerites of centipeds, both now sometimes considered to derive from basal leg segments subsumed into the body wall (Kukalová-Peck, 1993).

Permian milliped fossils are rare and not well-studied. The few Mesozoic records (Dzik, 1975, 1981) can be accommodated in living orders, as can all Cenozoic milliped fossils.

### 16.2.4 ARTHROPLEURIDS

Members of the Class Arthropleuridea range in age from Late Silurian to Upper Carboniferous (Shear and Selden, 1995) (The spelling of Arthropleuridea has been changed here from Arthropleurida to avoid duplication of the ordinal name, Arthropleurida.) The class is defined by diplosomy (Briggs and Almond, 1994; Shear and Selden, 1995), paranotal tergal lobes separated from the axis by a suture, and by a unique system of sclerotized plates buttressing the leg insertions (Rolfe and Ingham, 1967). Currently two orders are recognized in the class, and a third is in the process of being described (Shear and Selden, 1995; W. Shear and J. Almond, manuscript in preparation).

The Order Arthropleurida (Upper Carboniferous) contains the largest land arthropods of which we have knowledge. Extrapolating from incomplete specimens suggests a length of well over 2 m for some species (Rolfe, 1969; Hahn *et al.*, 1986). The spiny, three-part tergites were conspicuously flattened, perhaps a habitus like that of modern polydesmid millipeds. Rolfe and Ingham (1967) suggested that the legs consisted of up to 10 podomeres, each with strong ventral spines, but also offered alternative interpretations, some more consistent with a seven-segmented leg. Briggs and Almond (1994) demonstrated that *Arthropleura* must have been diplopodous, each tergite covering two pairs of legs. Each leg pair was accompanied by a median sternite, and at the base of each leg were three complex plates, designated K, B, and rosette plates by Rolfe and Ingham (1967). The rosette plate was divided into five or six lobes. The function of these plates is not clear, but Rolfe and Ingham (1967) suggested that they served as pleural sclerites which buttressed the massive legs' junctions with the body. They may be basal leg segments subsumed into the body wall; a similar arrangement is seen in centipeds. Little is known of the head; Rolfe (1969) described a poorly preserved head, from the single specimen on which it appeared, to be a rounded, oval structure, possibly bearing mandibles. Briggs and Almond (1994) described the dorsal surface of the head as much wider than long, and with shallow lateral embayments, from which projected rounded, narrow lobes. Subsequent examination of the material suggests that this may in fact be the first tergite, which completely covers the head (Figure 16.4), and that the projecting lobes are actually head structures.

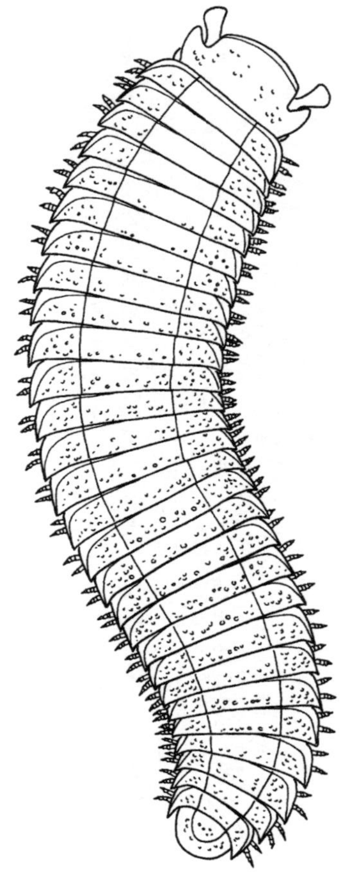

**Figure 16.4** Reconstruction of *Arthropleura armata* Jordan, from the Upper Carboniferous of the Saar region (modified from Hahn *et al.*, 1986). The living animal would have been more than 1 m long. The number of segments is conjectural, as is the head, which is based on studies of other arthropleurids and well-preserved specimens of another species from Montceau-les-Mines, France (see also Briggs and Almond, 1994).

The Order Eoarthropleurida (Late Silurian–Late Devonian) was established by Shear and Selden (1995) for the genus *Eoarthropleura* Størmer. Members of this genus are known only from fragments, usually isolated tergites, sternites and plates, but Shear and Selden (1995) found evidence in sternal and tergal proportions for diplosomy. The legs (recently discovered in Late Devonian rocks from North America) appear to have had the typical seven podomeres, with strong ventral spines like those found in *Arthropleura.* The plates at the base of the leg are difficult to homologize with the same structures in *Arthropleura*, but there are at least two per leg. Heads are unknown, as is the number of trunk segments, but length of the largest known specimen has been estimated at about 15 cm.

Specimens of an undescribed species of arthropleurid deserving ordinal distinction have been found as a part of the well-studied Middle Devonian Gilboa fauna. These animals diverged in morphology from the larger forms described above, being no more than 7–8 mm in length, consisting of a head, nine tergites and an anal cone. There are no traces of sutures between the body axis and the paranotal lobes. They, too, were diplopodous, with median sterna and only a single plate at the base each leg. The legs were long, gracile, and fitted with disproportionately large ventral spines. Several well-preserved heads are known and include mandibles and a gnathochilarium-like structure, but no sign of eyes, antennae or antennal sockets. Peculiar trumpet-like structures protruded from the sides of the head, possibly homologous to the 'rounded lobes' found in *Arthropleura.* There appears to have been strong sexual dimorphism in the tergites and legs of the last few segments.

The relative abundance and wide distribution of the remains of these creatures suggest that they were major participants in terrestrial ecosystems from the Silurian at least to the Late Carboniferous, but it must be remembered that no evidence of any respiratory structure is known, and the assertion of their terrestriality is based entirely upon analogy. Briggs *et al.* (1991) are sure that *Arthropleura* could not have moulted on land. Not enough is known about arthropleurids to find a place for them in atelocerate phylogeny, if indeed they are atelocerates, but the presence of diplosomites and what little is known of the head suggest a relationship with Dignatha.

### 16.2.5 KAMPECARIDA

A second extinct taxon probably worthy of recognition at the level of class is Kampecarida (Figure 16.5), treated as an order of uncertain position in Almond's (1986) unpublished thesis. Kampecarids are known only from the Late Silurian and Early Devonian of Scotland and the Anglo-Welsh Basin. Why they had not been intensively studied before the work of Almond is hard to understand, since they are the most abundant of any myriapod-like fossils from that time period, and new finds continue to be made. Specimens occur, sometimes in groups, in lacustrine or fluviatile sediments associated with finely divided plant material. There is insufficient evidence to establish their habitus as terrestrial or fresh-water (Almond, 1986).

The kampecarid head was subglobular and probably bore a single pair of antennae; there is some evidence for a mandible. The dorsal part of the head may have been covered by two plates, or by a single plate with a transverse suture, or have been followed by a legless collum segment, as in millipeds. At least the following two trunk segments appear to have been haplosegments, with a single pair of legs. The next trunk segments are clearly diplosegments, with deep strictures between prozonites and metazonites, the latter being strongly arched and rounded. The prozonites fit snugly into the preceding metazonite, producing a ball-and-socket joint. There were no paratergal folds or free pleurites, and the legs articulated with the sterna. Most peculiarly, the last diplosomite(?) was inflated and legless, enclosing dorsal and ventral anal valves with small, terminal processes. In a few specimens, the second trunk segment bears a pair of dorsal horns; this may have been an expression of sexual dimorphism (Almond, 1986). Again, while the presence of diplosegments, the articulation of legs with the sterna, and a possible collum segment implies a relationship with Diplopoda and Pauropoda, too few data are available to fully incorporate these animals into our phylogenetic schemes.

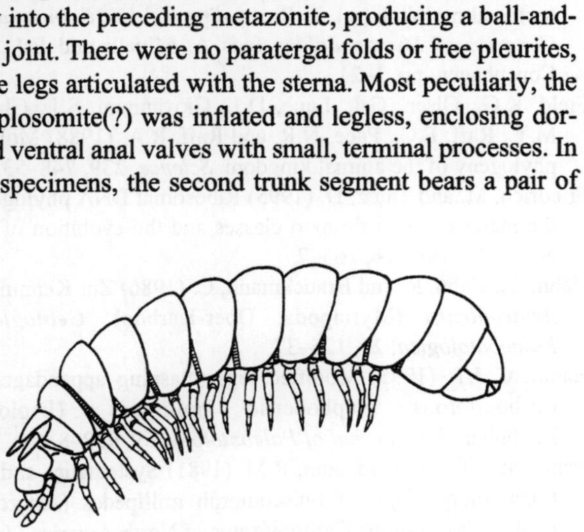

**Figure 16.5** Reconstruction of a generalized kampecarid, based mostly on *Kampecaris dinmorensis*, and modified from Almond (1986). The living animal probably would have been about 20 mm long.

### 16.2.6 EUTHYCARCINOIDS

Recently, a third fossil group, the Euthycarcinoidea, has been suggested as belonging to the Atelocerata (Schram and Rolfe, 1982; McNamara and Trewin, 1993; Wills *et al.*, 1995). In particular, it has even been proposed that these animals are more closely related to insects than any myriapod class (McNamara and Trewin, 1993; Wills *et al.*, 1995), a conclusion based on very weak evidence from poorly preserved fossils, and which is incongruent with the numerous well-established cladograms based on living forms. An examination of the published evidence and new specimens of *Kottyxerxes gloriosus* Schram leads me to question these conclusions. This is not the place for a detailed analysis of the evidence, but I would simply state that specimens of isopod crustaceans, with their uniramous limbs, highly reduced first antennae, and clear tagmatization, in a state of preservation similar to that of even the best preserved euthycarcinoids, let alone the poorly preserved *Kalbarria* (McNamara and Trewin, 1993), could as easily be considered atelocerates.

In the majority rule consensus cladogram of Wills *et al.* (1995), euthycarcinoids (*Kalbarria*) are nested paraphyletically among the insects, a clearly unacceptable result. They state, 'The hexapods, myriapods and the euthycarcinoid *Kalbarria* are united in all of the fundamentals by features including the absence of second antennae, the modification of the appendages of the third segment as whole-limb jaws, the presence of tracheae, ectodermal Malpighian tubules, and specialized ommatidial structures' (Wills *et al.*, 1995, p. 210). One pair of antennae is **preserved** in a few euthycarcinoids (though not in *Kalbarria*), but if the euthycarcinoids are isopod-like crustaceans, these could be the second antennae. It cannot be confirmed from the fossils that euthycarcinoids had whole-limb mandibles (indeed, the hypothesis of whole-limb mandibles in any arthropod is rapidly being dropped), there is no evidence of tracheae in *Kalbarria* and, of course, fossils can tell us nothing of the embryological derivation of internal organs or the ultrastructure of eyes. Edgecombe and Ramsköld (1996) have strongly criticized Wills *et al.* (1995; see also Wills, 1996) for their method of scoring inapplicable characters, as well as for the presence of a great deal of homoplasy and the many reversals treated as synapomorphies in their data matrix. In my opinion the euthycarcinoids are at best Atelocerata of very uncertain position, and quite possibly are more closely related to

Crustacea, as originally supposed. Certainly their many specializations, seen even in the Silurian *Kalbarria*, would seem to exclude them from the direct ancestry of insects.

## 16.3 POSTSCRIPT

What we have learned so far suggests that few, if any, questions about the phylogeny of the atelocerate classes will be answered by the fossil record, barring the discovery of truly remarkable levels of preservation, even better than those now being exploited by the HF maceration method (Shear *et al.*, 1984). What the fossils *do* tell us is that our living fauna samples only a fraction of the past high-level systematic diversity of this ancient group. Fossils must be taken into account, however, in any phylogenetic analysis; this is what is commendable in the approach of Wills *et al.* (1995). Nonetheless, at the same time, characters used to place living groups must be carefully checked by palaeontologists, and neontologists must come to recognize the serious limitations of fossil evidence.

## ACKNOWLEDGEMENTS

Profound thanks are due to John Almond for permission to use unpublished material from his thesis, the photograph used as Figure 16.1, and for many hours of discussion on fossil myriapods. I am also grateful to Joseph Hannibal for suggestions on the manuscript and for the use of photographs and drawings in the lecture on which this paper is based. Suggestions by Paul A. Selden and Asa Kreevitch substantially improved the manuscript.

## REFERENCES

Almond, J.E. (1985) The Silurian-Devonian fossil record of the Myriapoda. *Philosophical Transactions of the Royal Society of London, B*, **309**, 227–38.

Almond, J.E. (1986) *Studies on Palaeozoic Arthropoda*, Unpublished doctoral thesis, University of Cambridge.

Boore, J.L., Collins, T.M., Stanton, D., Daehler, L.L. and Brown, W.M. (1995) Deducing the pattern of arthropod phylogeny from mitochondrial DNA rearrangements. *Nature*, 376,163–5.

Boudreaux, H.B. (1979) *Arthropod Phylogeny, with Special Reference to Insects*, Wiley-Interscience, New York.

Briggs, D.E.G. and Almond, J.E. (1994) The arthropleurids from the Stephanian (Late Carboniferous) of Montceau-les-Mines (Massif Central - France), in *Quand le Massif Central était sous l'équateur: un écosystème Carbonifére à Montceau-les-Mines* (eds C. Poplin and D. Heyler), Editions du Comité des Traveaux Historiques et Scientifiques, Paris, pp. 127–35.

Briggs, D.E.G., Dalingwater, J.E. and Selden, P.A. (1991) Biomechanics of locomotion in fossil arthropods, in *Biomechanics and Evolution* (eds J.M.V. Rayner and R. J. Wootton), Cambridge University Press, Cambridge, pp. 37–56.

Brusca, R.C. and Brusca, G.J. (1990) *Invertebrates*, Sinauer Associates, Sunderland, Massachusetts.

Burke, J.J. (1973) Notes on the morphology of *Acantherpestes* (Myriapoda, Archipolypoda), with the description of a new species from the Pennsylvanian of West Virginia. *Kirtlandia*, **17**, 1–33.

Burke, J.J. (1979) A new millipede genus, *Myriacantherpestes* (Diplopoda, Archipolypoda) and a new species, *Myriacantherpestes bradebirksi*, from the English coal measures. *Kirtlandia*, **30**, 1–23.

Dohle, W. (1965) Uberr die Stellung der Diplopoden in System. *Zoologischer Anzeiger, Supplement*, **28**, 597–606.

Dohle, W. (1974) The origin and inter-relationships of the myriapod groups. *Symposia Zoological Society of London*, **32**, 191–8.

Dohle, W. (1980) Sind die Myriapoden eine monophyletische Gruppe? Ein Diskussion der Verwandtschaftsbeziehung der Antennaten. *Abhandlungen der Naturwissenschaften vereins Hamburg*, **23**, 45–104.

Dohle, W. (1985) Phylogenetic pathways in the Chilopoda. *Bijdragen tot de Dierkunde*, **55**, 55–66.

Dohle, W. (1988) Myriapoda and the ancestry of insects, Charles H. Brookes Memorial Lecture, published by The Manchester Polytechnic, Manchester, UK.

Dzik, J. (1975) Results of the Polish–Mongolian Palaeontological Expeditions – Part VI. Spiroboloid millipeds from the Late Cretaceous of the Gobi Desert, Mongolia. *Palaeontologia Polonica*, **33**, 17–24, plate VII.

Dzik, J. (1981) An Early Triassic milliped from Siberia and its evolutionary significance. *Neues Jahrbuch für Geologie und Paläontologie Monatshefte*, 1981, 395–404.

Edgecombe, G.D. and Ramsköld, L. (1996) Classification of the arthropod *Fuxianhuia. Science*, **272**, 747–8.

Enghoff, H. (1984) Phylogeny of millipeds – a cladistic analysis. *Zeitschrift für Zoologie, Systematik und Evolutionsforschung*, **22**, 8–26.

Enghoff, H. (1990) The ground-plan of chilognathan millipeds (external morphology), in *Proceedings of the 7th International Congress of Myriapodology* (eds A. Minelli and E.J. Brill), Copenhagen, pp. 1–21..

Field, K.G., Olsen, G.J., Lane, D.J., Giovannoni, S.J., Ghiselin, M.T., Raff, E.C., Pace, N.R. and Raff, R.A. (1988) Molecular phylogeny of the animal kingdom. *Science*, **239**, 748–53.

Friedrich, M. and Tautz, D. (1995) Ribosomal DNA phylogeny of the major extant arthropod classes and the evolution of myriapods. *Nature*, **376**, 165–7.

Hahn, G., Hahn, R. and Brauckmann, C. (1986) Zur Kenntnis von *Arthropleura* (Myriapoda; Ober-Karbon). *Geologica et Palaeontologica*, **20**, 125–37.

Hannibal, J.T. (1995) Modified legs (clasping appendages?) of Carboniferous euphoberiid millipeds (Diplopoda: Euphoberiida). *Journal of Paleontology*, 69, 932–8.

Hannibal, J.T. and Feldmann, R.M. (1981) Systematics and functional morphology of oniscomorph millipedes (Arthropoda: Diplopoda) from the Carboniferous of North America. *Journal of Paleontology*, **55**, 730–46.

Hannibal, J.T. and Feldmann, R.M. (1988) Millipeds from late Paleozoic limestones at Hamilton, Kansas, in *Regional Geology and Paleontology of Upper Paleozoic Hamilton Quarry Area*, Kansas Geological Society Guidebooks, Series 6, pp. 125–31.

Hilken, G. and Kraus, O. (1994) Struktur und Homologie de Komponenten des Gnathochilarium der Chilognatha (Tracheata, Diplopoda). *Verhandlungen des naturwissenschaftlichen Vereins in Hamburg*, **34**, 35–50.

Hoffman, R.L. (1963) New genera and species of upper Paleozoic Diplopoda. *Journal of Paleontology*, **37**,167–74.

Hoffman, R.L. (1969) Myriapoda, exclusive of Insecta, in *Treatise on Invertebrate Paleonotology* (ed. R.C. Moore), Geological Survey of America and University of Kansas, Lawrence, R572–606.

Hopkin, S. and Read, H. (1992) *The Biology of Millipedes*, Oxford University Press, Oxford.

Jeram, A.J., Selden, P.A. and Edwards, D. (1990) Land animals in the Silurian: arachnids and myriapods from Shropshire, England. *Science*, **250**, 658–61.

Johnson, E.W., Briggs, D.E.G., Suthern, R.J., Wright, J.L. and Tunnicliff, S.P. (1994) Non-marine arthropod traces from the subaerial Borrowdale Volcanic Group, English Lake District. *Geological Magazine*, **117**, 395–406.

Kraus, O. (1974) On the morphology of Palaeozoic diplopods. *Symposium of the Zoological Society of London*, **32**, 13–22.

Kraus, O. and Kraus, M. (1994) Phylogenetic system of the Tracheata (Mandibulata): on 'Myriapoda'–Insecta interrelationships, phylogenetic age and primary ecological niches. *Verhandlungen der naturwissenschaftlichen Vereins in Hamburg*, **34**, 5–31.

Kukalová-Peck, J. (1993) Origin of the insect wing and leg articulation from the arthropodan leg. *Canadian Zoologist*, **61**, 1618–69.

Lewis, J.G.E. (1981) *The Biology of Centipedes*, Cambridge University Press, New York.

Manton, S.M. (1977) *The Arthropoda: Habits, Functional Morphology, and Evolution*, Clarendon Press, Oxford.

McNamara, K.J. and Trewin, N.H. (1993) A euthycarcinoid arthropod from the Silurian of Western Australia. *Palaeontology*, **36**, 319–36.

Mikulic, D.G., Briggs, D.E.G. and Klussendorf, J. (1985) A new exceptionally preserved biota from the Lower Silurian of Wisconsin, USA. *Philosophical Transactions of the Royal Society of London, B*, **311**, 75–85, plates 1, 2.

Mundel, P. (1979) The centipedes (Chilopoda) of the Mazon Creek, in *Mazon Creek Fossils* (ed. M. Nitecki), Academic Press, New York, pp. 72–100.

Paulus, H.F. (1979) Eye structure and the monophyly of the Arthropoda, in *Arthropod Phylogeny* (ed. A.P. Gupta), Van Nostrand Reinhold, New York, pp. 299–383.

Peach, B.N. (1882) On some fossil myriapods from the Lower Old Red sandstone of Forfarshire. *Proceedings of the Royal Physical Society of Edinburgh*, **7**, 178–88.

Robison, R.A. (1990) Earliest known uniramous arthropod. *Nature*, **343**, 163–4.

Rolfe, W.D.I. (1969) Arthropleurida, in *Treatise on Invertebrate Paleontology* (ed. R.C. Moore), Geological Society of America, Lawrence, Kansas, pp. R607–20.

Rolfe, W.D.I. and Ingham, J.K. (1967) Limb structure, affinity and diet of the Carboniferous 'centipede' *Arthropleura. Scottish Journal of Geology*, **3**, 118–24.

Schram, F.R. and Rolfe, W.D.I. (1982) New euthycarcinoid arthropods from the Upper Pennsylvanian of France and Illinois. *Journal of Paleontology*, **56**, 1434–50.

Shear, W.A. (1994) Myriapodous arthropods from the Viséan of East Kirkton, West Lothian, Scotland. *Transactions of the Royal Society of Edinburgh*, **84**, 309–16.

Shear, W.A. and Bonamo, P.M. (1988) Devonobiomorpha, a new order of centipeds from the Middle Devonian of Gilboa, New York State, USA, and the phylogeny of centiped orders. *American Museum Novitates*, **2927**, 1–30.

Shear, W.A. and Selden, P.A. (1995) *Eoarthropleura* (Arthropoda, Arthropleurida) from the Silurian of Britain and the Devonian of North America. *Neues Jahrbuch Geologie und Paläontologie Abhandlungen*, **196**, 347–75.

Shear, W.A., Bonamo, P.M., Grierson, J.D., Rolfe, W.D.I., Smith, E.L. and Norton, R.A. (1984) Early land animals in North America: evidence from Devonian age arthropods. *Science*, 224, 492–4.

Simonetta, A. and Delle Cave, L. (1991) Early Palaeozoic arthropods and problems of arthropod phylogeny, with some notes on taxa of doubtful affinities, in *Early Evolution of Metazoa and the Significance of Problematic Taxa* (eds A. Simonetta and S. Conway Morris), Cambridge University Press, Cambridge, pp. 189–244.

Snodgrass, R.E. (1938) Evolution of the Annelida, Onychophora and Arthropoda. *Smithsonian Miscellaneous Collections*, **97**,1–159.

Štys, P. and Zrzavý, J. (1994) Phylogeny and classification of extant Arthropoda: review of hypotheses and nomenclature. *European Journal of Entomology*, **91**, 257–75.

Tesakov, A.S. and Alekseev, A.S. (1992) Myriapod-like arthropods from the Lower Devonian of central Kazakhstan. *Paleontologicheskii zhurnal*, **26**, 15–19.

Turbeville, J.McC., Pfeifer, D.M., Field,K.G. and Raff, R.A. (1991) The phylogenetic status of arthropods, as inferred from 18S rRNA sequences. *Molecular Biology and Evolution*, **8**, 669–86.

Verhoeff, K.W. (1926) Fossile Diplopoden, PP. 330–59 in *Bronns' Klassen und Ordnungen des Tier-Reichs*, 5, II, 2.

Wheeler, W.C., Cartwright, P. and Hayashi, C.V. (1993) Arthropod phylogeny: a combined approach. *Cladistics*, **9**, 1–39.

Wills, M.A. (1996) Classification of the arthropod *Fuxianhuia. Science*, **272**, 746–7.

Wills, M.A., Briggs, D.E.G., Fortey, R.A. and Wilkinson, M. (1995) The significance of fossils in understanding arthropod evolution. *Verhandlungen der Deutschen Zoologischer Gesellschaft*, **88**, 203–15.

# 17 The early history and phylogeny of the chelicerates

*J.A. Dunlop and P.A. Selden*

*Department of Earth Sciences, University of Manchester, Oxford Road, Manchester, M13 9PL, UK*
*email: jdunlop@fs1.ge.man.ac.uk*
*Department of Earth Sciences, University of Manchester, Oxford Road, Manchester, M13 9PL, UK*
*email:Paul.selden@man.ac.uk*

## 17.1 INTRODUCTION

Chelicerata is one of the major arthropod groups, characterized by a body divided into two tagmata, prosoma and opisthosoma, while the name Chelicerata refers to the chelicerae, the chelate first pair of appendages. Chelicerates comprise the arachnids (spiders, scorpions, mites, etc.), the extinct eurypterids (sea scorpions) and the xiphosurans (horseshoe crabs). Among Recent groups the limits of the Chelicerata are well defined, i.e. arachnids and xiphosurans, though the position of pycnogonids (sea spiders) remains uncertain (King, 1973). Phylogenetic relationships within the Chelicerata remain unresolved (Weygoldt and Paulus, 1979; van der Hammen, 1989; Shultz, 1990). When fossil taxa are considered, the limits of the Chelicerata become less well constrained. This is especially true of various problematic arthropods: fossils such as aglaspidids (Raasch, 1939), chasmataspids (Caster and Brooks, 1956), *Sanctacaris* Briggs and Collins, 1988 and other Burgess Shale-type arthropods (see Conway Morris, 1992 for a review). Different authors at different times have expanded the definition of Chelicerata to include some of these problematic groups (e.g. Bergström, 1979) and a range of higher taxon nomenclature has been generated.

The history of chelicerate phylogeny can be traced back to Lankester's (1881) classic paper, 'Limulus, *an arachnid*' in which it was proposed that xiphosurans were related to arachnids, not crustaceans as had previously been thought. Lankester (1881) was not the first to propose this theory, but he argued the case most convincingly. To Lankester, Arachnida embraced arachnids, xiphosurans and eurypterids. Similarities between eurypterids and xiphosurans had already been identified (Nieszkowski, 1859) and these two groups were united in the taxon Merostomata (Woodward, 1866). Similarities between xiphosurans and trilobites were noted by, among others, Lockwood (1870), and similarities between scorpions and eurypterids were summarized by Woods (1909). However, the general consensus around the turn of the century was that xiphosurans were closely related to eurypterids in the Merostomata, which were aquatic, and that together these were related to the arachnids, which were terrestrial (see Clarke and Ruedemann (1912) and Størmer (1944) for reviews of early work on chelicerate phylogeny). Heymons (1901) introduced the name Chelicerata for merostomes and arachnids, and this name became widely adopted.

Walcott (1912) described a number of unusual Cambrian fossils from the Burgess Shale and referred them to one of the four main arthropod groups (i.e. chelicerates, trilobites, crustaceans and atelocerates). Walcott (1912) placed some of these fossils in Merostomata and constructed an evolutionary tree showing how these taxa may have evolved into xiphosurans. Other problematic arthropod fossils which resembled merostomes, e.g. aglaspidids (Raasch, 1939) were also described in the early part of this century. Størmer (1944) reviewed these problematic arthropods and recognized a broad taxon, Arachnomorpha Heider, 1913, which comprised chelicerates (including aglaspidids), trilobites and the Trilobitomorpha Størmer, 1944; the latter comprising most of the problematic Burgess Shale-type Palaeozoic arthropods. Størmer (1944) concluded that chelicerates evolved from among the trilobitomorphs and subsequently Størmer (1959) recognized a taxon, Merostomoidea, for the most merostome-like of the trilobitomorphs. Other authors attempted to derive chelicerates from olenellid trilobites (Raw, 1957; Lauterbach, 1980). The Burgess Shale arthropods were subsequently reinterpreted as animals with unique body plans of uncertain affinity, unrelated to any of the four main arthropod groups. The Merostomoidea were redescribed by Bruton (1981), Whittington (1981) and Bruton and Whittington (1983), who argued on morphological grounds that none of these merostomoids could be classed as a chelicerate or a direct chelicerate ancestor. These revisions, and Briggs *et al.*'s (1979) removal of aglaspidids from Chelicerata (see below), left no Cambrian chelicerates, other than the dubi-

*Arthropod Relationships*, Systematics Association Special Volume Series 55, Edited by R.A. Fortey and R.H. Thomas. Published in 1997 by Chapman & Hall, London. ISBN 0 412 75420 7

ous *Eolimulus* (see below) until Briggs and Collins (1988) described *Sanctacaris* as a chelicerate.

Kraus (1976) argued that Merostomata represented an aquatic grade of chelicerate, rather than a discrete taxon and Bergström (1979) further suggested that Arachnida could simply represent terrestrial chelicerates. A number of evolutionary models for the chelicerates and/or the arachnomorphs have been proposed (Fedotov, 1925; Beklemishev, 1944; Zakhvatkin, 1952; Dubinin, 1962; Sharov, 1966; Bristowe, 1971; Savory, 1971; Grasshoff, 1978; Bergström, 1979; Simonetta and Delle Cave, 1981; Starobogatov, 1990). A discussion of all these evolutionary models is beyond the scope of this present work, but most were based on general resemblances of the various fossils rather than distinct synapomorphies. The most widely accepted recent model was the cladistic analysis of Weygoldt and Paulus (1979). These authors included Aglaspidida in the Chelicerata as sister group of a taxon, Euchelicerata, which itself had a basic phylogeny of (Xiphosura (Eurypterida (Scorpiones + other arachnids))).

Recent authors (Wheeler *et al.*, 1993; Wills *et al.*, 1995) have rejected treating all the problematic groups as unrelated arthropods of uncertain affinity, the latter recognizing broad clades of arthropods: Crustacea, Marellomorpha, Atelocerata and Arachnomorpha, which include both the problematic fossils and extant groups. In these models the Chelicerata clearly evolved from among the Arachnomorpha (alternatively called Arachnata), a taxon comprising chelicerates, trilobites and the non-marellomorph problematic arthropods. However, a consensus on the phylogenetic position of chelicerates and what their immediate sister groups are among the arachnomorphs has proved difficult to establish (Briggs and Fortey, 1989; Briggs *et al.*, 1992; Wills *et al.*, 1994, 1995). In this paper we present a phylogenetic analysis and evolutionary model of the major chelicerate groups and those fossil taxa we consider their immediate outgroups. Based on this we attempt a new diagnosis of Chelicerata.

## 17.2 THE LIMITS OF CHELICERATA

Living chelicerates are represented by xiphosurans and arachnids. The extinct eurypterids are also clearly chelicerates. It is still unclear whether the pycnogonids (sea spiders) are chelicerates (see King, 1973) and their meagre fossil record does not help clarify the situation (Bergström *et al.*, 1980); though these authors suggested pycnogonids could be derived from primitive merostomes. Wheeler *et al.* (1993) placed pycnogonids as sister group to the chelicerates on a combination of morphological and ribosomal DNA evidence, though for now we regard pycnogonids as Arthropoda *incertae sedis* and have not included them in our analysis (see below). Other extant arthropods have in the past been referred to Chelicerata (Dahl, 1913), though of these, pentastomids (tongue worms) are probably crustaceans (Abele *et al.*, 1989) and tardigrades are certainly not chelicerates but may have originated between the lobopods and stem group arthropods (Dewel and Dewel, 1997, this volume). Among fossil taxa, the Silurian–Triassic euthycarcinoids, originally thought to be crustaceans (Handlirsch, 1914), were suggested as being chelicerates by Riek (1964) based on their general resemblance to xiphosurans. However, the presence of antennae and apparently mandibulate mouthparts suggests that euthycarcinoids are atelocerates (McNamara and Trewin, 1993; Wills *et al.*, 1995).

Among arachnomorphs, the limits of Chelicerata have proved harder to define. Aglaspidids were for a long time included as chelicerates, either in Xiphosura (Størmer, 1944), Merostomata (Eldredge, 1974) or as sister group to the other chelicerates (Bergström, 1979; Weygoldt and Paulus, 1979). Briggs *et al.* (1979) removed aglaspidids from Chelicerata on the grounds they had only four or five cephalic appendages, none of which were chelicerae. *Sanctacaris* was described as a chelicerate by Briggs and Collins (1988), who placed it as the sister group to all other chelicerates, redefining Chelicerata to accommodate it. However, subsequent phylogenetic analyses (Briggs *et al.*, 1992; Wills *et al.*, 1994, 1995) (Figure 17.1), while placing *Sanctacaris* in the Arachnomorpha, do not place it as the sister group to other chelicerates (Wills, 1996), but as sister group to *Yohoia* (Wills *et al.*, 1995). Thus, *Sanctacaris* cannot be referred to Chelicerata without including the other Arachnomorpha, including trilobites.

Chen *et al.* (1995) described *Fuxianhuia* from the Lower Cambrian Chengjiang fauna as a basal euarthropod. Wills (1996) coded this fossil into his (Wills *et al.*, 1995) dataset and concluded that *Fuxianhuia* was a chelicerate, specifically sister group of arachnids and eurypterids on the synapomorphy of opisthosomal differentiation, a character we regard as synapomorphic for all chelicerates (see below). Edgecombe *et al.*'s reply to Wills rejected this interpretation (a view with which we concur), questioning Wills's (1996) coding of *Fuxianhuia* characters and pointing out that his model requires re-evolution of antennae and exopods on the trunk appendages and does not fit the known chelicerate fossil record. Babcock (1996) also proposed a Chengjiang arachnomorph as an early Cambrian chelicerate, though this fossil does not fit Chelicerata as we define it. In response to this trend of 'chelicerate spotting' among the early arthropod-bearing Lagerstätten, a new look at the limits of Chelicerata is clearly warranted.

## 17.3 PHYLOGENETIC ANALYSIS

We investigated the phylogeny and early history of the chelicerates to determine the boundary between chelicerates and non-chelicerates and hence diagnostic characters for the Chelicerata. To this end we undertook a cladistic analysis

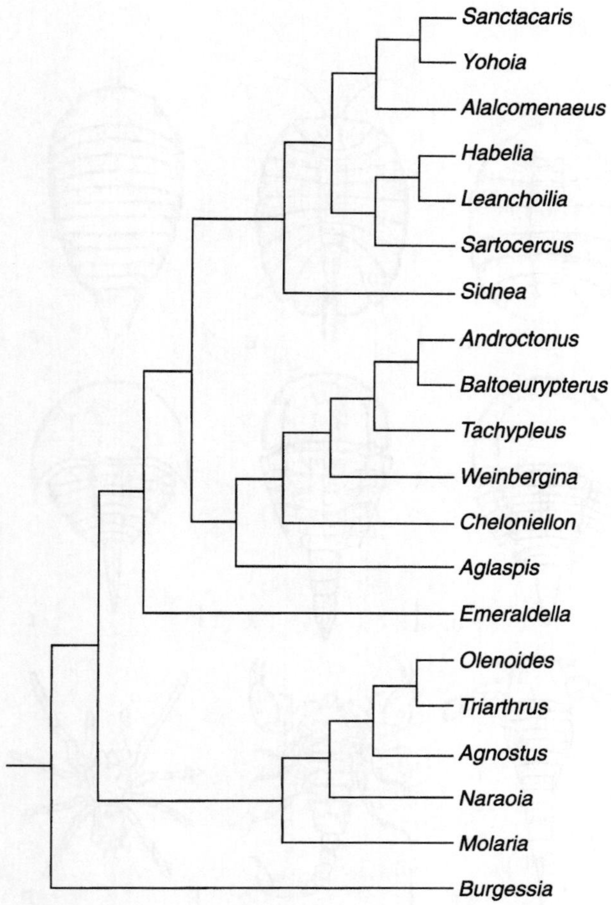


**Figure 17.1** Phylogeny of the Arachnomorpha clade of the Palaeozoic arthropods. (After Wills *et al.*, 1995, Figure 1A.)

using MacClade 3.04 and PAUP 3.1.1 of those taxa known to be chelicerates, i.e. xiphosurans, eurypterids and arachnids. Wills *et al.* (1995) placed Chelicerata + *Cheloniellon* + *Aglaspis* as a clade within Arachnomorpha and in our analysis we investigated this clade in more detail. We included the major chelicerate groups, aglaspidids *sensu stricto* and other genera referred to Aglaspidida (treated as separate genera), *Cheloniellon* and similar forms, and a number of problematic genera such as *Chasmataspis*. *Sanctacaris* is omitted from the analysis since in most recent phylogenies it is placed as sister group to other non-chelicerate Arachnomorpha (see above). Taxa known only from isolated carapaces, such as *Eolimulus,* were also excluded. Characters and character states are listed below and the data matrix is given in Table 17.1.

### 17.3.1 TAXA INCLUDED IN THE ANALYSIS

#### (a) *Neostrabops* (Figure 17.2(A))

The Upper Ordovician arthropod *Neostrabops* Caster and Macke, 1952, is a poorly preserved fossil from Stonelick Creek, Ohio. It is 4 cm long and is one of a group of fossils which resemble *Cheloniellon* (see below) in overall appearance, though many of its characters are unknown. The telson inferred by Caster and Macke (1952) in *Neostrabops* is not present in the fossil and appendages are absent.

#### (b) *Duslia* (Figure 17.2(B))

The Upper Ordovician arthropod *Duslia* Jahn, 1913 from the Barrandian area of the Czech Republic was originally interpreted as a polyplacophoran. This 9- to 12-cm-long arthropod is distinguished by a characteristic fringe of spines around the body. Appendages are not known and no eyes are preserved. *Duslia* has been referred to trilobites and cheloniellid arthropods (see Chlupač, 1988 for a review). Chlupač (1988) redescribed *Duslia* and concluded that it probably belonged in the Trilobitomorpha, rather than Trilobita, and that its systematic position was probably analogous to *Cheloniellon*.

#### (c) *Triopus* (Figure 17.2(C))

*Triopus* Barrande, 1872 is an Ordovician arthropod, about 4 cm long, possibly from the same horizon in the Czech Republic as *Duslia* (Chlupač, 1988). It was originally thought to be a trilobite. The holotype was reported lost (Chlupač, 1965), but was rediscovered and figured by Chlupač (1988). *Triopus* has been compared with aglaspidids (Chlupač, 1965), while Bergström (1968) interpreted it as a poorly preserved xiphosuran. Though incomplete and lacking appendages, *Triopus* resembles *Cheloniellon* and is therefore included in this analysis.

**Table 17.1** Data matrix used in the phylogenetic analysis.

| | | 1 | | 2 | |
|---|---|---|---|---|---|
| *Characters* | *12345* | *67890* | *12345* | *67890* | *12345* |
| *Neostrabops* | 11000 | 00??? | ?1000 | 0???? | ????0 |
| *Duslia* | 11000 | 001?? | ?1100 | 01?4? | ????1 |
| *Triopus* | 11??? | ????? | ??100 | 0???? | ????0 |
| *Pseudarthron* | 111?? | ????? | ??100 | 0???? | ????0 |
| *Cheloniellon* | 11100 | 00000 | 01100 | 01040 | 000?0 |
| *Paleomerus* | 10000 | 000?? | ?2000 | 00?0? | ????0 |
| *Strabops* | 10000 | 00??? | ?2000 | 00?0? | ????0 |
| Aglaspidida | 10010 | 000?0 | 02000 | 00110 | ?00?0 |
| *Lemoneites* | 1000? | ????? | ?2000 | 10?0? | ????0 |
| *Weinbergina* | 10001 | 1?010 | 02110 | 10001 | 100?0 |
| *Bunodes* | 10001 | 110?? | ?1111 | 10?0? | ????0 |
| *Kasibelinurus* | 10001 | 110?? | ?0110 | 10?1? | ????0 |
| Xiphosurida | 10011 | 11010 | 00113 | 10012 | 10000 |
| *Diploaspis* | 10000 | 010?? | ?3002 | 30?4? | ?00?0 |
| *Chasmataspis* | 10010 | 010?? | ?3002 | 3001? | ?00?0 |
| Eurypterida | 10000 | 01011 | 13000 | 20013 | ?1000 |
| Scorpiones | 10000 | 01011 | 13000 | 20034 | 20110 |
| Other arachnids | 10000 | 01011 | 13000 | 10024 | 30010 |

0–4 = alternative character states. ? = character state uncertain. See text for details.

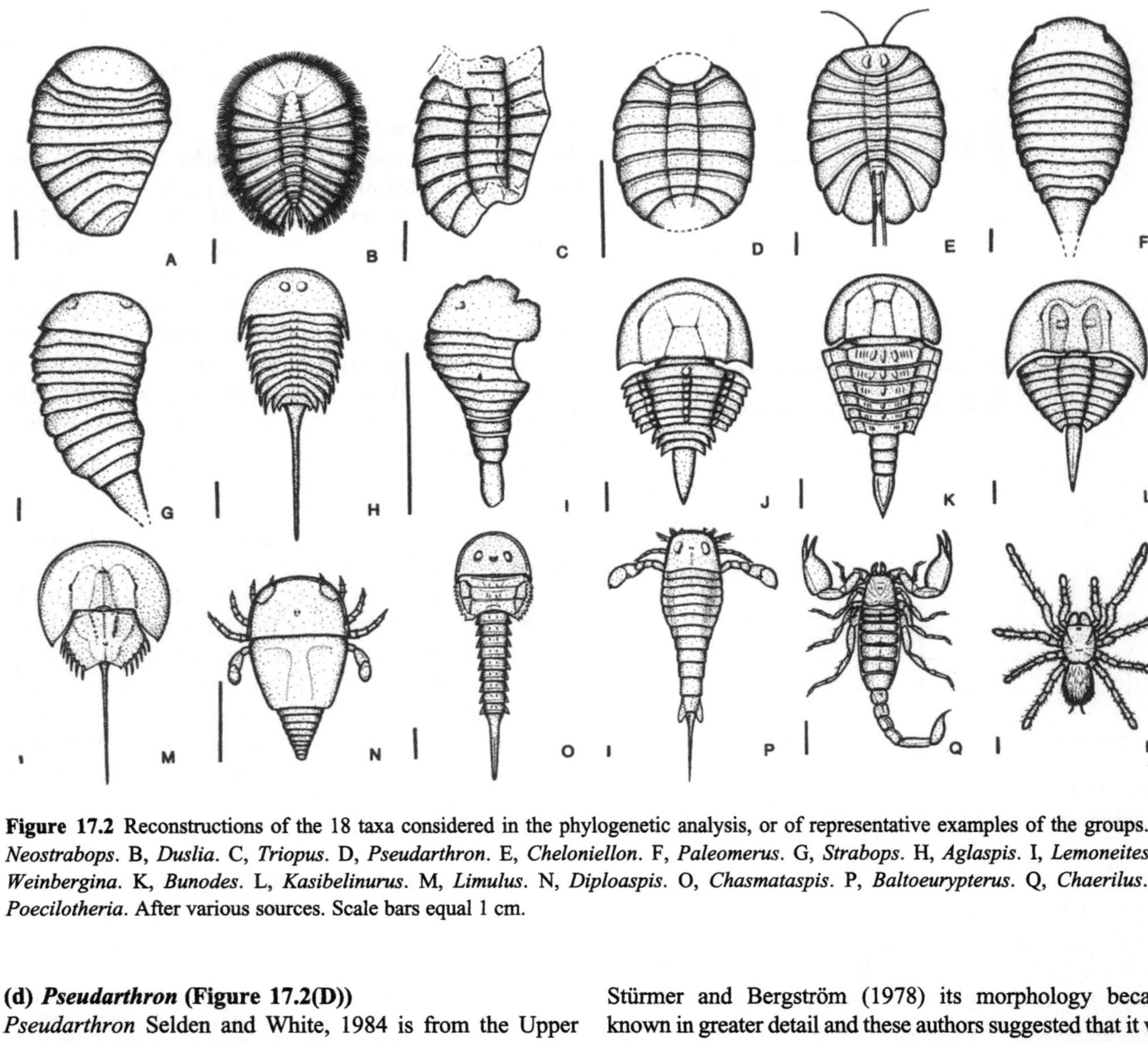


**Figure 17.2** Reconstructions of the 18 taxa considered in the phylogenetic analysis, or of representative examples of the groups. A, *Neostrabops*. B, *Duslia*. C, *Triopus*. D, *Pseudarthron*. E, *Cheloniellon*. F, *Paleomerus*. G, *Strabops*. H, *Aglaspis*. I, *Lemoneites*. J, *Weinbergina*. K, *Bunodes*. L, *Kasibelinurus*. M, *Limulus*. N, *Diploaspis*. O, *Chasmataspis*. P, *Baltoeurypterus*. Q, *Chaerilus*. R, *Poecilotheria*. After various sources. Scale bars equal 1 cm.

**(d) *Pseudarthron* (Figure 17.2(D))**

*Pseudarthron* Selden and White, 1984 is from the Upper Silurian of Lesmahagow, Scotland. It is about 16 mm long, and resembles *Cheloniellon*, but the prosoma is poorly preserved and the appendages are absent. These authors discussed its phylogenetic affinities with extant groups, including isopods, myriapods and chelicerates and a number of fossil genera, including the Devonian *Oxyuropoda* Carpenter and Swain, 1908, and the Carboniferous *Camptophyllia* Gill, 1924. Selden and White (1984) suggested that resemblances between *Pseudarthron*, *Oxyuropoda* and *Camptophyllia* were due to convergence, hence the latter two genera were omitted from our analysis.

**(e) *Cheloniellon* (Figure 17.2(E))**

The Lower Devonian arthropod *Cheloniellon* Broili, 1932, from the Hunsrück Shale, Germany, was originally described as a crustacean. Following a detailed re-description by Stürmer and Bergström (1978) its morphology became known in greater detail and these authors suggested that it was a late representative of a group of trilobitomorphs which gave rise to Chelicerata. Wills *et al.* (1995) included *Cheloniellon* in their phylogenetic analysis and placed it among the Arachnomorpha, as sister group to the xiphosurans, eurypterids and arachnids, with aglaspidids as sister group to *Cheloniellon* + Chelicerata. *Cheloniellon* is 6–10 cm long and is not particularly merostome-like in appearance; it has antennae, lacks a telson and the prosoma-opisthosoma boundary is poorly defined. However, it has gnathobasic appendages and, bearing in mind Wills *et al.*'s (1995) placement, we included *Cheloniellon*, and the similar forms: *Neostrabops*, *Pseudarthron*, *Duslia* and *Triopus*, in our analysis.

**(f) *Paleomerus* (Figure 17.2(F))**

The Lower Cambrian *Paleomerus* Størmer, 1956 (see also Størmer 1955) from Sweden is one of the oldest arthropods and is one of the most primitive-looking merostome-like arthropods known. *Paleomerus* is about 8 cm long, though none of the three *Paleomerus* fossils is well preserved and appendages are unknown.

**(g) *Strabops* (Figure 17.2(G))**

The Cambrian *Strabops* Beecher, 1901, is about 10 cm long and known from a single specimen from Missouri, USA, which shows the dorsal surface only. It was described by Clarke and Ruedemann (1912) as a eurypterid, based on a count of 12 opisthosomal segments (in fact it has 11), who interpreted it as the most primitive member of that group. Bergström (1971) removed *Strabops*, *Neostrabops* and *Paleomerus* (see above) from the Aglaspidida and referred them to an unnamed order of the Merostomoidea, though Briggs *et al.* (1979) doubted whether this was justified. These three genera show neither a typical aglaspidid carapace nor strongly curved margins of the tergites; we treat *Strabops*, *Neostrabops* and *Paleomerus* as separate taxa in our analysis. Note that the appendages figured in Clarke and Ruedemann's (1912, plate 1) reconstruction of *Strabops* are hypothetical. It proved difficult to find any significant characters to distinguish *Strabops* from *Paleomerus*.

**(h) Aglaspidida (Figure 17.2(H))**

Aglaspidids are primarily Cambrian fossils, first described by Hall (1862) who interpreted them as crustaceans, but are best known from the monograph of Raash (1939), who referred them to Merostomata. Aglaspidids have recently been recorded from the upper Lower Ordovician (D. Siveter, personal communication). Aglaspidids are 2–10 cm long and have been regarded as a taxon within Xiphosura (Størmer, 1955), a separate taxon of the Merostomata (Eldredge, 1974) or as a separate taxon of Chelicerata (Weygoldt and Paulus, 1979; Bergström, 1979). However, Briggs *et al.* (1979) removed Aglaspidida from Chelicerata, preferring not to assign them to any higher taxon within Arthropoda. Aglaspidida has become something of a 'bucket taxon' for non-trilobite arthropods which resemble xiphosurans (Størmer, 1955; Chlupač and Havlicek, 1965). Some taxa were removed from Aglaspidida by Bergström (1971). One aglaspidid genus, *Beckwithia* Resser, 1931, was described as having a fused tail plate and was thus raised to a family separate from other aglaspidids by Raasch (1939); it was placed in a monotypic order by Bergström (1979). *Beckwithia* has been interpreted as central to the evolution of the chelicerates (Starobogatov, 1990) and/or the arachnids in particular (Bergström, 1979). Hesselbo (1989) redescribed *Beckwithia* and found no evidence for the tail plate, and thus returned the genus to the order Aglaspidida. Another problematic genus is *Kodymirus* Chlupač and Havlicek, 1965, which was described as having 12 opisthosomal segments, as opposed to the 11 seen in typical aglaspidids such as *Aglaspis*. However, their photographs of the specimen suggest that even more segments may be present in the fossil. A supposed Carboniferous aglaspidid (Youchong and Shaowu, 1981) does not resemble typical aglaspidids, having fewer than 11 segments and lacking a carapace with genal spines and we regard it as an arthropod of uncertain affinity. The aglaspidids form a distinct group of arachnomorph arthropods and a revision of their constituent genera is clearly warranted. For the purposes of this analysis we regard Aglaspidida as a morphologically distinct group characterized by an 11-segmented opisthosoma, postanal plates, a tail spine, and a carapace bearing genal spines. We relied on the descriptions of Briggs *et al.* (1979) and Hesselbo (1989, 1992) for determining characters and character states.

**(i) Xiphosura (Figures 17.2(I–M))]**

The Xiphosura were traditionally seen as the most primitive chelicerates and incorporated the Aglaspidida in some schemes (see above). Significant accounts of xiphosuran phylogeny and systematics include Størmer (1952, 1955), Novojilov (1962), Eldredge (1974), Bergström (1975) and Selden and Siveter (1987). A recent analysis by Anderson and Selden (1997) concluded that class Xiphosura comprise the order Xiphosurida, supported by distinct autapomorphies (see below), and a paraphyletic group of genera, the synziphosurines. This differs from previous interpretations in which the synziphosurines were regarded as a distinct taxon, Synziphosurida, of equal rank to Xiphosurida (Eldredge, 1974). It has been suggested that arachnids evolved from among the synziphosurines (Beall and Labandeira, 1990) and this hypothesis merits consideration. For the purpose of this analysis we included the Xiphosurida (Figure 17.2(M)) as one taxon and selected four of the synziphosurines to include as separate taxa: *Lemoneites* Flower, 1968 (Figure 17.2(I)) is the oldest and smallest of the synziphosurines and which shows the most plesiomorphic character states. *Weinbergina* Richter and Richter, 1929 (Figure 17.2(J)) is one of the most intensively studied synziphosurines (Stürmer and Bergström, 1981), also included by Wills *et al.* (1995) in their analysis. *Bunodes* Eichwald, 1854 (Figure 17.2(K)) is perhaps the most arachnid-like of the synziphosurines. *Kasibelinurus* Pickett, 1993 (Figure 17.2(L)) emerged as the sister group of the Xiphosurida in Anderson and Selden's analysis. The supposed Cambrian xiphosuran *Eolimulus* (Möberg, 1892) is only known from a carapace and is not included in the analysis. From an analysis of these taxa it should be possible to test whether any of the other chelicerate groups is sister taxon to all Xiphosura, or to a particular synziphosurine.

**(j) *Diploaspis* (Figure 17.2(N))**

Størmer (1972) described two Devonian fossils, *Diploaspis* and *Heteroaspis*, from Germany and referred them to Chasmataspida. These small (2 cm) fossils were described as having a 12-segmented opisthosoma with three fused anterior segments and a postabdomen of nine ring-like segments similar to *Chasmataspis* (see below). The postabdomens of *Diploaspis* and *Heteroaspis* are short, not elongate as in *Chasmataspis*. The posteriormost prosomal appendage of *Diploaspis* is a eurypterid-like paddle. The preservation of some of this Devonian material is patchy and there is a suspicion that these taxa might represent a composite of a number of different fossils. The characters used to distinguish *Diploaspis* from *Heteroaspis* are of no phylogenetic significance and since *Diploaspis* is better preserved, it is used in our analysis. Eldredge (1974) placed Chasmataspida as the sister group of Eurypterida on account of the possession of paddles in some eurypterids and some chasmataspids. Bergström (1979) placed Chasmataspida as the sister group of Xiphosurida on the basis of the following characters: chilaria (unknown in *Chasmataspis*), chelate legs, and lamellate book-gills. *Diploaspis* was figured on a separate lineage leading to the arachnids. For the purposes of this analysis *Chasmataspis* and *Diploaspis* are treated as separate taxa.

**(k) *Chasmataspis* (Figure 17.2(O))**

The Ordovician fossil *Chasmataspis* Caster and Brooks, 1956 from Tennessee is about 5 cm long, with an aglaspidid-like carapace and an opisthosoma consisting of a 'buckler' of three segments, fused both dorsally and ventrally, and an elongate postabdomen of nine reduced, ring-like segments. Caster and Brooks (1956) described *Chasmataspis* as a 'merostomaceous arachnomorph' and erected a new order, Chasmataspida, which they referred to Merostomata. The affinities of Chasmataspida are discussed above. An unpublished Cambrian trace fossil from Texas may have been made by a chasmataspid.

**(l) Eurypterida (Figure 17.2(P))**

The eurypterids were traditionally allied to the Xiphosura in the Merostomata (Størmer, 1944, 1955), though more recently they have generally been placed as the sister group of the arachnids (Weygoldt and Paulus, 1979; Shultz, 1990). The relationship between eurypterids and scorpions is controversial (see below). Significant accounts of eurypterid phylogeny and systematics include Clarke and Ruedemann (1912), Størmer (1955, 1974), Dubinin (1962), Caster and Kjellesvig-Waering (1964), Waterston (1979) and Tollerton (1989). A number of genera, originally placed in the Cyrtoctenida Størmer and Waterston, 1968, were removed from Eurypterida by Tollerton (1989). Cyrtoctenids have a large, domed, carapace, a reduced genital appendage and a cleft metastoma. However, since they show the two eurypterid autapomorphies of a genital appendage and metastoma (albeit modified) we prefer to retain Cyrtoctenida as a taxon within Eurypterida. In this analysis Eurypterida is regarded as a single taxon, with stylonurids interpreted as the most primitive eurypterids (Størmer, 1955) in the determination of characters and character states.

**(m) Scorpiones (Figure 17.2(Q))**

Scorpions are the oldest arachnids and are widely perceived as the most primitive members of the group (Pocock, 1893). The interpretation of the earliest fossil scorpions as aquatic (Pocock, 1901) is a strong argument against the traditional split of chelicerates into the aquatic Merostomata and terrestrial Arachnida. The obvious similarities between scorpions and eurypterids (see Clarke and Ruedemann (1912), Kjellesvig-Waering (1986) and Sissom (1990) for reviews), led some authors to propose that eurypterids evolved from scorpions (Versluys and Demoll, 1920), that scorpions evolved from the eurypterids (Beklemeshev, 1944; Sharov, 1966; Bristowe, 1971) and/or that scorpions are more closely related to eurypterids than the rest of the arachnids (Grasshoff, 1978; Starobogatov, 1990). Other authors have regarded scorpions as true arachnids (Petrunkevitch, 1949; Shultz, 1990). Fossil scorpion phylogeny was extensively revised by Kjellesvig-Waering (1986), though many of his higher taxa, based on the nature of the respiratory organs, have not been adopted by subsequent authors (Stockwell, 1989; Jeram, 1994). The scorpions are treated as a separate taxon from the non-scorpion arachnids in this analysis, which allows us to test the hypotheses that scorpions are either sister group of the eurypterids (Grasshoff, 1978) (in which case Arachnida is paraphyletic), or sister group of the non-scorpion arachnids (Weygoldt and Paulus, 1979), (in which case Arachnida remains monophyletic). We cannot test Shultz's (1990) model in which Arachnida are monophyletic and opilionids are the sister group of (scorpions + (pseudoscorpions + solifugeds)) without expanding the analysis to include all the arachnid orders, which is beyond the scope of this analysis.

**(n) Other arachnids (Figure 17.2(R))**

The phylogeny of the arachnid orders (including scorpions) has been investigated by a number of authors and significant accounts of arachnid phylogeny include: Pocock (1893), Petrunkevitch (1949), Zakhvatkin (1952), Dubinin (1962), Savory (1971), Firstman (1973), Grasshoff (1978), Weygoldt and Paulus (1979), Lindquist (1984), van der Hammen (1989) and Shultz (1990). For the purpose of this analysis we regard all the non-scorpion arachnids as a single taxon, termed Lipostena by Weygoldt and Paulus (1979). There is a general consensus about which characters are primitive for arachnids (Shultz, 1990, Figure 3, modelled plesiomorphic states in arachnids, though this model included scorpions as arachnids) such as an undivided carapace bearing median and lateral eyes, a 12-segmented opisthosoma plus a telson and such conventions were used when coding non-scorpion arachnid characters.

### 17.3.2 CHARACTERS AND CHARACTER STATES

For the purpose of this analysis all characters are assumed to be unordered and polarity has not been argued, though autapomorphies and synapomorphies have been identified where appropriate. The alternative character states are noted in the text.

*Character 1 (Body tagmosis)*
A body divided into two distinct tagmata, a prosoma and opisthosoma, is present in all taxa under consideration and is coded as (1).

*Character 2 (Carapace margin)*
A distinctly procurved posterior margin of the carapace is only seen in the genera *Neostrabops*, *Triopus*, *Duslia*, *Pseudarthron* and *Cheloniellon* and is synapomorphic for these taxa (1), compared with the straight or slightly recurved posterior carapace margin seen in all other taxa (0).

*Character 3 (Carapace size)*
A highly reduced carapace, less than half the width of the opisthosoma, is synapomorphic for *Pseudarthron* (where its size can be estimated from the preserved tergites) and *Cheloniellon* (1), compared with the broad carapaces of all other taxa under consideration (0). This character is uncertain in *Triopus*.

*Character 4 (Genal spines)*
Strongly developed genal spines (1) are present in aglaspidids, *Chasmataspis* and xiphosurids. The carapaces of all other taxa lack genal spines (0). This character is unknown in *Triopus* and *Pseudarthron*.

*Character 5 (Ophthalmic ridges)*
An ophthalmic ridge, a discrete ridge on the carapace bearing the eyes, is seen in all xiphosurids and the synziphosurines *Weinbergina*, *Bunodes* and *Kasibelinurus*, and is synapomorphic for these taxa (1). The carapaces of all other taxa lack ophthalmic ridges (0). The carapaces of *Lemoneites*, *Triopus* and *Pseudarthron* are poorly preserved or unknown and this character is coded as uncertain in these taxa.

*Character 6 (Cardiac lobes)*
A cardiac lobe in the centre of the carapace is seen in *Weinbergina*, *Bunodes*, *Kasibelinurus* and xiphosurids and is synapomorphic for these taxa (1). The carapaces of all other taxa lack a cardiac lobe (0). The carapaces of *Lemoneites*, *Triopus* and *Pseudarthron* are poorly preserved or unknown and this character is coded as uncertain in these taxa.

*Character 7 (Median eyes)*
Median eyes, or ocelli, are present in xiphosurans, *Chasmataspis*, *Diploaspis* [where their presence is inferred in comparison to the related and better preserved carapace of *Heteroaspis* in which median eyes are clearly present (Størmer, 1972, plate 5)], eurypterids, scorpions and other arachnids. Median eyes are synapomorphic for these taxa (1). Median eyes are absent in the other taxa under consideration (0). The carapaces of *Lemoneites*, *Triopus* and *Pseudarthron* are poorly preserved or unknown and this character is coded as uncertain in these taxa.

*Character 8 (Lateral eyes)*
Lateral eyes are absent in *Duslia* where this is interpreted as an autapomorphic condition (1). Lateral eyes are present (0) in all other taxa where the carapace is sufficiently well preserved. The carapaces of *Neostrabops*, *Strabops*, *Lemoneites*, *Triopus* and *Pseudarthron* are poorly preserved or unknown and this character is coded as uncertain in these taxa.

*Character 9 (Prosomal appendage 1)*
The first prosomal appendage of *Cheloniellon* is antenniform (0). This appendage was interpreted as antenniform in aglaspidids (Hesselbo, 1992), though the evidence from his plates is not convincing and this character is coded as uncertain. A chelate first prosomal appendage, (i.e. a chelicera) is present in *Weinbergina*, xiphosurids, eurypterids, scorpions and arachnids (1). The first appendage of the remaining taxa is unknown.

*Character 10 (Divided femur)*
Shultz (1990) regarded a walking leg femur divided into two podomeres to be the plesiomorphic state in arachnids, including scorpions (and retained in taxa such as solifuges and ricinuleids) and also to be present in eurypterids (Selden, 1981). A divided femur (0) is coded as absent in *Cheloniellon*, Aglaspidida, *Weinbergina* and Xiphosurida and a divided femur (1) is present in eurypterids, scorpions and the other arachnids. This character is coded as unknown in all other taxa. Appendages are known from both chasmataspid taxa and while they appear not to have a divided femur, their preservation makes it is difficult to be certain which podomeres should be homologized with the chelicerate femur and this character is coded as uncertain.

*Character 11 (Basitarsus)*
The tarsus of prosomal appendages 3–6 in eurypterids, scorpions and other arachnids is subdivided into a basitarsus and telotarsus. A basitarsus on these appendages is absent in *Cheloniellon*, Aglaspidida, *Weinbergina* and Xiphosurida (0) and present in eurypterids, scorpions and other arachnids (1). This character is coded as uncertain in all other taxa. Chasmataspids may possess a basitarsus, but as noted above there are problems homologizing podomeres and this character is coded as uncertain.

*Character 12 (Opisthosomal segment number)*
The taxa under consideration show a range of visible opisthosomal segment counts. *Kasibelinurus* and xiphosurids have nine segments (0), a synapomorphy of these taxa. *Neostrabops, Cheloniellon, Duslia* and *Bunodes* have 10 segments (1). *Paleomerus, Strabops, Lemoneites, Weinbergina* and aglaspidids have eleven segments (2). *Diploaspis, Chasmataspis*, eurypterids, scorpions and other arachnids have twelve segments (3), a synapomorphy of these taxa. A transient 13th segment is cited by some authors as being present during scorpion embryology. Total opisthosomal segment number in *Triopus* and *Pseudarthron* is unknown and this character is coded as uncertain in these taxa.

*Character 13 (Opisthosomal axial region)*
A discrete opisthosomal axial region is present in *Cheloniellon, Pseudarthron, Duslia, Triopus, Bunodes, Weinbergina, Kasibelinurus* and all xiphosurids (1). This axial region was interpreted as an adaptation for enrolment in xiphosurans (Eldredge, 1974). An opisthosomal axial region is absent in the remaining taxa (0).

*Character 14 (Opisthosomal segment 1)*
Opisthosomal segment 1 is highly reduced in all xiphosurids and in *Weinbergina, Bunodes* and *Kasibelinurus* and all xiphosurids (1) and is synapomorphic for these taxa. Opisthosomal segment 1 is fully expressed in the remaining taxa (0).

*Character 15 (Opisthosomal tergite fusion)*
Fusion of tergites 2 and 3 into a diplotergite is autapomorphic for *Bunodes* (1) in this analysis. Fusion of the first three opisthosomal segments into a discrete buckler is synapomorphic for *Diploaspis* and *Chasmataspis* (2). Fusion of the opisthosomal tergites into a thoracetron is autapomorphic for Xiphosurida (3). Some derived arachnids have fused opisthosomal segments, but here they do not form a thoracetron. The tergites of all remaining taxa are coded as unfused (0).

*Character 16 (Opisthosomal differentiation)*
The last three opisthosomal segments of *Lemoneites, Weinbergina, Bunodes* and many non-scorpion arachnids are ring-like and form a discrete postabdomen (1). The last five opisthosomal segments of eurypterids and scorpions form a postabdomen (2). The last nine opisthosomal segments of *Diploaspis* and *Chasmataspis* form a postabdomen (3). The remaining taxa show no differentiation of the opisthosoma (0).

*Character 17 (Furcal rami)*
Paired furcal rami at the posterior end of the opisthosoma are present in *Duslia* and are also present in *Cheloniellon* as elongate structures (1). Furcal rami are not seen in the remaining taxa (0). The posterior opisthosoma of *Neostrabops, Triopus* and *Cheloniellon* is not preserved and this character is coded as uncertain in these taxa.

*Character 18 (Postanal plates)*
Postanal plates are autapomorphic for aglaspidids (1) and not recorded in other taxa where the ventral surface is known, i.e. *Cheloniellon, Weinbergina*, xiphosurids, *Chasmataspis*, eurypterids and both scorpion and non scorpion arachnids (0). This character is coded as uncertain in the remaining taxa.

*Character 19 (Telson shape)*
The telson is broad, but tapering in *Paleomerus, Strabops, Lemoneites, Weinbergina* and *Bunodes* (0). The telson (or tail spine incorporating opisthosomal segments in xiphosurans) is longer and more styliform in Aglaspidida, *Kasibelinurus*, xiphosurids, *Chasmataspis* and eurypterids (secondarily modified into a broad structure in some eurypterids) (1). The telson is subdivided in those non-scorpion arachnids which retain a telson, such as palpigrades and uropygids (an autapomorphy) (2) and is modified into a sting in scorpions (an autapomorphy) (3). The telson is absent in *Cheloniellon, Duslia* and *Diploaspis* (4). The posterior opisthosoma is unknown in *Neostrabops, Triopus* and *Pseudarthron* and this character is coded as uncertain in these taxa.

*Character 20 (Opisthosomal appendages)*
Pediform appendages running the length of the opisthosoma are present in aglaspidids and *Cheloniellon* (0). Pediform appendages are restricted to opisthosomal segment 1 in *Weinbergina* (1). We believe this appendage to be opisthosomal (Anderson's 1973 review of xiphosuran embryology suggests the chilaria belong to opisthosomal segment 1), not prosomal as suggested by Stürmer and Bergström (1981). This appendage is modified into chilaria in xiphosurids (an autapomorphic state) (2), may be modified into the metastoma of eurypterids (3), though an alternative interpretation of the metastoma is that they are the fused epicoxa of prosomal appendage 6 (S. Braddy, personal communication), (an autapomorphic state) and is lost in scorpions, though a transitory limb bud appears during scorpion embryology (Anderson, 1973) and all other arachnids (4). The appendages of the remaining taxa are unknown and this character is coded as uncertain in these taxa.

*Character 21 (Respiratory organs)*
A series of pairs of gills running the length of the opisthosoma is present in *Cheloniellon* (0). Five pairs of respiratory organs are seen in *Weinbergina* and xiphosurids (1), four pairs are seen in scorpions (2) and two pairs of lamellate respiratory organs are primitively present in non-scorpion

arachnids (3). Eurypterids have five pairs of branchial chambers, but it is unclear whether they all contained book-gills. The number of gills in eurypterids and the remaining taxa is coded as uncertain.

*Character 22 (Genital appendage)*
A genital appendage on the underside of the opisthosoma is autapomorphic for eurypterids (1) and is absent in all remaining taxa where the ventral surface is known (0), i.e. *Cheloniellon*, aglaspidids, *Weinbergina*, xiphosurids, scorpions and other arachnids. This character is coded as uncertain in the remaining taxa.

*Character 23 (Pectines)*
Pectines, paired, probably sensory, structures on the underside of the opisthosoma, are autapomorphic for scorpions (1) and are absent in all remaining taxa where the ventral surface are known (0), i.e. *Cheloniellon*, aglaspidids, *Weinbergina*, xiphosurids, eurypterids and other arachnids. This character is coded as uncertain in the remaining taxa.

*Character 24 (Slit sensilla)*
Slit sensilla are sensory cuticular strain gauges which are synapomorphic for scorpions and non-scorpion arachnids (1). They are absent in xiphosurids and eurypterids (0) and unknown in all remaining taxa.

*Character 25 (Marginal spines)*
A fringe of spines on the prosoma and opisthosoma is autapomorphic for *Duslia* (1) and is absent in all remaining taxa (0).

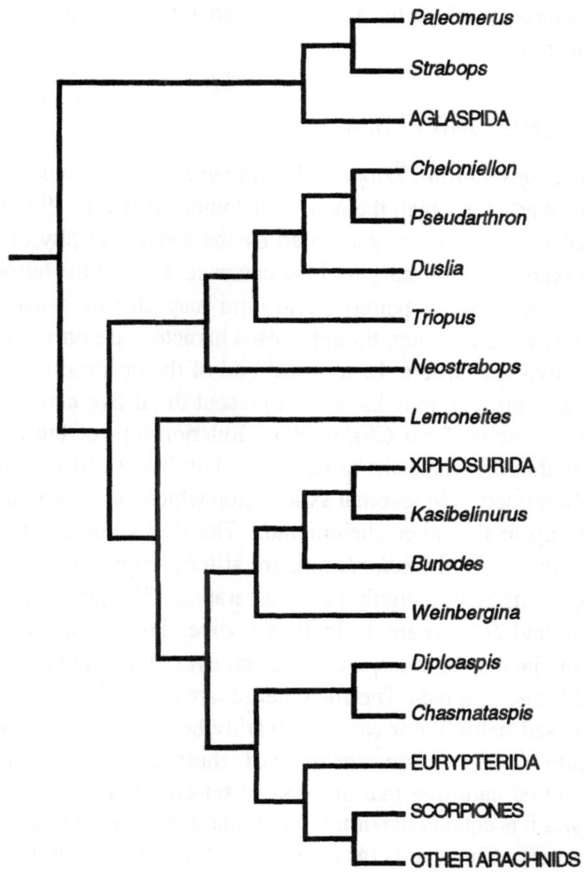


**Figure 17.3** Cladogram of the major chelicerate groups and their outgroups. The cladogram has a tree length of 46 and a consistency index of 0.83 and is one of a number of equally parsimonious trees.

## 17.4 DISCUSSION

Using these 18 taxa and 25 characters we have produced a cladogram (Figure 17.3) with a tree-length of 46 and a consistency index of 0.83. The cladogram is one of a number of equally parsimonious trees. This analysis has identified a number of well supported clades and highlights problems in resolution, particularly with respect to poorly preserved taxa.

### 17.4.1 *PALEOMERUS AND STRABOPS*

These two Cambrian genera are the most primitive-looking of the taxa considered in this analysis. Bergström (1971) removed these genera from Aglaspidida, which is supported by our analysis since they lack aglaspidid apomorphies (e.g. genal spines). It proved impossible to find meaningful characters to separate *Paleomerus* and *Strabops* and we regard them as sister groups in the analysis (Figure 17.3), though there are no synapomorphies to support this. In the evolutionary model below these taxa are treated as part of the same lineage, a primitive stem group from which Chelicerata, Aglaspidida and Cheloniellida can be derived. Despite the spectacular Chengjiang finds, *Paleomerus* is perhaps the best model of a primitive arachnomorph.

### 17.4.2 AGLASPIDIDA

The aglaspidids form a distinct group with an autapomorphy of postanal plates. The development of genal spines and an elongate telson is probably convergent with the chasmataspids and later xiphosurids, presumably reflecting a similar mode of life. Aglaspidids represent a distinct taxon within the Arachnomorpha, though as noted above, a number of problematic arthropods have been referred to Aglaspidida and the group as a whole needs revision. The position of Aglaspidida relative to the other taxa in the analysis is difficult to resolve. It is equally parsimonious to place aglaspidids as sister group to either *Paleomerus/Strabops* or the Chelicerata and aglaspidids share no synapomorphies with any of these groups. Since they originate in the Cambrian and lack opisthosomal differentiation we prefer to place aglaspidids as sister group to the

*Paleomerus/Strabops* lineage, rather than as sister group of the Chelicerata.

### 17.4.3 CHELONIELLIDA

*Neostrabops*, *Duslia*, *Triopus*, *Pseudarthron* and *Cheloniellon* form a clade for which the name Cheloniellida Broili, 1932 is available. This clade is supported by the synapomorphy of a procurved posterior margin of the carapace. Loss of the telson and the presence of paired furcal rami may also be synapomorphic for this group, though these characters are only seen in the two taxa where the terminal end of the opisthosoma is known. Antennae may have been present in all five taxa, but are only known from *Cheloniellon*. Relationships of the five taxa within this clade are harder to resolve; *Neostrabops* lacks a well-defined opisthosomal axial region which places it as sister group to the other cheloniellids. The two youngest taxa, *Pseudarthron* and *Cheloniellon*, are sister groups sharing the synapomorphy of a highly reduced carapace. The positions of *Duslia* and *Triopus* are difficult to resolve, though *Duslia* has well-defined autapomorphies of lateral eyes absent and a marginal fringe of spines. The cheloniellids are not chelicerates (as diagnosed below), nor can they readily be envisaged as chelicerate ancestors on account of their synapomorphies (procurved carapace margin, loss of telson, ?furcae). In this analysis it is equally parsimonious to place cheloniellids as sister group to Chelicerata [as in Wills *et al.*'s (1995) model], sister group to aglaspidids and *Paleomerus/Strabops* or sister group to all other taxa in the analysis. Since cheloniellids are currently known from the Ordovician (*Paleomerus*, *Strabops* and most Aglaspidida being known from the Cambrian) we follow Wills *et al.* (1995) and adopt the sister group to Chelicerata model (Figures 17.3 and 17.4). In any case, cheloniellids appear to represent a distinct taxon within the Arachnomorpha.

### 17.4.4 CHELICERATA

Chelicerata, restricted to *Lemoneites*, xiphosurans, chasmataspids, eurypterids and arachnids in this analysis, is a monophyletic group. We question the referral of certain Cambrian taxa to Chelicerata (Briggs and Collins, 1988; Babcock, 1996; Wills, 1996). The presence of chelicerae, the traditional diagnostic character of chelicerates, is difficult to prove in many of the fossil taxa under consideration. However, in this analysis we recognize alternative diagnostic characters for Chelicerata: (1) presence of median eyes and/or median ocular tubercle; and (2) differentiation of the opisthosoma into a preabdomen and postabdomen (mesosoma and metasoma are alternative terms, but usually used with reference to the presence of opisthosomal appendages which cannot be seen in all taxa). This opisthosomal differentiation is not seen in Xiphosurida, but is present in the synziphosurines and as the pygidium of the arachnid orders Palpigradi, Trigonotarbida, Araneae, Amblypygi, Uropygi, Acari and Ricinulei (Shultz, 1990).

### 17.4.5 *LEMONEITES*

These small, problematic, Ordovician fossils were originally referred to Aglaspidida, but show some synziphosurine features (Flower, 1968). Anderson and Selden (in press) excluded *Lemoneites* from Xiphosura. In our analysis it is equally parsimonious to place *Lemoneites* as sister group to chasmataspids + eurypterids + arachnids, to the Xiphosura or to all other chelicerates. *Lemoneites* shows opisthosomal differentiation, supporting its referral to Chelicerata, but its poor preservation makes determining diagnostic characters such as the xiphosuran autapomorphy of ophthalmic ridges impossible. With respect to its simple morphology, Ordovician age and lack of explicit xiphosuran synapomorphies, we prefer to place *Lemoneites* as sister group to all other chelicerates, and it is probably a reasonable model for a common ancestor of Chelicerata.

### 17.4.6 XIPHOSURA

In this analysis the Xiphosura emerge as a monophyletic group within Chelicerata united on the synapomorphies of ophthalmic ridges, a cardiac lobe and a reduced opisthosomal segment 1. An axial region of the opisthosoma (convergent with cheloniellids) also characterizes xiphosurans. The presence of these synapomorphies makes it difficult to derive any of the remaining chelicerate groups from among the Xiphosura (Beall and Labandeira, 1990); the status of *Lemoneites* was discussed above. Our analysis of the Xiphosura is essentially the same as that of Anderson and Selden (1997) where the synziphosurines form a series of plesions with Xiphosurida as a crown group characterized by the autapomorphies of chilaria and opisthosomal segments fused into a thoracetron.

### 17.4.7 CHASMATASPIDA

*Chasmataspis*, with its genal spines and elongate postabdomen, and the smaller, squatter *Diploaspis/Heteroaspis* group (the two genera may be synonymous with differences resulting from preservation) appear quite different. However, they emerge as sister groups, sharing the synapomorphies of fusion of the first three opisthosomal segments and a postabdomen of nine ring-like segments. This supports the referral of both groups to a separate chelicerate taxon, Chasmataspida. The position of the Chasmataspida within Chelicerata is harder to resolve and we place chasmataspids as the sister group of (Eurypterida (Scorpiones + other arachnids)) with which they share the synapomorphies of 12 opisthosomal segments. We find no synapomorphies in favour of Eldredge's (1974) placement of chasmataspids as sister group of eurypterids alone,

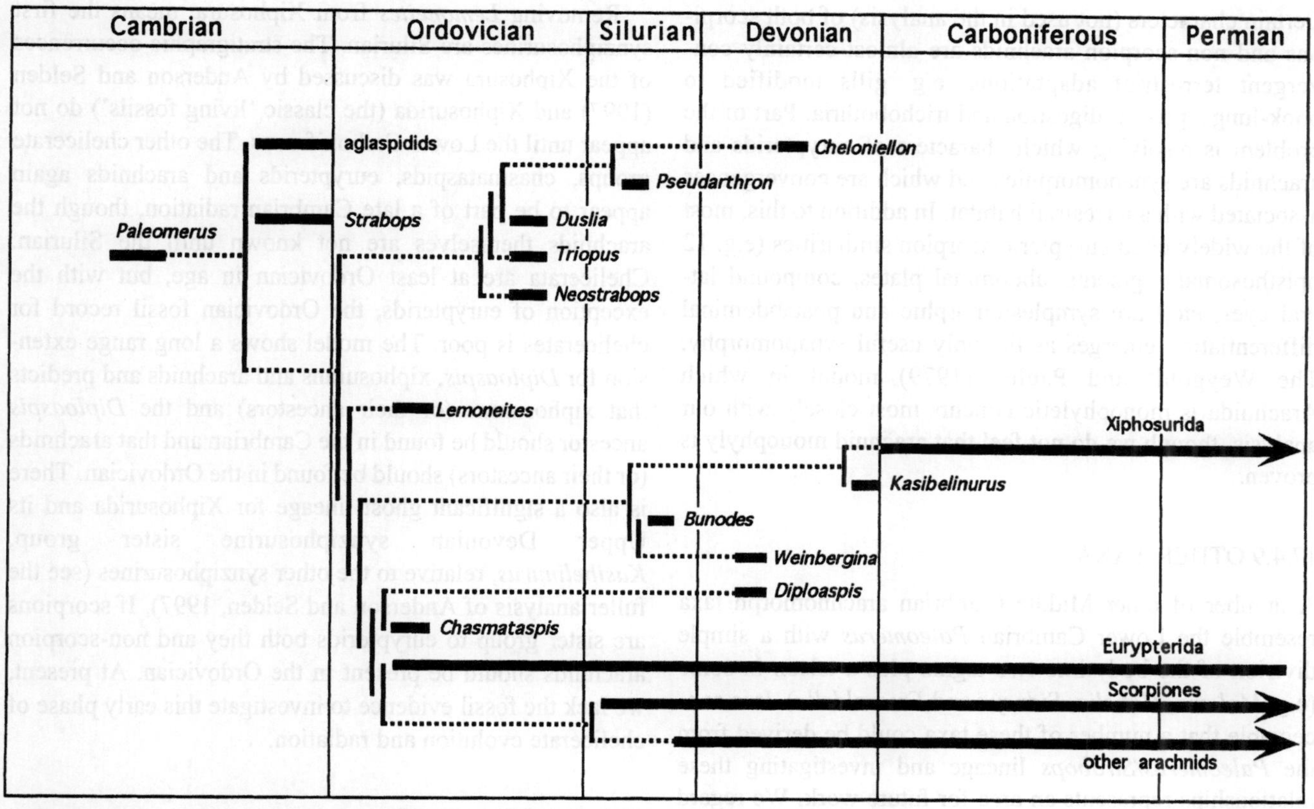


**Figure 17.4** Evolutionary tree of the major chelicerate groups and their outgroups derived from the cladogram and the stratigraphic record. Vertical lines show inferred relationships, solid horizontal lines the fossil record, dashed lines are ghost lineages.

and since chasmataspids lack ophthalmic ridges their inclusion in Xiphosura (Selden and Siveter, 1987) is unsupported. However, on the current characters it is equally parsimonious to place chasmataspids as sister group to any one of eurypterids, scorpions or other arachnids. Chasmataspids' nine-segmented postabdomen could be interpreted as part of a trend towards increasing postabdominal differentiation (a scorpion/eurypterid character), while their three-segmented preabdomen suggests they may only have had two or three pairs of gills which is closer to the non-scorpion arachnid condition. Since a number of characters are unknown in chasmataspids the model in Figure 17.3 is most parsimonious. Restudy of the original chasmataspid material would be welcome to resolve some of these uncertain characters and their phylogenetic position. Chasmataspids are a group of rare, Palaeozoic chelicerates; their synapomorphy of opisthosomal differentiation suggest they are not ancestral to any other chelicerate group, but may represent an offshoot of an early arachnid/eurypterid radiation.

## 17.4.8 EURYPTERIDA AND ARACHNIDA

Eurypterida is a monophyletic group within Chelicerata, characterized by a metastoma and a genital appendage. Eurypterids share no unique synapomorphies with xiphosurans, supporting Kraus's (1976) interpretation that the traditional concept of Merostomata (i.e. Xiphosura + Eurypterida) is artificial. Scorpions have autapomorphies of pectines and a sting, supporting the interpretation that they are not ancestral to any other arachnid order, while the other arachnids have a primitive condition of two pairs of respiratory organs and a segmented telson. Eurypterids, scorpions and arachnids form a well-supported clade with the synapomorphies of a divided femur and a basitarsus (though either of these characters may be present in chasmataspids). Resolving the interrelationships of these three taxa is difficult since scorpions share a number of characters with both eurypterids and other arachnids. Both eurypterids and scorpions have a postabdomen of five segments and this is a potential synapomorphy of these groups. In this respect, the other arachnids with a postabdomen (primitively) of three segments are closer to the xiphosurans. The models of Weygoldt and Paulus (1979) and Shultz (1990) in which scorpions and non-scorpion arachnids are placed together to form a monophyletic Arachnida is supported by the synapomorphies of slit sensilla and lack of appendages on opisthosomal segment 1 [though these are present as transient limb buds during scorpion embryology (Anderson, 1973)].

Certain characters (not used in the analysis) of both scorpions and non-scorpion arachnids are almost certainly convergent terrestrial adaptations, e.g. gills modified to book-lungs, preoral digestion and trichobothria. Part of the problem is resolving which characters of eurypterids and arachnids are synapomorphies and which are convergences associated with a terrestrial habitat. In addition to this, most of the widely cited eurypterid/scorpion similarities (e.g. 12 opisthosomal segments, abdominal plates, compound lateral eyes, etc.) are symplesiomorphic and postabdominal differentiation emerges as the only useful synapomorphy. The Weygoldt and Paulus (1979) model in which Arachnida is monophyletic concurs most closely with our analysis, though we do not feel that arachnid monophyly is proven.

### 17.4.9 OTHER TAXA

A number of other Middle Cambrian arachnomorph taxa resemble the Lower Cambrian *Paleomerus* with a simple division of the body into two tagma plus a telson in some (e.g. *Molaria*, *Habelia*, *Sidneya* and *Emeraldella*). It is conceivable that a number of these taxa could be derived from the *Paleomerus/Strabops* lineage and investigating these relationships represents an area for future work. We regard this present study as a framework, identifying three major arachnomorph clades: cheloniellids, aglaspidids and chelicerates. These three taxa are probably not a monophyletic group (Figure 17.3), rather part of a Cambrian radiation of arachnomorph arthropods whose relationships remain to be resolved.

## 17.5 EVOLUTIONARY TREE

The cladogram (Figure 17.3) was superimposed on the stratigraphic ranges (using both published and unpublished data) of these taxa to produce an evolutionary model (Figure 17.4). The *Paleomerus/Strabops* lineage (the two taxa cannot be separated in the analysis) is the oldest group, predating the Burgess Shale-type faunas; a number of arachnomorph taxa could be derived from this morphologically simple lineage. Aglaspidids can be derived from this group, and underwent a late Cambrian radiation, surviving to the Ordovician. Cheloniellids are relatively young, appearing in the late Ordovician through to the Lower Devonian, though the model predicts their occurrence in the Cambrian. Chelicerates can also be derived from a *Paleomerus/Strabops*-like ancestor. The occurrence of *Lemoneites*, *Chasmataspis* and the oldest eurypterids indicates that chelicerates had appeared by at least the early Ordovician. There is an unpublished trace fossil which may have been made by an Upper Cambrian chasmataspid and would therefore extend the range of Chelicerata back into the Cambrian.

Removing *Lemoneites* from Xiphosura, means the first synziphosurines are Silurian. The stratigraphic occurrences of the Xiphosura was discussed by Anderson and Selden (1997) and Xiphosurida (the classic 'living fossils') do not appear until the Lower Carboniferous. The other chelicerate groups, chasmataspids, eurypterids and arachnids again appear to be part of a late Cambrian radiation, though the arachnids themselves are not known until the Silurian. Chelicerata are at least Ordovician in age, but with the exception of eurypterids, the Ordovician fossil record for chelicerates is poor. The model shows a long range extension for *Diploaspis*, xiphosurans and arachnids and predicts that xiphosurans (or their ancestors) and the *Diploaspis* ancestor should be found in the Cambrian and that arachnids (or their ancestors) should be found in the Ordovician. There is also a significant ghost lineage for Xiphosurida and its Upper Devonian synziphosurine sister group, *Kasibelinurus*, relative to the other synziphosurines (see the fuller analysis of Anderson and Selden, 1997). If scorpions are sister group to eurypterids both they and non-scorpion arachnids should be present in the Ordovician. At present, we lack the fossil evidence to investigate this early phase of chelicerate evolution and radiation.

## 17.6 SYSTEMATICS

### CHELICERATA HEYMONS, 1901

*Diagnosis*

Arthropods with body divided into two tagmata: prosoma, covered by a primitively undivided carapace bearing both median and lateral eyes, and bearing six pairs of appendages, first of which (chelicerae) are chelate; and opisthosoma of 12 or fewer segments in which opisthosomal appendage 1 may be present and pediform, but in which subsequent opisthosomal appendages are plate-like and bear respiratory organs. Some degree of opisthosomal differentiation into a pre- and postabdomen developed. Post-anal telson primitively present.

*Remarks*

This diagnosis restricts Chelicerata to arachnids (including scorpions), eurypterids and xiphosurans (including synziphosurines) and is essentially a restatement of Heymons's (1901) concept of Chelicerata, but now including the chasmataspids. It excludes *Sanctacaris* (which lacks opisthosomal differentiation) and *Fuxianhuia* (which has biramous opisthosomal limbs), the remaining Arachnomorpha and pycnogonids (though pycnogonids have median eyes and their highly reduced opisthosoma would not show differentiation and is clearly derived). We note the following provisos: The chelicerae in some arachnids (e.g. spiders) are modified from the chelate state, into 'clasp-knife' structures. The arachnid orders Palpigradi and

Solifugae have a divided carapace. In our analysis this would represent a derived state permitting prosomal flexibility, rather than a primitive character, as implied in the phylogenies of Savory (1971) and Grasshoff (1978). Scorpions have been interpreted as having 13 opisthosomal segments (van der Hammen 1989), partly based on a reported transitory pregenital embryological segment-bearing appendages. However, Anderson's (1973) detailed review of scorpion embryology noted transitory limb buds on the pregenital segment which are later reabsorbed. These limb buds were still part of a groundplan of a seven-segmented preabdomen and a five-segmented postabdomen, supporting the interpretation that scorpions have 12, not 13, opisthosomal segments.

## 17.7 CONCLUSIONS

1. Chelicerata Heymons, 1901 belongs to a broad clade of arthropods, the Arachnomorpha Heider, 1913.
2. Chelicerata is monophyletic, diagnosed as above, and is restricted to *Lemoneites*, Xiphosura, Chasmataspida, Eurypterida and Arachnida.
3. Eleven opisthosomal segments appears to be primitive for Chelicerata, as occurs in *Lemoneites*, Xiphosura and outgroups such as *Paleomerus*. Chasmataspida are not xiphosurans and are placed as the sister group of Eurypterida + Arachnida, with which they share the synapomorphy of 12 opisthosomal segments.
4. Xiphosura, Eurypterida, Chasmataspida and Scorpiones all possess autapomorphies and thus none is ancestral to any other chelicerate group. Arachnida emerges as a monophyletic taxon while Merostomata is paraphyletic.
5. The Ordovician *Lemoneites* appears very close to the origins of the chelicerates, but is too poorly preserved to resolve whether it is a primitive synziphosurine, sister group to chasmataspids, eurypterids and arachnids, or a sister group to all other chelicerates.
6. Presence of chelicerae is difficult to determine in many Lower Palaeozoic fossils and the presence of median eyes (or their tubercle) and some degree of differentiation in the opisthosoma into a preabdomen and postabdomen are better characters for recognizing a fossil chelicerate.
7. *Neostrabops*, *Duslia*, *Triopus*, *Pseudarthron* and *Cheloniellon* form a monophyletic group, Cheloniellida, placed as sister group to Chelicerata in this analysis, and is united on the character of a procurved posterior carapace margin.
8. (*Paleomerus*, *Strabops* + Aglaspidida) represents the sister group to Chelicerata and Cheloniellida. Aglaspidids represent a poorly defined group which warrants revision and it is difficult to resolve the position of aglaspidids relative to chelicerates and the *Paleomerus/Strabops* lineage.
9. Cheloniellida, Aglaspidida and Chelicerata represent three distinct arachnomorph taxa, all of which could be derived from a *Paleomerus/Strabops*-like lineage, though their relationships to the other Arachnomorpha remain uncertain.

ACKNOWLEDGEMENTS

We thank Lyall Anderson, Simon Braddy, Andy Jeram, Otto Kraus, Bill Shear, Jeff Shultz and Derek Siveter for useful discussions and correspondence and Ruth Dewel for information on tardigrades. J.A.D. acknowledges a NERC postdoctoral fellowship into the origins and early radiation of the Chelicerata.

## REFERENCES

Abele, L.G., Kim, W. and Felgenhauer, B.E. (1989) Molecular evidence for inclusion of the phylum Pentastomida in the Crustacea. *Molecular Biology and Evolution*, **6**, 685–91.

Anderson, D.T. (1973) *Embryology and Phylogeny in Annelids and Arthropods*, Pergamon Press, Oxford.

Anderson, L.I. and Selden P.A. (1997) Opisthosomal fusion and phylogeny of Palaeozoic Xiphosura. *Lethaia*, **30**, 19–31.

Babcock, L.E. (1996) Early Cambrian chelicerate arthropod from China, in *Sixth North American Paleontological Convention. Abstracts of Papers.* (ed. J.E. Repetski), The Palaeontological Society, Special Publication No. 8, p. 17.

Beall, B.S. and Labandeira, C.C. (1990) Macroevolutionary patterns of the Chelicerata and Tracheata, in *Arthropod Palaeobiology, Short Courses in Palaeontology*, No. 3 (ed. S.J. Culver), The Palaeontological Society, pp. 257–84.

Bekleminshev, V.N. (1944) Osnovy sravnitelinoi anatomii bespozvonochnykh (Fundamentals of the comparative anatomy of invertebrates), Sovetskaya nauka, 1st edition.

Bergström, J. (1968) *Eolimulus*, a Lower Cambrian xiphosurid from Sweden. *Geologiska Förenigens i Stockholm Förhandlingar*, **90**, 489–503.

Bergström, J. (1971) *Paleomerus* – merostome or merostomoid. *Lethaia*, **4**, 393–401.

Bergström, J. (1975) Functional morphology and evolution of xiphosurids. *Fossils and Strata*, **4**, 291–305.

Bergström, J. (1979) Morphology of fossil arthropods as a guide to phylogenetic relationships, in *Arthropod Phylogeny* (ed. A.P. Gupta), Van Nostrand Reinhold Co., New York, pp. 3–56.

Bergström, J., Stürmer, W. and Winter, G. (1980) *Palaeoisopus*, *Palaeopantopus* and *Palaeothea*, pycnogonid arthropods from the Lower Devonian Hunsrück Slate, West Germany. *Paläontologishe Zeitschrift*, **54**, 7–54.

Briggs, D.E.G. and Collins, D.A. (1988) A Middle Cambrian chelicerate from Mount Stephen, British Columbia. *Palaeontology*, **31**, 779–98.

Briggs, D.E.G. and Fortey, R.A. (1989) The early radiation and relationships of the major arthropod groups. *Science*, **246**, 241–3.

Briggs, D.E.G., Bruton, D.L. and Whittington, H.B. (1979) Appendages of the arthropod *Aglaspis spinifer* (Upper

Cambrian, Wisconsin) and their significance. *Palaeontology*, **22**, 167–80.

Briggs, D.E.G, Fortey, R.A. and Wills, M.A. (1992) Morphological disparity in the Cambrian. *Science*, **256**, 1670–3.

Bristowe, W.S. (1971) *The World of Spiders*, 2nd edn, Collins, London.

Bruton, D.L. (1981) The arthropod *Sidneya inexpectans*, Middle Cambrian, Burgess Shale, British Columbia. *Philosophical Transactions of the Royal Society, London, B*, **295**, 619–56.

Bruton, D.L. and Whittington, H.B. (1983) *Emeraldella* and *Leancholia*, two arthropods from the Burgess Shale, British Columbia. *Philosophical Transactions of the Royal Society, London, B*, **300**, 553–85.

Caster, K.E. and Brooks, H.K. (1956) New fossils from the Canadian-Chazyan (Ordovician) hiatus in Tennessee. *Bulletins of American Paleontology*, **36**, 153–99.

Caster, K.E. and Kjellesvig-Waering, E.N. (1964) Upper Ordovician eurypterids of Ohio. *Palaeontographica Americana*, **4**, 301–58.

Caster, K.E. and Macke, W.B. (1952) An aglaspid merostome from the Upper Ordovician of Ohio. *Journal of Paleontology*, **26**, 753–57.

Chen, J.Y., Edgecombe, G.D., Ramsköld, L. and Zhou, G.Q. (1995) Head segmentation in early Cambrian *Fuxianhuia*: implications for arthropod evolution. *Science*, **268**, 1339–43.

Chlupač, I. (1965) Xiphosuran merostomes from the Bohemian Ordovician. *Sbornúk Geologickych Ved Praha*, Rada P, **5**, 7–38.

Chlupač, I. (1988) The enigmatic arthropod *Duslia* from the Ordovician of Czechoslovakia. *Palaeontology*, **31**, 611–20.

Chlupač, I. and Havlúcek, V. (1965) *Kodymirus* n. g., a new aglaspid merostome of the Cambrian of Bohemia. *Sborník Geologickych Ved Praha*, Rada P, **6**, 7–20.

Clarke, J.M. and Ruedemann, R. (1912) *The Eurypterida of New York.* Memoirs of the New York State Museum, Memoir 14.

Conway Morris, S. (1992) Burgess Shale-type faunas in the context of the 'Cambrian explosion': a review. *Journal of the Geological Society of London*, **149**, 631–6.

Dahl, F. (1913) *Vergleichende Physiologie und Morphologie der Spinnentiere*, Theil I, Jena.

Dubinin, V.B. (1962) Class Scorpionomorpha, in *Fundamentals of Paleontology, Volume 9, Arthropoda, Tracheata, Chelicerata* (ed. B.B. Rohdendorf), *Osnovy paleontologii*, 1991 English translation, Smithsonian Institution Libraries, Amerind Publishing Co., New Delhi, pp. 614–77.

Eldredge, N. (1974) Revision of the suborder Synziphosurina (Chelicerata, Merostomata), with remarks on merostome phylogeny. *American Museum Novitates*, **2543**, 1–41.

Fedotov, D. (1925) On the relations between the Crustacea, Trilobita, Merostomata and Arachnida. *Izvestiya Rossiiskoi Akademii Nauk*, (1924), 383–408.

Firstman, B. (1973) The relationship of the chelicerate arterial system to the evolution of the endosternite. *Journal of Arachnology*, **1**, 1–54.

Flower, R. (1968) Merostomes from a Cotter horizon of the El Paso group. *New Mexico State Bureau of Mines and Mineral Resources Memoir*, **22** (4), 35–44.

Grasshoff, M. (1978) A model of the evolution of the main chelicerate groups. *Symposium of the Zoological Society of London*, **42**, 273–84.

Hall, J. (1862) A new crustacean from the Potsdam Sandstone. *Canadian Naturalist*, **7**, 443–5.

Hammen, L. van der (1989) *An Introduction to Comparative Arachnology*, SPB Academic Publishing, The Hague.

Handlirsch, A. (1914) Eine interessante Crustaceenform aus der Trias der Vogesen. *Verhandlung der Kaiserlich Konlichen Zoologish–Botanischen Gesellschaft in Wien*, **64**, 1–8.

Hesselbo, S.P. (1989) The aglaspidid arthropod *Beckwithia* from the Cambrian of Utah and Wisconsin. *Journal of Paleontology*, **63**, 636–42.

Hesselbo, S.P. (1992) Aglaspidida (Arthropoda) from the Upper Cambrian of Wisconsin. *Journal of Paleontology*, **66**, 885–923.

Heymons, R. (1901) Die Entwicklungsgeschichte der Scolopender. *Zoologica*, **13**(33), 2–5.

Jeram, A.J. (1994) Scorpions from the Viséean of East Kirkton, West Lothian, Scotland, with a revision of the infraorder Mesoscorpionina. *Transactions of the Royal Society of Edinburgh: Earth Sciences*, **84**, 283–99.

King, P.E. (1973) *Pycnogonids*, Hutchinson and Co., Ltd., London.

Kjellesvig-Waering, E.N. (1986) A restudy of the fossil Scorpionida of the world. *Palaeontographica Americana*, **55**, 1–287.

Kraus, O. (1976) Zur phylogenetischen Stellung und Evolution der Chelicerata. *Entomologica Germanica*, **3**, 1–12.

Lankester, E.R. (1881) *Limulus*, an arachnid. *Quarterly Journal of Microscopical Science*, **23**, 504–548, 609–49.

Lauterbach, K.-E. (1980) Schlüsselerignisse in der Evolution des Grundplans der Arachnata (Arthropoda). *Abhandlungen des Naturwissenschaftlichen Vereins in Hamburg*, **23**, 163–327.

Lindquist, E.E. (1984) Current theories on the evolution of major groups of Acari and their relationships with other groups of Arachnida, with consequent implications for their classification, in *Acarology VI* (eds D.A. Griffiths and C.E. Bowman), Ellis Horwood Ltd, Chichester, pp. 28–62.

Lockwood, S. (1870) The horse foot crab. *American Naturalist*, **4**(5), 257–74.

McNamara, K.J. and Trewin, N.H. (1993) A euthycarcinoid arthropod from the Silurian of Western Australia. *Palaeontology*, **36**, 319–35.

Möberg, J.C. (1892) Om en nyupptäckt fauna i blok af kambrisk sandsten insamlade af Dr N.O. Holst. *Geologiska Föreningens i Stockholm Förhandlingar*, **14**, 103–20.

Nieszkowski, J. (1859) Der *Eurypterus remipes* aus den obersihurischen Schichten der Insel Oesel. *Archiv für Naturkunde der Liv- Est- und Kurlands* Series 1, **2**, 299–344.

Novojilov, N. (1962) Class Merostomata, in *Fundamentals of Paleontology, Volume 9, Arthropoda, Tracheata, Chelicerata* (ed. B.B. Rohdendorf), *Osnovy paleontologii*, 1991 English translation, Smithsonian Institution Libraries, Amerind Publishing Co., New Delhi, pp. 591–613.

Petrunkevitch, A.I. (1949) A study of Palaeozoic Arachnida. *Transactions of the Connecticut Academy of Arts and Sciences*, **37**, 69–315.

Pickett, J.W. (1993) A Late Devonian xiphosuran from near Parkes, New South Wales. *Memoirs of the Association of Australian Palaeontologists*, **15**, 279–87.

Pocock, R.I. (1893) On some points in the morphology of the Arachnida (s. s.) with notes on the classification of the group. *Annals and Magazine of Natural History*, **61**, 1–19.

Pocock, R.I. (1901) The Scottish Silurian scorpion. *Quarterly Journal of Microscopical Science*, **44**, 291–311.

Raasch, G.O. (1939) *Cambrian Merostomata*, Special Papers of the Geological Society of America, No. 19.

Raw, F. (1957) Origin of chelicerates. *Journal of Paleontology*, **31**, 139–92.

Richter, R., and Richter, G. (1929) *Weinbergina opitzi* n.g., n. sp., ein Schwerttrager (Merost. Xiphos.) aus dem Devon (Rheinland). *Senckenbergiana*, **11**, 193–209.

Riek, E.F. (1964) Merostomoidea (Arthropoda, Trilobitomorpha) from the Australian Middle Triassic. *Records of the Australian Museum*, **26**, 327–32.

Savory, T.H. (1971) *Evolution in the Arachnida*, Merrow Publishing Co., Watford.

Selden, P.A. (1981) Functional morphology of the prosoma of *Baltoeurypterus tetragonophthalamus* (Fischer) (Chelicerata, Eurypterida). *Transactions of the Royal Society of Edinburgh: Earth Sciences*, **72**, 9–42.

Selden, P.A. and Siveter, D.J. (1987) The origin of the limuloids. *Lethaia*, **20**, 383–92.

Selden, P.A. and White, D.E. (1984) A new Silurian arthropod from Lesmahagow, Scotland. *Special Papers in Palaeontology*, **30**, 43–9.

Sharov. A.G. (1966) *Basic Arthropodan Stock with Special Reference to Insects*, Pergamon Press, Oxford.

Shultz, J.W. (1990) Evolutionary morphology and phylogeny of Arachnida. *Cladistics*, **6**, 1–38.

Simonetta, A. and Delle Cave, L. (1981) An essay in the comparative and evolutionary morphology of Palaeozoic arthropods. *Atti dei Convengni Lincei*, **49**, 389–439.

Sissom, W.D. (1990) Systematics, biogeography and paleontology, in *The Biology of Scorpions* (ed. G.A. Polis), Stanford University Press, Stanford, pp. 64–160.

Starobogatov, Y.I. (1990) Sistema i filogeniya nizshikh khelitserovykh (analiz morfologii paleozoysikh grupp). (The systematics and phylogeny of the lower chelicerates (a morphological analysis of the Paleozoic groups)). *Paleontologicheskii Zhurnal*, **1**, 4–17 [English translation].

Stockwell, S.A. (1989) Revision of the phylogeny and higher classification of scorpions (Chelicerata). PhD Thesis, University of California, Berkeley. Published by University Microfilms International, Ann Arbour, Michigan.

Størmer, L. (1944) On the relationships and phylogeny of fossil and Recent Arachnomorpha. *Skrifter utgitt av det Norske Videnskaps Academi i Oslo, I. Matematisk-Naturvidenskapelig Klasse 1944 I*, **5**, 1–158.

Størmer, L. (1952) Phylogeny and taxonomy of fossil horseshoe crabs. *Journal of Paleontology*, **26**, 630–40.

Størmer, L. (1955) Merostomata, in *Treatise on Invertebrate Paleontology, Part P, Arthropoda 2* (ed. R.C. Moore), Geological Society of America and University of Kansas Press, Lawrence, Kansas, pp. 4–41.

Størmer, L. (1959) Trilobitomorpha, in *Treatise on Invertebrate Palaeontology, Part O, Arthropoda 1* (ed. R.C. Moore), Geological Society of America and University of Kansas Press, Lawrence, Kansas, pp. 22–37.

Størmer, L. (1972) Arthropods from the Lower Devonian (Lower Emsian) of Alken an der Mosel, Germany, Part 2, Xiphosura. *Senckenbergiana lethaea*, **53**, 1–29.

Størmer, L. (1974) Arthropods from the Lower Devonian (Lower Emsian) of Alken an der Mosel, Germany. Part 4, Eurypterida, Drepanopteridae and other groups. *Senckenbergiana lethaea*, **54**, 359–451.

Stürmer, W. and Bergström, J. (1978) The arthropod *Cheloniellon* from the Devonian Hunsrück late. *Paläontologiche Zeitschrift*, **52**, 57–81.

Stürmer, W. and Bergström, J. (1981) *Weinbergina*, a xiphosuran from the Devonian Hunsrück late. *Paläontologiche Zeitschrift*, **55**, 237–55.

Tollerton, V.P. (1989) Morphology, taxonomy and classification of the Order Eurypterida Burmeister, 1843. *Journal of Paleontology*, **63**, 642–57.

Versluys, J. and Demoll, R. (1920) Die Vervandschaft der Merostomata mit den Arachnida und der anderen Abteilungen der Arthropoda. *Proceedings Akademie Wetensschappen, Amsterdam*, **23**, 739–65.

Walcott, C.D. (1912) Middle Cambrian Branchiopoda, Malacostraca, Trilobita and Merostomata. Cambrian Geology and Palaeontology, II. *Smithsonian Miscellaneous Collections*, **57**, 145–288.

Waterston, C.D. (1979) Problems of the functional morphology and classification in stylonuroid eurypterids with observations on the Scottish Silurian Stylonuroidea. *Transactions of the Royal Society of Edinburgh: Earth Sciences*, **70**, 251–322.

Weygoldt, P. and Paulus, H.F. (1979) Untersuchungen zur Morphologie, Taxonomie und Phylogeny der Chelicerata. *Zeitschrift für Zoologie, Systematik und Evolutionsforschung*, **17**, 85–116, 177–200.

Wheeler, W.C., Cartwright, P. and Hayashi, C.Y. (1993) Arthropod phylogeny: a combined approach. *Cladistics*, **9**, 1–39.

Whittington, H.B. (1981) Rare arthropods from the Burgess Shale, Middle Cambrian, British Columbia. *Philosophical Transactions of the Royal Society, London, B*, **292**, 329–57.

Wills, M.A. (1996) Classification of the Arthropod *Fuxianhuia*. *Science*, **272**, 746–7.

Wills, M.A., Briggs, D.E.G. and Fortey, R.A. (1994) Disparity as an evolutionary index: a comparison of Cambrian and Recent arthropods. *Paleobiology*, **20**, 93–130.

Wills, M.A., Briggs, D.E.G., Fortey, R.A. and Wilkinson, M. (1995) The significance of fossils in understanding arthropod evolution. *Verhandlungen der Deutschen Zoologischen Gesellschaft*, **88**, 203–15.

Woods, H. (1909) The Crustacea and arachnids, in *The Cambridge Natural History* (eds S.F. Harmer and A.E. Shipley), Cambridge, pp. 283–94.

Woodward, H. (1866–1878) *Monograph of the British fossil Crustacea belonging to the order Merostomata*. Palaeontographical Society, London.

Youchong, H. and Shaowu, N. (1981) Discovery of new marine family-Sinaglaspidae (Aglaspida) in Shanxi Province. *Kexue Tongbao*, **26**(10), 911–14.

Zakhvatkin, A.A. (1952) Razdeleniye kleshchei (Acarina) na otryady i ikh polozheniye v sisteme Chelicerata (Division of ticks (Acarina) into orders and their position in the system of Chelicerata). *Parazitologiya Sbornik Instut Zoologii Akademiya Nauk SSSR*, **14**, 5–46.

# 18 Problem of the basal dichotomy of the winged insects

*A.P. Rasnitsyn*

*Arthropoda Laboratory, Paleontological Institute, Russian Academy of Science, Profsoyuznaya Str. 123, Moscow, 117647, Russia*
*email: rasna@glasnet.ru*

## 18.1 INTRODUCTION

There are two main competing hypotheses concerning the basal pterygote dichotomy. The first was proposed by Martynov (1925) and has been since widely supported (Rohdendorf, 1962; Boudreaux, 1979; Kristensen, 1981; Arnett, 1985; Carpenter, 1992; CSIRO, 1992). Currently, it is being most vigorously championed by Kukalová-Peck (1978–1992). This hypothesis infers that the main pterygote dichotomy separates Palaeoptera and Neoptera. The alternative hypothesis is mine (Rasnitsyn, 1976, 1980; Rohdendorf and Rasnitsyn, 1980). It claims that the basic pterygote dichotomy divides the Subclass Scarabaeona (= Pterygota) into Infraclasses Scarabaeones and Gryllones [the reasons for employing these names for higher taxa are explained in Rohdendorf (1977) and Rasnitsyn (1982, 1996)].

## 18.2 FIRST HYPOTHESIS

Martynov's (1938) hypothesis (Figure 18.1) infers that palaeodictyopteroids (Superorder Dictyoneuridea), mayflies (Superorder Ephemeridea), and dragonflies (Superorder Libellulidea) are symplesiomorphic in that their wings are unfoldable (Figure 18.2), lack the jugal area, and show advanced fluting due to convex MA (Figure 18.3) (wing morphology definitions as in CSIRO, 1992). Neoptera are synapomorphic in possessing an alternative set of character states: wings are foldable backward over the abdomen (Figure 18.4), bear a differentiated jugal area that rocks down in the folded wing (Figure 18.5), and have MA, if differentiated, neutral in its position in relation to the surrounding membrane, and not convex, that is, positioned on a top of a longitudinal convexity of the membrane (Figure 18.6).

This inference, however, is in conflict with various morphological and functional observations. Most important of these is that the permanently outstretched pro-wings and wing pads dramatically limit the mobility of an immature insect. It is particularly so because the wing development was very gradual in ancient insects, resulting in fairly long wing pads in later, but still non-flying ontogenetic stages. To cope with this problem, some of the immature stages of ancient palaeopterans had their wing pads bent rearward (Figure 18.7), while others acquired the foldable wing pads (Figure 18.8). In a flying, adult insect, permanently outstretched wings obviously hinder it when entering narrow spaces - and hence make the insect permanently exposed to predators, winds and rains. These obstacles are easy to cope with for a good flier, but it can be questioned whether this way of life applied to ancestral pterygotes whose flight necessarily should be weak (Figure 18.9).

Later, Kukalová-Peck (1992) modified her previous claim (Figure 18.10): she agreed that permanently outstretched wings were not a pterygote symplesiomorphy but a synapomorphy of Palaeoptera other than Diaphanopterida (= Diaphanopterodea, the order that is closely related to other dictyoneurideans but unlike them possesses the foldable wings). However, this hypothesis is still open to criticism. One reason is that the putative palaeopterous apomorphies are prone to independent origin. Unfoldable wings appear in butterflies and various moths (Alucitidae, Geometridae, etc.), in many dipterans (e.g. Bombyliidae), in some hymenopterans (e.g. Cephidae), and so on. The convex MA appears independently from palaeopterans, for instance, in the highly plesiomorphic Permian Homoeodictyon (Figure 18.11), in bittacid mecopterans (Figure 18.12) and in all dipterans (Shcherbakov *et al.*, 1995). In bittacids and dipterans, this vein is called RS4 but represents a branch of MA, because the mecopteroid and neuropteroid RS combines both true RS and MA (Figure 18.13). In addition, Shcherbakov *et al.* (1995) showed that the most archaic Triassic dipteran family, Vladipteridae, has convex developments of both the suggested MA homologues, RS3 and RS4.

Still more significant is the existence of a transition that connects palaeodictyopterans (Figure 18.2 and 18.3) to Paoliidae, which have the least advanced pterygote wings (Figure 18.14). The transition is formed by the order Diaphanopterida (Figure 18.15), and further by

*Arthropod Relationships*, Systematics Association Special Volume Series 55, Edited by R.A. Fortey and R.H. Thomas. Published in 1997 by Chapman & Hall, London. ISBN 0 412 75420 7

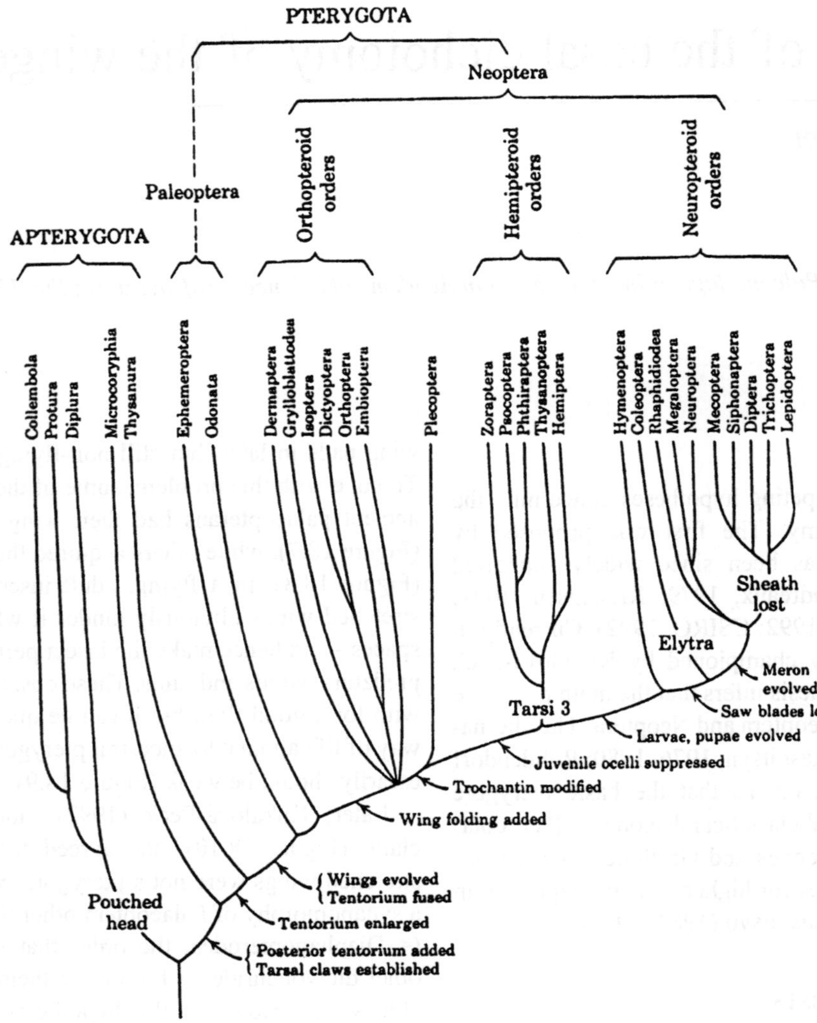


**Figure 18.1** Insect phylogeny by Ross (1965) displaying essentially Martynov's view on the basic insect dichotomy.

*Heterologopsis* Brauckmann (Figure 18.16). At first glance, *Heterologopsis* seems to be a normal diaphanopterid, with the typically palaeodictyopteroid sucking beak and prothoracic winglets, and with foldable wings. The wings, however, are typical of Cacurgidae, a group related to Paoliidae and venationally the second least advanced group among the winged insects (Figure 18.17).

The above series of taxa may well represent an example of an evolutionary transition from stem pterygotes to the palaeodictyopterans. The problem is that in this series the sucking beak had already originated at the Cacurgid stage (Figure 18.16), that is, long before than the palaeopteran venational synapomorphies appeared. In contrast, in the mayfly–dragonfly clade, the groundplan mouthparts are close to the pterygote groundplan mouthparts (Boudreaux, 1979). Apparently, the dragonflies and mayflies gained their palaeopteran apomorphies independently of dictyoneurideans.

## 18.3 SECOND HYPOTHESIS

The opposing hypothesis suggests rather that the basic pterygote dichotomy separates the Infraclass Gryllones (= Polyneoptera, or Orthopteroidea in the widest sense) from its ancestral (paraphyletic) Infraclass Scarabaeones. The latter group comprises all the rest of the winged insects, that is, Martynov's Oligoneoptera and Paraneoptera. The first of these two taxa, known also as Holometabola (insects with complete metamorphosis), is typified as the Cohors Scarabaeiformes. Paraneoptera embraces Hemiptera and Psocopteroidea and is typified, jointly with several extinct groups, as the Cohors Cimiciformes (Rasnitsyn, 1980; Rohdendorf and Rasnitsyn, 1980). This hypothesis states that Scarabaeones are symplesiomorphic in the roof-like repose position of their wings (Figure 18.18), and that in their hind wing the jugal area (Figure 18.5) is small, with the folding line running behind (usually well behind) the vein

**Figure 18.2** *Goldenbergia* sp. (Dictyoneurida) from the Upper Carboniferous of Siberia. Reconstruction by A.G. Ponomarenko. (From Rohdendorf and Rasnitsyn 1980.)

2A (using conservative vein nomenclature: otherwise, if 1A is re-defined as PCu, this vein should be termed 1A). In contrast, Gryllones are synapomorphic in acquiring horizontal wing repose position (Figure 18.4), with their jugal area enlarged in the hind wing, and rocking down before 2A at rest (Figure 18.19).

Under these assumptions, the palaeopteran apomorphies are considered to appear homoplastically on several distinct occasions within the Scarabaeones. One such occasion happened during the course of evolution of the palaeodictyopteroid insects (Superorder Dictyoneuridea), namely, at the appearance of the sister group of Diaphanopterida. This sister group is hypothesized to comprise the orders Dictyoneurida (= Palaeodictyoptera; Figure 18.2) and Mischopterida (= Megasecoptera; Figure 18.20).

The second major independent acquisition of palaeopterous apomorphies defines the Cohors Libelluliformes, the clade covering mayflies and dragonflies. Unlike my previous opinion (Rasnitsyn, 1980; Rohdendorf and Rasnitsyn, 1980), I now accept the sister group relationships between the Superorders Libellulidea and Ephemeridea based upon synapomorphies in wing morphology. These include the very basal separation of R and RS veins, and the pronounced fluting of their wing surface, so as in the abundantly branching convex vein systems (MA, CuA), the convex primary veins alternate with the concave intercalaries (longitudinal veins with a double, forking base), while in the corresponding concave systems (RS, MP) the concave primary veins alternate with the convex intercalaries. This inference is supported by the wing morphology of the ancient, Carboniferous representatives of both mayflies and dragonflies that lack several of their autapomorphies and thus are overall rather similar (Figures 18.21 and 18.22).

Because of their thoracic morphology, which is the most plesiomorphic within known Scarabaeones (pterothoracic sterna are not invaginated along the segment midline: no discrimen is developed), dragonflies and mayflies should be considered to be a sister group of the rest of the Scarabaeones. This inference is not yet secure, however, because the body structure is completely unknown for some ancient groups that are highly plesiomorphic in other respects. This applies to the Paoliidae, a group virtually lacking any apomorphy above the pterygote groundplan, but unfortunately known only from their wings (Figure 18.14). Paoliidae have been formerly supposed to represent the ancestral pterygote taxon, Protoptera (Sharov, 1966) or Paoliida (Rohdendorf and Rasnitsyn, 1980). However, as described above, there is a ‘smooth’ transition connecting them with Diaphanopterida via Cacurgidae (*Heterologopsis*). Diaphanopterida possess the invaginated pterothoracic sterna, the same should be hypothesized to *Heterologopsis* and hence to Cacurgidae because of their profound similarity in the body structure. Paoliidae and Cacurgidae are similar enough venationally to deserve attribution to one and the same order (Paoliida). Hence, we should assume that they had similar body morphology, at least until the contrary is proven. However, the discovery of the body of any representative of the Paoliidae may overturn this assumption. The Paoliidae may prove to be an early off-shoot of the pterygote ancestor: either a sister group, or even the ancestral taxon of the rest of the winged insects.

## 18.4 CONCLUSION

The basal pterygote dichotomy of Palaeoptera and Neoptera can be rejected in favour of that of Scarabaeones and Gryllones. This is based upon functional considerations, and upon fossils whose morphology is congruent with neither monophyly nor paraphyly of Palaeoptera.

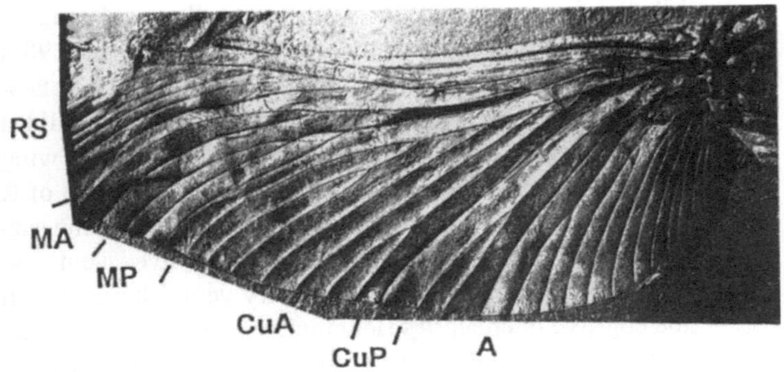


**Figure 18.3** Hind wing of *Dunbaria quinquefasciata* (Martynov) (Dictyoneurida) from the Upper Permian (Kungurian) of Tshekarda, Ural Mountains.

**Figure 18.4** *Sylvaphlebia tuberculata* Martynov (Grylloblattida) from the Upper Permian (Kungurian) of Tshekarda, Ural Mountains.

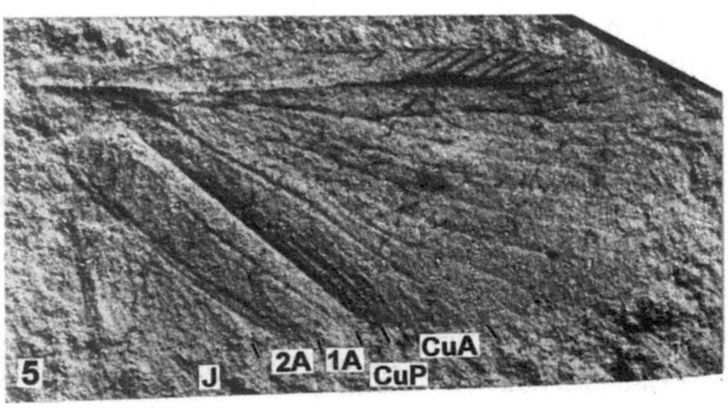


**Figure 18.5** *Glaphyrophlebia subcostalis* Martynov (Blattinopseida), the hind wing with unfolded jugum (J), from the Upper Permian (Kazanian) of Soyana, Arkhangelsk Region of Northern Russia.

**Figure 18.6** Fore wing of *Holasicia vetula* Kukalová (Paoliida) from the lowermost Middle Carboniferous (Namurian C). (From Kukalová, 1958.)

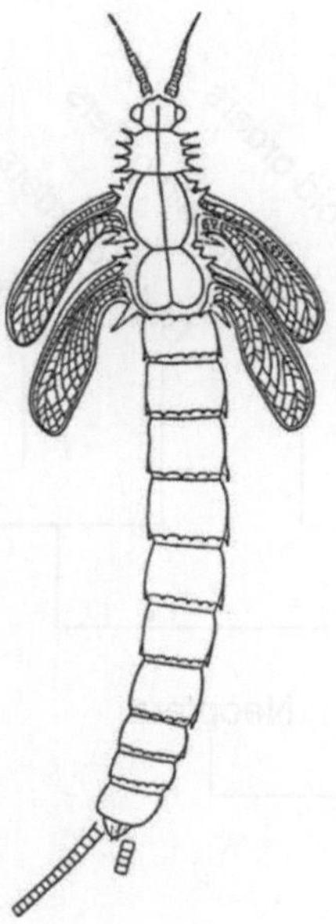

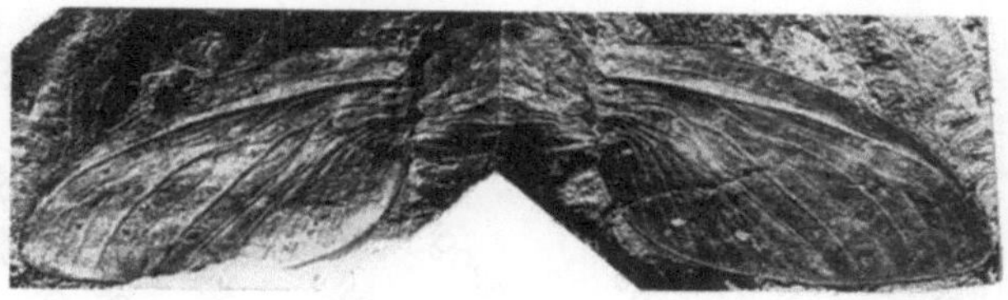

**Figure 18.8** Thoracic segment of a nymphal *Tchirkovaea guttata* G. Zalessky (Dictyoneurida) from the Upper Carboniferous of Siberia: aligned photographs of the obverse and reverse parts of the fossil. (From Rasnitsyn, 1980.)

**Figure 18.7** Nymphal *Mischoptera douglassi* Carpenter and Richardson (Mischopterida) from the Lower Permian (Artinskian) of Illinois, USA. (From Carpenter and Richardson, 1968.)

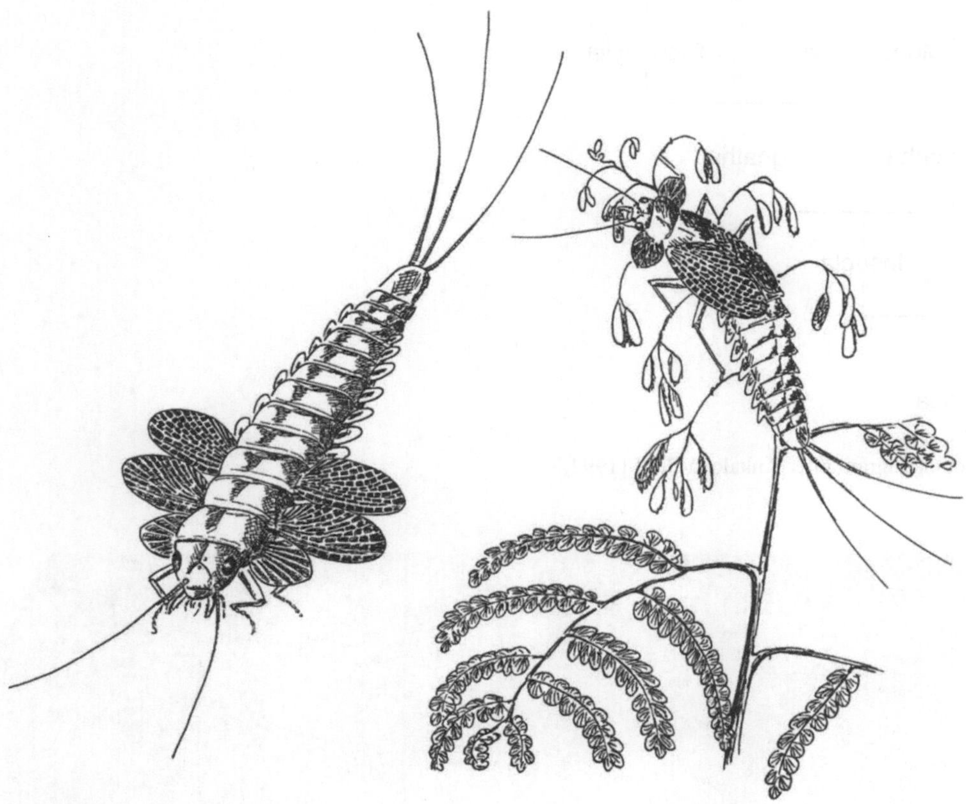

**Figure 18.9** Ancestral winged insects. Reconstruction by A.G. Ponomarenko. (From Rohdendorf and Rasnitsyn, 1980.)

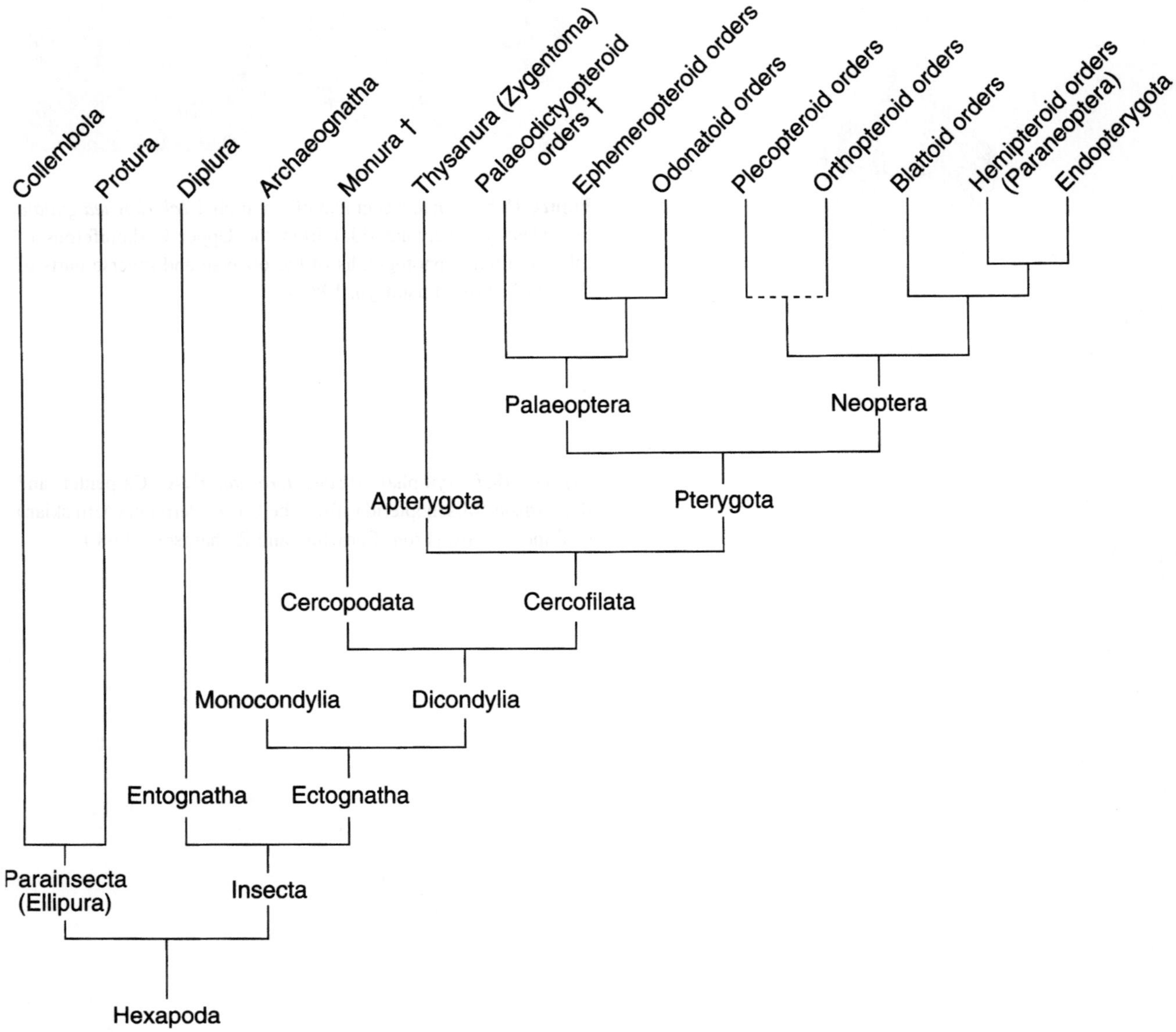


**Figure 18.10** Insect cladogram after Kukalová-Peck (1992).

**Figure 18.11** Fore wing of *Homoeodictyon elongatum* Martynov (?Hypoperlida) from the Upper Permian of Kargala, Southern Ural Mountains, showing convex MA stem.

**Figure 18.12** Fore wing of the Bittacid scorpionfly *Neorthophlebia robusta* Martynov (Panorpida) from the Lower Jurassic of Kirghizia (vicinity of Shurab), the fore wing with convex hind MA branch (= RS4). (From Martynov, 1937.)

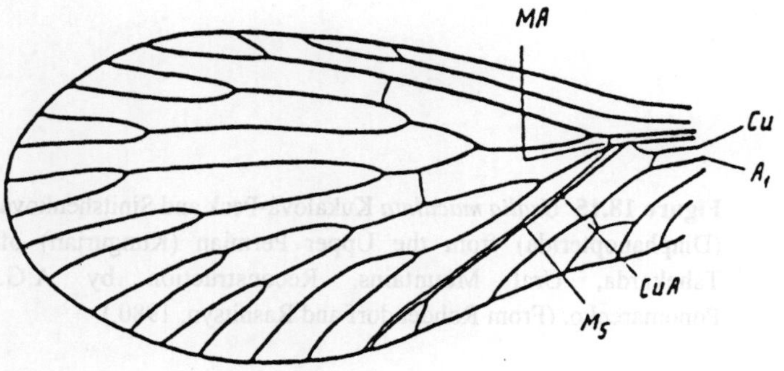


**Figure 18.13** Hind wing of the Kaltanid scorpionfly *Altajopanorpa pilosa* O. Martynova (Panorpida) from the Upper Permian (Ufimian) of Kuznetsky Basin, S. Siberia, showing MA base. (From Novokshonov, 1992.)

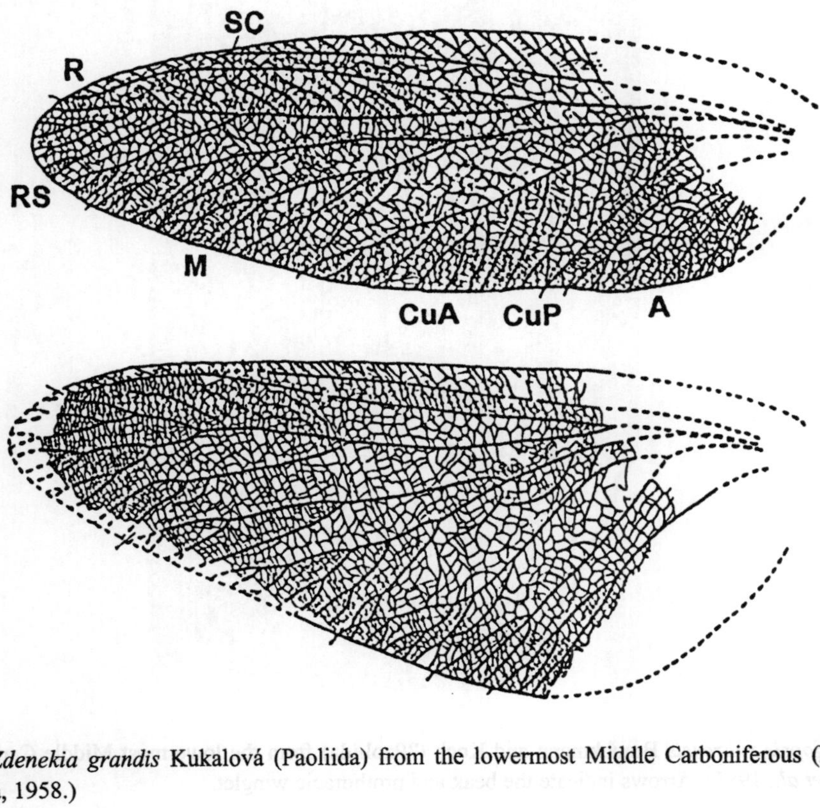


**Figure 18.14** Wings of *Zdenekia grandis* Kukalová (Paoliida) from the lowermost Middle Carboniferous (Namurian C) of Czech Republic. (From Kukalová, 1958.)

**Figure 18.15** *Uralia maculata* Kukalová-Peck and Sinitshenkova (Diaphanopterida) from the Upper Permian (Kungurian) of Tshekarda, Ural Mountains. Reconstruction by A.G. Ponomarenko. (From Rohdendorf and Rasnitsyn, 1980.)

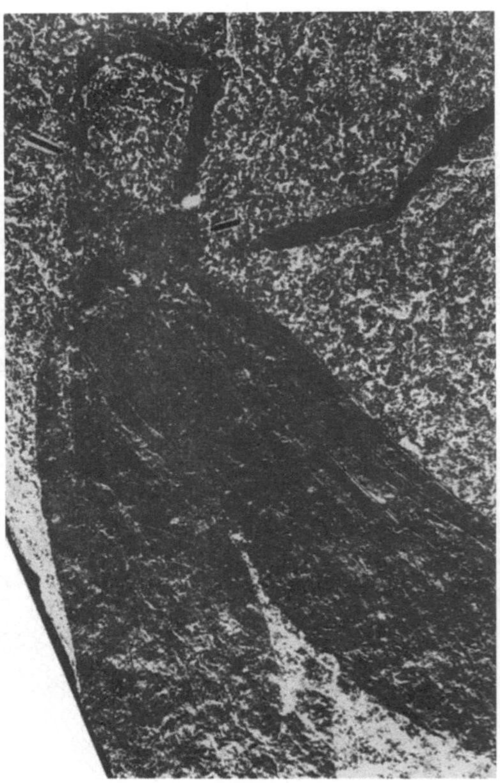

**Figure 18.16** *Heterologopsis ruhrensis* Brauckmann and Koch (?Paoliida) from the lowermost Middle Carboniferous (Namurian) of Germany (Brauckmann *et al.*, 1985). Arrows indicate the beak and prothoracic winglet.

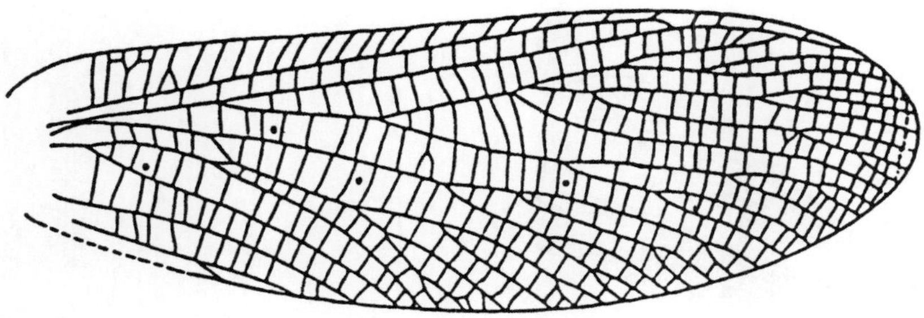

**Figure 18.17** Fore wing of *Heterologus langfordorum* Carpenter (?Paoliida) from the Lower Permian (Artinskian) of Illinois, USA. (From Carpenter, 1944.)

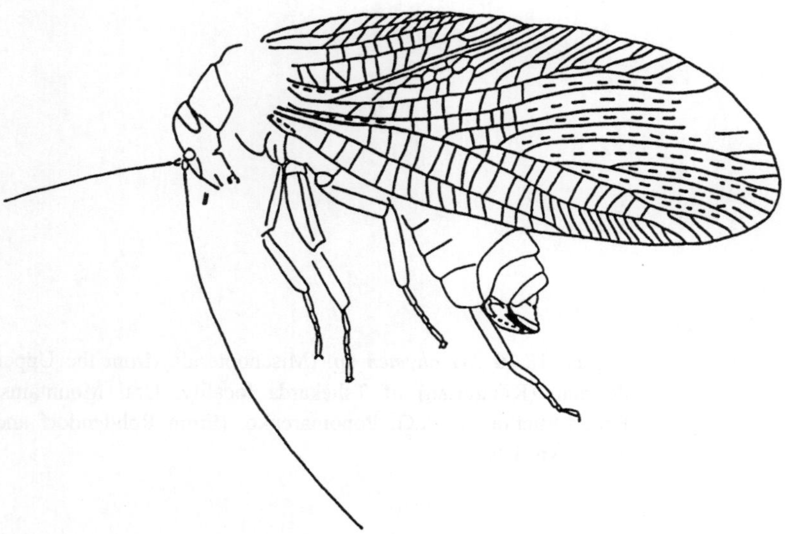

**Figure 18.18** *Glaphyrophlebia uralensis* (Martynov) (Blattinopseida) from the Upper Permian (Kungurian) of Tshekarda, Ural Mountains. (Rasnitsyn, 1980), habitus drawing to show the roof-like wing repose position.

**Figure 18.19** Hind wing of the roach *Phyloblatta*? sp. (Blattida) from the Upper Permian (Kazanian) of Soyana, Arkhangelsk Region of Northern Russia, to show semicircular jugum (J) bent down along the flexion line behind 1A.

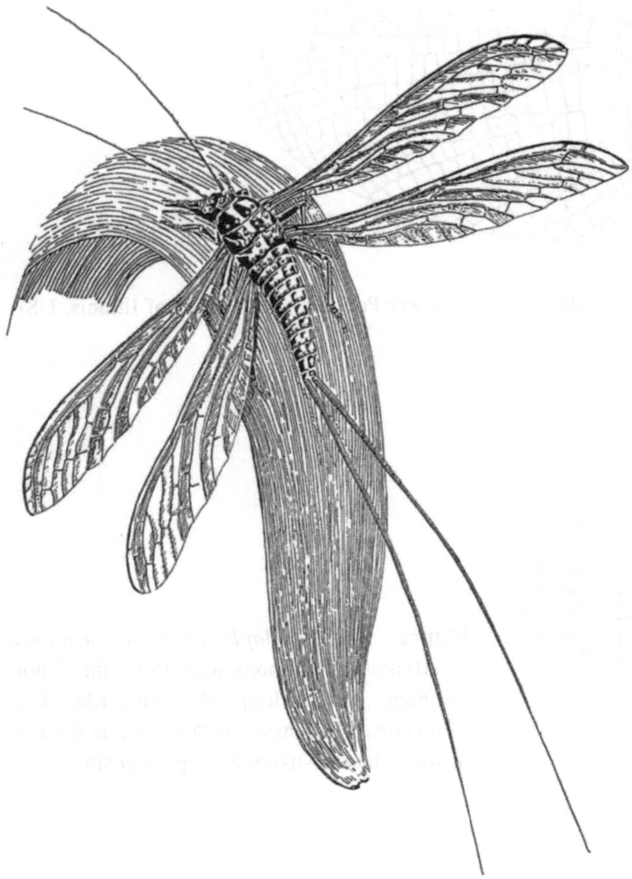

**Figure 18.20** *Sylvohymen* sp. (Mischopterida) from the Upper Permian (Kungurian) of Tshekarda locality, Ural Mountains. Reconstruction by A.G. Ponomarenko. (From Rohdendorf and Rasnitsyn, 1980.)

**Figure 18.21** Fore wing of the archaic mayfly *Lithoneura mirificum* Carpenter (Syntonopterida) from the Lower Permian (Artinskian) of Illinois, USA. (From Carpenter, 1944.)

**Figure 18.22** Fore wing of the archaic dragonfly *Eogeropteron arcuatum* Riek (Meganeurida) from the Carboniferous (Malanzan Formation) in Paganzo Basin, Argentina (Riek and Kukalová-Peck, 1984). The counterpart of the fossil shows the wing from below, with convex and concave veins being mirrored compared with the usual appearance.

## REFERENCES


Arnett, R.H. (1985) *American Insects. A Handbook of the Insects of America North of Mexico*, Van Nostrand Reinhold Co., New York.

Boudreaux, H.B. (1979) *Arthropod Phylogeny: with Special Reference to Insects*, John Wiley & Sons, New York.

Brauckmann, C., Koch, L. and Kemper, M. (1985) Spinnentiere (Arachnida) und Insekten aus den Vorhalle-Schichten (Namurium B; Ober-Karbon) von Hagen-Vorhalle (West-Deutschland) *Geologie und Paläontologie in Westfälen*, **3**, 1–132.

Carpenter, F.M. (1944) Carboniferous insects from the vicinity of Mazon Creek, Illinois. *Illinois State Museum, Scientific Papers*, **3**, 1–20.

Carpenter, F.M. (1992) *Treatise on Invertebrate Paleontology*. Pt. R. Arthropoda 4. Vol. 3, 4. Superclass Hexapoda. Geological Society of America, Boulder, Colorado, and University of Kansas, Lawrence, Kansas.

Carpenter, F.M. and Richardson, E.S. (1968) Megasecopterous nymphs from the Pennsylvanian concretions of Illinois. *Psyche*, **75**, 295–309.

CSIRO (ed.) (1992) *The Insects of Australia*, 2nd edn, Melbourne University Press, Melbourne.

Kristensen, N.P. (1981) Phylogeny of insect orders. *Annual Review of Entomology*, **26**, 135– 57.

Kukalová, J. (1958) Paoliidae Handlirsch (Insecta – Protorthoptera) aus dem Öberschlesischen Steinckohlenbecken. *Geologie*, **7**, 935–59.

Kukalová-Peck, J. (1978) Origin and evolution of insect wings and their relation to metamorphosis, as documented by the fossil record. *Journal of Morphology*, **15**, 53– 126.

Kukalová-Peck, J. (1983) Origin of insect wing and wing articulation from the arthropodan leg. *Canadian Journal of Zoology*, **61**, 1618–69.

Kukalová-Peck, J. (1992) Fossil history and the evolution of hexapod structure, in *The Insects of Australia*, 2nd edn. (ed. CSIRO), Melbourne, Melbourne University Press, pp. 144–82.

Martynov, A.V. (1925) aüber zwei Grundtypen der Flügel bei den Insekten und ihre Evolution. *Zeitschrift für Ökologie und Morphologie der Tiere*, **4**, 465–501.

Martynov, A.V. (1937) Liassic insects of Shurab and Kizil-Kiya. *Transactions of the Paleontogical Institute, Academy of Sciences of USSR*, 7, no. 1, Izdatelstvo AN SSSR, 232 pp. (in Russian, with English summary).

Martynov, A.V. (1938) An essay on the geological history and phylogeny of the insect orders (Pterygota). Pt. I. Palaeoptera and Neoptera-Polyneoptera. *Transactions of the Paleontological Institute, Academy of Sciences of USSR*, **7**, no. 4, Izdatelstvo AN SSSR, 148 pp. (in Russian, with French summary).

Novokshonov, V.G. (1992) Early evolution of caddisflies (Trichoptera). *Zoologicheskiy Zhurnal*, **71**, 58–67 (in Russian).

Rasnitsyn, A.P. (1976) On the early evolution of insect and the origin of Pterygota. *Zhurnal Obshchey Biologii*, **37**, 543–55 (in Russian with English summary).

Rasnitsyn, A.P. (1980) Origin and evolution of Hymenoptera. *Transactions of the Paleontological Institute, Academy of Sciences of USSR*, 174, Nauka Press, Moscow, 192 pp. (in Russian).

Rasnitsyn, A.P. (1982) Proposal to regulate the names of taxa above the family group. *Bulletin of Zoological Nomenclature*, **39**, 200–7.

Rasnitsyn, A.P. (1996) Conceptual issues in phylogeny, taxonomy, and nomenclature. *Contributions to Zoology*, **66**, 3–41.

Riek, E.F. and Kukalová-Peck, J. (1984) A new interpretation of dragonfly wing venation based upon Early Upper Carboniferous fossils from Argentina (Insecta: Odonatoidea) and basic character states in pterygote wings. *Canadian Journal of Zoology*, **62**, 1150–66.

Rohdendorf, B.B. (ed.) (1962) Osnovy paleontologii. Chlenistonogiye. Trakheinye i khelitserovaye, Izdatelstvo Akademii Nauk SSSR, Moscow, 560 pp [in Russian, translated into English as Rohdendorf B.B. (ed.) (1991) *Fundamentals of Paleontology*. Vol. 9, Arthropoda, Tracheata, Chelicerata. New Delhi, Amerind Co., 894 pp.]

Rohdendorf, B.B. (1977) The rationalization of names of higher taxa in zoology. *Paleontologicheskiy Zhurnal*, no. **3**, 11–21 [in Russian, translated into English in *Paleontological Journal*, **11**, 149–55 (1977).]

Rohdendorf, B.B. and Rasnitsyn A.P. (eds). (1980) Historical development of the class Insecta. *Transactions of the Paleontological Institute*, Academy of Sciences of USSR, **175**, Nauka Press, Moscow, 269 pp. (in Russian).

Ross, H.H. (1965) *A Textbook of Entomology*, 3rd edn, John Wiley & Sons, New York.

Sharov, A.G. (1966) On the position of the orders Glosselytrodea and Caloneurodea in the system of Insecta. *Paleontologicheskiy Zhurnal*, no. **3**, 84–93 (in Russian).

Shcherbakov, D.E., Lukashevich, E.D. and Blagoderov, V.A. (1995) Triassic Diptera and initial radiation of the order. *Dipterological Research*, **3**, 75–115.

# 19 Arthropod phylogeny and 'basal' morphological structures

*J. Kukalová-Peck*

*Department of Earth Sciences, Carleton University, 1125 Colonel By Drive, Ottawa, K1S 5B6, Canada*
*email: jkpeck@ccs.carleton.ca*

## 19.1 INTRODUCTION

The objective of this paper is to show that the relationships between arthropodan higher taxa are most reliably revealed by 'basal' arthropodan structures, such as head and body segmentation, and leg-derived appendages (mouthparts, legs, genitalia, wings, and their articulations). For reliable assessment, the character transformations should be compared throughout all arthropodan groups, extinct and extant. Phylogeny of the higher taxa (Subphyla, Classes, Orders, and Superfamilies) has been mainly approached in two different ways. In this paper I loosely term these approaches the 'restrictive' and 'historical', respectively.

### 19.1.1 THE 'RESTRICTIVE' APPROACH

This examines the relationships between higher taxa by research methods similar to those used for modern species (Kristensen, 1975, 1981). Each character is considered without broad analysis in numerous extant and fossil representatives, and adequate weighting. The 'restrictive' approach provides plentiful data for cladistic analysis and shows morphological similarities, but without deeper understanding why they are considered shared derived similarities (synapomorphies). The questions how morphological characters evolved in all related orders, and what are their exact homologues, are either minimized, or not pursued, believed to be impossible to find, considered unreliable, and/or irrelevant. However, unlike species, higher taxa are based on a small number of ancient, complex, multi-layered character transformations almost always involving 'basal' structures. Homoplasies are rampant and often indistinguishable from genuine synapomorphies. Direct comparisons, which at the species level bring more than 50% accuracy in character evaluations, can be much more unreliable at higher taxonomic levels, undermining the use of cladistic analysis. This has been interpreted as an incapacity of morphology effectively to analyse and resolve the relationships between the present higher taxa. My own view is that the 'restrictive' approach has serious limitations, and because of them, it is unable to deliver a reliable evaluation of higher characters. Examples will be shown below.

In the past 10 years, many systematists hoped that either molecular analysis, or advanced character processing of huge character sets, the so-called 'total evidence', would deliver a reliable phylogenetic tree. However, they have been disappointed. DNA and small subunit RNA sequencing is erratic in portraying phylogenetic events which happened 240 Myr ago and before. With the rise of complex morpho-developmental control systems, there has been homoplasy among sequences, many duplications of both structural and control genes, and gene complexes which can become dormant and then reactivated. It is extremely difficult to select molecular sequences that record the origin of the present day higher taxa. 'Total evidence' currently uses computer cladistics employing characters that are too often only weakly supported by proper comparative anatomy. The phylogenetic results are not reliable.

### 19.1.2 THE 'HISTORICAL' APPROACH

The historical approach to arthropod phylogeny, as advocated by Snodgrass (1935, 1938), is not currently fashionable. It is predicated on the assumption that higher level characters are unravelled by a deep understanding of all 'basal' structures, by cross-comparison between all arthropodan groups. The present higher taxa show change and divergence of 'basal' structures, which took place roughly between the Vendian and the end of the Palaeozoic era. If each structural character is examined simultaneously in all higher taxa, the probability of interpretative error is diminished, because: (i) each character is subject to numerous cross-checks; (ii) interlocking reveals previous misinterpretations much more readily than single characters; and (iii) all details are tested for correct homology. After homologization comes the reconstruction of the phylogeny, groundplan reconstruction, and finally development of adaptive scenarios. The groundplan must not contain char-

*Arthropod Relationships*, Systematics Association Special Volume Series 55, Edited by R.A. Fortey and R.H. Thomas. Published in 1997 by Chapman & Hall, London. ISBN 0 412 75420 7

acters which are unrecorded on actual specimens. In practice, leg podites are defined by individual muscular attachments and by articulation. If two podites fuse (an apomorphy), they continue to show their muscular attachment and, frequently, a discrete separation by a suture or by a sulcus. Fusion, subsegmentation, and flattening of a cylindrical podite into a pleuron, is derived. Such clues are used for homologization of arthropodan appendages in all groups, for phylogenetics, and for construction of the groundplan. The 'historical approach' is labour-intensive and slow, but ultimately effective in resolving phylogenetic puzzles. The basal, all-important, information is the exact number of leg podites and rami in the Arthropoda leg groundplan (Snodgrass, 1935; Sharov, 1966; Boudreaux, 1979; Hennig, 1981; Kukalová-Peck, 1983, 1991, 1992). The 'historical approach' pursues detailed homologization of all parts of fossil and modern examples, so as to understand all leg-derived appendages (mouthparts, genitalia, wings, wing venation and articulation) at all higher taxonomic levels, before the systematic study even starts. The next aim is to find out, through multiple criss-cross comparisons, which derived changes (fusions, reductions, subsegmentations, and the like) are shared by all members of a higher taxon. The resulting complex of characters defines the higher taxon. Comparisons of groundplans then usually reveal the relationships between the higher taxa. Insects have retained a number of primitive, 'basal' characters, some of which are already modified in trilobites, modern chelicerates and crustaceans: evidence from Insecta is thus critical. Each groundplan is extremely influential. For example, an erroneous basal number of leg podites in the 'limb base' (one podite instead of three, see below) obscured monophyletic, class level relationships and caused separation of Myriapoda + Insecta into 'Uniramia' by Manton (1972, 1977) and her followers. It is regrettable that detailed morphology, indispensable to achieve a reliable phylogenetic synthesis, is now so rarely taught in biology departments in our universities.

Presently, the use of the 'historical' approach is most advanced among the pterygote insects. It has so far documented relationships within the following groups: Coleoptera, at the order, suborder, and partly superfamily level (Kukalová-Peck and Lawrence, 1993); Blattoneoptera, order level, including Zoraptera (Kukalová-Peck and Peck, 1993); Orthoneoptera: work in progress with D.F.C. Rentz, order, suborder and partial superfamily level; Hymenoptera: work with M. Sharkey involves primitive superfamilies. Additional phylogenetic studies include the following: Coleoptera, Scarabaeoidea, Browne (1991a,b, 1993), Browne and Scholtz (1994, 1995, 1996), Browne *et al.* (1993), Scholtz *et al.* (1994), Scholtz and Browne (1996); together with PhD studies on Dermaptera, Odonatoptera, Ephemeroptera, and Hymenoptera. Developmental geneticists have found evidence supporting the origin of wings from mobile leg appendages (Kukalová-Peck, 1983; Štys and Zrzavý, 1994; Averof and Cohen, 1997).

## 19.2 ARTHROPODAN MONOPHYLY: THE MANDIBULAR APPENDAGE AND THE JAW

The leg segments of Hexapoda, as homologized with the basal arthropodan leg, contain the following podites: 1, subcoxa (SCX); 2, coxa (CX); 3, trochanter (TR) (these podites plus coxal and trochanteral endite compose the coxopodite); 4, prefemur (PFE); 5, femur (FE); 6, patella (PAT); 7, tibia (TI); 8, basitarsus (BT); 9, eutarsus (ET) (with 2 subsegments, by autapomorphy); and 10, pretarsus (PT) (with two curved ungues, by autapomorphy) (podites 4–10 compose the telopodite) (see Figures 19.3(e) and 19.4(c,d)). Based on three decades of comparative morphological research by E. L. Smith (California Academy of Sciences, San Francisco) and myself, the basal arthropodan leg contains 10 cylindrical, articulated podites. I propose that the 11th podite, the epicoxa (ECX), probably entered pleural membrane at the Arthropoda Phylum level, and changed into flat archipleuron placed between the tergum and the (cylindrical) subcoxa (see Figure 19.5(a,c,d)).

The arthropod coxopodite and telopodite have a different movement and function (Manton, 1977): the coxopodite (composed of three podites) is suspended and its motion is more or less rotary, but the telopodite (composed of seven podites) moves mainly forwards and backwards.

The principal reason why Manton (1972, 1977) saw Arthropoda as polyphyletic and separated Myriapoda + Hexapoda (= Atelocerata) into a special phylum Uniramia, was based on an erroneous interpretation in the use and structure of the mandibular appendage. Manton correctly observed that the part of the leg used for masticating (= jaw) in Trilobita, Chelicerata and Crustacea was the 'limb base'. However, she incorrectly assumed that the 'limb base' is formed by a single, leg podite, termed the 'coxa'. Myriapoda and Hexapoda have jaws demonstrably composed of **several** leg podites, which convinced Manton that they ate with the tip of a multisegmented leg, like onychophorans. This explanation was countered by Kukalová-Peck (1983, 1991, 1992; also Smith, 1970a,b, 1988, and *Atlas of Arthropods*, unpublished), as follows: (i) the arthropodan jaw is **always formed by the entire coxopodite** composed of three podites and two endites (subcoxa, coxa, trochanter, with subcoxal and trochanteral endites serving as two teeth); (ii) coxopodal podites are very frequently secondarily fused and give an impression of being only 'one' podite; (iii) the leg shaft or telopodite is frequently lost and never used in eating.

Thus, all Arthropoda originally ate with the segmented limb base, which may, or may not, show segmentation. The homoplasic fusion of podites in the mandibular jaw confused Manton (1972, 1977), and produced 'Uniramia'.

In Myriapoda (considered here as a clade) mandibular jaw podites are articulated and still partly movable in Diplopoda (Figure 19.1(a)), and they are fully fused, but still visibly separated by sutures and sulci in primitive Insecta (Archaeognatha, jumping bristletails (Figure 19.1(b,c); Ephemeroptera, mayflies; Figure 19.1(e)). Compared with this, the jaw podites of most, even Cambrian, Arthropoda (trilobites), derived Insecta (e.g. grasshoppers) and Crustacea (Figure 19.2), became completely fused together – a poly-convergence. However, they

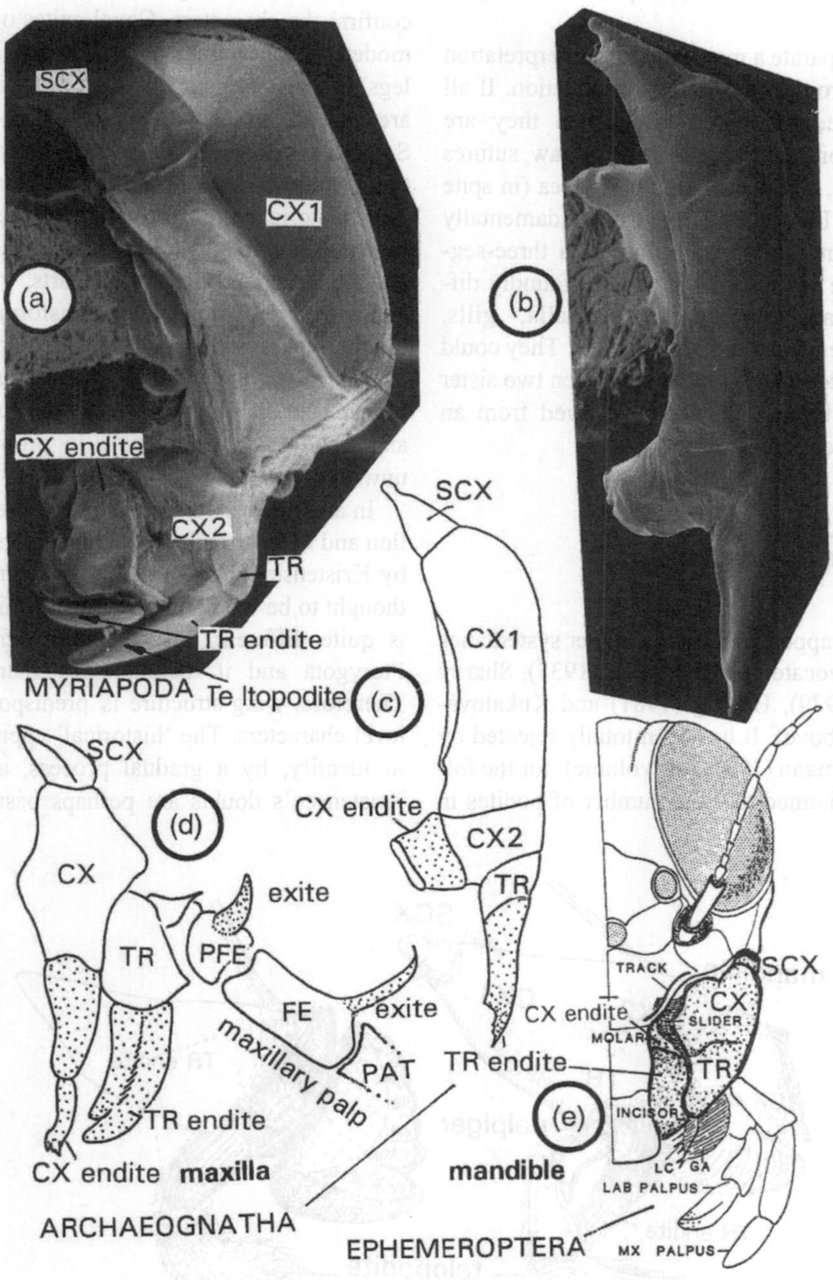


**Figure 19.1** Evidence that in Atelocerata (Myriapoda + Hexapoda) mandible and maxilla is formed by coxopodite (= three leg podites + two endites), and tolopodite is lost.. (a) Myriapoda: Diplopoda, articulation of podites is partly preserved. (b–d) Wingless Insecta: Archaeognatha. (b,c) Mandibular podites fused, derived, but still separated by sutures. (d) Maxillary podites are not fused, articulated, primitive. (e) Pterygota: Ephemeroptera, mandibular podites fused, but still outlined by sulci. [Sources: (a,b), scanning electron micrograph by Lew Ling; (c,d), originals; (e) redrawn after Kukalová-Peck, 1985.]

are often still marked by shallow sulci (Figure 19.2) and by muscular attachments.

Both the three-segmented jaw of bristletails and the seemingly 'one-segmented' jaw of grasshoppers, is followed by a serially homologous maxilla, which clearly shows a three-segmented coxopodite (Figures 19.1(d), 19.3(a,e) and 19.4(c,d)). This proves that the jaw has the same composition as the maxilla (compare with Figure 19.2).

It is impossible to separate a morphological interpretation of a 'basal' structure from its systematic implication. If all Arthropoda share three-segmented jaws, then they are monophyletic. If trilobitomorphs (in spite of jaw sutures shown by Cisne, 1975), Chelicerata and Crustacea (in spite of residual jaw sulci, Figure 19.2) have a fundamentally one-segmented jaw, but Myriapoda + Insecta a three-segmented jaw, then these groups also have a profoundly different leg groundplan, mouthparts, genitalia, gills, motion-promoting appendages and articulations. They could **not** be either monophyletic, or biphyletic, or even two sister group phyla, because they could not be derived from an immediate common ancestor.

## 19.3 ARTHROPODAN LEG EVIDENCE INDISPENSABLE FOR HIGHER PHYLOGENETICS

The use of leg-derived appendages in the higher systematics of Arthropoda was advocated by Snodgrass (1935), Sharov (1966), Boudreaux (1979), Hennig (1981) and Kukalová-Peck (see references above). It has been totally rejected by Meier (1993) and Willmann (1997, this volume), for the following reasons: it is claimed that the number of podites in arthropodan legs is unstable, and that podites can be added or reduced in an unpredictable way; that the occurrence and number of exites and endites (rami) is also unstable and cannot be predicted; and that exites are smooth, rather than primitively annulated (which distinguishes them from kinetodontia, the movable spines). The flaw in these objections is the neglect of evidence provided by the fossil record: those who are willing to study the material will be able to confirm the characters. Coxal exites on the thoracic legs of modern Archaeognatha, coxal exites called 'swimming legs' in Crustacea, and all other primitive exites and endites, are, indeed, annulated (Figure 19.3(c–e)) (Manton, 1977). Smooth exites, without flexibility and sites for muscle insertions, could not have been used for underwater movement. The 'historical' transformation of the arthropod leg has a long tradition in higher systematics; these compose the head and body appendages, mouthparts, genitalia, cerci, wings and wing articulations. Such transformations are the basis of the present system. If these are claimed by Meier and Willmann as meaningless, then presumably arthropodan higher taxa should be based in the future only on tagmata and upon internal organs – a proposition which seems unwise.

In a controversial criticism, the use of insect wing venation and wing articulation in higher systematics was rejected by Kristensen (1995, 1997, this volume), because they are thought to be too homoplasious for this purpose. The reality is quite different. Insectan wing structure is 'basal' for Pterygota and it distinguishes sharply all higher taxa. Therefore, wing structure is predisposed to contain higher level characters. The 'historical' approach has the capacity to identify, by a gradual process, all such homoplasies. Kristensen's doubts are perhaps best countered by the 10

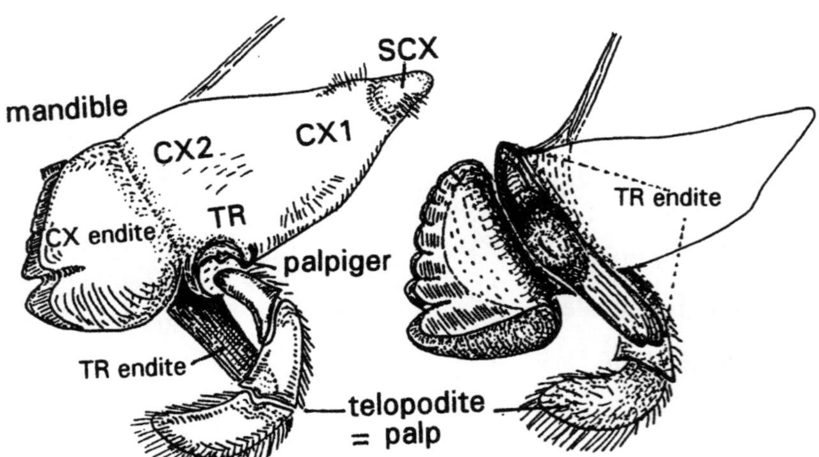


**Figure 19.2** Crustacean mandible is formed by coxopodite (= three leg podites + two endites, all fused), telopodite (palp) is present. Left, dorsal view with subcoxa and coxal endite delimited by shallow sulci. Right, ventral view showing long and narrow trochanteral endite. Malacostraca: Decapoda, Cambaridae, *Cambarus* sp. Original.

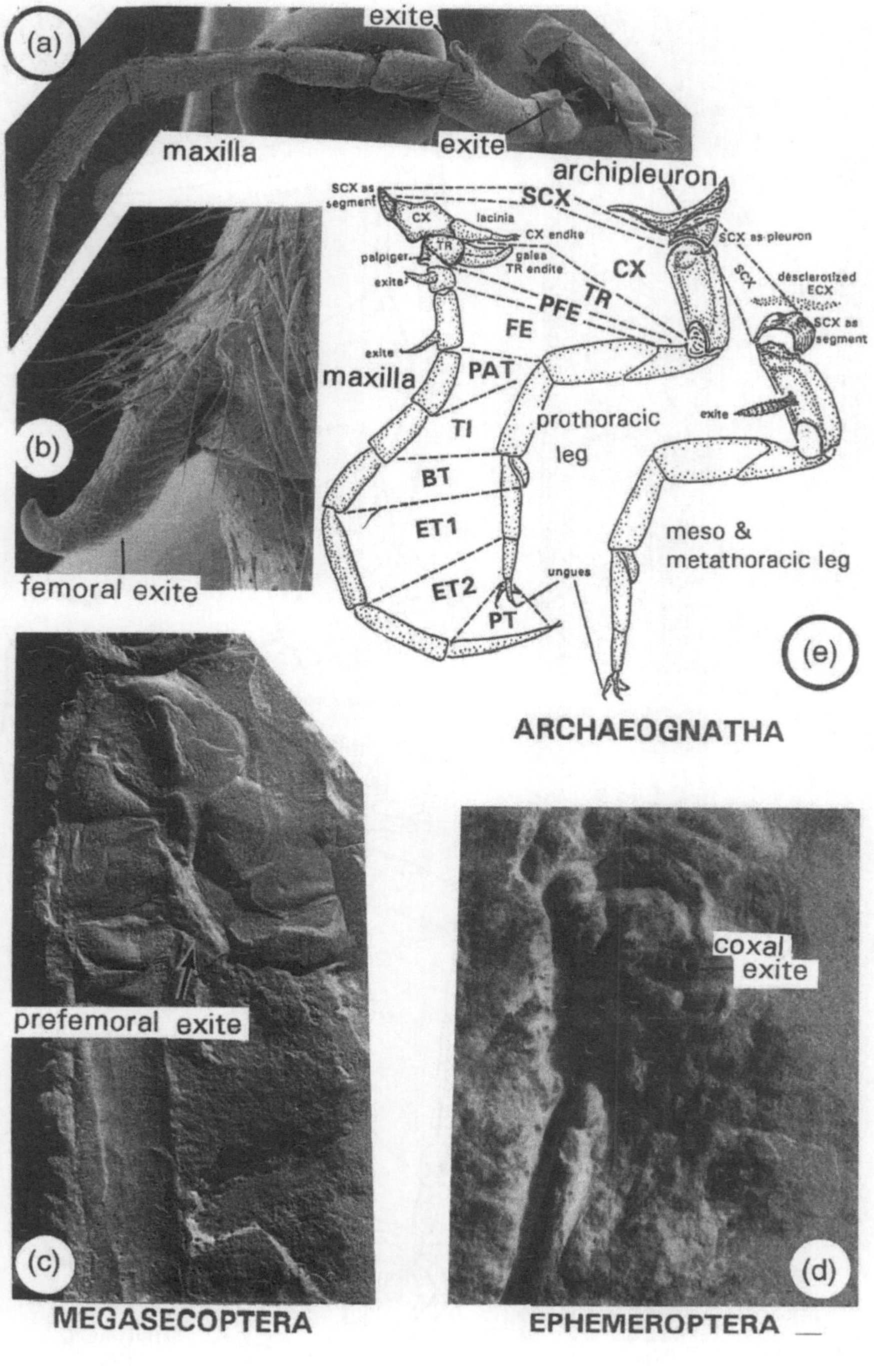


**Figure 19.3** Insecta: exites on maxillae and thoracic legs. Recent Archaeognatha. (a) Maxilla with prefemoral and femoral exites. (b) Femoral exite, embayment in femur, enlarged. (c) Permian Megasecoptera (order emended): *Monsteropterum moravicum* Kukalová-Peck, 1972, holotype, middle leg with annulated prefemoral exite. (d) Permian Ephemeroptera: Protereismatidae, ?hind leg with annulated coxal exite, specimen no. 8585, Museum of Comparative Zoology, Harvard University, Cambridge. (e) The most primitive modern arthropodan leg appendage. Archaeognatha: maxilla, coxopodite with three podites fully articulated, telopodite with seven podites (= full number), prefemoral and femoral exite present, patella articulated at both ends as in spiders, eutarsus bears two subsegments, and posttarsus two large sensilla. Prothoracic leg has large archipleuron and subcoxal pleuron; meso- and metathoracic legs have cylindrical subcoxa instead of a pleuron, and coxa with an annulated coxal exite. This shows that Insect thoracic pleuron is formed by flattened subcoxa. [Sources: (a,b), scanning electron micrograph by Lew Ling; (c,d), after Kukalová-Peck, 1983, relabelled; (e), original.]

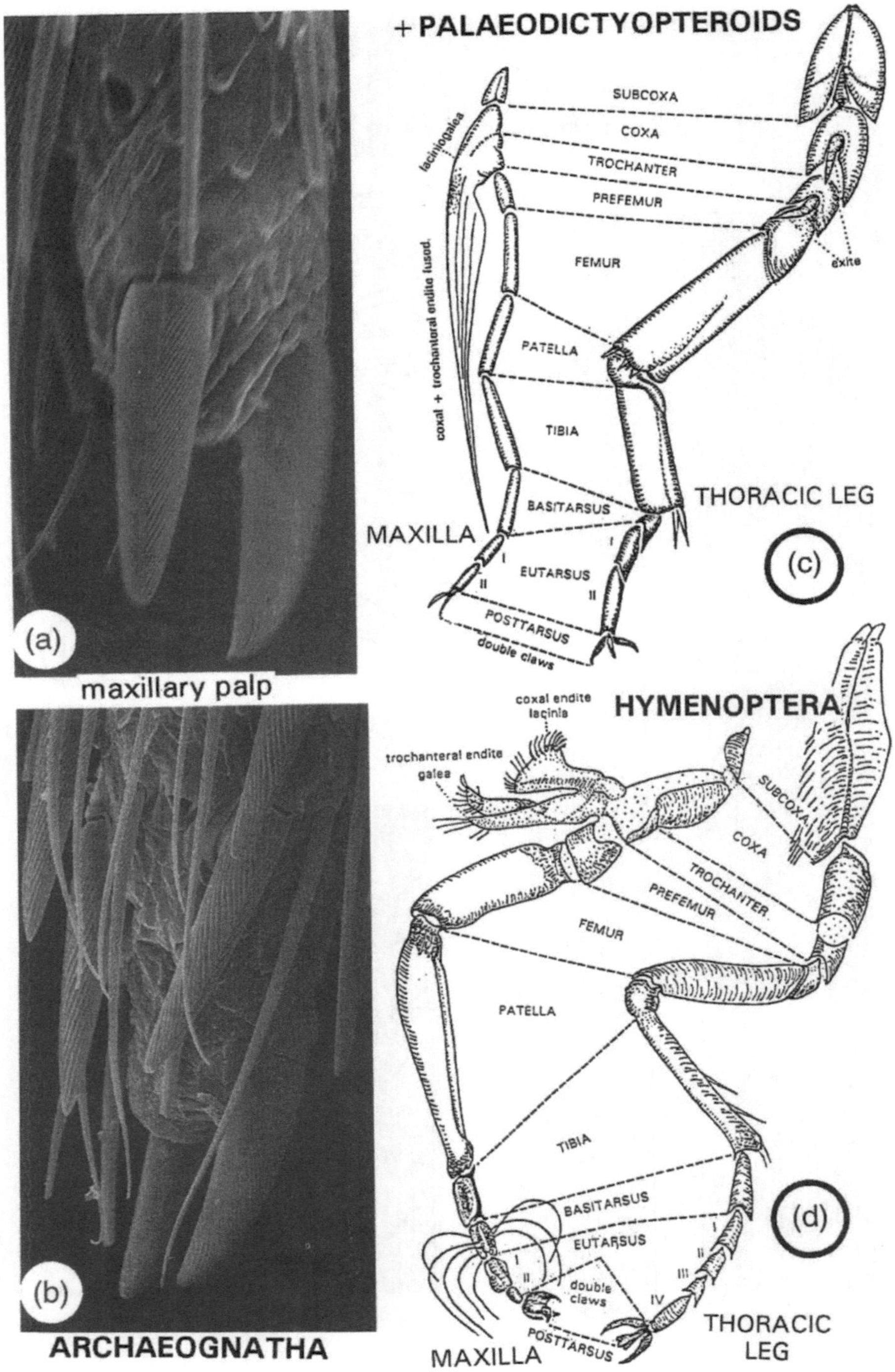


**Figure 19.4** (a,b) Archaeognatha: maxillary palps with two terminal sensilla. (c,d) Pterygota: Maxilla showing all 10 basal arthropodan leg podites: coxopodite with three podites + two endites, and telopodite with seven podites and terminal ungues. Thoracic leg is highly derived, with only 6–8 apparent podites. (c) Paleoptera: maxilla has coxal + trochanteral endite always fused (autapomorphy); Permian Diapnahopterodea retained fully segmented maxilla and the most primitive hexapodan thoracic leg. (d) Hymenoptera: Xyelidae: *Pleroneura* sp.: maxilla shows identical podites as Archaeognatha and Diaphanopterodea. Thoracic leg is highly derived, patella is 'lost', eutarsus is subdivided into derived four subsegments. [Source: (a,b), scanning electron micrograph by Lew Ling; (c), after Kukalová-Peck, 1992, relabelled; (d), original.]

recent publications, as well as PhD programmes referred to above. In my view, homoplasies appear confusing particularly to those who favour the 'restrictive' approach.

It is timely that the contribution of the 'historical approach' to higher systematics (Arthropoda are monophyletic; Mandibulata = Crustacea + (Myriapoda + Hexapoda)) is at last regaining acceptance. Over recent years, many changes based upon the 'restrictive' approach have proved to be disappointing. Sharov's (1966) massive contribution to Mandibulata was originally harshly criticized, but is now viewed as progressive. The account below is concerned with the examples of the 'historical' approach, several new discoveries, and with rebutting some recent criticisms. Morphological progress on the leg-derived appendages during the past two decades has been summarized by Kukalová-Peck (1978, 1983, 1985, 1987, 1991, 1992), Kukalová-Peck and Brauckmann (1990, 1992), Kukalová-Peck and Lawrence (1993), Kukalová-Peck and Peck (1993), Riek and Kukalová-Peck (1984) and Shear and Kukalová-Peck (1990).

### 19.3.1 THE BASAL ARTHROPODAN LEG AND LEG RAMI

The two articulated inner rami of the coxopodite (two endites, the coxal and trochanteral) are working appendages performing eating in the head and copulation in the abdomen; the outer rami of the coxopodite (maximum three exites, frequently lost) are engaged in respiration (gills and plate gills of trilobites, chelicerates (*Limulus*), primitive malacostracan crustaceans (*Anaspides*)), or movement (swimming 'legs' of crustaceans, flying wings of pterygote insects), or both (abdominal winglets of mayfly nymphs), or in tactile sensing (conical, annulated thoracic exites of modern Archaeognatha and fossil pterygote insects) (Figures 19.1(d), 19.3(a–e) and 19.4(c,d)).

The telopodite (composed of seven podites) secures movement in contact with the substrate and its terminal podite is either pointed, adapted for pushing back against the substrate (Crustacea, Myriapoda), or it bears two curved claws (ungues; adapted terminal spines) for taking hold while clinging to objects (Hexapoda). Telopodites with seven podites (the full number according to comparisons throughout arthropods) occur frequently in trilobites (Kukalová-Peck, 1983, 1991, 1992; Figure 19.3(e)). Fusions between several terminal podites are extremely frequent as adaptations for vertical pushing against the substrate. In the clinging–climbing legs of Hexapoda, evolution went in the opposite direction, towards the subsegmentation of podite 9 (eutarsus) to gain a curved, flexible, 'foot'-like grip. This singular adaptation explains why Hexapoda, high up the phylogenetic tree, nevertheless, preserved the basal arthropodan leg pattern better than any other living arthropodan group. The most primitive extant hexapodan (also, modern arthropod) appendage, closest to the basal arthropodan groundplan leg, is the archaeognathan maxilla, with **all three** coxopodal podites fully articulated (Figures 19.1(d) and 19.3(a,d)); the same are partly fused but still recognizable e.g. in extinct palaeodictyopteroids and modern Hymenoptera (Figure 19.4(c,d)); podite 6 (patella) in the insect maxilla is articulated at both ends as in trilobites and spiders; the eutarsus is subdivided into two subsegments; and the pretarsus bears two curved ungues (Figure 19.4(c,d)). The last two are autapomorphies for the Hexapoda, as shown by their presence in modern and fossil maxillary palps (Figures 19.4(c,d)) and in a hundred or so Palaeozoic fossils in abdominal leglets, in gonostyli, and in cercal leglets (Figure 19.6(a)) [Kukalová-Peck, 1985 (Figure 24); 1987, 1992 (Figures 1–5, 7, 21, 30, 35)]. Ancestral hexapods bore on the head (in maxillary and labial palps), the thorax (on three pairs of thoracic legs) and the abdomen (on eight pairs of abdominal leglets, one pair of gonostyli, and one pair of cercal leglets) no less than 15 pairs of telopodites with the eutarsus subdivided only once, terminated by post-tarsus with double-claws. Except for the thoracic legs, all remaining 12 pairs of legs have a patella articulated by both ends, as in spiders. Neither Bitsch (1994) nor Kristensen (1995) recognize the patella on the hexapod thoracic leg, thus making it non-homologous with legs of spiders, and insectan palps. The documentation for these interpretations, as presented by Bitsch (1994), is very poor, relying exclusively on the number of circles in the published figure of a leg anlage in one insect, *Drosophila*, as the sole information for the leg groundplan of all Arthropoda. The anlage is miniscule, and the fruitfly is both highly derived and very weakly sclerotized. Such an extremely influential decision cannot rest on a single piece of evidence, in my view quite unreliable. Willmann proposed that the primitive insectan eutarsus bears four subsegments, without taking into consideration that they occur only in some, derived thoracic legs (Figure 19.4(d)) and nowhere else. Legs of the most primitive fossil insects, and all complete abdominal and head telopodites (palps, gonostyli, leglets), always bear two subsegmented eutarsi (Figures 19.3(e) and 19.4(c)) (also Kukalová-Peck, 1992).

### 19.3.2 EXITES AND ENDITES

The initial presence and subsequent parallel loss of most or all rami in most Arthropodan groups, is easily understood in a 'historical' context. It is an unresolvable problem using the 'restrictive' approach, probably because conspicuous functional changes are rated arbitrarily, often more highly than their phylogenetic worth. All primitive arthropod rami are rhythmically constricted into annuli, which may resemble podites. Therefore, exites (swimming 'legs') of crustaceans are sometimes confused with legs, and overrated. Trilobitomorphs and crustaceans have independently, in

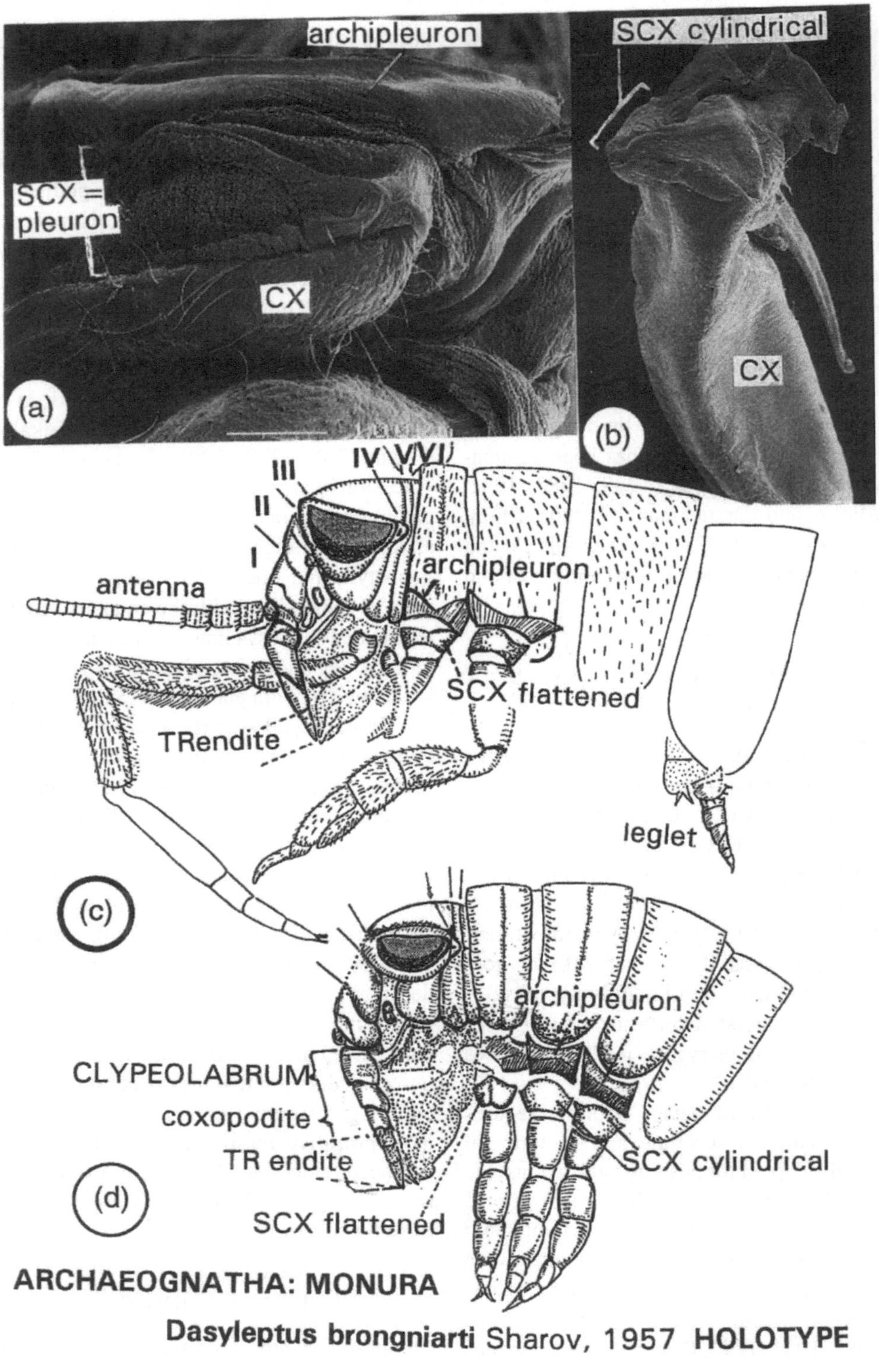


**Figure 19.5** Archaeognatha. (a) Prothoracic leg with archipleuron and subcoxal pleuron. (b) Mesothoracic leg with cylindrical subcoxa. (c) The most primitive heads of Atelocerata. Permian suborder Monura, undescribed gen.et sp., and *Dasyleptus brongniarti* Sharov, 1957, holotype. Heads are composed of six distinct segments, and an acron between ocelli, eyes, and tergum IV; tergum I and II flanks the acron, tergum III is under the eye which is pushed deep into segment IV. Clypeolabrum (= two coxopodites laterally fused) bears terminally annulated trochanteral endite. Strong archipleuron above all thoracic legs. Prothoracic subcoxa forms solid pleuron with pleural sulcus; meso- and metathoracic subcoxa is cylindrical. Abdominal leglet bears seven-segmented (= complete) telopodites. [Source: (a,b), scanning electron micrograph by Lew Ling; (c), originals.]

parallel, enlarged one of their limb base rami, and used it in respiration or swimming; other rami are frequently lost (but not always, as clearly shown by Manton, 1977, Figure 4.5a). This parallel adaptation has been interpreted not as multiple homoplasy, and simple reduction, but as Class, or Subphylum level diversification, the basal 'biramy' and 'uniramy' which supposedly divides all Arthropoda.

The total number of exites in the basal arthropodan leg is not yet known. In insects, the following five exites on and below the coxopodite (limb base) are on record: the epicoxal (= wing), subcoxal, coxal, prefemoral and femoral. In modern wingless insects, Zygentoma (silverfish, the sister group of winged insects) epicoxal (= wing) exite is changed into sidelobes, but it still contains the wing-type of tracheation (Sulc, 1927); coxal, prefemoral and femoral exites are still present in modern Archaeognatha (Figure 19.3(a,b,e)). This shows that Kristensen's statement (1996, p. 12) in the Programme and Abstracts of the Arthropod Symposium, that 'all extant apterygote hexapods are devoid of limb-base exites' is untrue. Annulated exites are commonplace in insects: in thoracic legs of Palaeozoic Palaeoptera (Figure 19.3(c,d)), ancestral hemipteroids (Kukalová-Peck and Brauckmann, 1992, their Figures 14 and 15), and in modern Archaeognatha (Figure 19.3(a,b,e); typical presence of setae, traces of annulation, sinusoid wave around exites). The prothoracic legs of 10 primitive families of modern orthopteroids bear stunted conical exites on their coxae (Kukalová-Peck, 1983, 1992; Kukalová-Peck and Brauckmann, 1990). It may prove that the exites (gills) in Trilobitomorpha may be subcoxal exites rather than coxal exites. A subcoxal exite occurs in chelicerates (prosomal limb of *Limulus*, Manton, 1977, Figure 4.4a); in Crustacea, *Anaspides* thoracic legs have retained more than one flattened and annulated exite (Manton, 1977, Figure 4.5). Hence, all arthropod subphyla have retained at least some leg rami. This shows clearly that Arthropoda share a **polyramous** basal leg. The primitive condition is neither biramous, nor uniramous, and the loss of rami is always secondary. The name 'Uniramia', or 'Unirama', created for a phylum comprising Atelocerata + Onychophora, and indicating the absence of rami in Atelocerata with four rami present in the modern fauna, is inappropriate and should be dropped.

In all higher arthropodan taxa exites are evaginations of membrane between two podites articulated and mobilized by musculature from both podites. Exites often more or less lose annulation, and move upwards into the proximal podite, causing a sinusoid wave in its distal margin (Figure 19.3(a,b)). Endites are also evaginations of the membrane and articulated, but are mobilized by musculature from only one segment (Kukalová-Peck, 1983, 1992). There is fundamentally one exite and one endite per podite, no more. Exites and endites frequently bear several tooth-like projections (as in jaws, maxillae and labium) (Figures 19.1(a) and 19.2). Movable kinetodontia are spines, not rami.

Manton (1972, 1977), Meier (1993), Bitsch (1994), Kristensen (1995) and Willmann (1997, this volume) represent hexapods by the highly derived, modern thoracic leg. This has caused much confusion. 'Historical' approaches showed that the insectan thoracic leg 'lost' the subcoxa (podite 1 has become the pleuron); it 'lost' either the trochanter or prefemur (podites 2 and 3 are usually fused together (Figures 19.5(c,d) and 19.6(a,b)), but in Diaphanopterodea (Figure 19.4(c)), Caloneurodea and Hymenoptera (Figure 19.4(d)), podite 3 is instead fused with podite 4); the patella is 'lost' [podite 6 is always fused with 7, the tibia, with (Figure 19.4(c)) or without (Figure 19.4(d)) a suture]; and the eutarsus (podite 9) may subdivide into four subsegments, or may even lose several or all subsegments (compare Figures 19.3(e), 19.4(c) and 19.4(d)). Recently, yet another podite-deficient arthropod groundplan was proposed by Bitsch (1994). This is because an average insect thoracic leg seemingly shows six to seven segments (Figures 19.3(e) and 19.4(d)), but in my view this was not the original number. Nevertheless, it convinced many arthropodologists that the 'limb base' (coxopodite) is composed only of the 'coxa'.

Arthropodan appendages can be correctly homologized only by employing a groundplan with one archipleuron and 10 freely articulated podites (Kukalová-Peck, 1983, 1991, 1992; Riek and Kukalová-Peck, 1984). All 10 articulated arthropodan leg podites are fully retained and can be verified by dissections, in the maxilla of primitive fossil and modern Insecta (Figures 19.3(a,e) and 19.4(c,d)). The documentation is again repeated here and above. In several arthropod groups, homologous podites were given very confusing, synonymous names, often to avoid a decision on the homologue. Short, well-documented hexapodan names would significantly facilitate the homologization of all arthropodan legs, rami, mouthparts and genitalia, and their use in higher systematics.

### 19.3.3 EPICOXA, ARCHIPLEURON AND WING ARTICULATION

While struggling to explain the origin of insect wings and wing articulation, I encountered the challenging task of accommodating the following facts: (i) the wing is an articulated, flattened, motion-promoting appendage of the leg; (ii) the wing is mobilized the same way as exites between two podites (by muscles attached above and below); and (iii) the wing articulation above the wing is derived in development from the leg anlage and not from the tergum (Stenzhorn, 1974). I proposed that the wing articulation above the wing originated from an undescribed leg podite, the **epicoxa** (Kukalová-Peck, 1983), which became flattened, and entered the pleural membrane as an archipleuron (i.e. wall support) in the arthropodan ancestor. In the Pterygota, the archipleuron became fragmented around

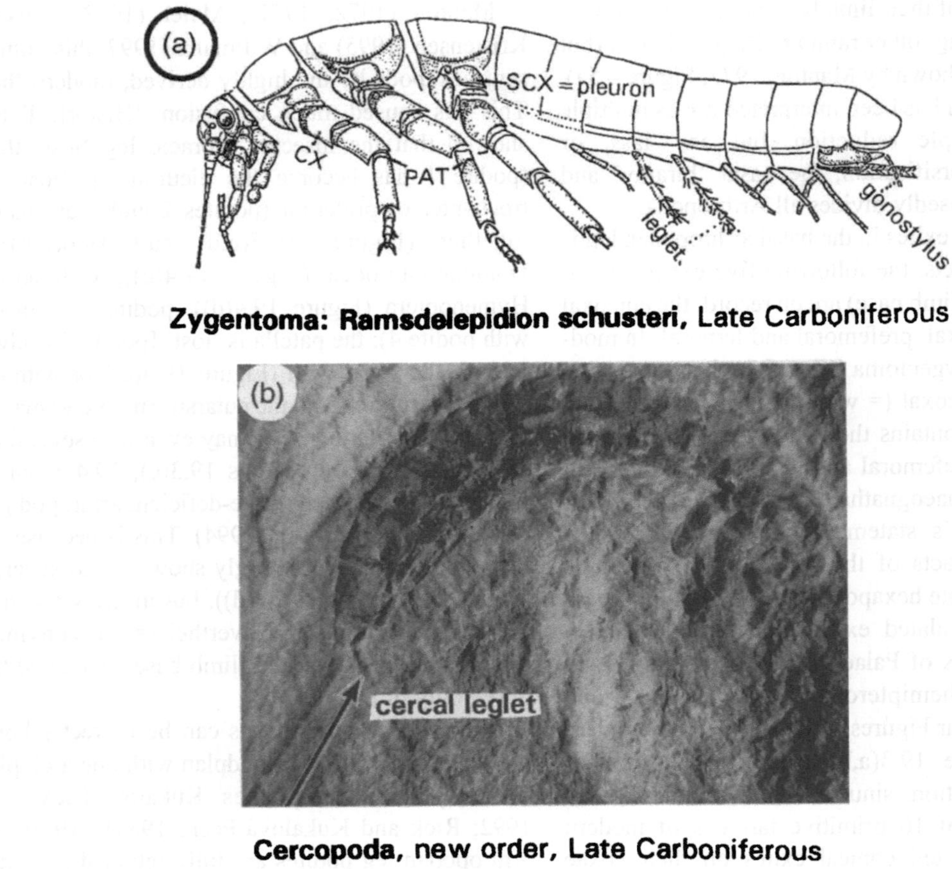


**Figure 19.6** (a) Carboniferous Zygentoma (silverfish) bears subcoxal pleuron identical to Pterygota, but fragmented and desclerotized (autapomorphy showing the monophyly of all members). (b) 'Cercopoda' is the only hexapod group retaining cercal legs (= predecessors of cerci). Cercopoda share with Zygentoma + Pterygota subcoxal pleura on three thoracic segments and ovipositor with clearly defined gonangulum. [Source: (b), original; (a), after Kukalová-Peck, 1985, Figure 23, relabelled.]

muscle attachments, and changed into the wing articulation. To test the hypothesis by the 'historical' approach, it was necessary to find the flattened archipleuron in other arthropods. A flat sclerite in the right place is mentioned in trilobites (Cisne, 1975); a flat sclerite was adjacent above the (subcoxal) pleuron in Carboniferous Myriapoda (Kukalová-Peck, 1983, Figure 6); a flat sclerite above the subcoxa was described in modern Archaeognatha (Manton, 1977, Figure 9.9 b,c,d); and also in *Limulus* (Manton, 1977, Figure 10.4f on leg III). In Crustacea, its occurrence should be pursued.

Recently, the archipleuron was found in Palaeozoic Monura (now categorized as an extinct suborder of Archaeognatha: Carpenter, 1992). Figure 19.5(c,d) shows large, flat, well-sclerotized sclerites above all thoracic legs as in modern Archaeognatha, but much larger and better sclerotized (compare Figures 19.3(e), 19.5(a) and 19.5(c,d)). It now seems quite probable that the epicoxa was, indeed, the 11th basal podite of the proto-arthropodan leg (the archipleuron probably served in Arthropoda to anchor the leg articulation to the tergum). Further support for this hypothesis would be if either fossil Tardigrada, or some other Cambrian arthropodoids, were found bearing legs with 11 cylindrical leg podites. A broad comparative study of the oldest leg segmentation in fossils is necessary.

### 19.3.4 PLEURON AND LEG PODITES

In arthropods, leg podites have become flattened, entered the pleural membrane, and changed into the body wall reinforcement – the pleuron – to create the sides of the box-like body segments. Snodgrass (1935) was aware of the problem that the pleuron may have a different composition in various orders **and** segments (e.g. in Myriapoda). In Hexapoda the situation is complex, because the thoracic and abdominal pleuron is composed differently (Kukalová-Peck, 1983, 1987, 1991, 1992). The abdominal pleura in modern arthropods (e.g. winged insects) may be seamless and look as if derived from a single flattened podite, but fossils show a different construction. Instead of one, there are three pleural plates in the abdomen of many fossil Diaphanopterodea (Kukalová-Peck, 1992, Figures 27–37). The number of podites in the free appendage below (if preserved) can show how many podites above them became fused into the pleuron. Abdominal leglets found in many fossil Pterygota consistently show podites 4 to 10 – the complete telopodite; this again testifies, consistently and independently from Diaphanopterodea, that the pterygote abdominal pleuron is composed of three flat podites (Kukalová-Peck, 1987, 1992).

The origin of the insectan thoracic pleuron was vigorously debated for many decades (Matsuda, 1970) until E. L. Smith found a simple, elegant solution to the problem: modern Archaeognatha have a primitive, cylindrical subcoxa on the 2nd and the 3rd thoracic leg, but on the 1st prothoracic leg it is replaced by a subcoxal pleuron (a flattened segment which entered the pleural membrane and serves as wall support) (compare Figures 19.3(e), 19.5(a) and 19.5(b)). Prothoracic subcoxa (pleuron), next to meso- and metathoracic subcoxa (= both cylindrical podites) are best seen by wiggling thoracic legs of a wet specimen. This provides proof that: (i) in basal Hexapoda the thoracic pleuron is absent on at least two (or, originally on all three?) segments; and (ii) if formed, it is represented by flattened subcoxa. The early, firm subcoxal pleuron (in Archaeognatha, 'Cercopoda', Pterygota) also shows that the membranized and fragmented thoracic pleuron of Zygentoma (Figure 19.6(b)) is secondary. Since Myriapoda have the pleuron always on all segments, this is very probably an autapomorphy. In addition, the pleuron of Crustacea is very probably also an autapomorphy. This example shows that one successful 'historical' interpretation can reveal a chain of character evaluations in the neighbouring taxa.

## 19.4 GROUNDPLAN OF THE HEXAPOD HEAD AND THE POSITION OF MYRIAPODA

### 19.4.1 MANDIBULATA AND ATELOCERATA

The composition of the head of insects has a long history of controversy (Rempel, 1975; Kukolová-Peck, 1992). The following outer features are now supported by the 'historical' approach: (i) bilobed acron is between ocelli, eyes and tergum IV; (ii) there are six pairs of leg-derived head appendages, and they are articulated to six head segments (shared in Crustacea, Hexapoda, Myriapoda); (iii) legs on segment II are changed into antennae (shared in Crustacea, Hexapoda and Myriapoda); (iv) leg appendages on segment III are invaginated, fused and combined with inner cranium as hypopharynx (shared in Hexapoda and Myriapoda; they are changed into 2nd antennae in Crustacea); (v) the mandible is on segment IV (shared in Crustacea, Hexapoda and Myriapoda); (vi) the mandible lost its telopodite (in Hexapoda, mandible retained its telopodite in many Crustacea); (vi) sternal projections (furcae) meet inside the head to form a complex, subdivided, cranial inner skeleton called the tentorium (in Hexapoda and Myriapoda). These features represent very ancient – and hence reliable – synapomorphies showing the monophyly of the Mandibulata and of the Atelocerata as (Crustacea + (Myriapoda + Hexapoda)).

The monophyly of Atelocerata and the position of the Myriapoda has been recently questioned (Popadić *et al.*, 1996). Developmental genetics shows that the insectan and crustacean jaw is similarly formed from a 'limb base' (= coxopodite), but that the myriapod jaw is formed from the 'entire limb' (= coxopodite plus telopodite). However, this means that a long list of arthropodan–'atelocerate' characters must have evolved twice in precisely the same way and in two distantly related groups. This seems unparsimonious, in cladistic parlance.

I infer that there must be some other explanation for the data of Popadić *et al.* (1996). Nobody doubts that the arthropod mandible is formed by the entire leg, but this is not the issue. The pertinent problem is which part of leg is used for eating. All arthropods eat exclusively with the entire 'limb base', called the coxopodite (i.e. subcoxa, coxa, trochanter, with coxal and trochanteral endite serving as two jagged teeth). The embryonic distal 'limb shaft', called the telopodite, may actually be present (Figures 19.1 and 19.2), but it is not used in adult mastication. In ontogeny, the telopodite may be partly or completely reduced, but it may still be expressed at the embryo level (as, possibly, in the Myriapoda examined by Popadić *et al.*, 1996); or, reduced and no longer expressed at the embryo level (as possibly in Insecta); or, not reduced at all but not examined (as possibly in Crustacea). In any case, myriapodan and insectan jaws share the same use in mastication, *pace* developmental evidence (Popadić *et al.*, 1996).

The common mandibulate ancestor probably had coxopodal segments at least partly movably articulated, and the coxa (as is often the case in mouthparts) was subsegmented. Such a mandibular 'limb base' was **more primitive** than that found in trilobites. In Crustacea, the coxopodal segments became fully fused into a 'single' 'limb base', and the telopodite was retained (Figure 19.2). In the common atelocerate ancestor (Hexapoda + Myriapoda), coxopodal segments retained their mobility, but the telopodite was diminished. Myriapoda inherited a plesiotypic jaw with some residual articulation. In Hexapoda, all coxopodal segments became fully fused (an autapomorphy) and the telopodite became completely lost. The sutures clearly retained in the jaws of Archaeognatha (Figure 19.1(b,c)) are reminiscent of the atelocerate ancestor, but the jaws of other insects show a 'single' segment, an apparent convergence with Crustacea (Figure 19.2).

The history of 'basal' structures shows that the Myriapoda are monophyletic and proves their sister group relationship with Hexapoda. Shared characters in the eye structure, etc. found in Crustacea and Insecta (Osorio and Bacon, 1994; Nilsson and Osorio, 1997, this volume), are thus interpreted as symplesiomorphies which, as an interesting twist, imply the monophyly of Myriapoda. All the above evidence sustains the original phylogenetic tree as proposed by Snodgrass (1935), Sharov (1966), Boudreaux (1979), Hennig (1981) and Kukalová-Peck (1978, 1991):

Mandibulata = (Crustacea + (Hexapoda + Myriapoda))

The taxa Mandibulata and Atelocerata, which were once supported by somewhat deficient 'historical' evidence, were deconstructed by superficially convincing 'restrictive' evidence. The improved 'historical' evidence briefly summarized here has restored them, so that these taxa are now much better understood than before. Molecular biologists and some developmental geneticists, could well benefit from an input from comparative morphology, the better to focus their research strategies.

### 19.4.2 INSECT HEAD: RECENT ADVANCES

While the presence of an acron on the surface of the insect cranium, as well as the number of head segments (six), is now established, the relationship between outer head segmentation and inner morphology is not yet understood (Kukalová-Peck, 1992). The heads of modern insects, except in primitive silverfish (Zygentoma, *Tricholepidion*), have lost the intersegmental sutures, but the sutures are often preserved in Palaeozoic insects. The problem is that the boundaries of the segments on the cranial surface do not match with the inner organs, and they are not reconciled with some developmental events. By far the most promising breakthrough in these two directions is the work of Dewel and Dewel (1996, 1997, this volume).

Carpenter (1992) recognized Sharov's holotypes of Monura as representing an extinct suborder of Archaeognatha. However, the collection also contains several members of wingless, but broad-jawed (= dicondylous) insects with gonangulum and cercal leglets (Kukalová-Peck, 1985, 1987) which closely resemble Archaeognatha, but which must be referred to an undescribed order informally designated 'Cercopoda'. Monura and 'Cercopoda' are found mixed together in Palaeozoic freshwater deposits, often as cast cuticles preserved *in situ*.

The restudied holotype of Monura: *Dasyleptus brongniarti* Sharov, 1957 (Figure 19.5(d)) and a related, undescribed species (Figure 19.5(c)) show some very interesting and previously unknown features. The head of the holotype is unquestionably the most primitive in all Hexapoda. Bilobed acron is above tergum I, II and II is pushed deeply into segment IV; segment III (which bears two invaginated leg appendages is expressed only under the eye and does not reach the ventral cranial margin. The shift of the eye from its original adult position between segments II and III (in Myriapoda and Parainsecta) into segment IV (in Insecta) was previously known (Kukalová-Peck, 1991, 1992), but a small embayment at the eye level of IV into V, and of V into VI, was not seen before. The clypeolabrum appears to be two coxopodites of head segment I, fused together along their lateral margins, with fused trochanteral endites at the distal end (Kukalová-Peck, 1992). This interpretation has received strong support from the clypeolabrum bears a fused, segmented appendage, possibly a telopodite and the cuticle of clypeolabrum, because of its special composition, is preserved in matrix in exactly the same way as the (leg-derived) mouthparts and thoracic legs, but very differently from that of all terga, on the cranium as well as on the thorax. The monuran head does not show any archipleuron fused with the cranium, a possibility proposed by Kukalová-Peck (1987). It seems that cephalic archipleura are rather adjacent to the cranium and inconspicuous, or perhaps completely reduced.

## 19.5 INSECTA, HIGHER PHYLOGENY AND THE 'HISTORICAL' APPROACH

The past 5 years have seen an array of dissimilar cladograms of insect orders, and not much stability of branching order (see Štys and Zrzavý, 1994 and Kristensen, 1995, for reviews). The alternative, 'historical approach' is based upon my evidence from 200 entries on Archaeognatha and Zygentoma, over 500 detailed figures of the wing articulations of modern insects, about 3000 wing venation studies, and some 2000 figures of fossil and extant leg appendages. This is now sufficient to maintain with confidence that 'basal' structures (leg-derived appendages, wings) do contain the evidence needed to resolve the higher level phylogenetics of Pterygota. Even if classification is, and must be, based only on synapomorphies, the intellectual foundation of higher classification is the full understanding of successive changes between the ancient common ancestor and the present-day higher taxa. For this, knowledge of **both** apomorphic and plesiomorphic characters is essential. The hierarchy of classification is, after all, built by the discrimination of the former from the latter.

### 19.5.1 ORDER ZYGENTOMA (SILVERFISH, SISTER GROUP OF PTERYGOTA)

According to Kristensen (1995, 1997, this volume), this group is quite probably not monophyletic, and only the advanced Zygentoma are the sister group of Pterygota. The following 'historical' characters argue against this. All subgroups of Zygentoma have the pleuron on all thoracic segments as in the sister group Pterygota. However, the pleuron (flattened subcoxa, wall support, as above) in Zygentoma is desclerotized and fragmented. Since the Palaeozoic, silverfish have used running from predators as their escape strategy, employing a unique stride, which requires leg articulation to a flexible pleuron (analysed by Manton, 1977). The pleuron is subdivided into concentric rings along the ends of muscle attachments. The oldest Carboniferous Zygentoma show a less fragmented pleura, with an easily-recognizable remnant of pleural sulcus (Figure 19.6(b)). The more primitive Archaeognatha and sister group Pterygota have a solid pleuron with pleural sulcus (Figures 19.3(e) and 19.5(a,c,d)). This shows that a solid pleuron is plesiomorphic, and the desclerotized, fragmented pleuron is an autapomorphy at the Zygentoma level. Archeognatha have their **solid** pleuron only on the prothoracic legs. Therefore, the presence of the pleura on all three thoracic segments is a synapomorphy shared by Dicondylia ('Cercopoda' + (Zygentoma + Pterygota)) (Kukalová-Peck, 1987, 1991). Thus, pleural desclerotization and fragmentation is autapomorphic and reveals Zygentoma as a monophyletic group.

### 19.5.2 THE USE OF WING STRUCTURE IN HIGHER PTERYGOTE PHYLOGENY

The Pterygota commenced at the origin of articulated, mobile protowings, which probably evolved from flattened epicoxal exites with repeatedly forked cuticular ridges supplied with blood, much as in the motion-promoting telsonal plates of Crustacea (Kukalová-Peck, 1983). All major evolutionary lines are clearly distinguished by wing structure. It is generally believed that protowings, their articulation, and the wing vein pattern originated only once, and that the entire wing structure is monophyletic and homologous (Boudreaux, 1979; Hennig, 1981; Kukalová-Peck, 1978, 1983, 1991; CSIRO, 1991; Carpenter, 1992). This casts the wing structure as a primary source of higher level characters for the Pterygota. But, all attempts to use wing venation at order level have failed (Ragge, 1955), and Kristensen (1995, 1997, this volume) considers further efforts futile because of frequent homoplasies. This is far from the truth. Past failure was caused, not by homoplasies, but by the mistaken belief that the vein nomenclature was correctly applied to all orders. Instead, each insect order has acquired an idiosyncratic (non-homologous) system of symbols for veins, and phylogenetic relations have become obscured. I have shown that **none** of the vein and articulation systems could be derived from any shared groundplan, because the systems themselves are flawed. However, now that the homologization is on a proper phylogenetic basis, wing structure does seem to be of use in higher systematics; for example, Brown and Scholtz (1995, p. 146, 1996) found wing characters superior to conventional ones.

#### (a) Wing articulation

The insect wing articulation perhaps had a more complex origin than the one I proposed in 1983. The first column of articular sclerites, called proxalaria (P), which is expressed only in the dorsal membrane, indeed, originated from the archipleuron. But, most or all sclerites in the second and third column, called axalaria (AX) and fulcalaria (F), are occasionally expressed in both the dorsal and ventral membrane. These were probably formed from the extended wing base (Figure 19.8(a)). The so-called pleurites, the basalaria (BA) and subalaria (SA) under the wing, may be of combined origin, perhaps partly from the wing base, and partly from pleural (= subcoxal) fragments. A detailed study is necessary to verify the hypothesis.

The groundplan of pterygote articulation (Kukalová-Peck, 1983) is based on the following evidence. The articular sclerites provide protection for the blood channels, which then continue into the wing veins (Arnold, 1964). The primitive condition is one of eight wing veins, the first two of which (precosta and costa, PC and C) are close to each other and form the anterior margin. There are eight rows of articular sclerites aligned with eight wing veins preserved in the articulation of Palaeozoic Palaeoptera (Figures 19.7(a,c) and 19.8(a)) and modern Odonata (Figure 19.8(b)). Sclerites near the wing evolved from fragmented veins, which explains why the protoarticulation (groundplan) has eight rows of sclerites. The rows apparently drifted apart (Figure 19.8(a)) and were subdivided by hinges. The hinges divided the articulation anteroposteriorly into three columns of sclerites, proxalar (P), axalar (AX), and fulcalar (F). Fulcalaria were articulated to sclerotized bases of wing veins, called basivenalia (B).

Fragmentation of veinal bases explains two curious facts: (i) that the sclerites are numerous (not less than 32); and (ii) that the underlying pattern is nearly regular (Kukalová-Peck, 1983, 1991 had already showed that proto-wing venation also showed an unexpected regularity). The groundplan has proved its worth in Coleoptera, Orthoneoptera, Hymenoptera, Blattodea, Dermaptera, and Odonatoptera + Ephemeroptera (co-workers and specialists are C. Brauckmann, J. F. Lawrence, D. C. F. Rentz, C. H. Scholtz, D. J. Brown, M. Sharkey and R. J. Wootton; PhD programmes involve D. J. Brown, F. Haas, G. Bechly, A. Relin and A. C. Roy). The several differences between the 'restrictive' phylogeny and that preferred by the author, are summarized below.

*Palaeoptera*

The first division of Pterygota is into Palaeoptera and Neoptera. Palaeoptera (comprising the extinct palaeodictyopteroid orders + (Ephemeroptera + Odonatoptera)) are the sister group of Neoptera (= Pleconeoptera ?+ Orthoneoptera) + (Blattoneoptera + (Hemineoptera + Endoneoptera))) (Kukalová-Peck, 1991). In contrast,

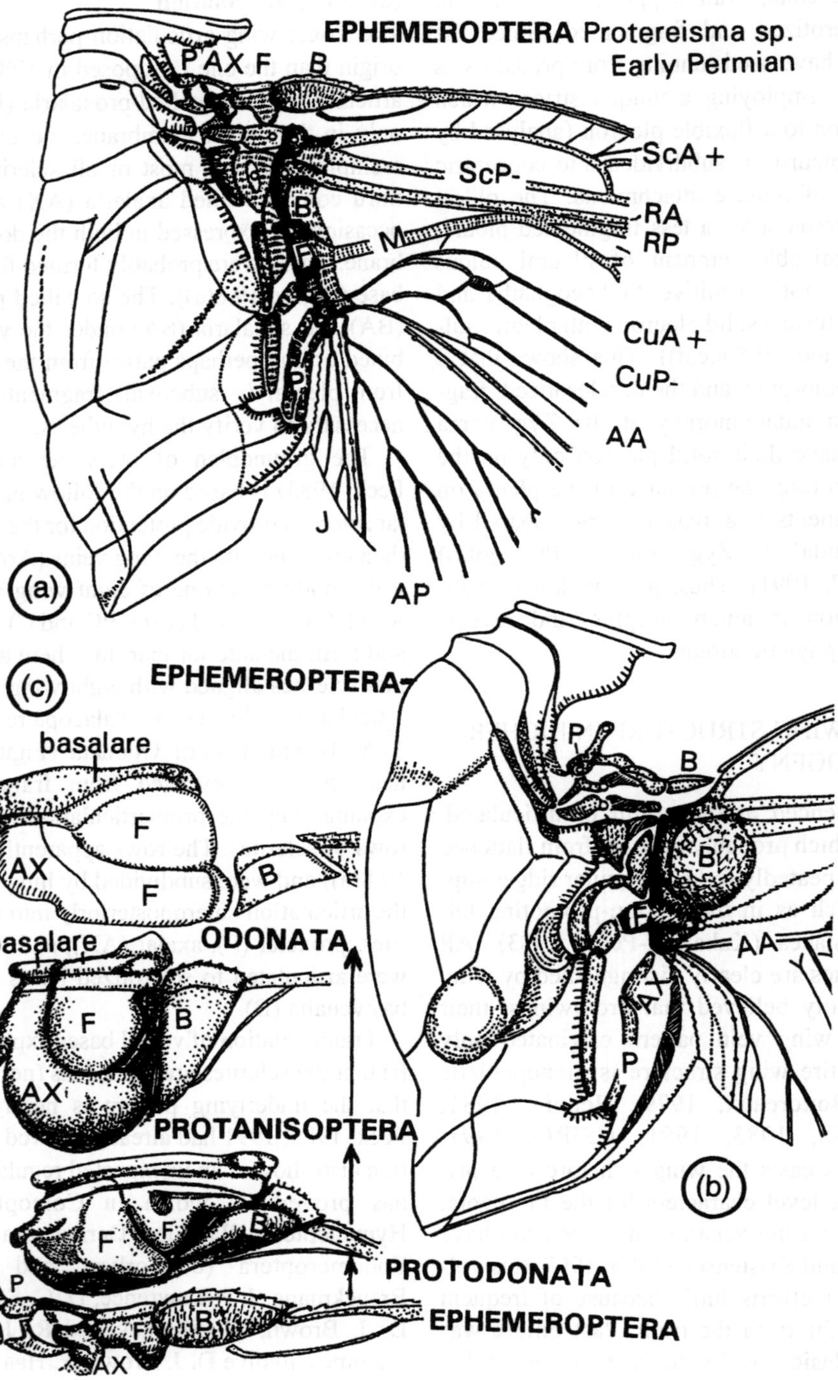


**Figure 19.7** Palaeoptera: articular sclerites are always arranged in columns. (a) Permian Ephemeroptera bear a primitive platform (darkened sclerites F and B). (b) Recent Ephemeroptera bear much more consolidated platform (darkened), and desclerotized sclerites (delineated). (c) Ephemeroptera and Odonatoptera share a unique articular plate (associated costal B and F and AX) (strong synapomorphy supporting the sister group relationship). (ascending) Permian Ephemerida, *Protereisma* sp., specimen no. 3411, Museum of Comparative Zoology, Harvard University, Cambridge, original. Modern siphlonurid Ephemerida, after Kukalová-Peck, 1983, relabelled. Protodonata: *Oligotypus tillyardi*, specimen no. 4804, original; Protanisoptera: *Ditaxineura* sp., specimen no. 3045, original; both specimens at Museum of Comparative Zoology, Harvard University, Cambridge; modern Odonata: *Uropetala carovei*, after Kukalová-Peck, 1983, interpreted.

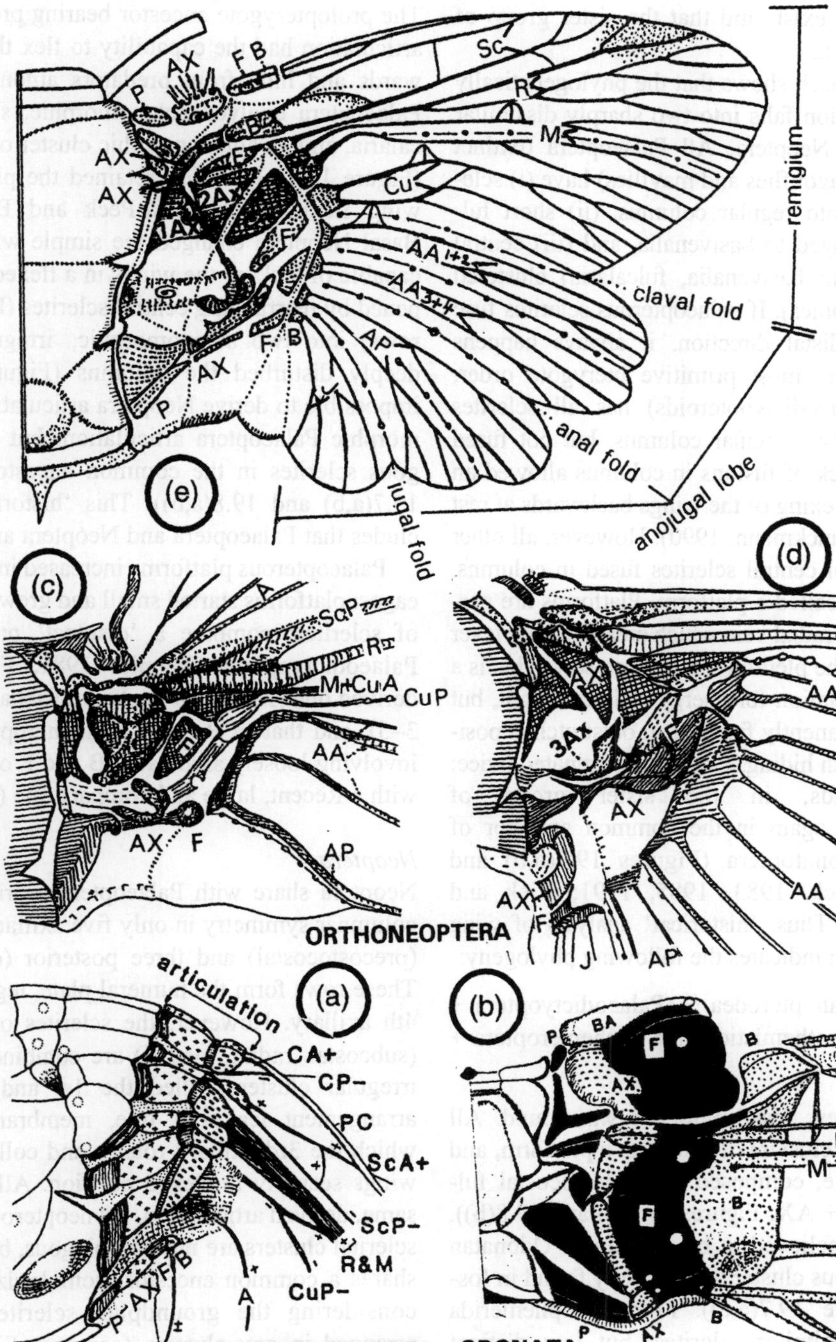


**Figure 19.8** Pterygote wing articulation, homologized. (a,b) Palaeoptera show derived articulation. (a) Diaphanopterodea, the most primitive member, have central basivenalia broadly hinged with each other and with fulcalaria (Paleoptera, autapomorphies), P and F sclerites very short, jugal row separated, and cubitojugal area sclerotized (palaeodictyopteroids, autapomorphies). (b) Recent Odonata have two columns fused into an articular plate and a platform, but the proxalar column is free. (c–e) Neoptera have sclerites of Sc, R, M rows fused into two hook-like, irregular clusters (1Ax and 2Ax), but PC, C, Cu, A, J rows are arranged in columns as in Paleoptera. (c,d) Orthoneoptera: Ensifera: Tettigoniidae: *Capnobotes fuliginosus*: (c) Forewing; (d) Hindwing. (e) Neoptera, groundplan reference scheme, improved. Comparing orthopteroid articulation with (e), groundplan, brings out character transformations at the Orthoneoptera and at the Ensifera, level. [Source: (a), after Kukalová-Peck and Brauckmann, 1990, relabelled; (b), after Kukalová-Peck, 1983, relabelled; (c–e), originals.]

Kristensen (1995) and Willmann (1997, this volume) argue that Palaeoptera do not exist, and that the sister group of Neoptera are the Odonata.

The 'historical' approach shows that the phylogenetically important wing articulation falls into two sharply dissimilar types: Palaeoptera and Neoptera. All Palaeoptera (extinct palaeodictyopteroids, dragonflies and mayflies) have (i) sclerites always arranged into regular columns, (ii) short fulcalaria very broadly hinged to basivenalia, and (iii) central sclerites (Sc, R, M, Cu: basivenalia, fulcalaria) clustered together (all autapomorphies). If palaeopterous sclerites fuse together in the anterodistal direction, it always happens exactly in columns. The most primitive pterygote order, Diaphanopterodea (palaeodictyopteroids) has all sclerites hinged in rows and in two regular columns, but not fused (Figure 19.8(a)). The lack of fusions in columns allowed an ultra-primitive, simple flexing of the wings backwards at rest (Kukalová-Peck and Brauckmann, 1990). However, all other Palaeoptera have several central sclerites fused in columns. The resulting cluster is called a platform. Platforms are permanently attached to the radial vein which creates a long lever resting and rocking on the pleuron – like a see-saw. This is a very advantageous adaptation for energy-efficient flight, but the wings become permanently fixed in an outstretched position which interferes with hiding. Platforms originated twice: in palaeodictyopteroids, in the sister group of Diaphanopterodea; and again in the common ancestor of Ephemeroptera + Odonatoptera (Figures 19.7(a,b) and 19.8(b)) (Kukalová-Peck, 1983, 1985, 1991; Riek and Kukalová-Peck, 1984). Thus, 'historical' analysis of wing articulation and venation indicates the following phylogeny:

Palaeoptera = {[Diaphanopterodea + (Palaeodictyoptera + (Megasecoptera + Permothemistida))] + [Ephemeroptera + Odonatoptera]}.

Recently, important, new fossil evidence was found. All Odonatoptera have two clusters of sclerites: a platform, and a smaller, anterior plate, composed of precostocostal fulcalaria and axalaria (F + AX; Figures 19.7(c) and 19.8(b)). An anterior plate was believed to be a uniquely Odonatan feature, but a homologous cluster was recently found in fossil Ephemerida (Figure 19.7(a,c)). Modern Ephemerida (Figure 19.7(b)) bear similar sclerites, but less distinct because they are desclerotized. This discovery further supports Odonata + Ephemerida as sister groups and Palaeoptera as monophyletic.

Palaeopteran monophyly is further supported by wing venation (below), and by the following autapomorphies in basal structures: the maxillary inner rami, lacinia (coxal endite) and galea (trochanteral endite), are fused; trochantins (sternal fragments near thoracic legs) are incompletely separated from sterna; all abdominal sterna are expanded (instead of triangular, as in Archaeognatha and in basal Neoptera) (Kukalová-Peck, 1991).

*Wing flexing*

The protopterygote ancestor bearing protowings and protoarticulation had the capability to flex the protowings backwards and hide from predators among vegetation. Basal Palaeoptera evolved autapomorphic, short and broad fulcalaria, and an autapomorphic cluster of central basivenalia (Figure 19.8(a)), which retained the plesiomorphic simple wing flexing (Kukalová-Peck and Brauckmann, 1990). Basal Neoptera changed the simple wing flexing into one capable of locking the wings in a flexed position. This happened by rearranging central sclerites (P, AX, F in Sc, R, M rows) into two autapomorphic, irregular clusters, which deeply disturbed the columns (Figure 19.8(c–e)). It is impossible to derive Neoptera articulation from the autapomorphic Palaeoptera articulation; but both share homologous sclerites in the common ancestor (compare Figures 19.7(a,b) and 19.8(a,b)). This 'historical' argument concludes that Palaeoptera and Neoptera are sister groups.

Palaeopterous platforms increased in all palaeopteran lineages: platforms started small and grew larger by apposition of sclerites: compare a 'disjunct' primitive platform of Palaeodictyoptera (Hennig, 1981, Figure 25B) with a derived one (Kukalová Peck and Richardson, 1983, Figures 3–5); and that of Permian Ephemeroptera (Figure 19.7(a); involving loosely associated B and F of veins Sc, R, M, A) with a Recent, large and compact one (Figure 19.7(b)).

*Neoptera*

Neoptera share with Palaeoptera a primitive adherence to columnar symmetry in only five axillary rows: two anterior (precostocostal) and three posterior (cubital, anal, jugal). These rows form the humeral plate, tegula, and the 3rd and 4th axillary. However, the sclerites of three central rows (subcostal, radial, medial) are combined obliquely, in two irregular clusters called the 1st and 2nd axillary. This arrangement leaves a free, membranous 'window' into which the 3rd axillary rotates and collapses, and locks the wings securely in flexed position. All Neoptera share the same, derived articulation. As neopterous and palaeopterous sclerital clusters are **not** homologous, but the wing structure shares a common ancestor, homologization is possible by considering the groundplan sclerites which were not arranged in any clusters (compare Figures 19.7(a,b) and 19.8(a–e)) (Kukalová-Peck, 1983). Previous comparisons on a one-to-one basis to homologize palaeopterous platforms with neopterous axillaries were not successful, because they left some sclerites uninterpreted. They never generated phylogenetic conclusions. My 1983 model of homology (Pterygota level) has been tested by myself and co-authors, so far on 700 modern specimens. Phylogenetic conclusions were pursued at the Superorder, Order, Suborder, and Superfamily level. The new reference scheme of Neoptera (Figure 19.8(e)) is an improvement of the 1991 model.

**(b) Wing venation**

Wing venation contains eight veins (precosta, PC; costa, C; subcosta, Sc; radius, R; media, M; cubitus, Cu, anal, A; jugal, J; see Kukalová-Peck, 1983). The criterion for judging the relatively primitive character state is based upon the well-tested principle which states that all fusions and losses of principal veins and branches and of articular sclerites, are derived and would not revert. Hence, based on over 3000 fossils (Kukalová-Peck, 1983, 1991), the wing venation of all pterygote lineages can be derived from a protowing (or, from prothoracic wings of fossils) that includes: eight principal veins; a precosta forming a strip fully adjacent to the costa; all other veins start from a basivenale diverged into two independent sectors (anterior convex and posterior concave; A and P); and, each sector is divided repeatedly and dichotomously into principal branches. No fusions and no braces (= important cross-veins) are shared by all pterygote orders. Because aerial flapping flight is impossible without fusions or braces (Brodsky, 1994), this absence has an important implication: the protowing could not have been capable of flapping flight. This deduction explains why major pterygote lines differ in veinal fusions and braces near the base, i.e. in the very features which enable flapping flight. The multiple origin of insect flight was also suspected by Brodsky (1994) and Marden (1995), based on functional observations. According to the most probable hypothesis, Pterygota diverged at a point when the protowings were flapping to propel insects away from predators, but not yet flying; the legs and abdomens of insects may have been resting on the water surface (Marden and Kramer, 1994).

Basal divergence was explained as follows:

- Palaeoptera: fore and hindwings are homonomous, in shape as well as in venational pattern; the stem of M is always present; the stem of Cu is always present; the anal brace is formed by stiffened membrane in conjunction with the basal portions of AA (the last three are autapomorphies).
- Neoptera: fore and hindwings are primitively heteronomous, forewings are mainly protective, narrower and more heavily sclerotized, hindwings are mainly flying, broader and thinner (all autapomorphies). Forewings: anal branches basally are asymmetrically distributed, AA1+2 and AA3+4 are well distant from each other, AA3+4 and AP are closely associated; jugal veins and area are membranized and largely reduced, folded under the wing (all autapomorphies). Hindwings: anojugal lobe enlarged, with claval, anal, and jugal flexion lines/folds; anal brace is formed by anal basivenale and first anal branch (all autapomorphies).

The palaeopterous and the neopterous sets of interlocked veinal characters cannot be derived one from another, without contravening the polarity in the evolution of veins. Again, this supports the monophyly of both Palaeoptera and Neoptera, and their sister group relationship. This conclusion gets additional support from the wing venation of Palaeozoic Ephemeroptera and Odonatoptera (Riek and Kukalová-Peck, 1984; Bechly, 1996, figures 67–70): the superorders share seven strong synapomorphies: the stem of M, the stem of Cu, simple CuP; three important braces: rp-ma, m-cu, and cup-aa1; a long curved anal brace. These veinal characters are interlocked and unique to Ephemerida + Odonata.

Willmann (1997, this volume) and Kristensen (1991, 1997, this volume) proposed that Ephemeroptera and palaeodictyopteroids diverged first, and that Neoptera and Odonata are sister groups. The supposed synapomorphies of this relationship were listed by Kristensen (1991): absence of a moult in the imaginal stage (in my view convergence, since all pterygote lines had imaginal moults in the Palaeozoic); absence of long terminal filament (probably a multiple reduction since also absent in palaeodictyopteroids); and, in the mouthparts, a mandible with a permanent anterior articulation; similar hypopharynx; loss of all tentoriomandibular muscle bundles (except one) and of the tentoriolacinial muscle. The apparently persuasive mandibular synapomorphies can be easily explained as homoplasies based on convergent feeding adaptation. 'Historical' character research shows that Odonata (for predation) and basal Neoptera (for crushing vegetable matter) needed strong, broadly articulated mandibles with a permanent anterior condyle. This, in Odonata, could easily have formed from the outer socket/groove present in basal Dicondylia, (known in 'Cercopoda': Kukalová-Peck, 1987, Figures 11 and 13), presumably present in protopterygotes, and present in all Palaeoptera except Odonatoptera. In Neoptera, it evolved from the same ancestral outer socket, presumably present in the protopterygote ancestor. The hypopharynx in Odonata and Neoptera is similar, because it is likewise pushed back in the head to make space for the broad, shearing mandibles. Similar mandibles required convergent adjustments (reductions) in muscles. But, maxilla confirms Palaeoptera. This example shows that allegedly convincing higher-level 'synapomorphies' may prove less robust if examined in their 'historical' context.

## 19.6 STREPSIPTERA: COULD THEY POSSIBLY BE FLIES?

It is believed that this parasitic, early diverged, but very highly specialized order with short elytra and broad hindwings, is related to Coleoptera. The hindwing venation shares six higher-level synapomorphies with the Coleoptera, and indicates their sister group relationship (Kukalová-Peck and Lawrence, 1993). This conclusion was challenged by Whiting and Wheeler (1994) who argued that Strepsiptera are rather the sister group of Diptera, in which the wing pairs became reversed by a developmental mutation (halteres became elytra and the forewings became hindwings).

This evidence was accepted and maintained as 'strong', by Akam *et al.* (1994). However, the fly relationship was based on a hypothetical genetic event, and presented without properly checking its admissibility on morphological grounds. Whiting and Kathirithamby (1995) chose to support their case by a pugnacious criticism of a paper on Coleoptera phylogeny (Kukalová-Peck and Lawrence, 1993) for employing a 'historical' approach, and for using the wing groundplans of Neoptera and Endopterygota. To prove their point, Whiting and Kathirithamby presented four scanning electron microscopy (SEM) and three light microscopy photos showing that in Strepsiptera the costal wing margin is broad (if dry, it pulls back up) and the subcostal vein (ScP-) is hidden in an adjacent groove. It apparently eluded them, that this unusual feature was noted and described by Kukalová-Peck and Lawrence (1993, p. 196) in Coleoptera. Thus, Whiting and Kathirithamby strengthened the Coleoptera argument by their new documentation. They claimed that we negligently 'overlooked' that Strepsiptera wings start with a 'convex' subcosta; in fact, in insect wings the anterior margin is convex, but the subcosta (= ScP-) is always concave. Furthermore, Whiting and Kathirithamby (1995) apparently did not check some morphological data in the classical literature on Strepsiptera. For example, they missed the information that the forewings have two radial sectors (RA and RP) not fused into a radial stem (Kinzelbach, 1971, p.40, p. 202), precisely as in Coleoptera forewings (elytra) (Kukalová-Peck, 1991). This ultra-primitive, plesiomorphic feature occurs in Endoneoptera **only** in Strepsiptera and Coleoptera, because the rest of Endoneoptera share a long radial stem (veinal stems are always recognized as derived). Also, Kinzelbach (1971, p. 203, Figure 63; p. 205, Figure 67) described another unusual and important morphological feature: in Strepsiptera, direct flight muscle 118 (= M 71) starts from the pleuron and is inserted in the membrane, just behind, or at the very end of, the 3rd axillary (3Ax). Larsén (1966) described in Coleoptera a homologous muscle with the same function in the same place. Kukalová-Peck and Lawrence (1993) described the feature again in detail, and figured it on their Figures 73, 75, 77, 78, 79, 80 and 81. Since the feature is old, complex, and unique, it represents a convincing synapomorphy of Strepsiptera + Coleoptera. Whiting and Kathirithamby's (1995) concentration on the philosophical criticism of comparative morphology may have obscured the real questions of the interpretation of synapomorphies, which are available in the literature.

I recently found two additional, strong synapomorphies with Coleoptera: radial and medial basivenale are separated by membranous groove rather than associated; and, all elements of the 3rd axillary (body and arms) are fused into a single sclerite (as described in Coleoptera by Larsén, 1966). These two features enable unique flexing of the wing apex. Also, a fan-like arrangement of veins determines Strepsiptera wings clearly as hind wings, not fore wings. The doubt about the endopterygote nature of Strepsiptera (Kristensen, 1991) can now be put to rest. Combined with the equally special muscle insertion in the membrane or at the edge of 3Ax (see above), this complex apparatus enables coleopteroids to pull in the wing apex and fold it under the elytra (Kukalová-Peck and Lawrence, 1993). The ancestor of Strepsiptera evidently did flex and fold the wing tips under the elytra in the same way as the present-day Coleoptera. The very close (sister group) Order level relationship between Strepsiptera and Coleoptera is as reliably documented as any in Pterygota.

## 19.7 CLADISTIC ANALYSIS IN HIGHER CLASSIFICATION

Cladistic methods are a great help in systematics, but correct evaluation of characters is a serious concern. In higher taxa, with few, extremely old characters, the selection of the correct outgroup is all-important as well as very difficult. Some modern higher taxa cannot be compared directly without errors, because anciently they diverged from a common ancestor in profoundly different ways. For example, Palaeoptera and Neoptera share an unknown protopterygote ancestor. Their higher-level wing characters defied recognition for 150 years, but they started unravelling immediately after the reconstruction of the protowing and of protoarticulation. I believe that the 10-segmented arthropodan leg groundplan can be equally helpful in unravelling the relationships of the higher subtaxa of Chelicerata and Crustacea, both extinct and extant.

## 19.8 SUMMARY

The 'historical' approach has yielded the following results in higher systematics:

1. Mandibulata = Crustacea + (Hexapoda + Myriapoda).
2. Myriapoda are monophyletic; Myriapoda and Hexapoda are sister groups.
3. Diplura are monophyletic; Diplura and Insecta (s.s.) are sister groups.
4. Zygentoma are monophyletic; Zygentoma and Pterygota are sister groups.
5. Pterygota = (Palaeoptera + Neoptera); Palaeoptera are monophyletic.
6. Odonata and Ephemerida are sister groups.
7. Strepsiptera are closely related to, and are probably the sister group of Coleoptera.

ACKNOWLEDGEMENTS

Dr R.A. Fortey reviewed this paper in several versions. Dr S.B. Peck (Carleton University), Dr R.A. Dewel (Appalachian State University, Boone), Dr P. Svacha

(Entomological Institute CAV), F. Haas (Jena University) and G. Bechly (University of Tübingen) commented on the manuscript. Mr Lew Ling (Carleton University) is thanked for SEM photographs. The research is supported by NSERC Canada, and facilities provided by Department of Earth Sciences, Carleton University.

## REFERENCES

Akam, M., Averof, M., Castelli-Gair, J., Dawes, R., Falciani, F. and Ferrier, D. (1994) The evolving role of *Hox* genes in arthropods. *Development*, Supplement, 209–15.

Arnold, J.W. (1964) Blood circulation in insect wings. *Memoirs of Entomological Society of Canada*, **38**, 5–48.

Averof, M. and Cohen, M. (1977) Evolutionary origin of insect wings from ancestral gills. *Nature*, **385**, 627–630.

Bechly, G. (1996) Morphologische Untersuchungen an Flügelgeäder der rezenten Libellen und deren Stammgruppenvertreter (Insecta; Pterygota; Odonata). *Petalura*, **2**, 1–402.

Bitsch, J. (1994) The morphological groundplan of Hexapoda: critical review of recent concepts. *Annales de la Société entomologique de France* (N.S.), **30**, 103–29.

Boudreaux, H.B. (1979) *Arthropod Phylogeny: with Special Reference to Insects*, John Wiley & Sons, New York.

Brodsky, A.K. (1994) *The Evolution of Insect Flight*, Oxford University Press, Oxford.

Browne, D.J. (1991a) Phylogenetic significance or wing characters in the Geotrupini (Coleoptera: Scarabaeoidea), MSc Thesis, University of Pretoria.

Browne, D.J. (1991b) Wing structure of the genus *Eucanthus* Westwood; confirmation of the primitive nature of the genus (Scarabaeoidea: Geotrupidae: Bolboceratinae). *Journal of the Entomological Society of Southern Africa*, **54**, 221–30.

Browne, D.J. (1993) Phylogenetic significance of the hind wing basal articulation of the Scarabaeoidea (Coleoptera), PhD Thesis, University of Pretoria.

Browne, D.J. and Scholtz, C.H. (1994) The morphology and terminology of the hindwing articulation and wing base of the Coleoptera, with specific reference to the Scarabaeoidea. *Systematic Entomology*, **19**, 133–43.

Browne, D.J. and Scholtz, C.H. (1995) Phylogeny of the families of Scarabaeoidea (Coleoptera) based on characters of the hindwing articulation, hindwing base and wing venation. *Systematic Entomology*, **20**, 145–73.

Browne, D.J. and Scholtz, C.H. (1996) The morphology of the hind wing articulation and wing base of the Scarabaeoidea (Coleoptera) with some phylogenetic implications, *Bonner Zoologische Monographien*, **40**, 1–200.

Browne, D.J., Scholtz, C.H. and Kukalová-Peck, J. (1993) Phylogenetic significance of wing characters in the Trogidae (Coleoptera: Scarabaeoidea). *African Entomology*, **1**, 195–206.

Carpenter, F.M. (1992) Superclass Hexapoda, in *Treatise on Invertebrate Paleontology* (eds R.C. Moore and R.L. Kaesler), Arthropoda 4, (4), The Geological Society of America and The University of Kansas, pp. 1–655.

Cisne, J.L. (1975) Anatomy of *Triarthrus* and the relationships of the Trilobita. *Fossils Strata*, **4**, 45–63.

CSIRO (1991) *The Insects of Australia*, 2nd edn (eds I.E. Naumann and CSIRO), Melbourne University Press, Melbourne.

Dewel, R.A. and Dewel, W.C. (1996) The brain of *Echiniscus viridissimus* Peterfi, 1956 (Heterotardigrada): a key to understanding the phylogenetic position of tardigrades and the evolution of the arthropod head. *Zoological Journal of the Linnean Society*, 116, 35–49.

Hennig, W. (1981) *Insect Phylogeny*, John Wiley and Sons, New York.

Kinzelbach, R.K. (1971) Morphologische Befunde an Fächerflügelrn und ihre phylogenetische Bedeutung (Insecta: Strepsiptera). *Zoologica*, **41**(119), (1–2), 1–128, 129–256.

Kristensen, N.P. (1975) The phylogeny of hexapod 'orders'. A critical review of recent accounts. *Zoologische Systematische Evolution-forschungen*, **13**, 1–44.

Kristensen, N.P. (1981) Phylogeny of insect orders. *Annual Review of Entomology*, **26**, 135–57.

Kristensen, N.P. (1991) Phylogeny of extant hexapods, in *CSRIO The Insects of Australia*, 2nd edn (eds I.E. Naumann and CSIRO), Melbourne University Press, Melbourne, pp. 125–40.

Kristensen, N.P. (1995) Forty years' insects phylogenetic systematics. *Zoologische Beiträge, NF*, **36**(1), 83–124.

Kukalová-Peck, J. (1978) Origin and evolution of insect wings and their relation to metamorphosis, as documented by the fossil record. *Journal of Morphology*, **156**, 53–126.

Kukalová-Peck, J. (1983) Origin of the insect wing and wing articulation from the arthropodan leg. *The Canadian Journal of Zoology*, **61**, 1618–69.

Kukalová-Peck, J. (1985) Ephemeroid wing venation based upon new gigantic Carboniferous mayflies and basic morphology, phylogeny, and metamorphosis of pterygote insects (Insecta, Ephemerida). *The Canadian Journal of Zoology*, **63**, 933–55.

Kukalová-Peck, J. (1987) New Carboniferous Diplura, Monura, and Thysanura, the hexapod groundplan, and the role of thoracic lobe in the origin of wings (Insecta). *The Canadian Journal of Zoology*, **65**, 2327–45.

Kukalová-Peck, J. (1991) Fossil history and the evolution of hexapod structures, in *CSIRO The Insects of Australia*, 2nd edn (eds I.E. Naumann and CSIRO), Melbourne University Press, Melbourne, pp. 141–79.

Kukalová-Peck, J. (1992) The 'Uniramia' do not exist: the groundplan of Pterygota as revealed by the Permian Diaphanopterodea from Russia (Insecta: Paleodictyopteroidea). *The Canadian Journal of Zoology*, **70**, 236–55.

Kukalová-Peck, J. and Brauckmann, C. (1990) Wing folding in pterygote insects, and the oldest Diaphanopterodea from the early Late Carboniferous of West Germany. *The Canadian Journal of Zoology*, **68**, 1104–11.

Kukalová-Peck, J. and Brauckmann, C. (1992) Most Paleozoic Protorthoptera are ancestral hemipteroids: major wing braces as clues to a new phylogeny of Neoptera (Insecta). *The Canadian Journal of Zoology*, **70**, 2452–73.

Kukalová-Peck, J. and Lawrence, J.F. (1993) Evolution of the hind wing in Coleoptera. *The Canadian Entomologist*, **125**, 181–258.

Kukalová-Peck, J. and Peck, S.B. (1993) Zoraptera wing structures: evidence for new genera and relationship with the blattoid orders (Insecta: Blattoneoptera). *Systematic Entomology*, **18**, 333–50.

Kukalová-Peck, J. and Richardson, E.S., Jr (1983) New Homoiopteridae (Insecta: Paleodictyoptera) with wing articulation from Upper Carboniferous strata of Mazon Creek, Illinois. *The Canadian Journal of Zoology*, **61**(7), 1670–87.

Larsén, O. (1966) On the morphology and function of the locomotor organs of the Gyrinidae and other Coleoptera. *Opuscula Entomologica Supplementum*, **30**, 1–40.

Marden, J.H. (1995) Flying lessons from a flightless insect. *Natural History*, **2** (95), 6–8.

Marden, J.H. and Kramer, M.G. (1994) Surface–skimming stoneflies: a possible intermediate stage in insect flight evolution. *Science*, **266**, 427–30.

Manton, S.M. (1972) The evolution of arthropodan locomotory mechanisms. Part 10. Locomotory habits, morphology and evolution of the hexapod classes. *Journal of the Linnean Society, Zoology*, **51**, 203–400.

Manton, S.M. (1977) *The Arthropoda. Habits, Functional Morphology and Evolution*, Clarendon Press, Oxford.

Matsuda, R. (1970) Morphology and evolution of the insect thorax. *Memoirs of the Entomological Society of Canada*, **76**, 1–431.

Meier, R. (1993) [Book review] *The Insects of Australia; A Textbook for Students and Research Workers*, 2nd edn, *Systematic Biology*, **42**, 588–91.

Osorio, D. and Bacon, J.P. (1994) A good eye for arthropod evolution, *BioEssays*, **16**(6), 419–24.

Popadić, A., Rusch, D., Peterson, M., Rogers, B.T. and Kaufman, T.C. (1996) Origin of the arthropod mandible. *Nature*, **380**, 395.

Ragge, D.R. (1955) The wing venation of the Orthoptera–Saltatoria with notes on Dictyopteran wing-venation. *Bulletin of the British Museum, Natural History*, **6**, 1–159.

Rempel, J.G. (1975) The evolution of the insect head: the endless dispute. *Questiones Entomologicae*, **11**, 7–25.

Riek, E.F. and Kukalová-Peck, J. (1984) A new interpretation of dragonfly wing venation based upon Early Upper Carboniferous fossils from Argentina (Insecta: Odonatoidea) and basic character states in pterygote wings. *The Canadian Journal of Zoology*, **62**, 1150–66.

Scholtz, C.H. and Browne, D.J. (1996) Polyphyly in the Geotrupidae (Coleoptera: Scarabaeoidea): a case for a new family. *Journal of Natural History*, **30**, 597–614.

Scholtz, C.H., Browne, D.J. and Kukalová-Peck, J. (1994) Glaresidae, archaeopteryx of the Scarabaeoidea (Coleoptera). *Systematic Entomology*, **19**, 259–77.

Sharov, A.G. (1966) *Basic Arthropodan Stock*, Pergamon Press, New York.

Shear, W.A. and Kukalová-Peck, J. (1990) The ecology of Paleozoic terrestrial arthropods: the fossil evidence. *The Canadian Journal of Zoology*, **68**, 1807–34.

Smith, E.L. (1970a) Biology and structure of some California bristletails and silverfish. *Pan-Pacific Entomologist*, **46**, 212–25.

Smith, E.L. (1970b) Biology and structure of the dobsonfly *Neohermes californicus* (Walker) (Megaloptera: Corydalidae). *Pan-Pacific Entomologist*, **46**(2), 142–50.

Smith, E.L. (1988) Morphology and evolution of the hexapod head, 81. Insect reproductive system and external genitalia, 73. Main lines of arthropod evolution: Morphological evidence, 91, *Proceedings of the XVIII. International Congress of Entomology, Vancouver*, pp. 1–499.

Snodgrass, R.E. (1935) *Principles of Insect Morphology*, McGraw-Hill Book Company, New York.

Snodgrass, R.E. (1938) Evolution of Annelida, Onychophora and Arthropoda. *Smithsonian Miscellaneous Collection*, **138**, 1–77.

Stenzhorn, H.J. (1974) Experimentelle Untersuchungen zur Entwicklung des Flügelgelenkes von *Lymantria dispar* L. (Lepidoptera). *Wilhelm Roux Archive der Entwicklungsmechanischen Organe*, **175**, 65–86.

Štys, P. and Zrzavý, J. (1994) Phylogeny and classification of extant Arthropoda: Review of hypotheses and nomenclature. *European Journal of Entomology*, **91**, 257–75.

Sulc, K. (1927) Das Tracheensystem von *Lepisma* (Thysanura) und Phylogenie der Pterygogenea. *Acta Societatis de Sciencias Naturales de Moravia*, **4**(7/39), 1–108.

Whiting, M.F. and Kathirithamby, J. (1995) Strepsiptera do not share hind wing venational synapomorphies with Coleoptera: a reply to Kukalová-Peck and Lawrence. *Journal of the New York Entomological Society*, **103**(1), 1–14.

Whiting, M.F. and Wheeler, W.C. (1994) Insect homeotic transformations. *Nature*, **368**, 696.

# 20 Advances and problems in insect phylogeny

*R. Willmann*

*II. Zoologisches Institut der Universität, Georg August Universität Göttingen, Berliner Strasse 28, D-37073 Göttingen, Germany*

## 20.1 INTRODUCTION

The insects were one of the first groups whose relationships were investigated using the theory and methods of phylogenetic systematics. Today, almost 50 years after Hennig's first consideration of insect phylogeny, it might be expected that the relationships among the major insect groups would be clear. Figure 20.1 shows these relationships as currently accepted by many entomologists. With only a few alterations, this is almost the same view as was held by Hennig in 1953, a view that was largely derived from comparative morphological analyses, for example by Weber (1933) and Snodgrass (1935 and subsequent papers). However, the fact that it has prevailed over decades does not imply that it is well-founded. The cladogram (Figure 20.1) shows numerous doubtful or weakly supported sister group relationships.

There are several reasons for this situation. Most studies on insect phylogeny are based on comparative morphology, and many of them have centred either upon the question whether or not the terminal taxa indicated in Figure 20.1 are monophyletic or upon resolving sister group relationships **within** these taxa. Relationships **among** the major taxa are not commonly addressed. Another source of many uncertainties is that the Insecta is both a very old and rapidly evolving group and that the amount of morphological restructuring has been enormous. Accordingly, synapomorphies of some taxa may have become veiled, while plesiomorphies have been preserved in other lineages.

Progress in our knowledge of insect phylogeny can still be expected from detailed morphological studies, but these are time-consuming. An example is a recent, monumental work on the Dictyoptera (Klass, 1995). The Dictyoptera comprises the cockroaches, termites and praying mantises (Blattodea, Isoptera and Mantodea), and there has been a debate for many years on the sister group relationships among the three. Now, Klass has found that the termites are in fact a subgroup of the cockroaches and that the wingless roach *Cryptocercus* is the sister taxon of the termites. The result is an entirely new phylogenetic system embracing the groups involved.

It is interesting to note that Hennig had already developed the idea that *Cryptocercus* and the termites could be sister groups. In 1969, in his book *Die Stammesgeschichte der Insekten* (translated 1981), he reviewed the similarities shared by *Cryptocercus* and the termite *Mastotermes*. Among them was the fact that *Mastotermes* digests wood in the same way as *Cryptocercus*, and that it has a fauna of flagellates that is very similar to that of *Cryptocercus* (Hennig, 1969, 1981). Phylogenetic reasoning was not well developed in the 1960s, and therefore, nobody drew any far-reaching conclusions from these observations. Only Hennig stated that, provided that the similarities shared by *Cryptocercus* and the Isoptera are neither symplesiomorphies nor the result of convergence, *Cryptocercus* is more closely related to the termites than the rest of the Blattaria (Hennig, 1981).

Kristensen (1991, 1995) has reviewed current knowledge of the relationships among the major insect groups. In the following section, I am going to focus on a few topics that go beyond his papers.

## 20.2 SEGMENTATION OF THE HEXAPOD LEG

Several relationships among the insects appear quite clear. This has implications for some current theories of their evolution. The Ellipura (= Collembola and Protura) and possibly also the Diplurans (together: 'Entognatha') are the sister group of the rest of the Hexapods (Ectognatha, Insecta s. str.). The archaeognaths are probably the sister group of the Dicondylia (= Zygentoma and Pterygota) (Figure 20.1).

In the groundplans of all these groups, the leg consists of six free podomeres – coxa, trochanter, femur, tibia, tarsus and pretarsus (Figure 20.2). Kukalová-Peck (1987, 1991) assumes that the hexapodan leg originally consisted of many more segments and that even the pterygote insects primarily had nine free leg-segments. This implies the

*Arthropod Relationships*, Systematics Association Special Volume Series 55, Edited by R.A. Fortey and R.H. Thomas. Published in 1997 by Chapman & Hall, London. ISBN 0 412 75420 7

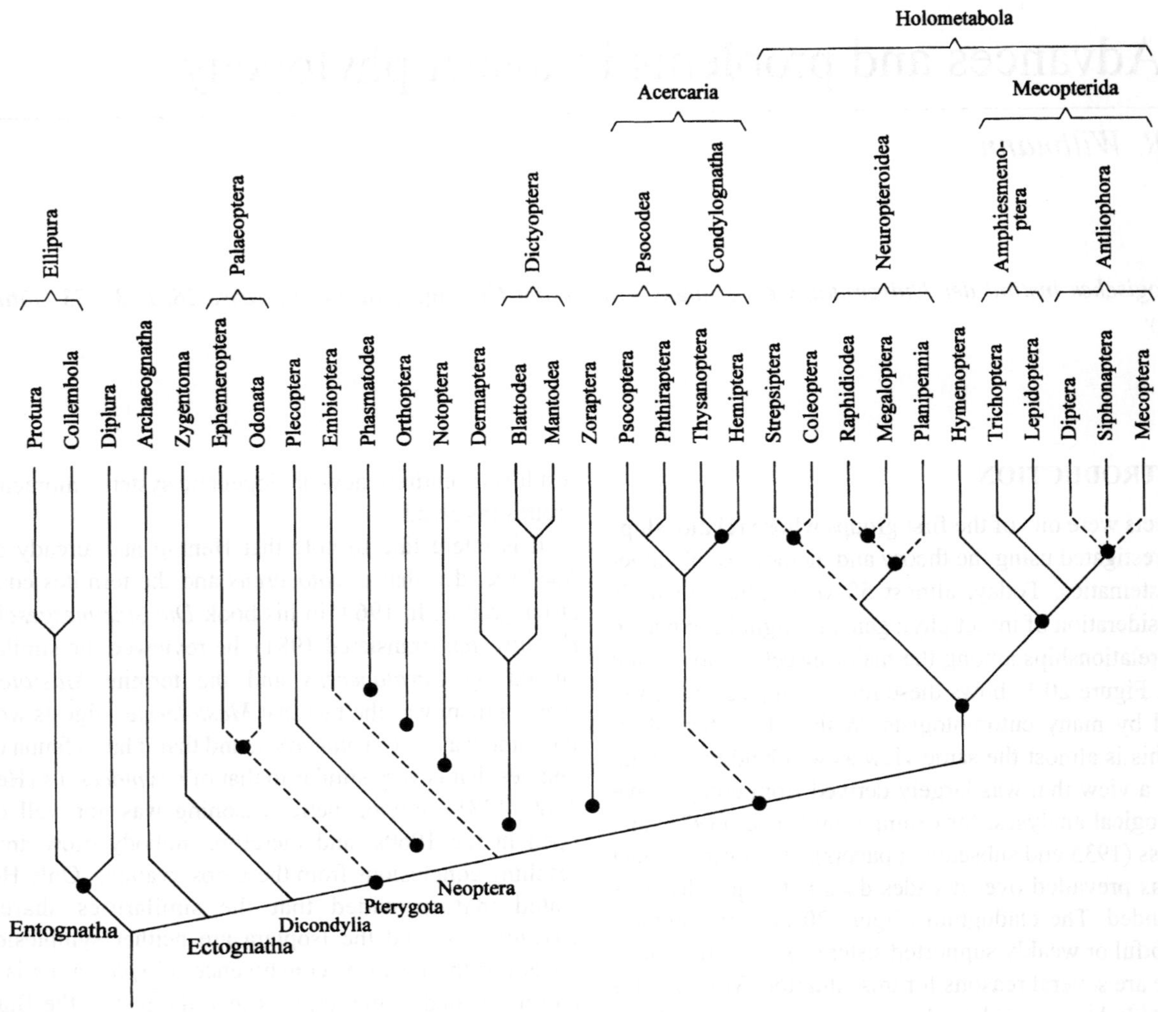


**Figure 20.1** Relationships among the major insect taxa. Black dots and broken lines indicate various degrees of uncertainty with respect to the position of the groups. Only extant taxa are considered. The Mantodea include the termites.

convergent loss of the same number of leg segments – three – in the Ellipura, Diplura, Archaeognatha, Zygentoma and at least once within the pterygotes, that is, five times. It appears more parsimonious to assume that the original number of podomeres in the hexapods was six and that this number has been retained by most taxa. In the Collembola as a derived group of the Ellipura, however, only five podomeres have remained, since tibia and tarsus have become fused (Figure 20.2(a)). Secondary subdivision of leg segments is not unusual. In the Odonata, an additional segment has been separated from the trochanter. The two trochanteral segments are not, however, moveable on each other (Snodgrass, 1935). In the Hymenoptera, the proximal part of the femur is separated from the main portion by a suture line. The Palaeozoic Diaphanoptera, a pterygote taxon, has, according to Kukalová-Peck, far more than six leg segments. If this is plesiomorphic, it requires the assumption of additional reductions to the number of six within the pterygotes.

In the other major tracheate groups, the Progoneata and Chilopoda, additional podomeres occur (Figure 20.2(e)), and therefore it is possible that early ancestors of the hexapods had more than six. However, the number had already been reduced in the last stem species of the hexapods.

## 20.3 COXAL STYLI

Kukalová-Peck (1987) described leg exites (= exopodites) from several fossil insect taxa, adding (Kukalová-Peck, 1987, p. 2333) that 'coxal exites are still preserved in some Recent machilids'. From various insects, Kukalová-Peck described exites from more than one podomere. She noted that thoracic leg exites were present 'in the arthropodan and insectan groundplan and occur in trilobites, chelicerates, crustaceans, arthropleurids, and myriapods'.

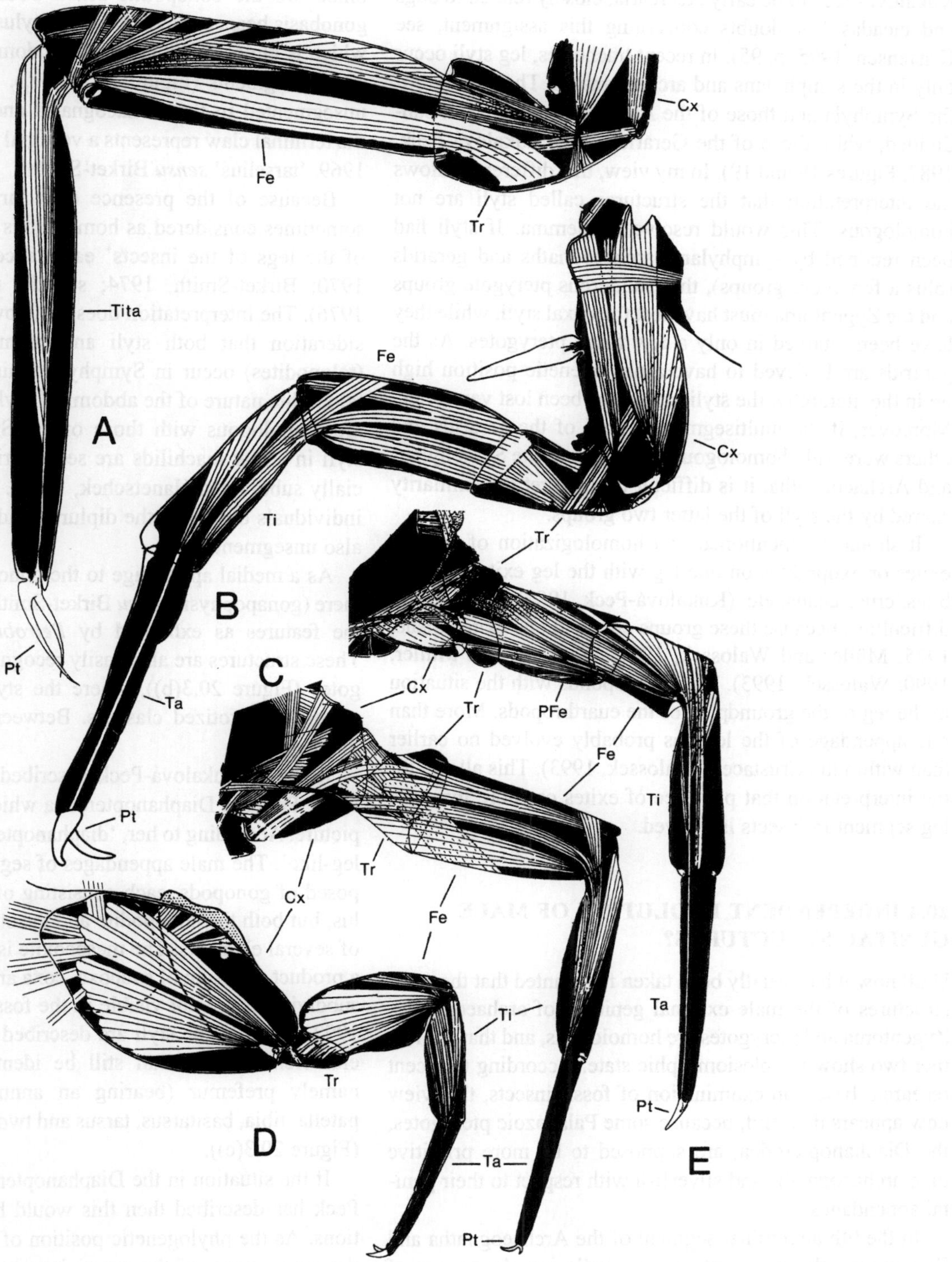


**Figure 20.2** Legs of (A) *Tomocerus* (Collembola), tibia and tarsus fused to form a tibiotarsus. (B) *Petrobius* (Archaeognatha); (C) *Campodea* (Diplura); (D) *Lepisma* (Zygentoma); and (E) for comparison, *Lithobius* (Chilopoda) with an additional podomere, the prefemur. Six podomeres as in (B–D) are here considered as plesiomorphic for the hexapods. Cx, Coxa; Fe, Femur; PFe, prefemur; Pt, pretarsus; Ta, tarsus; Ti, Tibia; Tita, tibiotarsus; Tr, trochanter. (After Manton, 1977.)

Among the taxa described as bearing leg exites are some Carboniferous gerarids. The gerarids are believed by Kukalová-Peck to be early Acercaria, closely related to bugs and cicadas (for doubts concerning this assignment, see Kristensen, 1995, p. 95). In recent tracheates, leg styli occur only in the symphylans and archaeognaths. The leg styli of the Symphyla and those of the Archaeognatha are not subdivided, while those of the Geraridae are (Kukalová-Peck, 1987, Figures 18 and 19). In my view, this difference allows the interpretation that the structures called styli are not homologous. This would resolve a dilemma. If styli had been retained by symphylans, archaeognaths and gerarids (plus a few more groups), then numerous pterygote groups and the Zygentoma must have lost the coxal styli, while they have been retained in only a few fossil pterygotes. As the gerarids are believed to have a phylogenetic position high up in the hierarchy, the styli must have been lost very often. Moreover, if the multisegmented styli of the gerarids and others were truly homologous with those of the Symphylans and Archaeognatha, it is difficult to interpret the similarity shared by the styli of the latter two groups.

It should be mentioned that homologization of **several** exites or exopodites on one leg with the leg exites in trilobites, crustaceans, etc. (Kukalová-Peck, 1987) faces several difficulties, because these groups only had one exite (Cisne, 1975; Müller and Walossek, 1987; Walossek and Müller, 1990; Walossek, 1993). This corresponds with the situation in the leg of the groundplan of the euarthropods. More than one appendage of the leg has probably evolved no earlier than within the Crustacea (Walossek, 1993). This allows for the interpretation that presence of exites on more than one leg segment in insects is derived.

## 20.4 INDEPENDENT EVOLUTION OF MALE GENITAL STRUCTURES?

Until now it has usually been taken for granted that the main structures of the male external genitalia of archaeognaths, Zygentoma and pterygotes are homologous, and that the former two show the plesiomorphic state. According to recent research, based on examination of fossil insects, this view now appears doubtful, because some Palaeozoic pterygotes, the Diaphanopterodea, are supposed to be more primitive than archaeognaths and silverfish with respect to their genital appendages.

In the 9th abdominal segment of the Archaeognatha and Zygentoma, the male external genitalia mainly consist of structures that are usually regarded as derivatives of abdominal legs, and therefore they are often termed gonopods. The homology of the various parts of the gonopods is unclear, and therefore they are named differentially. A proximal unpaired sclerite is sometimes called the 'sternite' ('basal sclerite' *sensu* Birket-Smith, 1974). Distally the gonobases follow (gb in Figure 20.3(a,b); interpreted as a divided sternum by Birket-Smith, 1974). Gonobasis and sternite combined are the coxopodite *sensu* Snodgrass (1935). Each gonobasis bears a movable (gono)stylus (Figure 20.3(a), gs) which is almost identical to the abdominal styli of the preceding segments (Figure 20.3(a), $s_8$). Abdominal styli are unsegmented in the archaeognaths and zygentoma, unless the terminal claw represents a vestigial 2nd segment (Smith, 1969, 'tarsellus' *sensu* Birket-Smith).

Because of the presence of a tarsellus, the styli are sometimes considered as homologous with the telopodites of the legs of the insects' early ancestors (Smith, 1969, 1970; Birket-Smith, 1974; see the review in Matsuda, 1976). The interpretation does not, however, take into consideration that both styli and segmented walking legs (telopodites) occur in Symphyla. This would exclude the telopodite nature of the abdominal styli of hexapods if they are homologous with those of the Symphyla. Annulated styli in a few machilids are secondarily and only superficially subdivided (Janetschek, 1957). They occur in a few individuals only, and the dipluran and symphylan styli are also unsegmented.

As a medial appendage to the gonobasis there is a paramere (gonapophysis *sensu* Birket-Smith). Figure 20.3 shows the features as exhibited by *Petrobius* (Archaeognatha). These structures are also easily recognizable in many pterygotes (Figure 20.3(b)), where the styli may develop into strongly sclerotized claspers. Between the gonobases lies the penis.

In 1992, Kukalová-Peck described male genitalia from lower Permian Diaphanopterodea which do not fit into this picture. According to her, 'diaphanopterid male genitalia are leg-like'. The male appendages of segment 9 are also composed of gonopods, each consisting of coxopodite and stylus, but both the coxopodite and the styli are said to consist of several elements. The mesocoxite is interpreted by her as a product of fusion of subcoxa, coxa and trochanter, but 'the subdivisions are not visible in the fossils' (Kukalová-Peck, 1992, p. 245). The styli are described as consisting of several elements that can still be identified as podomeres, namely prefemur (bearing an annulated exite), femur, patella, tibia, basitarsus, tarsus and two claws (the pretarsus) (Figure 20.3(c)).

If the situation in the Diaphanopterodea is as Kukalová-Peck has described then this would have serious implications. As the phylogenetic position of the archaeognaths as the sister group of the remaining ectognaths (Dicondylia) and the position of the Zygentoma as the sister group of the pterygotes are not a matter of debate (Kristensen, 1991, 1995), Kukalová-Peck's descriptions leave two possibilities: either the subdivision of the coxopodites in the Diaphanopterodea and possession of two claws are secondary (an interpretation which she obviously does not

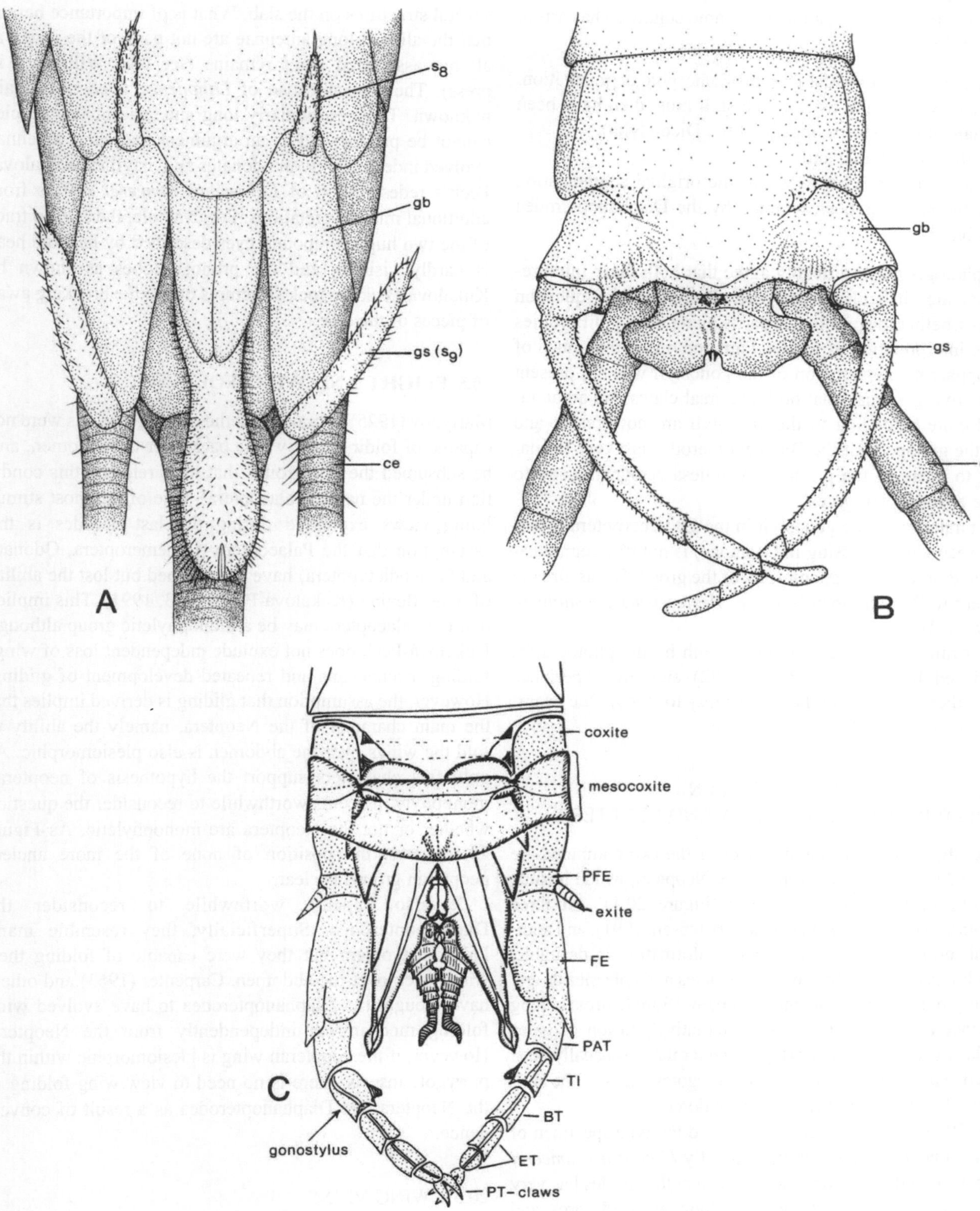


**Figure 20.3** Male genitalia of (A) *Petrobius brevistylis* (Archaeognatha), ventral view. Gonostyli flexed cephalad. (B) *Ecdyonurus* sp. (Ephemeroptera), ventral view, gonostyli secondarily annulated. (c) Reconstruction of the genitalia of the Permian *Uralia* (Diaphanopterodea; after Kukalová-Peck, 1992). FE, femur; PAT, patella; PFE, prefemur; TI, tibia; BT, basitarsus; ET, tarsal segments; PT, pretarsus; ce, cerci; gb, gonobsis; gs, gonostylus; $s_8$, styli.

share) or if the state in the Diaphanopterodea is plesiomorphic within the insects, almost the same condition has arisen at least three times:

1. In the Archaeognatha (the plesiomorphic segmentation, the claws and the exite of the styli must then have been retained in the stem lineage of the Dicondylia).
2. In the Zygentoma; and
3. Within the pterygotes, where the original segmentation of the styli has been retained by the Diaphanopterodea only.

Morphologists will probably agree that both these interpretations are difficult to imagine. Therefore, the question arises whether the problem is partly generated by difficulties in the interpretation of fossils. I am convinced that much of the apparent segmentation of the gonostyli was not present in the living animal, that both pretarsal claws and an annulated exite belonging to the gonostyli are not present and that the gonopods of the Diaphanopterodea are very similar both to those of the primarily wingless ectognaths and to those of most pterygotes.

Annulation of the gonostyli in the Ephemeroptera (probably secondarily missing in Caenidae) is not true segmentation and probably an apomorphy of the group. It was already present in Permian mayflies (e.g. *Protereisma permianum*, Tillyard, 1932).

I regard my view as supported both by the photographs published by Kukalová-Peck (1992) and my experience with other fossils I had the opportunity to study. One example is given in the following section.

### 20.4.1 RELATIONSHIPS BETWEEN EPHEMEROPTERA, ODONATA AND NEOPTERA

Either the mayflies or the odonates or the two combined are believed to be the sister group of the Neoptera, which include the rest of the pterygote insects (Figure 20.1). All three hypotheses have their supporters (Kristensen, 1991), and additional morphological or molecular information is desirable. Possible synapomorphies of the odonates and ephemerids are, among others, short antennae (taxon 'Subulicornia', long antennae are the plesiomorphic alternative), fusion of galea and lacinia and possibly a wing character that has usually been regarded as plesiomorphic for the pterygotes – lack of the ability to fold back the wings over the abdomen.

In 1985, Kukalová-Peck redescribed the type specimen of the syntonopterid Carboniferous mayfly *Lithoneura lameerei* Carpenter, and one of her results was that the species has very long antennae (being at least as long as head, pro- and mesothorax combined). An early mayfly possessing long antennae would certainly rule out the character 'short antennae' as a synapomorphy of mayflies and odonates.

Upon the request of the late Prof. Carpenter and together with him, I restudied the specimen and it became immediately clear that Kukalová-Peck (1985) had misinterpreted several structures on the slab. What is of importance here is that the alleged long antennae are not parts of the insect at all but associated plant remains (see also Willmann, in press). The true antennae of *Lithoneura lameerei* remain unknown. Thus, particularly long antennae in this species cannot be put forward as an argument that short antennae evolved independently in odonates and mayflies. Kukalová-Peck's redescription of *Lithoneura lameerei* suffers from additional misinterpretations. For example, there is no trace of the two huge compound eyes described by her (the head is hardly visible), and the pronotal lobes as shown by Kukalová-Peck are rock structures due to the breaking away of pieces of the slab.

## 20.5 FLIGHT AND WING FOLDING

Martynov (1925) assumed that the early pterygotes were not capable of folding their wings back over the abdomen, and he subsumed the few groups that have retained this condition under the name 'Palaeoptera'. One of the most stimulating views expressed during the last decades is the assumption that the Palaeoptera (Ephemeroptera, Odonata and Palaeodictyoptera) have not retained but lost the ability of wing-flexing (Kukalová-Peck, 1983, 1991). This implies that the Palaeoptera may be a monophyletic group although Kukalová-Peck does not exclude independent loss of wing-folding mechanisms and repeated development of gliding. However, the assumption that gliding is derived implies that the main character of the Neoptera, namely the ability to fold the wings over the abdomen is also plesiomorphic. As only few characters support the hypothesis of neopteran monophyly, it seems worthwhile to reconsider the question whether or not the Neoptera are monophyletic. As Figure 20.1 shows, the position of none of the more ancient neopteran groups is clear.

It also appears worthwhile to reconsider the Diaphanopterodea. Superficially, they resemble many Palaeodictyoptera but they were capable of folding their wings back over the abdomen. Carpenter (1963) and others have thought the Diaphanopterodea to have evolved wing folding mechanisms independently from the Neoptera. However, if the neopteran wing is plesiomorphic within the pterygote insects, there is no need to view wing folding in the Neoptera and Diaphanopterodea as a result of convergence.

### 20.5.1 WING VEINS

Another hypothesis can be seen in a light different from that currently in vogue. According to Kukalová-Peck (1983, 1991) numerous longitudinal wing veins and in particular an additional precosta, an additional costa (costa posterior) and an additional subcosta (subcosta anterior) are plesiomorphic within the pterygotes. She has also put forward the idea that the articulation of the wing consisted originally of 32 and not of only 11 or so sclerites. She has based her views particularly on the examination of exceptionally large species of Carboniferous homoiopterids, members of the Palaeodictyopterodea with a wing span of up to 40 cm.

It is perhaps probable that the homoiopterids were derived with respect to their exceptional size, and they may not, therefore, have a particularly primitive position within the pterygotes. If so, it also seems possible that their wings do not display plesiomorphic, but rather numerous derived character states. Their venation could be the result of a well-known correlation: increase in wing size often leads to an increase in the number of veins. If this is so, usage of some of the wing terminology that Kukalová-Peck has proposed appears premature. The most primitive pterygotes were probably of small size, say 2 cm in body length, or even less.

## 20.6 PLECOPTERA

Many authors consider the Plecoptera to be the sister group of the remaining Neoptera, which implies that the stone flies are a very old group (Figure 20.1). However, in 1992, Carpenter listed very few Palaeozoic Plecoptera, as did Sinitshenkova (1987). It is obviously difficult to recognize the earliest stone flies, and the reasons may be that the extant Plecoptera have but few apomorphic characters that may have been preserved in fossils, or that their autapomorphies have not yet been worked out.

I believe that the following characters of the Plecoptera are apomorphic:

1. The nymphs are aquatic.
2. The nymphs have segmental abdominal gills (Zwick, 1980).
3. The media has only two or three branches.
4. In the forewing, CuA has two or three branches (a few species that are probably derived may have four, e.g. *Brachyptera trifasciata*).
5. The tarsi are three-segmented (this character has been considered as plesiomorphous by Hennig, 1986).

There is a species-rich group of Palaeozoic and Mesozoic insects whose affinities are unclear: the 'Protorthoptera', an assemblage of taxa which may be only distantly related. Within the Protorthoptera there is a small group that deserves the attention of those interested in plecopteran evolution, the Palaeozoic Lemmatophoridae (Figure 20.4). In the lemmatophorids: (i) the nymphs were aquatic (they have been found in considerable numbers in Permian limnic sediments); (ii) on each abdominal segment, the nymphs had a pair of gills; (iii) CuA has more than one branch; and (iv) except for *Lecorium*, the number of branches in the media is reduced to three (*Lecorium* has four branches to M, but here, this is possibly not pleisiomorphic as the first branch of CuA

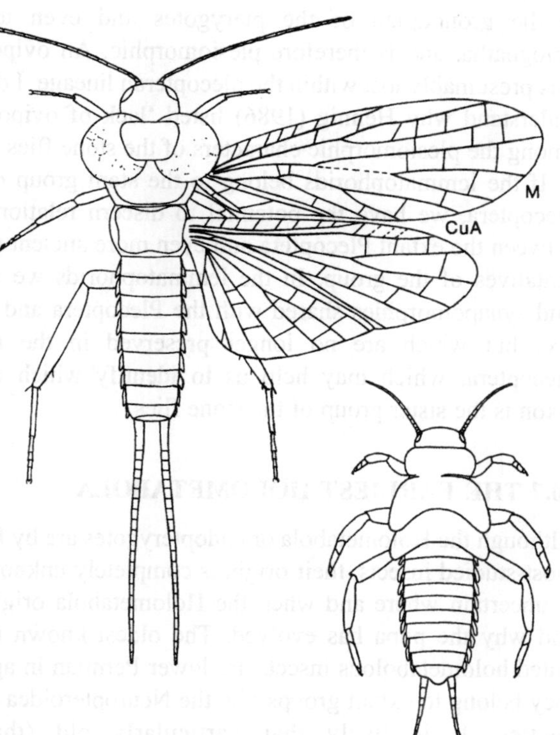


**Figure 20.4** Lemmatophoridae, *Lemmatophora*, nymph and imago. Lower Permian, Kansas. (After Carpenter, 1935, 1992.)

seems to have shifted to M. The primitive condition is probably exhibited by *Liomopterum* with four long branches to M, while CuA has retained two branches).

Tillyard – and also Carpenter in his earlier papers – considered these forms as the stem group of the Plecoptera, and Tillyard (1928) even named them 'Protoperlaria'. However, Carpenter later thought that the Protoperlaria were better placed within the Protorthoptera. At the same time, he did not mention further any phylogenetic relationships between the lemmatophorids and the Plecoptera. I assume that this is unjustified and that the similarities mentioned above are synapomorphies. There may be one more. In the Plecoptera, both adults and nymphs have three-segmented tarsi, while adult lemmatophorids have five (Figure 20.4). The nymphs of the lemmatophorids had, however, reduced the number of tarsal segments to four or three (Carpenter, 1935) (Figure 20.4). Possibly, the reduction in number began in the nymphs and was later transferred to the adults. If so, this means that the Lemmatophoridae are the stem group or the sister group of the recent stone flies and that Plecoptera were quite common during the Permian.

There is one remarkable difference between the Lemmatophoridae and the Plecoptera. The Lemmatophoridae had an ovipositor, while recent stone flies do not. However, this is no argument to exclude the lemmatophorids from the Plecoptera, as an ovipositor belongs to the groundplan of the pterygotes and even to the Ectognatha, and is therefore plesiomorphic. An ovipositor was presumably lost within the plecopteran lineage. I do not understand why Hennig (1986) listed 'lack of ovipositor' among the plesiomorphic characters of the stone flies.

If the lemmatophorids belong to the stem group of the Plecoptera, we have the potential to discern relationships between the extant Plecoptera and even more ancient representatives of the group. In the lemmatophorids we might find synapomorphies shared with the Plecoptera and other taxa but which are no longer preserved in the recent Plecoptera, which may help us to identify which extant taxon is the sister group of the stone flies.

## 20.7 THE EARLIEST HOLOMETABOLA

Although the Holometabola or endopterygotes are by far the most-studied insects, their origin is completely unknown. It is uncertain where and when the Holometabola originated and why the pupa has evolved. The oldest known undisputed holometabolous insects are lower Permian in age but they belong to extant groups like the Neuropteroidea or the beetles. It is likely that particularly old (that is, Carboniferous) Holometabola have already been discovered, but that we are unable to recognize them for what they are. We have no idea about the groundplan of the holometabolous insects, and the reason for this is lack of knowledge about the sister group relationships **within** the Holometabola, just as we cannot identify their sister group. Hence, ancient holometabolan remains are of outstanding interest, and a candidate example is discussed below.

### 20.7.1 A CARBONIFEROUS HOLOMETABOLAN LARVA?

A few years ago, a Carboniferous fossil was described as being a holometabolan larva (Kukalová-Peck, 1991); however, I consider it debatable whether the specimen (Figure 20.5) is an endopterygote, because many features of the specimen can hardly be expected from juvenile Holometabola. If it were insisted that the specimen was indeed an immature endopterygote, a whole list of well-founded views on insect evolution would no longer be tenable. For example, it would then be necessary to view both the evolution of the thorax and abdomen, and the loss of segmented abdominal walking legs, as a result of repeated convergence.

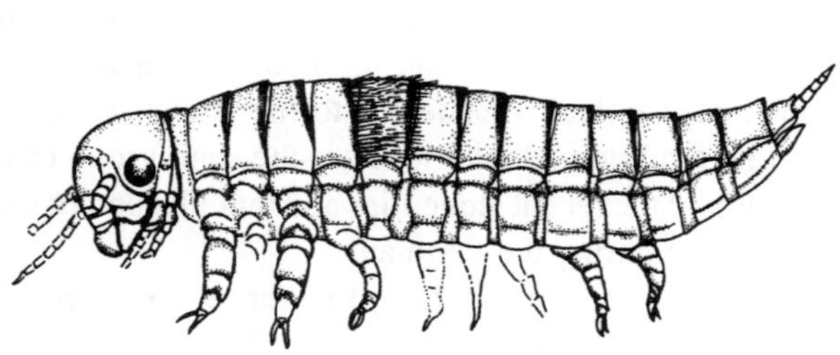

**Figure 20.5** Alleged Carboniferous endopterygote larva. Mazon Creek area, Illinois. (After Kukalová-Peck, 1991, Figure 6.26.)

In the following, I am going to discuss some of the features of the alleged Carboniferous larva from a more traditional standpoint, accepting that the insects are monophyletic, that the archaeognaths are the sister group of the Dicondylia and that the Zygentoma are the sister group of the pterygotes. Under this view it has always been assumed that abdominal appendages functioning as walking legs had already disappeared from the groundplan of the Neoptera and probably much earlier (Snodgrass, 1941). What was possibly left in immature stages of early endopterygote insects were abdominal vesicles and unsegmented styli similar to those of the Archaeognatha. Thus, all one can expect in holometabolous larvae are these structures or derivations of them, but not entire abdominal walking arthropodia bearing pairs of claws.

Snodgrass (1931), for example, assumed that retractile gill-bearing vesicles in larval Megaloptera and the abdominal legs of hymenopterous larvae and lepidopteran caterpillars are possibly homologous, adding (p. 82) that they are 'highly suggestive of the eversible sacs of the thysanuran appendages' (see also Smith 1969).

The specimen described as a Carboniferous holometabolan larva has, however, segmented abdominal walking legs, which terminate in two claws. This alone excludes the specimen from belonging to the endopterygotes.

Again, the alleged Carboniferous larva is described by Kukalová-Peck (1991) as having 'ocelli, [and] probably compound eyes' in holometabolous larvae ocelli are, however, reduced, and if this is not due to convergence ocelli were not present in the groundplan of larval endopterygotes. Furthermore, in immature Holometabola, compound eyes are replaced by a few single eyes, making the interpretation of Kukalová-Peck even more doubtful. Compound eyes as well as ocelli may, however, still have been present in early members of the stem group of the Holometabola.

With the exception of Coleoptera, immature holometabolans are devoid of annulated cerci. The homology of the cerci-like appendages of coleopteran larvae with true cerci is unclear (Snodgrass, 1931, 1935, Smith, 1969). Weidner (1982), however, considers them as being true cerci. The specimen as figured by Kukalová-Peck has annulated cerci, making its position within the endopterygotes unlikely. It must be admitted, however, that current knowledge does not rule out the presence of cerci in the earliest holometabolan larvae. Thus, the cerci only exclude the specimen from being a member of an extant holometabolan taxon with the exception of the Coleoptera.

Possibly, larvae of the groundplan representative of the Holometabola had appendages belonging to segment 11, a terminal pair of abdominal legs, the so-called 'pygopods' or 'postpedes'. According to Smith (1969) they are possibly derivatives of the vesicles of the respective segment. The structures are well-known from hymenopteran larvae and lepidopteran caterpillars, but they are present in most holometabolan groups. The alleged endopterygote Carboniferous larva has, however, no such structures.

If segmented abdominal walking legs had not already excluded the specimen from the Holometabola, possession of ocelli, compound eyes and cerci would point to a position within the stem group of the endopterygotes. For such stem group representatives one should expect many other features that are plesiomorphic relative to the remaining holometabolans. One might expect, for example, a clear distinction between thorax and abdomen. Such a distinction has been lost in several derived larvae, like those of some beetles or the larvae of the Diptera and fleas. A clear distinction between thorax and abdomen is, however, not visible in the figure given by Kukalová-Peck. One would also not expect a thorax in wingless arthropods carrying legs in almost all segments, as the insect thorax was developed to function as the tagma that is solely responsible for locomotion. Thus it seems possible that lack of a clear distinction between thorax and abdomen excludes the fossil from being a hexapod.

To summarize, the evidence that the 'oldest known endopterygote larva' (Kukalová-Peck, 1991) is, indeed, an endopterygote is equivocal; it can even be doubted that it belongs to the pterygote insects. Any view on insect evolution which puts this fossil in a central position should be viewed with caution.

## 20.7.2 EARLY HOLOMETABOLAN WINGS

For reconstructing endopterygote phylogeny, it would be most helpful to know whether the Holometabola originally had a rich venation, similar to those of many Neuroptera, or wings with only few veins, similar to those of the groundplan of the Antliophora and Mecoptera (Willmann, 1989). Many fossil insects are preserved as wings only, and as wings are complicated structures, they can contribute much to our understanding of insect phylogeny. In 1990, Kukalová-Peck and I described several insect species from the lowermost Permian of the Czech Republic (Kukalová-Peck and Willmann, 1990). Superficially, their wings look mecopteroid-like, which means that they resemble the wings of Mecoptera or Trichoptera (Figure 20.6).

There was an important difference, however. It seemed that the wings had at least weak indications of one or two short veinlets near the wing base and that these veins were remnants of what Kukalová-Peck had called the costa posterior and the subcosta anterior (Figure 20.6(a)). These veins originally had been described from non-holometabolous insects, and they are unknown from any other fossil or extant holometabolous insect. Thus, the specimens from the Czech Republic could be plesiomorphic Holometabola, still possessing the posterior costa and anterior subcosta, which were lost by the rest of the Endopterygotes. Another interpretation is that they are non-holometabolous insects with endopterygote-like wings.

Unfortunately, weak veins are often poorly-preserved in the material we have studied, and although our figures (Kukalová-Peck and Willmann, 1990, Figures 1–11) show that the respective veins are the posterior costa and the anterior subcosta, we could not reach a definite conclusion whether or not this is the case. The alleged veins may also be slight folds due to wing deformation that occurred during or after sedimentation.

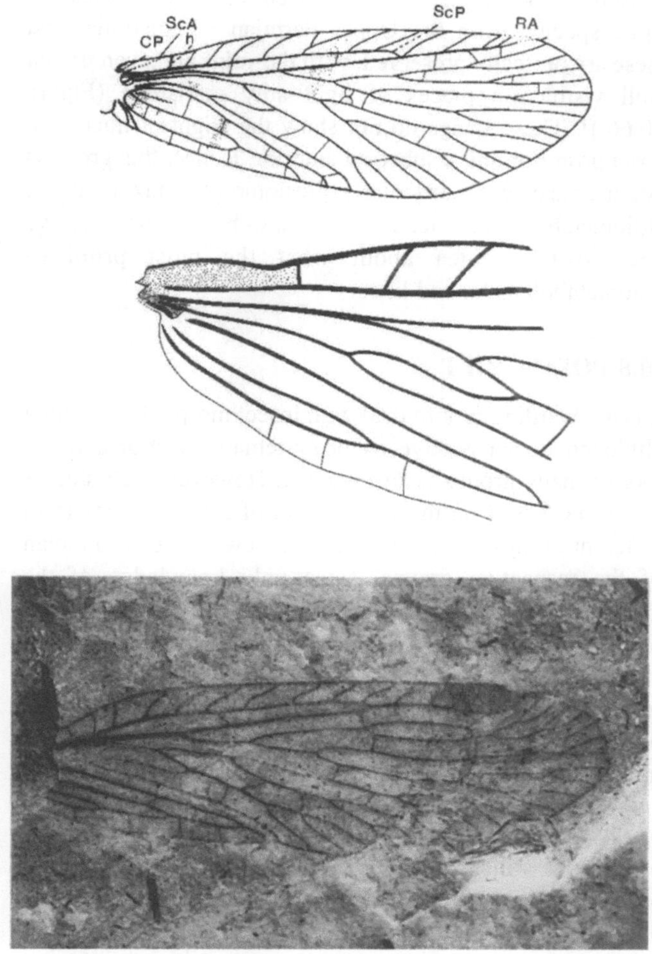


(a)

(b)

(c)

**Figure 20.6** Permian endopterygote wings. (a) *Pseudomerope mareki*, lower Permian of Obora, Bohemia. (After Kukalová-Peck and Willmann, 1990.) (b,c) *Pseudomerope* sp. from the Upper Permian of Belmont-Warner's Bay, Queensland, Australia. (b) wing base; (c) wing showing the excellent state of preservation of another wing. The Australian material shows that there is not the slightest trace of a costa posterior and subcosta anterior and the veins can now be considered as also absent from the Bohemian species. CP, costa posterior; ScA, subcosta anterior; h, humeral vein; RA, radius.

Subsequently, I have studied a collection of almost the same species from the Upper Permian of Australia, and these are perfectly preserved, with the veins and even trichia well visible on pieces of light silty marlstone (Figure 20.6(c)). These wings do not show the slightest trace of a costa posterior and a subcosta anterior. Hence, this group is not necessarily a particularly plesiomorphic taxon of the Holometabola. So, once more it has to be admitted that we have still no idea about what the most primitive holometabolans looked like.

## 20.8 POSTSCRIPT

It is a rewarding task to deal with insect morphology from a phylogenetic perspective, as the systematic position and history of many groups is still unclear. However, such studies cannot be based on the comparison of a few species from different groups. A prerequisite is a view on the groundplan of the respective groups, since only knowledge of the groundplan of a taxon allows for objective comparison with other taxa. Knowledge of the groundplan also requires an awareness of the sister group relationships within the respective group. Hence thorough morphological studies are bound to be time-consuming. In spite of this, they are of immense value: we not only need them as independent data against which trees provided from molecular evidence can be tested, but also they provide biologists with the identity of the taxa which will provide the most fruitful reference base for further studies. This in turn feeds back to phylogenetic analyses based on molecular data. Morphological studies provide us with a scenario for the evolution of body structures, and hence with an image of insect evolution in terms of adaptive character change.

There are severe obstacles, however. Many insect groups are very old, and they may have lost many of the similarities they originally shared with other taxa because of restructuring in their anatomy over many millions of years. In this case, fossils may be of importance, because they may possess autapomorphies of the recent members of a clade while still retaining synapomorphies with their closest relatives. The hundreds of localities with fossil insects now known have provided palaeoentomology with a huge amount of perfectly preserved material which awaits further research.

## REFERENCES


Birket-Smith, S.J. (1974) On the abdominal morphology of Thysanura (Archaeognatha and Thysanura s. str.). *Entomologica Scandinavica, Supplement*, **6**, 1–67.

Carpenter, F.M. (1935) The Lower Permian insects of Kansas. Part 7. The order Protoperlaria. *Proceedings of the American Academy of Arts and Sciences*, **70**, 103–46.

Carpenter, F.M. (1963) Studies of Carboniferous insects from Commentry, France. Part V. The genus *Diaphanoptera* and the order Diaphanopterodea. *Psyche*, **70**, 240–56.

Carpenter, F.M. (1992) Superclass Hexapoda, in *Treatise on Invertebrate Paleontology, Part R*, Arthropoda 4, Volumes 3 and 4. Boulder, Colorado and Lawrence, Kansas.

Cisne, J. (1975) Anatomy of *Triarthrus* and the relationships of the Trilobita. *Palaeontographica Americana*, **9**, 99–142.

Hennig, W. (1953) Kritische Bemerkungen zum phylogenetischen System der Insekten. *Beiträge zur Entomologie*, **3**, Sonderheft, 1–85.

Hennig, W. (1969) *Die Stammesgeschichte der Insekten*, Verlag W. Kramer, Frankfurt.

Hennig, W. (1981) *Insect Phylogeny*, J. Wiley and Sons, New York.

Hennig, W. (1986) *Wirbellose II. Gliedertiere*, Harri Deutsch Verlag, Thun, Frankfurt.

Janetschek, H. (1957) Über die mögliche phylogenetische Reversion eines Merkmals bei Felsenspringern mit einigen Bemerkungen über die Natur der Styli der Thysanuren (Ins.). *Broteria*, **26**, 1–22.

Klass, K.-D. (1995) *Die Phylogenie der Dictyoptera*, Cuvillier Verlag, Göttingen.

Kristensen, N. (1991) Phylogeny of extant hexapods, in *The Insects of Australia*, I. Melbourne University Press, Melbourne, pp. 125–40.

Kristensen, N. (1995) Forty years insect phylogenetic systematics. Hennig's 'Kritische Bemerkungen...' and subsequent developments. *Zoologische Beiträge NF*, **36**, 83–124.

Kukalová-Peck, J. (1983) Origin of the insect wing and wing articulation from the arthropodan leg. *Canadian Journal of Zoology*, **61**, 1618–69.

Kukalová-Peck, J. (1985) Ephemeroid wing venation based upon new gigantic Carboniferous mayflies and basic morphology, phylogeny, and metamorphosis of pterygote insects (Insecta, Ephemerida). *Canadian Journal of Zoology*, **63**, 933–55.

Kukalová-Peck, J. (1987) New Carboniferous Diplura, Monura, and Thysanura, the hexapod groundplan, and the role of thoracic lobe in the origin of wings (Insecta). *Canadian Journal of Zoology*, **65**, 2327–45.

Kukalová-Peck, J. (1991) Fossil history and the evolution of hexapod structures, in *The Insects of Australia*, I, Canberra, pp. 141–79.

Kukalová-Peck, J. (1992) The 'Uniramia' do not exist: the groundplan of the Pterygota as revealed by Permian Diaphanopterodea from Russia (Insecta: Paleodictyopteroidea). *Canadian Journal of Zoology*, **70**, 236–55.

Kukalová-Peck, J. and Willmann, R. (1990) Lower Permian 'mecopteroid-like' insects from central Europe (Insecta, Endopterygota). *Canadian Journal of Earth Sciences*, **27**, 459–68.

Manton, S.M. (1977) *The Arthropoda. Habits, Functional Morphology and Evolution*, Clarendon Press, Oxford.

Martynov, A. (1925) Über zwei Grundtypen der Flügel bei Insecten und ihre Evolution. *Zeitschrift für Morphologie und Ökologie der Tiere*, **4**, 465–501.

Matsuda, R. (1976) *Morphology and Evolution of the Insect Abdomen*, Pergamon Press, Oxford.

Müller, K.J. and Walossek, D. (1987) Morphology, ontogeny, and life habit of *Agnostus pisiformis* from the Upper Cambrian of Sweden. *Fossils and Strata*, **19**, 1–124.

Sinitshenkova, N.D. (1987) Istoricheskoe Razvitie Vesnyanok. *Trudy Paleontologicheskii Instituta*, **221**, 127–36.

Smith, E.L. (1969) Evolutionary morphology of external insect genitalia. 1. Origin and relationships to other appendages. *Annals of the Entomological Society of America*, **62**, 1051–79.

Smith, E.L. (1970) Biology and structure of some California bristletail and silverfish. *Pan-Pacific Entomologist*, **46**, 212–25.

Snodgrass, R. (1931) Morphology of the insect abdomen. Part I. General structure of the abdomen and its appendages. *Smithsonian Miscellaneous Collections*, **85**(6), 1–127.

Snodgrass, R. (1935) *Principles of Insect Morphology*, McGraw-Hill, New York.

Snodgrass, R. (1941) The male genitalia of Hymenoptera. *Smithsonian Miscellaneous Collections*, **99**(14), 1–86.

Tillyard, R.J. (1928) Kansas Permian insects. Part 10. The new order Protoperlaria. *American Journal of Science*, **216**, 185–220.

Tillyard, R.J. (1932) Kansas Permian insects. Part 15. The order Plectoptera. *American Journal of Science*, **223**, 97–134, 237–72.

Walossek, D. (1993) The upper Cambrian *Rehbachiella* and the phylogeny of Branchiopoda and Crustacea. *Fossils and Strata*, **32**, 1–202.

Walossek, D. and Müller, K.J. (1990) Upper Cambrian stem-lineage crustaceans and their bearing upon the monophyly of Crustacea and the position of *Agnostus*. *Lethaia*, **23**, 409–27.

Weber, H. (1933) *Lehrbuch der Entomologie*, G. Fischer Verlag, Jena.

Weidner, H. (1982) Morphologie, Anatomie und Histologie. *Handbuch der Zoologie* IV: Arthropoda (2) Insecta 1/11. Berlin.

Willmann, R. (1989) Evolution und phylogenetisches System der Mecoptera (Insecta: Holometabola). *Abhandlungen der Senckenbergischen Naturforschenden Gesellschaft*, **544**, 1–153.

Willmann, R. Phylogeny and the consequences of phylogenetic systematics, in *Proceedings of the 8th International Conference on Ephemeroptera and the 12th International Conference on Plecoptera*, Lausanne, 1995 (eds P. Landolt and M. Sartori) (in press).

Zwick, P. (1980) Plecoptera (Steinfliegen). *Handbuch der Zoologie*, IV, Insecta 26, Berlin.

# 21 The groundplan and basal diversification of the hexapods

*N.P. Kristensen*

*Zoological Museum, University of Copenhagen, Universitetsparken 15, DK-2100, Copenhagen Ø, Denmark*
*email: npkristens@zmuc.ku.dk*

## 21.1 INTRODUCTION

It is well known that at the present time the Hexapoda comprise the majority of the arthropods, and indeed of all living organisms. The really species-rich hexapod lineages, however, are all cladistically very subordinate. Therefore, while an elucidation of the groundplan and basal diversification of the Hexapoda would be of great interest in its own right, the nested sets of apomorphies thereby disclosed would not suffice to explain 'why there are so many different kinds of insects'. The full answer to that question will require explanation of the selective advantages of apomorphies originating in many later splitting events in insect evolution.

As here delimited the extant Hexapoda comprise the primarily wingless, hence soil-bound, taxa Collembola (>6000 described species), Protura (ca. 500 species), Diplura (ca. 800 species), Archaeognatha (>300 species), Zygentoma (= Thysanura *s.str.* >300 species) as well as the winged insects, the Pterygota (>900 000 described species). There has long been a consensus about the strongly founded monophyly of an entity comprising the three last mentioned taxa, and in compliance with one frequent practice (which I have repeatedly supported myself; e.g. Kristensen, 1975, 1981, 1989, 1991, 1995) the name Insecta is here reserved for that entity; 'Ectognatha' is a strict synonym. The extinct Monura (non-monophyletic?) undoubtedly belong to the Insecta lineage.

While a close relationship of the Diplura to the Insecta has appeared uncontroversial, such relationships of the phenetically overall more distinctive Collembola and Protura have been questioned from time to time, in the case of the Protura even quite recently (Dallai, 1991). However, there is reasonable evidence supporting the monophyly of an entity Ellipura (= Parainsecta) comprising these two disparate taxa (see below and Kraus, 1997, this volume) and, importantly, there seems to be no evidence of close relationship between one or both of them with taxa outside the hexapod assemblage. Currently, the question of the basal split within the Hexapoda is identical with the question about the status and affinities of the Diplura. Diplurans, like the Ellipura, have an 'entognathan' head structure, and both will here be talked of collectively as 'entognathans', but it must be emphasized that while the monophyly of the entognathan assemblage remains a real possibility, this monophyly has been repeatedly challenged. There are recent suggestions that the sister group of the Insecta is the Diplura only (Kukalová-Peck, 1991), or even just a subgroup of the latter (Štys *et al.*, 1993) – the monophyly of the Diplura themselves having so far never been strongly supported (Figure 21.1).

There has long been near-consensus that the closest affinities of the Hexapoda are with the myriapod assemblage, though it has remained unsettled whether the two were sister groups, or the latter was paraphyletic in terms of the former; the myriapods and hexapods together are variably known as the Atelocerata (preferred by me), Tracheata or Antennata. Recently, however, the old idea of the hexapods being the sister group of, or subordinate within, the crustacean assemblage has been revived on the basis of both structural (see Kraus, 1997, this volume; Dohle, 1997, this volume) and molecular (Friedrich and Tautz, 1995) characters.

## 21.2 THE HEXAPOD GROUNDPLAN

A comprehensive reconstruction of the groundplan of all body regions and major organ systems of the Hexapoda is beyond the scope of the present contribution. Particular attention will be paid here to characters which in some states have recently been ascribed significance as groundplan autapomorphies of the Hexapoda as a whole or of their basal subordinate lineages.

The uncertainty about the closest relatives and primary internal splits of the Hexapoda clearly impedes the feasibility of ascertaining the groundplan state of some of the characters which are variable among the basal hexapod clades. States may straightforwardly be ascribed to the hexapod groundplan if they occur in basal (apterygote) Insecta as well as within the Ellipura. States with a more

*Arthropod Relationships*, Systematics Association Special Volume Series 55, Edited by R.A. Fortey and R.H. Thomas. Published in 1997 by Chapman & Hall, London. ISBN 0 412 75420 7

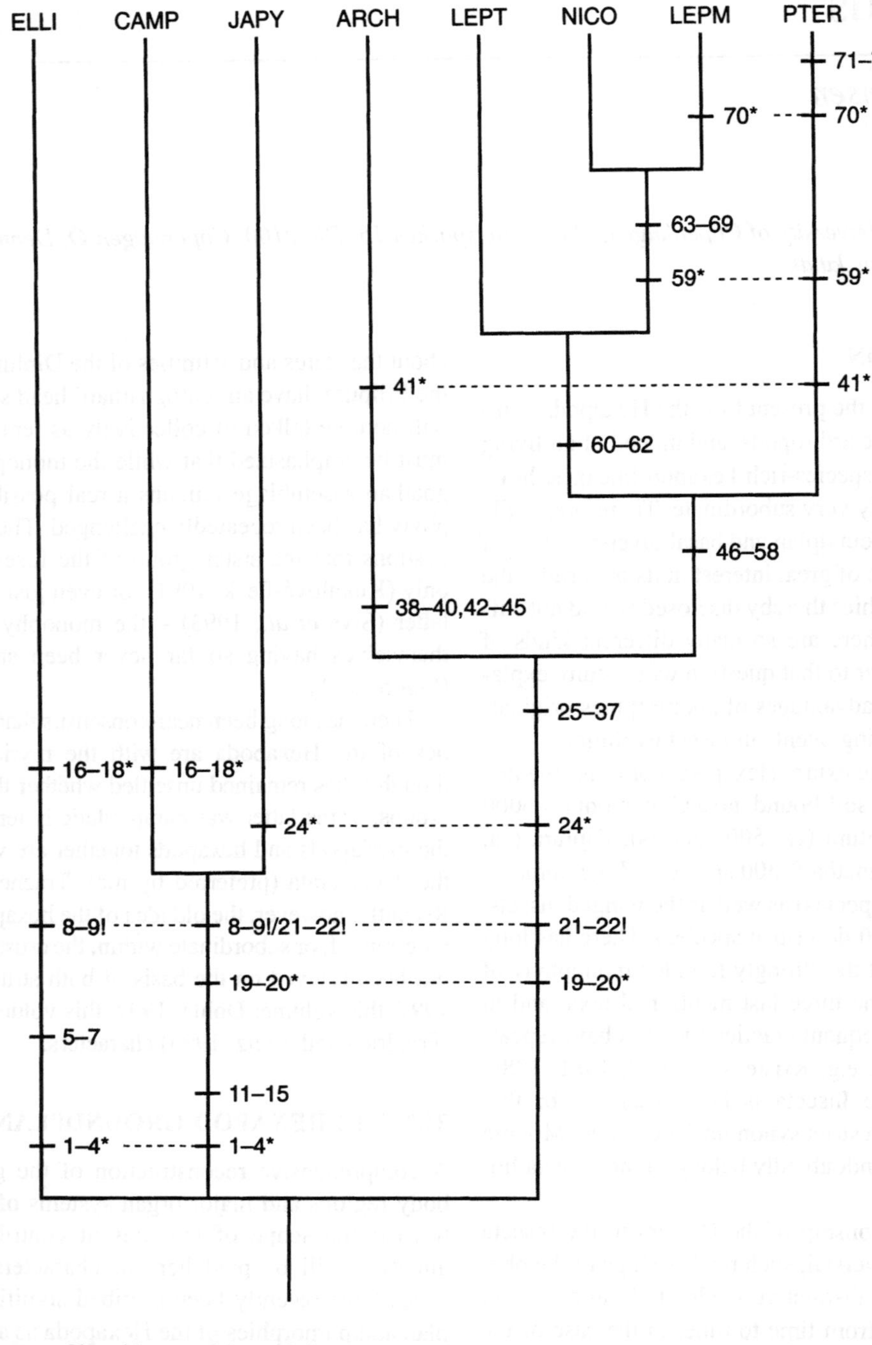


**Figure 21.1** Cladogram of basal hexapod (extant) lineages. For character numbers see text. Apomorphies marked with * are homoplasious under the preferred phylogeny, or (1–4) the monophylum they support is here left unindicated. Characters marked with ! are used with alternative interpretations of state polarity. ELLI, Ellipura (= Collembola + Protura); CAMP, Diplura–Campodeoidea; JAPY, Diplura–Japygoidea; ARCH, Archaeognatha; LEPT, Zygentoma–Lepidothrichidae; NICOL, Zygentoma–Nicoletiidae (*s.lat.*); LEPM, Zygentoma–Lepismatidae; PTER, Pterygota.

restricted occurrence among basal hexapods may be ascribed to the hexapod groundplan if present also in not-too-subordinate myriapod (or crustacean?) taxa; inevitably decisions on such matters will often appear debatable. A somewhat detailed, and largely innovative reconstruction of the groundplan of hexapod integumental structure has been worked out by Kukalová-Peck (1991, 1994, 1997, this volume). Bitsch (1994) has advocated more cautious interpretations of many traits, and I find myself in overall agreement with his views.

### 21.2.1 THE HEAD

More than 20 years ago Rempel (1975) characterized the dispute over the composition of the hexapod head as 'endless', a statement not contradicted by subsequent developments. However, it does seem unquestioned that all hexapod heads are derivable from a single type. The cuticular coverings of the head segments have fused into a solid head capsule with few external traces of segmental boundaries. There is consensus about the identity of a 'gnathocephalon' component of the head, comprising the mandibular, maxillary and labial segments; the debate is concerned with the composition of the pre-mandibular 'procephalon', as well as with the exact location of the cephalothoracic boundary.

The anteriormost head region comprises the clypeolabrum immediately above the mouth. A transverse 'epistomal suture' demarcating the clypeolabrum from the frons does not seem to pertain to the hexapod groundplan; it is indicated in Snodgrass' textbook (1935, 1952) drawings of archaeognathan heads, but is actually absent in these insects (Denis and Bitsch, 1963) as well as in entognathan hexapods.

In the hexapod groundplan the cephalic endoskeleton was largely a 'ligamentous' (non-cuticular) formation, as it still is in the entognathan taxa (review by François and Chaudonneret, 1982). However, the anterior component of the true tentorium has also been ascribed to this groundplan, and Boudreaux (1979) considered it a groundplan autapomorphy of the Atelocerata. There has been much confusion about the interpretation of so-called 'tentoria' in atelocerates, and for instance none of the formations in entognathan hexapods called tentorial in Manton's writings (1964, 1977) are homologous with the tentorium in Insecta. Their cuticular components are the hypopharyngeal 'fulturae' (Francois and Chaudonneret, 1982), and I believe this will prove to be the case also with at least some cuticular 'tentoria' in myriapods. The Protura do have cuticular rods, which are contiguous with the anterior cranial wall and in that respect comparable with the anterior tentorial arms of the Insecta; however, external pits clearly demonstrating the origin of the anterior tentorial arms as invaginations from the outer cuticle are restricted to the latter. Further comparative morphological and developmental work is obviously needed to assess the homology of the 'anterior tentorial arms' in myriapods, Protura and Insecta.

All visual organs (Paulus, 1979) are usually ascribed to the preantennary segment ('prosocephalon', the first genuine segment). The hexapod groundplan possessed, in addition to the compound eyes, three pairs of ocelli: median, lateral and dorsal. Only the Collembola have retained the dorsal pair and a distinctly paired (with two separate nerves) configuration of the median pair; however, even in some pterygotes the paired origin of the latter may remain detectable because of the paired nerve roots. Extant 'Diplura' and Protura lack visual organs altogether, and ocelli are lost in most Zygentoma and many subordinate pterygote taxa. *The ommatidia of the compound eyes had two corneageneous cells transformed into 'primary pigment cells* (note that hexapod groundplan autapomorphies are italicized); according to Paulus' (1979) comparisons with conditions in other mandibulate arthropods this trait is a groundplan autapomorphy of the Hexapoda. Kukalová-Peck believes all cephalic segments had visual organs in ancestral hexapods: Ocelli on segments 1 (her 'labral segment') and 2 are represented by the unpaired and paired ocelli respectively, the compound eyes belong to the fused segments 2+3 'but have shifted to segment 4 (the mandibular segment, NPK) in Insecta'; small eyes reported from the maxillary and labial segments in Palaeozoic Monura have no counterparts in extant hexapods. I consider the palaeontological evidence for this theory less than convincing, and I have also not found it supported by the innervation pattern of the visual organs in extant hexapods.

The antennary (deuterocerebral) segment had well-developed segmental appendages, the antennae, movable by muscles between adjacent segments as in most other Mandibulata.

The postantennary (tritocerebral) segment was devoid of (at least externally identifiable) segmental appendages. This regressive trait is one of the classical putative hexapod–myriapod synapomorphies, as reflected in the name Atelocerata for the composite entity.

The hexapod mandible is usually a single-piece appendage, but Kraus and Kraus (1996) interpret a weakness zone on some archaeognathan mandibles as a primitive podomere joint; in any case the morphological composition of the mandible remains debatable (Denis and Bitsch, 1973). Like the myriapod mandible it is devoid of a palp (telopodite) and had, in the groundplan, a single articulation with the head capsule: a mandibular condylus fitting a cranial ginglymus. A passively movable 'prostheca' or 'lacinia mobilis' on the inner mandibular surface has a very scattered occurrence within the Hexapoda (some 'Diplura', Ephemeroptera, some higher pterygotes) and probably cannot be attributed to their groundplan.

The maxilla comprises a basal piece (cardo) followed by the stipes which bears two endite lobes (galea, lacinia) and a multi-segmented palp whose serial homology with the telopodites of the locomotory limbs is generally accepted. It seems straightforward to consider the very long and somewhat leg-like maxillary palp present in Archaeognatha (and the extinct Monura) to be the closest approximation to the hexapod groundplan configuration seen in extant taxa. However, like Bitsch (1994) I believe the question of the primitive segment number in the palp still remains open, and I am hesitant to accept Kukalová-Peck's assertion that apical claws on the palp can be ascribed to the groundplan of the hexapod crown group. As in the non-chilopod myriapods (according to Kraus, 1997, this volume) the appendages of the following segment are proximally fused, forming an unpaired lower lip, the labium. The extrinsic musculature of the gnathal appendages originated partly on the cranium, partly (ventral adductors) on the ligamentous endoskeleton.

Exocrine glands (putative segmental organ derivatives) opening on or near the gnathal appendage bases are generalized arthropod traits. Glands of this kind can presumably be ascribed to all three gnathal segments in the hexapod groundplan. However, while those of the mandibular segment are widespread and those of the labial segment almost universally present, those pertaining to the maxillary segment are retained only in the Ellipura among the basalmost hexapods; their reported occurrence in scattered higher pterygotes (Richards and Davies, 1977) may represent cases of autapomorphic character reversals.

The medioventral walls of some of the postoral segments constitute the tongue-shaped hypopharyngeal lingua which in ancestral hexapods presumably bore laterodorsal lobes, the superlinguae. Kukalová-Peck interprets the superlinguae to be vestiges of the postantennary segmental limbs. However the embryonic development and innervation of these formations (Denis and Bitsch, 1973; Chaudonneret, 1987) indicate that they pertain to a more posterior segment, viz. the mandibular segment, or perhaps a highly reduced 'tetrocerebral' segment located between the postantennary and mandibular segments. I am unaware of the basis for Kukalová-Peck's statement (1991, p. 147) that the superlinguae 'have intrinsic endite muscles'. The lingual wall was reinforced by sclerotized braces, the suspensorium, the ground pattern of which within the Hexapoda remains to be worked out. One of these braces, located at the maxillary–labial segmental boundary, was continued into a prominent proximally directed rod, the fulcrum (= fultura *auct.*).

Homologues of the myriapod 'Tömösváry's organ' were probably retained in the hexapod groundplan, and are represented by the 'postantennal organs' in the Collembola and the 'pseudoculus' in Protura.

The cephalic aorta (Pass, 1991, 1996) was equipped with a pair of lateral branches which extended around the oesophagus to anastomose into a longitudinal trunk below the latter; a similar configuration occurs in centipedes. Antennary vessels were present, branching off from the aorta itself (Pass, 1991).

### 21.2.2 THE THORAX

The most immediately conspicuous groundplan autapomorphy of the Hexapoda is the trunk tagmatization pattern: the *thorax comprising the three postcephalic segments bearing locomotory appendages*, and the abdomen comprising the remaining segments which are devoid of such appendages.

The debate over the ancestral structure of the hexapod thoracic segments is certainly no less 'endless' than that over the head composition. A comprehensive literature review was given by Matsuda in 1970 (the 1979 version is more succinct and not much updated). Subsequent work by Manton (1972, 1977) insisted on extensive non-homology of body-wall components across the hexapods/atelocerates, while Kukalová-Peck (1983, 1987, 1991, 1994, 1997, this volume), derived a sizeable proportion of the segmental body wall (including the pterygote wing) from limb-base components. Particularly important are the detailed accounts of thoracic morphology in basal hexapods presented in numerous studies by Carpentier and Barlet from the mid-1940s onward; a succinct review (Barlet, 1988) addresses selected issues concerning the apterygote grade. A believable reconstruction of the hexapod thoracic groundplan will certainly require a detailed compilation of the information presented in the literature, together with descriptive work on additional taxa and comparisons with relevant myriapod/crustacean outgroups.

The soft-walled cervical region behind the head capsule most probably pertains mainly to the prothorax, but may also contain elements pertaining to the labial segment; the evidence from different criteria for segmental assignment (muscle attachments, innervation, morphogenesis) is contradictory (Bitsch, 1969; Bitsch and Ramond, 1970). Discrete laterocervical sclerites are not a hexapod groundplan feature.

The thoracic terga and sterna of ancestral hexapods probably bore distinctly delimited pretergites and presternites respectively (Barlet, 1988).

The locomotory limb originates laterad from or above the sternal plate and is separated from the tergum by a complement of pleural sclerotizations. It is commonplace to interpret the latter as being (at least largely) derived from one or two basal limb segments which have been incorporated into the trunk wall following desclerotization of their medial or ventral components (and/or fusion of the latter with the sternal sclerotization). Following the works of Carpentier and Barlet (also Matsuda, 1970) the ancestral hexapod had two arched pleural sclerotizations: the lower catapleuron and the upper anapleuron. The two together correspond to what Kukalová-Peck believes represents the lower ('subcoxa') of two basal limb segments incorporated above the free appendage; the upper segment, the 'epicoxa', is allegedly represented by a paratergal area on which the pterygote wing is borne (see below).

Deuve (1994) proposed that the 'subcoxa' *sensu* Kukalová-Peck comprises two discrete components, but the most dorsal of these, the 'epipleurite' apparently corresponds to a much smaller area than the anapleurite of Carpentier/Barlet (whose work is unmentioned).

*The 'free' limb comprises only six universally recognized, tubular segments: coxa, trochanter, femur, tibia, tarsus, pretarsus* (= post-tarsus *auct.*, = transtarsus *auct.*). Kukalová-Peck recognizes another three segments in the free limb, viz. a 'prefemur', a (pretibial) 'patella' and a 'basitarsus'; however, like Bitsch (1994) I am hesitant to accept that these leg sections really correspond to genuine limb segments in the hexapod groundplan. It also remains uncertain whether paired pretarsal claws pertain to the hexapod, or even atelocerate groundplan. Are the so-called paired claws in Chilopoda–Notostigmophora and Symphyla (to which Hennig, 1969, 1981, referred) really homologous with those present in Diplura and Insecta? Boudreaux (1979) even believed these to be non-homologous, and the apparently unpaired claws in the Palaeozoic Monura [as described by Sharov, see section 21.3.2(c)] could support this view. If dipluran and insect claws pertain to the same transformation series, it seems natural to assume that the condition in the former (Figure 21.2(h)), with the claws articulating on a full-sized pretarsal annulus (see below) is the plesiomorphic one. A single (flexor) muscle inserts on

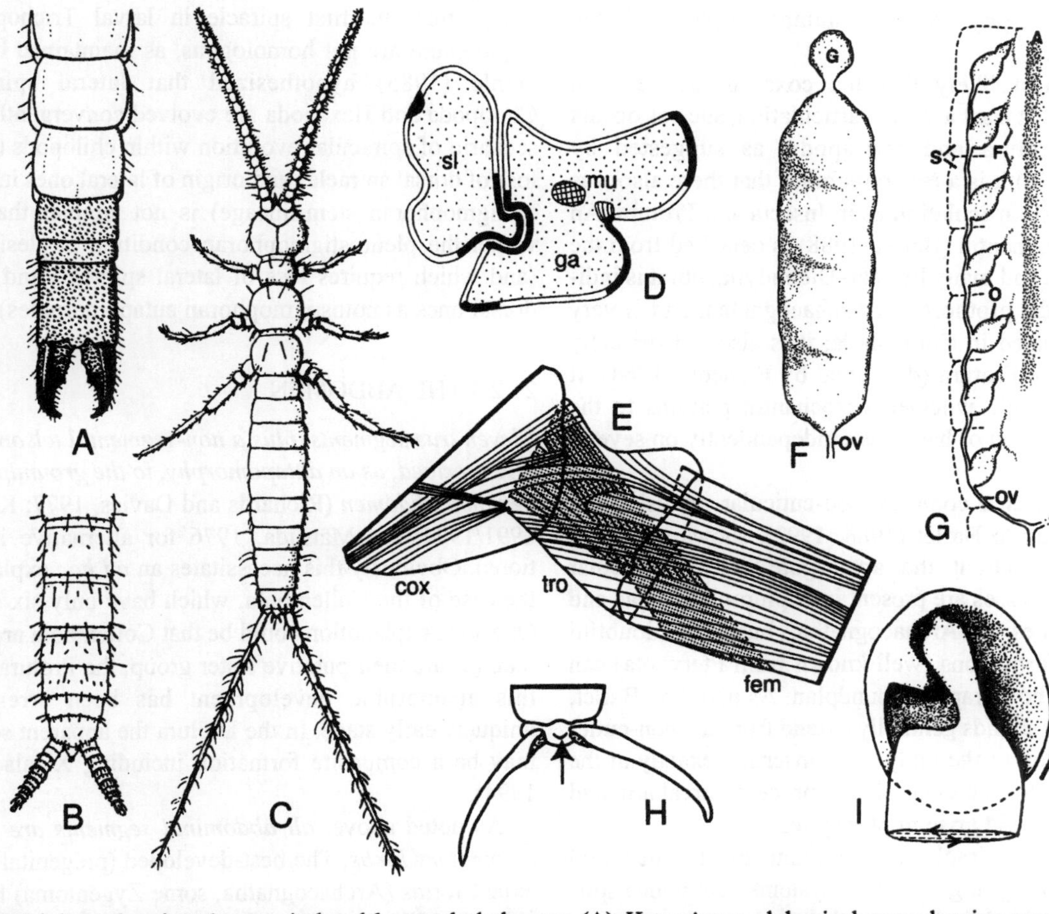


**Figure 21.2** Diplura, the pivotal group in basal hexapod phylogeny. (A) *Heterojapyx*, abdominal apex showing secondarily unsegmented, forcipate cerci. (B) *Anajapyx*, abdominal apex showing short, stout cerci. (C) *Campodea*, habitus. (D) *Campodea*, transverse section of interlocking superlingua (sl) and maxillary galea (ga); mu, galea muscles. (E) *Campodea*, thoracic leg base, showing specialized trochanteral femur-twisting muscle (shaded); cox, coxa; fem, femur; tro, trochanter. (F) *Campodea*, sac-like ovary. (G) *Catajapyx*, ovary with 'metameric' ovariole arrangement; A, gut; F, fat body; o, ovariole; ov, oviduct; s, outline of abdominal segments. (H) Campodeid, apex of leg, showing complete pretarsus (arrow) on which claws articulate. (I) *Heterojapyx*, femurotibial pivot joint, end view, posterior is right; right hand arrow shows direction of rotation of femur caused by contraction of muscle shaded in (E). [Source: (A–C) after Richards and Davies (1977); (D) redrawn from François (1970); (E) redrawn from Manton (1972); (F,G) after Bilinski and Szklarzewicz (1992); (H) redrawn from Boudreaux (1979); (I) after Manton (1972).]

the pretarsus; its antagonist must be haemolymph pressure. The absence of an antagonistic (levator) pretarsal muscle is one of the potential synapomorphies of myriapods and hexapods. It is straightforward to assume that antagonistic muscles are a plesiomorphic feature of arthropod podomeres, and such muscles do occur in Crustacea and Arachnida.

A crucial feature of Kukalová-Peck's groundplan limb model is the presence of exites arising just beyond the distal border of each of the basal limb segments; the pterygote wing is believed to represent the epicoxal exite. The model is based upon morphology derived from Palaeozoic fossils, and its veracity depends upon these having been correctly interpreted. The well-known 'stylus' on the coxa of Archaeognatha is considered a homologous exite, but it is a difficulty, then, that this formation arises at mid-length on the segment in question, not from its apex. I consider it an open question whether this stylus is a unique (at the hexapod level) plesiomorphy or an autapomorphy of the Archaeognatha.

It seems most likely that the coxa in the hexapod groundplan had at least a sternal articulation, such as occurs in Diplura, Protura and myriapods; as suggested by Kristensen (1975) it is a real possibility that there also was a dorsal (pleural) articulation as in Insecta and Protura. An anterolateral trochantin (Barlet, 1988) is detached from the coxa in Diplura and many Insecta–Dicondylia, but it is indicated only on the prothorax in Archaeognatha, and is very weakly demarcated in Collembola; it is also not typically developed in the Protura (described by François, 1964). It is thus questionable whether a trochantin pertains to the hexapod groundplan or has arisen independently on several occasions.

The thorax has a complex non-cuticular endoskeleton which, according to Barlet (1988, 1992), shows important similarities throughout the apterygote grade. Cuticular spinasternal processes are present in Diplura–Japygidae and pleural apodemes in Archaeognatha, but it is doubtful whether such formations (well known from Pterygota) can be ascribed to the hexapod groundplan. As noted by Barlet, however, tissue strands generally extend from the non-cuticular endoskeleton to the body wall, inserting exactly at the positions where the cuticular pleural process, furcal arm and spinal process would arise in pterygotes.

The thoracic spiracle complement in the hexapod groundplan is intriguing. Tracheal systems, and hence spiracles, have evolved independently on several occasions among terrestrial arthropods, and within the Hexapoda the cervical spiracle in Collembola–Symphypleona is certainly a formation *sui generis*. The maximum number of thoracic spiracles in hexapods is found in Diplura–Japygidae, which have four of them. Thoracic spiracles have been classified, according to their relations to adjacent sclerites, into three categories (within which there is presumably serial homology): 'postsegmentary', 'suprapleural' (homologized with those of Chilopoda–Pleurostigmophora) and 'intersternal/intersegmentary'; a fourth, 'presegmentary' category is reported from Zygentoma (Barlet and Carpentier, 1962; Barlet, 1988). Barlet and Carpentier suggest that absence of representatives of one or more of the four spiracular categories in hexapods is likely to be due to secondary reduction. It is a corollary of the theory that the first thoracic spiracle, the segmental assignment of which has been so controversial, is not necessarily homologous in different hexapod taxa; for instance, it is claimed that it is 'postsegmentary' in Diplura–Japygidae and 'intersegmentary' in Diplura–Campodeidae. The suggested presence of more than one spiracle series in the hexapod groundplan is an interesting theory which certainly deserves more attention, though I believe that the criteria for distinguishing between some of the series need further scrutiny, not least in the higher insects. I am, for instance, extremely reluctant to accept that the first spiracle in larval Trichoptera and Lepidoptera are not homologous, as maintained by Barlet. Dohle (1985) hypothesized that lateral spiracles in Chilopoda and Hexapoda are evolved convergently, but his scenario of spiracular evolution within chilopods (requiring loss of dorsal spiracles and origin of lateral ones in the pleurostigmophoran stem-lineage) is not simpler than one in which the pleurostigmophoran condition is plesiomorphic (and which requires loss of lateral spiracles and origin of dorsal ones as notostigmophoran autapomorphies).

### 21.2.3 THE ABDOMEN

*Eleven true segments, plus a non-segmental telson, are usually ascribed, as an autapomorphy, to the groundplan of the hexapod abdomen* (Richards and Davies, 1977; Kristensen, 1991/1994; see Matsuda, 1976 for alternative interpretations). Obviously this necessitates an *ad hoc* explanation in the case of the Collembola, which have only six segments. One such explanation could be that Collembola are anamorphic (as are their putative sister group, the Protura) and that this anamorphic development has been arrested at a uniquely early stage. In the Diplura the apparent segment X may be a composite formation including XI also (Bitsch, 1994).

As noted above, *all abdominal segments are devoid of locomotory limbs*. The best-developed (pregenital) limbs in extant forms (Archaeognatha, some Zygentoma) have sizeable plate-like bases ('coxites') which form part of the ventral segmental plate, but may be distinctly demarcated from a median area, which perhaps corresponds to the primitive sternum; there are indications that the coxite includes more than one limb base segment (Bitsch, 1979; Kukalová-Peck, 1987). The coxite bears an unsegmented telopodite 'stylus'. However, such extensive limb reduction must have taken place independently on more occasions if the fossil evidence is correctly interpreted: there are records of Palaeozoic apterygotes, and even pterygotes, with multiarticulated, sometimes claw-bearing, abdominal telopodite 'leglets' (Kukalová-Peck, 1983, 1987, 1991, 1994). Segment X is apparently consistently devoid of limbs in the Hexapoda.

'Eversible vesicles' (membranous, water-absorbing protuberances) on the limb base mediad from the telopodite or stylus are one of the potential synapomorphies of hexapods and 'dignathan/labiophoran' myriapods (but note that similar structures have evolved also in Onychophora and some Arachnida). In the hexapod groundplan they were probably present on abdominal segments I–VII, but only the Archaeognatha retains the full complement. Some members of this group even have two vesicles on each coxite in some segments; Kukalová-Peck (1987) considers this to be the plesiomorphic condition, reflecting the incorporation of two limb-base segments in the 'coxite'.

Cerci, elongate multiarticulated appendages of segment XI, are likely groundplan attributes of the Diplura and Insecta, though they are apparently completely absent in some Monura [as described by Sharov, see section 21.3.2(c)]; the shorter, thicker cerci found in non-campodeid Diplura and numerous higher pterygotes are generally considered secondary. According to Kukalová-Peck (1985, 1987, 1991, 1994) an apterygote Carboniferous insect [referred to the Monura, but see section 21.3.2(c) below] has retained 'normal', i.e., leg-like, even claw-bearing, limbs on XI. However, the alleged claw apart, the cercus in question (Kukalová-Peck, 1985, Figures 24 and 26) does not seem particularly aberrant; it is certainly much shorter than the terminal filament, but so is the cercus in many extant Archaeognatha (juveniles in particular). I am hesitant to accept the fossil evidence against considering clawless elongate cerci a groundplan autapomorphy of the least inclusive monophylum comprising the Diplura and Insecta. A secondary character reversal is a possible alternative explanation of the 'monuran' claw-bearing cercus, if the latter has really been correctly interpreted.

The non-cuticular abdominal endoskeleton in apterygotegrade hexapods has been less extensively studied than that of the thorax.

The largest complement of abdominal spiracles in hexapods is one on each of segments I–VIII, and this may be the groundplan configuration. Spiracle VIII occurs in Insecta only.

The gonopores were probably located on X in males and behind VII in females (Bitsch, 1979). It is uncertain which kinds of body-wall modifications (if any) were present adjacent to the gonopores in ancestral hexapods.

While it seems straightforward to assume that metameric arrangements of gonads represent the hexapod groundplan condition, this is not necessarily correct. Thus, outgroup comparisons lend support to the notion that metameric ovarioles in **adult** (not embryonic) hexapods is apomorphic (paedomorphosis) (Štys *et al.*, 1993). In the sperm cells the axoneme configuration was likely of the '9+9+2' type.

### 21.2.4 ALIMENTARY AND RESPIRATORY SYSTEM

Apterygote hexapods suggest that the alimentary canal in the hexapod groundplan probably had an overall simple stomodaeum without cuticular teeth, a midgut with a few anterior caeca, and six Malpighian tubules. The tracheal system was without longitudinal connectives or transverse commissures between branches arising from individual spiracles.

### 21.2.5 EMBRYOGENESIS

The total egg cleavage in Collembola was long considered to be very aberrant for hexapods. However, a brief total cleavage phase preceding blastoderm formation is now known to be widespread among Archaeognatha (Machida *et al.*, 1990) and it may well prove to represent the hexapod groundplan mode.

## 21.3 THE BASAL PHYLOGENETIC HIERARCHY WITHIN THE HEXAPODA

The following inquiry into the evidence for the hypothesized basal hexapod clades is largely based on characters discussed in more detail in other publications (most recently Kristensen, 1991, 1994 and 1995) where additional information and references may be found. The character numbers (in bold) refer to the cladogram in Figure 21.1.

### 21.3.1 THE ENTOGNATHAN ASSEMBLAGE

The problems with elucidating the basal splits within the Hexapoda hinge primarily on the status and affinities of the 'Diplura', i.e., the assemblage comprising the 'campodeine' and 'japygine' lineages of primarily apterous, entognathan hexapods.

#### (a) Evidence for the monophyly of Entognatha *s.lat.*

The time-honoured principal argument for recognizing a monophylum Entognatha comprising the Ellipura (= Collembola + Protura) **plus** the Diplura is **(1)** Entognathy itself: the overgrowth of the mouthparts (at least the mandibles and maxillae) by integumental 'oral folds', so that they come to be located inside deep pockets; the maxillary and labial palps become greatly shortened; as for *Testajapyx*, see below. Other characters supporting entognathan monophyly are regressions, which obviously carry modest weight only. **(2)** Reduction of compound eyes (in extant taxa represented only by the dispersed ommatidia present in Collembola; as for *Testajapyx*, see below). **(3)** Reduction of Malpighian tubules. **(4)** Perhaps loss of centriole adjunct in sperm.

#### (b) Evidence for the monophyly of Ellipura (= Parainsecta)

While the Collembola and Protura both are very distinctive lineages, the notion that they constitute together a monophylum is supported by **(5)** Entognathy more advanced than in Diplura, oral folds almost or entirely meeting in posterior/ventral midline. **(6)** Presence of an integumental specialization, the 'linea ventralis' in this posterior/ventral mid-line. **(7)** Enlarged 'epipharyngeal ganglia' (aggregations of sensory neurones). **(8)** Perhaps the unpaired pretarsal claws. If the Diplura are indeed members of a monophyletic Entognatha, then paired claws are most parsimoniously ascribed to the groundplan of this entity, as to the Hexapoda as a whole. If they are not, the unpaired claws in Collembola and Protura could as well be plesiomorphic. **(9)** Perhaps the loss of cerci. This is another potential ellipuran autapomorphy which hinges on the affinities of the Diplura in much the same way as the unpaired claws.

### (c) Evidence for the monophyly of Diplura

While the 'order' Diplura is recognized in all entomology texts, little evidence for the monophyly of this entity was available. A classical, but trivial, regressive apomorphy of extant diplurans is **(10)** complete loss of all visual organs; as for *Testajapyx*, see below. Another loss character concerns **(11)** the anterior tentorial arms, but as noted earlier it is not established beyond doubt that this formation was indeed a hexapod groundplan character. More impressive, 'positive' synapomorphies of campodeids and japygoids later identified include **(12)** a unique position of the gonopore which in **both** sexes is located between VIII and IX (Boudreaux, 1979 said 'on the venter of the ninth segment'), in repose more or less hidden by the hind border of sternum VIII (Pagés, 1989), **(13)** a unique trochanteral femur-twisting muscle (Figure 21.2(e)) and **(14)** a unique femur-tibia pivot joint (Figure 21.2(i)), to which can now be added **(15)** a peculiar interlocking of galea and superlingua (Figure 21.2(d)). The allegedly distinctive dipluran cercal base (E.L. Smith, in Kukalová-Peck, 1987) remains undocumented.

Some **caveats** are in order here. While much is known about the familiar Campodeidae (Figure 21.2(c)) and the Japygoidea (with the characteristic 'earwig-type' cerci; Figure 21.2(a)) none of characters 10, 12–14 has been investigated in the small families Anajapygidae (postabdomen in Figure 21.2(b)), Projapygidae and Procampodeidae, whose structure is very incompletely known. It is commonplace to group the non-japygoid Diplura into a taxon Rhabdura (or 'Campodeina'), but while the monophyly of a taxon Campodeoidea (Procampodeidae + Campodeidae) seems a reasonable working hypothesis (Bareth *et al.*, 1989), anajapygid and projapygid affinities remain very questionable. Moreover, the evidence for the monophyly of the Diplura must be weighted against indications of a closer relationship between one of their subgroups and another hexapod taxon. Thus, the Campodeoidea share some potential synapomorphies with the Ellipura: **(16)** Absence of abdominal spiracles. **(17)** Non-metameric **embryonic** ovaries – in adults this condition is now considered plesiomorphic at the hexapod level (Štys *et al.*, 1993). **(18)** Ovariole with linear meroism (Štys and Bilinski, 1990). Similarities with Insecta are listed below (characters 23–24).

### (d) Difficulties with entognathan monophyly

Arguments against the monophyly of the Entognatha come in two forms. The definitive head configuration in the Diplura and Ellipura is so different that it has been questioned whether the entognathy types in the two groups pertain to the same transformation series (Kraus, 1997, this volume). Another line of argument is furnished by potential synapomorphies of 'Diplura', or a subgroup thereof, and the Insecta; such arguments have led Kukalová-Peck to include the Diplura in the Insecta. Characters of this kind include: **(19)** Loss of Tömösváry's organ homologue; **(20)** loss of antennary ligamentous endoskeleton ('pseudotentorium'), judging from the account of François and Chaudonneret (1982); **(21)** perhaps the paired claws, cf. (8) above; and **(22)** perhaps the cerci, cf. (9) above.

Kukalová-Peck's principal argument for uniting the Diplura with the Insecta is that both have the abdominal limb bases **inclusive of the trochanter** incorporated in the trunk wall; in Ellipura the trochanter allegedly remains non-incorporated. I have (Kristensen, 1995) questioned the feasibility of making decisions on this issue.

Each of the two principal dipluran lineages shows similarities with true insects not shown by the other. Thus, the Campodeidae share with the Archaeognatha **(23)** a peculiar sperm modification, viz. a 4+5 grouping of the peripheral singlets in the axoneme; this condition does not occur in Zygentoma or basal pterygotes, so it is homoplasious in any case. The Japygidae share with the Insecta **(24)** the probably apomorphic (paedomorphic!) 'metameric' arrangement of ovarioles in the adult (cf. 17 above).

Yet another problem with entognathan monophyly is presented by the Carboniferous *Testajapyx thomasi* Kukalová-Peck, 1987. This large (ca. 5 cm) apterygote hexapod is recorded as having well-developed compound eyes and largely exposed mouthparts with rather long, multisegmented maxillary and labial palps; abdominal segments I–VI bore rather long, annulated limbs. The cerci, however, are forcipate and strongly reminiscent of those of Japygoidea. If *Testajapyx* really belongs to the japygoid lineage, and if its structure is correctly interpreted, then the principal support for the Entognatha collapses, and extensive homoplasy must be accepted. Bitsch (1994) assumes it is the *Testajapyx*/japygoid cercal similarity which is homoplasious and this solution, of course, remains a real possibility.

In conclusion, any phylogenetic arrangement of the entognathans will have a lot of homoplasy to account for. I do believe that the present support for a monophyletic Diplura (comprising at least the Campodeoidea + Japygoidea) suffices for it to be considered a reasonable working hypothesis. On the other hand, while the monophyly of a taxon Entognatha certainly should not be ruled out, I believe that an unresolved basal trichotomy in the hexapod cladogram (as in Figure 21.1 here) remains the preferable arrangement at present.

21.3.2 THE INSECTA *S. STR*

Several potential, and partly strong, groundplan autapomorphies support the monophyly of the Insecta as here circumscribed: (**25**) no intrinsic antennal muscles beyond scapus; (**26**) antennal pedicellus with Johnston's organ; (**27**) antennal flagellum multiarticulated; (**28**) antennal vessels separate from aorta; (**29**) posterior tentorium present; forming transverse bar; (**30**) no ventral coxal articulation; (**31**) tarsi subsegmented; (**32**) pretarsal annulus reduced, paired claws articulating on distal tarsomere (Boudreaux, 1979); (**33**) only two thoracic spiracles; (**34**) female abdomen with long ovipositor formed by gonapophyses on VIII and IX; (**35**) prominent 'terminal filament' borne on segment XI; (**36**) perhaps sperm axoneme with 1–3 'accessory bodies ' developed from centriole adjunct; (**37**) amniotic cavity formed during embryogenesis.

**(a) The basal hierarchy within the Insecta**

There is good evidence that, as far as extant insects are concerned, the basal dichotomy lies between the Archaeognatha (the 'machiloid' component of the old 'Thysanura') and all other Insecta, known collectively as the Dicondylia. The status and affinities of a Palaeozoic taxon Monura remains problematic.

**(b) Evidence for the monophyly of Archaeognatha (= Microcoryphia)**

While archaeognathans are overall generalized insects their monophyly is adequately supported by the following suite of apomorphies: (**38**) compound eyes enlarged, contiguous in mid-line; (**39**) perhaps hypertrophy of maxillary palps. While, as noted above, it seems straightforward to assume that leg-like palps are plesiomorphic (cf. 49 below), those of the archaeognathans are even larger than the thoracic locomotory limbs; (**40**) labial endites bilobed; (**41**) very strong development of pleural apodeme in meso- and metathorax (paralleled in Pterygota); anapleurite largely desclerotized in older instars (Barlet, 1988); (**42**) trochantin undeveloped in meso- and metathorax (Barlet, 1988); (**43**) no delimitation between thoracic basisternum and furcasternum. Such delimitations are present in Diplura –'Y'apodeme – and Zygentoma (Barlet, 1988); (**44**) spiracle I lost; (**45**) specialized trunk endoskeleton/musculature, related to jumping mechanism.

**(c) A note on the Monura**

Sharov's description (1966) of the Palaeozoic apterygote 'order' Monura has been considerably emended by Kukalová-Peck (1985, 1987, 1991, 1994), drawing on subsequently collected material. The insects described by Sharov had very large maxillary palps which together with the strong terminal filament conferred upon them a strong phenetic resemblance to Archaeognatha. Their mandibles were narrow (in lateral view) and apparently monocondylous. The compound eyes were well developed, though not extending to the mid-line. The tarsi were not subsegmented, the pretarsal claws apparently unpaired and the abdominal 'styli' at least two-segmented. A most unexpected feature was the apparently complete absence of cerci. In Kukalová-Peck's textbook account (1991, 1994) the Monura are ascribed dicondylous mandibles, double claws, 'fully segmented' abdominal limbs and, as noted above, leg-like and claw-bearing cerci. It is not clear to which extent the emendations are due to re-study of Sharov's material, or to the fact that Monura as circumscribed by Kukalová-Peck comprises very disparate taxa. Insects as different as the reconstructions shown in Sharov's Figure 33 and Kukalová-Peck's Figure 6.9 would definitely not appear to be each other's closest relatives. Decisions on monuran affinities, and on their significance for interpretations of structure and relationships of extant taxa should await clarification of these issues.

**(d) Evidence for the monophyly of Dicondylia**

The Dicondylia, then, comprise the winged insects plus the Zygentoma (the 'lepismatoid' component of the old Thysanura; they are sometimes still referred to as Thysanura, or Thysanura *s.str.*). Groundplan autapomorphies of the Dicondylia include: (**46**) enlarged mandibular base with secondary (anterior) cranial articulation; this articulation (a mandibular concavity and a small cranial condylus) is initially rather loose; (**47**) postoccipital suture complete dorsally; (**48**) anterior tentorial arms united posteriorly; (**49**) maxillary palps distinctly smaller than thoracic legs; (**50**) hypopharyngeal fultura obliterated; (**51**) loss of postcephalic spina (this formation is present in a suite of machilids and japygids; Barlet, 1988); (**52**) perhaps tarsi with five subsegments; (**53**) eversible vesicles lost on II; (**54**) ovipositor base with 'gonangulum' sclerite fully developed, articulating with tergum IX and attached to '1st valvula/valvifer' (Rousset, 1970); (**55**) abdomen with tracheal commissures and connectives; (**56**) perhaps sperm with outer singlets regularly distributed (cf. 23 above), and mitochondrial derivative with 'crystallomitin'; (**57**) amniopore at least temporarily closed; (**58**) superficial egg cleavage.

**(e) What is the sister group of the winged insects?**

There is now general agreement on the monophyly of the winged insects, the Pterygota. In contrast, the monophyly of the Zygentoma (comprising the families Lepidotrichidae, Lepismatidae and Nicoletiidae) has so far not been strongly supported, and their possible paraphyly in terms of the Pterygota has repeatedly been pointed out (Kristensen, 1981, 1991, 1994, 1995). The principal basis for this suggestion is the statement (which has remained undocumented) by Boudreaux (1979) that the family Lepidothrichidae (with a single, Nearctic, species *Tricholepidion gertschi*) has retained parts of the ligamentous cephalic endoskeleton, while all other Dicondylia (**59**) have only the cuticular tentorium; *Tricholepidion* therefore might be the sister group of all other Dicondylia.

I have examined serial sections of the *Tricholepidion* head, and I can confirm that a 'lamellar tendon' or 'ligament' pertaining to zygomatic mandibular and maxillary muscles is indeed retained, immediately below the true anterior tentorium (Figure 21.3(b)). On the other hand, my examination of the female postabdomen of this insect has shown that its ovipositor base does conform with the typical dicondylian configuration (cf. 54), with an elongate–triangular gonangulum articulating with tergum XI and its anterior corner reaching the base of the 1st valvula; Wygodzinsky's illustration (1961, Figure 47) showing the *Tricholepidion* gonangulum as a rectangular plate far separated from the 1st valvula is therefore misleading. I have also confirmed that the previously undescribed hypopharynx of *Tricholepidion* is indeed devoid of fulturae, as in Dicondylia generally (cf. 50 above).

In my evaluation, the retention in *Tricholepidion* of the 'lamellar tendon' cannot outweigh the evidence for the inclusion of this taxon in a monophyletic Zygentoma, which can now be supported by the following synapomorphies of its constituent taxa: (**60**) markedly widened apical segment of labial palp. This unusual feature is generally very distinctive throughout the Nicoletiidae and Lepismatidae, and it is prominent in *Tricholepidion* (Figure 21.3(a)); (**61**) obliteration of superlinguae. Superlinguae are still well developed in Ephemeroptera (mayflies - in the nymphs, which contrary to the adults have functional mouthparts). They must, therefore, be ascribed to the dicondylian groundplan, and their absence in Zygentoma must be an autapomorphy; I have checked that they are indeed also absent in *Tricholepidion*. (**62**) Sperm conjugation; Wygodzinsky (1961) described this unusual phenomenon from *Tricholepidion*, and it is also known from both Lepismatidae and Nicoletiidae (Jamieson, 1987).

Therefore, it is suggested that the sister group of the winged insects is, after all, most likely to be the taxon Zygentoma as conventionally delimited. Within the latter there is most likely a sister group relationship between *Tricholepidion* and Nicoletiidae + Lepismatidae. Apparent synapomorphies of the two last-mentioned families (which may not both be monophyla) include (Wygodzinsky, 1961), in addition to the loss (cf. 59) of the cephalic lamellar tendon: (**63**) loss of ocelli, (**64**) reduction of compound eyes, (**65**) reduced number of 'pectinate appendages' on lacinia, (**66**) fewer than five tarsomeres (if that number – here ascribed to the dicondylian groundplan, cf. 52 above – is not an autapomorphy of *Tricholepidion*, as believed by Wygodzinsky), (**67**) size-reduction of abdominal pregenital

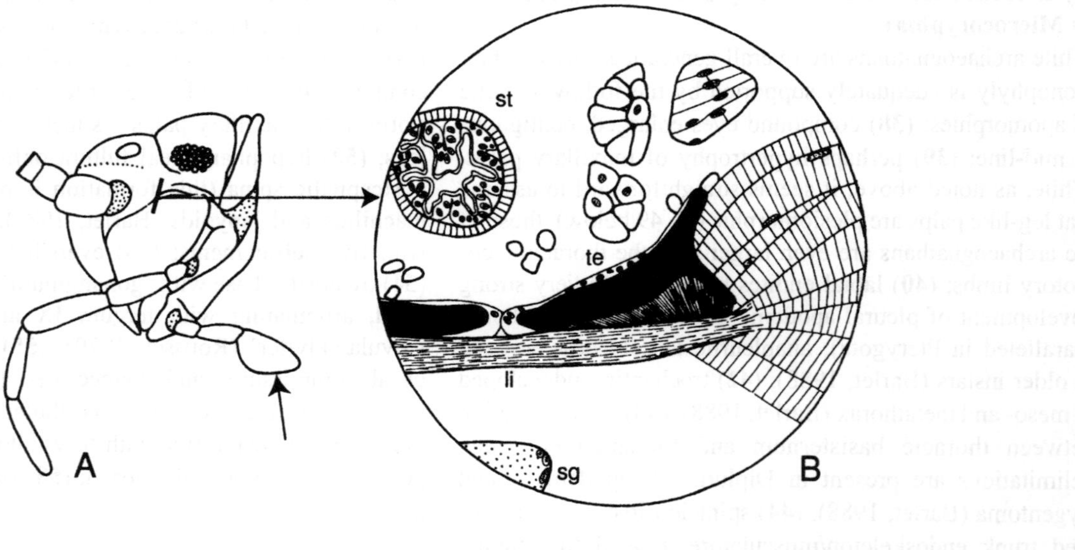


**Figure 21.3** *Tricholepidion gertschi*, the single extant lepidothrichid zygentoman. (A) Head (redrawn after Wygodzinsky, 1961), showing prominently widened apical segment of labial palp (arrow), an apparent zygentoman autapomorphy. (B) Diagrammatic transverse section of head (as indicated in (A)), showing transverse 'lamellar tendon' or 'ligament' (Hi) with insertions of 'zygomatic' mandibular muscle immediately below anterior tentorial arms (te) and linked to the latter by fibrous tissue; the tentorial arms will fuse a short distance behind the level of this section. sg, suboesophageal ganglion; st, stomodaeum (original).

sterna (paralleled in Archaeognatha), **(68)** sternum and coxopodite on male VIII and IX non-discrete (paralleled in Archaeognatha), **(69)** fewer than seven ovarioles per side.

As illustrated in Figure 21.3(b), the zygomatic mandibular muscles in *Tricholepidion* run extremely close to their tentorial counterparts, and a loss of the transverse ligament (accompanied by loss of the muscle fibres in question – or shift of their origins onto the tentorium) would be a minor rearrangement only. A loss of the ligament can therefore easily have taken place independently within the Zygentoma and in the groundplan of the Pterygota. A homoplasious complete loss **(70)** of the eversible vesicles in Zygentoma–Lepismatidae and Pterygota can similarly easily have happened.

## 21.4 THE PTERYGOTA, AND PROBLEMS IN THEIR BASAL EVOLUTION

Any detailed account of the problems in pterygote phylogeny is outside the scope of this contribution. The monophyly of the Pterygota is, of course, primarily indicated by the entire character complex **(71)** comprising the wing and flight musculature. The evolution of the flight motor was accompanied by a major restructuring of the meso- and metathoracic segments **(72)**, including development of high pleura stiffened by a pleural suture, and development of invaginated apodemes from pleural (parallelism with Archaeognatha, cf. 41) and sternal sclerites. Pterygote monophyly is also supported by characters unrelated to flight; suffice it here to note that in the pterygote head **(73)** the anterior and posterior tentorium is fused, and **(74)** circum- and suboesophageal blood vessels lost (Pass, 1991).

It is commonplace to recognize three basal groups among extant Pterygota: the Ephemeroptera, Odonata, and the Neoptera which comprise all other winged insects. The Neoptera are characterized primarily by their ability to flex the wings backwards over the abdomen through contraction of a pleural muscle inserting on the so-called '3rd axillary' in their (overall distinctive and uniform) complement of wing-base sclerotizations. Admittedly it cannot be known for certain that this ability is apomorphic at the pterygote level (one lineage among Palaeozoic non-neopterans – the palaeodictyopteroid Diaphanopterodea – could flex their wings), and neopteran monophyly is being challenged by Rasnitsyn (1997, this volume). Non-alar potential neopteran autapomorphies are few: perhaps the full-length '3rd-valvula-sheath' in the ovipositor, and the pretarsal arolium.

While the monophyly of the Neoptera may still be a reasonable working hypothesis, the basal dichotomy among the extant lineages remains unclarified. Is it between a taxon 'Palaeoptera' (= Ephemeroptera + Odonata) and Neoptera, or is it between Ephemeroptera and (Odonata + Neoptera)? The principal arguments are presented in the companion chapters by Kristensen and Kukalová-Peck (1991, 1994). There is a great need for additional evidence bearing on this issue, in particular for evidence from non-alar characters, which can be confidently polarized by out-group criteria.

A final remark on the morphological basis of wing formation, which rightly has attracted so much attention: the plausibility of Kukalová-Peck's exciting 'wing-from-leg-exite-theory' is dependent on the number of splitting events that are inferred to have occurred 'below' the pterygotes among extant Hexapoda/Atelocerata, **all** of which are devoid of free exites basad from coxa. Each of these splits therefore has a 'cost' in terms of homoplasious elimination of the free exite (be it through loss or fusion). On this basis the theory may appear quite 'costly', but it certainly deserves continued scrutiny. I suggest that additional developmental studies focusing on the thorax of nymphal mayflies, dragonflies and stoneflies may hold the greatest promise for elucidating the origin of that remarkable piece of machinery which, when overlayered by additional suites of apomorphies, has mediated the greatest bursts of speciation in the living world.

## ACKNOWLEDGEMENTS

I am grateful to Professor R. Elofsson (Lund) for loan of some of the *Tricholepidon* sections examined, to Dr G. Pass (Vienna) and Professor J. Bitsch (Toulouse) for helpful discussions, and to Professor H. Enghoff (Copenhagen) and Dr R.A. Fortey for pertinent comments on the manuscript. Ms B. Rubæk and Mr S. Langemark provided greatly appreciated assistance with the illustrations.

## REFERENCES

Bareth, C., Conde, B. and Pages, J (1989) Les procampodeides, des diploures peu connus, in *3rd International Seminar on Apterygota* (ed. R. Dallai), University of Sienna, pp. 137–44

Barlet, J. (1988) Considérations sur le squelette thoracique des insectes Aptérygotes. *Bulletin et Annales de la Societé Royale Belge d'Entomologie*, **124**, 171–87.

Barlet, J. (1992) L'endosquelette thoracique des insectes. *Mémoires de la Societé Royale Belge d'Entomologie*, **35**, 701–5.

Barlet, J. and Carpentier, F. (1962) Le thorax des Japygides. *Bulletin et Annales de la Societé Entomologique de Belge*, **98**, 95–123.

Bilinski, S.M and Szklarzewicz, T. (1992) The ovary of *Catajapyx aquilonaris* (Insecta, Entognatha): ultrastructure of germarium and terminal filament. *Zoomorphology*, **112**, 247–51.

Bitsch, J. (1969) Evolution de la morphologie prothoracique chez les insectes apterygotes et les pterygotes polyneopteres. *I Simposio International de Zoofilogenia, Facultad de Ciencias, Universidad de Salamanca*, pp. 317–21.

Bitsch, J. (1979) Morphologie abdominale des Insects, in *Traité de Zoologie*, Vol. 8, 2 (ed. P.-P. Grassé), Masson, Paris, pp. 291–578.

Bitsch, J. (1994) The morphological groundplan of Hexapoda: critical review of recent concepts. *Annales de la Societé Entomologique de France, N.S.*, **30**, 103–29.

Bitsch, J. and Ramond, S. (1970) Étude du squelette et de la musculature prothoraciques d'*Embia ramburi* R.-K. (Insecta Embioptera). *Zoologische Jahrbücher, Anatomie*, **87**, 63–93.

Boudreaux, H.B. (1979) *Arthropod Phylogeny: with Special Reference to Insects*, John Wiley & Sons, New York.

Chaudonneret, J. (1987) Evolution of the insect brain with special reference to the so-called tritocerebrum, in *Arthropod Brain. Its Evolution, Development, Structure and Functions* (ed. A.P. Gupta), John Wiley and Sons, New York, pp. 3–26.

Dallai, R. (1991) Are Protura really insects? in *The Early Evolution of Metazoa and the Significance of Problematical Taxa* (eds A.M. Simonetta and S.C. Morris), Cambridge University Press, Cambridge, pp. 263–9.

Deuve, T. (1994) Sur la présence d'un 'épipleurite' dans le plan de base du segment des Hexapodes. *Bulletin de la Societé Entomologique de France*, **99**, 199–210.

Denis, J.R. and Bitsch, J. (1963) Morphologie de la tête des Insects, in *Traité de Zoologie*, Vol. 8 (ed. P.-P. Grassé), Masson, Paris, pp. 1–593.

Dohle, W. (1985). Phylogenetic pathways in the Chilopoda. *Bijdragen tot de Dierkunde*, **55**, 55–66.

François, J. (1964) Le squelette thoracique des Protoures. *Travaux du Laboratoire de Zoologie et de la Station Aquicole Grimaldi de la Faculté des Sciences de Dijon*, **55**, 1–18.

François, J. and Chaudonneret, J. (1982) Les formations endo-squellettiques céphaliques des Collemboles et autres Aptérygotes. *Bulletin de la Societé Zoologique de France*, **107**, 537–43.

Friedrich, M. and Tautz, D. (1995) Ribosomal DNA phylogeny of the major extant arthropod classes and the evolution of myriapods. *Nature*, **376**, 165–7.

Hennig, W. (1969) *Die Stammesgeschichte der Insekten*. Waldemar Kramer, Frankfurt am Main. [Revised English edition (1981): *Insect Phylogeny* (eds A.C. Pont and D. Schlee), John Wiley & Sons, Chichester.]

Jamieson, B.G.M. (1987) *The Ultrastructure and Phylogeny of Insect Spermatozoa*, Cambridge University Press, Cambridge.

Kraus, O. and Kraus, M. (1996). On myriapod/insect interrelationships, in *Acta Myriapodologica: Mémoires du Muséum National d'Histoire Naturelle* (eds J.J. Geoffroy, J.P. Mauriès and M. Nguyen Duy-Jaquemin), Vol. 169, pp. 283–90.

Kristensen, N.P. (1975) The phylogeny of hexapod 'orders'. A critical review of recent accounts. *Zeitschrift für Zoologische Systematik und Evolutionsforschung*, 13, 1–44.

Kristensen, N.P. (1981) Phylogeny of insect orders. *Annual Review of Entomology*, **26**, 135–57.

Kristensen, N.P. (1989) Insect phylogeny based on morphological evidence, in *The Hierarchy of Life. Molecules and Morphology in Phylogenetic Analysis* (eds B. Fernholm *et al.*), Elsevier, Amsterdam, pp. 295–306.

Kristensen, N.P. (1991) Phylogeny of extant hexapods, in *The Insects of Australia*, 2nd edn (ed. CSIRO), Melbourne University Press, Carlton, Victoria, pp. 125–40. [Reprinted with minor alterations in 1994 as *Systematic and Applied Entomology: An Introduction* (ed. I.D. Naumann), Carlton, Victoria, Melbourne University Press.]

Kristensen, N.P. (1995) Forty years insect phylogenetic systematics. Hennig's 'Kritische Bemerkungen...' and subsequent developments. *Zoologische Beiträge NF*, **36**, 83–124.

Kukalová-Peck, J. (1983) Origin of the insect wing and the wing articulation from the arthropodan leg. *Canadian Journal of Zoology*, **61**, 1618–69.

Kukalová-Peck, J. (1985) Ephemeroid wing venation based upon new gigantic Carboniferous mayflies and basic morphology, phylogeny, and metamorphosis of pterygote insects (Insecta, Ephemerida). *Canadian Journal of Zoology*, **63**, 933–55.

Kukalová-Peck, J. (1987) New Carboniferous Diplura, Monura, and Thysanura, the hexapod groundplan and the role of thoracic side lobes in the origin of wings (Insecta). *Canadian Journal of Zoology*, **65**, 2327–45.

Kukalová-Peck, J. (1991) Fossil history and the evolution of hexapod structures, in *The Insects of Australia*, 2nd edn (ed. CSIRO), Melbourne University Press, Carlton, Victoria, pp. 141–79. [Reprinted with minor alterations in 1994 as *Systematic and Applied Entomology: An Introduction* (ed. I.D. Naumann), Carlton, Victoria, Melbourne University Press.]

Machida, R., Nagashima, T. and Ando, H. (1990) The early embryonic development of the jumping bristletail *Pedetontus unimaculatus* Machida (Hexapoda: Microcoryphia, Machilidae). *Journal of Morphology*, **206**, 181–95.

Manton, S.M. (1964) Mandibular mechanisms and the evolution of arthropods. *Philosophical Transactions of the Royal Society, B*, **247**, 1–183.

Manton, S.M. (1972) The evolution of arthropodan locomotory mechanisms. Part 10. Locomotory habits, morphology and evolution of the hexapod classes. *Journal of the Linnean Society of London, Zoology*, **51**, 203–400.

Manton, S.M. (1977) *The Arthropoda: habits, functional morphology and evolution*, Oxford University Press, London.

Matsuda, R. (1970) Morphology and evolution of the insect thorax. *Memoirs of the Entomological Society of Canada*, **76**, 1–431.

Matsuda, R. (1976) *Morphology and Evolution of the Insect Abdomen*, Pergamon Press, Oxford.

Pagés, J. (1989) Sclérites et appendices de l'abdomen des Diploures (Insecta, Apterygota). *Archives des Sciences* (Geneva), **42**, 509–51.

Pass, G. (1991) Antennal circulatory organs in Onychophora, Myriapoda and Hexapoda: Functional morphology and evolutionary implications. *Zoomorphology*, **110**, 145–64.

Pass, G. (1996) Morphology and evolution of circulatory organs in the Tracheata, in *Acta Myriapodologica: Mémoires du Muséum National d'Histoire Naturelle* (eds J.- J. Geoffroy, J.P. Mauriès and M. Nguyen Duy-Jaquemin), Vol. 169, pp. 291–2.

Paulus, H.F. (1979) Eye structure and the monophyly of the Arthropoda, in *Arthropod Phylogeny* (ed. A.P. Gupta), Van Nostrand, New York, pp. 299–383.

Rempel, J.G. (1975) The evolution of the insect head: the endless dispute. *Quaestiones Entomologicae*, **11**, 7–24.

Richards, O.W. and Davies, R.G. (1977) *Imms' Textbook of Entomology* 1–2. Chapman & Hall, London.

Rousset, A. (1973) Squelette et musculature des régions génitales et postgénitales de la femelle de *Thermobia domestica* (Packard). Comparaison avec la région génitale de *Nicoletia* sp. (Insecta: Apterygota: Lepismatidae). *International Journal of Insect Morphology and Embryology*, **2**, 55–80.

Sharov, A.G. (1966) *Basic Arthropodan Stock*, Pergamon, Oxford.

Snodgrass, R.E. (1935) *Principles of Insect Morphology*. McGraw-Hill, New York.

Snodgrass, R.E. (1952) *A Textbook of Arthropod Anatomy*, Comstock Publishing Association, Ithaca, New York.

Štys, P. and Bilinski, S. (1990) Ovariole types and the phylogeny of Hexapods. *Biological Reviews*, **65**, 401–29.

Štys, P., Zrzavý, J. and Weyda, F. (1993) Phylogeny of the Hexapoda and ovarian metamerism. *Biological Reviews*, **68**, 365–79.

Wygodzinsky, P. (1961) On a surviving representative of the Lepidotrichidae (Thysanura). *Annals of the Entomological Society of America*, **54**, 621–7.

# 22 Phylogenetic relationships between higher taxa of tracheate arthropods

*O. Kraus*

*Zoologisches Institut und Zoologisches Museum, Universität Hamburg, Martin-Luther-King-Platz 3, D-20146 Hamburg, Germany*

## 22.1 INTRODUCTION

The monophyletic origin of the Tracheata (synonyms: Antennata; Atelocerata) has never been questioned seriously by previous authors (Snodgrass, 1938; Hennig, 1966, 1982; Manton, 1977; Hennig, 1981; Ax, 1987; Boudreaux, 1987; Kraus and Kraus, 1994, 1996). Apparently, the Tracheata form the sister taxon of the Crustacea (or a part of them). Recent work by developmental biologists also points in this direction: for example, this is shown by details of the segmental composition of the crustacean head as compared with insects (Scholtz, 1995).

Recent developmental genetic data (Akam and Averof, 1993; Akam, 1995; Averof and Akam, 1995; Panganiban *et al.*, 1995) has called the monophyly of the Tracheata into question, and suggests that only the Insecta form the sister group to the Crustacea. If this were true, one would have to regard the Tracheata as a paraphyletic assemblage. In an analysis based on nuclear ribosomal gene sequences, Friedrich and Tautz (1995) concluded 'that the crustaceans and not the myriapods should be considered to be the sister group of the insects'. In their cladogram, the 'Myriapoda' are presented as the sister group of the Chelicerata. Remarkably different trees were derived from ribosomal and ubiquitin protein sequence data (Wheeler *et al.*, 1993).

From the viewpoint of the morphologist, conclusions such as a presumed myriapod–chelicerate relationship (with consequences for the phylogenetic position of the insects) seem to be highly improbable. They are in conflict with much other facts, most of them drawn from comparative anatomy and functional morphology. Some ideas of this kind could have been the result of selection of a particular taxon as outgroup. This is especially true for the assemblage of non-insects among the Tracheata commonly called 'Myriapoda'. Their existence as a natural unit has been, and is still generally accepted by a majority of authors (Hennig, 1981; Boudreaux, 1987; Kukalová-Peck, 1991). Various workers have been misled by this concept, which was simply based on typology. Some of them picked just one representative from this assemblage of 'myriapods' for comparison – frequently a juliform diplopod. This is true of studies in molecular biology (Ballard *et al.*, 1992). Ubiquitin sequence data (Wheeler *et al.*, 1993) may have reduced value, because relatively high mutation rates could superimpose 'signals' for branching events which happened in Cambrian (at the latest in Early Silurian) times (H.-J. Bandelt, personal communication). The presentation of a 'consensus cladogram' (Wheeler *et al.*, 1993, their Figure 9) based on molecular and morphological data may be criticized for several reasons, even though the relationships assumed are close to the views presented here. Other reasons why molecular phylogenies tend to differ considerably from those obtained by traditional approaches are discussed by Wägele (1995a,b). The assumed dichotomy of the Hexapoda was questioned by Kukalová-Peck (1991). Could it be that the Entognatha/Ectognatha concept, as proposed by Hennig (1981) and others, is no longer valid?

Classifications at different levels hitherto believed to be sound are now questioned, leading to a high degree of uncertainty. Here, I will try to deal with three main questions: (i) monophyly and sister group relationships of the Tracheata as a whole; (ii) phylogenetic analysis of the so-called Myriapoda; and (iii) phylogenetic position and early phylogeny of the Hexapoda. Some more general views on phylogenetic age, terrestrialization and functional morphology will be presented.

## 22.2 METHODS

Reconstructions of phylogenetic relationships were deduced using Hennig's original approach to phylogenetics. Details of construction and functional morphology were investigated and taken into consideration (for example, see characters 6 and 8 in Table 22.1). Different types of construction were analysed, including interdependencies between the acquisition of niches and the origin of evolutionary novelties. It should be emphasized that this approach is primarily

*Arthropod Relationships*, Systematics Association Special Volume Series 55, Edited by R.A. Fortey and R.H. Thomas. Published in 1997 by Chapman & Hall, London. ISBN 0 412 75420 7

based on peculiarities of the functioning phenotype – the real subject of natural selection.

Like other characters, data derived from ontogeny need careful interpretation. Adaptations may occur in the modes how and when various structures are realized in onotogeny (Gould, 1977). In diplopods, the so-called thoracic segments are a good example of potentially misleading ontogenic information. Each of these anterior rings of the postcephalic trunk is composed of two segments (Kraus, 1990), but this composition is not expressed, and hence remains invisible in ontogenetic stages (Dohle, 1964, 1974). The realization of certain other structures may be postponed to later, free-living stages. This is illustrated by an observation by Enghoff (1993). He found that certain diplopods hatch and even moult as haplopodous stages until the regular diplopody is finally acquired. This demonstrates how misleading it could be to rely directly on what is visible (or invisible) in ontogenetic stages.

## 22.3 PHYLOGENETIC ANALYSIS

### 22.3.1 THE TRACHEATA: SISTER GROUP RELATIONSHIPS, MONOPHYLY AND TERRESTRIALIZATION

At present, there is no sound reason for questioning the view that Crustacea and Tracheata form sister groups. However, the arguments available are not numerous: (i) there is the fact that mandibles are present in both groups; they are homologous insofar as they originated by transformations of the 4th pair of the appendages of the head capsule; (ii) further, one could perhaps refer to resemblances in the formation and composition of the head capsule (Lauterbach, 1980).

The following peculiarities confirm the monophyletic origin of the Tracheata.

*Appendage loss*

Loss of the appendages of the 3rd metamere of the head capsule, homologous to the 2nd pair of antennae in crustaceans.

*'Telognathic' mandibles*

A second peculiarity was telognathic mandibles – of course without palps [in principle, Manton (1977, p. 132) was correct, but see Kraus and Kraus, 1994)]. However, it is questionable whether this type of mandible really represents the relevant appendage as a whole ('whole limb mandible'), with the number of podomeres reduced. The components of the mandibles in different tracheate taxa never exceed three. This can be observed in diplopods (Manton, 1977, p. 102, Figure 3.17), Pauropoda (Kraus and Kraus, 1996, Figure 3), Symphyla (Kraus and Kraus, 1996, Figure 5) (see also Kraus and Kraus, 1994, p. 20, Figures 16–17), Archaeognatha (ibid, Figure 6), and even in the silverfish *Lepisma saccharina* (Zygentoma). Mandibles of this type may well represent the proximal components of this pair of appendages (in crustaceans: praecoxa/coxa/basis, or, according to Walossek and Müller (1990) and Walossek (1995), proximal endite/endopod/basipod (with further distal elements remaining unrealized)).

*Molar hooks*

Another feature of the mandibles provides a third autapomorphy: the presence of molar hooks ('Molar-Höcker') (Figures 22.1(a,b) and 22.2). If this is the case, one would have to assume that hooks of this kind were reduced in the branch formed by the Diplura and in the Dicondylia. This aspect needs more detailed investigation.

Originally, tracheate mandibles had two different functions: mastication, and transporting food from the oral cavity towards the pharynx. The latter function is correlated with the maintenance and increasing functional adaptation of movable, succeeding mandibular components. They are interpreted as specialized vestiges of the original (proximal?) podomeres of the relevant appendages. This is correlated with details of their internal musculature – even in chilopods (Kraus and Kraus, 1994, their Figure 22). Vestiges of former articulations are discernible in representatives of all major tracheate taxa, except for the Protura, Diplura (?) and Pterygota. The evolution of stiff, inflexible mandibles – at least in the Dicondylia – was apparently correlated with the transfer of one of the original mandibular functions, i.e. transportation, to the 2nd maxillae. Thus, the function of the mandibles was finally reduced to mastication alone, correlated with an increase in efficiency of the whole.

Some other characters are frequently mentioned as tracheate autapomorphies, but they should no longer be considered to be reliable.

*Malpighian tubules*

All representatives of the Tracheata lack midgut glands. Their reduction was apparently correlated with the transition to terrestrial life. Instead, Malpighian tubules were acquired. However, only a few studies on the tubules are available so far (Hopkin and Read, 1989; Wenning, 1989; Wenning et al., 1991). It remains uncertain whether Malpighian tubules in different tracheate subgroups are homologous or not. This feature needs further, comparative investigation.

*Tracheae*

As implied by their name, tracheates are frequently characterized by the presence of tracheae. However, respiratory organs of this type also exist in the Onychophora and in various subtaxa of the Arachnida (Hilken, 1997).

There are sound reasons for believing that the arachnid tracheal systems originated more than once. In araneomorph

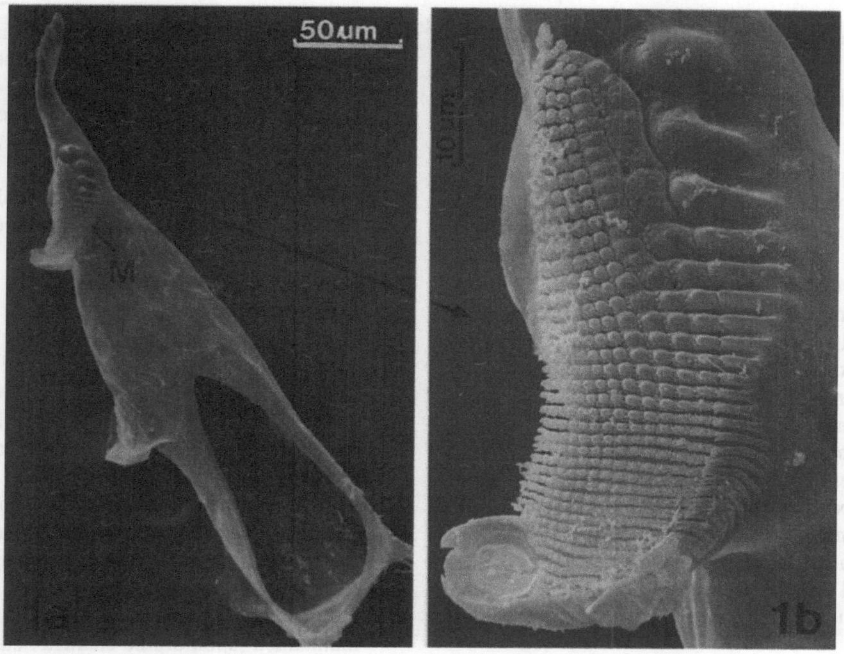


**Figure 22.1** Molar hooks (M) on mandibles of Tracheata. *Tomocerus* sp. (Hexapoda, Collembola).

spiders (Araneae) multiple transformation of book lungs into true tracheae apparently correlates with parallel reductions of the body size (Levi, 1967; Weygoldt and Paulus, 1979a,b). This illustrates that tracheae may have originated independently when required.

In a recent comparative study on respiratory organs carried out in my laboratory, Hilken (1997) investigated respiratory systems of representatives of the Tracheata and also of other arthropods. For the tracheates, he demonstrated that tracheae originated independently at least four and perhaps even five times (Figure 22.3, ro): twice in centipedes, in the Dignatha (= Diplopoda + Pauropoda), in the Hexapoda, and (?) in the Symphyla. I refer to Hilken's work as the results and conclusions cannot be given in detail in the present review. However, differences may be briefly illustrated by mentioning some of the most important facts.

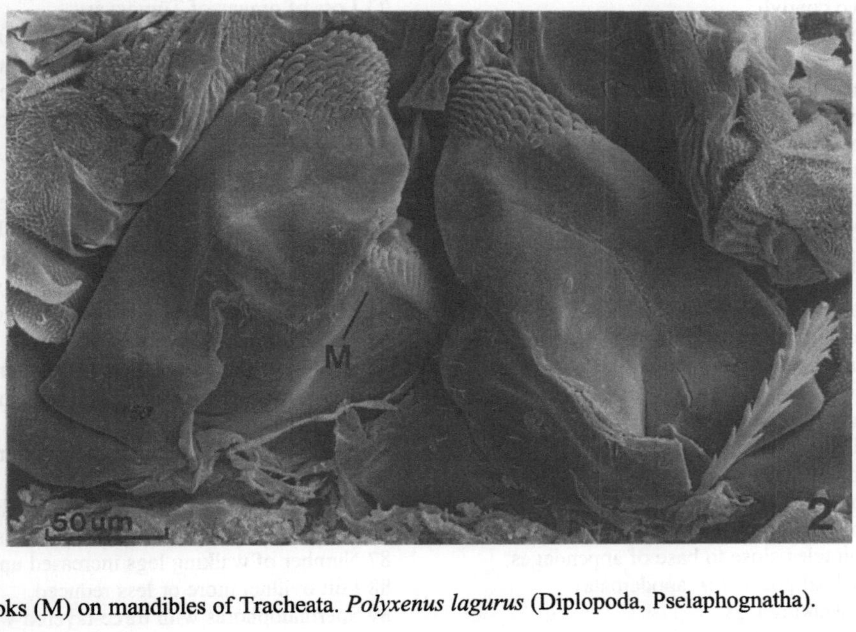


**Figure 22.2** Molar hooks (M) on mandibles of Tracheata. *Polyxenus lagurus* (Diplopoda, Pselaphognatha).

**Table 22.1** The cladogram presented in Figure 22.3 is based on the following apomorphic character states. (Not all characters available are included; for details see Borucki, 1996; Hilken, 1996; Koch, 1995; Kraus and Kraus, 1994).

01 Nauplius eye: median eyes closely associated to form a functional unit
02 Mandibles without palps, formed by three components ('telognathic')
03 Intercalary segment: complete loss of appendages of 2nd segment of head capsule
04 Loss of midgut glands; instead, presence of Malpighian tubules (homology of tubules in different higher taxa may be questionable)
05 Presence of coxal vesicles and styli
06 Maxillary plate: ventral side of mouth cavity bordered by basal parts of maxillae II
07 Maxillipedes: appendages of 1st postcephalic segment transformed into forcipules
08 Ventral side of mouth cavity bordered by 1st (!) maxillae
09 Stemmata: compound eyes transformed accordingly
10 Loss of median eyes
11 Formation of thorax: head capsule followed by three thoracomeres, locomotory in function
12 11 abdominal segments (in ground pattern)
13 Progoneaty: genital opening follows 3rd pair of legs (1st pair of legs may be reduced; in this case: genital opening behind 2nd pair of legs)
14 Bothriotrichia with special enlargement ('bulbs') close to their base
15 Maxillae I without palps
16 Loss of median eyes
17 Epimorphosis
18 Special type of entognathy combined with presence of linea ventralis on ventral side of head capsule
19 Loss of terminal filaments or cerci, respectively
20 Antennae composed of three parts: scapus, pedicellus and (major) terminal segment, subdivided into antennomeres
21 Tarsus of walking legs secondarily subdivided; presence of two tarsal claws
22 Special type of entognathy, different from (18)
23 Presence of genital papilla
24 Only 10 abdominal segments visible
25 Loss of unpaired terminal filament
26 Mandibles with two condyli
27 Walking legs: five tarsomeres
28 Ontogeny: closed amniotic cavity
29 Insertion of antennae approximated; compound eyes adjoining in dorsal median line of head capsule
30 Loss of stigmata on 1st abdominal segment
31 Paired wings on meso- and metathorax
32 Loss of styli on pregenital segments of abdomen
33 Loss of coxal vesicles
34 Reduction of compound eyes to a few facets (value of character doubtful)
35 Complete loss of antennae and eyes; appendages of prothoracic segment elongated, with tactile function
36 Suboesophageal and prothoracal ganglia fused
37 Genital chamber (with porus) between 11th abdominal segment and telson
38 Abdomen reduced: six segments only
39 Specialized form of appendages on 1st, 3rd and 4th abdomimal segments
40 Walking legs: formation of a tibiotarsus
41 Special tracheal spiracles close to base of appendages, associated with tracheal pockets (= apodemata)
42 Loss of 1st pair of walking legs
43 Presence of 'penes': appendages behind 2nd (morphologically 3rd) pair of legs with male genital opening on tip
44 Genital opening unpaired
45 Head spiracles: one single pair of tracheal stigmata on lateral sides of head capsule
46 Loss of all eyes
47 Maxillae II fused to form a functional labium
48 Female spermathecae formed by paired lateral pockets in mouth cavity
49 Paired terminal spinnerets
50 Diplopody
51 Distal segment of antennae with four sensory cones
52 Unique type of antennae with special sensory organ (globulus)
53 Maxillae II reduced to an unpaired triangular plate
54 Paired pseudoculi on lateral sides of head capsule
55 Exsertile coxal vesicles on ventral side of 1st postcephalic segment
56 Compound eyes transformed into stemmata
57 Maxillae I and II fused to form a 'perfect' gnathochilarium
58 Compound eyes reduced to a few single isolated 'ocelli' (= ommatidia!)
59 No gnathochilarium; ground pattern: maxillae with telepodites maintained (apomorphic character state questionable)
60 Special arrangement and types of bristles, including two pairs of terminal brushes
61 Maxillae II: loss of trochanter
62 Head capsule secondarily flattened
63 Male gonads secondarily unpaired
64 Maxillipedes: formation of a tarsungulum
65 Tracheal spiracles on tergites, unpaired, special tracheal lungs
66 Pseudocompound eyes (secondarily evolved?)
67 Tarsi and antennae secondarily subdivided into numerous tarso-/antennomeres
68 Paired maxillary organs on 1st maxillae
69 Epimorphosis (limited)
70 Brood care: female body bent ventrally
71 Formation of coxopleurites on last leg-bearing segment
72 Loss of maxillar nephridia
73 Loss of organs of Tömösváry
74 Male gonad subdivided into numerous vesicles
75 Female 'gonopods' with special spines and terminal claws (character not yet reliable as an autapomorphy)
76 Maxillipedes: direct lateral articulation of tarsungulum with trochantero-praefemur
77 Epimorphosis complete
78 Tracheal system with longitudinal stems bridging several segments, and anastomoses between both sides of trunk
79 Segment of 15th walking legs with ring-like sclerite
80 Special ano-genital capsule, shape bivalvular
81 Maxillae I: telopodites reduced to one segment
82 Number of antennomeres: 18
83 Complete loss of heterotergy: segments of trunk equal in shape; number of leg-bearing segments considerably increased (from 31 up to 181)
84 Acquisition of complete series of spiracles on trunk segments; tracheal system with chiasmata
85 Brood care: female body bent dorsally
86 Tergites of 1st and 2nd postcephalic segments fused
87 Number of walking legs increased up to 21 (or 23)
88 Left oviduct more or less reduced
89 Spermatophores with three-layered wall

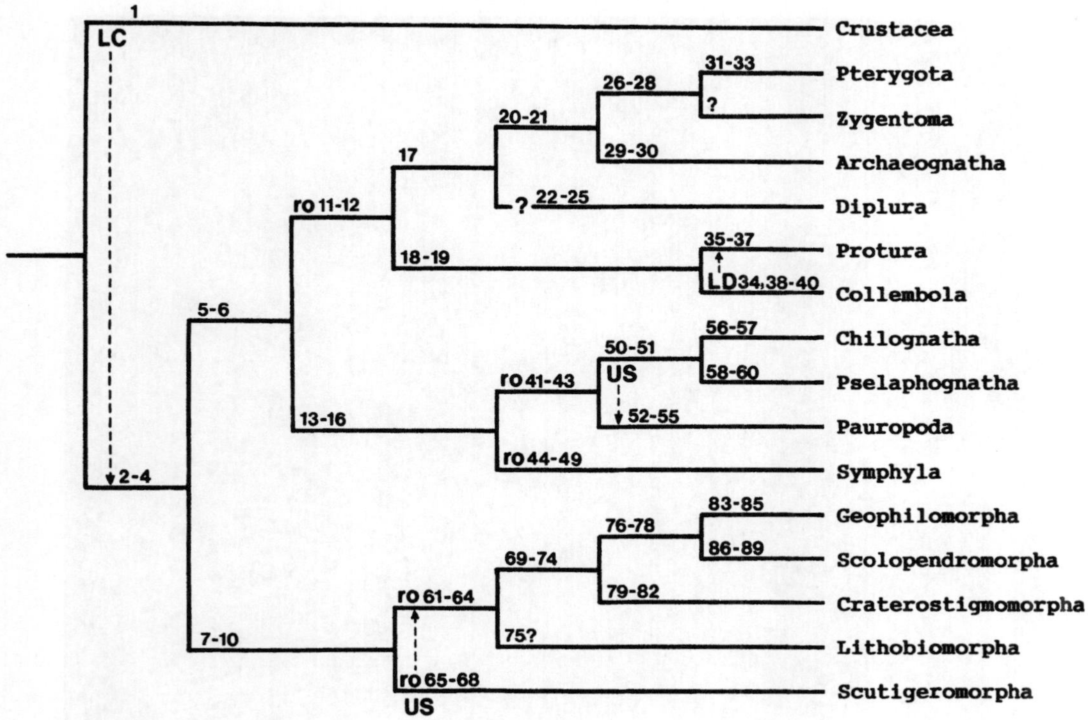


**Figure 22.3** Cladogram of phylogenetic relationships of higher taxa of the Tracheata, with Crustacea as sister group. Abbreviations: LC, Lower Cambrian; US, Upper Silurian; LD, Lower Devonian. Dashed vertical lines (with arrows) indicate equivalent minimum age according to oldest known fossils; ro: independent acquisition of respiratory organs. Numbers refer to characters listed in Table 22.1.

The Notostigmophora are equipped with unpaired dorsal lungs, whereas all other chilopods have lateral, paired spiracles. Each notostigmophoran spiracle communicates with an internal atrium just underneath the tergites (Figure 22.4). Multiple tracheae originate from the wall of each of these air sacs. Tracheae of this kind are immersed in the haemolymph of the body cavity, and the animals make use of haemocyanin for transporting oxygen. Accordingly, tracheae of this type do not enter at all into the musculature and internal organs, there are no tracheoles at all.

Quite another type of tracheae evolved in the Dignatha: in Diplopods, paired spiracles occur in a strictly ventral position. They are located on the sternites, next to the base of the appendages. Spiracles of this kind open into the lumen of solid internal tracheal pockets. Whole bunches of fairly simple, delicate tracheae originate from these pockets; they are not interconnected at all. Simultaneously, the tracheal pockets serve as apodemata for the musculature of the appendages. The same is true for certain pauropods (Hexamerocerata), which also maintained various other plesiomorphic features (Remy, 1953). The position of the spiracle and the presence of tracheal pockets with strictly segmental tracheae form one of the sound autapomorphies of the Dignatha.

I am reluctant to interpret the somewhat enigmatic single pair of spiracles on the lateral sides of the head capsule of the Symphyla.

Tracheal systems of pleurostigmophoran chilopods and hexapods are regarded as being of independent origin. They differ in various details (Hilken, 1997).

### 22.3.2 CONCLUSIONS

1. Because of the lack of information on details of their ultrastructure, the presence of Malpighian tubules in all tracheates cannot be used yet as an autapomorphy of the taxon.
2. The presence of tracheal systems should no longer be regarded as an autapomorphy of the Tracheata.
3. There is no longer any reason to believe that terrestrialization was already completed somewhere in the stem lineage of the Tracheata. As tracheal systems of different construction and function originated in different phylogenetic lineages (Figure 22.3, ro), it is much more probable that tracheate terrestrialization happened several times independently. The assumption of one single terrestrialization event (as previously indicated by Kraus and Kraus (1994, Figure 2)) is no longer supported by the facts presented by Hilken (1997), but this also depends on the definition of terrestrialization.

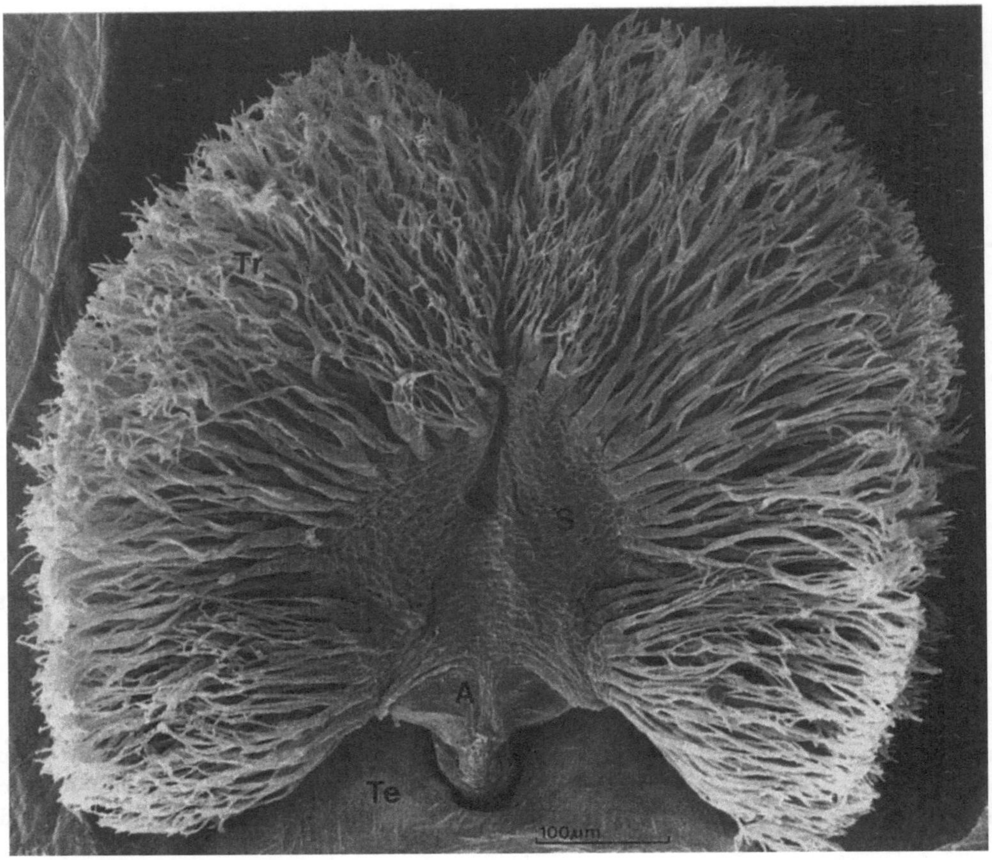


**Figure 22.4** Tracheal lung of *Scutigera coleoptrata* (Chilopoda, Notostigmophora), ventral view on tergite (Te) with Atrium (A), paired air sacs (S), and blind-ending Tracheae (Tr). (Courtesy of G. Hilken.)

## 22.4 BASAL BRANCHING EVENTS IN THE TRACHEATA

The question of whether the major subgroups of the Tracheata are monophyletic units, and their phylogenetic relationships have been studied by Kraus and Kraus (1994, 1996). The 1994 paper includes an analysis of numerous characters and should be consulted for additional details. Arguments for the inferred phylogenetic relationships of the higher taxa concerned were summarized in a cladogram (Kraus and Kraus, their Figure 2); for reasons of convenience, most of them are repeated in the present paper (Figure 22.3; Table 22.1). The analysis revealed that the Chilopoda represent the sister taxon to all other tracheates. These form a monophyletic unit called Labiophora. Chilopods transformed the 1st maxillae to border the mouth cavity ventrally (or posteriorly, respectively). In the Labiophora, instead, the same functional need was matched by a transformation of the 2nd maxillae – a fundamentally different construction.

The interrelationships between the higher chilopod taxa were quite recently restudied in detail by Borucki (1996). His results support the following phylogenetic system (see Figure 22.3):

Chilopoda
- Notostigmophora
- Pleurostigmophora
  - Lithobiomorpha
  - Epimorpha *s. lat.*
    - Craterostigmomorpha (syn. Devonobiomorpha ?)
    - Epimorpha *s. str.*
      - Scolopendromorpha
      - Geophilomorpha

The Labiophora branched into the monophyletic units Progoneata (with Symphyla and Dignatha [= Diplopoda + Pauropoda]), and the Hexapoda. According to Kraus and Kraus (1994, their Figure 2) (see also Figure 22.3 and Table 22.1), the following phylogenetic system reflects the relationships within the labiophores:

Labiophora
  Progoneata
  Symphyla
  Dignatha
    Diplopoda
      Chilognatha
      Pselaphognatha
    Pauropoda
Hexapoda (= Insecta *s. lat.*)

However, one detail of the basic branching events within the Hexapoda is different from Kraus and Kraus (1994): the traditional 'Entognatha concept' is no longer valid. Most previous authors, including Hennig (1981), believed in the 'Entognatha'. Manton (1977, p. 400) was already taking issue with this concept. Later, Kukalová-Peck (1991) maintained a taxon Ellipura (= Collembola + Protura). She called this unit Parainsecta and classified it as the sister group to all other hexapods – including the entognathous Diplura. Hence, the designation Insecta was restricted to the Diplura plus Entognatha.

In order to clarify the crucial question of the relationships between the primarily wingless, entognathous hexapods, studies my laboratory by Koch (1997) found that, in principle, Manton (1977) was right. There are different kinds of entognathy in the Ellipura on the one hand, and in the Diplura on the other. This result is not surprising considering that entognathous conditions were also achieved by representatives of the Diplopoda (Colobognatha) and Pauropoda (Figure 22.5). Kraus and Kraus (1994) had reasons to introduce the term 'semientognathy' for the configuration and function of the mouth parts in various other progoneate groups (Symphyla, Pselaphognatha). Accordingly, entognathy, as such, no longer constitutes an autapomorphy.

In accordance with Kristensen's view (1997, this volume), Koch confirmed the Ellipura as a monophyletic unit. Several somewhat weak arguments put forward by various authors in favour of the Ellipura concept (Hennig, 1981, p. 103) are now supplemented by additional facts. They primarily concern details of the mouth parts. One character among others is the presence of a median rim on the ventral side of the 2nd maxillae (Figures 22.6 and 22.7); it is called *linea ventralis*. This feature is present in the Collembola as well as in the Protura. Hence, it is regarded as an autapomorphy of the Ellipura. Koch (1997) confirmed a special kind of ellipuran entognathy with mandibles and 1st maxillae in paired, succeeding gnathal pockets; furthermore, he refers to polytroph–meroistic ovarioles.

Until now, there has been no sound reason to question the monophyletic origin of the Diplura. The phylogenetic position of this taxon has not yet been definitively settled. At present, I tend to favour a sister group relationship to the Ectognatha (Figure 22.3). If this is true, the phylogenetic system of the hexapods should read as follows:

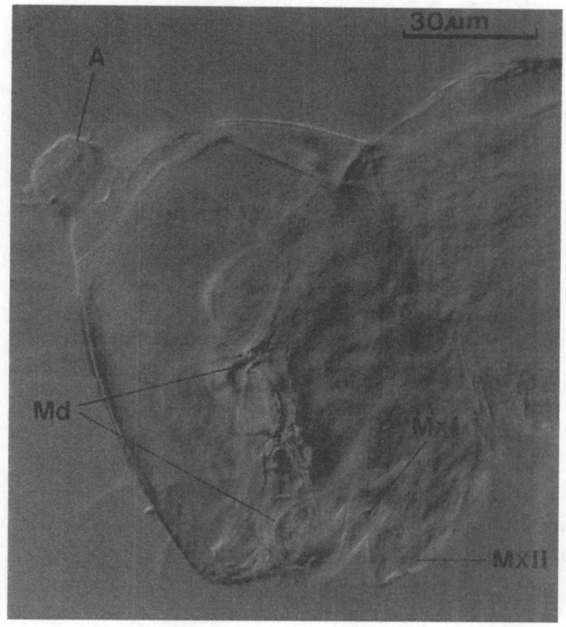


**Figure 22.5** *Millotauropus silvestrii* (Symphyla, Hexamerocerata), head, lateral view, with three-partite mandibles in a semi-entognathous position. Abbreviations: A, base of antennae; Md, mandibles; MxI, MXII, first and second maxillae. (From Kraus and Kraus, 1994.)

Hexapoda (= Insecta *s. lat.*)
  Ellipura
    Collembola
    Protura
  Insecta *s. str.*
    Diplura
    Ectognatha
      Archaeognatha
      Dicondylia
        Zygentoma
        Pterygota

## 22.5 GENERAL VIEWS OF EVOLUTION

The major subgroups of the Tracheata are geologically ancient. The occurrence of the oldest known fossils is indicated in Figure 22.3 by capital letters. Arrows refer to relevant sister taxa; equal age has to be inferred. According to the fossil record, the basic branching events happened in the late Cambrian to early Silurian. Non-predators presumably fed primarily on fungi and/or green algae by penetrating or damaging cell walls – similar to tardigrades. Accordingly, early tracheates were remarkably small animals. Their primary food niche coincides with various independent origins of semi-endognathy and with the endognathous condition in two basic branches of the Hexapoda (Ellipura, Diplura) as well as

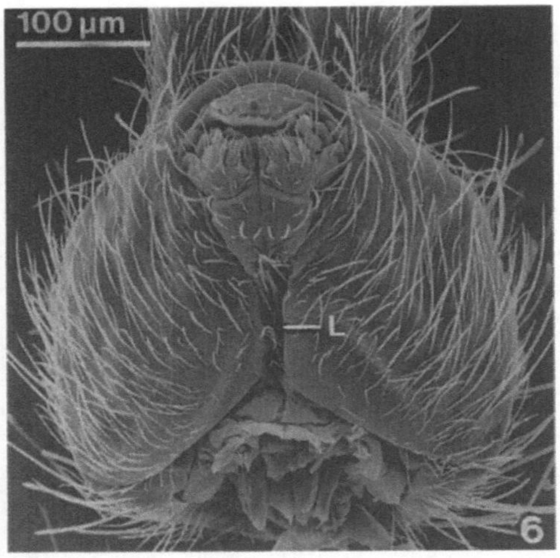


**Figure 22.6** Linea ventralis (L) in representatives of the Ellipura. *Entomobrya* sp. (Collembola). (Courtesy of M. Koch.)

in the Colobognatha (Diplopoda). In principle, only representatives of the Chilognatha (among diplopods) and of many subtaxa of the Pterygota (among hexapods) successfully entered new food niches. This was correlated with the increase in body size of such non-predatory animals. The unusually large size of the Carboniferous Arthropleurida remains enigmatic. In the majority of the higher taxa (e.g. Symphyla, Pauropoda, Pselaphognatha, Ellipura), however,

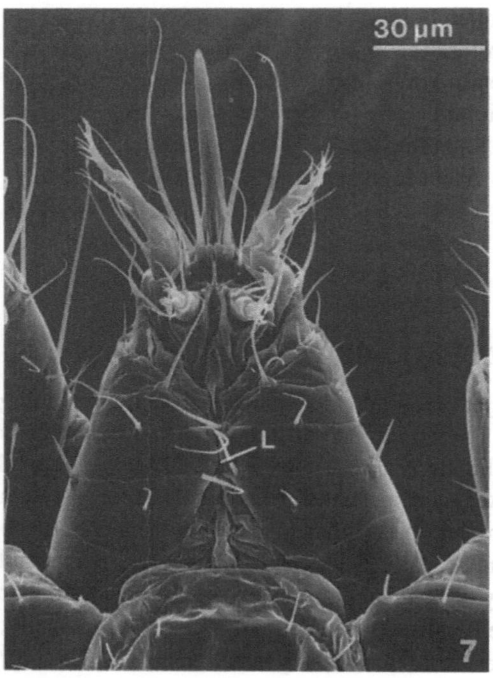


**Figure 22.7** Linea ventralis (L) in representatives of the Ellipura. *Acerentomon* sp. (Protura). (Courtesy of M. Koch.)

non-predators more or less maintained the original small size – body length approximately 1–2 mm. This correlates with the fact that the original, primary food niches continue to exist. This concept is not valid for predators. The digestion of protein materials is much easier than the digestion and uptake of nutrients from algae, plants or even rotten leaves. This is probably accounts for the relatively large size of some scolopenders and japygids, including representatives of the latter taxon from the Carboniferous (see Shear and Kukalová-Peck, 1990; Kraus and Kraus, 1994, pp. 25–27).

## REFERENCES


Akam, M. (1995) *Hox* genes and the evolution of diverse body plans. *Philosophical Transactions of the Royal Society of London*, B, **349**, 313–19.

Akam M. and Averof, M. (1993) HOM/Hox genes of *Artemia:* implications for the origin of insect and crustacean body plans. *Current Biology*, **3**, 73–8.

Averof, M. and Akam, M. (1995) Insect–crustacean relationships: insights from comparative developmental and molecular studies. *Philosophical Transactions of the Royal Society of London, B*, **347**, 293–303.

Ax, P. (1987) *The Phylogenetic System. The Systematization of Organisms on the Basis of their Phylogenies*, Springer Verlag.

Ballard, J.W.O., Olsen, G.J., Faith, D.P., Odgers, W.A., Rowell, D.M. and Atkinson, P.W. (1992) Evidence from 12S ribosomal RNA sequences that onychophorans are modified arthropoda. *Science*, **258**, 1345–8.

Borucki, H. (1996) Evolution und Phylogenetisches System der Chilopoda (Tracheata). *Verhandlungen des naturwissenschaftlichen Vereins in Hamburg, NF*, **35**, 95–226.

Boudreaux, H.B. (1987) *Arthropod Phylogeny. With special reference to insects*, John Wiley & Sons, New York.

Dohle, W. (1964) Die Embryonalentwicklung von *Glomeris marginata* (Villers) im Vergleich zur Entwicklung anderer Diplopoden. *Zoologische Jahrbücher, Abteilung Anatomie*, **81**, 241–310.

Dohle, W. (1974) The segmentation of the germ band of Diplopoda compared with other classes of arthropods. *Symposia of the Zoological Society of London*, **32**, 143–61.

Enghoff, H. (1993) Haplopodous diplopods: a new type of millipede body construction in cambalopsid juveniles (Diplopoda, Spirostreptida). *Acta Zoologica*, **74**(3), 257–61.

Friedrich, M. and Tautz, D. (1995) rDNA phylogeny of the major extant arthropod classes and the evolution of Myriapods. *Nature*, **376**, 165–7.

Gould, S.J. (1977) *Ontogeny and Phylogeny*, Harvard University Press, Cambridge.

Hennig, W. (1966) *Phylogenetic Systematics*, University of Illinois Press, Urbana.

Hennig, W. (1981) *Insect Phylogeny.* Translation of Hennig (1969).

Hennig, W. (1982) *Phylogenetische Systematik*, Paul Parey, Hamburg.

Hilken, G. (1997) Vergleich von Tracheensystemen unter phylogenetischem Aspekt. *Verhandlungen des naturwissenschaftlichen Vereins in Hamburg, NF*, **37**.

Hopkin, S.P. and Read, H.J. (1989) *The Biology of Millipedes*, Oxford University Press, Oxford.

Koch, M. (1997) Verwandtschafts-Verhältnisse hochrangiger Taxa bei primär flügellosen Insekten. Thesis, University of Hamburg (to be published in *Verhandlungen des naturwissenschaftlichen Vereins in Hamburg, NF*, **37**.)

Kraus, O. (1990) On the so-called thoracic segments in Diplopoda, in *Proceedings of 7th International Congress of Myriapodology* (ed. A. Minelli), E.J. Brill, Leiden, pp. 63–8.

Kraus, O. and Kraus, M. (1994) Phylogenetic system of the Tracheata (Mandibulata): on 'Myriapoda'–Insecta interrelationships, phylogenetic age and primary ecological niches. *Verhandlungen des naturwissenschaftlichen Vereins in Hamburg, NF*, **34**, 5–31.

Kraus, O. and Kraus, M. (1996) On myriapod/insect interrelationships, in *Acta Myriapodologica: Mémoires de la Musée National d'Histoire Naturelle Paris* (ed. J.-J. Geoffroy *et al.*), Vol. 169, pp. 283–90.

Kukalová-Peck, J. (1991) Fossil history and the evolution of hexapod structures, in *The Insects of Australia* (eds I.D. Naumann *et al.*), Vol. 1, pp. 141–79.

Lauterbach, K.-E. (1980) Schlüsselereignisse in der Evolution des Grundplans der Mandibulata (Arthropoda). *Abhandlungen des naturwissenschaftlichen Vereins in Hamburg, NF*, **23**, 163–327.

Levi, H.W. (1967) Adaptations of respiratory systems of spiders. *Evolution*, **21**, 571–83.

Manton, S.M. (1977) *The Arthropoda. Habits, Functional Morphology, and Evolution*, Oxford University Press, Oxford.

Panganiban, G., Sebring, A., Nagy, L. and Carroll S. (1995) The development of crustacean limbs and the evolution of arthropoda. *Science*, **270**, 1363–6.

Remy, P.A. (1953) Description de nouveaux types de pauropodes: *Millotauropus* et *Rabadauropus*. *M,moires de l'Institut Scientifique de Madagascar* (A), **8**, 24–41.

Scholtz, G. (1995) Head segmentation in Crustacea – an immunocytochemical study. *Zoology*, **98**, 104–14.

Shear, W.S. and Kukalová-Peck, J. (1990) The ecology of Paleozoic terrestrial arthropods: the fossil evidence. *Canadian Journal of Zoology*, **68**, 1807–34.

Snodgrass, R.E. (1938) Evolution of the Annelida, Onychophora and Arthropoda. *Smithsonian Miscellaneous Collections*, **97**(6), 1–159.

Wägele, J.W. (1995a) On the information content of characters in comparative morphology and molecular systematics. *Journal of Zoological Systematics and Evolutionary Research*, **33**(1), 42–7.

Wägele, J.W. (1995b) Arthropod phylogeny inferred from partial 12S rRNA revisited: monophyly of the Tracheata depends on sequence alignment. *Journal of Zoological Systematics and Evolutionary Research*, **33**(2), 75–80.

Walossek, D. (1995) The Upper Cambrian *Rehbachiella*, its larval development, morphology and significance for the phylogeny of Branchiopoda and Crustacea. *Hydrobiologia*, **298**, 1–13.

Walossek, D. and Müller, K.J. (1990) Upper Cambrian stem-lineage crustaceans and their bearing upon the monophyletic origin of Crustacea and the position of *Agnostus*. *Lethaia*, **23**, 409–27.

Wenning, A. (1989) Transporteigenschaften der Malpighischen Gefässe von *Lithobius forficatus* L. (Myriapoda, Chilopoda). *Verhandlungen der deutschen zoologischen Gesellschaft*, **82**, 215–16.

Wenning, A., Greisinger, U. and Proux J.P. (1991) Insect characteristics of the Malpighian tubules of a non insect: fluid secretion in the centipede *Lithobius forficatus* (Myriapoda, Chilopoda). *Journal of Experimental Biology*, **158**, 165–80.

Weygoldt, P. and Paulus, H.F. (1979a,b) Untersuchungen zur Morphologie, Taxonomie und Phylogenie der Chelicerata. I: Morphologische Untersuchungen; II: Cladogramme und die Entfaltung der Chelicerata. *Zeitschrift für zoologische Systematik und Evolutionsforschung*, **17**, (a) 85–116; (b) 177–200.

Wheeler, W.C., Cartwright, P. and Hayashi, C.Y. (1993) Arthropody phylogeny, a combined approach. *Cladistics*, **9**(1), 1–39.

# 23 Myriapod–insect relationships as opposed to an insect–crustacean sister group relationship

*W. Dohle*

*Institut für Zoologie, Freie Universität Berlin, Königin-Luise-Strasse 1–3, D-14195 Berlin, Germany*

## 23.1 INTRODUCTION

Most textbooks on invertebrate zoology combine the myriapods and the insects and treat them under the heading: Antennata, Tracheata or Atelocerata. Characters often cited in favour of such an assembly include: loss of second antennae, formation of Malpighian tubules, postantennal organs, and tracheae. Kristensen (1991, pp. 126–7) states: 'For several decades there has been near-universal agreement that the closest relatives of the Hexapoda are to be sought among the myriapods'.

However, there are characters which are astoundingly similar in insects and at least some crustaceans and which cannot be found in equivalent form in myriapods, such as neuroblasts, patterns of axonogenesis in early differentiating neurons, the fine structure of ommatidia and expression patterns of segmentation genes. The resultant conflicting evidence has not yet been thoroughly discussed, partly because some of these characteristics have only recently been revealed. Indeed, in October 1994, when I was asked to propose a title for my presentation, few facts were known which supported the insect–crustacean alternative. Since then, many new facts have been published which give additional support for a close insect–crustacean relationship. The facts come mostly from neurobiology and molecular genetics. As these are fields of very active research it is probable that the evidence for an insect–crustacean relationship will grow further in the near future. However, the counter-arguments are also worth reviewing.

This paper assesses the critical characters which play a role in the discussion of myriapod–insect–crustacean relationships and focuses on the following questions:

1. Are the similarities between two or more characters so detailed that homology and thus a single evolutionary origin has to be postulated? Or must convergent evolution be taken into consideration?
2. If homology is postulated and the characters in question are confined to but a few taxa, does this indicate synapomorphy and thus support a recent common ancestry of these taxa? Or is it possible that the characters have evolved longer ago and have been lost or altered in other closely related groups?

## 23.2 MONOPHYLY OF THE HEXAPODA

If relationships between species-rich and diverse groups are debatable it is first essential to ascertain that these groups are reliably monophyletic taxa, and that there is a reasonable conception of the ground pattern of their stem species. The monophyly of Hexapoda has been questioned several times. It can be taken for granted that the Ectognatha are monophyletic. If, as Hennig proposed as early as 1953, one group of 'thysanurans', namely the Lepismatida, has a more recent common ancestor with the Pterygota than the Machilida this would be a fortunate situation. As rockhoppers and silverfish share many plesiomorphic characters the common ancestor of the Ectognatha must have had the appearance of a 'thysanuran'. It would be a tempting task to make a detailed reconstruction of this ectognathan ancestor, not only of its cuticular morphology, but also of its inner anatomy, fine structure and physiology.

The monophyly of the entognathans is a matter of recent debates. It is possible that the common ancestor had the appearance of a 'dipluran'. However, the synapomorphies of ectognathans and entognathans are less numerous. I suspect that there is a number of details of neurobiology and hormone physiology which speak in favour of the monophyly of Hexapoda. However, they have not yet been summarized or been demonstrated for all hexapod groups. I hold the common belief that Hexapoda are monophyletic on the ground of their tagmatization (Kristensen, 1991) but I admit that this has not been established beyond doubt.

## 23.3 MONOPHYLY OF THE DIFFERENT 'MYRIAPOD' GROUPS

Each of the four groups which are numbered among the 'myriapods' can easily be shown to be monophyletic.

*Arthropod Relationships*, Systematics Association Special Volume Series 55, Edited by R.A. Fortey and R.H. Thomas. Published in 1997 by Chapman & Hall, London. ISBN 0 412 75420 7

1. **Diplopoda** have eight antennomeres with four sensorial cones on the end-plate of the distal segment (Enghoff, 1984). In the trunk, two segments are covered by only one tergite – with the exception of the most anterior ones. I am quite confident that it will soon be confirmed using molecular techniques that my assertion is correct that the first four trunk segments of the Diplopoda are simple segments and that the 5th and 6th segments are covered by the first diplotergite (Dohle, 1964, 1974). Enghoff (1990) has given a good picture of the putative ancestor of the subgroup Chilognatha. The ancestor of all Diplopoda must have had several traits which we only find in the extant Polyxenida: non-calcified cuticle, trichobothria with a bulb-like basal extension, and indirect sperm transfer.
2. **Pauropoda** are clearly monophyletic. Their ramified antennae, number and distribution of the trichobothria at the margins of the tergites as well as the peculiar pseudoculi are apomorphic.
3. **Symphyla** are a rather homogeneous group. All species have 12 pairs of legs. They have but one pair of spiracles which are situated in the head above the mandibles, a pair of cercus-like spinnerets and terminal trichobothria.
4. **Chilopoda** vary in their appearance and life-style. All subgroups have some characters in common which cannot be found elsewhere: first trunk appendages transformed into poison-claws, egg-tooth on embryonic cuticle of the second maxillae, similar fine structure of the spermatozoa, with a striated cylinder around the axonema and a helicoidal 'mantle'. In addition the characteristics of the chilopod ancestor have been worked out in detail (Dohle, 1985).

## 23.4 DIGNATHA: A WELL-FOUNDED MONOPHYLETIC TAXON

Millipedes and pauropods can be put together to form the well-supported monophyletic taxon Dignatha. The arguments have already been discussed at length (Dohle, 1980, 1988). Some of the arguments have been questioned since (Kraus and Kraus, 1994). I consider the following characters as synapomorphic for Diplopoda and Pauropoda: lack of appendages in second maxillary segment, formation of a 'lower lip' by the appendages of the first maxillary segment and the intervening sternite, genital pores at the base of the second trunk leg pair, sternal spiracles which open into a tracheal pouch giving rise to an apodema and to tracheae, pupoid stage, first free-living juvenile with three pairs of legs.

The second maxillary segment (= postmaxillary segment) lacks appendages (Figure 23.1). I have investigated several millipede embryos (*Polyxenus*, *Glomeris*, *Polydesmus*, *Oxidus*, *Ommatoiulus*) from germ band formation up to eclosion. In no phase of development has a vestige of a limb bud been found in this segment (Dohle, 1964, 1974). According to Tiegs (1947) the same is true for pauropods: the segment behind the maxillary segment, the so-called collum segment, is devoid of appendages.

The lack of second maxillae may be convergent. In contrast, the convergent formation of a 'lower lip' out of the appendages of the first maxillary segment is highly improbable. I take it as a convincing synapomorphy. In millipedes, the sternal part of the first maxillary segment grows out to form a plate connecting the two limb buds (Figure 23.2). Good markers for identifying the structures during their transformation are the ganglion grooves and the pores of the maxillary nephridia. Exactly the same formation has been described by Tiegs (1947) for *Pauropus*, the only difference being that the appendages and the sternal plate (= intermaxillary plate) do not coalesce. By not accepting the relevance of these facts, Kraus and Kraus (1994) and Hilken and Kraus (1994) repeat the speculations of Verhoeff (1911–1914) that the gnathochilarium of millipedes is composed of two pairs of maxillae. They also disregard the fact that the gnathochilarium is innervated by but one pair of ganglia and that the muscles are those of one segment only (Fechter, 1961).

The genital pores in millipedes and pauropods are found at the base of the second leg pair. In polyxenid males, the vas deferens opens on the tip of a cone-like movable penis, as in pauropods (Figure 23.3). In derived cases the vas deferens can pierce the coxa of the leg. In polyxenids as well as in pauropods, the males fabricate a web of threads and deposit sperm droplets on the meshes.

The sternal spiracles are unique and they are regarded as novel acquisitions of the Dignatha. In millipedes, they each open into a tracheal pouch which elongates into a long, curved apodema. From the tracheal pouch a few tracheae ramify (Figure 23.4). Most pauropods lack tracheae; the Hexamerocerata, however, are endowed with spiracles, tracheal pouches, apodemata and tracheae in exactly the same combination on the first leg pair segment (Rémy, 1953).

Millipede and pauropod embryos pass through a motionless pupoid stage which is surrounded by the split chorion. The first emerging juvenile has only three pairs of legs (Enghoff *et al.*, 1993).

## 23.5 PROGONEATA: PROBABLY MONOPHYLETIC

The most plausible candidate for sister group to the Dignatha is Symphyla. Though several characters can be named as possible synapomorphies between these groups they are less convincing than those put forward for the Dignatha. Characters in favour of a taxon Progoneata (Dignatha + Symphyla) are: gonopore situated in anterior trunk segment; midgut developing within the yolk, the lumen thus being devoid of yolk; formation of fat body out

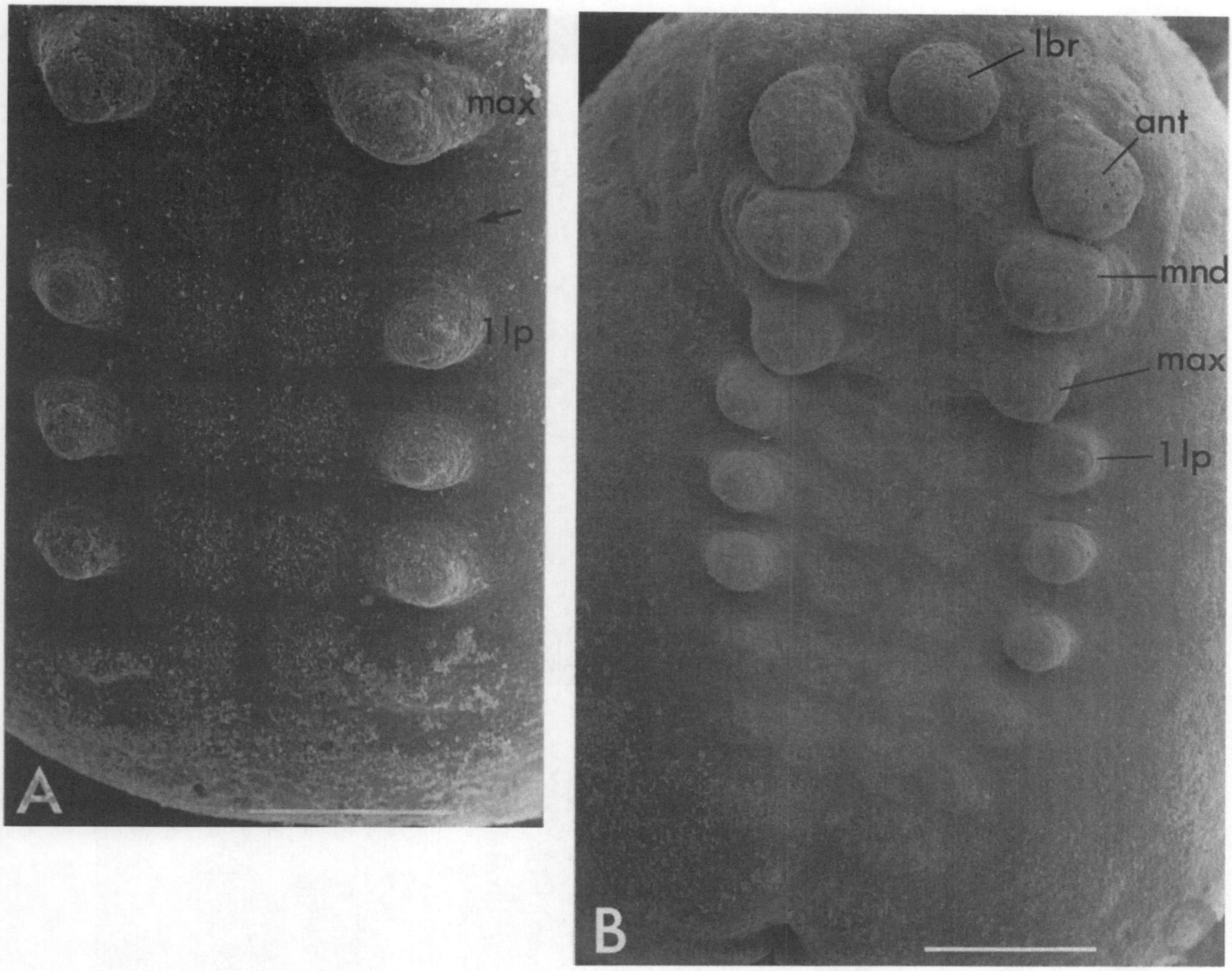


**Figure 23.1** Scanning electron micrographs of germ bands of *Glomeris marginata* (Diplopoda). (A) Early germ band. Appendage buds of first maxillae and the first three leg pairs are formed. The postmaxillary segment (= second maxillary segment) is devoid of appendage buds. (B) Slightly older germband. A furrow is formed at the site of the postmaxillary segment, but there are no appendage buds. Limb buds on the 4th and 5th segments are vestigial. ant, antenna; lbr, labrum; 1lp, limb bud of first leg; mnd, mandible; max, maxilla. Bar = 0.1 mm.

of vitellophages; trichobothria with a basal bulb; sternal apodemata; formation of a clypeo-labral ridge and others (Dohle, 1980).

## 23.6 THE QUESTION OF THE MONOPHYLY OF MYRIAPODA

There have been debates about the question whether the progoneates and the centipedes constitute a monophyletic unit called Myriapoda or whether one or the other group has closer relationships with the insects, the myriapods thus remaining a paraphyletic group (Dohle, 1980, 1988; Kraus and Kraus, 1994). I have discussed this at length and have come to the conclusion that although the question cannot be answered definitely, nothing points to a monophyly of the myriapods. Other authors maintain that the myriapods are monophyletic (Boudreaux, 1979; Ax, 1987). Some characters which have been put forward are: absence of median eyes, absence of scolopidia, absence of typical ommatidia. Absence of a character should not be considered a valid argument in phylogenetic reasoning. Absence can be primary or secondary. Primary absence means that a character which cannot be found in a species or species group has never existed in their ancestral lineages. It is therefore of no value whatsoever for phylogenetic reasoning. Secondary absence means that a character once present in an ancestral lineage has been lost in

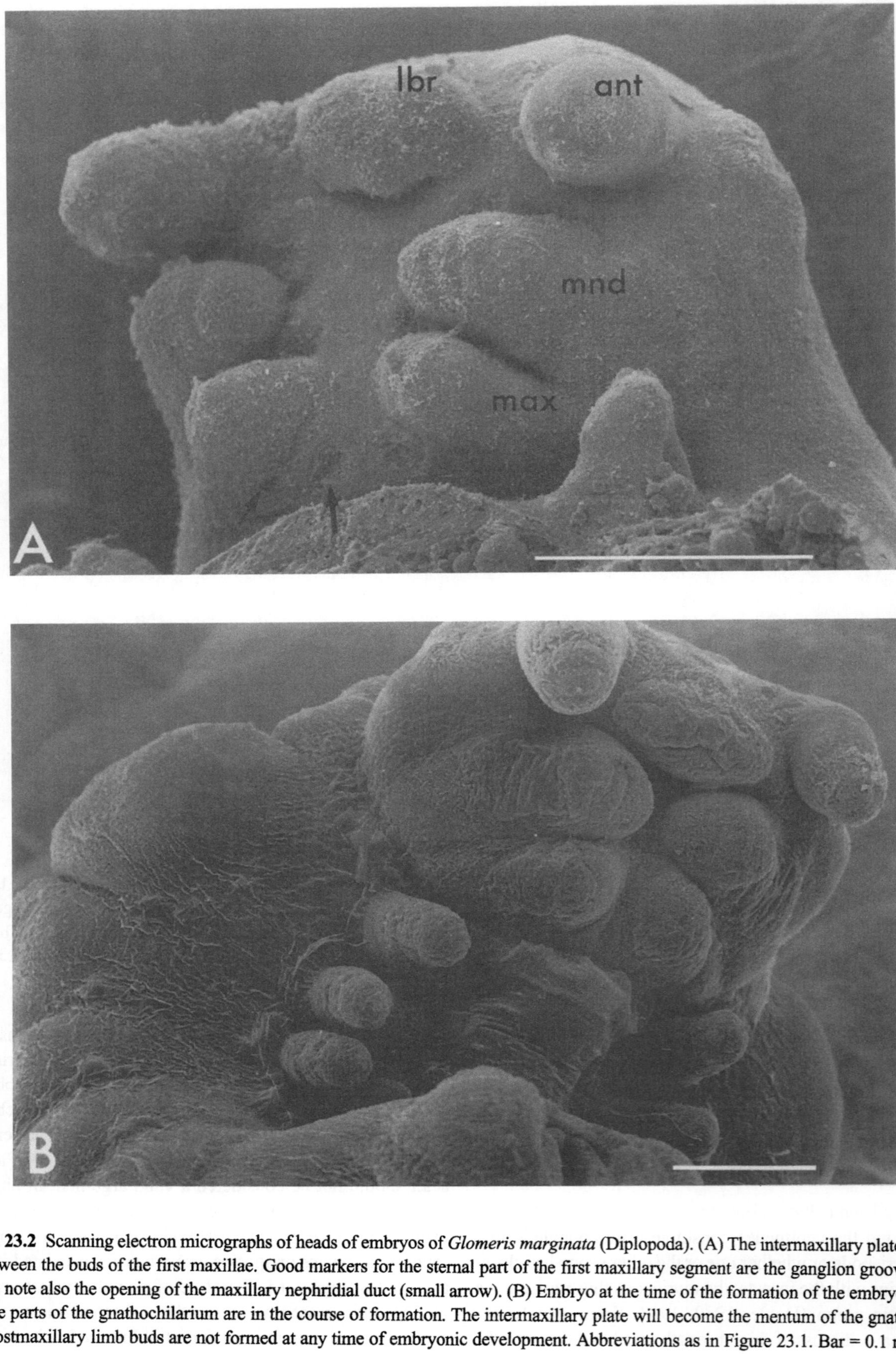


**Figure 23.2** Scanning electron micrographs of heads of embryos of *Glomeris marginata* (Diplopoda). (A) The intermaxillary plate is forming between the buds of the first maxillae. Good markers for the sternal part of the first maxillary segment are the ganglion grooves (large arrow); note also the opening of the maxillary nephridial duct (small arrow). (B) Embryo at the time of the formation of the embryonic cuticle. The parts of the gnathochilarium are in the course of formation. The intermaxillary plate will become the mentum of the gnathochilarium. Postmaxillary limb buds are not formed at any time of embryonic development. Abbreviations as in Figure 23.1. Bar = 0.1 mm.

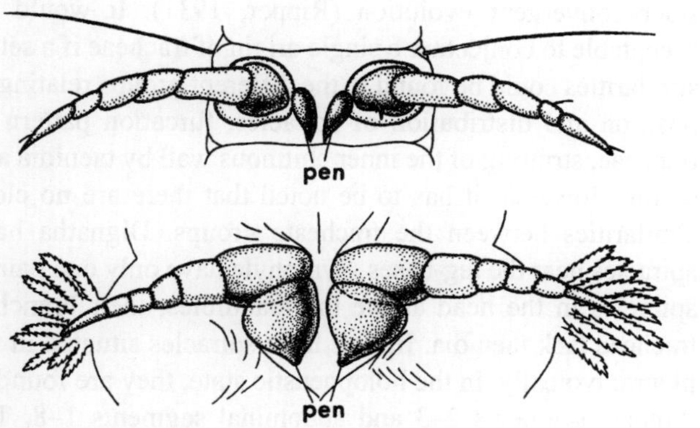


**Figure 23.3** Formations of penes (pen) at the second leg pair segment in pauropods and millipedes. (A) *Pauropus silvaticus* (after Tiegs, 1947). (B) *Polyxenus lagurus* (after Reinecke, 1910).

the descendants. This can only be inferred if one has already a clear notion of the cladogram. In addition, loss can easily be explained as convergence. If it is obvious that a character is secondarily absent in a well-established monophyletic group it is perhaps most parsimonious to postulate a single loss in the stem species. However, to claim loss of a character as the prime argument for the establishment of a relationship which cannot otherwise be demonstrated is clearly circular reasoning. I conclude that no positive character can be found in favour of the Myriapoda.

## 23.7 CHARACTERS IN FAVOUR OF ANTENNATA

Even if the mutual relationships between the myriapod groups and the insects are not clear, the question can be posed whether we are entitled to regard myriapods plus insects as a monophyletic unit. Kristensen (1991) and Wägele and Stanjek (1995) have summarized the characters supporting the Antennata. The main characters will be discussed here: absence of second antennae, formation of Malpighian tubules, postantennal organs, and tracheae.

### 23.7.1 ABSENCE OF SECOND ANTENNAE

In insects and myriapods, appendages are lacking at the segment between antennae and mandibles which is equivalent to the second antennal segment in crustaceans and is called the intercalary segment. Provided the view of an original homonomy of all segments is correct, absence of second antennae can be taken as a loss and a derived feature. Whether it has been lost once or several times in the groups and what the appearance of the appendage before the loss in the ancestral lineage was like cannot, of course, be specified. Except for the secondary absence of this character, I know of no other similarities in the intercalary segment of the myriapods and insects which are specific for both these groups. Absence of second antennae is therefore a very weak argument.

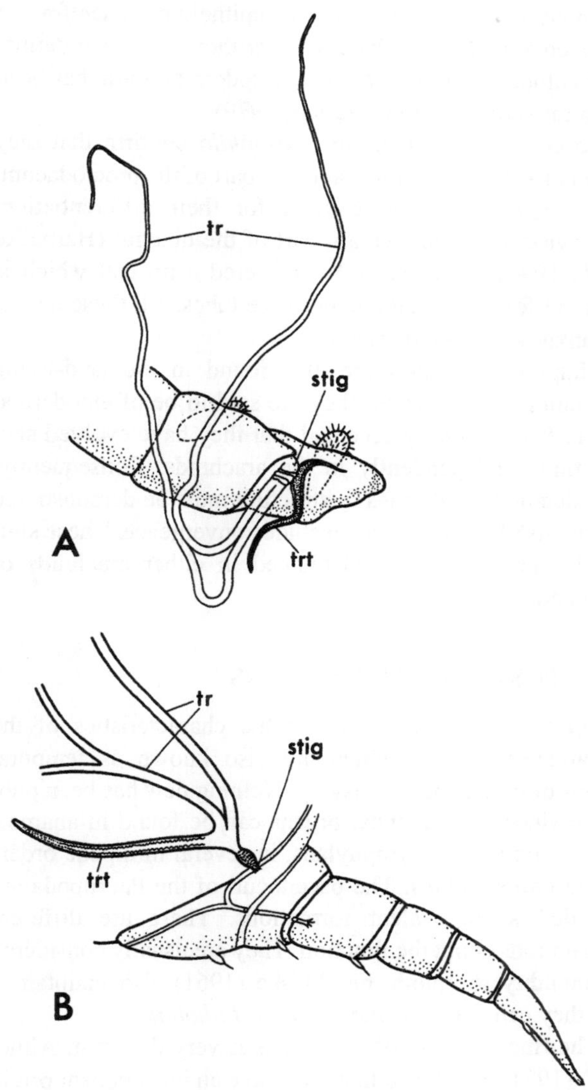


**Figure 23.4** Tracheae in pauropods and millipedes. (A) *Millotauropus latiramosus*, base of first leg (after Rémy, 1953). (B) *Polyxenus lagurus*, first leg (after Reinecke, 1910). stig, spiracle; tr, tracheae; trt, tracheal pouch.

### 23.7.2 MALPIGHIAN TUBULES

Malpighian tubules are common to myriapods and the ectognathous insects. In Entognatha they are either lacking (Collembola) or only represented by papillae between midgut and hindgut (Protura and Diplura). It has never been seriously discussed whether these papillae have to be regarded as reduced tubules. In ectognathous insects, Malpighian tubules open into the gut near the junction of the posterior midgut and hindgut. Their cells resemble hindgut cells more than midgut cells. However, they do not have a chitinous intima like the hindgut epithelium. Therefore, it has been a matter of debate whether they are of ectodermal or of entodermal origin. Even a mesodermal origin has been taken into consideration (Seifert, 1979).

Recent investigations on *Drosophila* confirm that they are outgrowths of the most anterior part of the proctodaeum. The *Krüppel* gene is required for their differentiation. Otherwise their cells remain part of the hindgut (Harbecke *et al.*, 1996). Skaer (1989) discovered a tip cell which is essential for the development of the tubes. All these details are unknown in the myriapods.

Malpighian tubules are also found in the land-living Arachnida. In this group they are said to be of entodermal origin. It is generally accepted that they have evolved several times independently in the arachnids. Consequently, very detailed similarities in structure must be demonstrated to establish homology and exclude convergence. These similarities have not been elucidated in either arachnids or antennates.

### 23.7.3 POSTANTENNAL ORGANS

Haupt (1979) has summarized the characteristics of the postantennal organs which are also known as temporal organs or organs of Tömösváry. Nothing new has been published since. Postantennal organs can be found in anamorphic centipedes, in symphylans, in several millipede orders and in collembolans. The pseudoculi of the Pauropoda are regarded as equivalent formations. There are different assumptions about the function. They are usually considered as humidity receptors, but Meske (1961) also maintained that they can detect sound waves in *Lithobius*.

The fine structure of the organs is very different. Altner *et al.* (1971) concluded that they have an independent origin in collembolans. Hoffman (1969) even doubted the single origin in millipedes. If the monophyly of the antennates were firmly established it would be no contradiction to assume a single origin of the postantennal organs, but these organs are not sufficient to support the hypothesis of a single origin of the antennates.

### 23.7.4 TRACHEAE

The assembly of myriapods and insects is often called Tracheata. This name suggests homology of tracheae in the different groups and thus a single origin in a common ancestor. Some recent authors maintain this view (Wägele and Stanjek, 1995). If this were true another important question would have been answered: the common ancestor must have been an air-breathing species.

Tracheae are defined as ectodermal tubes with a chitinous intima and with respiratory function. Such structures are found in many other land-living arthropods, e. g. arachnids and even onychophorans. Therefore, we have to consider convergent evolution (Ripper, 1931). It would be acceptable to conjecture a single origin of tracheae if a set of similarities could be found in the different groups relating to position and distribution of spiracles, furcation pattern of tracheae, strutting of the inner chitinous wall by taenidia and so on. However, it has to be noted that there are no close similarities between the tracheate groups. Dignatha have spiracles near the leg-bases. Symphyla have only one pair of spiracles in the head above the mandibles; their branched tracheae lack taenidia. Insects have spiracles situated in the pleura; typically, in the holopneustic state, they are found in thoracic segments 2–3 and abdominal segments 1–8. The collembolan ancestor must have lost the tracheae and one group within the Collembola, the Symphypleona, has gained tracheae as a new acquisition. Their tracheae radiate from a single pair of spiracles in the 'neck' region between head and thorax.

Pleurostigmophoran chilopods have paired lateral spiracles and branching tracheae strutted with taenidia as in insects. Nevertheless, this situation is regarded as secondary (Dohle, 1985). All evidence points to the conclusion that the respiratory system of the Notostigmophora (main genus *Scutigera*) with unpaired median dorsal spiracles represents the ancestral state in centipedes. Centipedes have the most complicated arterial system in all tracheate arthropods. It comprises 'aortic' and rectal arches between the dorsal heart and a ventral supraneural vessel as well as arteries to gnathal segments, trunk appendages and gut. It is nearly identical in notostigmophorans and pleurostigmophorans (Fahlander, 1938, Figures 13, 15 and 16). Such a system is functionally necessary in animals with localized and concentrated respiratory organs. In notostigmophorans, the dorsal median slit of the spiracle leads into a chamber from which short and sparsely branched tracheae emanate (Dubuisson, 1928). They only extend into the pericardial cavity. The aerated blood from this cavity must be distributed in the body via a complicated arterial system. This is a situation analogous to arachnids with localized book lungs. In pleurostigmophorans, the task of bringing oxygen to the organs has been taken over by the tracheal system with long branches which form anastomoses in scolopenders and geophilomorphs. Such a system normally leads to a profound reduction of the arterial system, as is the case in insects and also in arachnids with an equivalent status of the tracheal system (Solpugida, Opiliones). In contrast, the complicated arterial system of the centipede ancestor has been conserved in the pleurostigmophorans (Dohle, 1985). The alternative assumption, namely an evolution of the pleurostigmophoran state into the notostigmophoran condition would be incomprehensible in functional terms.

In conclusion, a six-fold convergent appearance of spiracles and tracheae must be postulated (Dohle, 1988). In Dignatha there are sternal spiracles; in Symphyla, a pair of head spiracles; in insects, pleural spiracles; in collembolans loss and reacquistion of tracheae; in the ancestral centipede, a condition as in *Scutigera* with unpaired dorsal spiracle and short tracheae; in pleurostigmophorans, evolution of a new tracheal system with pleural spiracles and long branched tracheae.

## 23.8 CHARACTERS COMMON TO INSECTS AND CRUSTACEANS

When all the arguments in favour of a close relationship between myriapods and insects are frail, as has been shown in the preceding sections, the question arises whether characters common to insects and crustaceans and lacking in the myriapods can be taken as synapomorphies. In this context, the problem of the monophyly of the crustaceans must be assessed. Whereas the different groups of crustaceans (e.g. Copepoda, Anostraca, Cirripedia, Malacostraca) can be easily shown to be monophyletic, their mutual relationships are far from clear. In addition there is no one character which gives convincing evidence that the crustaceans are monophyletic (Lauterbach, 1983; Ax, 1987). However, several of the characters in favour of a close insect–crustacean relationship have only been described in malacostracans. These characters accordingly only corroborate a close insect–malacostracan relationship, and further research is necessary to establish whether they can be found in other crustacean groups.

### 23.8.1 BRAINS

On the basis of extensive comparative research on the brains of different arthropod and other invertebrate groups, Hanström (1926) came to the conclusion that insects and malacostracans are most closely related (Figure 23.5). This conjecture has been substantiated recently through many very detailed investigations on single definable neurons in the brains of these two groups. Osorio *et al.* (1995) have summarized some of these similarities which cannot easily be explained as convergent. In addition, motor neurons in the ventral nervous system have very similar characteristics in the grasshopper and the crayfish (Wiens and Wolf, 1993). I refer the reader to the article by Nilsson and Osorio (1997, this volume).

### 23.8.2 AXONOGENESIS IN EARLY DIFFERENTIATING NEURONS

A strong case for a relatively recent common origin of insects and crustaceans can be made by referring to the early differentiation and axonogenesis of central neurons of the ventral nervous system. Axonogenesis in identified neurons was first elucidated in *Drosophila* and other insects. Thomas *et al.* (1984) postulated a common plan for neuronal development not only in insects, but also in crustaceans. As they had discovered equivalent patterns in the crayfish, they extended their claim to this group and speculated that it could be valid for all arthropods. Whitington and co-workers made an attempt to test this hypothesis in a centipede (Whitington *et al.*, 1991), in two crustaceans (Whitington *et al.*, 1993) and an apterygote insect (Whitington *et al.*, 1996). Based on the evidence of cell body position, axon morphology, and relative timing of axonal outgrowth, Whitington *et al.* (1996) found that eight neurons 'are almost certainly homologous' in winged insects and silverfish, whereas two further neurons are possibly homologous. In two crustaceans, a woodlouse and a crayfish, examined by Whitington *et al.* (1993), there were apparent differences from insects. However, at least six neurons common to both crustacean embryos appear to be homologous to equivalent insect neurons. Interestingly enough, some of the crustacean features which cannot be found in the winged insects, are present in the silverfish. Whitington *et al.* (1996: 280) concluded that 'the insects and crustaceans share a common Bauplan for the construction of central axonal pathways'. However, the pattern of neurons involved in pioneering axon growth in the centipede *Ethmostigmus rubripes* bears no obvious similarities to the insect pattern (Whitington *et al.*, 1991).

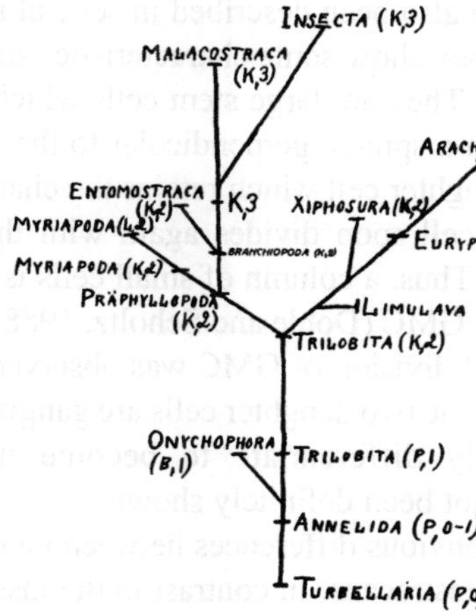


**Figure 23.5** Phylogenetic tree after Hanström (1926). P,0 = pigment cup ocelli, no vision centres; P,1 = pigment cup ocelli, formation of one vision centre; B,1 = vesicle eye, 1 vision centre; L,2 = ocelli with lens, 2 vision centres; K,2 = complex eyes, 2 vision centres; K,3 = complex eye, 3 vision centres.

### 23.8.3 NEUROBLASTS

Neuroblasts in insects were first described by Wheeler (1893) and Heymons (1895) at the turn of the century. It remained for Bate (1976) to demonstrate that they have a fixed number and arrangement in the ventral nerve cord. Subsequent studies showed that the generation of ganglion mother cells (GMC) and the differentiation of the neurons also follows a defined pattern.

Neuroblasts have also been described in several malacostracan embryos. They show some characteristics similar to those in the insects. They are large stem cells which divide asymmetrically with a spindle perpendicular to the surface. The larger distal daughter cell which retains the characteristics of the mother cell soon divides again with the same spindle orientation. Thus, a column of small cells is formed which constitute the GMC (Dohle and Scholtz, 1988). As no more than one equal division of GMC was observed, it has been concluded that the two daughter cells are ganglion cells and will eventually differentiate to become neurons. However, this has not been definitely shown.

There are some obvious differences between neuroblasts in insects and malacostracans. In contrast to the insects, the neuroblasts of the postnaupliar segments in malacostracans are generated by a stereotyped sequence of divisions. They remain at the surface and are not surrounded by sheath cells. In at least one case it has been demonstrated that a neuroblast, after having generated GMC, divides symmetrically to give rise to an epidermal cell and another neuroblast (Dohle, 1976). This is not known to happen in insects. Because of these differences, and also because at that time no neuroblasts had been found either in entomostracans or in myriapods, I concluded that neuroblasts have been evolved convergently in insects and malacostracans (Dohle, 1976). It may be time to revise this opinion.

Neuroblasts with the above characteristics are lacking in the myriapods. The ganglia are formed by a kind of invagination process. In centipedes and millipedes, a ganglion groove is formed. As the groove sinks deeper, the edges of the ganglion groove eventually coalesce. The closed groove can be observed as a hole within the developing ganglion of late embryos for a long period (Heymons, 1901; Dohle, 1964). As onychophorans and chelicerates show ganglion formation which resembles the mode in the myriapods, it seems plausible to regard it as the original mode, which cannot, therefore, play a role in the discussion about sister group relationships within the arthropods.

In entomostracan crustaceans, prior investigations could not demonstrate the presence of neuroblasts (summarized by Weygoldt, 1994). However, in the cladoceran *Leptodora*, Gerberding (1997) found cells with the characteristics of neuroblasts. Without further information and additional evidence of similarities in the mode of formation, arrangement, fate of GMC and neurone genealogy, the question whether neuroblasts in insects and crustaceans can be traced back to a common origin or whether they are the result of convergent evolution cannot be answered. With regard to the homology of the pattern of neurons dealt with in the preceding section it is, however, important to bear in mind that these two questions are not inextricably interconnected. Whitington (1996, p. 611) rightly states 'that even if crustacean neuroblasts are ultimately shown not to be homologous to insect neuroblasts, this would not necessarily invalidate the hypothesis that the early differentiating neurons shared between insects and crustaceans are true homologues'. Nearly identical and most certainly homologous patterns can be generated by cells which have a different genealogical origin (Dohle, 1989; Scholtz and Dohle, 1996).

### 23.8.4 OMMATIDIA

There are very detailed structural similarities in the ommatidia of insects and crustaceans. These 'cannot be due to convergence' (Paulus, 1979, p. 364). Myriapods are either blind (Symphyla, Pauropoda) or have lateral fields of ocelli which are completely different from the ommatidia in insects (Bähr, 1974; Spies, 1981). Only in *Scutigera* we find a faceted eye with cone-shaped ommatidia. However, these have a fine structure different from hexapodan and crustacean ommatidia and have been regarded as the result of parallel evolution (Paulus, 1979). In the light of the new considerations it is possible that the lateral ocelli of myriapods may have a different origin and must not necessarily be regarded as reduced or altered ommatidia of the hexapodan type.

### 23.8.5 SEGMENTATION GENES

It is to be expected that the growing knowledge on distribution, structure and function of genes affecting segmentation and differentiation in arthropods will enhance or modify our present opinions on relationship and phylogeny. Since methods of molecular genetics have been extended to other arthropods (Patel *et al.*, 1989; Averof and Akam, 1995; Panganiban *et al.*, 1995; Popadić *et al.*, 1996) several unexpected similarities have been revealed between insects and crustaceans.

In *Drosophila*, the segmentation gene *engrailed* is first expressed in one transversal row of cells. These cells mark the anterior compartment of a parasegment. The distance to the next *engrailed*-positive row is approximately two to three cells wide. The *engrailed* stripe widens and eventually defines the posterior margin of the embryonic segment. Posterior cells of a clone may lose *engrailed* expression (Vincent and O'Farrell, 1992). Therefore, the commitment of cells as *engrailed*-positive is not exclusively specified by their lineage.

In several higher crustaceans, *engrailed* expression can first be detected on the postnaupliar germ band in every second row of cells (Scholtz *et al.*, 1993; Patel, 1994). This is the anterior row descendant from a row of cells which have divided with longitudinal spindles. When this row divides once more the posterior row loses *engrailed* expression so that each *engrailed* stripe is separated from the next one by exactly three cells (Scholtz and Dohle, 1996). This shows that expression of *engrailed* is not mainly dependent on cell lineage.

The monoclonal antibody which binds to the *engrailed*

gene product in *Drosophila* as well as in other insects and higher crustaceans (Patel *et al.*, 1989) does not bind in the millipede *Glomeris*. The situation in the germ band of millipedes shortly before visible segmentation is completely different from insects and higher crustaceans. There are many small and tightly packed ectodermal cells, at least 15 in longitudinal sections between two just forming intersegmental furrows (Dohle, 1964).

There are several other hints that the expression of genes in insects and crustaceans is comparable. Osorio *et al.* (1995) reported that Patel and co-workers have found an expression pattern of *even-skipped* in isopod ganglion cells which is similar to insects. These and other correspondences remain to be described in detail before further conclusions can be drawn.

## 23.9 CONCLUSIONS

For several decades the close relationship between insects and myriapods has not seriously been questioned. However, if a critical attitude is adopted, the arguments in favour of this relationship are far from being persuasive. On the contrary, an increasing number of specific similarities between insects and crustaceans, especially higher crustaceans, have been revealed which cannot easily be dismissed as haphazard or as convergent. These similarities comprise features of brain morphology, specification of single-defined neurons, fine structure of ommatidia and certain points of expression of segmentation genes. As some of these features cannot be found in equivalent form in the myriapods they can either be explained as synapomorphies of crustaceans and insects or as ancient inheritances from the mandibulate ancestor which have vanished in the myriapods. The latter viewpoint becomes progressively more unlikely.

I have not yet mentioned several phylogenetic trees inferred from molecular data which also point to a close proximity of insects and crustaceans and which exclude the myriapods (Ballard *et al.*, 1992; Friedrich and Tautz, 1995). They can provide an additional stimulus to check carefully on the correspondence and distribution of all relevant characters in order to establish the framework of a well-founded and stable phylogenetic tree.

## ACKNOWLEDGEMENTS

Gerhard Scholtz and Stefan Richter made valuable comments on the manuscript. The referee, Dr Vane-Wright, induced some revisions and clearer formulations. Richard Thomas and Richard Fortey are not only thanked for the organization of the Symposium and for their kind invitation, but also for many valuable suggestions. Mrs Cooper-Kovacs corrected the English text. To these persons I offer my sincere thanks.

## REFERENCES

Altner, H., Karuhize, G. and Ernst, K.-D. (1971) Untersuchungen am Postantennalorgan der Collembolen (Apterygota) II. Cuticulärer Apparat und Dendritenendigung bei *Onychiurus* spec. *Revue d'Écologie et de Biologie du Sol*, **8**, 31–5.

Averof, M. and Akam, M. (1995) Insect–crustacean relationships: insights from comparative developmental and molecular studies. *Philosophical Transactions of the Royal Society of London B*, **347**, 293–303.

Ax, P. (1987) *The Phylogenetic System. The systematization of organisms on the basis of their phylogenesis*, John Wiley and Sons, Chichester.

Ballard, J.W.O., Olsen, G.J., Faith, D.P., Odgers, W.A., Rowell, D.M. and Atkinson, P.W. (1992) Evidence from 12S ribosomal RNA sequences that onychophorans are modified arthropods. *Science*, **258**, 1345–8.

Bähr, R.R. (1974) Contribution to the morphology of chilopod eyes. *Symposia of the Zoological Society of London*, **32**, 383–404.

Bate, C.M. (1976) Embryogenesis of an insect nervous system I. A map of the thoracic and abdominal neuroblasts in *Locusta migratoria*. *Journal of Embryology and Experimental Morphology*, **35**, 107–23.

Boudreaux, H.B. (1979) *Arthropod Phylogeny with Special Reference to Insects*, John Wiley and Sons, New York.

Dohle, W. (1964) Die Embryonalentwicklung von *Glomeris marginata* (VILLERS) im Vergleich zur Entwicklung anderer Diplopoden. *Zoologische Jahrbücher, Abteilung für Anatomie und Ontogenie der Tiere*, **81**, 241–310.

Dohle, W. (1974) The segmentation of the germ band of Diplopoda compared with other classes of arthropods. *Symposia of the Zoological Society of London*, **32**, 143–61.

Dohle, W. (1976) Die Bildung und Differenzierung des postnauplialen Keimstreifs von *Diastylis rathkei* (Crustacea, Cumacea) II. Die Differenzierung und Musterbildung des Ektoderms. *Zoomorphologie*, **84**, 235–77.

Dohle, W. (1980) Sind die Myriapoden eine monophyletische Gruppe? Eine Diskussion der Verwandtschaftsbeziehungen der Antennaten. *Abhandlungen des naturwissenschaftlichen Vereins Hamburg, NF*, **23**, 45–104.

Dohle, W. (1985) Phylogenetic pathways in the Chilopoda. *Bijdragen tot de Dierkunde*, **55**, 55–66.

Dohle, W. (1988) *Myriapoda and the Ancestry of Insects*, The Manchester Polytechnic, Manchester, UK.

Dohle, W. (1989) Zur Frage der Homologie ontogenetischer Muster. *Zoologische Beiträge, NF*, **32**, 355–89.

Dohle, W. and Scholtz, G. (1988) Clonal analysis of the crustacean segment: the discordance between genealogical and segmental borders. *Development*, **104** Supplement, 147–60.

Dubuisson, M. (1928) Recherches sur la ventilation trachéenne chez les Chilopodes et sur la circulation sanguine chez les Scutigères. *Archives de Zoologie expérimemtale et génerale*, **67**, 49–63.

Enghoff, H. (1984) Phylogeny of millipedes – a cladistic analysis. *Zeitschrift für zoologische Systematik und Evolutionsforschung*, **22**, 8–26.

Enghoff, H. (1990) The ground-plan of the chilognathan millipedes (external morphology), in *Proceedings of the 7th*

*International Congress of Myriapodology* (ed. A. Minelli), E. J. Brill, Leiden, pp. 1–21.

Enghoff, H., Dohle, W. and Blower, J.G. (1993) Anamorphosis in millipedes (Diplopoda) – the present state of knowledge with some developmental and phylogenetic considerations. *Zoological Journal of the Linnaean Society*, **109**, 103–234.

Fahlander, K. (1938) Beiträge zur Anatomie und sytematischen Einteilung der Chilopoden. *Zoologiska Bidrag fran Uppsala*, **17**, 1–148; plates 1–18.

Fechter, H. (1961) Anatomie und Funktion der Kopfmuskulatur von *Cylindroiulus teutonicus* (Pocock). *Zoologische Jahrbücher, Abteilung Anatomie und Ontogenie der Tiere*, **79**, 479–582.

Friedrich, M. and Tautz, D. (1995) Ribosomal DNA phylogeny of the major extant arthropod classes and the evolution of myriapods. *Nature*, **376**, 165–7.

Gerberding, M. (1997) Germ band formation and early neurogenesis of *Leptodora kindti* (Cladocera): first evidence for neuroblasts in the entomostracan crustaceans, *Invertebrate Reproduction and Development*, **32**, 63–73.

Hanström, B. (1926) Eine genetische Studie über die Augen und Sehzentren von Turbellarien, Anneliden und Arthropoden (Trilobiten, Xiphosuren, Eurypteriden, Arachnoiden, Myriapoden, Crustaceen und Insekten). *Kungl. Svenska Vetenskapsakademiens Handlingar*, Tredje Serien **4** (1), pp. 1–176.

Harbecke, R., Meise, M., Holz, A., Klapper, R., Naffin, E., Nordhoff, V. and Janning, W. (1996) Larval and imaginal pathways in early development of *Drosophila*. *The International Journal of Developmental Biology*, **40**, 197–204.

Haupt, J. (1979) Phylogenetic aspects of recent studies on myriapod sense organs, in *Myriapod Biology* (ed. M. Camatini), Academic Press, London, pp. 391–406.

Hennig, W. (1953) Kritische Bemerkungen zum phylogenetischen System der Insekten. *Beiträge zur Entomologie*, **3**, Sonderheft, 1–85.

Heymons, R. (1895) *Die Embryonalentwicklung von Dermapteren und Orthopteren unter besonderer Berücksichtigung der Keimblätterbildung*. Gustav Fischer, Jena.

Heymons, R. (1901) Die Entwicklungsgeschichte der Scolopender. *Zoologica*, **33**, 1–244, plates I–VIII.

Hilken, G. and Kraus, O. (1994) Struktur und Homologie der Komponenten des Gnathochilarium der Chilognatha (Tracheata, Diplopoda). *Verhandlungen des naturwissenschaftlichen Vereins Hamburg, NF*, **34**, 33–50.

Hoffman, R.L. (1969) Myriapoda, exclusive of Insecta, in *Treatise on Invertebrate Palaeontology* (ed. R.C. Moore), Arthropoda 4, Kansas, pp. 572–606.

Kraus, O. and Kraus, M. (1994) Phylogenetic system of the Tracheata (Mandibulata): on 'Myriapoda'–Insecta interrelationships, phylogenetic age and primary ecological niches. *Verhandlungen des naturwissenschaftlichen Vereins Hamburg, NF*, **34**, 5–31.

Kristensen, N.P. (1991) Phylogeny of extant hexapods, in *The Insects of Australia* (ed. Csiro), 2nd edn., Vol. 1. Cornell University Press, Ithaca, NY, pp. 125–40

Lauterbach, K.-E. (1983) Zum Problem der Monophylie der Crustacea. *Verhandlungen des naturwissenschaftlichen Vereins Hamburg, NF*, **26**, 293–320.

Meske, C. (1961) Untersuchungen zur Sinnesphysiologie von Diplopoden und Chilopoden. *Zeitschrift für vergleichende Physiologie*, **45**, 61–77.

Osorio, D., Averof, M. and Bacon, J.P. (1995) Arthropod evolution: great brains, beautiful bodies. *Trends in Evolution and Ecology*, **10**, 449–54.

Panganiban, G., Sebring, A., Nagy, L. and Carroll, S. (1995) The development of crustacean limbs and the evolution of arthropods. *Science*, **270**, 1363–6.

Patel, N.H. (1994) The evolution of arthropod segmentation: insights from comparisons of gene expression patterns. *Development*, **1994** Supplement, 201–7

Patel, N.H., Kornberg, T.B. and Goodman, C.S. (1989) Expression of *engrailed* during segmentation in grasshopper and crayfish. *Development*, **107**, 201–12

Paulus, H.F. (1979) Eye structure and the monophyly of the arthropods, in *Arthropod Phylogeny* (ed. A.P. Gupta), Van Nostrand Reinhold, New York, pp. 299–383.

Popadic, A., Rusch, D., Peterson, M., Rogers, B.T. and Kaufman, T.C. (1996) Origin of the arthropod mandible. *Nature*, **380**, 395.

Reinecke, G. (1910) Beiträge zur Kenntnis von *Polyxenus*. *Jenaische Zeitschrift für Naturwissenschaften*, **46** (NF39), 845–96, plates 31–5.

Rémy, P.A. (1953) Description de nouveaux types de Pauropodes: 'Millotauropus' et 'Rabaudauropus'. *Mémoires de l'Institut Scientifique de Madagascar, série A*, **8**, 25–41.

Ripper, W. (1931) Versuch einer Kritik der Homologiefrage der Arthropodentracheen. *Zeitschrift für wissenschaftliche Zoologie*, **138**, 303–69.

Scholtz, G. and Dohle, W. (1996) Cell lineage and cell fate in crustacean embryos – a comparative approach. *The International Journal of Developmental Biology*, **40**, 211–20.

Scholtz, G., Dohle, W., Sandeman, R.E. and Richter, S. (1993) Expression of *engrailed* can be lost and regained in cells of one clone in crustacean embryos. *The International Journal of Developmental Biology*, **37**, 299–304.

Seifert, G. (1979) Considerations about the evolution of excretory organs in terrestrial arthropods, in *Myriapod Biology* (ed. M. Camatini), Academic Press, London, pp. 353–72.

Skaer, H. (1989) Cell division in Malpighian tubule development in *D. melanogaster* is regulated by a single tip cell. *Nature*, **342**, 566–9.

Spies, T. (1981) Structure and phylogenetic interpretation of diplopod eyes (Diplopoda). *Zoomorphology*, **98**, 241–60.

Thomas, J.B., Bastiani, M.J., Bate, M. and Goodman, C.S. (1984) From grasshopper to *Drosophila*: a common plan for neuronal development. *Nature*, **310**, 203–7.

Tiegs, O.W. (1947) The development and affinities of the Pauropoda, based on a study of *Pauropus silvaticus*. *Quarterly Journal of Microscopical Science*, **88**, 165–267, 275–336, Plates I–XI.

Verhoeff, K.W. (1911–1914) Die Diplopoden Deutschlands, zusammenfassend bearbeitet, zugleich eine allgemeine Einführung in die Kenntnis der Diplopoden-Systematik, der Organisation, Entwicklung, Biologie und Geographie, Winter, Leipzig.

Vincent, J.-P. and O'Farrell, P.H. (1992) The state of *engrailed* expression is not clonally transmitted during early *Drosophila* development. *Cell*, **68**, 923–31.

Wägele, J.W. and Stanjek, G. (1995) Arthropod phylogeny inferred from 12S rRNA revisited: monophyly of the Tracheata depends on sequence alignment. *Journal of Zoological Systematics and Evolutionary Research*, **33**, 75–80.

Weygoldt, P. (1994) Le développement embryonnaire, in *Traité de Zoologie*, Tome VII, Fasc. 1. (ed. P.-P. Grassé), Masson, Paris, pp. 807–89.

Wheeler, W.M. (1893) A contribution to insect embryology. *Journal of Morphology*, **8**, 1–160.

Whitington, P.M. (1996) Evolution of neural development in the arthropods. *Seminars in Cell and Developmental Biology,* (**7**, 605–14

Whitington, P.M., Meier, T. and King, P. (1991) Segmentation, neurogenesis and formation of early axonal pathways in the centipede, *Ethmostigmus rubripes* (Brandt). *Roux's Archives of Developmental Biology*, **199**, 349–63.

Whitington, P.M., Leach, D. and Sandeman, R. (1993) Evolutionary change in neural development within the arthropods: axonogenesis in the embryos of two crustaceans. *Development*, **118**, 449–61.

Whitington, P.W., Harris, K.-L. and Leach, D. (1996) Early axonogenesis in the embryo of a primitive insect, the silverfish *Ctenolepisma longicaudata*. *Roux's Archives of Developmental Biology*, **205**, 272–81.

Wiens, T.J. and Wolf, H. (1993) The inhibitory motor neurons of crayfish thoracic limbs: identification, structures, and homology with insect common inhibitors. *Journal of Comparative Neurology*, **336**, 261–78.

# 24 Cleavage, germ band formation and head segmentation: the ground pattern of the Euarthropoda

G. Scholtz

*Institut für Biologie, Vergleichende Zoologie Humboldt-Universität zu Berlin, Philippstrasse 13, D-10115 Berlin, Germany*
*email: gerhard=scholtz@rz.hu-berlin.de*

## 24.1 INTRODUCTION

### 24.1.1 EMBRYOLOGY AND ARTHROPOD PHYLOGENY

Comparative embryological studies have always played an important role in the interpretation and understanding of arthropod origins and phylogeny (Bowler, 1994). Embryonic data have been used to support both arthropod monophyly (Weygoldt, 1979, 1986) as well as polyphyly (Anderson, 1973). Now there is convincing and increasing evidence from palaeontological, embryological, morphological, and molecular studies that arthropods are monophyletic (Weygoldt, 1986; Wägele, 1993; Walossek, 1993; Wheeler *et al.*, 1993; Friedrich and Tautz, 1995; Wills *et al.*, 1995; Wägele and Stanjek, 1995; Garey *et al.*, 1996). This is especially clear for the 'true' arthropods, the Euarthropoda, which include the chelicerates, crustaceans, myriapods, and insects. The method of phylogenetic systematics (Hennig, 1950, 1966) allows not only the analysis of the phylogenetic relationships between taxa but it also offers a tool for the reconstruction of the ground pattern of a given monophyletic group (Ax, 1987). This ground pattern is inferred on the basis of the phylogenetic systematics and the character distribution in that monophyletic group. It represents the set of characters which were present in the group's stem species. The ground pattern is a mixture of apomorphic and plesiomorphic characters of the taxon under investigation. The knowledge of the ground pattern provides us with the starting point for the analysis of evolutionary alterations (Lauterbach, 1980; Bitsch, 1994; Sandeman and Scholtz, 1995; Scholtz, 1995a,b). The phylogenetic relationships between the higher euarthropod taxa are not settled. In particular, the close relationship between insects and myriapods has been challenged by recent investigations (Averof and Akam, 1995; Friedrich and Tautz, 1995; Osorio *et al.*, 1995; see other contributions in this volume). Nevertheless, it is possible and worthwhile to reconstruct the ground pattern of euarthropods with regard to the cleavage type, the mode of germ band formation, and the segmentation of the head. In addition, some embryonic characters are discussed that provide further arguments for arthropod and euarthropod monophyly.

### 24.1.2 THE SISTER GROUP OF ARTHROPODS

It is a widespread view that annelids and arthropods are closely related or, depending on annelid monophyly, even sister groups (Lauterbach, 1980; Weygoldt, 1986; Brusca and Brusca, 1990; Wheeler *et al.*, 1993; Nielsen, 1995; Westheide, 1996). On the other hand, this close relationship has been disputed and a monophyletic origin of animals with a trochophora larva (Eutrochozoa) has been suggested (Eernisse *et al.*, 1992, Winnepenninckx *et al.*, 1995; Eernisse, 1997, this volume). However, the assumption of a close relationship between annelids and arthropods is supported by a number of derived characters shared by annelids and arthropods: an ectodermal and mesodermal growth zone anterior to a terminal body portion bearing the anus, segment formation in an anteroposterior sequence with the ventral side showing a higher degree of differentiation than the dorsal side, segmental paired ganglia forming a ladder-like central nervous system, segmental paired coelomic sacs, segmental paired metanephridia, external annulation, a long dorsal tube-like heart, and mushroom bodies characterized by a certain cell type in a characteristic arrangement situated in the anterior part of the brain (Strausfeld *et al.*, 1995). The characters related to segmentation, in particular, form a complex that cannot be found in any other protostome group. On the molecular level this is supported by the similar mode of dual *engrailed* gene expression during early segmentation and neurogenesis in annelids and arthropods (Weisblat *et al.*, 1993). Thus, segmentation in annelids and arthropods is homologous and most likely an apomorphy for the Articulata. Alternatively, according to the eutrochozoan hypothesis one has to assume that either segmentation of annelids and arthropods is convergent (which is unlikely) or the common ancestor of eutrochozoans and arthropods was already segmented (Eernisse *et al.*, 1992).

*Arthropod Relationships*, Systematics Association Special Volume Series 55, Edited by R.A. Fortey and R.H. Thomas. Published in 1997 by Chapman & Hall, London. ISBN 0 412 75420 7

The latter assumption needs to explain the independent loss of segmentation in most spiralian taxa and it seems more plausible to interpret the serially arranged characters in some molluscs as either a first step towards the articulate-like segmentation complex (Nielsen, 1995) or as an independent development (Lauterbach, 1983, Haszprunar and Schaefer, 1996)).

## 24.2 CLEAVAGE

### 24.2.1 THE CLEAVAGE MODES IN EUARTHROPODS

Euarthropods exhibit a great variety of different cleavage types (Siewing, 1969; Fioroni, 1970; Anderson, 1973) (Figures 24.1–24.4). The common mode of superficial cleavage is characterized by yolky eggs with intralecithal cleavage divisions with the cleavage products (energids) not being separated by membranes. Later, the energids migrate to the egg surface and form the blastoderm with a central yolk mass and the germ becomes cellular (Figure 24.4). In some taxa we find a holoblastic cleavage type with a coeloblastula (Figure 24.3) and an invagination gastrula. In addition, there are cleavage modes that are intermediate between these two extremes (mixed cleavage, see Fioroni, 1970), e.g. they start with total cleavage and switch in later stages to the superficial mode showing a blastoderm stage. This variety of cleavage types is found in each of the major euarthropod groups: chelicerates, crustaceans, myriapods and insects. Even in closely related taxa several cleavage types occur and it is apparently not a big step to alter a superficial cleavage into holoblastic cleavage and vice versa. This character distribution makes it difficult to reconstruct the ground pattern of the early development of euarthropods. Contradictory views include the opinion that superficial cleavage in combination with a yolky egg is a synapomorphy of onychophorans and euarthropods (Weygoldt, 1986), while it has also been suggested that the plesiomorphic condition for arthropods is holoblastic cleavage with eggs which are poor in yolk (Siewing, 1969). The claim that this holoblastic cleavage is still a type of spiral cleavage or that it at least shows traces of the spiral cleavage pattern found in annelids (Anderson, 1973; Nielsen, 1995) has been challenged by Siewing (1969) and Dohle (1989).

The variety of combinations of developmental processes evident in the mixed cleavage types shows that early development is a sequence of stages where each step can be evolutionarily altered without affecting subsequent stages (Scholtz and Dohle, 1996). For instance, early intralecithal divisions are changed towards total divisions but the blastoderm stage is retained. Thus, it is not possible to infer cleavage mode as a whole from individual stages. We cannot conclude that a large amount of yolk leads to intralecithal divisions and that these lead to a blastoderm stage. Accordingly, these cleavage modes are subdivided into different stages which will be discussed separately.

#### (a) Early cleavage – intralecithal or total, radial or spiral?

The problem of reconstructing the original arthropod pattern of early cleavage is due to the difficulties in homologizing intralecithal cleavage as such. One needs certain distinct characters such as position or lineage of the division products to claim homology; otherwise, it is just the absence of membranes separating the blastomeres that unifies intralecithal cleavage. There are very few studies which have tried to trace the lineage of the early cleavage products in intralecithal cleavages (Dohle, 1970). In several euarthropod lineages intralecithal cleavage has been altered towards total cleavage and vice versa. Therefore, a different path is pursued here – cases of total cleavage are compared in order to look for more detailed similarities. With the help of additional data the question of the original euarthropod cleavage type is readdressed.

The early total cleavage of representatives of chelicerates (Pantopoda: Dogiel, 1913), crustaceans (Cladocera: Kühn, 1913; Cirripedia: Anderson, 1965; Copepoda: Fuchs, 1914; Ostracoda: Weygoldt, 1960; Euphausiacea: Taube, 1909; Decapoda: Zilch, 1978; Hertzler and Clark, 1992; Isopoda: Strömberg, 1971; Amphipoda: Rappaport, 1960), myriapods (Symphyla: Tiegs, 1940; Pauropoda: Tiegs, 1947; Diplopoda: Dohle, 1964), and insects (Collembola: Claypole, 1898) show some distinct similarities (Figures 24.1 and 24.2). They undergo an early 'radial' cleavage with mostly no oblique spindle orientation, regardless of whether the cleavages are equal or unequal (Figures 24.1 and 24.2). Subsequent divisions show mainly spindle directions orthogonal to the preceding ones, although there is some variation in this. In addition, the relative positions of blastomeres vary between individual eggs. Thus, cleavage appears irregular and no traceable cell lineage exists in advanced stages. This has been reported for several crustaceans (Weygoldt, 1960; Benesch, 1969; Scheidegger, 1976) and myriapods (Tiegs, 1940, 1947; Dohle, 1964) (Figure 24.1). The same pattern can be seen in clear cases of secondary early total cleavage as in parasitic isopods (Strömberg, 1971) (Figure 24.1) and even in intralecithally cleaving eggs (Dohle, 1970). The general pattern of early cleavage allows the conclusion that 'radial' cleavage is part of the arthropod ground pattern. The term 'radial' cleavage is used in the meaning of a radially oriented position of the cleavage products (energids or cells); it is not meant in the strict definition for the radial cleavage type given by Siewing (1969, 1979). Since arthropods are part of the Spiralia, the 'radial' cleavage is an apomorphy of arthropods. This conclusion relates only to the relative position of the division products of early cleavage. It does not clarify the question as to whether the radially oriented nuclei and the cytoplasm were separated by membranes or not.

Several authors have claimed that spiral cleavage occurs in some crustaceans with mixed or holoblastic cleavage modes (Taube, 1909; von Baldass, 1941; Anderson, 1969, 1973;

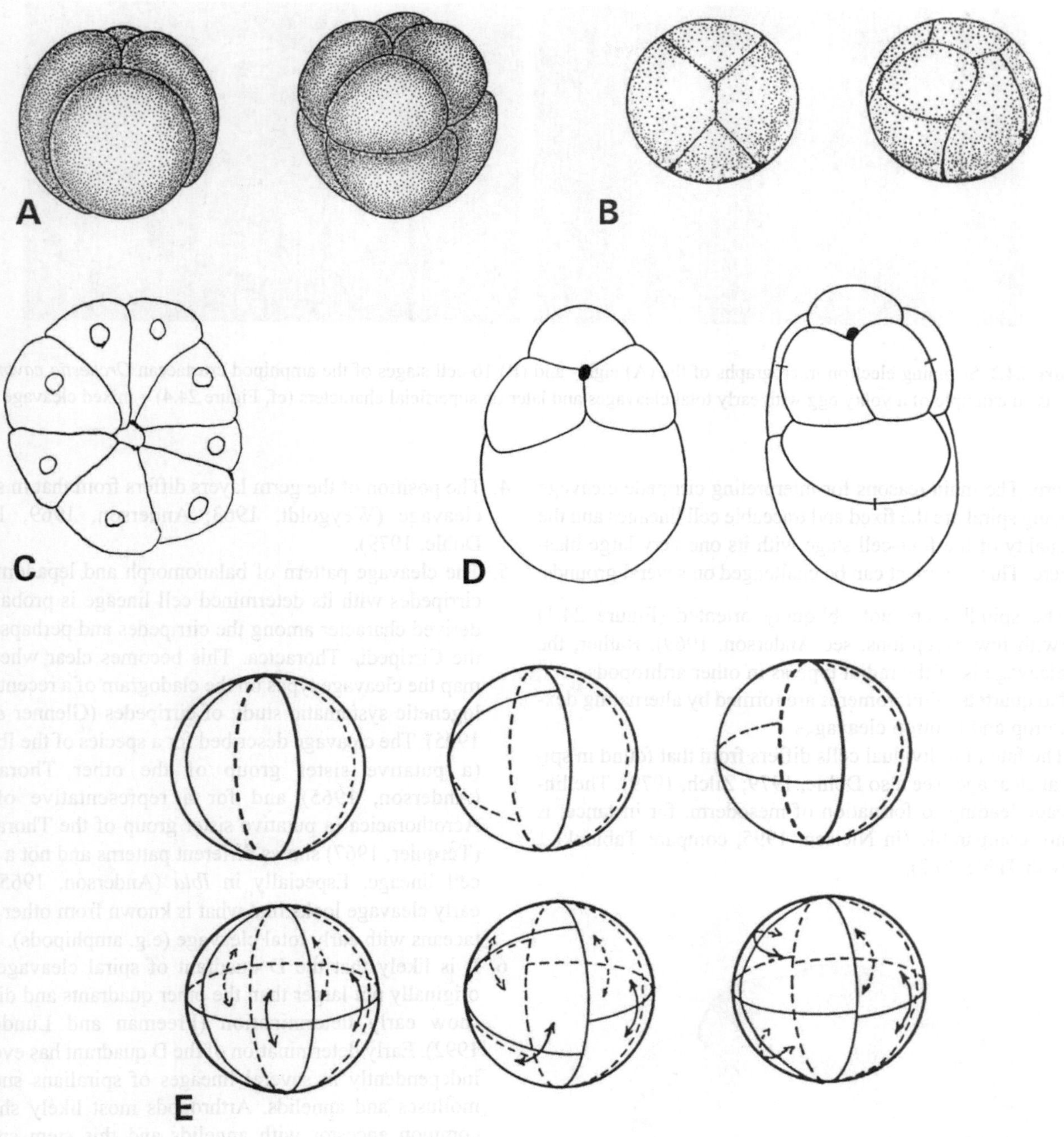


**Figure 24.1** The early total cleavage of some arthropods showing the pattern of irregular 'radial' cleavage. (A) Insecta, Collembola, four- and eight-cell stages. (Modified from Anderson, 1973, after Claypole, 1898.) (B) Myriapods, Diplopoda, four- and irregular eight-cell stages. (After Dohle, 1964.) (C) Chelicerata, Pantopoda, section through a 32-cell stage. (After Dogiel, 1913.) (D) Crustacea, Cirripedia, four- and eight-cell stages. (Modified from Nielsen, 1995, after Delsman, 1917.) (E) Three different arrangements of the blastomeres in the four- and eight-cell stages of the parasitic isopod *Bopyroides.* (After Strömberg, 1971.)

Nielsen, 1995). This has been disputed by Siewing (1979), Dohle (1979, 1989), Zilch (1979) and Hertzler and Clark (1992). In particular, the cleavage patterns found in some species of Cirripedia have been interpreted in this way (Bigelow, 1902; Delsman, 1917; Anderson, 1973; Costello and Henley, 1976; Nielsen, 1995). However, these interpretations differ considerably with regard to the application of spiral nomenclature and fate of individual cells (compare Anderson, 1973; Costello and Henley, 1976; Nielsen, 1995) and cell lineages have never been analysed up to the formation of germ layers. Nielsen (1995) even considered the cirripede cleavage pattern as part of the euarthropod ground

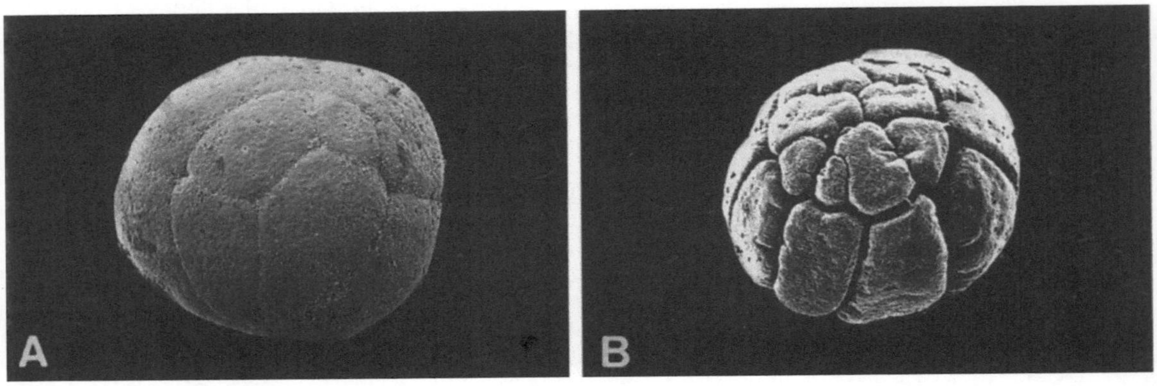

**Figure 24.2** Scanning electron micrographs of the (A) eight- and (B) 16-cell stages of the amphipod crustacean *Orchestia cavimana*. This is an example of a yolky egg with early total cleavages and later on superficial characters (cf. Figure 24.4) – mixed cleavage type.

pattern. The main reasons for interpreting cirripede cleavage as being spiral are the fixed and traceable cell lineages and the inequality of the four-cell stage with its one very large blastomere. This argument can be challenged on several grounds:

1. The spindles are not obliquely oriented (Figure 24.1) (with few exceptions, see Anderson, 1969). Rather, the cleavage is of the radial type as in other arthropods.
2. No quartets of micromeres are formed by alternating dexiotrop and leiotrop cleavages.
3. The fate of individual cells differs from that found in spiral cleavage (see also Dohle, 1979; Zilch, 1979). The lineage leading to formation of mesoderm, for instance, is not comparable (in Nielsen, 1995, compare Table 11.1 with Table 20.2).

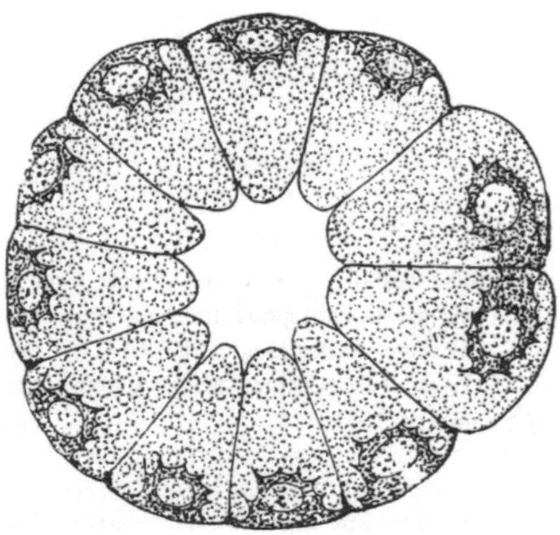

**Figure 24.3** Coeloblastula of a penaeid crustacean. (After Zilch, 1979.) Although this is a holoblastic cleavage mode, the nuclei and the cytoplasm migrate towards the periphery as in the blastoderm stage of superficial cleavages (cf. Figure 24.4).

4. The position of the germ layers differs from that in spiral cleavage (Weygoldt, 1963; Anderson, 1969, 1973; Dohle, 1979).
5. The cleavage pattern of balanomorph and lepadomorph cirripedes with its determined cell lineage is probably a derived character among the cirripedes and perhaps only the Cirripedia Thoracica. This becomes clear when we map the cleavage types on the cladogram of a recent phylogenetic systematic study of cirripedes (Glenner *et al.*, 1995). The cleavage described for a species of the Iblidae (a putative sister group of the other Thoracica) (Anderson, 1965) and for a representative of the Acrothoracica (a putative sister group of the Thoracica) (Turquier, 1967) shows different patterns and not a fixed cell lineage. Especially in *Ibla* (Anderson, 1965) the early cleavage looks like what is known from other crustaceans with early total cleavage (e.g. amphipods).
6. It is likely that the D quadrant of spiral cleavage was originally not larger than the other quadrants and did not show early determination (Freeman and Lundelius, 1992). Early determination of the D quadrant has evolved independently in several lineages of spiralians such as molluscs and annelids. Arthropods most likely share a common ancestor with annelids and this stem species was then devoid of a large D quadrant. Thus, the yolky cell in cirripedes cannot be derived from an annelid-like spiral cleavage.

However, accepting that arthropods and annelids share an ancestor which was a spiralian it is sensible to look for traces of spiral cleavage in arthropod development. Representatives of the arthropod ancestral lineage must have had spiral cleavage, but in the cleavage patterns of recent arthropods there have to be at least some similarities to claim homology with spiral cleavage and this does not seem to be the case in the cirripede and other arthropod cleavage patterns (see also Dohle, 1979, 1989).

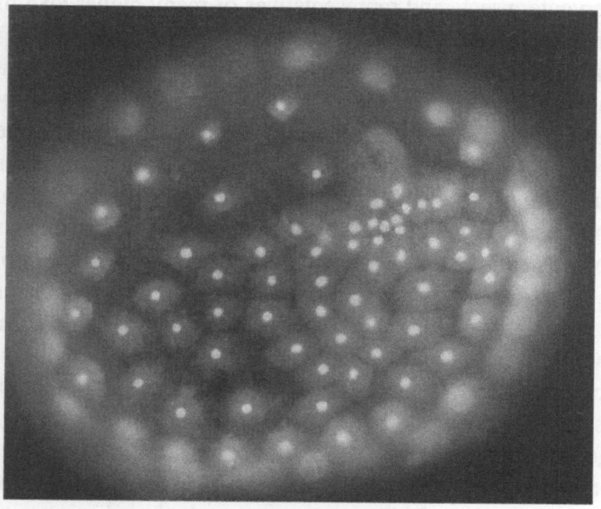

**Figure 24.4** Late blastoderm stage and beginning aggregation of germ band cells in the egg of an amphipod crustacean. Note the central yolk mass. Fluorescent dye: bisBenzimide.

**(b) The blastoderm stage**

In superficial and mixed cleavage types the result of the cleavage process is the blastoderm stage defined by a central yolk mass surrounded by a layer of cells (energids) – the blastoderm (Figure 24.4). Although there are some differences in the way the yolk is distributed and the yolk mass is separated from the blastoderm cells, the general final pattern is very much alike (Fioroni, 1970; Anderson, 1973). One phenomenon found in euarthropod eggs with the holoblastic cleavage type and a coeloblastula is that the nuclei migrate towards the periphery (Tiegs, 1940; Zilch, 1978; Hertzler and Clark, 1992) – this resembles processes of the superficial cleavage type and indicates a derivation from a true blastoderm stage (Figure 24.3). In animals with an original holoblastic cleavage mode, such as many cnidarians, echinoderms or spiralians, the nuclei remain in a central position in the cells during blastula and gastrula stages (Siewing, 1969).

### 24.2.2 THE GROUND PATTERN OF EUARTHROPOD CLEAVAGE

One can reconstruct the ground pattern of early cleavage in euarthropods as follows: it was a 'radial' cleavage with some irregularities concerning spindle orientation and the relative position of blastomeres and no stereotyped cell lineage. Its widespread occurrence among euarthropods suggests that early cleavage was intralecithal. This suggestion is also supported by the fact that onychophorans originally show intralecithal cleavages (Anderson, 1973). However, the possibility cannot be ruled out that early cleavage in euarthropods was total, and I do not see a way to settle this question beyond doubt. Originally the cleavage process (intralecithal or total) led to a blastoderm with a central yolk mass. The blastoderm stage suggests a relatively yolky egg in the ground pattern of euarthropods. This is again supported by the blastoderm stages of the yolky onychophoran eggs (Anderson, 1973). Furthermore, new findings of putative trilobite eggs from the Middle Cambrian exhibiting a blastoderm stage are indicative of yolky eggs in early euarthropods (Zhang and Pratt, 1994). In summary, it is obvious that early development of the euarthropod stem species followed the superficial or mixed cleavage mode. Holoblastic cleavage modes with eggs poor in yolk are apomorphic among euarthropods.

## 24.3 THE GERM BAND

### 24.3.1 SHORT AND LONG GERM BANDS

The formation and characteristics of the early germ band have been described for representatives of all higher euarthropod taxa (Anderson, 1973). Besides several similarities there are some distinct differences in the appearance and the development of germ bands. For instance, two extreme types of germ band have been discriminated in insects – the 'short germ' (e.g. *Schistocerca*) and the 'long germ' type (e.g. *Apis*) of development (Krause, 1939). In the extreme short germs the material for only the head lobes is formed and the rest is successively budded by a posterior growth zone. Segments are differentiated in a general anteroposterior sequence. In long germs, the material for the whole length of the embryo appears from the onset of germ band formation and segments are formed almost simultaneously along the whole germ band (Krause, 1939; Sander, 1983). Between these extremes all transitions are found, the intermediate or semi-long germs. Similar phenomena have been described for other euarthropod taxa (Scholtz, 1992) and even for onychophorans (Walker, 1995).

### 24.3.2 THE GROUND PATTERN OF THE EUARTHROPOD GERM BAND AND SOME EVOLUTIONARY ALTERATIONS

From the distribution of germ band characters among euarthropods the following ground pattern is reconstructed. The early germ band is formed by aggregation of blastomeres on the prospective ventral side of the embryo (Figure 24.4). The anterior end is characterized by the paired semicircular head lobes that give rise to the lateral eyes and the lateral parts of the protocerebrum (Figures 24.5 and 24.6). Only the material for a few anterior segments is present. The rest is produced by the activity of a growth zone. It is likely that the original euarthropod germ band was neither of the extreme short type as in *Schistocerca*, where only the head lobes and the growth zone are present, nor a long germ type. This is deduced from the following observations. An intermediate germ has been suggested as the original state for insects because this is found in

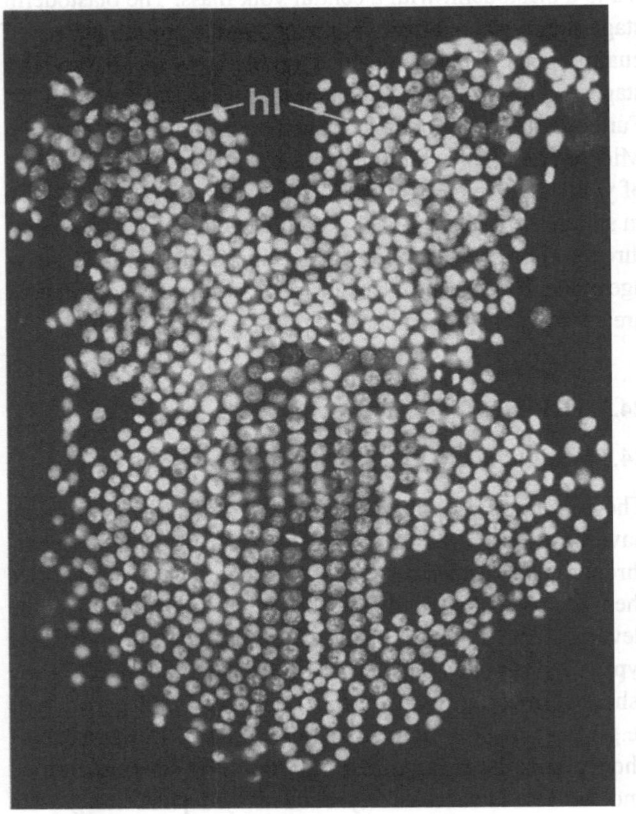


**Figure 24.5** Early germ band of an amphipod crustacean (*Gammarus roeselii*) showing the anterior head lobes (hl). Fluorescent dye: bisBenzimide.

apterygote insects and representatives of many higher taxa of pterygote insects (Tautz *et al.*, 1994). Futhermore, neither crustaceans nor chelicerates possess an early germ band that comprises only the head lobes and a growth zone – the crustacean nauplius larva, for instance, consists of at least three segments and the posterior growth zone. In contrast to what is found in annelids, the growth zone of the original euarthropod germ band was not characterized by specialized cells such as teloblasts (Figure 24.7). It seems likely that the elongation of the germ band was caused by scattered irregular divisions along the germ band since Patel (1993) describes this for a short germ insect and Gerberding (1997) for a non-malacostracan crustacean (Figure 24.7). Originally, the differentiation of segments such as limb buds and ganglion formation follows a general anteroposterior sequence (Figures 24.6–24.8). This is inherited from the common ancestor of annelids and arthropods and vestiges of this process can even be found in extreme long germs like *Drosophila* (Karr *et al.*, 1989). However, the anteriormost segment is not necessarily the one in which differentiation starts ('Differenzierungszentrum'; Seidel, 1975). It can be either the mandibular segment, or the maxillary segment or the first trunk segment (Patel *et al.*, 1989; Fleig, 1990; Scholtz *et al.*, 1994; Manzanares *et al.*, 1996). It is not clear what the situation was in the ground pattern. The segment-polarity gene *engrailed* showed a characteristic pattern in the germ band of the euarthropod stem species (Figures 24.8–24.11). In the germ band of insects and crustaceans, it is expressed in iterated transverse stripes in the posterior portion of each developing segment comprising the neurogenic region and limb buds (DiNardo *et al.*, 1985; Patel *et al.*, 1989; Scholtz

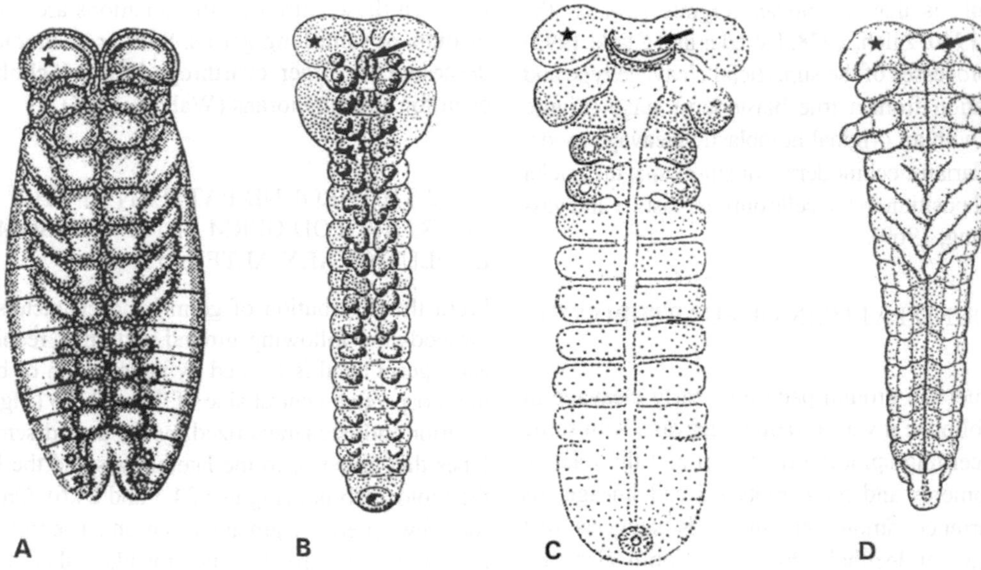


**Figure 24.6** Comparison of advanced germ bands of (A) Scorpion (chelicerate) (after Brauer, 1895); (B) Isopod (crustacean) (modified from Kaestner, 1967 after Silvestri); (C) Chilopod (myriapod) (after Hertzel, 1984); (D) Caddisfly (insect) (after Kobayashi and Ando, 1990). Note the overall similarity. All germ bands are characterized by the head lobes (star) the paired labral rudiment (arrow), and the anteroposterior decline of differentiation.

*et al.*, 1993, 1994; Brown *et al.*, 1994; Patel 1994; Schmidt-Ott *et al.*, 1994; Manzanares *et al.*, 1996; Dohle, 1997, this volume) (Figures 24.8–24.11). A corresponding *engrailed* expression pattern occurs in annelids (Lans *et al.*, 1993). It is likely that *engrailed* already played a role in segmentation in the common ancestor of annelids and arthropods. The similarity of the *engrailed* expression in insects and crustaceans conflicts with the theory that crustacean segments and biramous appendages are the result of the fusion of two adjacent original segments (Emerson and Schram, 1990; Schram and Emerson, 1991). There is now enough evidence to show that this 'duplo-segment' hypothesis is not well founded (Zrzavý and Štys, 1994; Panganiban *et al.*, 1995; Scholtz, 1995c).

Various aspects of this ground pattern have been altered in several euarthropod lineages. A long germ has been independently evolved in different taxa (Scholtz, 1992). In malacostracan crustaceans (perhaps also in cirripedes, Anderson, 1973) the growth zone underwent changes that are convergently similar to the situation in annelids (clitellates). In this group, teloblasts have evolved that lie in front of the telson and give rise to the material for posterior segments by unequal divisions (Dohle and Scholtz, 1988; Scholtz and Dohle, 1996) (Figure 24.7). This teloblastic growth is correlated with a unique stereotyped cell division pattern during germ band growth and differentiation (Dohle and Scholtz 1988) (Figure 24.7). In some spiders (Anderson, 1973), scolopendromorph centipedes (Whitington *et al.*, 1991), and amphipod crustaceans (G. Scholtz, unpublished results) the germ band is split along its longitudinal axis. The two lateral halves separate and fuse again in a later stage. In some crustaceans it is questionable whether there is an early germ band stage at all. In penaeid decapods, for instance, the germ

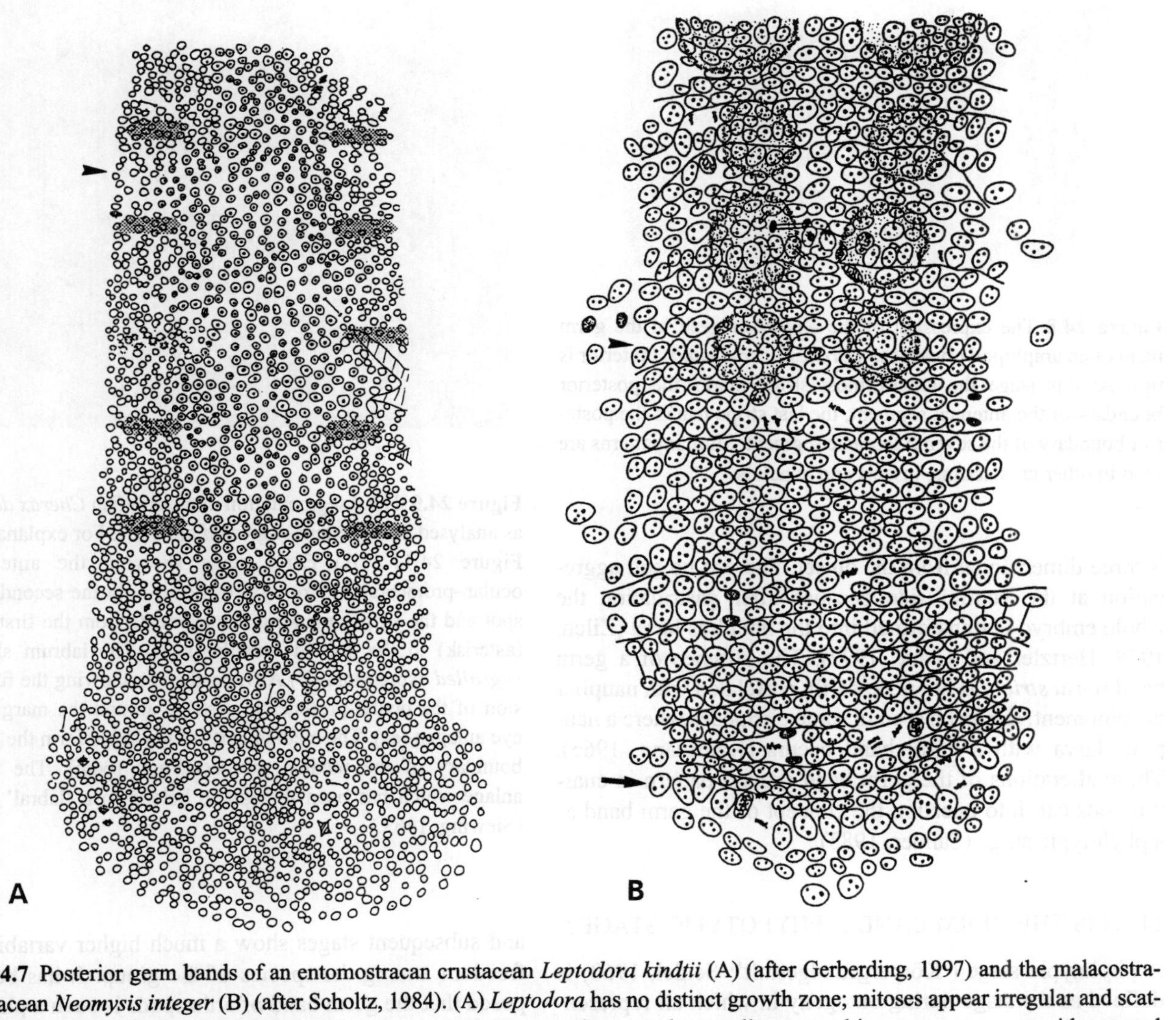


**Figure 24.7** Posterior germ bands of an entomostracan crustacean *Leptodora kindtii* (A) (after Gerberding, 1997) and the malacostracan crustacean *Neomysis integer* (B) (after Scholtz, 1984). (A) *Leptodora* has no distinct growth zone; mitoses appear irregular and scattered throughout the germband. (B) *Neomysis* shows a teloblastic growth zone – large cells arranged in a transverse row with unequal divisions (arrow). Note the regular division pattern in the entire germband (for details, see Dohle and Scholtz, 1988). Arrowheads mark the position of the first thoracic segment.

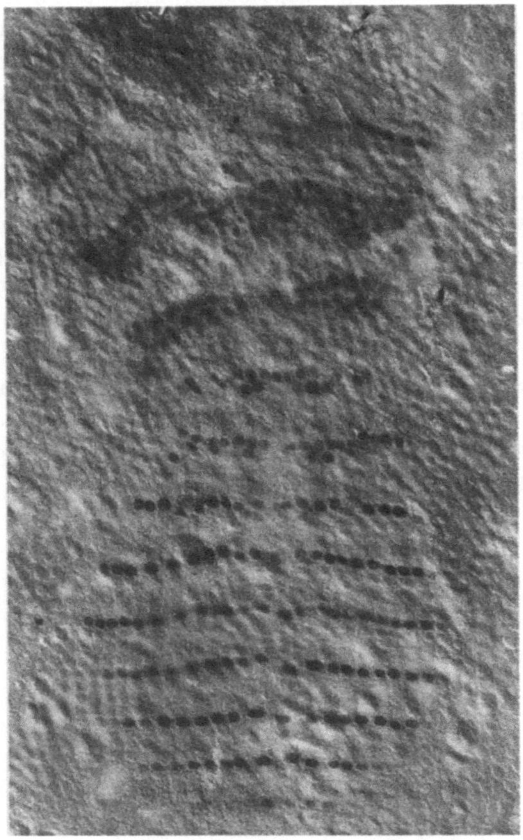

**Figure 24.8** The expression of the *engrailed* gene in the germ band of an amphipod crustacean (*Orchestia cavimana*) (anterior is up). At this stage the anteriormost stripe marks the posterior boundary of the antennal segment, the last stripe marks the posterior boundary of the seventh thoracic segment. Similar patterns are seen in other crustaceans, insects and annelids.

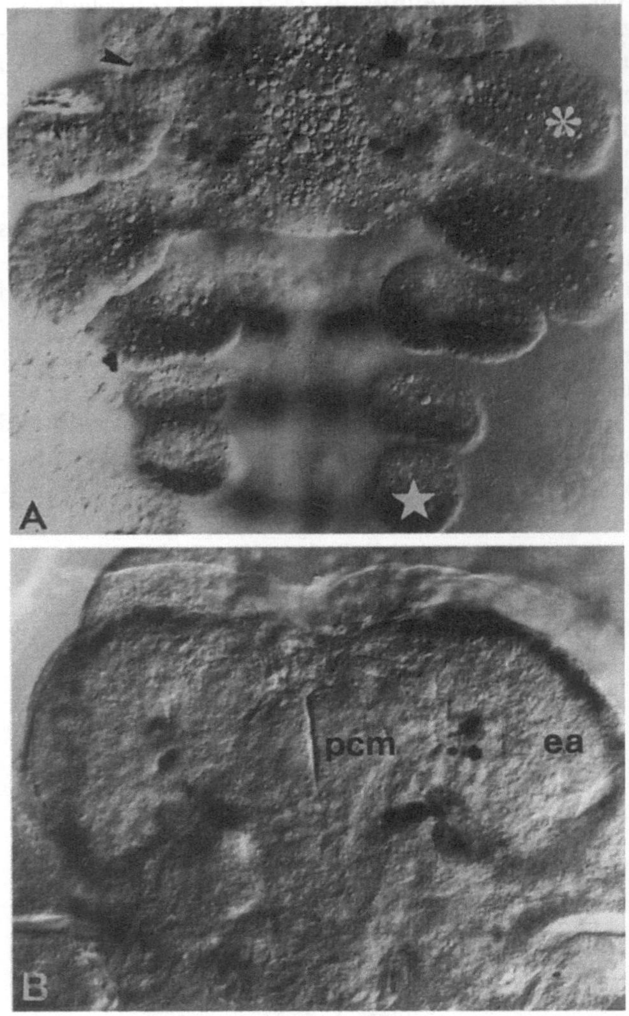


**Figure 24.9** Head segmentation in the crayfish *Cherax destructor* as analysed with the anti-*engrailed* antibody. For explanation, see Figure 24.11. (A) Early stage showing the anteriormost ocular–protocerebral stripe (arrowhead) with the secondary head spot and the stripes marking the segments from the first antenna (asterisk) to the second maxilla (star). The labrum shows no *engrailed* expression. (B) Advanced stage showing the full extension of the ocular-protocerebral stripe around the margin of the eye anlagen (ea). The median parts of the stripe form the posterior boundary of the median protocerebrum (pcm). The ganglion anlage which has been thought to belong to a 'labral' segment (Siewing, 1963.)

is three-dimensional from the outset. There is no cell aggregation at the ventral side because after gastrulation the whole embryo is transformed into the nauplius larva (Zilch, 1978; Hertzler and Clark, 1992). The absence of a germ band *sensu stricto* is not necessarily correlated with naupliar development, as can be seen in some cirripedes where a nauplius larva with a germ band occurs (Kaufmann, 1965). These alterations of the germ band ground pattern in euarthropods call into question the concept of the germ band as a phylotypic stage (Sander, 1983).

### 24.3.3 IS THE GERM BAND A PHYLOTYPIC STAGE?

A phylotypic stage ('Körpergrundgestalt'; Seidel, 1960) is defined as the stage during ontogeny in which all representatives of a given animal group most resemble each other (Sander, 1983). The phylotypic stage is thought to represent a developmental constraint – a bottleneck – whereas earlier and subsequent stages show a much higher variability and freedom among the species of the group. It has been proposed that the germ band of arthropods is such a phylotypic stage (Sander, 1983). According to this view, the germ band would be the stage in arthropod development with the fewest differences between taxa and a necessary step in the

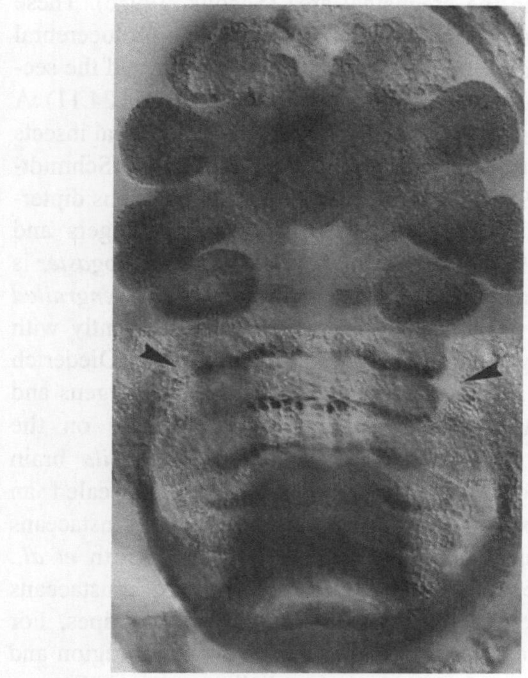

**Figure 24.10** The expression of *engrailed* in the margin of the carapace of *Cherax destructor*. The stripe of the first maxillary segment continues into the circular carapace anlage (arrowheads). The stripes of the subsequent segments fuse also with the carapace margin.

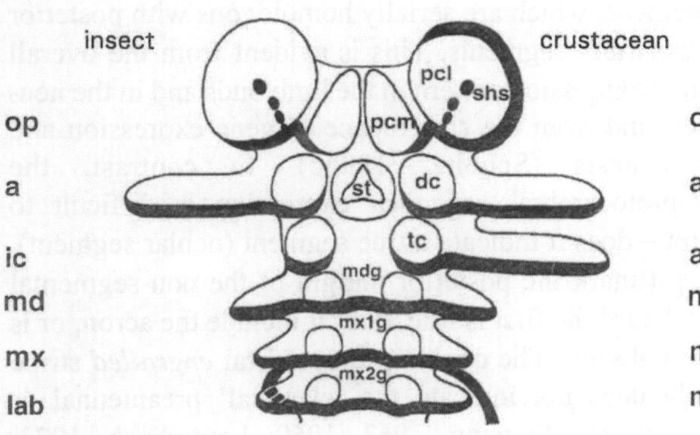


**Figure 24.11** Schematic representation of the pattern of *engrailed* expression in the head of insects (left) and crustaceans (right). In insects, the ocular stripe is not as distinct as in crustaceans, the second antennal segment is devoid of an appendage, and the stripes of the maxillary and the labial segments are connected. The carapace stripe in the crustacean is also indicated. a, antenna; dc, deuterocerebrum; ic, intercalary segment; lab, labium; md, mandible; mdg, mandibular ganglion; mx, maxilla; mxg, maxillary ganglion; op, ocular–protocerebral region; pcl, lateral protocerebrum; pcm, median protocerebrum; shs, secondary head spot; st, stomodaeum; tc, tritocerebrum.

development of arthropods. Although it is clear that the arthropod stem species went through a germ band stage during embryonic development, this does not necessarily mean that this character is conserved throughout all recent arthropod species. In contrast, the concept of the phylotypic stage does not allow exceptions because of the constraints which stabilize this stage. It is well-documented that on each developmental level from the genes to organogenesis homologous structures can arise, despite alterations in preceding developmental stages (Sander, 1983; Dohle, 1989; Wagner and Misof, 1993; Scholtz and Dohle, 1996). Ontogenetic stages and processes can be altered without affecting the resulting product. There is no reason to believe that single developmental steps should be an exception to this rule, and the various modes of germ band formation and differentiation point in that direction.

## 24.4 HEAD SEGMENTATION

### 24.4.1 THE PROBLEMS OF ANALYSING HEAD SEGMENTATION

There are numerous theories concerning head segmentation in euarthropods (Weber, 1952; Manton, 1960; Siewing, 1963, 1969; Rempel, 1975; Bitsch, 1994; Scholtz, 1995c). There is no doubt that the euarthropod head is composed of segments which are serially homologous to trunk segments. However, serial homology and segment identity in the head are obscured by fusion, loss and alteration of structures, and by morphogenetic movements and displacement during ontogeny. The real issue then becomes the nature and the number of segments that make up the euarthropod head. There are two aspects to this question: (i) how many segments can be identified in the head of extant arthropods?; and (ii) how many head segments existed in the euarthropod stem species and its predecessors? The problem we are faced with is the definition of a segment and of the nature of structures that might indicate the former existence of a full segment. For instance, is the occurrence of coelomic cavities enough evidence to postulate the existence of a complete segment in the past? It seems plausible that the clue for solving the problem of segment number in the euarthropod head lies in early embryology. Here, the serial arrangement of structures is more obvious and fusion has not yet occurred. Unfortunately, morphological–embryonic data have not provided us with unambiguous interpretations. Molecular markers now offer a new basis for a different approach to the question of head segmentation. However, it is unlikely that the 'endless dispute' (Rempel, 1975) has come to an end or that the 'field of mental exercise' (Snodgrass, 1960) no longer exists.

Head segmentation involves two particular problematical issues:

1. The number and identity of segments in the anterior pregnathal area.
2. The posterior limit of the original euarthropod head.

In the anterior region, we face the problem of identification of putative segmental vestiges and of the serial homologization of segmental structures. Traditionally, these structures have been coelomic sacs, appendages, ganglia, sutures and muscles. In addition, we face the problem of whether arthropods possess an anteriormost non-segmental part, the acron. The existence of an acron has been deduced from the annelid prostomium which is formed by the larval episphere (Siewing, 1963) and which bears the eyes.

The problem of the posterior boundary of the arthropod head is of a different kind. It is related to the criteria that make up a head segment as opposed to a trunk segment. These criteria can be the extension of the head shield, the functional transformation of appendages into feeding or sensory appendages, the fusion of ganglia with anterior head ganglia, and differences in ontogenetic and genetic aspects of head versus trunk segment formation. For instance, malacostracan crustaceans show different cell division patterns in the naupliar and the post-naupliar regions (Dohle and Scholtz, 1988; Scholtz, 1990). The anteriormost segments up to the intercalary segment of *Drosophila* differ from more posterior segments in the mode of regulatory gene interactions (Cohen and Jürgens, 1991).

The possibility of using molecular markers in a variety of animals offers new ways to readdress questions of head segmentation. The expression pattern of the segment-polarity gene *engrailed*, for instance, has been used to analyse head segmentation patterns in various insects and crustaceans (Patel *et al.*, 1989; Schmidt-Ott and Technau, 1992; Fleig, 1994; Schmitt-Ott *et al.*, 1994; Scholtz, 1995c; Rogers and Kaufman, 1996). Unfortunately no data are available on chelicerates, myriapods, and representatives of onychophorans but some data exist on *engrailed* expression in the anterior region of annelids (Dorresteijn *et al.*, 1993; Lans *et al.*, 1993).

### 24.4.2 THE PREGNATHAL REGION

#### (a) The *engrailed* expression pattern as an analytical tool

*engrailed* provides us with a useful marker for embryonic segment anlagen before they are morphologically present. All *engrailed* stripes in the head region share several characteristic features which they also share with the more posterior trunk stripes (Scholtz *et al.*, 1994; Scholtz, 1995c): all are formed in a mediolateral progression, the distance between two adjacent stripes is the same, *engrailed* is expressed in cells on the surface first and later in neurogenic cells in the interior. In malacostracan crustaceans it has been shown that there are three distinct regions of *engrailed* expression in the pregnathal area (Scholtz, 1995c). These are in anteroposterior sequence, the ocular–protocerebral stripe, the first antennal/deuterocerebral stripe, and the second antennal/tritocerebral stripe (Figures 24.9 and 24.11). A corresponding pattern has been described for several insects such as beetles (Fleig, 1994; Brown *et al.*, 1994; Schmidt-Ott *et al.*, 1994), locusts (Patel *et al.*, 1989), various dipterans (Schmidt-Ott *et al.*, 1994) and others (Rogers and Kaufman, 1996). The pattern in *Drosophila melanogaster* is somewhat more complicated and the various *engrailed* stripes and patches have been interpreted differently with regard to segment sequence, number and identity (Diederich *et al.*, 1991; Schmidt-Ott and Technau, 1992; Jürgens and Hartenstein, 1993). However, a recent study on the metameric subdivision of the anterior *Drosophila* brain using confocal laser scanning microscopy revealed an *engrailed* expression pattern identical to that of crustaceans and other insects (three pregnathal stripes) (Hirth *et al.*, 1995). The similarity between insects and crustaceans extends even to characteristics of individual stripes. For instance, the stripes of the ocular–protocerebral region and the first antennal segment are medially separated (Figures 24.9 and 24.11). Furthermore, a secondary head spot occurs in the eye region which has been described in *Drosophila*, the beetle *Tribolium* and crustaceans (Schmidt-Ott and Technau, 1992; Brown *et al.*, 1994; Scholtz, 1995c) (Figures 24.9 and 24.11).

#### (b) The nature of metameres

The *engrailed* stripes in the first antennal/deuterocerebral and the second antennal/tritocerebral regions clearly indicate segments which are serially homologous with posterior head and trunk segments. This is evident from the overall *engrailed* expression pattern in the limb buds and in the neuromeres, and from the congruence of gene expression and morphogenesis (Scholtz, 1995c). In contrast, the ocular–protocerebral *engrailed* expression is difficult to interpret – does it indicate a true segment (ocular segment), or does it mark the posterior margin of the non-segmental acron? And if the first is true, does it include the acron, or is an acron absent? The ocular–protocerebral *engrailed* stripe certainly does not indicate the 'classical' preantennal or labral segment (Siewing, 1963, 1969; Lauterbach, 1973; Jürgens and Hartenstein, 1993) which has been thought to lie between the eye-bearing acron and the (first) antennal segment. This is because the ocular–protocerebral stripe comprises the eye anlagen, which have been thought to indicate the acron together with the brain parts attributed to the labral segment (Siewing, 1963) (Figs 9, 11). This continuous stripe shows that these structures form **one** unit and not two subsequent ones. Furthermore a mutant analysis in *Drosophila* shows that there is no evidence for a labral segment situated between the ocular region and the antennal segment (Schmidt-Ott *et al.*, 1995). If the ocular–protocerebral *engrailed* stripe does indicate a segment, it would be an ocular segment which has been postulated by several authors (Reichenbach, 1886; Sharov, 1966; Schmidt-Ott and Technau, 1992; Schmidt-Ott *et al.*, 1995; Rogers and Kaufman, 1996). The acron would then either be included in this segment or absent. However, there are arguments in favour of the acron nature of the body part marked by the ocular–protocerebral *engrailed* stripe. The posterior margin of the episphere of the trochophora larva of the polychaete *Platynereis* seems to express *engrailed* (Dorresteijn *et al.*, 1993). Since this is the anteriormost *engrailed* stripe in the polychaete it might correspond to the anteriormost stripe in arthropods. The leech homologue of the *Deformed* gene of *Drosophila*, *Lox 6*, is expressed in the third suboesophageal ganglion of *Hirudo medicinalis* (Aisemberg *et al.*, 1995). In insects, *Deformed* is expressed in the mandibular and maxillary segments (Diederich *et al.*, 1991; Fleig *et al.*, 1992). Thus, the supraoesophageal ganglion of annelids would correspond to the protocerebrum of arthropods and again the first *engrailed* stripe would mark the acron in arthropods (insects and crustaceans).

### (c) The nature of the labrum

The labrum originating from a bilobed anlage is an apomorphy of the euarthropods. A corresponding structure occurs in neither onychophorans nor annelids. The labrum has been interpreted as being a ventral outgrowth of the acron anterior to the mouth (Walossek and Müller, 1990), as being the medially fused appendages of a preantennal (labral) segment (Siewing, 1969; Lauterbach, 1973; Weygoldt, 1979) or as the anteriormost body segment (Sharov, 1966; Schmidt-Ott and Technau, 1992). The lack of *engrailed* expression in the labrum of all crustaceans investigated (Scholtz, 1995c; Manzanares *et al.*, 1996) (Figure 24.9) and in several insect species (locust: Patel *et al.*, 1989; beetle, nematocerans: Schmidt-Ott *et al.*, 1994) contradicts the assumption of an origin of the labrum from appendages and the segmental nature of the labrum in general (Rogers and Kaufman, 1996). On the other hand, *engrailed* expression in the labrum of other insects (*Drosophila*: Diederich *et al.*, 1991; nematocerans, brachycerans: Schmidt-Ott *et al.*, 1994; beetle: Fleig, 1994) seems to support the appendiculate nature of the labrum. The *Distal-less* gene is expressed in the tips of limb buds in representatives of all higher euarthropod groups (Panganiban *et al.*, 1995). The *Distal-less* expression in the labrum of some crustaceans and insects (Panganiban *et al.*, 1995) seems to argue even more strongly for a possible homology between the labrum and segmental appendages. However, if the labrum is considered as being leg-like in its nature the question arises as to which segment it belongs. In addition, from the distribution of the labral *engrailed* expression amongst insects it has been concluded that this is a secondary feature in some insects (Scholtz, 1995c). Morever, the correlation between *engrailed* expression and morphogenesis is different between legs and the labrum – in the labrum *engrailed* expression does not precede morphogenesis as is the case in limb bud areas, the labrum is a morphologically well defined structure before *engrailed* expression occurs (see Scholtz, 1995c). This makes the homology between the labrum and legs doubtful (see Dickinson 1995; Bolker and Raff, 1996). The interpretation of the pattern of *Distal-less* expression is also problematic. Although it is clearly expressed in the tips of limbs of insects, myriapods, chelicerates and crustaceans it is also expressed in the telson of crustaceans (Panganiban *et al.*, 1995). The telson is clearly no limb derivative. Claims that *Distal-less* expression is indicative of a limb character of the labrum are, therefore, doubtful. Also since *Distal-less* expression is not restricted to tips of limb buds it seems likely that the *Distal-less* expression in the labrum is comparable with that in the telson and that labrum and telson mark the extreme ends of the body axis.

The evidence for the labrum being the fused appendages of a preantennal segment situated between the acron and the (first) antennal segment appears weak. It is more likely that the labrum is the tip of the acron which moved ventrally and posteriorly during evolution of the euarthropods.

## 24.4.3 THE POSTERIOR BOUNDARY OF THE HEAD

*engrailed* is expressed in the gnathal region of crustaceans and insects in the segments of the mandibles, (first) maxillae and second maxillae (labium). There is no trace of an additional segment between the second antennal (intercalary) segment and the mandibular segment such as postulated by Chaudonneret (1987) (Figures 24.9 and 24.10). There is also no distinct separation from the trunk segments. Nevertheless, the pattern of *engrailed* expression might contribute to the recognition of the original posterior boundary of the euarthropod head. In embryos of the decapod crustacean *Cherax destructor* the cells of the margin of the developing carapace express *engrailed* (Figure 24.10). This *engrailed* expression continues into the *engrailed* stripe of the first maxillary segment (Figure 24.10). It is concluded that the carapace margin is the extended posterior margin of the segment of the first maxillae. The *engrailed* stripes of the subsequent second maxillary segment and the trunk segments fuse laterally with this circular *engrailed* region (Figure 24.10). These findings speak against an origin of the carapace from the second maxillary segment (Lauterbach, 1974; Newman and Knight, 1984) or the antennal segment (Casanova, 1993) – at least ontogenetically. Moreover, the results presented here might point towards the ground pattern of the euarthropod head. In several insects, the *engrailed* stripes of the maxillary and the labial segments are connected by a bow-like region of *engrailed* positive cells (Patel *et al.*, 1989; Diederich *et al.*, 1991; Fleig *et al.*, 1992; Brown *et al.*, 1994; Rogers and Kaufman, 1996) (Figure 24.11). These comparable expression patterns found in the crustacean *Cherax* and in insects can be interpreted as indicating the original posterior margin of their heads. This supports the idea that in the stem species of the mandibulates only the first maxillary segment was included in the head (Lauterbach, 1980; Walossek and Müller, 1990), a hypothesis based on the fact that the cephalocarid crustaceans possess only one pair of maxillae. The 'second maxillae' do not differ from the thoracic appendages (Sanders, 1963). The pattern of maxillary muscle attachment in other crustaceans (Pilgrim, 1973) and the embryonic development of the head shield in centipedes (Dunger, 1993) also support this hypothesis. Furthermore, trilobites show a head shield that covers the antennal segment and three postantennal segments (Cisne, 1975; Walossek, 1993) (Figure 24.12). The posterior head margin in embryonic insects and crustaceans as indicated by the (first) maxillary *engrailed* expression would correspond, as a recapitulation, to the posterior margin of the trilobite head as indicated by the head shield.

### 24.4.4 THE GROUND PATTERN OF THE EUARTHROPOD HEAD

To summarize, the original euarthropod head probably comprised an acron, an antennal segment and three postantennal segments, all covered by a head shield. This is at least true for the crown-euarthropods (the descendants of last common ancestor of extant euarthropods; see Jefferies, 1980). In the stem lineage of euarthropods there are certainly representatives with fewer cephalic segments (e.g. *Sidneyia*; Bruton, 1981). The acron bears the compound eyes, the protocerebrum and the labrum. The deuterocerebrum is the ganglion of the antennal segment. The tritocerebrum is the ganglion of the first postantennal segment. Originally, it lay posterior to the mouth as can be seen in crustacean and insect embryos (Weygoldt, 1979; Boyan *et al.*, 1995) and some adult crustaceans such as branchiopods (Hanström, 1928). The subdivision of the head into pregnathal and gnathal areas is an apomorphy of the mandibulates. Within the mandibulates the 'second maxillary (labial in insects) segment' became fused to the head and its appendages altered their function into head limbs in different lineages independently. Insects and myriapods lost the appendages of the original first postantennal segment, but retained the tritocerebrum and the corresponding *engrailed* stripe (unknown for myriapods). As mentioned above, there are no *engrailed* data available for chelicerates. The general opinion is that chelicerates reduced the antennal segment (Weygoldt, 1985). However, most embryonic studies in chelicerates have been biased by the idea that there is a preantennal (labral) segment (Pross, 1966; Winter, 1980; Weygoldt, 1985). This led to some confusions concerning the coelomic cavities and ganglion anlagen (Pross, 1977; Weygoldt, 1985). Within the concept of euarthropod head segmentation discussed here, the characteristics of chelicerate embryos can be more easily interpreted. One has to look for vestiges of only one precheliceral segment. Nevertheless, a study using molecular markers to analyse the segmentation pattern in chelicerates is badly needed.

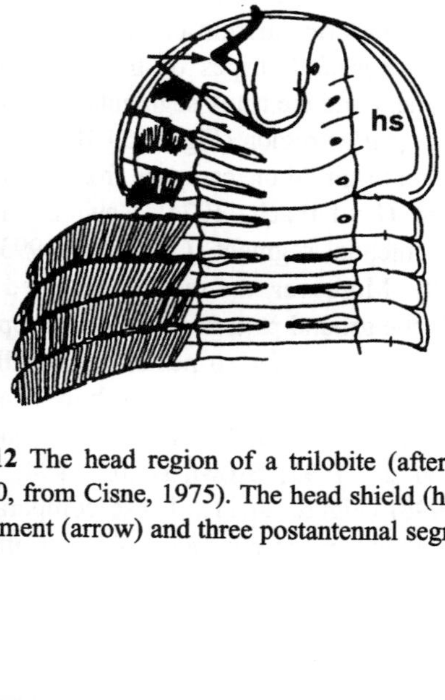


**Figure 24.12** The head region of a trilobite (after Brusca and Brusca, 1990, from Cisne, 1975). The head shield (hs) covers the antennal segment (arrow) and three postantennal segments.

## 24.5 SOME APOMORPHIES SUPPORTING ARTHROPOD AND EUARTHROPOD MONOPHYLY

### CLEAVAGE

1. The radial position of the cleavage products is an apomorphy for arthropods. Plesiomorphically there was spiral cleavage.
2. The superficial cleavage type or mixed cleavage type might be an arthropod apomorphy. Comparable developments do not occur in annelids.
3. The blastoderm stage with a central yolk mass is an arthropod apomorphy. In spiral cleavage there is originally a coeloblastula.

### GERM BAND

1. The formation of a germ band by aggregation of blastomeres on the ventral side of the germ is an apomorphy of arthropods. Polychaetes have originally no germ bands [exceptions can be found in species with a large amount of yolk (A. Dorresteijn, personal communication)]. The whole germ is transformed into the trochophora larva and the later worm. Clitellates possess germ bands. These are, however, formed on the dorsal side of the germ and migrate and fuse ventrally (Penners, 1924; Smith *et al.*, 1996). This germ band type is a clitellate apomorphy.
2. The head lobes are an arthropod apomorphy since they are also found in onychophorans (Anderson, 1973; Walker, 1995). There is no corresponding structure in annelids.
3. A non-teloblastic growth zone is an arthropod apomorphy, whereas in annelids the mesoderm buds from mesoteloblasts which are derivatives of the 4d cell.

## HEAD SEGMENTATION

1. A labrum originating from a bilobed anlage is an apomorphy of euarthropods. A corresponding structure can be found neither in onychophorans nor in annelids.
2. A head consisting of an antennal and three postantennal segments is an apomorphy of crown-group euarthropods. The onychophoran head is difficult to interpret but it probably consists of the eye bearing acron, the antennal segment, the jaw segment, and the segment of the oral papilla.

## ACKNOWLEDGEMENTS

I thank Richard Fortey and Richard Thomas for inviting me to the arthropod meeting. I also thank Stefan Richter and Paul Whitington for critically reading the manuscript.

## REFERENCES

Aisemberg, G.O., Wong, V.Y. and Macagno, E.R. (1995) Genesis of segmental identity in the leech nervous system, in *The Nervous Systems of Invertebrates – an Evolutionary and Comparative Approach* (eds O. Breidbach and W. Kutsch), Birkhäuser Verlag, Basel, pp. 77–87.

Anderson, D.T. (1965) Embryonic and larval development and segment formation in *Ibla quadrivalvis* (Cuv.) (Cirripedia). *Australian Journal of Zoology*, **13**, 1–15.

Anderson, D.T. (1969) On the embryology of the cirripede crustaceans *Tetraclita rosea* (Krauss), *Tetraclita purpurascens* (Wood), *Chtamalus antennatus* (Darwin) and *Chamaesipho columna* (Spengler) and some considerations of crustacean phylogenetic relationships. *Philosophical Transactions of the Royal Society, B*, **256**, 183–235.

Anderson, D.T. (1973) *Embryology and Phylogeny in Annelids and Arthropods*, Pergamon Press, Oxford.

Averof, M. and Akam, M. (1995) Insect–crustacean relationships: insights from comparative developmental and molecular studies. *Philosophical Transactions of the Royal Society, B*, **347**, 293–303.

Ax, P. (1987) *The Phylogenetic System*, Wiley and Sons, Chichester.

Baldass, F. von (1941) Die Entwicklung von *Daphnia pulex*. *Zoologische Jahrbücher, Abteilung Anatomie und Ontogenie der Tiere*, **67**, 1–60.

Benesch, R. (1969) Zur Ontogenie und Morphologie von *Artemia salina* L. *Zoologische Jahrbücher, Abteilung Anatomie und Ontogenie der Tiere*, **86**, 307–458.

Bigelow, M.A. (1902) The early development of *Lepas*. A study of cell lineage and germ layers. *Bulletin of the Museum of Comparative Zoology*, Harvard, **40**, 61–144.

Bitsch, J. (1994) The morphological groundplan of Hexapoda: critical review of recent concepts. *Annales de la Société Entomologique de France* (N.S.), **30**, 103–29.

Bolker, J.A. and Raff, R.A. (1996) Developmental genetics and traditional homology. *BioEssays*, **18**, 489–94.

Bowler, P.J. (1994) Are the Arthropoda a natural group? An episode in the history of evolutionary biology. *Journal of the History of Biology*, **27**, 177–213.

Boyan, G.S., Williams, J.L.D. and Reichert, H. (1995) Morphogenetic reorganization of the brain during embryogenesis in the grasshopper. *Journal of Comparative Neurology*, **361**, 429–40.

Brauer, A. (1895) Beiträge zur Kenntnis der Entwicklungsgeschichte des Skorpions. II. *Zeitschrift für Wissenschaftliche Zoologie*, **59**, 351–433.

Brown, S.J., Patel, N.H. and Denell, R.E. (1994) Embryonic expression of the single *Tribolium engrailed* homolog. *Developmental Genetics*, **15**, 7–18.

Brusca, R.C. and Brusca, G.J. (1990) *Invertebrates*, Sinauer, Sunderland, Massachusetts.

Bruton, D.L. (1981) The arthropod *Sidneyia inexpectans*, Middle Cambrian, Burgess Shale, British Columbia. *Philosophical Transactions of the Royal Society, B*, **295**, 619–53.

Casanova, B. (1993) L'Origine protocéphalique de la carapace chez les Thermosbaenacés, Tanaidacés, Cumacés et Stomatopodes. *Crustaceana*, **65**, 144–50.

Chaudonneret, J. (1987) Evolution of the insect brain with special reference to the so-called tritocerebrum, in *Arthropod Brain* (ed. A.P. Gupta), Wiley, New York, pp 3–26.

Cisne, J.L. (1975) Anatomy of *Triarthrus* and the relationships of the Trilobita. *Fossils and Strata*, **4**, 45–63.

Claypole, A.M. (1898) The embryology and oogenesis of *Anurida maritima*. *Journal of Morphology*, **14**, 219–300.

Cohen, S. and Jürgens, G. (1991) *Drosophila* headlines. *Trends in Genetics*, **7**, 267–72.

Costello, D.P. and Henley, C. (1976) Spiralian development: a perspective. *American Zoologist*, **16**, 277–91.

Delsman, H.C. (1917) Die Embryonalentwicklung von *Balanus balanoides* Linn. *Tijdschrift nederlandse dierkunde Vereen* Ser. 2, **15**, 419–520.

Dickinson, W.J. (1995) Molecules and morphology: where is the homology? *Trends in Genetics*, **11**, 119–21.

Diederich, R.J., Pattatucci, A.M. and Kaufmann, T.C. (1991) Developmental and evolutionary implications of *labial*, *Deformed* and *engrailed* expression in the *Drosophila* head. *Development*, **113**, 273–81.

DiNardo, S., Kuner, J.M., Theis, J. and O'Farrell, P.H. (1985) Development of embryonic pattern in *D. melanogaster* as revealed by accumulation of the nuclear *engrailed* protein. *Cell*, **43**, 59–69.

Dogiel, V. (1913) Embryologische Studien an Pantopoden. *Zeitschrift für wissenschaftliche Zoologie*, **107**, 575–741.

Dohle, W. (1964) Die Embryonalentwicklung von *Glomeris marginata* (Villers) im Vergleich zur Entwicklung anderer Diplopoden. *Zoologische Jahrbücher, Abteilung Anatomie und Ontogenie der Tiere*, **81**, 241–310.

Dohle, W. (1970) Die Bildung und Differenzierung des postnauplialen Keimstreifs von *Diastylis rathkei* (Crustacea, Cumacea) I. Die Bildung der Teloblasten und ihrer Derivate. *Zeitschrift für Morphologie der Tiere*, **67**, 307–92.

Dohle, W. (1979) Vergleichende Entwicklungsgeschichte des Mesoderms bei Articulaten. *Fortschritte in der zoologischen Systematik und Evolutionsforschung*, **1**, 120–40.

Dohle, W. (1989) Zur Frage der Homologie ontogenetischer Muster. *Zoologische Beiträge, NF*, **32**, 355–89.

Dohle, W. and Scholtz, G. (1988) Clonal analysis of the crustacean segment: the discordance between genealogical and segmental borders. *Development*, **104** Supplement, 147–60.

Dorresteijn, A.W.C., O'Grady, B., Fischer, A., Porchet-Hennere, E. and Boilly-Marer, Y. (1993) Molecular specification of cell lines in the embryo of *Platynereis* (Annelida). *Roux's Archives of Developmental Biology*, **202**, 260–9.

Dunger, W. (1993) 1. Klasse Chilopoda, in *Lehrbuch der Speziellen Zoologie, Band I, Teil 4* (ed. H.-E. Gruner), Fischer Verlag, Jena, pp. 1047–94.

Eernisse, D.J., Albert, J.S. and Anderson, F.E. (1992) Annelida and Arthropoda are not sister taxa: a phylogenetic analysis of spiralian metazoan morphology. *Systematic Biology*, **41**, 305–30.

Emerson, M.J. and Schram, F.R. (1990) The origin of crustacean biramous appendages and the evolution of Arthropoda. *Science*, **250**, 667–9.

Fioroni, P. (1970) Am Dotteraufschluss beteiligte Organe und Zelltypen bei höheren Krebsen; der Versuch zu einer einheitlichen Terminologie. *Zoologische Jahrbücher, Abteilung Anatomie und Ontogenie der Tiere*, **87**, 481–522.

Fleig, R. (1990) *Engrailed* expression and body segmentation in the honeybee *Apis mellifera. Roux's Archives of Developmental Biology*, **198**, 467–73.

Fleig, R. (1994) Head segmentation in the embryo of the Colorado beetle *Leptinotarsa decemlineata* as seen with anti-*en* immunostaining. *Roux's Archives of Developmental Biology*, **203**, 227–9.

Fleig, R., Walldorf, U., Gehring, W.J. and Sander, K. (1992) Development of the *Deformed* protein pattern in the embryo of the honeybee *Apis mellifera* L. (Hymenoptera). *Roux's Archives of Developmental Biology*, **201**, 235–42.

Freeman, G. and Lundelius, J.W. (1992) Evolutionary implications of the mode of D quadrant specification in coelomates with spiral cleavage. *Journal of Evolutionary Biology*, **5**, 205–47.

Friedrich, M. and Tautz, D. (1995) Ribosomal DNA phylogeny of the major extant arthropod classes and the evolution of the myriapods. *Nature*, **376**, 165–7.

Fuchs, F. (1914) Die Keimblätterentwicklung von *Cyclops viridis* Jurine. *Zoologische Jahrbücher, Abteilung Anatomie und Ontogenie der Tiere*, **38**, 103–56.

Garey, J.R., Krotec, M., Nelson, D.R. and Brooks, J. (1996) Molecular analysis supports a tardigrade–arthropod association. *Invertebrate Biology*, **115**, 79–88.

Gerberding, M. (1997) Germ band formation and early neurogenesis of *Leptodora kindti* (Cladocera) first evidence for neuroblasts in the entomostracan crustaceans. *Invertebrate Reproduction and Development*, **32**, 63–73.

Glenner, H., Grygier, M.J., Hoeg, J.T. Jensen, P.G. and Schram, F.R. (1995) Cladistic analysis of the Cirripedia Thoracica. *Zoological Journal of the Linnean Society*, **114**, 365–404.

Hanström, B. (1928) *Vergleichende Anatomie des Nervensystems der wirbellosen Tiere*, Springer, Berlin.

Haszprunar, G. and Schaefer, K. (1996) Anatomy and phylogenetic significance of *Micropilina arntzi* (Mollusca, Monoplacophora, Micropilinidae Fam. Nor.) *Acta Zoologica*, 77, 315–34.

Hennig, W. (1950) *Grundzüge einer Theorie der phylogenetischen Systematik*, Deutscher Zentralverlag, Berlin.

Hennig, W. (1966) *Phylogenetic Systematics*. University of Illinois Press, Urbana.

Hertzel, G. (1984) Die Segmentation des Keimstreifens von *Lithobius forficatus* (L.) (Myriapoda, Chilopoda). *Zoologische Jahrbücher, Abteilung Anatomie und Ontogenie der Tiere*, **112**, 369–86.

Hertzler, P.L. and Clark, W.H. Jr (1992) Cleavage and gastrulation in the shrimp *Sicyonia ingentis*: invagination is accompanied by oriented cell division. *Development*, **116**, 127–40.

Hirth, F., Therianos, S., Loop, T., Gehring, W.J., Reichert, H. and Furukubo-Tokunaga, K. (1995) Developmental defects in brain segmentation caused by mutations of the homeobox genes *orthodenticle* and *empty spiracles* in *Drosophila. Neuron*, **15**, 769–78.

Jefferies, R.P.S. (1980) Zur Fossilgeschichte des Ursprungs der Chordaten und Echinodermen. *Zoologische Jahrbücher, Abteilung Anatomie und Ontogenie der Tiere*, **23**, 285–353.

Jürgens, G. and Hartenstein, V. (1993) The terminal regions of the body pattern, in *The Development of Drosophila melanogaster* (eds M. Bate and A. Martinez-Arias), Cold Spring Harbor Laboratory Press, New York, pp. 687–746.

Kaestner, A. (1967) *Lehrbuch der Speziellen Zoologie*, I, 2. Fischer Verlag, Stuttgart.

Karr, T.L., Weir, M.P., Ali, Z. and Kornberg, T. (1989) Patterns of *engrailed* protein in early *Drosophila* embryos. *Development*, **105**, 605–12.

Kaufmann, R. (1965) Zur Embryonal- und Larvalentwicklung von *Scalpellum scalpellum* L. (Crustacea, Cirripedia) mit einem Beitrag zur Autökologie dieser Art. *Zeitschrift für Morphologie und Ökologie der Tiere*, **55**, 161–232.

Kobayashi, Y. and Ando, H. (1990) Early embryonic development and external features of developing embryos of the caddisfly, *Nemotaulis admorsus* (Trichoptera: Limnephilidae). *Journal of Morphology*, **203**, 69–85.

Krause, G. (1939) Die Eitypen der Insekten. *Biologisches Zentralblatt*, **59**, 495–536.

Kühn, A. (1913) Die Sonderung der Keimesbezirke in der Entwicklung der Sommereier von *Polyphemus pediculus* de Geer. *Zoologische Jahrbücher, Abteilung Anatomie und Ontogenie der Tiere*, **35**, 243–340.

Lans, D., Wedeen, C.J. and Weisblat, D.A. (1993) Cell lineage analysis of the expression of an *engrailed* homolog in leech embryos. *Development*, **117**, 857–71

Lauterbach, K.-E. (1973) Schlüsselereignisse in der Evolution der Stammgruppe der Euarthropoda. *Zoologische Beiträge, NF*, **19**, 251–99.

Lauterbach, K.-E. (1974) Über die Herkunft des Carapax der Crustaceen. *Zoologische Beiträge, NF*, **20**, 273–327.

Lauterbach, K.-E. (1980) Schlüsselereignisse in der Evolution des Grundplans der Mandibulata (Arthropoda). *Abhandlungen des naturwissenschaftlichen Vereins Hamburg, NF*, **23**, 105–61.

Lauterbach, K.-E. (1983) Erörterungen zur Stammesgeschichte der Mollusca, insbesondere der Conchifera. *Zeitschrift für zoologische Systematik und Evolutionsforschung*, **21**, 201–16.

Manton, S.M. (1960) Concerning head development in the arthropods. *Biological Reviews*, **35**, 265–82.

Manzanares, M., Williams, T.A., Marco, R. and Garesse, R. (1996) Segmentation in the crustacean *Artemia*: engrailed staining studied with an antibody raised against the *Artemia* protein. *Roux's Archives of Developmental Biology*, **205**, 424–31.

Newman, W.A. and Knight, M.D. (1984) The carapace and crustacean evolution – a rebuttal. *Journal of Crustacean Biology*, **4**, 682–7.

Nielsen, C. (1995) *Animal Evolution*, Oxford University Press, Oxford.

Osorio, D., Averof, M. and Bacon, J.P. (1995) Arthropod evolution: great brains, beautiful bodies. *Trends in Ecology and Evolution*, **10**, 449–54.

Panganiban, G., Sebring, A., Nagy, L. and Carroll, S. (1995) The development of crustacean limbs and the evolution of arthropods. *Science*, **270**, 1363–6.

Patel, N.H. (1993) Evolution of insect pattern formation: a molecular analysis of short germ band segmentation, in *Evolutionary Conservation of Developmental Mechanisms* (ed. A.C. Spradling), Wiley-Liss, New York, pp. 85–110.

Patel, N.H. (1994) The evolution of arthropod segmentation: insights from comparisons of gene expression patterns. *Development*, Supplement, 201–7.

Patel, N.H., Kornberg, T.B. and Goodman, C.S. (1989) Expression of *engrailed* during segmentation in grasshopper and crayfish. *Development*, **107**, 201–12.

Penners, A. (1924) Die Entwicklung des Keimstreifs und die Organbildung bei *Tubifex rivulorum* Lam. *Zoologische Jahrbücher, Abteilung Anatomie und Ontogenie der Tiere*, **45**, 251–308.

Pilgrim, R.L.C. (1973) Axial skeleton and musculative in the thorax of the hermit crab. *Pagurus bernhardus* (Anomura Paguridae). *Journal of the Marine Biological Association of the UK*, **53**, 363–96.

Pross, A. (1966) Untersuchungen zur Entwicklungsgeschichte der Araneae (*Pardosa hortensis* (Thorell)) unter besonderer Berücksichtigung des vorderen Prosomaabschnittes. *Zeitschrift für Morphologie und Ökologie der Tiere*, **58**, 38–108.

Pross, A. (1977) Diskussionsbeitrag zur Segmentierung des Cheliceraten-Kopfes. *Zoomorphologie*, **86**, 183–96.

Rappaport, R. Jr (1960) The origin and formation of blastoderm cells of gammarid Crustacea. *Journal of Experimental Zoology*, **144**, 43–60.

Reichenbach, H. (1886) Studien zur Entwicklungsgeschichte des Flusskrebses. *Abhandlungen der senckenbergischen naturforschenden Gesellschaft*, **14**, 1–137.

Rempel, J.G. (1975) The evolution of the insect head: the endless dispute. *Quaestiones Entomologicae*, **11**, 7–25.

Rogers, B.T. and Kaufman, T.C. (1996) Structure of the insect head as revealed by the EN protein pattern in developing embryos. *Development*, **122**, 3419–32.

Sandeman, D.C. and Scholtz, G. (1995) Groundplans, evolutionary changes, and homologies in decapod crustacean brains, in *The Nervous Systems of Invertebrates – an Evolutionary and Comparative Approach* (eds O. Breidbach and W. Kutsch), Birkhäuser Verlag, Basel, pp. 329–47.

Sander, K. (1983) The evolution of patterning mechanisms: gleanings from insect embryogenesis and spermatogenesis, in *Development and Evolution* (eds B.C. Goodwin, N. Holder and C.G. Wylie), Cambridge University Press, Cambridge, pp. 137–58.

Sanders, H.L. (1963) The Cephalocarida. Functional morphology, larval development, comparative external anatomy. *Memoirs of the Connecticut Academy of Arts and Science*, **15**, 1–80.

Scheidegger, G. (1976) Stadien der Embryonalentwicklung von *Eupagurus prideauxi* Leach (Crustacea, Decapoda, Anomura) unter besonderer Berücksichtigung der Darmentwicklung und der am Dotterabbau beteiligten Zelltypen. *Zoologische Jahrbücher, Abteilung Anatomie und Ontogenie der Tiere*, **95**, 297–353.

Schmidt-Ott, U. and Technau, G.M. (1992) Expression of *en* and *wg* in the embryonic head and brain of *Drosophila* indicates a refolded band of seven segment remnants. *Development*, **116**, 111–25.

Schmidt-Ott, U., Sander, K. and Technau, G.M. (1994) Expression of *engrailed* in embryos of a beetle and five dipteran species with special reference to the terminal regions. *Roux's Archives of Developmental Biology*, **203**, 298–303.

Schmidt-Ott, U., González-Gaitán, M. and Technau, G.M. (1995) Analysis of neural elements in head-mutant *Drosophila* embryos suggests segmental origin of the optic lobe. *Roux's Archives of Developmental Biology*, **205**, 31–44.

Scholtz, G. (1984) Untersuchungen zur Bildung und Differenzierung des postnauplialen Keimstreifs von *Neomysis integer* Leach (Crustacea, Malacostraca, Peracarida). *Zoologische Jahrbücher, Abteilung Anatomie und Ontogenie der Tiere*, **112**, 295–349.

Scholtz, G. (1990) The formation, differentiation and segmentation of the post-naupliar germ band of the amphipod *Gammarus pulex* L. (Crustacea, Malacostraca, Peracarida). *Proceedings of the Royal Society London, B*, **239**, 163–211.

Scholtz, G. (1992) Cell lineage studies in the crayfish *Cherax destructor* (Crustacea, Decapoda): germ band formation, segmentation, and early neurogenesis. *Roux's Archives of Developmental Biology*, **202**, 36–48.

Scholtz, G. (1995a) Ursprung und Evolution der Flusskrebse (Crustacea, Astacida). *Sitzungsberichte der Gesellschaft Naturforschender Freunde zu Berlin, NF*, **34**, 95–115.

Scholtz, G. (1995b) The attachment of the young in the New Zealand freshwater crayfish *Paranephrops zealandicus* (White, 1847) (Decapoda, Astacida, Parastacidae). *New Zealand Natural Sciences*, **22**, 81–9.

Scholtz, G. (1995c) Head segmentation in Crustacea – an immunocytochemical study. *Zoology*, **98**, 104–14.

Scholtz, G. and Dohle, W. (1996) Cell lineage and cell fate in crustacean embryos – a comparative approach. *International Journal of Developmental Biology*, **40**, 211–20.

Scholtz, G., Dohle, W., Sandeman, R.E. and Richter, S. (1993) Expression of *engrailed* can be lost and regained in cells of one clone in crustacean embryos. *International Journal of Developmental Biology*, **37**, 299–304.

Scholtz, G., Patel, N.H. and Dohle, W. (1994) Serially homologous *engrailed* stripes are generated via different cell lineages in the germ band of amphipod crustaceans (Malacostraca, Peracarida). *International Journal of Developmental Biology*, **38**, 471–8.

Schram, F.R. and Emerson, M.J. (1991) Arthropod pattern theory: a new approach to arthropod phylogeny. *Memoirs of the Queensland Museum*, **31**, 1–18.

Seidel, F. (1960) Körpergrundgestalt und Keimstruktur, Eine Erörterung über die Grundlagen der vergleichenden und experimentellen Embryologie und deren Gültigkeit bei phylogenetischen Überlegungen. *Zoologische Anzeiger*, **164**, 245–305.

Seidel, F. (1975) *Entwicklungsphysiologie der Tiere*, Walter de Gruyter, Berlin.

Sharov, A.G. (1966) *Basic Arthropodan Stock with Special Reference to Insects*, Pergamon Press, Oxford.

Siewing, R. (1963) Das Problem der Arthropodenkopfsegmentierung. *Zoologischer Anzeiger*, **170**, 429–68.

Siewing, R. (1969) *Lehrbuch der vergleichenden Entwicklungsgeschichte der Tiere*, Parey, Hamburg.

Siewing, R. (1979) Homology of cleavage types? *Fortschritte in der zoologischen Systematik und Evolutionsforschung*, **1**, 7–18.

Smith, C.M., Lans, D. and Weisblat, D.A. (1996) Cellular mechanisms of epiboly in leech embryos. *Development*, **122**, 1885–94.

Snodgrass, R.E. (1960) Facts and theories concerning the insect head. *Smithsonian Miscellaneous Collections*, **142**, 1–61.

Strausfeld, N.J., Buschbeck, E.K. and Gomez, R.S. (1995) The arthropod mushroom body: its functional roles, evolutionary enigmas and mistaken identities, in *The Nervous System of Invertebrates: An Evolutionary and Comparative Approach* (eds O. Breidbach and W. Kutsch), Birkhäuser, Basel, pp. 349–81.

Strömberg, J.O. (1971) Contribution to the embryology of bopyrid isopods with special reference to *Bopyroides*, *Hemiarthrus*, and *Pseudione* (Isopoda, Epicaridea). *Sarsia*, **47**, 1–46.

Taube, E. (1909) Beiträge zur Entwicklungsgeschichte der Euphausiden. I. Die Furchung des Eies bis zur Gastrulation. *Zeitschrift für Wissenschaftliche Zoologie*, **92**, 427–64.

Tautz, D., Friedrich, M. and Schröder, R. (1994) Insect embryogenesis – what is ancestral and what is derived? *Development*, Supplement, 193–9.

Tiegs, O.W. (1940) The embryology and affinities of the Symphyla, based on a study of *Hanseniella agilis*. *Quarterly Journal of Microscopical Science*, **82**, 1–225

Tiegs, O.W. (1947) The development and affinities of the Pauropoda, based on a study of *Pauropus sylvaticus*. *Quarterly Journal of Microscopical Science*, **88**, 165–267, 275–336.

Turquier, Y. (1967) L'embryogénèse de *Trypetesa nassarioides* Turquier (cirripède acrothoracique). Ses rapports avec celle des autres cirripèdes. *Archives Zoologie éxperimentelle et génerale*, **11**, 573–628.

Wagner, G.P. and Misof, B.Y. (1993) How can a character be developmentally constrained despite variation in developmental pathways? *Journal of Evolutionary Biology*, **6**, 449–55.

Wägele, J.W. (1993) Rejection of the 'Uniramia' hypothesis and implications on the mandibulate concept. *Zoologische Jahrbücher, Abteilung Systematik*, **120**, 253–88.

Wägele, J.W. and Stanjek, G. (1995) Arthropod phylogeny inferred from partial 12 SrRNA revisited: monophyly of the Tracheata depends on sequence alignment. *Journal for Zoological Systematics and Evolutionary Research*, **33**, 75–80.

Walker, M.H. (1995) Relatively recent evolution of an unusual pattern of early embryonic development (long germ band?) in a South African onychophoran, *Opisthopatus cinctipes* Purcell (Onychophora: Peripatopsidae). *Zoological Journal of the Linnean Society*, **114**, 61–75.

Walossek, D. (1993) The Upper Cambrian *Rehbachiella* and the phylogeny of Branchiopoda and Crustacea. *Fossils and Strata*, **32**, 1–202.

Walossek, D. and Müller, K.J. (1990) Upper Cambrian stem-lineage crustaceans and their bearing upon the monophyly of Crustacea and the position of *Agnostus*. *Lethaia*, 23, 409–27.

Weber, H. (1952) Morphologie, Histologie und Entwicklungsgeschichte der Articulaten II. Die Kopfsegmentierung und die Morphologie des Kopfes überhaupt. *Fortschritte der Zoologie*, **9**, 18–231.

Weisblat, D.A., Wedeen, C.J. and Kostriken, R. (1993) Evolutionary conservation of developmental mechanisms: comparison of annelids and arthropods, in *Evolutionary Conservation of Developmental Mechanisms* (ed. A.C. Spradling), Wiley-Liss, New York, pp. 125–40.

Westheide, W. (1996) Articulata, Gliedertiere, in *Spezielle Zoologie*, Teil 1 (eds W. Westheide and R. Rieger), Gustav Fischer, Stuttgart, pp. 350–2.

Weygoldt, P. (1960) Embryologische Untersuchungen an Ostracoden: Die Entwicklung von *Cyprideis litoralis* (G. S. Brady). Ostracoda, Podocopa; Cytheridae. *Zoologische Jahrbücher, Abteilung Anatomie und Ontogenie der Tiere*, **78**, 369–426.

Weygoldt, P. (1963) Grundorganisation und Primitiventwicklung bei Articulaten. *Zoologischer Anzeiger*, **171**, 363–76.

Weygoldt, P. (1979) Significance of later embryonic stages and head development in arthropod phylogeny, in *Arthropod Phylogeny* (ed. A.P. Gupta), Van Nostrand Reinhold, New York, pp. 107–35.

Weygoldt, P. (1985) Ontogeny of the arachnid central nervous system, in *Neurobiology of Arachnids* (ed. F.G. Barth), Springer, Heidelberg, pp. 20–37.

Weygoldt, P. (1986) Arthropod interrelationships – the phylogenetic–systematic approach. *Zeitschrift für zoologische Systematik und Evolutionsforschung*, **24**, 19–35.

Wheeler, W.C., Cartwright, P. and Hayashi, C.Y. (1993) Arthropod phylogeny: a combined approach. *Cladistics*, **9**, 1–39.

Whitington, P.M., Meier, T. and King, P. (1991) Segmentation, neurogenesis and formation of early axonal pathways in the centipede, *Ethmostigmus rubripes* (Brandt). *Roux's Archives of Developmental Biology*, **199**, 349–63.

Wills, M.A., Briggs, D.E.G., Fortey, R.A. and Wilkinson, M. (1995) The significance of fossils in understanding arthropod evolution. *Verhandlungen der Deutschen Zoologischen Gesellschaft*, **88**(2), 203–15.

Winnepenninckx, B., Backeljau, T., Mackey, L.Y., Brooks, J.M., De Wachter, R., Kumar, S. and Garey, J.R. (1995) 18S rRNA data indicate that Aschelminthes are polyphyletic in origin and consist of at least three distinct clades. *Molecular Biology and Evolution*, **12**, 1132–7.

Winter, G. (1980) Beiträge zur Morphologie und Embryologie des vorderen Körperabschnitts (Cephalosoma) der Pantopoda Gerstaecker, 1863. *Zeitschrift für zoologische Systematik und Evolutionsforschung*, **18**, 27–61.

Zhang, X. and Pratt, B.R. (1994) Middle Cambrian arthropod embryos with blastomeres. *Science*, **266**, 637–9.

Zilch, R. (1978) Embryologische Untersuchungen an der holoblastischen Ontogenese von *Penaeus trisulcatus leach (Crustacea, Decapoda) Zoomorphologie*, **90**, 67–100.

Zilch, R. (1979) Cell lineage in arthropods? *Fortschritte in der zoologischen Systematik und Evolutionsforschung*, **1**, 19–41.

Zrzavý, J. and Štys, P. (1994) Origin of crustacean schizoramous limb: re-analysis of the duplosegmentation hypothesis. *Journal of Evolutionary Biology*, **7**, 743–56.

# 25 Homology and parallelism in arthropod sensory processing

D.-E. Nilsson and D. Osorio

*Department of Zoology, University of Lund, Helgonavägen 3, S-223 62 Lund, Sweden*
*email: dan-e.nilsson@zool.lu.se*
*Biological Sciences, Sussex University, Brighton, BN1 9QG, UK*
*email: D.Osorio@sussex.ac.uk*

## 25.1 INTRODUCTION

Prominent eyes and antennae serving vision and olfaction are distinctive features of many arthropods, while underlying ganglia such as insect optic lobes, antennal lobes and mushroom bodies (Figure 25.1) occupy much of the nervous system. Well-developed sensory systems serving motor coordination, navigation and memory are characteristic of mobile animals, and so understanding their evolution promises insights into the emergence of arthropods from simpler ancestors.

Recent work on the eye and brain permits identification of cellular and genetic homologues by developmental and molecular criteria (Whitington and Bacon, 1997, this volume). However, the evolution of large and complex structures cannot be understood simply by comparing cellular components, or developmental genetics of two lineages. The difficulty is exemplified by the recent finding that the *Pax6* gene (as well as other homeotic genes) is required for normal eye development in mammals and *Drosophila* (Quiring *et al.*, 1994). The inference that eyes are developmental homologues in the two lineages here is inconsistent with the marked differences between invertebrate and vertebrate photoreceptor embryology, structure and biochemistry (Nilsson, 1996). In other cases where either cellular components or developmental programmes for a complex structure are shared by two phyla it may be easier to conclude, simplistically, that the structures themselves rather than their cellular components are apomorphic. In the light of this caution we review comparative work on eyes and brains in various lineages, but chiefly insects and decapod malacostracans, and ask what, if any, conclusions may be drawn about the evolutionary history of sensory processing in the arthropods.

Clearly, inferences about evolutionary relationships should be informed by an appreciation of function. Structures and physiology of sensory mechanisms are constrained by physical laws, natural stimuli and the behavioural requirements of an animal. It is not surprising therefore that parallel and convergent evolution are commonplace. For example, the various types of physiological optics, such as simple and compound eyes, have appeared independently on many occasions (Land and Fernald, 1992). Similarly, the columnar neural architecture of visual centres reflects the way in which optical stimuli are processed by parallel arrays of neural modules. Likewise, glomerular architecture of olfactory centres in diverse phyla probably reflects general requirements, but the functional principles underlying the neural architectures of olfactory centres are obscure (Hopfield, 1991; Laurent and Naraghi, 1994; Osorio *et al.*, 1994; Hildebrand, 1995).

### 25.1.1 BACKGROUND

Pioneered by workers such as Grenacher, Retzius, Zwarazin and Cajal, the comparative study of arthropod eyes and nervous systems enjoyed a golden age from the turn of the century to the second world war. Hanström (e.g. 1928) in particular produced many studies on arthropods, which given the scarcity of subsequent work on most taxa remain useful. However Hanström's pictures of cells may in some cases be diagrammatic (e.g. in crustacean and chelicerate optic ganglia) and interpretation needs to be made with caution. Fortunately (especially for English monoglots) the prewar comparative work was comprehensively summarized by Bullock and Horridge (1965), and for publications before 1950 we generally refer to this book, with a page number, rather than to original papers. Recent sources on comparative anatomy of eye and brain include Paulus' (1979) review on eye structures, Strausfeld's (1976) *Atlas of an insect brain*, and Strausfeld and Nässel's (1981) review of optic lobe anatomy in insects and crustaceans, and more recently volumes edited by Gupta (1987) and by Breidbach and Kutsch (1995).

*Arthropod Relationships*, Systematics Association Special Volume Series 55, Edited by R.A. Fortey and R.H. Thomas. Published in 1997 by Chapman & Hall, London. ISBN 0 412 75420 7

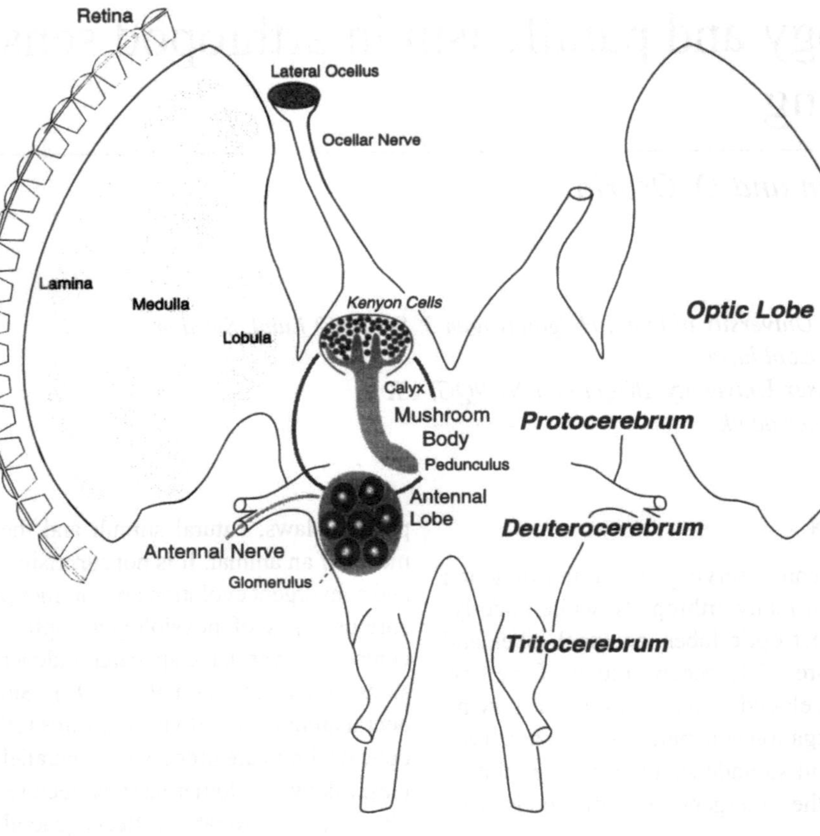


**Figure 25.1** Cephalic nervous system of a locust (*Schistocerca gregaria*) viewed from the front (technically the ventral surface relative to the axis of the ventral nerve chord). Three optic ganglia – lamina, medulla and lobula have a columnar organization reflecting the point-by-point visual processing found in early stages of the visual pathway in many animals. The antennal lobe comprises glomerular aggregations of neural process which are also seen in olfactory centres of mammals and molluscs. The insect mushroom body is primarily and olfactory centre, but in some taxa receives a substantial visual input. The intrinsic cells of the mushroom body are Kenyon cells. Structures which may be homologous to all these centres have been described in various arthropods and annelids.

## 25.2 OPTICS AND THE RETINA

### 25.2.1 DIVERSITY OF EYE TYPES

The presence or absence of eyes, or the type of eye, has frequently been used as part of the data in cladistic analyses of arthropod relationships. However, a scan through the literature on comparative eye morphology (Bullock and Horridge, 1965) should alert the reader to the surprising mixture of eye types within the major arthropod groups. An attempt to group arthropods according to eye type would show how misleading such characters are: scutigeromorph centipedes have compound eyes, whereas scolopendromorphs have single lens eyes; most crustacean groups have compound eyes but not copepods, which may have lens eyes instead; xiphosurids and eurypterids have large compound eyes, but terrestrial chelicerates invariably have lens eyes. With sensory organs, it is necessary to look at details to find plausible evidence for true homology hidden among similarities dictated by functional constraint.

An intriguing peculiarity within the Crustacea are the three pairs of single-lens eyes (camera-type eyes) in ampeliscid amphipods, which is discordant with the general occurrence of compound eyes in amphipods. Analysis of the retina of ampeliscid camera eyes, however, reveals typical amphipod ommatidia grouped under large common lenses (Elofsson *et al.*, 1980; Hallberg *et al.*, 1980). One explanation of such a transformation from compound eyes to camera eyes would be a temporary loss of selection favouring vision, causing first a reduction of the eye's optics and later, as vision becomes useful again, a re-invention of optics. Such a temporary loss of selection for vision could in the case of the ampeliscids have been the acquisition of a burrowing life-style which was abandoned before eye structure had been entirely lost (Nilsson, 1989).

### 25.2.2 CONVERGENT EVOLUTION OF COMPOUND EYES

The compound eyes of insects and crustaceans offer potentially useful information for phylogenetic analysis. Later, we discuss similarities between insect and crustacean compound eyes that extend beyond eye type and involve cellular organization and physiology of the eye and associated parts of the nervous system. However, it is instructive to look first at how convergent evolution in eye design can lead to deceptive similarities.

There are several types of compound eye which rely on different optical principles for image formation. The major division is between apposition and superposition eyes (Figure 25.2). In apposition eyes, which are likely to be the ancestral type, the photoreceptors of each ommatidium receive light only from their own facet. Superposition eyes collect light much more efficiently because a large number of ommatidia cooperate optically to make a bright superimposed image on the retina. However, this efficiency is not achieved easily; the optical system of superposition eyes is fundamentally different from that of apposition eyes. In fact, it is often hard to see how superposition eyes can evolve from apposition eyes without passing through non-functional intermediates. Plausible ways around these evolutionary obstacles do however exist and a common theme is that special life styles have made the apposition eyes preadapted for transformation into superposition eyes (Nilsson, 1983, 1988, 1989, 1990; Nilsson *et al.*, 1984).

Superposition imaging can be accomplished in different ways, using lenses, mirrors or a combination of both (Nilsson, 1989). The most common type, termed refracting superposition, is based on graded index lenses in the crystalline cones, effectively forming a small Kepplerian telescope in each ommatidium. Superposition optics require a precisely tuned gradient of refractive index in the crystalline cone, together with accurately matched eye geometry and alignment of the optical units. None of the foregoing are as critical in apposition optics. The position and design of the superposition retina, the eye's pigment shields and the pupil mechanism are moreover fundamentally different from those in apposition eyes.

With all these specialized and advanced traits, and the forbidding degree of precision required for the eye to work, one would expect refracting superposition eyes to be a rare invention of significance in phylogenetic discussions. However, the list of crustacean and insect groups that have these eyes tells a different story. Among crustaceans, they are found in anaspideaceans, mysids, euphausiids, one genus of decapod shrimp and some hermit crabs. In insects, refracting superposition eyes exist in one group of mayflies, most neuropterans, three separate groups of beetle, most moths, and all caddis flies. Undoubtedly, this heterogeneous crowd of creatures does not share superposition eyes because they descend from a common ancestor with this type of eye. Refracting superposition eyes must have evolved numerous times from apposition eyes because of their superiority in dim conditions.

It is fascinating to compare the internal structure of refracting superposition eyes from the various groups of insects and crustaceans. From histological sections of the eyes, it is virtually impossible to determine which group they belong to. This remarkable conformism demonstrates

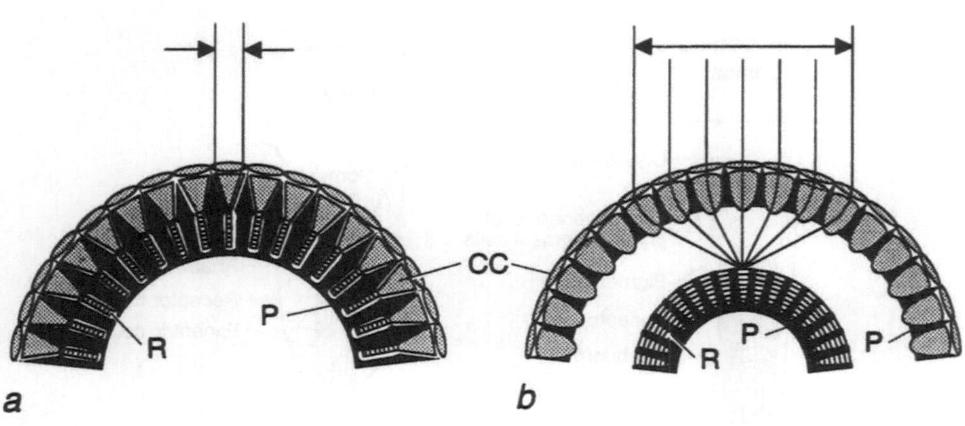


**Figure 25.2** The two most common optical types of compound eye, apposition (a) and refracting superposition (b). In both types of eye the optical image is transduced by an array of rhabdoms, each equivalent to an image pixel. A weakness of the apposition type is that each rhabdom receives only the light entering its own facet, whereas a superposition eye may admit light through several hundred facets to each rhabdom. Given that the number of available photons often limits visual performance, superposition eyes have a clear advantage. Two of the distinctive features of superposition eyes are the upright imaging of the ommatidial optics (inverted imaging in apposition eyes) and the clear zone separating the optics from the retina. Abbreviations: R, rhabdom; P, screening pigment; CC, crystalline cone.

how tight the functional constraints can be and how precisely evolution responds to them. However, the optical structures of eyes are unique, not for the essential mechanisms and precision of their evolution, but for our understanding of the functional constraints. The bases for optimization of biological structures are usually understood to a far smaller degree. That indistinguishable superposition eyes evolved independently in so many groups, emphasizes that we have to treat sceptically similarities which seem to be apomorphic, especially in structures where functional principles are less easily understood.

### 25.2.3 PHOTORECEPTORS

Arthropod photoreceptor cells are of the microvillar type (as opposed to the ciliary type of chordates), and their neurotransmitter is histamine (Callaway and Stuart, 1989; Hardie, 1989; Batelle *et al.*, 1991; Schmid and Dunker, 1993). These features may seem to unite arthropods, but the eyes in the head (cephalic eyes) of annelids, molluscs and other protostomes are always microvillar, but non-cephalic eyes, such as the tentacular eyes of sabellid polychaetes and the mantle eyes of bivalves are often of the ciliary type (Salvini-Plawen and Mayr, 1977). It is of course possible that the similarities reflect a common origin for the receptor cells themselves, but since they are present in animals with very simple eyespots (Land and Fernald, 1992) they say little about relationships of different arthropod groups. Meanwhile it is interesting that histamine is found in photoreceptors of flatworms (Panula *et al.*, 1995) but not in molluscs, even though it is widespread in their nervous system (Soinila *et al.*, 1990; Alkon *et al.*, 1993; Karhunen *et al.*, 1993).

### 25.2.4 ANATOMY OF COMPOUND EYE OMMATIDIA

While median eyes occur in most arthropod groups, they are simple organs without any striking similarities between specific groups, the cellular structure of compound eye ommatidia is of interest here. Insect and crustacean ommatidia are built on identical groundplans (Figure 25.3). Most distally in the ommatidium, just below the cuticular cornea, is a pair of cells termed primary pigment cells in insects and corneageneous cells in crustaceans. Surrounded by these is the crystalline cone which forms part of the ommatidium's optical system. In both insects and crustaceans the cone is generally composed of four cells, although a few variants exist. A group of eight receptor cells occupies the proximal part of the ommatidium. Some crustaceans have fewer receptor cells, and in both insects and crustaceans the number may exceed eight (Paulus, 1979; Meinertzhagen, 1991). Yet another similarity between insect and crustacean ommatidia is the division of receptor cells into two distinct classes: cells with axons terminating in the first optic ganglion, the lamina, and cells with axons continuing to the second optic ganglion, the medulla (Figures 25.4 and 25.5). The cells with short visual fibres (SVFs) are often green-sensitive and constitute the majority of receptors in the ommatidium, whereas the cells with long visual fibres (LVFs) are generally sensitive to short-wavelength light and only one (crustaceans), two or three (insects) are present in each ommatidium (Meinertzhagen, 1991). The dichotomy of receptor cell types appears to be ancient because the green photopigments in decapods and insects are more closely related than UV and green pigments are in flies (Gartner and Towner, 1995).

The affinity between insect and crustacean compound

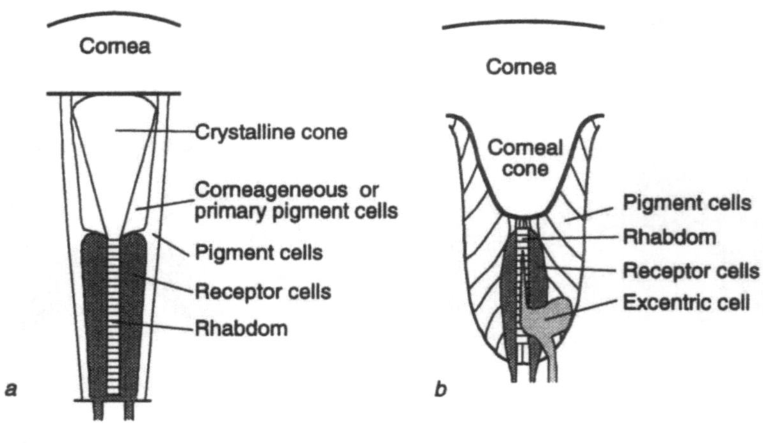


**Figure 25.3** The cellular composition of compound eye ommatidia in (a) insects and crustaceans, and (b) *Limulus*. In both cases the ommatidia are specialized parts of the epidermis. Insects and crustaceans have a determinate number of each cell type, notably two corneageneous cells, four cone cells and seven or eight receptor cells. The *Limulus* ommatidium contains a much larger and indeterminate number of cells. The eccentric cell, a specialized photoreceptor, has no analogue in the insect retina, its function being similar to that of the large monopolar cells in the insect lamina.

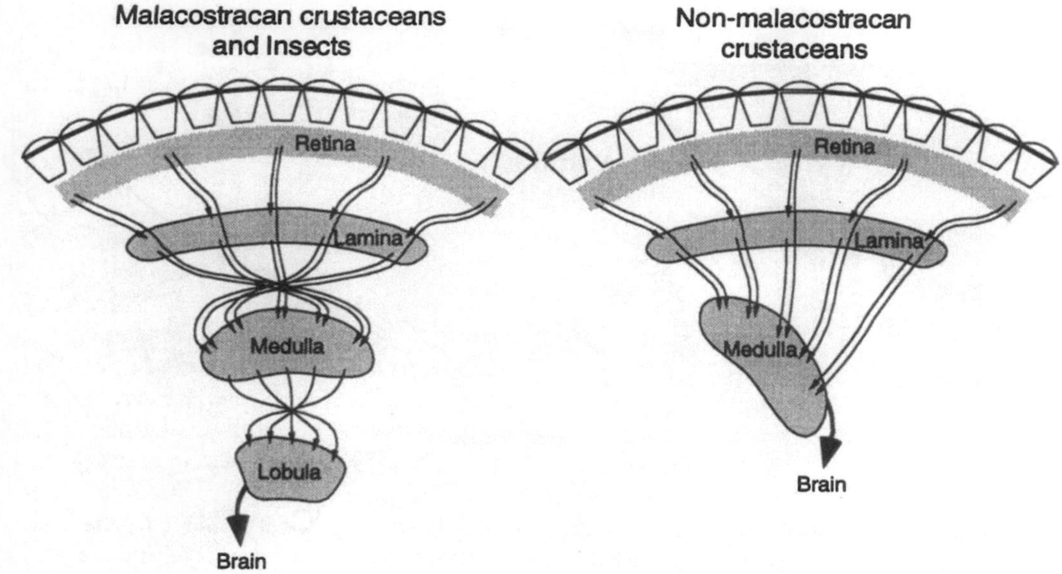


**Figure 25.4** The general lay-out of insect and crustacean optic lobes. The outer neural chiasm of malacostracans and insects make the image representation turn front to back in the medulla, but the second chiasm restores image orientation in the lobula. Non-malacostracan crustaceans, such as branchiopods, differ from malacostracan crustaceans and insects in having only two visual ganglia, the lamina and the medulla, and no chiasms. Common to all insects and crustaceans is the distinction between retinal photoreceptors with short axons terminating in the lamina and those with long axons terminating in the medulla.

eyes is also implied by strong labelling of their crystalline cones by an antibody to the crystallin 3G6 (Edwards and Meyer, 1990; Tomarev and Piatigorsky, 1996), which also gives 'very faint but specific' labelling in the eye of *Lithobius* (Chilopoda). These similarities between insect and crustacean ommatidia are unlikely to be due to convergence, for example the compound eyes of Myriapoda and of *Limulus* are built in a different way (Figure 25.3). The specific cell types and determinate number of cells common to insect and crustacean ommatidia contrast to the indeterminate and much larger number of cells forming the ommatidia in millipedes and xiphosurids. Ommatidia of scutigeromorph centipedes show a superficial similarity to those of insects but the number of cells are different and indeterminate (Paulus, 1979).

## 25.3 NERVOUS SYSTEM

The facility with which flies pursue one another, or bees learn the colour and odour of flowers and navigate in a three-dimensional landscape may leave the human observer unmoved because we share many similar abilities. However, for an engineer the performance of arthropod brains is impressive. A complex body plan with diverse articulated limbs would have co-evolved with the nervous system. Just as a car or a camera is easier to build than an autonomous robot, acquisition of a nervous system may be critical in the evolution of arthropods. If emergence of developmental mechanisms was a key event in metazoan phylogeny (Davidson *et al.*, 1995) the evolution of the elaborate architecture of the nervous system may have been of singular importance. In this context, an additional attraction of the nervous system is its conservatism compared with more peripheral structures. Comparative physiologists often note that central circuits are less diverse than sensory or motor periphery. Variations to serve diverse behavioural needs are chiefly at the level of synaptic connections rather than the cells themselves (Dumont and Robertson, 1986; Robertson and Olberg, 1988; Arbas *et al.*, 1991; Fullard and Yack, 1991; Katz, 1991). This conservatism offers the opportunity to find similarities between remotely related taxa. However, some parts of the nervous system are more variable than others, and sense organs together with underlying neural circuitry are quickly lost when not used. Consequently differences between taxa – for example, in the eyes of cursorial or aerial insects and the more fossorial myriapods – may well reflect lifestyle and history more than evolutionary relationships.

Recently, comparative work on arthropod nervous systems has been given impetus by Thomas *et al.*'s (1984) suggestion that several orders of insect and also a malacostracan share a developmental homologue, namely the G-neurone of the thoracic ganglia. Subsequent molecular and developmental studies have vindicated the contention that insect

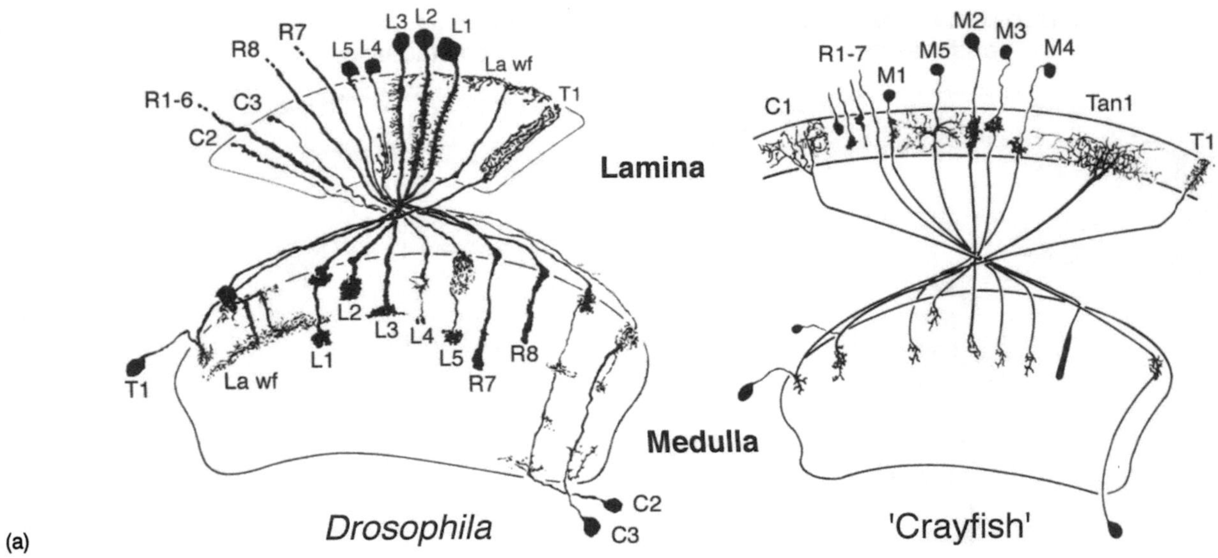


**Figure 25.5** (a) Principal columnar neurones which make up a lamina cartridge in the fly *Drosophila* (from Fischbach and Dittrich 1989), and decapods. (Modified from Strausfeld and Nässel, 1981 and Wang-Bennett and Glantz, 1987a,b.) The complement of cells in the lamina cartridges of the two orders are much alike. Amacrine cells whose processes are confined to the lamina are not illustrated. (b) Diagram of some of the anatomical pathways in the outer optic lobe common to flies and decapods. The nomenclature follows that for *Drosophila*. Whereas flies have two monopolar cells that do not receive direct photoreceptor inputs, the decapod probably has only one such monopolar cell which denotes the fly cell L4. Physiological responses of anatomical analogues of the long and short visual fibre receptors, large monopolar cells, L4 and the T-cell are very similar in decapods, fly and locust (Wang-Bennett and Glantz, 1987a,b; Bartels and Glantz, 1991; Osorio and Bacon, 1994; James and Osorio, 1996). Abbreviations: LVF, long visual fibre photoreceptor; SVF, short visual fibre photoreceptor; LMC, large monopolar cell.

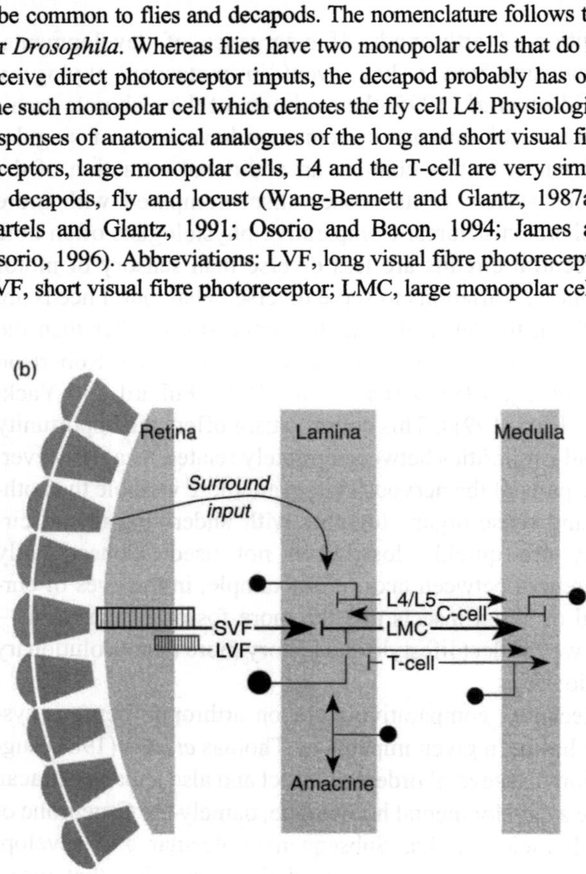


nervous systems share a common developmental plan, and there is good evidence that this extends to crustaceans (Osorio *et al.*, 1995; Elson, 1996; Whitington and Bacon, 1997, this volume). Although impressive as evidence for conservatism, similarities between insect and malacostracan segmental nervous systems do not inexorably point to the common ancestor being an arthropod. Just as ommatidial units of the compound eye retina probably formed as simple ocelli, the pattern of neural development seen in the ganglia of the ventral nerve chord could have emerged in an animal with little segmental specialization and subsequently evolved in parallel in insect and crustacean lineages.

By comparison, specialized neural circuits serving complex sensory–motor tasks are less likely to have emerged in a simple animal. Ganglia in the head that serve vision and olfaction are conspicuous in many arthropods (Figure 25.1), and comparisons of their neural circuitry promise insights into arthropod evolution. In insects, and crustaceans, the cephalic nervous system is divided into three principal parts: the protocerebrum, the deuterocerebrum and the tritocerebrum (Hirth *et al.*, 1995; Scholtz, 1997, this volume). The optic lobes may be derived from the segment anterior to the antenna (Schmidt-Ott *et al.*, 1995). The cephalic ganglia are probably derived from three or more sets of segmental neuroblasts resembling those in the ventral nerve chord (Zacharias *et al.*, 1993; Whitington and Bacon, 1997, this volume), but they are elaborated for sensory processing and contain over 80% of all neurones in large insects such as the fly *Calliphora* and the bee *Apis* (Strausfeld, 1976). Unlike the ganglia along the ventral nerve chord, detailed developmental data is lacking in the head. Nonetheless, similarities in the overall architecture and mature neural circuitry in the optic lobes warrants attention.

25.3.1 OPTIC LOBE

Description of the optic lobe is based on insects and malacostracans which are best known; subsequently, we consider the place of other arthropods.

The functional demands of vision enforce a columnar organization on the nervous system. Each facet in the compound eye corresponds to a point or pixel in the optical image. The insect optic lobe, including that of Archaeognatha, comprises three successive ganglia – the lamina, the medulla and the lobula – separated by two anteroposterior chiasms (Figure 25.4). Beneath each ommatidium a set of neurones is repeated to form a columnar unit in the lamina and medulla. The lobula, has a column of cells for every four facets in the eye. In flies, each of the columnar neurones of the lamina and medulla have been individually identified, there are about 60 for each facet (Strausfeld, 1976; Fischbach and Dittrich, 1989), giving 360 000 cells for the 6000 or so facets in the two eyes; a large proportion of the total number of cells in the insect brain. In addition to the columnar neurones, widefield or tangential neurones occur throughout the optic lobe, forming a conspicuous 'serpentine layer' in the proximal half of the medulla. These fibres integrate inputs across the visual field, make feedbacks from the brain and provide pathways for lateral signal flow.

Malacostracan crustaceans have an optic lobe resembling that of insects, with three columnar ganglia separated by two anteroposterior chiasms. Hanström called the outer two ganglia the lamina ganglionaris and medulla externa, and the third ganglion the medulla interna. Regardless of homology, the architecture of these ganglia are much like that of the insect, and we follow Strausfeld and Nässel (1981) who advocate use of insect nomenclature, referring to the ganglia as lamina, medulla and lobula respectively. As in insects, the malacostracan lamina and medulla are composed of columnar units, one for each ommatidium, and there is a serpentine layer in the medulla. Again, periodicity of the lobula is coarser than that in the outer ganglia, but the degree of convergence is unknown.

The optic lobes of non-malacostracan crustaceans lack the lobula and there are no optic chiasms (Figure 25.4). Information on non-malacostracan optic lobes is sparse (for an overview see Elofsson and Dahl, 1970) and details are known only from branchiopods where the lamina and medulla have a columnar organization much as in malacox stracan crustaceans (Strausfeld and Nässel, 1981; Elofsson and Hagberg, 1986).

25.3.2 COMPARATIVE STUDIES OF LAMINA

The insect and crustacean laminas provide a good basis for comparisons of neural anatomy and physiology in an entire ganglion. Whereas comparisons between mature nervous systems elsewhere are generally restricted to single cells (Whitington and Bacon, 1997, this volume), in the lamina each of the ten or so neurones that make up a cartridge, the columnar unit beneath each ommatidium, have analogues in the two mandibulate lineages. Even without developmental data, similarities of insect and malacostracan laminas both anatomical and physiological are striking; evidence either for common ancestry in neural processing beneath the compound eye, or for closely convergent/parallel evolution of neurones and physiological mechanisms (Osorio and Bacon, 1994; Osorio *et al.*, 1995). Here, we review the structure and function of the lamina (Figure 25.5), and then turn to consider what we can learn about evolution of vision in arthropods.

**(a) Insect lamina**

The insect lamina lies beneath the compound eye retina, and has been described in several orders (Kral, 1987). Each cartridge receives direct inputs from six or seven optically coaxial (i.e. viewing a single point in space) short visual fibre receptors, and so preserves the optical order of the eye. Two or three long visual fibre axons pass through the lamina to terminate in the coaxial medulla column. Each cartridge contains a set of about 10 neurones of six main anatomical types, with little or no variation across the eye (Figure 25.5). These were first described by Cajal and Sanchez (1915; Bullock and Horridge, p. 1079) whose nomenclature has since been retained for insects and extended to crustaceans. The lamina neurones which relay signals from eye to medulla are monopolar cells and a T-cell. The monopolar cell bodies lie between the lamina and retina, while those of the T-cells lie between the lamina and medulla. The large monopolars (LMCs) a subtype of monopolar, usually with fat axons, are directly driven by SVF receptors; there are three LMCs in Diptera and similar numbers in other orders. Other monopolars and the T-cell receive inputs from the LMCs. Meanwhile, C-cells – whose cell bodies lie between the medulla and lobula – are feedback neurones, sometimes projecting to multiple lamina columns. Small field tangential cells (Tan 1) resemble T-cells but their dendrites spread across several cartridges, while amacrine cells make lateral connections within the lamina.

Lamina physiology, centred on LMCs, has been studied in Odonata (Yang and Osorio, 1996), Orthoptera (James and Osorio, 1996) and Diptera (Jansonius and van Hateren, 1993a,b; Laughlin, 1994), with some work on Hymenoptera (deSouza *et al.*, 1992) and Lepidoptera (Horridge *et al.*, 1984). The LMCs are much the most easily recorded neurones. The primary response is a hyperpolarizing synaptic input from the photoreceptors. In addition, spatial and temporal antagonism fine-tunes the LMC responses. Differences between the complement of three LMCs in each lamina cartridge of flies are small and unimportant here (Laughlin, 1994).

Lamina neurones other than the LMCs are little known. In the locust, two monopolar cells are physiologically quite different from the LMCs (James and Osorio, 1996). One is depolarized by illumination and has comparatively strong spatial and temporal antagonism; the other has a receptive field much larger than a single facet and is briefly excited both by increases and decreases in intensity. These responses resemble those of unidentified afferent neurones in the fly lamina (Jansonius and van Hateren, 1993a,b) characterized by their 'transient' and 'sustaining' spiking responses (LMCs are non-spiking cells). One locust neurone may be an anatomical and physiological analogue of the fly monopolar L4/sustaining unit, while the transient locust neurone is analogous to the fly cell L5 (James and Osorio, 1996). In addition to three main types of monopolars, T- and C-cells have been recorded in locust (James and Osorio, 1996) but not in other insects. The T-cell response resembles that of the LMCs but is more sluggish and lacks spatial or temporal antagonism.

Given the similarities of locust and fly laminas, which may extend to crayfish, it is salutary to note that dragonfly monopolars are different in that all five monopolars have physiological responses superficially resembling those of the fly LMCs. Chromatic coding by dragonfly lamina is more complex than in calliphorid flies and locusts, and the differences partly reflect specialization for colour vision (Yang and Osorio, 1996).

**(b) Decapod lamina**

The cellular anatomy of decapod lamina is much like that of the insects (Figure 25.5; Strausfeld and Nässel, 1981). Seven SVF receptors terminate on one set of (four or five) monopolars (LMC type) called M1–4, while a single long visual fibre passes through the lamina to the medulla. A T-cell, and a monopolar M5 forms a secondary afferent pathway as in insects, while amacrine cells make lateral connections. There are also neurones with their cell bodies proximal to the medulla, as in insect C-cells. Intracellular records from several types of monopolar, the T-cell and amacrines in the crayfish *Procambarus* and *Pacifastacus* provide a useful basis for comparisons with insects, especially locust (Wang-Bennett and Glantz, 1987a,b; Glantz and Bartels, 1994). In each case where comparisons are possible there are direct anatomical and physiological analogues common to fly and locust (Figure 25.5): the LMC-type monopolars show weak temporal and spatial inhibition. In addition, M5 is a sustaining monopolar (Wang-Bennett and Glantz, 1987b) which is depolarized by light and with comparatively strong lateral inhibition. This resembles the locust-sustaining monopolar. Likewise, the crayfish T-cell has responses resembling those of the LMCs but lacking spatiotemporal inhibition, just like the locust T-cell.

Some similarities between insect and crustacean laminas are to be expected for functional reasons. Anatomically, the general layout with repeated columns of neurones connected by a horizontal network of amacrine cells is echoed by vertebrate retina. LMC physiology may be explained by engineering principles consistent with their function as being efficient encoding of the retinal signal (Laughlin, 1981, 1994). This is a general task useful to any visual system and LMCs have functional analogues in the vertebrate retinal bipolar cells, and also in the eccentric cells of *Limulus* lateral eye (Laughlin, 1981, 1994). However, in contrast to LMCs, the sustaining monopolar and the T-cells, both having analogues in locust and crayfish, do not have physiological analogues in vertebrate retina and their significance is obscure. Even though early visual processing should be similar in all visually guided animals, it seems clear that the task can be implemented in more than one way, making the close anatomical and physiological similarities between insect and crustacean visual systems phylogenetically important.

It is not physiological and anatomical similarities in any one class of neurone that make the lamina of particular interest, but the overall consonance of anatomy and physiology. The complement of cells in each lamina cartridge, and so far as is known their connectivity and physiology are common to the two groups. Thus, the neural circuitry of the lamina cartridge may be homologous in the insects and malacostracans (Figure 25.5). Overall, the lamina cartridge presents a picture of cellular homology consistent with that in the eyes themselves on the one hand and the segmental ganglia on the other.

**(c) Inner optic lobe ganglia**

It may reasonably be objected that the lamina is the most peripheral of the ganglia in the optic lobe and the circuitry is comparatively simple. Even if molecular studies were to confirm the homology of the neurones in lamina of insects and malacostracans, the conclusion that visual ganglia are apomorphic in insects and malacostracans should not be extended by default to the medulla and lobula. The general arrangement of the optic lobes of insects and malacostracans is indeed alike (Figure 25.4); three columnar ganglia are separated by two anteroposterior chiasms. However, this is comparatively weak evidence for these ganglia being apomorphic; for example, *Limulus* shares a similar ganglionic arrangement, while their eyes and neural connections are quite different from the insect/crustacean plan.

### 25.3.3 CHIASMS AND THE NON-MALACOSTRACAN ENIGMA

The detailed similarities between insect and malacostracan ommatidia, lamina neurones and general optic lobe lay-out seems to indicate that they share a common ancestor with good vision. However, non-malacostracan compound eyes and optic ganglia only partly fit into this pattern. The ommatidial design of branchiopods (Güldner and Wolff, 1970; Nilsson and Odselius, 1981; Nilsson *et al.*, 1983), ostracods (Land and Nilsson, 1990), barnacle larvae (Hallberg and Elofsson, 1983), ascothoracid larvae (Hallberg *et al.*, 1985) and the branchiuran *Argulus* (Hallberg, 1982) is not fundamentally different from that in malacostracans and insects. The optic lobes in non-malacostracans, do differ in having only two ganglia, lamina and medulla, and no chiasms (Figure 25.4; Elofsson and Dahl, 1970). Only for branchiopod crustaceans (Anostraca, Cladocera) do studies permit comparison of neuronal types. Golgi staining in the anostracan lamina reveals one to three types of monopolar cell, C- and small field tangential (Tan1) cells (Elofsson and Hagberg, 1986), while cladocerans have neurones resembling the LMCs and a T-cell (Bullock and Horridge, 1965, p. 1070; Strausfeld and Nässel, 1981). Apparent variation may arise because Golgi staining is capricious, and it is possible for a cell type to be entirely overlooked, especially if only one variant of the technique is used (Strausfeld and Nässel, 1981). There are comparatively few cell types in the brachiopod medulla (Strausfeld and Nässel, 1981; Elofsson and Hagberg, 1986), and the whole optic lobe is simpler than those of malacostracans and insects (see also Osorio, 1991; Osorio *et al.*, 1995).

Despite its comparative simplicity there is no obvious reason to see the branchiopod visual system as ancestral to the malacostracan. Moreover, it is hard to see how the branchiopod optic lobes without chiasms could gradually transform into the malacostracan type with chiasms. Equally, the reverse transformation, which would be a reduction, is just as difficult to explain with a gradual change (although compare with the eyes of amplescid amphipods described above). Put simply, chiasms cannot be untwisted without creating new chiasms in the orthogonal direction (turning a malacostracan medulla 180° around the eye stalk axis would resolve the two anteroposterior chiasms, but create two new dorsoventral chiasms). The embryological reason for presence or absence of chiasms in Crustacea was investigated by Elofsson and Dahl (1970). New units in the eye and the lamina are produced by proliferation in the same direction in all crustaceans. The proliferation of medulla units, however, is fundamentally different so that the sides of the ganglion which become distal and proximal are reversed between malacostracans and non-malacostracans. The medulla cartridge of course has cell bodies and axons attached to it, and depending on the direction of proliferation some of these cell bodies and connections will end up on one side of the medulla and others on the other side (Figure 25.6). The difference between malacostracans and non-malacostracans lies in the topology of proliferation in relation to cell bodies and axon paths. The original growth direction of the medulla is proximal in all crustaceans, but because of the difference in growth topology the medulla turns in different directions in the two crustacean subgroups (Figure 25.6). Our conjecture that growth directions cannot be altered gradually is supported by the observation that insects and malacostracan invariably have chiasms – the optic tracts seem never to have uncrossed. Consequently, we face a phylogenetic problem, with insects and malacostracans being alike, whereas non-malacostracans are profoundly different. It is possible that a single mutation may alter the direction of proliferation and effectively flip one topology into the other. It is even possible that reasonably correct neuronal connections could be established in the medulla, even when this deep and stratified neuropile develops upside down. However, it is very hard to believe that such a hypothetical mutation would ever become established in a natural population, since it would not improve the function, and probably disturb it.

A plausible explanation of the differences between malacostracans and non-malacostracan optic lobes is that their common ancestor had no compound eyes, only advanced ocelli built like insect and crustacean ommatidia and connected to small optical ganglia containing the basic neurone types of modern lamina and medulla cartridges. If proliferation to obtain spatial vision with compound eyes developed independently in different lineages, it would explain why ommatidia and neurones are so similar but the ganglionic topology so different.

An ancestor with only ocelli – and thus no spatial vision – cannot have been a large mobile creature, since arthropodization probably resulted from selection for more efficient locomotion, the common crustacean ancestor would date back to the initial stages of arthropod evolution, just pre-dating the Cambrian explosion (Figure 25.7). Insects, in this scenario, must either have branched off at the ocellar stage and elaborated along similar lines to the malacostracans. Alternatively, insects may be descendants of malacostracan crustaceans.

### 25.3.4 OPTIC LOBE IN OTHER ARTHROPODS

In comparison with the insects and crustaceans, optic lobes of other arthropods are too little known and disparate for much to be said with confidence; nonetheless, as 'outgroups' they deserve attention. Myriapods generally have reduced compound eyes, there is a small optic lobe in the diplopod *Julus* with no chiasms (Bullock and Horridge, p. 842); an observation which adds little to what can be deduced from comparisons of the eyes and segmental nervous system (Whitington and Bacon, 1997, this volume).

The chelicerate *Limulus* is of particular interest. The lateral (compound) eyes are very well known, and not only does the cellular anatomy of the retina differ from that of insects and crustaceans, but the mechanism of lateral inhibition used to sharpen the receptor signals (a function of the SVF to LMC synapse in insects) is mediated by quite a different cell type, the eccentric cell (Laughlin, 1981). This specialized photoreceptor is electrically coupled to the

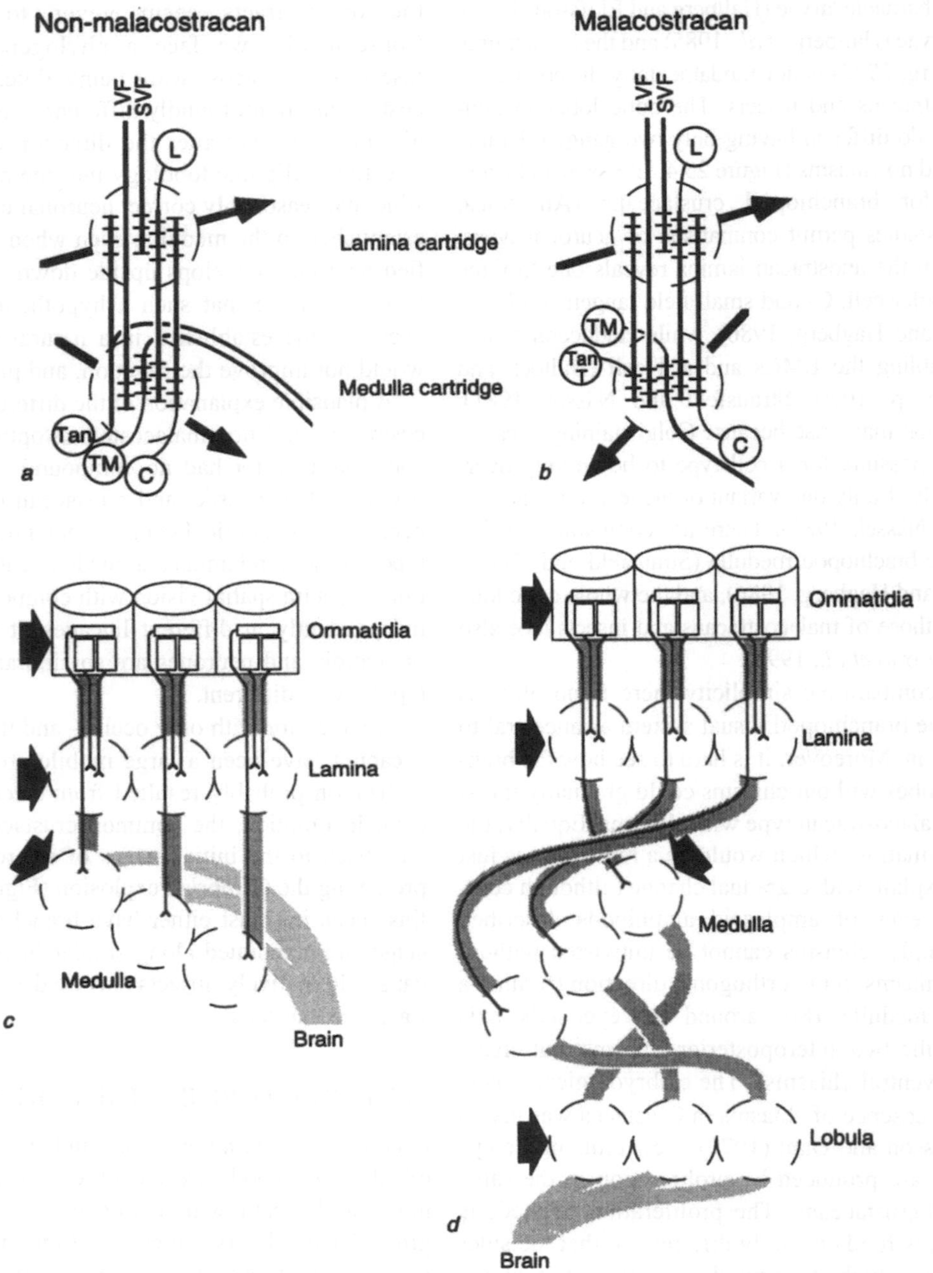


**Figure 25.6** Schematic diagrams illustrating the topology of proliferation in the optic lobes of malacostracan and non-malacostracan crustaceans. The arrangement of axons and cell bodies of some of the principal types of neurone in single lamina and medulla cartridges is shown in (a) and (b). The cell types are: receptor cells with long visual fibres (LVF) and short visual fibres (SVF), lamina monopolar cells (L), centrifugal cells (C), trans-medulla neurones (TM), T-cells (T) and tangential cells (Tan). Arrows indicate the direction of proliferation as new cartridges are added. Note the topological difference between malacostracans and non-malacostracans. The consequence of this difference in topology is illustrated in (c) and (d): when new ommatidia and neural cartridges are added, non-malacostracan crustaceans develop straight projections whereas two chiasms are formed in malacostracan crustaceans. The difference is phylogenetically important because there is no way of gradually transforming one system into the other.

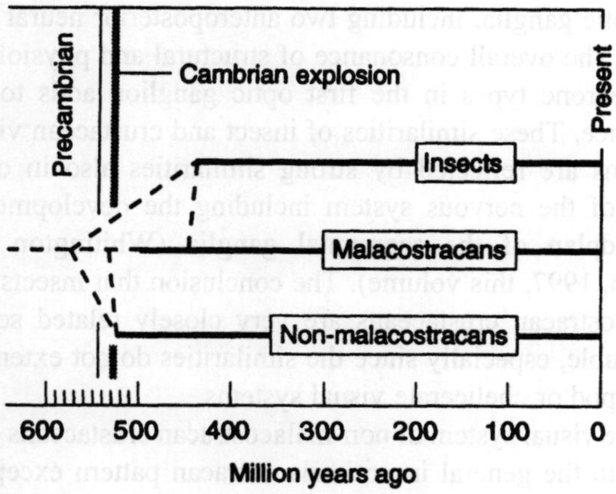


**Figure 25.7** Phylogeny of insects and crustaceans based on the compound eyes and optic lobes. The almost identical ommatidia and peripheral neural connections indicate strong affinities between insects and crustaceans, but the fundamentally different topologies of their optic lobes separate non-malacostracan crustaceans from malacostracans and insects. The differences indicate that the two crustacean subdivisions separated before compound eyes developed, indicating a last common ancestor with ocelli built like ommatidia. Insects and malacostracan crustaceans, which have virtually identical compound eyes and optic lobes, may independently have followed the same evolutionary path, or insects are in fact malacostracans.

principal receptors, and inhibited by itself and by lateral inputs from neighbouring eccentric cells, it projects to the first optic ganglion the lateral optic glomerulus (about which little is known). Hanström (cited in Bullock and Horridge, 1965, p. 845) illustrates the neurones from *Limulus* lateral optic glomerulus with neurones resembling monopolar and T-cells. This implies affinities with insect/malacostracan lamina anatomy which are surprising in view of differences between the eyes themselves; it would be interesting if Hanström's rather diagrammatic picture of affinities between *Limulus* and mandibulate cellular anatomy were confirmed. Spiders, which have simple not compound eyes, have optic neuropiles – inner and outer optic glomeruli. In spiders, chiasms separate the principal eyes from the first optic ganglion and the two ganglia, while cells in the first optic ganglion which receive receptor inputs (i.e. analogous to insect LMCs) have T-cell morphology in a ctenid spider (Strausfeld and Barth, 1993). Summing up, comparisons of lateral eyes of myriapods and chelicerates with the common insect/crustacean plan give us no clear grounds for proposing common arthropod ancestors or independent non-arthropod origins. The firm conclusion is that insect and crustacean visual systems reveal strong affinities, which are not challenged by any other combination of major arthropod groups.

## 25.3.5 THE OLFACTORY PATHWAY

In insects, olfactory neurones in the antennal nerve project to the antennal lobe which is part of the deuterocerebrum (Figure 25.1). The principal outputs of the antennal lobe are to the mushroom bodies (corporea pedunculata) which are part of the protocerebrum (Bullock and Horridge, 1965; Mobbs, 1982; Strausfeld *et al.*, 1995). An analogous pathway occurs in crustaceans, but here the protocerebral olfactory centre is called the hemiellipsoid body. Having emphasized the conservatism of neural anatomy it noteworthy that the size of the insect mushroom body and extent of non-olfactory inputs varies within insects. Meanwhile, decapods acquired *de novo*, and in some cases lost, a large deuterocerebral olfactory centre called the accessory lobe (Sandeman *et al.*, 1993; Sandeman and Scholtz, 1995).

The first ganglion on the olfactory pathway in insects and crustaceans is the antennal lobe. Whereas the optic lobe has a columnar architecture, the basic unit of the antennal lobe is the glomerulus, a dense aggregation of input and output neurones and intrinsic neural processes, connected to one another directly and via intrinsic interneurones (Hildebrand, 1995). Glomerular architecture is characteristic of olfactory centres and is seen also in the chilopod *Lithobius*, in the primary olfactory centres in terrestrial gastropods and mammals (Chase and Tolloczko, 1986; DeVries and Baylor, 1993; Sandeman *et al.*, 1993; Hildebrand, 1995; Strausfeld *et al.*, 1995).

### (a) Mushroom bodies

The major target of antennal lobe cells in insects are mushroom bodies which lie in the protocerebrum (Figure 25.1). Mushroom bodies have two parts, a calyx which receives inputs, and a pedunculus which is the principal output area (Mobbs, 1982). These two parts are connected by intrinsic neurones called Kenyon cells which are often very numerous; several hundred thousand are present in some insects (Strausfeld, 1976; Strausfeld *et al.*, 1995). Kenyon cells form a dense mass outside the mushroom body calyces, they are monopolar neurones belonging to a general class known as globuli cells characterized by large nuclei and little cytoplasm.

Insect mushroom bodies are olfactory centres but they vary in size – large in *Periplaneta* and social Hymenoptera but smaller in many Diptera, while in *Apis* but not Diptera they receive visual as well as olfactory inputs (Mobbs, 1982; Laurent and Naraghi, 1994; Strausfeld *et al.*, 1995). Genes involved in memory are expressed in *Drosophila* mushroom bodies, and their ablation abolishes olfactory learning (Davis, 1993; deBelle and Heisenberg, 1994). The plasticity of mushroom bodies is also seen in neurogenesis where, unusually in insects, Kenyon cells are not produced by a fixed number of cell divisions, but can be generated by an indeterminate number of divisions of ganglion mother cells under hormonal control (Cayre *et al.*, 1994; cf. Whitington and Bacon, 1997, this volume).

The presence of masses of globuli cells and other neuroanatomical features has led to structures being called mushroom bodies in several taxa. Structures named mushroom bodies (or corporea pedunculata) have been described from polychaetes, onychophorans, myriapods and chelicerates (Bullock and Horridge, 1965; Fahrenbach and Chamberlain, 1987; Schürmann, 1995; Strausfeld *et al.*, 1995, Table 1). Holmgren and Hanström (reviewed by Strausfeld *et al.*, 1995) argued that the mushroom bodies in these taxa are homologous, a view endorsed by Bullock and Horridge (1965, pp. 716 and 817) who, for example, say that the characteristic (insect-like) appearance of the mushroom bodies in some polychaetes 'strongly suggest homology with similar structures in arthropods'. Strausfeld *et al.* (1995) in their comparative study agree that mushroom bodies may be homologous in polychaetes, onychophorans and the mandibulates (perhaps including the crustacean hemiellipsoid body) and construct a cladogram based on various characters. However, as Strausfeld *et al.* point out, functional constraints in 'association centres' (i.e. brain regions whose function is obscure) may well influence their organization, and these are poorly understood (Hopfield, 1991; Osorio *et al.*, 1994; see also Schürmann, 1995). At present, we feel the term 'mushroom body' is probably best understood as applying to a neural architecture rather than homologous structures.

## 25.4 CONCLUSIONS

Where sensory organs or neural circuitry of arthropods are analysed in sufficient detail, they are found to be nearly perfectly optimized and tuned to the specific life style of the species (van Hateren and Nilsson, 1987; van Hateren, 1992; Laughlin and Weckström, 1993; Warrant and McIntyre, 1996). Moreover, it seems there are often only a small number of possible solutions to any given sensory task. Sensory systems in different animals may thus share many similar features because of functional constraints and optimization. Striking examples of this conformism are compound eyes of the refracting superposition type which have evolved in parallel to virtually identical organs in a large number of separate insect and crustacean groups. This is not a particularly desirable situation for analyses of phylogenetic relationships and there are only few cases where sensory systems provide compelling evidence for relationships between various arthropod groups.

The most important such evidence is the nearly identical compound eyes and optic ganglia of insects and malacostracan crustaceans. The two groups share the same cellular composition of the ommatidia of the eye, a unique antigen in the crystalline cone, the same two types of short- and long-axon photoreceptors, and the same general layout of the optic ganglia, including two anteroposterior neural chiasms. The overall consonance of structural and physiological neurone types in the first optic ganglion adds to the evidence. These similarities of insect and crustacean visual systems are reflected by strong similarities also in other parts of the nervous system including the developmental groundplan of the segmental ganglia (Whitington and Bacon, 1997, this volume). The conclusion that insects and malacostracan crustaceans are very closely related seems inevitable, especially since the similarities do not extend to myriapod or chelicerate visual systems.

The visual system of non-malacostracan crustaceans conform to the general insect/malacostracan pattern except in the topology of the optic ganglia. The differences are due to the profoundly different way in which new units are added in the growing eye. It is impossible gradually to derive one topology from the other, pointing to a common crustacean ancestor with only single ocelli connected to ganglia resembling the columnar cartridges beneath their modern compound eyes. Despite the undeniable and close affinities between insects and crustaceans, their common ancestor might thus have been a very primitive Precambrian creature.

## REFERENCES


Alkon, D.L., Anderson, M.J., Kuzirian, A.J., Rogers, D.F., Fass, D.M., Collin, C., Nelson, T.J., Kapetanovic, I.M. and Matzel, L.D. (1993) GABA-mediated synaptic interaction between the visual and vestibular pathways of *Hermissenda*. *Journal of Neurochemistry*, **61**, 556–66.

Arbas, E.A., Meinertzhagen, I.A. and Shaw, S.R. (1991) Evolution in nervous systems. *Annual Review of Neuroscience*, **14**, 9–38.

Battelle, B.A., Calman, B.G., Andrews, A.W., Grieco, F.D., Mleziva, M.B., Callaway, J.C. and Stuart, A.E. (1991) Histamine – a putative afferent neurotransmitter in *Limulus* eyes. *Journal of Comparative Neurology*, **305**, 527–42.

Breidbach, O. and Kutsch, W. (eds) (1995) *The Nervous Systems of Invertebrates: an Evolutionary and Comparative Approach*, Birkhäuser, Basel.

Bullock, T.H. and Horridge, G.A. (1965) *Structure and Function of the Nervous Systems of Invertebrates*, Freeman, San Francisco.

Cajal S.R. and Sanchez y Sanchez, D. (1915) Contribucion al conocimiento de los centros nerviosos de los insectos. Parte I. Retina y los centros opticos. *Trabajos del laboratorio de investigaciones biológicas de la Universidad Madrid*, **13**, 1–168.

Callaway C. and Stuart, A.E. (1989) Biochemical and physiological evidence that histamine is the transmitter of barnacle photoreceptors. *Visual Neuroscience*, **3**, 311–25.

Cayre, M., Strambi, C. and Strambi, A. (1994) Neurogenesis in an adult insect brain and its hormonal-control. *Nature*, **368**, 57–9.

Chase, R. and Tolloczko, B. (1986) Tracing neural pathways in snail olfaction: from the tip of the tentacles to the brain and beyond. *Microscopy Research and Technique*, **24**, 214–30.

Davidson, E.H., Peterson, K.J. and Cameron, R.A. (1995) Origin of bilaterian body plans – evolution of developmental regulatory mechanisms. *Science*, **270**, 1319–25.

Davis, R.L. (1993) Mushroom bodies and *Drosophila* learning. *Neuron*, **11**, 1–14.

deBelle, S. and Heisenberg, M. (1994) Associative odor learning of *Drosophila* abolished by chemical ablation of mushroom bodies. *Science*, **263**, 692–5.

DeSouza, J., Hertel, H., Ventura, D.F. and Menzel, R. (1992) Response properties of stained monopolar cells in the honeybee lamina. *Journal of Comparative Physiology A*, **170**, 267–74.

DeVries, S.H. and Baylor, D.A. (1993) Synaptic circuitry of the retina and olfactory bulb. *Cell*, **72** (suppl.), 139–49.

Dumont, J.P.C. and Robertson, R.M. (1986) Neuronal circuits – an evolutionary perspective. *Science*, **233**, 849–53.

Edwards, J.S. and Meyer, M.R. (1990) Conservation of antigen 3G6: a crystalline cone constituent in the compound eye of arthropods. *Journal of Neurobiology*, 21, 441–52.

Elofsson, R. and Dahl, E. (1970) The optic neuropiles and chiasmata of Crustacea. *Zeitschrift Zellforschung*, **107**, 343–60.

Elofsson, R. and Hagberg, M. (1986) Evolutionary aspects on the construction of the first optic neuropil (lamina) in Crustacea. *Zoomorphology*, **106**, 174–8.

Elofsson, R., Hallberg, E. and Nilsson, H.L. (1980) The juxtaposed compound eye and organ of Bellonci in *Haploops tubicola* (Crustacea: Amphipoda) – The fine structure of the organ of Bellonci. *Zoomorphology*, **96**, 255–62.

Elson, R.C. (1996) Neuroanatomy of a crayfish thoracic ganglion – sensory and motor roots of the walking-leg nerves and possible homologies with insects. *Journal of Comparative Neurology*, **365**, 1–17.

Fahrenbach, W.H. and Chamberlain, S.C. (1987) The brain of the horseshoe crab, *Limulus polyphemus*, in *Arthropod Brain: its Evolution, Development, Structure and Functions* (ed. A.P. Gupta), Wiley, New York, pp. 63–93.

Fischbach, K.-F. and Dittrich, A.P.M. (1989) The optic lobe of *Drosophila melanogaster*. I. A Golgi analysis of wild-type structure. *Cell and Tissue Research*, **258**, 441–75.

Fullard, J.H. and Yack, J.E. (1991) The evolutionary biology of insect hearing. *Trends in Ecology and Evolution*, **8**, 248–52.

Gartner, W. and Towner, P. (1995) Invertebrate visual pigments. *Photochemistry and Photobiology*, **62**, 1–16.

Glantz, R.M. and Bartels, A. (1994) The spatiotemporal transfer-function of crayfish lamina monopolar neurons. *Journal of Neurophysiology*, **71**, 2168–82.

Güldner, F.H. and Wolff, J.R. (1970) Über die Ultrastruktur des Komplexauges von *Daphnia pulex*. *Zeitschrift Zellforschung*, **104**, 259–74.

Gupta, A.P. (ed.) (1987) *Arthropod Brain: its Evolution, Development, Structure and Functions*. Wiley, New York.

Hallberg, E. (1982) The fine structure of the compound eye of *Argulus foliaceus* (Crustacea: Branchiura). *Zoologischer Anzeiger Jena*, **208**, 227–36.

Hallberg, E. and Elofsson, R. (1983) The larval compound eye of barnacles. *Journal of Crustacean Biology*, **3**, 17–24.

Hallberg, E., Nilsson, H.L. and Elofsson, R. (1980) Classification of amphipod compound eyes – the fine structure of the ommatidial units (Crustacea, Amphipoda). *Zoomorphology*, **94**, 279–306.

Hallberg, E., Elofsson, R. and Grygier, M.J. (1985) An ascothoracid compound eye (Crustacea). *Sarsia*, **70**, 167–71.

Hardie, R.C. (1989) A histamine-activated chloride channel involved in neurotransmission at a photoreceptor synapse. *Nature*, **339**, 704–6.

Hanström, B. (1928) *Vergleichende Anatomie des Nervensystems der wirbellosen Tiere*, Springer, Berlin.

Hateren, J.H. van (1992) Real and optimal natural images in early vision. *Nature*, **360**, 68–9.

Hateren, J.H. van and Nilsson D.-E. (1987) Butterfly optics exceed the theoretical limits of conventional apposition eyes. *Biological Cybernetics*, **57**, 159–68.

Hildebrand, J.G. (1995) Analysis of chemical signals by nervous systems. *Proceedings of the National Academy of Sciences, USA*, **92**, 67–74.

Hirth, F., Therianos, S., Loop, T., Gehring, W.J., Reichert, H. and Furukubotokunaga, K. (1995) Developmental defects in brain segmentation caused by mutations of the homeobox genes orthodenticle and empty spiracles in *Drosophila*. *Neuron*, **15**, 769–78.

Hopfield, J.J. (1991) Olfactory computation and object perception. *Proceedings of the National Academy of Sciences, USA*, **88**, 6462–6.

Horridge, G.A., Marcelja, L. and Jahnke, R. (1984) Color-vision in butterflies. 1. single color experiments. *Journal of Comparative Physiology*, **155**, 529–42.

James, A.C. and Osorio, D. (1996) Characterisation of columnar neurons and visual signal processing in the medulla of the locust optic lobe by system identification techniques. *Journal of Comparative Physiology A*, **178**, 183–99.

Jansonius, N.M. and van Hateren, J.H. (1993a) On-off units in the first optic chiasm of the blowfly 2. Spatial properties. *Journal of Comparative Physiology A*, **172**, 467–71.

Jansonius, N.M. and van Hateren, J.H. (1993b) On spiking units in the first optic chiasm of the blowfly. 3. The sustaining unit. *Journal of Comparative Physiology A*, **173**, 187–92.

Karhunen, T., Airaksinen, M.S., Tuomisto, L and Panula, P. (1993) Neurotransmitters in the nervous-system of *Macoma balthica* (Bivalvia). *Journal of Comparative Neurology*, **334**, 477–88.

Katz, P.S. (1991) Neuromodulation and the evolution of a simple motor system. *Seminars in the Neurosciences*, **3**, 379–89.

Kral, K. (1987) Organization of the first optic neuropil (or lamina) in different insect species, in *Arthropod Brain: its Evolution, Development, Structure and Functions* (ed. A.P. Gupta), Wiley, New York, pp. 181–201.

Land, M.F. and Fernald, R.D. (1992) The evolution of eyes. *Annual Review of Neuroscience*, **15**, 1–29.

Land, M. F. and Nilsson, D.-E. (1990) Observations on the compound eyes of the deep-sea ostracod *Macrocypridina castanea*. *Journal of Experimental Biology*, **148**, 221–33.

Laughlin, S.B. (1981) Neural principles in the peripheral visual systems of invertebrates, in *Handbook of Sensory Physiology* (ed. H.-J. Autrum), Vol. VII/6B, Springer, Berlin, pp. 133–280.

Laughlin, S.B. (1994) Matching coding, circuits, cells, and molecules to signals – general principles of retinal design in the fly's eye. *Progress in Retinal and Eye Research*,.**13**, 165–96.

Laughlin, S.B. and Weckström, M. (1993) Fast and slow photoreceptors – a comparative study of the functional diversity of coding and conductances in the Diptera. *Journal of Comparative Physiology A.*, **172**, 593–609.

Laurent, G. and Naraghi, M. (1994) Odorant-induced oscillations in the mushroom bodies of the locust. *Journal of Neuroscience*, **14**, 2992–3004.

Meinertzhagen, I.A. (1991) Evolution of the cellular organization of the arthropod compound eye and optic lobe. in *Vision and Visual Dysfunction, vol. II: Evolution of the Eye and Visual System* (eds J.R. Cronly-Dillon and R. Gregory), Macmillan, London, pp 341–63.

Mobbs, P.G. (1982) The brain of the honeybee *Apis mellifera* 1. The connections and spatial organisation of the mushroom bodies. *Philosophical Transactions of the Royal Society of London Series B Biological Sciences*, **298**, 309–54.

Nilsson, D.-E. (1983) Evolutionary links between apposition and superposition optics in crustacean eyes. *Nature*, **302**, 818–21.

Nilsson, D.-E. (1988) A new type of imaging optics in compound eyes. *Nature*, **332**, 76–8.

Nilsson, D.-E. (1989) Optics and evolution of the compound eye, in *Facets of Vision* (eds D.G. Stavenga and R.C. Hardie), Springer, Berlin Heidelberg, pp. 30–73.

Nilsson, D.-E. (1990) Three unexpected cases of refracting superposition eyes in crustaceans. *Journal of Comparative Physiology A*, **167**, 71–8.

Nilsson, D.-E. (1996) Eye ancestry – old genes for new eyes. *Current Biology*, **6**, 39–42.

Nilsson, D.-E. and Odselius, R. (1981) A new mechanism for light–dark adaptation in the *Artemia* compound eye (Anostraca, Crustacea). *Journal of Comparative Physiology*, **143**, 389–99.

Nilsson, D.-E., Odselius, R. and Elofsson, R. (1983) The compound eye of *Leptodora kindtii* (Cladocera): an adaptation to planktonic life. *Cell and Tissue Research*, **230**, 401–10.

Nilsson, D.-E., Land, M.F. and Howard, J. (1984) Afocal apposition optics in butterfly eyes. *Nature*, **312**, 561–3.

Osorio, D. (1991) Patterns of function and evolution in the arthropod optic lobe, in *Vision and Visual Dysfunction, vol. II: Evolution of the Eye and Visual System* (eds J.R. Cronly-Dillon and R. Gregory), Macmillan, London, pp. 203–29.

Osorio, D. and Bacon, J.P. (1994) A good eye for arthropod evolution. *BioEssays*, **16**, 419–26.

Osorio, D., Getz, W.M. and Rybak, J. (1994) Insect Vision and olfaction: different neural architectures for different kinds of sensory signal? in *Proceedings of the third international meeting on the Simulation of Adaptive Behaviour* (eds D. Cliff, P.H. Husbands, J.-A. Meyer and C. Wilson.), MIT, Boston, pp. 73–81.

Osorio, D., Averof, M. and Bacon, J.P. (1995) Arthropod evolution: great brains, beautiful bodies. *Trends in Ecology and Evolution*, **10**, 449–54.

Panula, P., Eriksson, K., Gustafsson, M. and Reuter, M. (1995) An immunocytochemical method for histamine – application to the planarians. *Hydrobiologica*, **305**, 291–5.

Paulus, H.F. (1979) Eye structure and monophyly of the Arthropoda, in *Arthropod Phylogeny* (ed. A.P. Gupta), van Nostrand Reinhold, New York, pp. 299–383.

Quiring, R., Walldorf, U., Kloter, U. and Gehring, W.J. (1994) Homology of the eyeless gene of *Drosophila* to the *Small eye* gene in mice and *aniridia* in humans. *Science*, **265**, 785–9.

Robertson, R.M. and Olberg, R.M. (1988) A comparison of flight interneurones in locusts, crickets, dragonflies and mayflies. *Experientia*, **44**, 735–8.

Salvini-Plawen, L. von and Mayr, E. (1977) On the evolution of photoreceptors and eyes. *Evolutionary Biology*, **10**, 207–63.

Sandeman, D. and Scholtz, G. (1995) Groundplans, evolutionary changes and homologies in decapod crustacean brains, in *The Nervous Systems of Invertebrates: an Evolutionary and Comparative Approach* (eds O. Breidbach and W. Kutsch), Birkhäuser, Basel, pp. 329–47.

Sandeman, D.C, Scholtz, G. and Sandeman, R.E (1993) Brain evolution in decapod Crustacea. *Journal of Experimental Zoology*, **265**, 112–33.

Schmid, A. and Duncker, M. (1993) Histamine immunoreactivity in the central-nervous-system of the spider *Cupiennius salei*. *Cell and Tissue Research*, **273**, 533–54.

Schmidt-Ott, U., González Gaitán, M. and Technau, G.M. (1995) Analysis of neural elements in head-mutant *Drosophila* embryos suggests segmental origin of the optic lobes. *Roux's Archive for Developmental Biology*, **205**, 31–44.

Schürmann, F.-W. (1995) Common and special features of the nervous system of Onychophora: a comparison with Arthropoda, Annelida and some other invertebrates, in *The Nervous Systems of Invertebrates: an Evolutionary and Comparative Approach* (eds O. Breidbach and W. Kutsch), Birkhäuser, Basel, pp. 139–58.

Soinila, S., Mpitsos, G.J. and Panula, P. (1990) Comparative-study of histamine immunoreactivity in nervous systems of *Aplysia* and *Pleurobranchaea*. *Journal of Comparative Neurology*, **298**, 83–96.

Strausfeld, N.J. (1976) *Atlas of an Insect Brain*, Springer, Berlin.

Strausfeld, N.J. and Barth, F.G. (1993) Two visual systems in one brain: neuropils serving the secondary eyes of the spider *Cupiennius salei*. *Journal of Comparative Neurology*, **328**, 43–62.

Strausfeld, N.J. and Nässel, D.R. (1981) Neuroarchitecture of brain regions that subserve the compound eyes of Crustacea and insects, in *Handbook of Sensory Physiology*, vol. VII/6B (ed. H.-J. Autrum), Springer, Berlin, pp. 1–132.

Strausfeld, N.J., Buschbeck, E.K. and Gomez, R.S. (1995) The arthropod mushroom body: its functional roles, evolutionary enigmas and mistaken identities, in *The Nervous Systems of Invertebrates: an Evolutionary and Comparative Approach* (eds O. Breidbach and W. Kutsch), Birkhäuser, Basel, pp. 349–81.

Thomas, J.B., Bastiani, M.J., Bate, M. and Goodman, C.S. (1984) From grasshopper to *Drosophila* – a common plan for neuronal development. *Nature*, **310**, 203–7.

Tomarev, S.I. and Piatigorsky, J. (1996) Lens crystallins of invertebrates – diversity and recruitment from detoxification enzymes and novel proteins. *European Journal of Biochemistry*, **235**, 449–65.

Wang-Bennett, L.T. and Glantz, R.M. (1987a) The functional organization of the crayfish lamina ganglionaris. I. Nonspiking monopolar cells. *Journal of Comparative Physiology A*, **161**, 131–45.

Wang-Bennett, L.T. and Glantz, R.M. (1987b) The functional organization of the crayfish lamina ganglionaris. II. Large field spiking and nonspiking cells. *Journal of Comparative Physiology A*, **161**, 147–60.

Warrant, E.J. and McIntyre, P.D. (1996) The visual ecology of pupillary action in superposition eyes. *Journal of Comparative Physiology A*, **178**, 75–90.

Yang, E.C. and Osorio, D. (1996) Spectral responses and chromatic processing in the dragonfly lamina. *Journal of Comparative Physiology A*, **178**, 543–50.

Zacharias, D., Williams, J.L.D., Meier, T. and Reichert, H. (1993) Neurogenesis in the insect brain – cellular identification and molecular characterization of brain neuroblasts in the grasshopper embryo. *Development*, **118**, 941–55.

# 26 The organization and development of the arthropod ventral nerve cord: insights into arthropod relationships

*P.M. Whitington and J.P. Bacon*

*Department of Zoology, University of New England, Armidale, NSW, 2351 Australia*
*email: pwhiting@metz.une.edu.au*
*Sussex Centre for Neuroscience, School of Biological Sciences, University of Sussex, Brighton, BN1 9QG, UK*
*email: J.P.Bacon@sussex.ac.uk*

## 26.1 INTRODUCTION

The central nervous system (CNS) of arthropods is a particularly suitable organ in which to search for characters to reconstruct evolutionary relationships between the major arthropod groups. Evolution has clearly wrought changes in nervous systems – indeed behavioural traits which are a prime target for natural selection are determined by the structure and function of this organ system. Despite this, the mature structure of the CNS and the developmental processes that generate it show a high degree of conservation compared with other, more malleable, features such as external body parts. Secondly, the CNS of arthropods is populated by an array of cells which can be recognized as individuals at a cellular level. Each neuron has a unique set of characters differentiating it from other neurons which can greatly assist the recognition of homologues between species, an essential first step when drawing inferences about evolutionary change.

This chapter reviews recent comparative studies of the development and mature structure of the ventral nerve cord (VNC), which have shed new light on the evolutionary relationships between three of the major arthropod groups: the insects, the crustaceans and the myriapods. Nilsson and Osorio (1997, this volume) consider arthropod phylogeny by examination of the brain. Both chapters focus at the level of identified neurons, since these allow the most definitive conclusions to be drawn. We begin with insects since our understanding of development and neural structure is more extensive for this group than for any other arthropod (and arguably for any other animal) group.

## 26.2 INSECTS

### 26.2.1 GROSS MORPHOLOGY OF THE VNC

The ladder-like organization of the insect VNC is familiar to even the most casual biology student. A chain of ganglia runs along the ventral side of the body. Each ganglion is joined to its neighbours by a pair of longitudinal connectives, consisting of the axonal processes of intersegmental neurons (Figure 26.1). Some ganglia are formed by the fusion of two or more segmental neural units, called neuromeres, while others consist of a single neuromere. There has been a trend in insect evolution for progressive neuromere fusion, reaching its climax in some Diptera where the VNC consists of a single compound ganglion (Bullock and Horridge, 1965). Body segments are supplied by nerves which run laterally from the neuromeres; these contain the axons of centrally located motorneurons and peripherally located sensory neurons. Typically, each body segment is innervated by one or more segmental nerves (SN) from one neuromere, and a single intersegmental nerve (ISN), with roots in the same and the adjacent anterior neuromere.

### 26.2.2 ORGANIZATION OF ADULT GANGLIA AT VARIOUS LEVELS: FIBRE TRACTS, GROUPS OF NEURONS AND IDENTIFIED NEURONS

The basic architecture of ganglia in the insect VNC is remarkably constant across orders as diverse as the hemimetabolous Orthoptera and the holometabolous Diptera (Boyan and Ball, 1993) (Figure 26.2). Each ganglion contains a cortex of neuron and glial cell bodies and an inner neuropile region, which is divided into various morphologically distinct areas. These regions include: nine longitudinal fibre tracts, consisting of anteroposteriorly running axons; six dorsal commissures and

*Arthropod Relationships*, Systematics Association Special Volume Series 55, Edited by R.A. Fortey and R.H. Thomas. Published in 1997 by Chapman & Hall, London. ISBN 0 412 75420 7

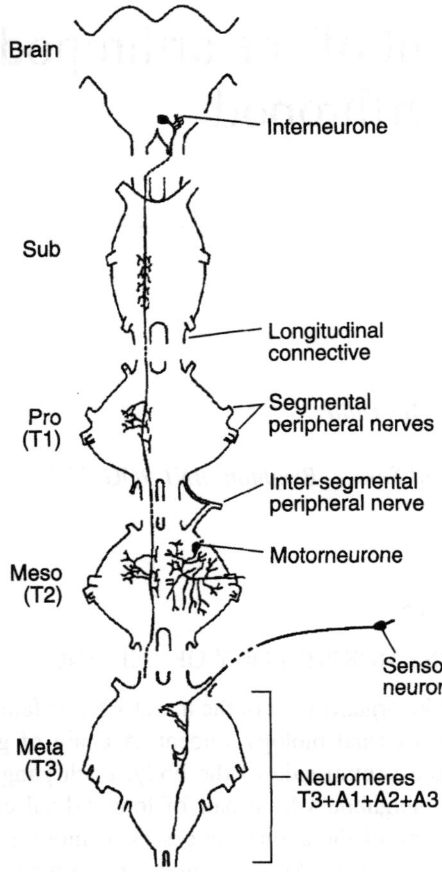


**Figure 26.1** Diagram of the adult grasshopper CNS in the head and thorax; the abdominal ganglia are not shown. The third thoracic ganglion consists of the metathoracic neuromere fused to the first three abdominal neuromeres. Typical morphology of the three major types of neuron in the nervous system – interneuron, motorneuron and sensory neuron – are drawn. Unlike the other two cell types, sensory neurons are born in the periphery and project into the CNS. (Modified from Tyrer and Gregory, 1982.)

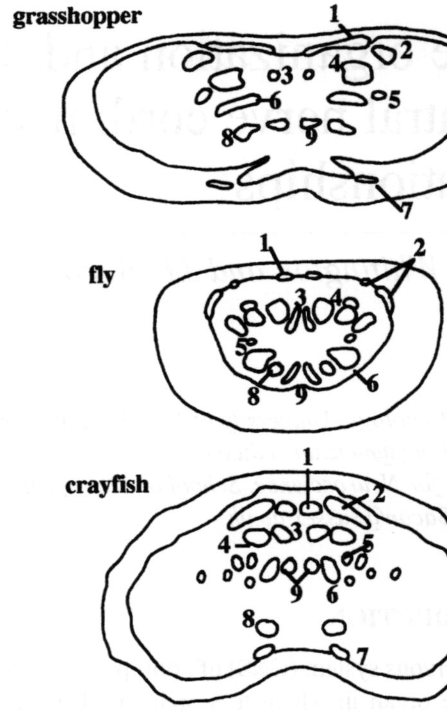


**Figure 26.2** Transverse sections through a thoracic ganglion of an orthopterous insect, a dipterous insect, and a crayfish show them to be remarkably similar in general morphology. Tracts of axon bundles: 1 = MDT, 2 = LDT, 3 = DMT, 4 = DIT, 5 = VLT, 6 = VIT, 7 = LVT, 8 = MVT, 9 = VMT. [Data modified from Les Williams (unpublished), Boyan and Ball (1993) and Elson (1996).]

four ventral commissures, containing neuron processes that cross the ganglion; some oblique tracts linking areas of neuropile; and a series of association areas, containing the arborizations of sensory axons and the dendrites of motorneurons and interneurons. This basic structure is conserved both between segments and between species. Homology of some of these tracts and neuropile areas between species is supported by the fact that they contain, respectively, the axons and arborizations of identified neurons that have been homologized using other, independent criteria (Figure 26.3(A,B)).

A particularly striking example of evolutionary conservation concerns the set of motorneurons (DLMs) that innervate the dorsal longitudinal muscles, which is very similar in insects as diverse as Diptera, Orthoptera and the apterygote Thysanura (Heckmann and Kutsch, 1995; reviewed in Kutsch and Heckmann, 1995). These motorneurons are situated in the ganglion ipsisegmental to the dorsal longitudinal muscle and the adjacent anterior ganglion (Figure 26.4). The posterior group consists of one or two contralateral motorneurons, one or two dorsal unpaired median (DUM) cells and a midline neuron with an ipsilaterally projecting axon. The anterior group consists of between four and eight ipsilateral, posterior neurons, one contralateral neuron, and one midline neuron. Several of these evolutionary conserved neurons, such as the large DLM–MN1, can be identified as individuals.

The soma location and axon morphology of neurons in the VNC which express particular classical neurotransmitters or neuropeptides show a high degree of consistency across different insect orders. Serotonin-like immunoreactivity is a useful neuronal marker for homology hunting because it is a relatively rare phenotype. Serotonin-like immunoreactive neurons in different insect orders show considerable interspecies homology in cockroach (Bishop and O'Shea, 1983), locust (Tyrer *et al.*, 1984), *Drosophila* (Vallés and White, 1988) and beetle (van Haeften and Schooneveld, 1992), particularly among neurons of the ventral cord, but also in the brain and optic lobes.

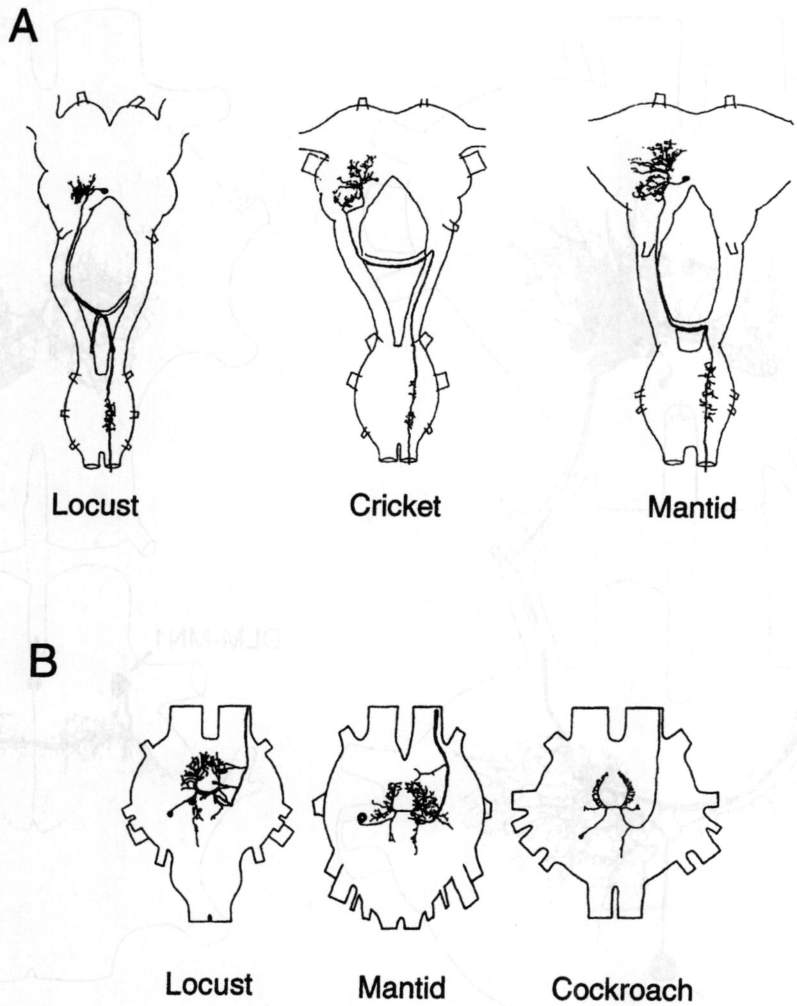


**Figure 26.3** Putative homologous neurons in different insect species arborizing in homologous neuropile area. (A) The tritocerebral giant neuron (TCG) in grasshopper (Bacon and Tyrer, 1978) and its putative homologues in cricket and mantid. TCG has its cell body and dendrites in the deuterocerebrum of the brain, its axon is the largest in the tritocerebral commissure through which it crosses the midline, and it grows posteriorly to the thoracic ganglia. One difference is that the axon of the cricket TCG runs in VIT in thoracic ganglia but in DIT in locust and mantid. Cricket and mantid data from J. Bacon (unpublished). (B) Putative neuronal homologues in the metathoracic ganglion of a locust, mantid and cockroach. In each insect, this neuron has a cell body in dorsolateral cortex, it crosses the midline in the SMC, has dendrites in each VIT and its axon in the LDT. (Figure modified from Boyan, 1995.)

A rare neuropeptide phenotype is vasopressin-like peptide in insects. Using antibodies to vertebrate vasopressin, a prominent pair of neuron cell bodies is stained on the ventral midline of the suboesophageal ganglion of locust (Thompson *et al.*, 1991), grasshopper (Tyrer *et al.*, 1993), cricket (Musiol *et al.*, 1990), beetle (Veenstra, 1984), mantid (Davis and Hildebrand, 1992) and cockroach (Davis and Hildebrand, 1992). The similarity of cell body position in this wide range of species, and their unusual immunoreactivity, suggest that these neurons are all homologues.

More detailed examination of these vasopressin-like immunoreactive (VPLI) neurons in Acrididae and Romaleidae provides additional characters to secure the homology; however, some aspects of their morphologies in individual species are strikingly different. The major difference is the presence of VPLI fibre bundles in most peripheral nerves and an array of fibres in the optic lobes of Oedipodine species, exemplified by *Locusta migratoria*, and the absence of any extensive peripheral branching in non-Oedipodine species, as exemplified by *Schistocerca gregaria* (Tyrer *et al.*, 1993) (Figure 26.5).

We infer the direction of evolutionary change in VPLI morphology by identifying character states shared by distantly related taxa (the 'outgroup') as primitive, while character states shared by only a group of closely related taxa (the 'ingroup') are derived (Maslin, 1952; Weston, 1988). The

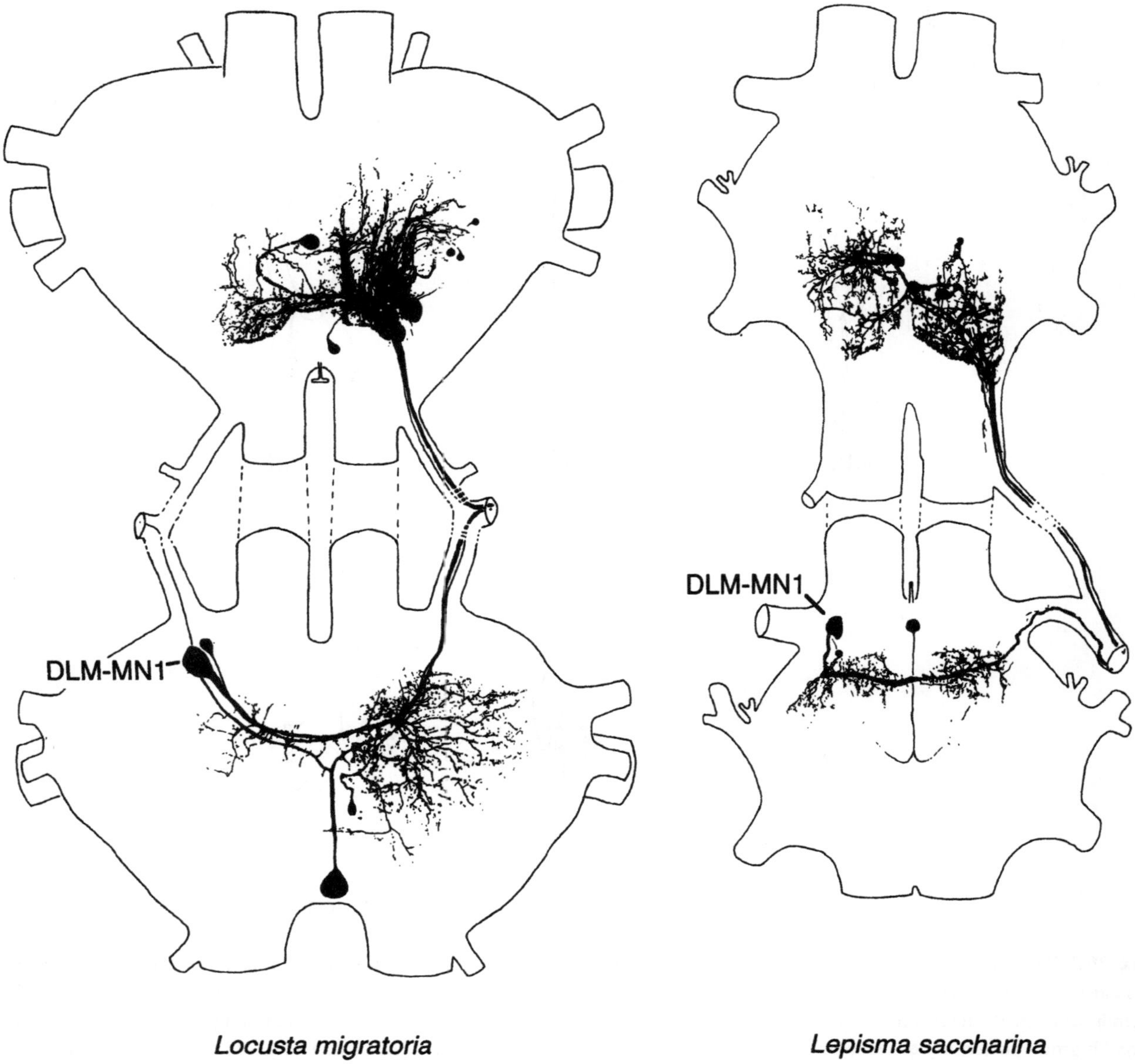


**Figure 26.4** Camera lucida drawings showing the strikingly similar arrangement and anatomy of motorneurons innervating the dorsal longitudinal muscles in the locust, *Locusta migratoria* and the silverfish, *Lepisma saccharina*. In both species, these motorneurons are situated in the ganglion ipsisegmental to the dorsal longitudinal muscle and the adjacent anterior ganglion. The posterior group consists of one or two contralateral motorneurons and a single DUM cell. The grasshopper has an additional midline neuron with an ipsilaterally projecting axon. The anterior group consists of between four and eight ipsilateral, posterior neurons, one contralateral neuron, and one midline neuron. (Modified from Kutsch and Heckmann, 1995.)

finding of no significant VPLI branching in peripheral nerves or optic lobes in any genus of the three acridid subfamilies Cyrtacanthacridinae, Gomphocerinae and Melanoplinae, or within the more distantly related Romaleinae (Tyrer *et al.*, 1993), or the even more distantly related Blattidae (order: Dictyoptera; Davis and Hildebrand, 1992) suggests that their '*Schistocerca*-type' VPLI morphology represents the ancestral form. In contrast, extensive branching by the VPLI neurons in the peripheral nerves and optic lobes of grasshoppers from only the six genera of the subfamily Oedipodinae suggests that the presence of peripheral VPLI fibres, the '*Locusta*-type' morphology, is a derived trait evolved early in the radiation of only the Oedipodinae. Nevertheless, this group has a very ancient history: the presence of representatives of this subfamily in North America, Africa and Eurasia suggests that radiation of Oedipodines began more than 100 million years ago, before the continents began to separate (Vickery, 1987). The presence of VPLI fibres in peripheral

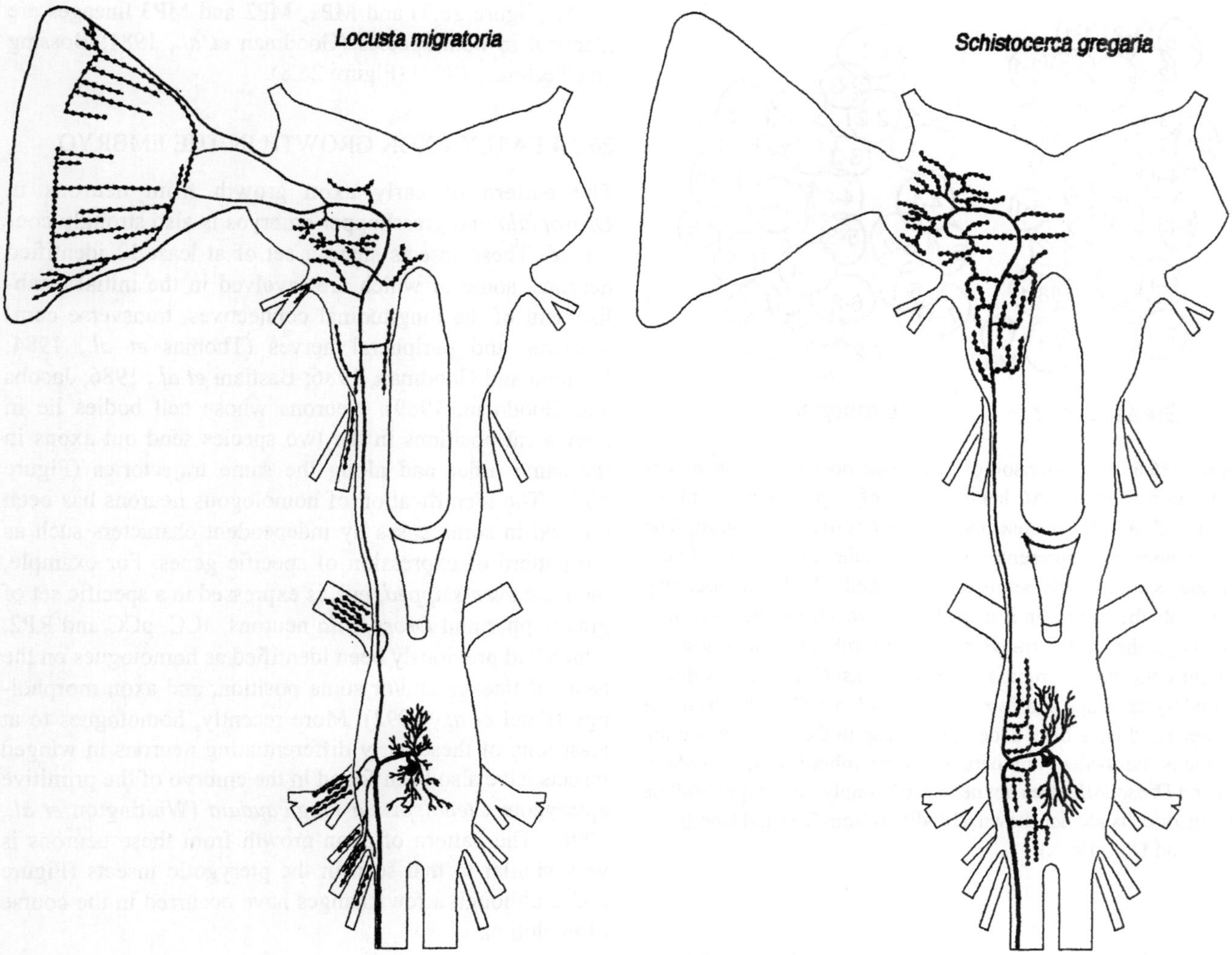


**Figure 26.5** Schematic morphology of the VPLI neurons in the brain and suboesophageal ganglion of two species of locust. The neuron has a cell body on the ventral midline of the suboesophageal ganglion and extensive arborization in the brain and throughout the VNC. VPLI anatomy is determined by phylogeny; in the subfamily Oedipodinae, to which *Locusta migratoria* belongs, VPLI neurons have bundles of fibres in the peripheral nerve roots and elaborate arborization in the optic lobes. In the other subfamilies examined, VPLI cells have no fibres in the peripheral nerve roots and little, or no arborization in optic lobes; this is typified by *Schistocerca gregaria.*

nerves and optic lobes is therefore congruent with other characters such as 'brightly-coloured hindwings' (Uvarov, 1966) used for the classification of the Oedipodines. It is reassuring that taxonomic groupings based almost entirely on cuticular structure can be confirmed in this way by using even single identified neurons of the nervous system. The functional consequence of this evolutionary modification of VPLI morphology remains a mystery (Tyrer *et al.*, 1993).

### 26.2.3 NEUROGENESIS

Conservation of the final structure of the nervous system within the insects is mirrored by a striking evolutionary conservation of the developmental processes by which it is assembled. A common mode of neurogenesis is found throughout the insects; in all species examined, neurons are produced by repeated, asymmetrical divisions of large neuron precursor cells, neuroblasts, which delaminate from the neuroectoderm early in embryogenesis. Each of the ganglion mother cells produced by the neuroblasts divides once symmetrically to generate neurons and/or glial cells.

The final number of neuroblasts per thoracic hemisegment is very similar in the embryos of insects as disparate as the grasshopper (Bate, 1976; Doe and Goodman, 1985), *Drosophila* (Hartenstein and Campos-Ortega, 1984; Doe, 1992), the phasmid *Carausius morosus* (Tamarelle *et al.*, 1985) and the silverfish *Ctenolepisma longicaudata* (J. Truman and E. Ball, personal communication). The 29–31 neuroblasts are arranged roughly in 6–7 rows and four columns, together with a single posterior unpaired median

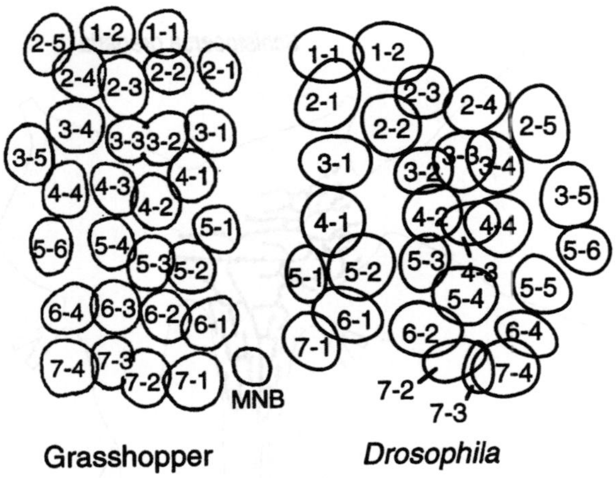


**Figure 26.6** A comparison of neuroblast positions, viewed ventrally, in a right A1–A6 hemisegment of a grasshopper embryo and a left A1–A6 hemisegment of a *Drosophila* embryo. The grasshopper map represents a stage when the final pattern of neuroblasts is evident and before any have died. The *Drosophila* map represents the pattern in a stage 11 embryo, chosen because after this stage the clear arrangement of neuroblasts into rows and columns becomes disrupted. One neuroblast (2-2) that has disappeared by this stage and five (1-2, 3-3, 4-3, 5-1, 7-3) that have not yet appeared have been added to the map in their previous/future positions. Equivalent numbering of a neuroblast in the grasshopper and *Drosophila* does not necessarily imply homology. Midline precursor cells are not shown. (Modified from Doe and Goodman, 1985, and Doe, 1992.)

neuroblast in each insect species examined (Figure 26.6). In addition, a similar set of midline precursor cells (MPs) exists in the grasshopper (Bate and Grunewald, 1981) and *Drosophila* (Doe, 1992; Bossing and Technau, 1994); each of these divides symmetrically to produce two neurons.

These similarities suggest that there is a conserved, ancestral programme for neuroblast production within the insects. Other similarities support the same conclusion. For example, the *Drosophila* genes *engrailed*, *fushi-tarazu*, *prospero* and *seven-up* and their grasshopper orthologues are expressed in particular subsets of neuroblasts and MPs found in similar positions in the two species (Doe *et al.*, 1988; Doe and Technau, 1993; Gutjahr *et al.*, 1993; Condron *et al.*, 1994; Dawes *et al.*, 1994; Broadus and Doe, 1995).

There are strong indications that the pattern of neuron production from individual neuroblasts has also been conserved during insect evolution. In a very few cases, it has been shown that some neurons meet the most rigorous definition of homology; that is, they are generated by identical lineages in different insect species. Thus, in both grasshopper and *Drosophila*, neurons aCC and pCC are products of the division of the first ganglion mother cell of neuroblast 1-1 (Goodman *et al.*, 1984; Udolph *et al.*, 1993; Broadus *et al.*, 1995) (Figure 26.7) and MP1, MP2 and MP3 lineages are identical in both species (Goodman *et al.*, 1981; Bossing and Technau, 1994) (Figure 26.8).

### 26.2.4 EARLY AXON GROWTH IN THE EMBRYO

The pattern of early axon growth from neurons in *Drosophila* and grasshopper embryos is also strongly conserved. These insects share a set of at least 12 identified neurons, some of which are involved in the initial establishment of the longitudinal connectives, transverse commissures and peripheral nerves (Thomas *et al.*, 1984; Bastiani and Goodman, 1986; Bastiani *et al.*, 1986; Jacobs and Goodman, 1989). Neurons whose cell bodies lie in equivalent positions in the two species send out axons in the same order and along the same trajectories (Figure 26.9). The identification of homologous neurons has been secured in some cases by independent characters such as the pattern of expression of specific genes. For example, the gene *even-skipped*, *eve*, is expressed in a specific set of grasshopper and *Drosophila* neurons, aCC, pCC and RP2, which had previously been identified as homologues on the basis of lineage and/or soma position, and axon morphology (Patel *et al.*, 1992). More recently, homologues to at least nine of these early differentiating neurons in winged insects have also been found in the embryo of the primitive apterygote *Ctenolepisma longicaudata* (Whitington *et al.*, 1996). The pattern of axon growth from these neurons is very similar to that seen in the pterygotic insects (Figure 26.9), although a few changes have occurred in the course of evolution.

### 26.2.5 CHANGE VERSUS CONSTANCY IN THE INSECT VNC AND PHYLOGENETIC IMPLICATIONS

Much of the final structure and development of the VNC is conserved across different insect orders, which suggests that a Bauplan for the construction of the central nervous system arose even before the divergence of the winged and apterygotic insects. Nonetheless, some evolutionary change has occurred. There are clear differences in the number of cells produced by homologous neuroblasts in different insects: the grasshopper MNB gives rise to glial cells, in addition to neurons, which are not present in the lineage of its *Drosophila* homologue. It also generates many more neurons than its *Drosophila* homologue (Goodman *et al.*, 1980; Thompson and Siegler, 1993; Bossing and Technau, 1994).

The relative timing of axon growth from certain neurons and patterns of fasciculation between axons in the connectives differ between *Drosophila*, grasshopper and silverfish (Figure 26.10) (Thomas *et al.*, 1984; Bastiani and Goodman, 1986; Bastiani *et al.*, 1986; Jacobs and Goodman, 1989; Whitington *et al.*, 1996). Differences are also seen in the final axon morphology of putative homologous neurons

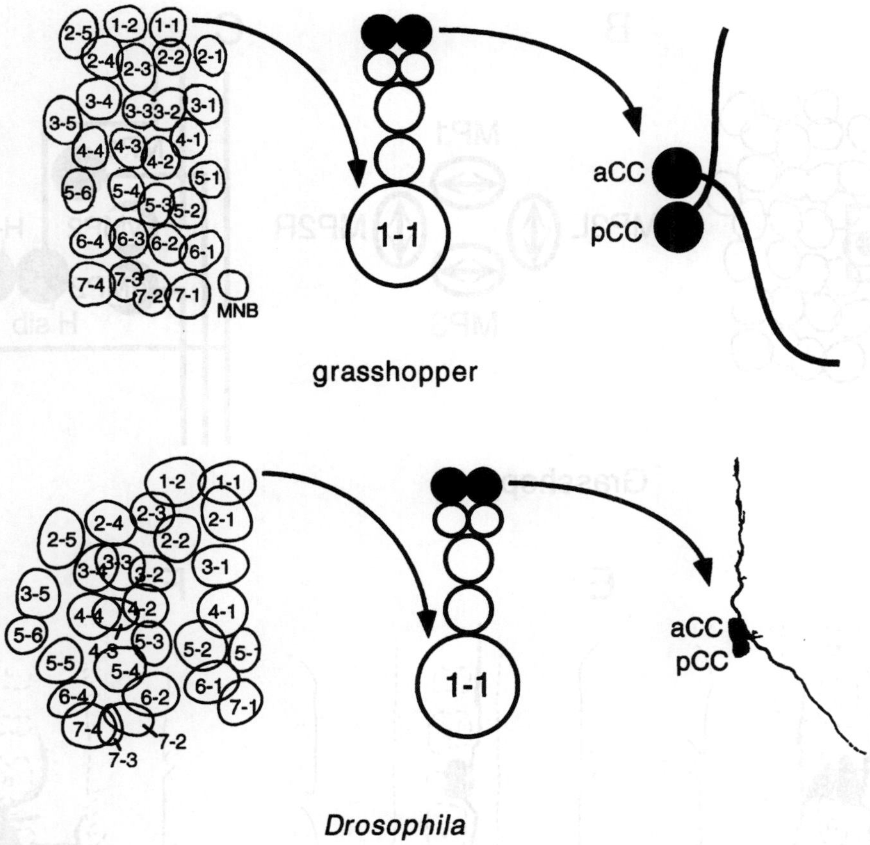


**Figure 26.7** aCC and pCC in the grasshopper and *Drosophila* fit the most stringent criterion of homology; they have identical lineages in both insects, arising from the symmetrical division of ganglion mother cell 1 of neuroblast 1.1. (Modified from Goodman *et al.*, 1984, and Udolph *et al.*, 1993.)

(Figure 26.9). For example, the axon of the silverfish dMP2 neuron projects anteriorly along the longitudinal connectives, whereas its putative grasshopper and *Drosophila* homologues run posteriorly (Thomas *et al.*, 1984; Whitington *et al.*, 1996). A null mutation in the *numb* gene in *Drosophila* generates a similar transformation of dMP2 axon morphology (Spana *et al.*, 1995), raising the intriguing possibility that this evolutionary change in axon morphology may be due to a change in the expression of this gene. Finally, there is a class of neurons which appears to be specific to one or the other insect species. For example, the silverfish neuron I, which pioneers a fascicle in the posterior commissure, does not appear to be present in the winged insects (Figure 26.9).

These differences between the insects provide an indication of the plasticity of developmental processes which will be useful when drawing comparisons between insects and non-insect arthropods. The relative timing of axon outgrowth is a feature which appears to be quite changeable, whereas axon morphology is more highly conserved. Our ability to analyse the system at the level of identified neurons has allowed us to identify homologous neurons with a high degree of confidence and thereby to recognize these conserved and changed features. When a wider spectrum of insect species has been examined, such characters should prove useful in cladistic analyses of the various insect orders.

## 26.3 CRUSTACEA

### 26.3.1 ORGANIZATION OF ADULT GANGLIA

The VNC of crustaceans closely resembles that of insects at a gross morphological level and it also displays the same evolutionary trend for ganglionic fusion (Bullock and Horridge, 1965; Sandeman, 1982). Furthermore, the arrangement of fibre tracts in ventral ganglia of decapod crustacea shows a strong similarity to the insect plan (Elson, 1996) (Figure 26.2). These neuroanatomical similarities may reflect convergent evolution, although it should be noted that tracts that have been homologized in the crayfish and insects do receive input from equivalent classes of neurons. In addition, a number of comparisons of the pattern

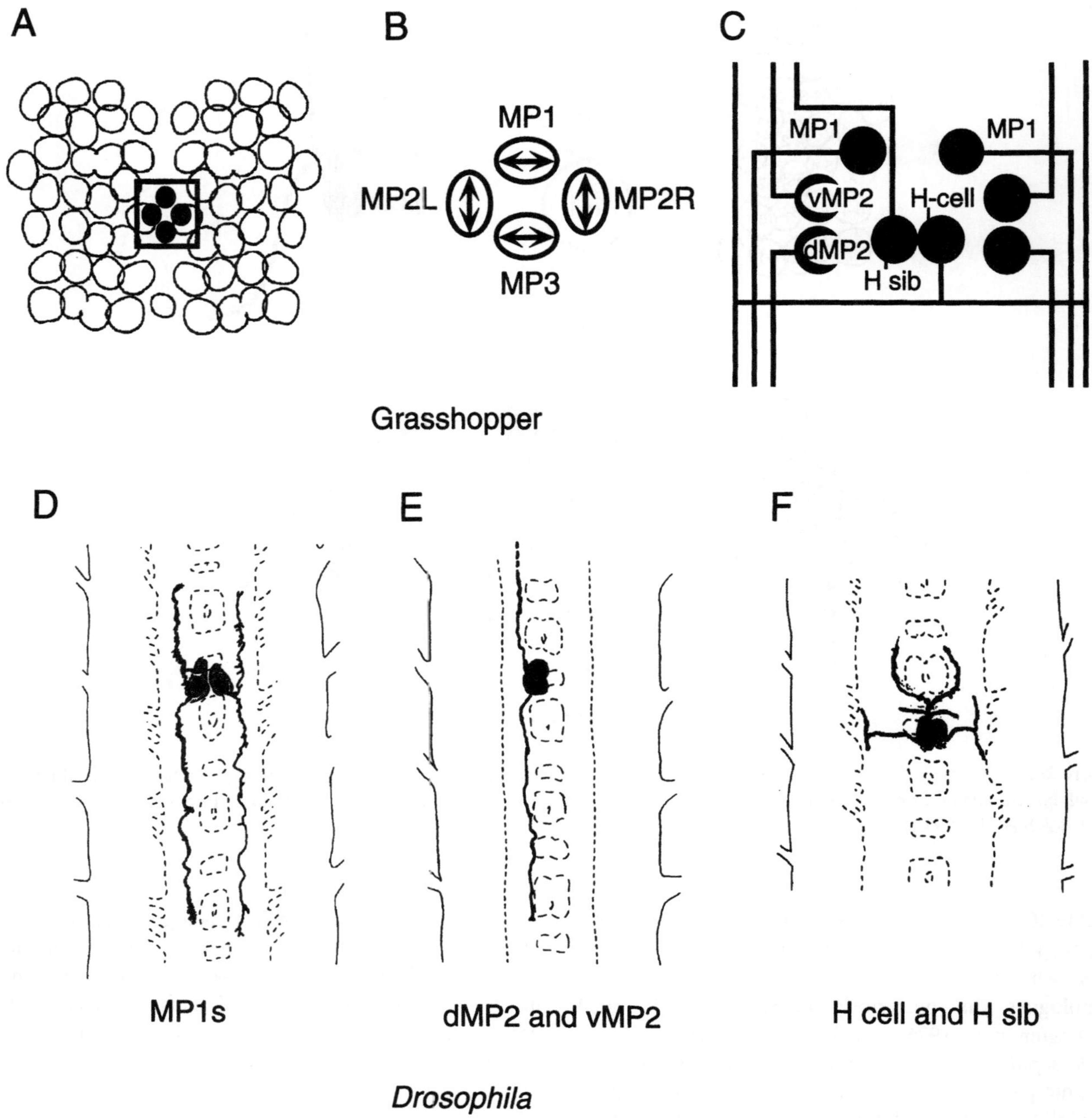


**Figure 26.8** MP1/2/3 lineages are identical in the grasshopper and *Drosophila*. (A) Position of the MP cells in the grasshopper neuroblast map. (B) Plane of mitotic division of the grasshopper MP cells. (C) Diagrammatic representation of the axon trajectories of MP progeny in the grasshopper embryo. (D–F) Axon morphologies of clones of neurons derived from single midline precursor cells in *Drosophila*. These neurons have similar axon morphologies to the grasshopper MP1s (D), dMP2 and vMP2 (E) and the H and H sibling (F) neurons. [Data in (D) and (F) are taken from Bossing and Technau (1994); (E) is from Bossing and Technau (personal communication).]

and axon morphology of individual identified neurons or groups of neurons in the adult crustacean VNC with similar neurons in the insect provide substantial support for the idea that there is a common Bauplan for the construction of the VNC in these two groups.

The pattern of motorneurons innervating the fast flexor muscles in the abdomen of *Procambarus clarkii* and *Homarus americanus* (Otsuka *et al.*, 1967; Mittenhall and Wine, 1978) and the dorsal extensor muscles in the thorax of the isopod *Idotea balthica* (Kutsch and Heckmann, 1995) shows several features in common with the set of dorsal longitudinal motorneurons in insects (Figure 26.11;

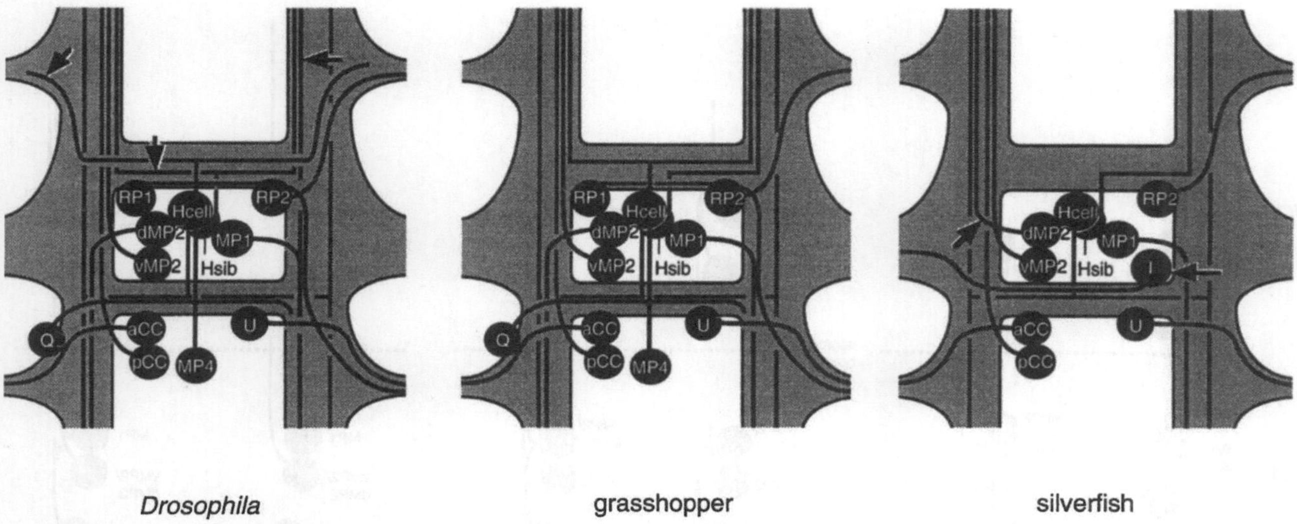


**Figure 26.9** Schematic diagrams of axon morphologies of early differentiating neurons in embryonic *Drosophila*, grasshopper and silverfish CNS. Nine putatively homologous neurons – aCC, pCC, dMP2, vMP2, MP1, H cell, H sib, RP2, U – are found in all three species. Some neurons – RP1, Q, MP4 – are shared by two species, while others are unique to a particular species [e.g. I in silverfish (arrow)]. Some putative homologues display minor differences in axon morphology: the H sib in *Drosophila* has a bifurcating axon; MP1 in *Drosophila* projects both anteriorly and posteriorly; MP4 in *Drosophila* projects into the periphery; the silverfish dMP2 neuron projects anteriorly. These differences are indicated by arrows.

cf. Figure 26.4). The motorneuron somata are distributed throughout two ganglia, one ipsisegmental to the muscles and the other anterior to the muscles' segment. The posterior ganglion contains a group of anterior, contralateral neurons and the anterior ganglion contains between 6 and 8 neurons which are mainly located in the posterior region of the ganglion.

Studies of serotonin-immunoreactive cells in decapods (Beltz and Kravitz, 1983), isopods (Thompson *et al.*, 1994) and a distantly related malacostracean, *Anaspides tasmaniae* (Harrison *et al.*, 1985) provide a few examples of tentative homology between serotonin-like immunoreactive neurons in crustacea and insects. However, as pointed out before for the insect VPLI cells, homologous neurons may change their morphology in evolution; more crucially, their transmitter phenotype may change rendering them invisible to particular antibodies. Therefore, any lack of apparent homology between serotonergic neurons of crustaceans and insects cannot be regarded as evidence against the view that arthropod nervous systems have a common Bauplan.

Clearer candidates for homologous neurons in insects and crustaceans are the inhibitory GABAergic motorneurons which innervate leg muscle. The soma position, axon trajectory and location of target muscle within the limb of the three GABAergic motorneurons in the crayfish is strikingly similar to insect GABAergic motorneurons (Wiens and Wolf, 1993) (Figure 26.12).

## 26.3.2 NEUROGENESIS

Decapods (Scholtz, 1992; Harzsch and Dawirs, 1994), isopods (Dohle and Scholtz, 1988) and amphipods (Scholtz, 1990) possess neuron progenitor cells which have many of the characteristics of insect neuroblasts. The crustacean neuroblasts bud off columns of smaller cells towards the interior of the embryo and each of these cells, like the ganglion mother cells in insects, subsequently undergoes an equal mitotic division to produce two neurons (Figure 26.13(A,B)). In the crayfish *Cherax destructor*, there is a total of 25–30 neuroblasts per hemisegment arranged in an insect-like pattern of 6–7 rows each containing 4–5 neuroblasts (Scholtz, 1992). Finally, the gene *engrailed* is expressed in a similar spatial pattern in insect and crustacean neuroectoderm (Patel *et al.*, 1989) and neuroblasts (N. Patel, personal communication).

However, crustacean neuroblasts do differ from those of insects. They do not delaminate from the surface layer of neuroectodermal cells during their divisions and they are not associated with the specialized sheath cells found in insects (Doe and Goodman, 1985; Scholtz, 1992). In addition, at least some crustacean neuroblasts can give rise to epidermal cells after they have begun to bud off neurons (Dohle and Scholtz, 1988), which is not described for any insect.

A very different pattern of neurogenesis is seen in non-malacostracan crustaceans. Cells with the morphological characteristics of insect or malacostracan neuroblasts have

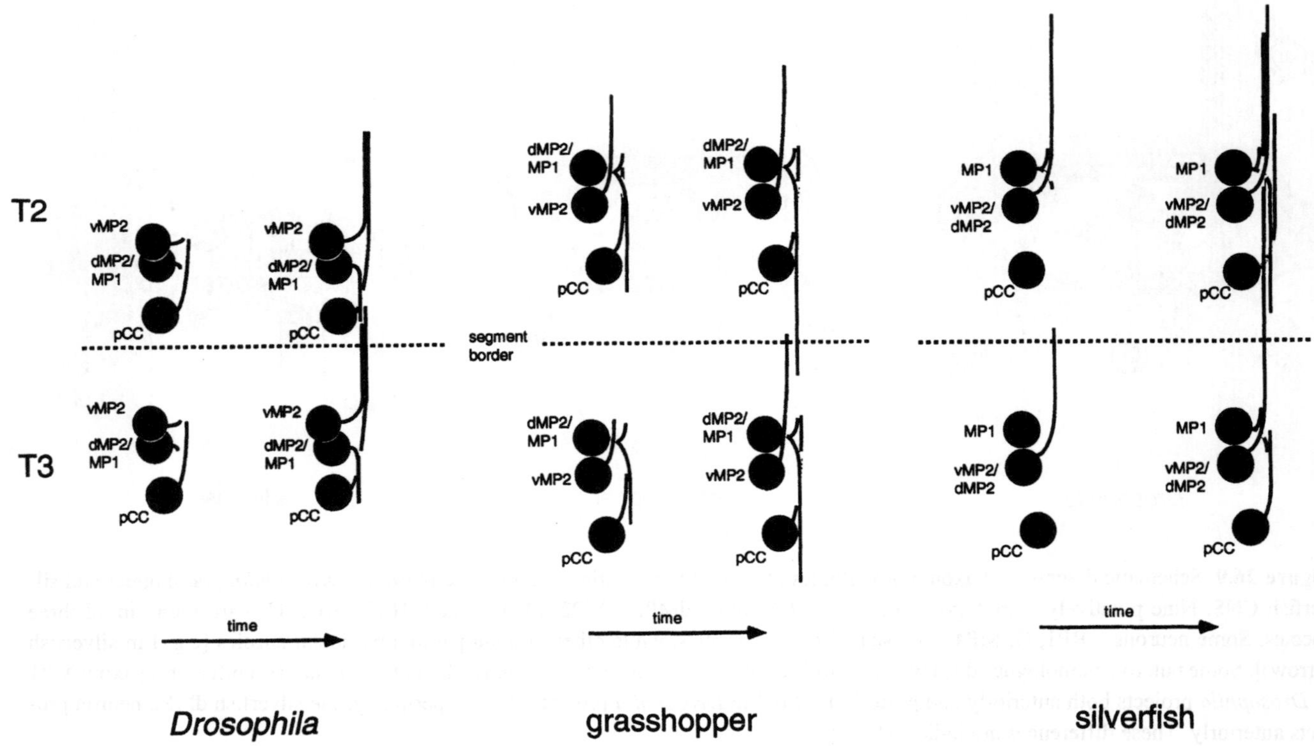


**Figure 26.10** Differences in timing and patterns of fasciculation between identified *Drosophila*, grasshopper and silverfish embryonic neurons. These differences include: silverfish pCC and MP1 begin axogenesis after the vMP2/dMP2 pair, whereas all four axons grow simultaneously in the winged insects; *Drosophila* pCC axon grows anteriorly alongside vMP2 whereas in the grasshopper its anterior advance is delayed until the MP1/dMP2 axons from the next anterior segment have arrived in its vicinity; grasshopper vMP2 axon pioneers its own fascicle whereas it fasciculates with pCC, MP1 and dMP2 in *Drosophila* and silverfish.

not been reported in these crustacean taxa. Rather, neurons arise by generalized, inwards proliferation of ectodermal cells (Anderson, 1973).

The evolutionary relationships between the various crustacean groups and the insects would be clarified if it were known whether neuron production via neuroblasts were a plesiomorphic condition for the Crustacea or an apomorphy of the Malacostraca, or even whether malacostracan neuroblasts were truly homologous to those of insects.

### 26.3.3 EARLY AXON GROWTH IN MALACOSTRACAN EMBRYOS

Is the conservative pattern of early axon growth seen in insect embryos also evident in crustacean embryogenesis? A set of nine early differentiating neurons conserved between the grasshopper and *Drosophila* embryo is also present in the embryo of the crayfish *Procambarus* (Thomas *et al.*, 1984). A more detailed study in two crustaceans, the decapod *Cherax destructor* and the isopod *Porcellio scaber*, uncovered a pattern of axon growth which has much in common with that seen in the insects (Whitington *et al.*, 1993). Both of these crustaceans possess at least five neurons which show a similar relative timing of axon outgrowth and axon morphology to insect neurons that lie in equivalent positions in the ganglion (Figure 26.14). These neurons are early contributors to the longitudinal connectives, the intersegmental nerve roots and the median fibre tracts.

However, there are a number of differences in the early axon growth pattern between these crustaceans and the insects. Some of these represent apomorphies of one or the other crustacean group. For example, *Porcellio* apparently lacks neurons with an axon morphology like the insect RP1 and U neurons, whereas *Cherax* has putative homologues to these neurons (V and E[1], respectively). If we consider only those neurons that are common to both crustaceans, we are then left with the insect–crustacean differences shown in Figure 26.15. A comparison of early axon growth between the two crustaceans (Figures 26.14–26.16) shows some neurons in common and some differences. In general, however, the pattern of axon growth is remarkably similar in these two relatively distantly related malacostracans.

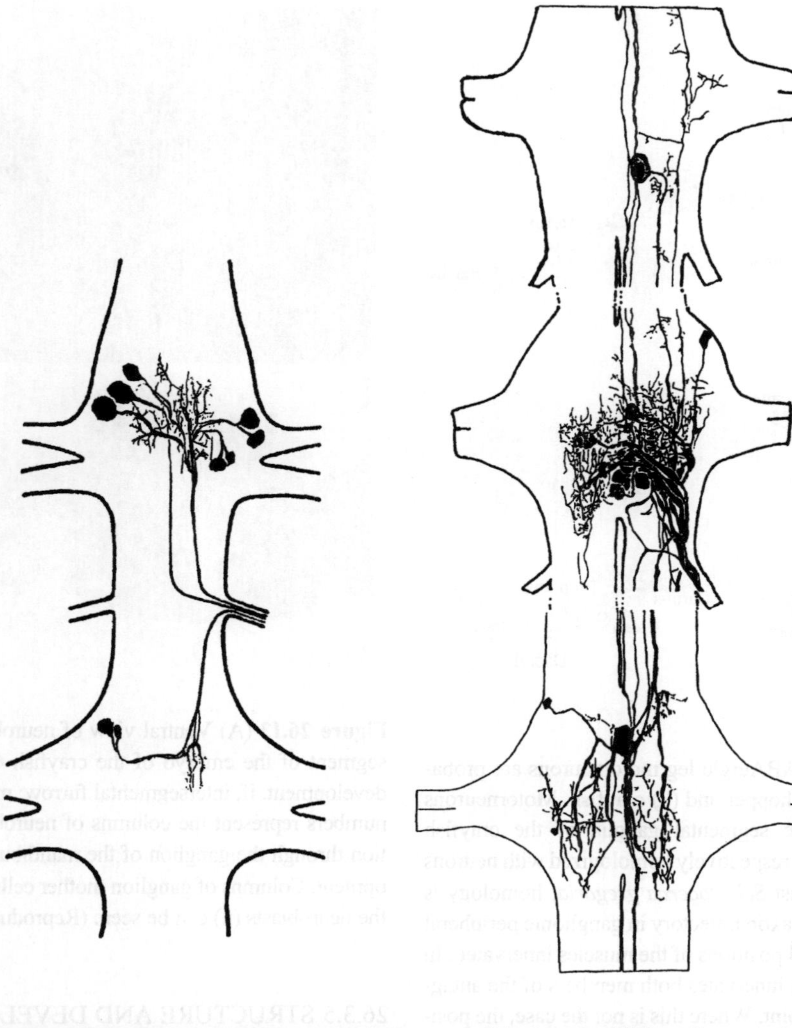


**Figure 26.11** Fast flexor motorneurons of the crayfish *Procambarus clarkii* and the dorsal extensor motorneurons of the isopod *Idotea balthica* have a similar appearance to insect dorsal longitudinal motorneurons in that the motorneuron somata are distributed throughout two ganglia, one ipsisegmental to the muscles and the other anterior to the muscles' segment. In addition, the posterior ganglion contains a group of anterior, contralateral neurons and the anterior ganglion contains between six and eight neurons which are mainly located in the posterior region of the ganglion. [Crayfish data modified from Mittenhall and Wine (1978); isopod data modified from Kutsch and Heckmann (1995).]

### 26.3.4 USING THE STRUCTURE AND DEVELOPMENT OF THE CRUSTACEAN VNC FOR PHYLOGENETIC PURPOSES

Overall, the similarities seen in the structure and development of the insect and malacostracan VNC strongly support the conclusion that these arthropods share a common Bauplan for the construction of their nervous systems. However, to what extent can we use individual characters seen in the development and mature structure of the VNC in these arthropods for the purposes of cladistic analysis? The Crustacea would be a useful outgroup when reconstructing evolutionary relationships between different insect groups. Our ability to carry out this sort of analysis depends critically upon our ability to establish homologies across the wide Insecta–Crustacea phylogenetic gap at the level of individual neurons or their precursors. Knowledge of the cell lineage of individual neurons and/or the expression of orthologous genes by neurons and their precursors will greatly assist in putting these homologies on a solid footing.

Leaving aside the question of homology of insect and crustacean neurons, there would appear to be little question about the homology of at least 11 early differentiating neurons (C, D, E, F, J, K, M, N, O, Q, S) in the isopod *Porcellio scaber* and

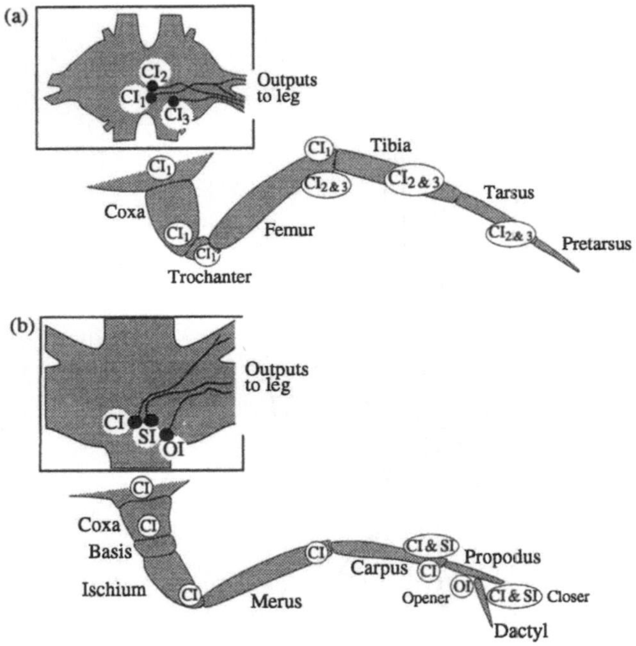


**Figure 26.12** Inhibitory GABAergic leg motorneurons are probably homologous in (a) grasshopper and (b) crayfish. Motorneurons CI, SI and OI in thoracic segmental ganglia of the crayfish *Pacifastacus leniusculus* are respectively homologized with neurons $CI_1$, $CI_2$ and $CI_3$ of the locust *Schistocerca gregaria*; homology is based on cell body position, axon trajectory in ganglionic peripheral nerves and the leg segmental positions of the muscles innervated. In most cases, the same neuron innervates both members of the antagonistic muscle pair at each joint. Where this is not the case, the position of the label indicates whether an extensor (above) or a flexor (below) is innervated by a given neuron. From Osorio et al. (1995).

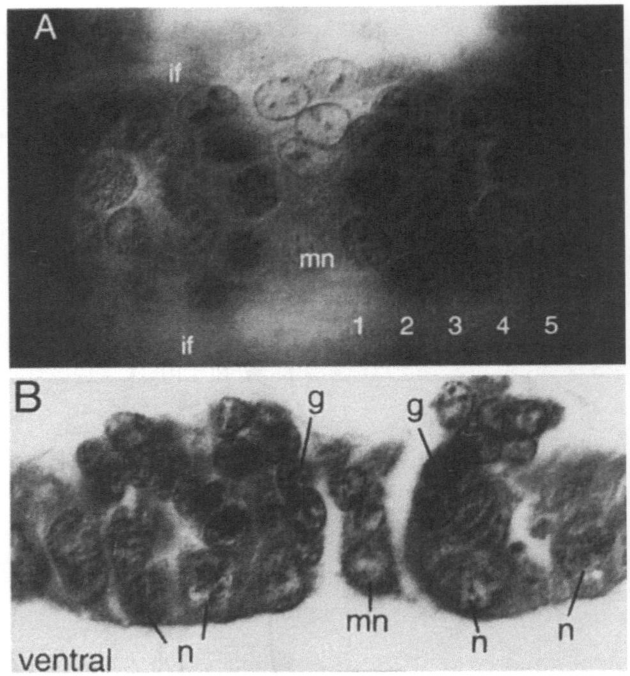


**Figure 26.13** (A) Ventral view of neuroblasts of the sixth thoracic segment of the embryo of the crayfish *Cherax destructor* at 45% development. if, intersegmental furrow; mn, median neuroblast. The numbers represent the columns of neuroblasts. (B) Transverse section through the ganglion of the mandibular segment at 40% development. Columns of ganglion mother cells (g) running dorsally from the neuroblasts (n) can be seen. (Reproduced from Scholtz, 1992.)

the decapod *Cherax destructor*. The axon morphology of these identified pioneering neurons will provide useful characters in an analysis of the relationships between the major crustacean groups. Unfortunately, we do not know the axon morphology of embryonic VNC neurons for any crustacean in a sister group to the Peracarida and Decapoda, and so we cannot yet say whether the pronounced similarities seen between these two groups represent eumalacostracan synapomorphies or Crustacean plesiomorphies. Finally, we may ask whether the similarities seen in the final structure and development of the VNC in insects and crustaceans provide evidence for a sister-group relationship between these two groups of arthropods. Clearly, information on the organization and development of the VNC in the other two major arthropod groups – the Chelicerata and the Myriapoda – is required to address this question. Unfortunately, both of these groups have been largely neglected.

### 26.3.5 STRUCTURE AND DEVELOPMENT OF THE VNC IN CHELICERATA

Descriptions of the gross morphology of the VNC in Chelicerata (Bullock and Horridge, 1965) reveal the familiar arthropod ladder-like arrangement of segmental ganglia with transverse commissures joined by longitudinal connectives, with a pronounced trend for ganglionic fusion in most groups. Wegerhoff and Breidbach (1995) have reviewed studies in various Chelicerata of the gross structural organization of the neuropile and the topology of identified neurons as revealed by immunohistochemical staining for various neurotransmitters and neuropeptides, including serotonin and CCAP. Perisulfakinin immunoreactivity has been described briefly by Agricola and Bräunig (1995). While revealing a conservative neuro-architecture within the arachnids, these studies do not reveal clear homologies with neural structures in the Insecta or Crustacea. For example, the somata of serotonin immunoreactive neurons in the harvestman (Arachnida; Opiliones) are located medially, rather than laterally as in insects and crustaceans and project axons ipsilaterally, rather than contralaterally (Breidbach and Wegerhoff, 1993).

Apart from a few classical accounts, neural development

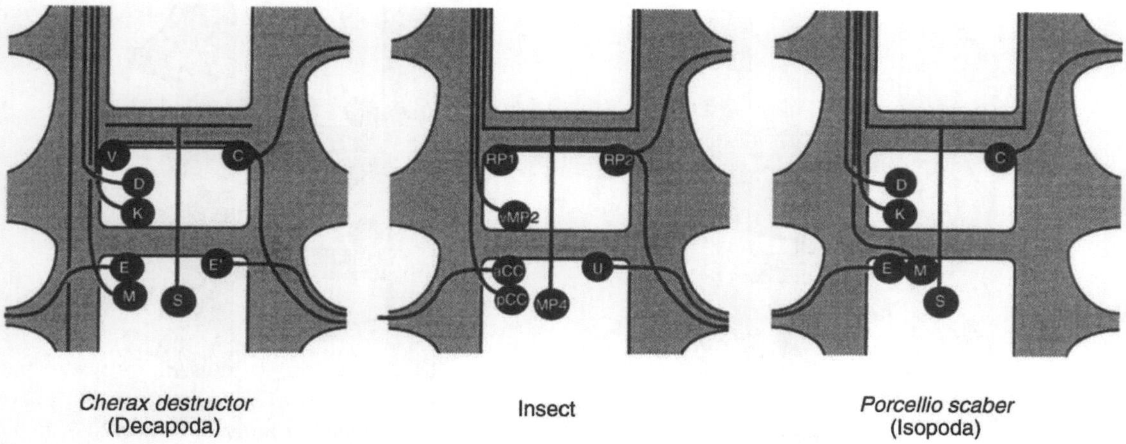


**Figure 26.14** Early embryonic axon growth in the CNS of insects and two malacostracan crustaceans, *Cherax destructor* and *Porcellio scaber*, shares a number of common features. Five insect neurons – vMP2, aCC, pCC, RP2 and MP4 – have putative homologues in both of the crustaceans (D/K, E, M, C, S, respectively). The insect neurons RP1 and U have putative homologues in the crayfish (V and $E^1$, respectively), but apparently not in the woodlouse. The insect neurons RP2, aCC and pCC express *even-skipped* (*eve*) during embryogenesis (Patel *et al.*, 1992) and so, in the absence of lineage data, a way to secure the putative homologies for these neurons would be to determine whether the C, E and M neurons express the crustacean orthologue of the *eve* gene.

in the Chelicerates has received little attention. In xiphosurans, and most scorpions and arachnids, neurogenesis occurs by a generalized proliferation of ventral ectoderm cells to produce paired segmental thickenings (Anderson, 1973). Neuroblasts have been reported in only one scorpion, *Heterometrus scaber* (Mathew, 1956) and one spider, *Heptathela kamurai* (Yoshikura, 1955). In a number of arachnid embryos, the left and right halves of the germband become widely separated as the embryo undergoes segmentation (Anderson, 1973). Correspondingly, the left and right halves of each ganglion rudiment are initially separate and only come together in the ventral midline later in embryogenesis. A similar behaviour is exhibited by the embryos of certain myriapods and onychophorans. No information is available concerning the pattern of axon growth from identified neurons in the VNC of chelicerate embryos to enable comparisons to be drawn with the Insecta or Crustacea.

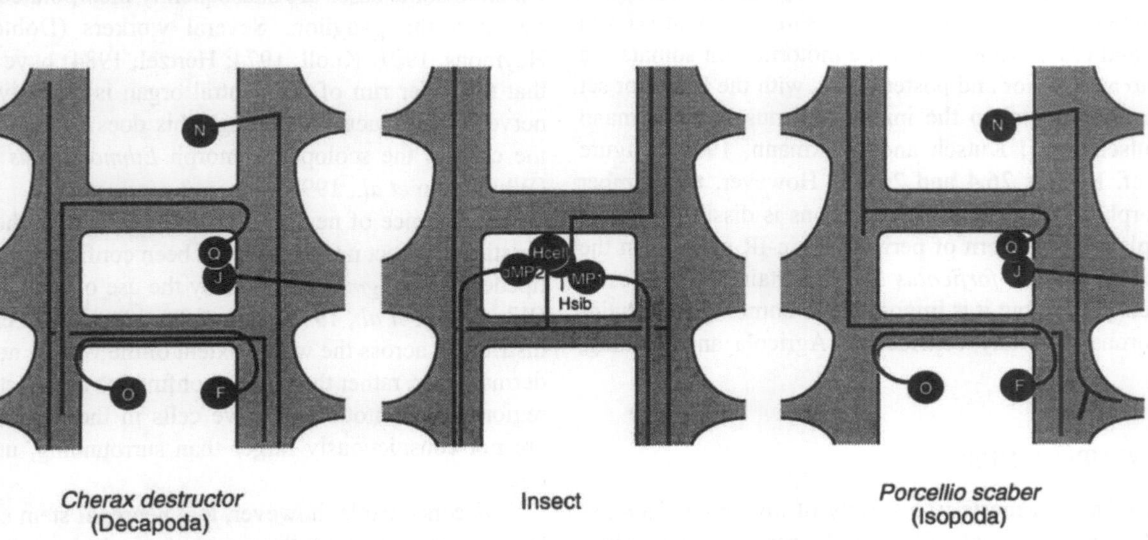


**Figure 26.15** Despite the overall conservation of the pattern of neuronal cell bodies and early axogenesis in insect and crustacean CNS, there are some notable differences between these two major arthropod groups. These diagrams show early differentiating neurons in insects (dMP2, MP1, H cell and H sib) that are not present in the two malacostracan crustaceans, *Cherax destructor* and *Porcellio scaber* and conversely, crustacean neurons (N, Q, J, O and F) that are not present in insects. The figure also shows some differences in axon morphology between the putative homologues Q, F and N in *Cherax* and *Porcellio*.

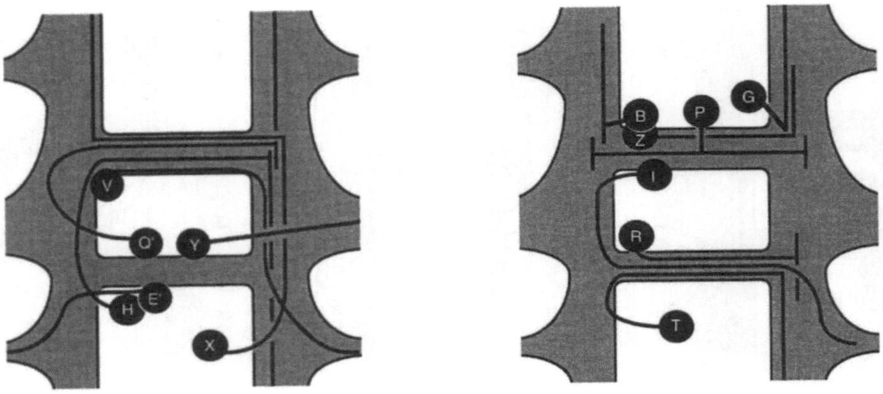


**Figure 26.16** Axon morphologies of early differentiating neurons in the embryo of *Cherax destructor* that are not apparently present in *Porcellio scaber* and vice-versa. The absence of a particular neuron in one species (e.g. R in *Cherax*) was judged by the failure to find a neuron of a similar axon morphology after an exhaustive search of the corresponding region of the ganglion. Clearly, we cannot exclude the possibility that a homologue to the *Porcellio* R neuron is present in *Cherax* but is located in a different region of the ganglion. Alternatively, the axon morphology of R may have been transformed during evolution such that its *Cherax* homologue cannot be recognized.

## 26.4 MYRIAPODS

### 26.4.1 ORGANIZATION OF ADULT GANGLIA

The VNC of myriapods conforms to the general arthropod plan in its external gross morphology (Bullock and Horridge, 1965). However, there are only scattered reports concerning its internal architecture. The arrangement of motorneurons innervating the dorsal longitudinal muscles in the centipede *Lithobius forficatus* is similar to that seen in insects and crustaceans in that the motorneuron somata are split into an anterior and posterior set, with the posterior set being ipsisegmental to the innervated muscle (Heckmann and Kutsch 1995; Kutsch and Heckmann, 1995) (Figure 26.17, cf. Figures 26.4 and 26.11). However, the number and morphology of these motorneurons is dissimilar to the insect plan. The pattern of perisulfakinin-IR neurons in the centipede *Lithobius forficatus* shows certain similarities to the insect pattern but it is impossible to homologize individual neurons with any confidence (Agricola and Bräunig, 1995).

### 26.4.2 NEUROGENESIS

Studies of neurogenesis in a variety of myriapods have, in general, failed to reveal cells with the morphological characteristics of insect neuroblasts (Chilopoda: Heymons, 1901; Symphyla, Pauropoda: Tiegs, 1940, 1947; Diplopoda: Dohle, 1964). Knoll (1974) claims that neuroblasts occur in the outer region of the neuroectoderm in the Chilopod *Scutigera coleoptrata* and that these generate vertical columns of neurons. The cells identified as neuroblasts are, however, only slightly larger than the overlying neurons. In all other cases, neurons are reported to be produced by a generalized proliferation of the ventral ectoderm. The formation of the ventral ganglia is associated with the formation of 'ventral organs'; these are shallow pits which develop within the ectoderm external to the ganglia and which in some cases are subsequently incorporated as cavities into the ganglion. Several workers (Dohle, 1964; Heymons, 1901; Knoll, 1974; Hertzel, 1984) have reported that the inner rim of the ventral organ is an active site of nerve cell production, although this does not appear to be the case in the scolopendromorph *Ethmostigmus rubripes* (Whitington *et al.*, 1991).

The absence of neuron progenitor cells with the characteristics of insect neuroblasts has been confirmed in the centipede *Ethmostigmus rubripes* by the use of BrdU labelling (Whitington *et al.*, 1991). Mitotic activity in this embryo is distributed across the whole extent of the ventral neuroectodermal layer, rather than being confined to the most external regions, and mitotically active cells in the neuroectoderm are not conspicuously larger than surrounding, unlabelled nuclei.

It is conceivable, however, that neuronal stem cells with the proliferative, if not the morphological, characteristics of insect neuroblasts exist in myriapods. Indeed, clusters of dye-coupled cells which are found within the centipede neuroectoderm may represent the progeny of single precursor cells, while BrdU-labelled cells which are often found near the middle of these clusters may be functional equivalents of

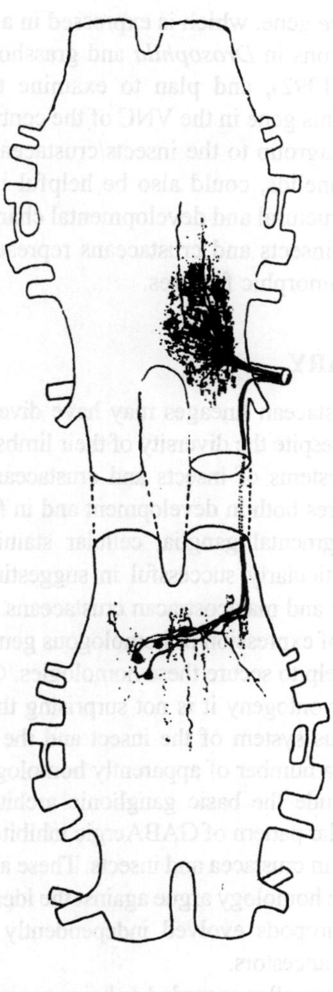


**Figure 26.17** The location and morphology of motorneurons innervating the dorsal longitudinal muscles in the centipede *Lithobius forficatus*. The pattern of motorneurons is similar to that seen in insects and crustaceans in that the motorneuron somata are split into an anterior and posterior set, with the posterior set being ipsisegmental to the innervated muscle. (Modified from Kutsch and Heckmann, 1995.)

insect neuroblasts (Whitington *et al.*, 1991). Studies of cell lineage within the centipede neuroectoderm should help to resolve this issue, together with descriptions of the expression of centipede homologues to insect genes involved in neurogenesis.

### 26.4.3 AXON GROWTH

Only one study has dealt with axon growth in the VNC of myriapod embryos at the level of identified neurons, although earlier workers have described the time of appearance of the first nerve fibres in the VNC. Using single neuron dye-labelling, Whitington *et al.* (1991) showed that the earliest central axon pathways in the embryo of the centipede *Ethmostigmus rubripes* arise by the posteriorly directed growth of axons originating from neurons located in the brain, rather than from neurons in each segmental neuromere, as in insects. This primary tract appears to act as a scaffold along which the longitudinal axons from segmental axons later grow.

This result might initially be taken to indicate a fundamental difference between insects and myriapods in the pattern of early axon growth. However, a similar difference was observed in the study of axon growth in isopod and decapod embryos: the first axons to appear in segmental ganglia of crayfish embryos also originate from brain neurons, whereas in the woodlouse they arise from neurons in the same segment. This difference has been attributed to the relative precocity of formation of head segments with respect to more posterior body segments in the crayfish (Scholtz, 1992). This myriapod–insect difference may also reflect a relatively minor change in the timing of common developmental events.

Nonetheless, the pattern of axon growth within segmental neuromeres of the centipede shows many other differences to the common pattern seen in insects and crustaceans. Firstly, axonogenesis begins within segmental ganglia when there is a much larger population of cells than that present in insect ganglia. This also appears to be the case in the pauropod *Pauropus sylvaticus* (Tiegs, 1947) and the symphylan *Hanseniella agilis* (Tiegs, 1940). Secondly, the neurons that participate in this early phase of axon growth in the centipede are widely separated from one another, whereas pioneering axons in the insect CNS arise from neurons in close proximity to one another. This may be a direct result of the delay in axonogenesis relative to neurogenesis. Thirdly, commissural axon growth in the centipede is substantially retarded relative to longitudinal growth, whereas in insect and crustacean embryos these pathways are laid down at around the same time. This difference may be an apomorphic trait 'forced' upon the scolopendromorph centipedes because the left and right hemiganglia are separated by a large expanse of extraembryonic ectoderm through much of embryogenesis. It will be interesting to see whether commissural axon growth is delayed in other myriapods which do not have a split germband. Finally, the pattern of neurons involved in pioneering axon growth in the centipede bears no obvious similarities to the insect pattern (Figure 26.18). It is, however, likely that the dye-injection technique used to stain these neurons sampled only a limited subset of the total population of neurons involved in pioneering axon growth. A more comprehensive sample of neurons might reveal a pattern of cells more closely comparable with that involved in early axon growth in insects and crustaceans.

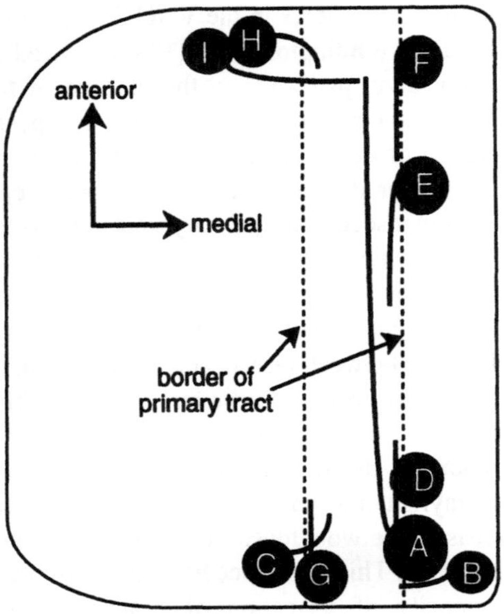


**Figure 26.18** Diagrammatic summary of the soma positions and axon morphology of early differentiating segmental neurons in the embryo of the centipede *Ethmostigmus rubripes*. The lines running from the somata represent the course taken by the axons while their lengths represent the relative degree of axon growth attained by the neurons.

### 26.4.4 STRUCTURE AND DEVELOPMENT OF THE VNC OF MYRIAPODS – PHYLOGENETIC IMPLICATIONS

The structure and development of the myriapod CNS shows little in common with the insects, especially when one considers the numerous similarities between the insect and crustacean VNC and brain (Nilsson and Osorio, 1997, this volume). These differences would at first sight appear to be at variance with the conventional view that myriapods are the sister group to insects. This is based on their common possession of such features as uniramous appendages and the lack of appendages on the segment of the tritocerebrum. It is possible, therefore, that the differences we see between the insect and myriapod CNS development represent relatively minor variations to an underlying common developmental Bauplan. We have already seen that some of the differences seen between insects and myriapods could be explained in terms of heterochronic change, a relatively minor and common evolutionary phenomenon.

It would be very helpful to obtain evidence for homologies between myriapods and insects at the identified neuron level. We would then be in a much better position to assess the significance of their developmental differences. Probes that reveal the expression of myriapod orthologues of genes that display highly conserved neural expression patterns within the insects could be especially useful in establishing such homologies. We have cloned the centipede homologue to the insect *eve* gene, which is expressed in a conserved set of central neurons in *Drosophila* and grasshopper embryos (Patel *et al.*, 1992), and plan to examine the pattern of expression of this gene in the VNC of the centipede embryo. Study of an outgroup to the insects/crustaceans/myriapods, such as the annelids, could also be helpful in determining whether the structural and developmental characters in common between insects and crustaceans represent plesiomorphic or synapomorphic features.

## 26.5 SUMMARY

Insect and crustacean lineages may have diverged over 500 Myr ago, but despite the diversity of their limbs and lifestyles, the nervous systems of insects and crustaceans share many common features both in development and in function. In the developing segmental ganglia, cellular staining techniques have been particularly successful in suggesting homologies between insect and malacostracan crustaceans. Future studies of the pattern of expression of homologous genes in identified neurons will help to secure these homologies. Given the similarities in their ontogeny it is not surprising that the postembryonic nervous system of the insect and the malacostracan CNS contains a number of apparently homologous structures. Examples include the basic ganglionic architecture and the strikingly similar pattern of GABAergic inhibitory innervation of leg muscles in crustacea and insects. These and other examples of putative homology argue against the idea that these two groups of arthropods evolved independently from separate non-arthropod ancestors.

Myriapods are often regarded as being more closely related to the insects than are the crustaceans. However, we find that myriapods show substantial differences, for example in the pattern of early axon growth, to the common pattern of early neural development seen in the Crustacea/Insecta. To determine whether these differences reflect a fundamentally different Bauplan for VNC construction or simply modifications to a common developmental plan, we shall need to examine orthologous gene expression in neurons of the developing VNC in these three groups.

## ACKNOWLEDGEMENTS

We thank Jim Truman, Eldon Ball, Nipam Patel, Torsten Bossing and Matthias Gerberding for permission to cite unpublished results and Gerhard Scholtz for helpful discussions. P.M.W. is supported by grants from the Australian Research Council. J.P.B. is supported by the BBSRC (UK).

## REFERENCES

Agricola, H. and Bräunig, P. (1995) Comparative aspects of peptidergic signalling pathways in the nervous systems of arthropods, in *Nervous Systems of Invertebrates: an Evolutionary and*

*Comparative Approach* (eds O. Breidbach and W. Kutsch), Birkhäuser Verlag, Basel, pp. 303–27.

Anderson, D.T. (1973) *Embryology and Phylogeny in Annelids and Arthropods*, Pergamon Press, Oxford.

Bacon, J.P and Tyrer, N.M. (1978) The tritocerebral commissure giant (TCG): a bimodal interneuron in the locust, *Schistocerca gregaria*. *Journal of Comparative Physiology*, **126**, 317–25

Bastiani, M.J. and Goodman, C.S. (1986) Guidance of neuronal growth cones in the grasshopper embryo. III Recognition of specific glial pathways. *Journal of Neuroscience*, **6**, 3542–51.

Bastiani, M.J., du Lac, S. and Goodman, C.S. (1986) Guidance of neuronal growth cones in the grasshopper embryo. I. Recognition of a specific axonal pathway by the pCC neuron. *Journal of Neuroscience*, **6**, 3518–31.

Bate, C.M. (1976) Embryogenesis of an insect nervous system: I. A map of the thoracic and abdominal neuroblasts in *Locusta migratoria*. *Journal of Embryology and Experimental Morphology*, **35**, 107–23.

Bate, C.M. and Grunewald, E.B. (1981) Embryogenesis of an insect nervous system: II. A second class of neuron precursor cell and the origin of the intersegmental connectives. *Journal of Embryology and Experimental Morphology*, **61**, 317–30.

Beltz, B.S. and Kravitz, E.A. (1983) Mapping of serotonin-like immunoreactivity in the lobster nervous system. *Journal of Neuroscience*, **3**, 585–602.

Bishop, C.A. and O'Shea, M. (1983) Serotonin immunoreactive neurons in the central nervous system of an insect (*Periplaneta americana*). *Journal of Neurobiology*, **14**, 251–64.

Bossing, T. and Technau, G.M. (1994) The fate of the CNS midline progenitors in *Drosophila* as revealed by a new method for single cell labelling. *Development*, **120**, 1895–906.

Boyan, G.S. (1995) Lineage analysis as an analytic tool in the insect central nervous system: bringing order to interneurons, in *Nervous Systems of Invertebrates: an Evolutionary and Comparative Approach* (eds O. Breidbach and W. Kutsch), Birkhäuser Verlag, Basel, pp. 273–301.

Boyan, G.S. and Ball, E.E. (1993) The grasshopper, *Drosophila* and neuronal homology (advantages of the insect nervous system for the neuroscientist). *Progress in Neurobiology*, **41**, 657–82.

Breidbach, O. and Wegerhoff, R. (1993) Neuroanatomy of the central nervous system of the harvestman, *Rilaena triangularis* (HERBST 1799) (Arachnida; Opiliones) – principal organization, GABA-like and Serotonin-immunohistochemistry. *Zoologischer Anzeiger*, **230**, 55–81.

Broadus, J. and Doe, C.Q. (1995) Evolution of neuroblast identity: *seven-up* and *prospero* expression reveal homologous and divergent neuroblast fates in *Drosophila* and *Schistocerca*. *Development*, **121**, 3989–96.

Broadus, J., Skeath, J.B., Spana, E.P., Bossing, T., Technau, G. and Doe, C.Q. (1995) New neuroblast markers and the origin of the aCC/pCC neurons in the *Drosophila* central nervous system. *Mechanisms of Development*, **53**, 393–402.

Bullock, T.H. and Horridge, G.A. (1965) *Structure and Function in the Nervous Systems of Invertebrates*, vol. 2, W.H. Freeman, San Francisco.

Condron, B.G., Patel, N.H. and Zinn, K. (1994) *engrailed* controls glial/neuronal cell fate decisions at the midline of the central nervous system. *Neuron*, **13**, 541–54.

Davis, N.T. and Hildebrand, J.G. (1992) Vasopressin-immunoreactive neurons and neurohaemal systems in cockroaches and mantids. *Journal of Comparative Neurology*, **320**, 381–93.

Dawes, R., Dawson, I., Falciani, F., Tear, G. and Akam, M. (1994) *Dax*, a locust Hox gene related to fushi-tarazu but showing no pair-rule expression. *Development*, **120**, 1561–72

Doe, C.Q. (1992) Molecular markers for identified neuroblasts and ganglion mother cells in the *Drosophila* central nervous system. *Development*, **116**, 855–63.

Doe, C.Q. and Goodman, C.S. (1985) Early events in insect neurogenesis. I Development and segmental differences in the pattern of neuronal precursor cells. *Developmental Biology*, **111**, 193–205.

Doe, C.Q. and Technau, G.M. (1993) Identification and cell lineage of individual neural precursors in the *Drosophila* CNS. *Trends in Neuroscience*, **16**, 510–14.

Doe, C.Q., Hiromi, Y., Gehring, W.J. and Goodman, C.S. (1988) Expression and function of the segmentation gene *fushi-tarazu* during *Drosophila* neurogenesis. *Science*, **239**, 170–5.

Dohle, V.W. (1964) Die Embryonalentwicklung von Glomeris marginata (Villers) im Vergleich zur Entwicklung anderer Diplopoden. *Zoologische Jahrbücher Anatomie Ontogenese*, **81**, 241–310.

Dohle, W. and Scholtz, G. (1988) Clonal analysis of the crustacean segment: the discordance between genealogical and segmental borders. *Development* (Supplement), **104**, 147–60

Elson, R.C. (1996) Neuroanatomy of a crayfish thoracic ganglion: sensory and motor roots of the walking-leg nerves and possible homologies with insects. *Journal of Comparative Neurology*, **365**, 1–17.

Goodman, C.S., Pearson, K.G. and Spitzer, N.C. (1980) Electrical excitability: a spectrum of properties in the progeny of a single embryonic neuroblast. *Proceedings of the National Academy of Sciences, USA*, **77**, 1676–80.

Goodman, C.S., Bate, M. and Spitzer, N.C. (1981) Embryonic development of identified neurons: origin and transformation of the H Cell. *Journal of Neuroscience*, **1**, 94–102.

Goodman, C.S., Bastiani, M.J., Doe, C.Q., du Lac, S., Helfand, S.L., Kuwada, J.Y. and Thomas, J.B. (1984) Cell recognition during neuronal development. *Science*, **225**, 1271–9.

Gutjahr, T., Patel, N.H., Li, X., Goodman, C.S. and Noll, M. (1993) Analysis of the gooseberry locus in *Drosophila* embryos: *gooseberry* determines the cuticular pattern and activates *gooseberry neuro*. *Development*, **118**, 21–31.

Harrison, P.J., Macmillan, D.L. and Young, H.M. (1995) Serotonin immunoreactivity in the ventral nerve cord of the primitive crustacean *Anaspides tasmaniae* closely resembles that of crayfish. *Journal of Experimental Biology*, **198**, 531–5.

Hartenstein, V. and Campos-Ortega, J.A. (1984) Early neurogenesis in wild-type *Drosophila* melanogaster. *Roux's Archives of Developmental Biology*, **193**, 308–25.

Harzsch, S. and Dawirs, R.R. (1994) Neurogenesis in larval stages of the spider crab *Hyas araneus* (Decapoda, Brachyura): proliferation of neuroblasts in the ventral nerve cord. *Roux's Archives of Developmental Biology*, **204**, 93–100.

Heckmann, R. and Kutsch, W. (1995) Motor supply of the dorsal longitudinal muscles. 2. Comparison of motorneuron sets in Tracheata. *Zoomorphology*, **115**, 197–211.

Hertzel, G. (1984) Die Segmentation des Keimstreifens von *Lithobius forficatus* (L.) (Myriapoda, Chilopoda). *Zoologische Jahrbücher Anatomie*, **112**, 369–86.

Heymons, R. (1901) Die Entwicklungsgeschichte der Scolopender. *Zoologica*, Stuttgart, **13**, 1–224.

Jacobs, J.R. and Goodman, C.S. (1989) Embryonic development of axon pathways in the *Drosophila* CNS. 2. Behavior of pioneer growth cones. *Journal of Neuroscience*, **9**, 2412–22.

Knoll, H.J. (1974) Untersuchungen zur Entwicklungsgeschichte von Scutigera coleoptrata L. (Chilopoda). *Zoologische Jahrbücher Anatomie*, **92**, 47–132.

Kutsch, W. and Heckmann, R. (1995) Homologous structures, exemplified by motorneurons of Mandibulata, in *Nervous Systems of Invertebrates: an Evolutionary and Comparative Approach* (eds O. Breidbach and W. Kutsch), Birkhäuser Verlag, Basel, pp. 221–48.

Maslin, T.P. (1952) Morphological criteria for phyletic relationships. *Systematic Zoology*, **1**, 49–70.

Mathew, A.P. (1956) Embryology of *Heterometrus scaber* (Thorell) Arachnida. *Zoological Memoirs. University of Travancore Research Institute*, **1**, 1–96.

Mittenthal, J.E. and Wine, J.J. (1978) Segmental homology and variation in flexor motorneurons of the crayfish. *Journal of Comparative Neurology*, **177**, 311–34.

Musiol, I.M., Jirikowski, G.F. and Pohlhammer, K. (1990) Immunocytochemical characterisation of a widely spread arg8-vasopressin-like neuroendocrine system in the cricket *Teleogryllus commodus* Walker (Orthoptera, Insecta). *Acta Histochemistry*, Supplement **40**, 137–42.

Osorio, D., Averof, M. and Bacon, J.P. (1995) Arthropod evolution: great brains, beautiful bodies. *Trends in Ecology and Evolution*, **10**, 449–54

Otsuka, M., Kravitz, E.A. and Potter, D.D. (1967) Physiological and chemical architecture of a lobster ganglion with respect to gamma-aminobutyrate and glutamate. *Journal of Neurophysiology*, **30**, 725–52.

Patel, N.H., Ball, E.E. and Goodman, C.S. (1992) Changing role of *even-skipped* during the evolution of insect pattern formation. *Nature*, **357**, 339–42.

Patel, N.P., Kornberg, T.B. and Goodman, C.S. (1989) Expression of *engrailed* during segmentation in grasshopper and crayfish. *Development*, **107**, 201–12.

Sandeman, D.C. (1982) Organization of the central nervous system, in *The Biology of Crustacea, vol. 3 Neurobiology: Structure and Function* (eds H.L. Atwood and D.C. Sandeman), Academic Press, New York, pp. 1–61.

Scholtz, G. (1990). The formation, differentiation and segmentation of the post-naupliar germ band of the amphipod *Gammarus pulex L.* (Crustacea, Malacostraca, Peracarida). *Proceedings of the Royal Society of London, B*, **239**, 163–211.

Scholtz, G. (1992) Cell lineage studies in the crayfish *Cherax destructor* (Crustacea, Decapoda) – germ band formation, segmentation, and early neurogenesis. *Roux's Archives of Developmental Biology*, **202**, 36–48.

Spana, E.P., Kopczynski, C., Goodman, C.S. and Doe, C.Q. (1995) Asymmetric localization of *numb* autonomously determines sibling neuron identity in the *Drosophila* CNS. *Development*, **121**, 3489–94.

Tamarelle, M., Haget, A. and Ressouches, A. (1985) Segregation, division, and early patterning of lateral thoracic neuroblasts in the embryos of *Carausius morosus Br.* (Phasmida: Lonchodidae). *International Journal of Insect Morphology and Embryology*, **14**, 307–17.

Thomas, J.B., Bastiani, M.J., Bate, M. and Goodman, C.S. (1984) From grasshopper to *Drosophila*: a common plan for neuronal development. *Nature*, **310**, 203–7.

Thompson, K.J. and Siegler, M.V.S. (1993) Development of segment specificity in identified lineages of the grasshopper CNS. *Journal of Neuroscience*, **13**, 3309–18.

Thompson, K.S.J., Tyrer, N.M., May, S.T. and Bacon, J.P. (1991) The vasopressin-like immunoreactive (VPLI) neurons of the locust, *Locusta migratoria:* I. Anatomy. *Journal of Comparative Physiology A*, **168**, 605–17

Thompson, K.S.J., Zeidler, M.P. and Bacon, J.P. (1994) Comparative anatomy of serotonin-like immunoreactive neurons in isopods: putative homologues in several species. *Journal of Comparative Neurology*, **347**, 553–69

Tiegs, O.W. (1940) The embryology and affinities of the Symphyla, based on a study of *Hanseniella agilis*. *Quarterly Journal of Microscopical Science*, **82**, 1–225.

Tiegs, O.W. (1947) The development and affinities of the *Pauropoda*, based on a study of *Pauropus silvaticus*. *Quarterly Journal of Microscopical Science*, **88**, 165–336.

Tyrer, N.M. and Gregory, G.E. (1982) A guide to the neuroanatomy of locust suboesophageal and thoracic ganglia. *Philosophical Transactions of the Royal Society of London, B*, **297**, 91–123.

Tyrer, N.M., Turner, J.D. and Altman, J.S. (1984) Identifiable neurons in the locust central nervous system that react with antibodies to serotonin. *Journal of Comparative Neurology*, **227**, 313–30.

Tyrer, N.M., Davis, N.T., Arbas, E.A., Thompson, K.S.J. and Bacon, J.P. (1993) The morphology of the vasopressin like immunoreactive (VPLI) neurons in many species of grasshopper. *Journal of Comparative Neurology*, **329**, 385–401.

Udolph, G., Prokop, A., Bossing, T. and Technau, G.M. (1993) A common precursor for glia and neurons in the embryonic CNS of *Drosophila* gives rise to segment-specific lineage variants. *Development*, **118**, 765–75.

Uvarov, B. (1966) *Grasshoppers and Locusts. A Handbook of General Acridology*, Vol. 1, Cambridge University Press, Cambridge.

Vallés, A.M. and White, K. (1988) Serotonin-containing neurons in *Drosophila melanogaster*: development and distribution. *Journal of Comparative Neurology*, **268**, 414–28.

van Haeften, T. and Schooneveld, H. (1992) Serotonin-like immunoreactivity in the ventral nerve cord of the Colorado potato beetle, *Leptinotarsa decemlineata*: identification of five different neuron classes. *Cell and Tissue Research*, **270**, 405–13.

Veenstra, J.A. (1984) Immunocytochemical demonstration of a homology in peptidergic neurosecretory cells in the suboesophageal ganglion of a beetle and a locust with antisera to bovine pancreatic polypeptide, FMRF amide, vasopressin and a-MSH. *Neuroscience Letters*, **48**, 185–90.

Vickery, V.R. (1987) The Northern Nearctic Orthoptera: their origins and survival, in *Evolutionary Biology of Orthopteroid Insects* (ed. B. Baccetti), Ellis Horwood Ltd, Chichester, pp. 581–91.

Wegerhoff, R. and Breidbach, O. (1995) Comparative aspects of the chelicerate nervous systems, in *Nervous Systems of Invertebrates: an Evolutionary and Comparative Approach* (eds O. Breidbach and W. Kutsch), Birkhäuser Verlag, Basel, pp. 159–79.

Weston, P.H. (1988) Indirect and direct methods in systematics, in *Ontogeny and Systematics* (ed. C.J. Humphries), Columbia University Press, New York, pp. 27–56.

Whitington, P.M., Meier, T. and King, P. (1991) Segmentation, neurogenesis and formation of early axonal pathways in the centipede, *Ethmostigmus rubripes* (Brandt). *Roux's Archives of Developmental Biology*, **199**, 349–63.

Whitington, P.M., Leach, D. and Sandeman, R. (1993) Evolutionary change in neural development within the arthropods – axonogenesis in the embryos of two crustaceans. *Development*, **118**, 449–61.

Whitington, P.M., Harris, K.-L. and Leach, D. (1996) Early axonogenesis in the embryo of a primitive insect, the silverfish *Ctenolepisma longicaudata. Roux's Archives of Developmental Biology*, **205**, 272–81.

Wiens, T.J. and Wolf, H. (1993) The inhibitory motorneurons of crayfish thoracic limbs – identification, structures, and homology with insect common inhibitors. *Journal of Comparative Neurology*, **336**, 261–78.

Yoshikura, M. (1955) Embryological studies on the liphistiid spider *Heptathela kumurai. II. Kumamoto Journal of Science*, **B2**, 1–86.

# Index

Page numbers in **bold** refer to illustrations, page numbers in *italics* refer to tables.

MIX
Papier aus verantwortungsvollen Quellen
Paper from responsible sources
FSC® C105338

If you have any concerns about our products,
you can contact us on
**ProductSafety@springernature.com**

In case Publisher is established outside the EU,
the EU authorized representative is:
**Springer Nature Customer Service Center GmbH**
**Europaplatz 3, 69115 Heidelberg, Germany**

Printed by Libri Plureos GmbH
in Hamburg, Germany